Mobile Radio Communications

SECOND EDITION

Mobile Radio Communications

SECOND EDITION

Second and Third Generation Cellular and WATM Systems

Edited by

Raymond Steele

Professor of Communications, University of Southampton
Chairman of Multiple Access Communications Ltd, Southampton, UK

and

Lajos Hanzo

Professor of Communications, University of Southampton
Consultant to Multiple Access Communications Ltd, Southampton, UK

JOHN WILEY & SONS, LTD
Chichester • New York • Weinheim • Brisbane • Singapore • Toronto

First published under the title *Mobile Radio Communications* in Great Britain in 1992 by
Pentech Press Limited, London. Copyright © Raymond Steele, 1992

Copyright © 1999 John Wiley & Sons Ltd
      Baffins Lane, Chichester,
      West Sussex, PO19 1UD, England

     National   01243 779777
     International (+44) 1243 779777

e-mail (for orders and customer service enquiries): cs-books@wiley.co.uk

Visit our Home Page on http://www.wiley.co.uk or http://www.wiley.com

Reprinted March 2000

Other Wiley Editorial Offices

John Wiley & Sons, Inc., 605 Third Avenue,
New York, NY 10158-0012, USA

Wiley-VCH Verlag GmbH, Pappelallee 3,
D-69469 Weinheim, Germany

Jacaranda Wiley Ltd, 33 Park Road, Milton,
Queensland 4064, Australia

John Wiley & Sons (Asia) Pte Ltd, 2 Clementi Loop #02-01,
Jin Xing Distripark, Singapore 129809

John Wiley & Sons (Canada) Ltd, 22 Worcester Road,
Rexdale, Ontario, M9W 1L1, Canada

British Library Cataloguing in Publication Data

A catalogue record for this book is available from the British Library

ISBN 0 471 97806 X

Produced from PostScript files supplied by the authors
Printed and bound in Great Britain by Bookcraft Ltd, Midsomer Norton
This book is printed on acid-free paper responsibly manufactured from sustainable
forestry in which at least two trees are planted for each one used for paper production.

Contents

v

Preface to the Second Edition

Second generation (2G) digital cellular mobile radio systems have taken root in many countries, untethering the telephone and enabling people to conduct conversations away from the home or office and while on the move. The systems are spectrally efficient with the frequency bands assigned by the regulatory bodies being reused repeatedly over countries and even continents. At the time of writing the standardisation of three third generation (3G) systems is also well under way in Europe, the United States and in Japan. This book aims to portray the evolutionary avenue bridging the second and third generation systems.

The fixed networks have also become digital, enabling the introduction of the integrated digital service network (ISDN). No longer are communications to be restricted to voice. Instead a range of services, such as fax, video conferencing and computer data transfer is becoming increasingly available. The second generation digital cellular networks have complex radio links, connecting the mobile users to their base stations. Mobile voice and data communications are supported by elaborate network protocols that support registration and location of mobile users, handovers between base stations as the mobiles roam, call initiation and call clear-down, and so forth. In addition there are management, maintenance, and numerous other functions unseen by the user that combine to facilitate high quality mobile communications. Some of these network issues are considered in the context of the Global System of Mobile (GSM) communications in Chapter 8 and in Wireless Asynchronous Transfer Mode (WATM) systems in Chapter 11, but this book principally addresses the so-called physical layer aspects of mobile communications.

Chapter 1 is a bottom-up approach to cellular radio. Commencing with the propagation environment of a single mobile communicating with a base station, Chapter 1 progresses via multiple access methods, first generation and second generation mobile systems, to cordless telecommunications and concludes with a discussion on the teletraffic aspects of mobile radio systems. The chapter is designed to equip the reader with a range of concepts that will prepare her or him for the more focused in-depth chapters which follow.

Chapter 2 considers mobile radio propagation in a quantitative manner, establishing the background material that is the backbone of mobile radio communications. A prerequisite to digital telephony is the selection of an appropriate speech encoder, converting the analogue speech signal into a digital format. Chapter 3 provides an in-depth discourse on analysis-by-synthesis codecs.

Having encoded the speech signal, forward error correction coding is applied together with interleaving of the coded speech bits, in order to combat the channel error bursts that occur due to the fading inflicted by the mobile radio channel. Chapter 4 addresses these issues. The interleaved data are transmitted via a suitable modulator over a mobile radio channel to a distant receiver which recovers the data. There are many different methods of modulation but we opted for describing those, which are particularly appropriate for mobile communications. In Chapter 5 we consider quaternary frequency shift keying (QFSK), which was a contending modem for the pan-European cellular network. Chapter 6 deals with a more complex family of modulation schemes, which are known as generalised phase modulation arrangements. In this chapter we consider Viterbi equalisation of wideband dispersive mobile radio channels.

Frequency hopping is an important technique in mobile radio communications, whereby a user's channel hops from one frequency carrier to another in order to avoid being in a deep fade for long periods of time. Chapter 7 is devoted to slow frequency hopping cellular systems, and an estimation of their spectral efficiency is presented. This is followed by a description of the pan-European mobile radio system in Chapter 8, which is now known as the Global System of Mobile communications, or GSM. This chapter guides the reader through the complexities of this mobile radio network, providing an overall system study and amalgamating the system components introduced in the preceding chapters.

Since the standardisation of the second generation systems, such as GSM, a decade has elapsed and the wireless community has been working towards the third generation of mobile systems. There have also been important evolutionary developments on the 2G scene, such as the definition of the half-rate Japanese Personal Digital Cellular (PDC) system's speech codec and that of the GSM half-rate speech-coding standard, the introduction of a new breed of enhanced full-rate speech codecs and the spread of advanced data, fax and email services. Further important developments have taken place in the area of high-speed wireless local area networks. Motivated by these trends and a range of other new developments in the field, **this second edition incorporates three new chapters**.

Chapter 9 presents a range of multimedia system components, which have the potential to provide attractive enhanced services in the context of both the existing 2G and the forthcoming 3G systems. Specifically, various video codecs and handwriting codecs are described, in order to support wireless video telephony and electronic 'white-board' services. Chapter 9 also provides an overview of the recent activities in the field of multi-level modulation schemes, which can be advantageously invoked in so-called intelligent multi-mode transceivers that are capable of re-configuring themselves on a burst-by-burst basis, supporting more robust transmissions in hostile propagation environments while transmitting an increased number of bits per symbol in benign propagation scenarios.

Chapter 10 provides an overview of the recently proposed 3G wideband Code Division Multiple Access (W-CDMA) standards. The systems considered are the so-called 'Intelligent Mobile Telecommunications in the year 2000' (IMT-2000), the 'Universal Mobile Telecommunications System' (UMTS) scheme and the pan-American cdma2000 arrangement. Despite the call for a common global standard, there are some differences in the proposed technologies, notably the chip rates and inter-cell operation. These differences are partly due to the 2G infrastructure already in use all over the world, specifically the GSM and the IS-95 systems; an issue elaborated in Chapter 10.

Our final chapter is rather different from the others in that it is concerned with network issues related to wireless asynchronous transfer mode (WATM) networks. With the aid of a WATM simulator numerous scenarios for the transport of multimedia traffic over cellular networks are addressed. The results verify the effectiveness of the WATM concept, successfully mixing real-time, non-real-time, constant bit rate, and variable bit rate services. A number of network control enhancements have been suggested. The simulations confirm that the medium access control protocols, data link control protocols, and network management schemes must be dynamic and intelligent, and should take into account the instantaneous traffic loading on each BS and in the surrounding network. Intelligent handover and call admission schemes can provide vast improvements in the Quality of Service (QoS). The rapid re-assignment of capacity over a wide area would be beneficial. It must be emphasised that, given current bandwidth availabilities, satisfying the QoS expected in the fixed ATM network is economically impractical in wireless networks. Therefore, acceptable mobile service grades should be defined, or the available radio spectrum increased.

To our original text dealing with many of the fundamentals of the physical aspects of mobile communications, we have added new chapters dealing with the exciting subjects of multimedia mobile communications, the proposed 3G CDMA systems, and WATM. It is our hope that you will find this second edition comprehensive, technically challenging, valuable and above all, enjoyable.

Raymond Steele
Lajos Hanzo

Acknowledgements

The book has been written by the staff in the Electronics and Computer Science Department at the University of Southampton and at Multiple Access Communications Ltd. The names of the authors of each chapter are presented at the beginning of their chapters. All of the contributors are indebted to our many colleagues who have enhanced our understanding of the subject. These colleagues and valued friends, too numerous all to be mentioned, have influenced our views concerning various aspects of wireless multimedia communications and we thank them for the enlightment gained from our collaborations on various projects, papers and books. We are grateful to J. Brecht, Jon Blogh, Marco Breiling, M. del Buono, Clare Brooks, Peter Cherriman, Stanley Chia, Byoung Jo Choi, Joseph Cheung, Peter Fortune, Lim Dongmin, D. Didascalou, S. Ernst, Eddie Green, David Greenwood, Hee Thong How, Thomas Keller, W.H. Lam, C.C. Lee, M.A. Nofal, Xiao Lin, Chee Siong Lee, Tong-Hooi Liew, Matthias Muenster, V. Roger-Marchart, Redwan Salami, David Stewart, Juergen Streit, Jeff Torrance, Spiros Vlahoyiannatos, William Webb, John Williams, Jason Woodard, Choong Hin Wong, Henry Wong, James Wong, Lie-Liang Yang, Bee-Leong Yeap, Mong-Suan Yee, Kai Yen, Andy Yuen and many others with whom we enjoyed an association.

We also acknowledge our valuable associations with Roke Manor Research, BT Laboratories, the Department of Trade and Industry and the Radiocommunications Agency. Our sincere thanks are also due to the EP-SRC, UK; the Commission of the European Communities, Brussels; and Motorola ECID, Swindon, UK for sponsoring some of our recent research.

Those authors who did not typeset their final manuscripts in Latex thank Jenny Clark, Debbie Sheridan and Denise Harvey for the work they did on their behalf in preparing the camera-ready copy. We are grateful to Phil Evans for the production of many of the drawings, and we feel particularly indebted to Peter Cherriman and Rita Hanzo for their skilful assistance with the final typesetting in Latex. Similarly, our sincere thanks are due to Juliet Booker, Mark Hammond and a number of other staff from John Wiley & Sons Ltd for their kind assistance throughout the preparation of this second edition.

Raymond Steele
Lajos Hanzo

Contributors

Editors:

R. Steele, BSc, PhD, DSc, FEng, FIEE, FIEEE,
Professor of Telecommunications, University of Southampton and
Chairman of Multiple Access Communications Ltd, UK

L. Hanzo, Dipl-Ing., PhD, MIEE, SMIEEE,
Professor of Telecommunications, University of Southampton and
Consultant to Multiple Access Communications Ltd, UK

Co-authors

Chapter 1: R. Steele
Chapter 2: D. Greenwood, L. Hanzo
Chapter 3: R.A. Salami, L. Hanzo, F.C.A. Brooks, R. Steele
Chapter 4: K.H.H. Wong, L. Hanzo
Chapter 5: I.J. Wassell, R. Steele
Chapter 6: I.J. Wassell, R. Steele
Chapter 7: Y.F. Ko, D.G. Appleby
Chapter 8: L. Hanzo, J. Stefanov
Chapter 9: L. Hanzo
Chapter 10: K. Yen, L. Hanzo
Chapter 11: P. Pattullo, R. Steele

Chapter **1**

Introduction to Digital Cellular Radio

R.Steele[1]

1.1 The Background to Digital Cellular Mobile Radio

Following the pioneering work of Hertz, the experiments of Marconi at the end of the 19th century demonstrated the feasibility that radio communications could take place between transceivers that were mobile and far apart. Henceforth telegraphic and voice communications were not inherently limited to users' equipment tethered by wires. Instead, a freedom to roam and yet still communicate was possible. However, only relatively few individuals were to enjoy this type of communication during the next several decades.

Morse-coded on-off keying was mainly used for mobile radio communications until the 1920s. It was not until 1928 that the first land mobile radio system for broadcasting messages to police vehicles was deployed. In 1933 a two-way mobile radio voice system was introduced by the Bayonne New Jersey Police Department. The early mobile radio transceivers were by today's standards very simple, noisy, bulky and heavy. They used power-hungry valves, operated in the lower frequency part of the VHF band, and had a range of some ten miles. Reference [1] has interesting photographs of mobile radio equipment in 1936.

The military embraced mobile communications in order to effectively

---

[1]University of Southampton and Multiple Access Communications Ltd

1

deploy their forces in battle. Vital mobile services, such as those of the police, ambulance, fire, marine, aviation and so forth, also introduced mobile communications to facilitate their operations. Early mobile radio communications were of poor quality. This was due to the radio propagation characteristics that resulted in the received signal being composed of many versions of the transmitted signal, each with different time delays, amplitudes and phases. The vector sum of these signals gave a received envelope with temporal and spatial variations. As the mobile station travelled, the received signal level often experienced large and rapid variations that caused considerable speech degradation. Of course these propagation characteristics exist today, but they had to be combatted then by a technology that was in a primitive evolutionary state. Whereas today's semiconductor technology can use millions of transistors to compensate for the propagation effects, early transceivers usually had less than ten valves.

The bandwidth that could be utilised by existing technology has always been a scarce commodity in radio communications. The long and medium wave bands were used for broadcasters, while the low frequency (LF) and high frequency (HF) bands were soon occupied with world wide communication services. The technology was inappropriate for good quality mobile communications in the VHF and UHF bands. The concept of frequency reuse was appreciated, but not its application to high user density mobile communications. Thus, for many years the quality of mobile communications was significantly worse than for wire communications as the technology was inadequate, and the operators were unable to utilise bandwidth in the higher frequency bands.

While fixed commercial analogue telephone networks were evolving into digital networks (thanks to the invention of low power, miniscule size, microelectronic devices) the mobile radio scene was also slowly altering. Private land mobile (PLM) radio systems came into use, catering for special groups, rather than individual members of the general public. Although Bell Laboratories conceived cellular radio in 1947, their parent company did not start to deploy a cellular network until 1979. The long gestation period was due to the waiting for necessary developments in technology. It was not until the arrival of custom designed integrated circuits, microprocessors, frequency synthesisers, high capacity fast switches, and so forth, that a cellular radio network could be realised. The 1980s witnessed the introduction of a number of analogue cellular radio systems, often known as public land mobile radio (PLMR) networks. Operating in the UHF band they represented a step-change in the complexity of civil communication systems. They enabled mobile users to have telephone conversations while on the move with any other users who were connected to the public switched telephone networks (PSTNs) or the integrated service digital networks (ISDNs). In the 1990s we witnessed another leap forward in mobile communications with the deployment of digital cellular networks and digital cordless telecommunication systems. These second-generation mobile radio systems provided a range of services in addition to telephony, such as data and short-message, as well as email services. Early in the new milleneum the third-generation networks will be launched, which are based on the so-called code division multiple access (CDMA) principle, delivering a whole host of further attractive services, including high-speed data transmission at rates up to 2 Mbit/s and also videotelephony.

This book is concerned with the principles and techniques of digital mobile radio transmissions, and only marginally with mobile radio network issues. This opening chapter gives an overview of digital mobile radio in a qualitative way, leaving subsequent chapters to treat the subject in greater depth. Rather than a top-down approach, we tackle the subject from the bottom upwards. This means we commence our discourse by considering the propagation phenomena that are responsible for much of the complexity that occurs in mobile radio communication systems. Armed with some understanding of this topic we describe in Section 1.3 how multiple users may be accommodated. Once we have introduced the principles of multiple access, we discuss in Section 1.4 the first-generation mobile radio systems which employ analogue modulation. The principles of digital cellular mobile radio transmissions are introduced in Section 1.5, paving the way for an overview of the second-generation cellular mobile radio systems in Section 1.6. Section 1.7 considers cordless communications, while Section 1.8 provides useful teletraffic equations.

1.2   Mobile Radio Propagation

Mobile radio communications in cellular radio take place between a fixed base station (BS) and a number of roaming mobile stations (MSs) [2-4]. The geographical area in which these communications occur is called a cell, and we may consider that the cell boundary marks the maximum distance that a MS can roam from the BS before the quality of communications becomes unacceptably poor. The cells in mobile radio communications vary substantially in size and shape. Traditionally their size is large, up to 30 km radius, when there is rarely a line-of-slight (LOS) between the BS and its MSs. More recently small cells of some 1 km radius have been used where LOS is more probable. Cells, known as micro-cells, have been proposed [5-7] whose size may be only 100 m along the side of a city block. In microe-ells LOS is often a feature. The presence of LOS has a profound effect on radio propagation, and this means that the characteristics of radio propagation are highly dependent on the cell size and shape.

In Section 1.4 we will describe how the cells are organised in clusters that use the entire spectrum assigned by the regulatory bodies, how these clusters are replicated and tessellated using the same radio frequency band to give coverage over an entire country, and how the base stations in the cells communicate with their mobiles and with other users via the PSTN [8].

However, for the present we will describe issues that relate to radio propagation of signals between a base station (BS) and a mobile station (MS).

Mobile radio propagation [2–4, 9] is considered indepth in Chapter 2. However, it is essential to introduce here some rudimentary notions on propagation in order to proceed to the wider system issues. As the distance between a BS and a MS increases the received mean signal level tends to decrease. The way this occurs is examined in Section 1.2.5. Over relatively short distances the received mean signal is essentially constant, but the received signal level can vary rapidly by amounts typically up to 40 dB. These rapid variations are known as fast fading and we deal with this phenomenon in this section and in Sections 1.2.2 to 1.2.4.

Let us consider a BS transmitting an unmodulated carrier which pervades the coverage area in which a MS is travelling. The MS does not receive one version of the transmitted carrier, but a number which have been reflected and diffracted by buildings and other urban paraphernalia. Indeed in most environments, each version of the transmitted signal received by the MS is subjected to a specific time delay, amplitude, phase and Doppler shift depending on its path from the BS to the MS. As a consequence the constant amplitude carrier signal transmitted may be substantially different from the signal the MS receives. When the signals from the various paths sum constructively at the MS antenna, the received signal level is enhanced. A serious condition occurs when the multipath signals, i.e., the transmitted signal arriving via many paths, vectorially sum to a small value. When this occurs the received signal is said to be in a fade, and the phenomenon is called multipath fading. As the MS travels it passes through an electromagnetic field that results in the received signal level experiencing fades approximately every half wavelength along its route. When a very deep fade occurs the received signal is essentially zero, and the receiver output is dependent on the channel noise, i.e., the channel signal-to-noise ratio (SNR) can be negative.

The above discussion relates to the transmission of an unmodulated carrier, an event that does not occur in practice. We are concerned here with digital mobile communications, where the propagation phenomena are highly dependent on the ratio of the symbol duration to the delay spread of the time variant radio channel. The delay spread may be considered as the length of the received pulse when an impulse is transmitted. We can see that if we transmit data at a slow rate the data can easily be resolved at the receiver. This is because the extension of a data pulse due to the multipaths is completed before the next impulse is transmitted. However, if we increase the transmitted data rate a point will be reached where each data symbol significantly spreads into adjacent symbols, a phenomenon known as intersymbol interference (ISI). Without the use of channel equalisers to remove the ISI the bit error rate (BER) may become unacceptably high.

Suppose that we continue to transmit at the high data rate which caused ISI, but move the MS closer to the BS while decreasing the radiated power to allow for the smaller BS to MS separation distance. If this distance is sufficiently small the delay spread will have decreased as delays of the multipaths components are, in general, smaller. The ISI will cease to be significant, removing the need for channel equalisation. The communications are still subjected to fading, but these fades may be very deep. The fading is said to be flat as it occurs uniformly across the frequency band of the channel. This is not so when ISI occurs as frequency selective fading is exhibited, i.e., some frequencies fade relative to others over the channel bandwidth.

The consequence is that cellular radio networks using large cells, where the excess delay spread may exceed 10 μs, need equalisers when the bit rate is relatively low, say 64kb/s, while cordless communications in buildings where the excess delay spread is often significantly below a microsecond may exhibit flat fading when the bit rate exceeds a megabit/sec (Mb/s). Small cells are not just smaller, they have different propagation features [7]. Very small cells, sometimes referred to as picocells may support many Mb/s without equalisation because the delay spread is only tens of nanoseconds.

1.2.1   Gaussian Channel

The simplest type of channel is the Gaussian channel. It is often referred to as the additive white Gaussian noise (AWGN) channel. Basically it is the noise generated in the receiver when the transmission path is ideal. The noise is assumed to have a constant power spectral density over the channel bandwidth, and a Gaussian amplitude probability density function (PDF). This type of channel might be considered to be unrealisable in digital mobile radio, but this is not so. In micro-cells it is possible to have a LOS with essentially no multipath, giving a Gaussian channel. Even when there is multipath fading, but the mobile is stationary and there are no other moving objects, such as vehicles, in its vicinity, the mobile channel may be thought of as Gaussian with the effects of fading represented by a local path loss.

The Gaussian channel is also important for providing an upper bound on system performance. For a given modulation scheme we may calculate, or measure in a laboratory, the BER performance in the presence of a Gaussian channel. When multipath fading occurs the BER will increase for a given channel SNR. By using techniques to combat multipath fading, such as diversity, equalisation, channel coding, data interleaving, and so forth, techniques to be described throughout the book, we can observe how close the BER approaches that for the Gaussian channel.

1.2.2   Rayleigh Fading Channel

If each multipath component in the received signal is independent then the PDF of its envelope is Rayleigh. A typical received signal's fading envelope

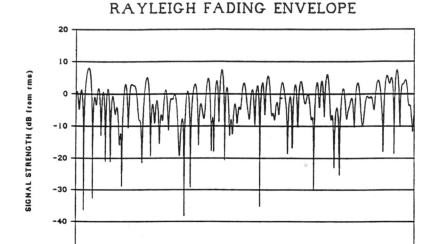

Figure 1.1: Typical profile of the received signal's Rayleigh fading envelope and phase at a vehicular speed of 30 mph and carrier frequency of 900 MHz.

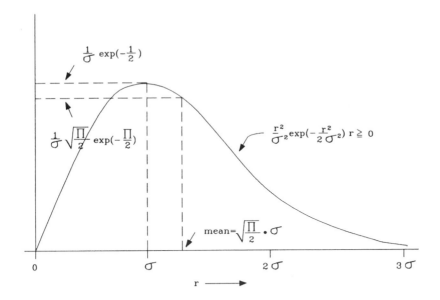

$$\frac{1}{\sigma} \exp(-\frac{1}{2})$$

$$\frac{1}{\sigma}\sqrt{\frac{\Pi}{2}} \exp(-\frac{\Pi}{2})$$

$$\frac{r^2}{\sigma^2}\exp(-\frac{r^2}{2\sigma^2}) \; r \geq 0$$

$$mean=\sqrt{\frac{\Pi}{2}} \cdot \sigma$$

0   σ 2σ 3σ

$r \longrightarrow$

Figure 1.2: Rayleigh PDF.

and phase as a function of time is shown in Figure 1.1. The Rayleigh PDF of the envelope is presented in Figure 1.2. The probability of experiencing a deep fade below, say, 3σ, is the area in the tail of the PDF between 3σ and infinity, where σ is the rms value of the received signal envelope $r(t)$. Notice that the Rayleigh PDF applies to positive values, namely the magnitude of the received signal envelope, and that σ and the mean of the distribution are similar.

The impulse response of the flat Rayleigh fading mobile radio channel consists of a single delta function whose weight has a Rayleigh PDF. This occurs because all the multipath components manifest themselves in a bunch with negligible delay spread between them, and when modelled as a single delta function they combine to have a Rayleigh PDF. Thus as the MS travels the received signal fades in a manner similar to that shown in Figure 1.1, while the weight of the delta function in the impulse response also changes according to a Rayleigh PDF. When the MS experiences a deep fade the weight of the delta function is small, and vice versa when the received signal is enhanced.

Notice that the Gaussian channel may be represented by an impulse response having a constant weight delta function, i.e., an ideal channel, to which is added an AWGN source.

Representation of mobile radio channels is required for both mathematical analysis and computer simulation of mobile radio systems. A Rayleigh fading profile channel can be modelled using the arrangement shown in

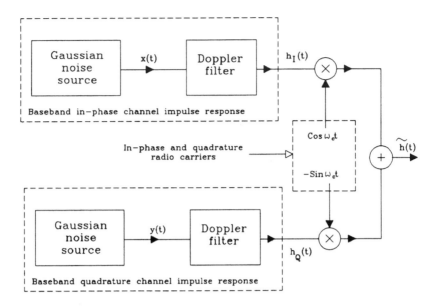

Figure 1.3: Model to generate a Rayleigh fading profile.

Figure 1.3. Observe that there are two quadrature channels in the model. The outputs from the Gaussian noise sources are applied to filters that represent the effects of Doppler frequency shifts. So before continuing with the description of the model we must say a few words regarding the Doppler changes to the transmitted signal as perceived by the MS. As throughout this chapter, we confine ourselves to basic concepts. Let us again consider the transmission of an unmodulated carrier from a BS. A MS travelling in a direction making an angle α_i with respect to the signal received on the i-th path has its carrier frequency f_c modified to $f_c + f_m \cos \alpha_i$, where $f_m = v/\lambda$ $= v f_c / c$, and v is the speed of the MS, $\lambda = c/f_c$ is the wavelength of the carrier, and c is the velocity of light. Notice that a Doppler frequency can be positive or negative depending on α_i, and that the maximum and minimum Doppler frequencies are $\pm f_m$. These extreme frequencies correspond to the $\alpha_i = 0°$ and $180°$, when the ray is aligned with the street that the MS is travelling along, and corresponds to the ray coming towards or from behind the MS, respectively. It is analogous to the change in the frequency of a whistle from a train perceived by a person standing on a railway line when the train is bearing down or receding from the person, respectively.

Assuming that α_i is uniformly distributed, the Doppler frequency has a random cosine distribution. The Doppler power spectral density $S(f)$ can be computed by equating the incident received power in an angle $d\alpha$ with the Doppler power $S(f)df$, where df is found by differentiating the Doppler frequency term $f_m \cos \alpha$ with respect to α. The incident received power at

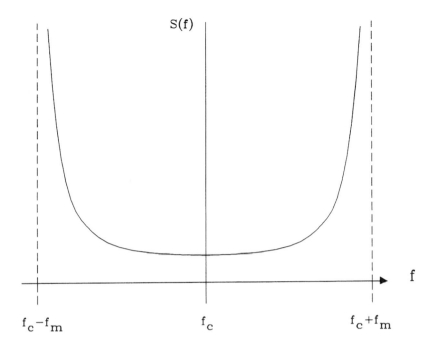

Figure 1.4: Sketch of the Doppler spectrum of an unmodulated carrier.

the MS depends on the power gain of the antenna and the polarisation used. Thus the transmission of an unmodulated carrier is received as a multipath signal whose spectrum is not a single carrier frequency f_c, but contains frequencies up to $f_c \pm f_m$. A typical spectrum is sketched in Figure 1.4. In general we can express the received RF spectrum $S(f)$ for a particular MS speed, antenna and polarisation as,

$$S(f) = \frac{A}{\sqrt{1 - (f/f_m)^2}} \tag{1.1}$$

where A is a constant. Observe that f_m depends on the product of the speed of the MS and the propagation frequency.

Let us now return to Figure 1.3. If the Doppler filters are absent such that $h_I(t) = x(t)$ and $h_Q(t) = y(t)$, then the output is

$$\tilde{h}(t) = x(t) \cos \omega_c t - y(t) \sin \omega_c t \tag{1.2}$$

where $x(t)$ and $y(t)$ are independent Gaussian random variables. This equation is simply that of band-limited white noise where

$$\tilde{h}(t) = R(t) \cos (\omega_c t + \psi(t)) \tag{1.3}$$

and we infer in Chapter 2 that

$$R(t) = \left(x^2(t) + y^2(t)\right)^{\frac{1}{2}} \tag{1.4}$$

is Rayleigh distributed, and

$$\psi(t) = \tan^{-1}\left(\frac{y(t)}{x(t)}\right) \tag{1.5}$$

has a uniform PDF. We observe that we have obtained a signal $\tilde{h}(t)$ having a Rayleigh envelope, and with a white power spectral density (PSD) as both $x(t)$ and $y(t)$ have flat PSDs. By introducing the Doppler filters in Figure 1.3 we do not change the Rayleigh envelope statistics of $\tilde{h}(t)$, but we do introduce the necessary correlation between frequency components in $\tilde{h}(t)$.

The Rayleigh fading signal profile $\tilde{h}(t)$ is composed of two quadrature channels, a feature we will utilise in connection with the transmission of continuous phase modulated (CPM) signals addressed in Chapter 6. We also notice that if the quadrature carriers are removed we are left with a quadrature baseband representation of the Rayleigh fading channel, a representation that is essential when doing computer simulations because to simulate the many radio frequency (RF) cycles per data symbol is impracticable. The signal $\tilde{h}(t)$ may be represented as

$$\tilde{h}(t) = Re\left[h(t)e^{j\omega_c t}\right] \tag{1.6}$$

where $h(t)$ is the complex baseband representation of $\tilde{h}(t)$,

$$h(t) = h_I(t) + jh_Q(t) \tag{1.7}$$

with $h_I(t)$ and $h_Q(t)$ marked on Figure 1.3, ω_c is the angular carrier frequency and $Re[(\cdot)]$ is the real part of $(\cdot)$. If the output from the transmitter is $\tilde{s}(t)$, the received RF signal is

$$\tilde{r}(t) = \tilde{h}(t) * \tilde{s}(t) + \tilde{n}(t) \tag{1.8}$$

where $\tilde{n}(t)$ is the additive receiver noise at RF, and $*$ means convolution. Observe that $\sim$ above a symbol indicates that it is an RF and not a baseband signal.

1.2.3   Rician Channel

In microcellular mobile radio a dominant path, which may be a line-of-sight (LOS) path, often occurs at the receiver, in addition to the many scattered paths. This dominant path may significantly decrease the depth of fading. The PDF of the received envelope is said to be Rician. We introduce a

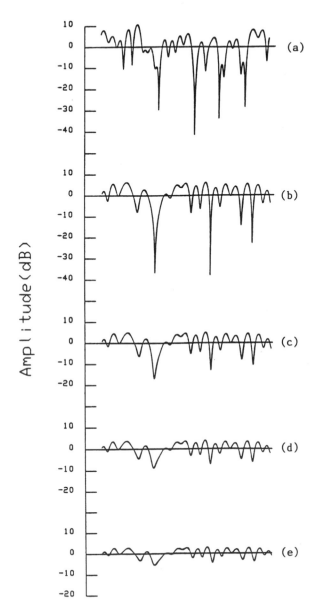

Figure 1.5: Rician fading profiles for a MS travelling at 30 mph. Sub-figures (a), (b), (c), (d) and (e) refer to a Rician K value of 0, 4, 8, 16 and 32, respectively.

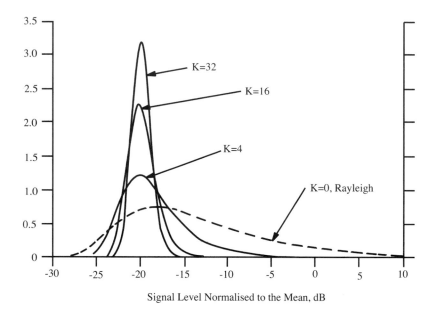

Figure 1.6: Rician PDFs normalised to their local means.

Rician parameter,

$$K = \frac{\text{power in the dominant path}}{\text{power in the scattered paths}} \tag{1.9}$$

and emphasise that sometimes this parameter is defined as the ratio of the power in the scattered path to the power in the dominant path. Notice that when K is zero the channel is Rayleigh, whereas if K is infinite the channel is Gaussian. There are mobile radio channels that do not conform to either Gaussian, Rayleigh or Rician fading statistics. However, usually one does apply, and the Rician channel may be considered as the general case.

Figure 1.5 shows a set of envelope fading profiles for different values of K recorded over an arbitrary interval. The fades have a high probability of being very deep when $K=0$ (Rayleigh fading) to being very shallow when $K = 32$ (approaching Gaussian). When the received signal is in a deep fade below the average level of channel noise an error burst occurs. However, the same average noise level will not cause as many error bursts when K is higher.

The Rician PDFs for different values of K normalised with respect to their local mean values are displayed in Figure 1.6. It is evident that the Rayleigh PDF has the highest probability of being in a deep fade below the mean, while the Gaussian PDF has the lowest. This is also apparent

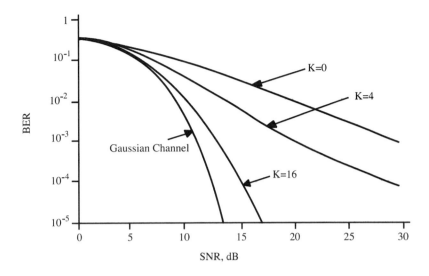

Figure 1.7: BER versus channel SNR for different Rician K values using non-coherent FSK.

in Figure 1.5. As an example of how Rician fading provides a superior performance to Rayleigh fading we show in Figure 1.7 the curves of BER as a function of signal-to-noise ratio (SNR) for different values of K [10]. The modulation, is non-coherent frequency shift keying (FSK). For a BER of 10^{-3} an SNR of 30 dB is required for a Rayleigh fading channel, because even for this low channel noise power there are occasionally deep fades that go below the noise floor of the receiver inducing an error burst. The Rayleigh fading channel is a feature of large cells. Macrocells of some 2 km diameter can exhibit both Rayleigh fading and Rician fading, but with usually low values of K. When microcells are used K can vary widely, but is often above five and values of 30 are not uncommon, see Section 1.2.6. When K is sufficiently high that it approaches a Gaussian channel an SNR of only 11 dB is required to achieve a BER of 10^{-3}.

Another by-product of the Rician channel is the improvement in cochannel interference performance. For the same FSK modulation, the BER performance in the presence of a single cochannel interferer is displayed in Figure 1.8 for different values of K. Again we observe that the higher the value of K the lower the signal-to-interference ratio (SIR) required to achieve a particular BER. This improvement in SIR performance due to the presence of a Rician mobile radio channel having a high K factor means that handover of communications to another BS due to an unacceptably high BER is required to be rapid as the MS needs only lower its SIR level by a few dBs for the BER to rapidly increase. This increasing BER with

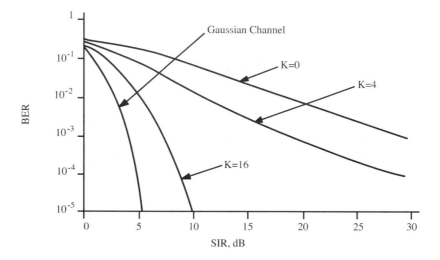

Figure 1.8: BER versus SIR for different Rician K values using non-coherent FSK.

decreasing SIR is much slower in Rayleigh fading channels.

The Rician channel representation is shown in Figure 1.9. The diagram is essentially that shown in Figure 1.3 for the Rayleigh channel, but with the dominant radio path represented by

$$\tilde{r}_D(t) = (I_D^2 + Q_D^2)^{1/2} \cos\left\{(\omega_C + \omega_D)t + \phi_D\right\} \qquad (1.10)$$

where I_D and Q_D are amplitudes of the quadrature components, ϕ_D is arctan (Q_D/I_D) and $\omega_C + \omega_D$ is the angular frequency of the dominant path. If this path intersects the MS at an angle of α_D when the MS is travelling at velocity v, the input RF spectrum shown in Figure 1.4 has an additional delta function at the frequency $f_c + (v/\lambda)\cos\alpha_D$, i.e., $\omega_D = (2\pi v/\lambda)\cos\alpha_D$. By calculating the ratio of the mean square value of $\tilde{r}_D(t)$ to the mean square value of $\tilde{h}(t)$ for the Rayleigh fading model shown in Figure 1.3, we obtain the value of K.

1.2.4   Wideband Channels

We have argued that the effect of multipath propagation is to spread the received symbols. In wideband channels the symbol rate is sufficiently high that each symbol is spread over adjacent symbols causing intersymbol interference (ISI). In order for the receiver to remove the ISI and regenerate the symbols correctly it must determine the impulse response of the mobile radio channel. This response must be frequently remeasured as the mobile channel may change rapidly both in time and space. Channel sounding

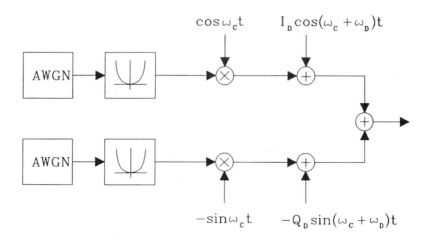

Figure 1.9: Model to generate a Rician fading channel.

is the means by which the impulse response of the time variant mobile radio channel is estimated. We will show in Chapter 6 that an estimate of the channel impulse response at the receiver is vital if equalisation of the channel is to be accomplished in order to operate the link at an acceptable BER.

Viewing the mobile radio channel as a time varying linear network we can measure its impulse response by a number of techniques. Suffice to say here is that if a pseudo random binary sequence (PRBS) modulates an RF carrier and is transmitted over an ideal channel to a receiver having a quadrature demodulator, then the use of correlators on each quadrature output will yield a narrow pulse whose width is approximately two bits wide. When the same PRBS is transmitted over a mobile radio channel the combined correlator outputs yield the impulse response of the channel, or more accurately, the response to the narrow pulse obtained for the ideal channel. Notice that as a quadrature demodulator is used the impulse response is determined in both magnitude and phase.

The magnitude of a typical impulse response is shown in Figure 1.10(a). If we partition the time delay axis as shown in Figure 1.10(b) into equal delay segments, usually called delay bins, then there will be, in general, a number of received signals in each bin corresponding to the different paths whose times of arrival are within the bin duration. These signals when vectorially combined can be represented by a delta function occurring in the centre of the bin having a weight that is Rayleigh distributed, see Figure 1.10(c). As the smaller impulses are of less significance we may introduce a threshold T and discard all components whose weight is below T. This leads to the simplified discrete impulse response shown in Fig-

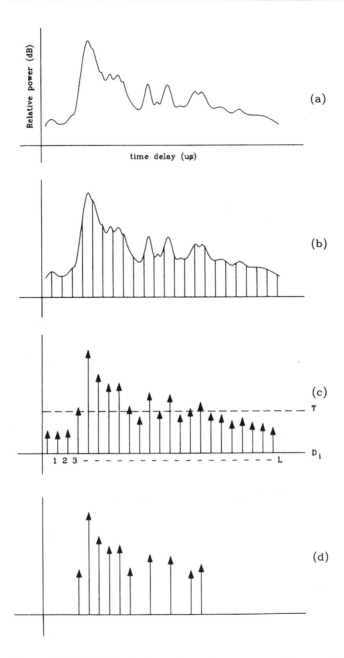

Figure 1.10: Wideband channel impulse response: (a) actual, (b) partitioned into equal delay segments, (c) discretised and (d) simplified discretised response.

ure 1.10(d). The discrete impulse in Figure 1.10(c) can be represented at
RF by

$$\tilde{h}(t) = \left\{ \sum_{i=1}^{L} \beta_i e^{j\phi_i} \delta(t - \tau_i) \right\} e^{j\omega_c t} \tag{1.11}$$

where β_i and ϕ_i are the weight and the phase of the component in the i-th
bin, respectively, which occurs at time $t = \tau_i$, and L is the last component
in the response. The weight of the delta function $\delta(t - \tau_i)$ is β_i, and the
complex baseband representation of $\tilde{h}(t)$ is

$$
\begin{aligned}
h(t) &= \sum_{i=1}^{L} \beta_i e^{j\phi_i} \delta(t - \tau_i) \\
&= \sum_{i=1}^{L} \beta_i \cos \phi_i \ \delta(t - \tau_i) + j \sum_{i=1}^{L} \beta_i \sin \phi_i \delta(t - \tau_i) \\
&= h_I(t) + j h_Q(t).
\end{aligned}
\tag{1.12}
$$

To model the discretised baseband impulse response $h(t)$ of the wideband
channel shown in Figure 1.10(c) we need to formulate L narrow-band baseband Rayleigh fading channels. Figure 1.3 represents a narrow-band bandpass Rayleigh fading model. By removing the quadrature carriers and the
adder we obtain the baseband inphase (I) and quadrature (Q) channels,
each consisting of an AWGN source in cascade with a filter representing
the effects of Doppler shifts. For the wideband channel we need L of these
baseband narrow band channels for both the I and Q components.

Suppose a data signal $\alpha(t)$ is applied to a modulator to give a high
frequency signal $\tilde{s}(\alpha, t)$, where again the $\sim$ above the symbol indicates
that it is a bandpass radio signal. In Section 6.1.1 we describe a continuous
phase modulation signal

$$\tilde{s}(\alpha, t) = A \cos \left(2\pi f_o t + \phi(t, \alpha)\right) \tag{1.13}$$

where A is a constant, f_o is the carrier frequency, and $\phi(t, \alpha)$ is the phase
angle of the phasor that carries $\alpha(t)$. We consider this signal as an example,
to show that by writing it as

$$
\begin{aligned}
\tilde{s}(\alpha, t) &= Re\left[A \exp\left(\phi(t, \alpha)\right) . \exp(2\pi f_o t)\right] \\
&= Re\left[s(t, \alpha) \exp(2\pi f_o t)\right]
\end{aligned}
\tag{1.14}
$$

we can identify the baseband signal component of $\tilde{s}(\alpha, t)$ as

$$s(t, \alpha) = s_I(t, \alpha) + j s_Q(t, \alpha) \tag{1.15}$$

where

$$s_I(t, \alpha) = A \cos \phi(t, \alpha) \tag{1.16}$$

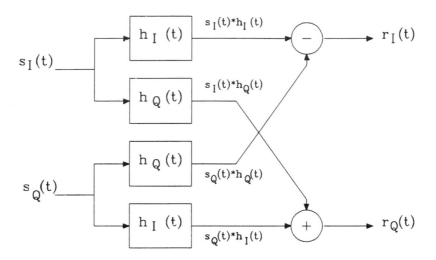

Figure 1.11: Quadrature representation of a baseband mobile radio channel, where $h_I(t)$ and $h_Q(t)$ are the in-phase and quadrature-phase channel impulse responses, respectively. (Additive noise is not shown.)

and

$$s_Q(t, \alpha) = A \sin \phi(t, \alpha). \tag{1.17}$$

When this signal $s(t, \alpha)$ is convolved with the channel impulse response $h(t, \alpha)$ we obtain the baseband received signal

$$r(t, \alpha) = s(t, \alpha) * h(t, \alpha) \tag{1.18}$$

where $*$ means convolution. Substituting in the complex values of $s(t, \alpha)$ and $h(t, \alpha)$ yields

$$
\begin{aligned}
r(t, \alpha) &= (s_I(t, \alpha) + j s_Q(t, \alpha)) * (h_I(t) + j h_Q(t)) \\
&= \{s_I(t, \alpha) * h_I(t) - s_Q(t, \alpha) * h_Q(t)\} \\
&\quad + j \{s_I(t, \alpha) * h_Q(t) + s_Q(t, \alpha) * h_I(t)\} \\
&= r_I(t, \alpha) + j r_Q(t, \alpha). \tag{1.19}
\end{aligned}
$$

This equation implies a block diagram of the baseband mobile radio channel of the form shown in Figure 1.11. To formulate each of the $r_I(t, \alpha)$ and $r_Q(t, \alpha)$ signals we use an L-stage shift register, whose delay D is equal to the duration between the delay bins. Consider the inphase modulated signal $s_I(t)$ applied to this register as shown in Figure 1.12. As $s_I(t)$ passes along the register, the outputs from each delay stage in the register are applied to two sets of multipliers. The signals $h_I(t)$ and $h_Q(t)$ are produced

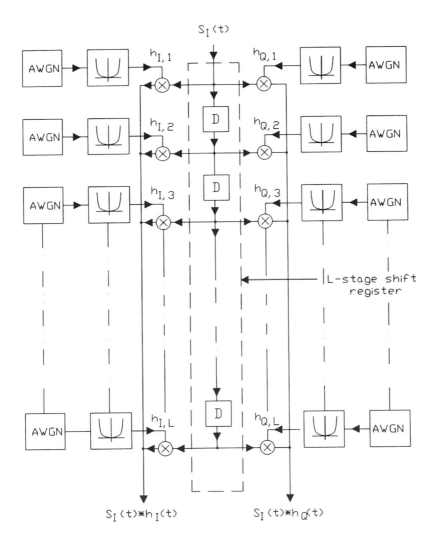

Figure 1.12: Generation of the baseband inphase channel output signals.

having components $h_{I,i}$ and $h_{Q,i}$; $i = 1, 2, ...L$, respectively, where each $h_{(),i}$ is a Doppler filtered AWGN signal, i.e., a signal having a Rayleigh envelope. Hence $h_{I,i}$, and $h_{Q,i}$ constitute the wideband channel impulse response where each component has a Rayleigh fading envelope. An identical arrangement to that in Figure 1.12 is used for $s_Q(t)$, and by formulating the convolutional terms in Equation 1.19 we obtain $r_I(t)$ and $r_Q(t)$ in Figure 1.11. Notice that all the Doppler filters in Figure 1.12 are the same, although they would all need to be modified for different MS speeds. The AWGN sources are statistically independent. This type of modelling is useful for computer simulation (with appropriate modifications).

In general the independent fading components in the wideband channel impulse response are not all Rayleigh distributed. However, Rayleigh is usually a reasonable assumption. When dealing with a specific type of environment, cell size, and so forth, it is better to use experimental results in modelling the channel. By averaging a set of experimental channel impulse responses an average channel impulse response is obtained and divided into delay bins. Again, each bin contains a coefficient ($h_{I,i}$ or $h_{Q,i}$) whose mean is adjusted to the value of the average impulse response in that bin.

1.2.4.1   GSM Wideband Channels

In formulating the specification for the pan-European digital mobile radio standard, the GSM Committee identified a number of channel impulse response models [11]. These models have impulse responses with either 6 or 12 components. In general the classical Doppler spectrum defined by Equation 1.1 is used, but they also employ the Rician Doppler spectrum as previously described. For rural areas (RA) the impulse response has 6 components, the first component h_1 has a Rician-type Doppler spectrum, while $h_2 \cdots h_6$ have the classical Rayleigh one. The other channel impulse responses specified by GSM have all their components subjected to the classical Doppler spectrum. We emphasise that measured Doppler spectra may be significantly different from the shape shown in Figure 1.4.

Chapter 8 is concerned exclusively with the GSM system, and Figure 8.25 displays the impulse responses specified for different environments. Notice that rural response has uniformly spaced components. The transmission bit rate of GSM is 271 kb/s, giving a bit duration of 3.7 μs which is long compared to the maximum delay of 0.5 μs. Consequently there is little dispersion with this channel. However, if the bit rate is, say, 2 Mb/s, then there would be significant ISI. The channel response for the hilly terrain (HT) is a group of components with a delay $< 0.5\mu s$, followed by other components at much greater delays that are the consequence of long paths resulting from reflection and diffraction from the hills. The typical urban channel (TU) impulse response spans 5 μs and, as for the HT response, equalisation is essential at transmitted rates of 271 kb/s. There is also a channel impulse response specified for testing equalisers having six compo-

input   output

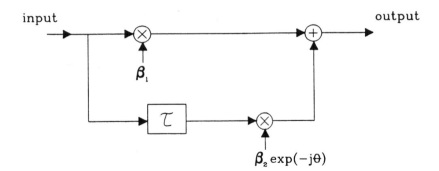

Figure 1.13: Dispersive two-ray channel model.

nents equally spaced by 3.2 μs and spanning 16 μs. This type of channel
response ensures considerable ISI and tests the equaliser's ability to remove
it.

1.2.4.2   The Two-ray Rayleigh Fading Channel

The block diagram of the two-ray channel model is shown in Figure 1.13.
The input signal is separated into two components or rays, one weighted by
β_1 and the other delayed by τ and weighted by $\beta_2\exp(-j\theta)$. Both β_1 and
β_2 are random variables from Rayleigh distributions, while θ is the relative
phase between the two rays and is uniformly distributed between 0 and 2π.
The outputs from the two rays are summed at the receiver input. In using
this model, τ is held constant and system performance parameters, such as
bit error rate (BER), determined. For a particular τ, the channel impulse
response is

$$h(t) = \beta_1\delta(t) + \beta_2\exp(-j\theta)\delta(t - \tau) \tag{1.20}$$

corresponding to a channel frequency response of

$$H(f) = \beta_1 + \beta_2\exp(-j\theta)\exp(-j2\pi f\tau). \tag{1.21}$$

We observe that $H(f)$ exhibits frequency selective fading due to the variations in β_1 and β_2. Further, the model is based on the assumption that
enough paths occur in the two bins to justify that only two rays need be
considered, and that both rays are subjected to independent Rayleigh fading.

   It is usual to determine the system parameters as a function of τ. For
example, if the receiver had an equaliser that operated by using a window
of W seconds of the estimated channel impulse response where the energy
was greatest, then when τ was zero the BER performance would be poor
as the channel would be a single path flat fading Rayleigh channel. As
no dispersion occurs there is nothing the equaliser can do to improve per-

CHAPTER 1. INTRODUCTION TO DIGITAL CELLULAR RADIO

formance. When τ is larger and both paths occur within W, the equaliser performs well, making use of the information in both paths. However, when τ is so large that only one ray resides within W, dispersion occurs, but the equaliser cannot exploit the diversity that inherently resides in the second path. Hence the graph of BER versus τ has a minimum when the equaliser can make best use of the two rays residing within W.

1.2.4.3   Real Channel Impulse Responses

The above wideband channel impulse responses are models. They relate to average conditions. What is often required are the sequences of actual channel impulse responses in circumstances that have a profound effect on system performance. One of these is when a MS enters a handover region. Handover is when the communications to the MS are switched from one BS to a more suitable one. Approaching a handover condition can be gradual when a MS is proceeding along a street or in a rural area. However, in a microcell the received signal level can rapidly decrease by some 20 dB due to diffraction losses as the MS turns a corner, coupled with a radically different impulse response. In these conditions the channel may rapidly deteriorate and fast hand-over is required. For those involved in studying handover procedures, it is not only generalised models of channels that are required in the simulation, but experimental soundings of worst case scenarios that occur near and at the cell boundaries. This means that soundings at a rapid rate are required at crucial locations.

1.2.5   Path Loss

So far we have discussed the fast fading phenomenon that occurs when a MS travels over distances where the mean signal level is approximately constant. In addition to these rapid fluctuations in the received signal level, less signal power is received as the BS to MS distance increases. Even in free space the power captured by an antenna decreases with increasing distance as there is less power on the surface of a larger than a smaller sphere whose radii are the transmitter to receiver separation. In cities there are losses in power due to reflection, diffraction around structures and refraction within them. We refer to this diminution of received power with distance as path loss(PL). There are many treatises dealing with the fundamentals of radio propagation and how to estimate PL [2,3,9]. They tend to start with PL for propagation in free space, and then in the presence of different types of surfaces, followed by diffraction and reflection around obstacles, and so forth. Techniques are available to replace the undulations in the terrain relative to a flat earth by the use of knife edges. For example [9,12], an equivalent knife edge can be positioned where the optical paths from each terminal to its horizon intersect. Allowing for many obstacles by knife edge diffraction techniques can become complex. Terrain-based techniques [13] determine the terrain profile between the BS and MS. A check is made to

see if there is a LOS path, and whether there is Fresnel-zone clearance over
the path implying 'free-space' propagation characteristics. If this is true the
free-space and plane-earth loss calculations are made and the higher one is
adopted. Should this situation not prevail multiple knife edge calculations
are performed (up to three edges) to find the PL.

The above approaches are used for BS antennas located at high elevations providing coverage over tens of kilometers. In urban environments
the BS antennas for large cells are located on tops of tall buildings. As the
MSs move along the streets, rarely within the line-of-sight (LOS) of their
BS antennas, the calculation of the PL for any particular MS is, in general,
analytically impossible. This means that network designers must site their
BSs in a sub-optimal way; disregarding the other legal and financial difficulties involved in acquiring cell site locations. They must site the BSs to
provide radio coverage over most of the geographical area they serve, which
results in some areas receiving considerable overlapping radio coverage by
a number of BSs. This increases the infrastructure costs.

However, the network designers do have a number of techniques for
wisely locating their BSs. These are usually company confidential, built
up with a mixture of published information, company experience and computer modelling. We will not stray into this specialised field. However,
the important contributions of Okumura et al [14] and Hata [15] should
be mentioned. Faced with intractable analytical calculations, an empirical
approach of estimating PL was used by Okumura *et al.* who carried out
a detailed propagation measurement programme in the Tokyo area. Their
results provide an estimate of the PL in dBs as the free-space path loss
in dBs for a BS to MS distance d, to which is added the urban loss in
dBs for a quasi-smooth terrain when the BS antenna is at $h_t = 200$ m
while the MS antenna height is $h_r = 3$ m. From this are subtracted correction factors in dBs to allow for the actual BS and MS antenna heights
in a particular situation. In addition, the PL calculations can be modified
for different terrains, types of urbanisation, and so forth. This excellent
work was presented in graphical form, and it was Hata [14] who turned
it into a set of readily usable equations. These equations are presented
in Chapter 2, and are defined in terms of the propagation frequency f in
MHz ($150 < $ f $ < 1500$), h_t in m($30 < h_t < 300$) and d in km ($1 < $ d $ < 20$).
We now know that the frequency range can be extended to at least 2 GHz
and that providing the BS antenna clears the local building line, h_t can be
reduced below 30 m, by adding a constant factor of some 6 dBs to PL. The
Hata equations are primarily concerned with estimating PL in large cells.
They yield straight-line PL(dB) versus log d curves.

When the received signal at the MS is averaged out to remove the effects
of fast fading the local mean signal level at a particular distance d is obtained. The statistics of the local means are log-normally distributed, and
seem to be dependent on the local environment. The standard deviations
are typically 6 to 9 dB, and may be as small as 3 dB. The signal received

at the MS is attenuated as the distance d increases due to increasing PL. At a particular d we may compute the PL using an empirical formula, such as Hata's, and then allow for the log-normally distributed mean, and the Rayleigh distributed fast fading. Identifying the receiver's noise floor level (NFL), we allow for there to be a low probability of the signal level rapidly fading below it, i.e., the NFL intersects the Rayleigh PDF such that there is only a small area in the tail of the PDF below NFL. Next we construct both the Rayleigh and log-normal PDFs. The mean of the log-normal PDF coincides with the path loss curve, while the mean of the Rayleigh PDF overlays the log-normal PDF at its two or three sigma point. Hata's straight-line curve enables the transmitted power to be estimated for the maximum cell distance d_{max}. Increasing the transmitter power moves the curve upwards pro rata and increases d_{max} accordingly.

1.2.6   Propagation in Microcells for Highways and City Streets

We will discuss microcellular mobile radio in later sections. Suffice to say here that microcells are small cells which may be a small segment of a highway, a street along the side of a city block, part of a park, an office floor and so forth. They are small areas where the teletraffic is high. Indeed, microcells are the most effective means of providing high user density mobile radio communications.

As we are still setting the propagation framework for system concepts, we consider in a qualitative way propagation in highway and street microcells [16]. Vehicular mobile stations (MSs) are applicable to both these types of microcells, but pedestrian MSs are expected to be more numerous in city street microcells than vehicular ones. To establish a microcell we must contain the radiation so that the frequency band allotted by the regulatory authority can be re-used far more frequently than in conventional large cells. Accordingly the BS antenna is not mounted at a high elevation to get wide area coverage. Instead it is some 5 to 12 m above the ground, and the radiated power is generally in the milliwatt range [7]. We will commence by considering the path loss characteristics of vehicular highway microcells.

1.2.6.1   Path Loss

Early conceptual and theoretical investigations into highway microcells were made by Steele and Prabhu [6]. Having no experimental data, and being concerned with interference from other highway microcells, they assumed that free space propagation applied. Chia *et al.* [17] later made measurements at 900 MHz along a number of highways in Southern England. An 18-element Yagi BS antenna having a gain of 15 dB and a front-to-back ratio of 25 dB was used. The power delivered to the antenna was 16 mW.

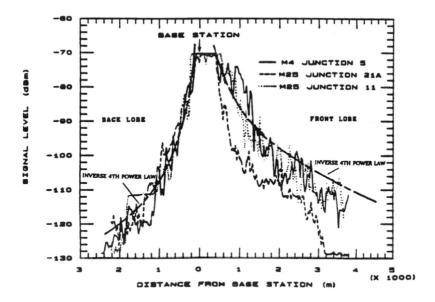

Figure 1.14: Received signal profiles at three different motorway locations (Chia *et al.* [17]).

During each experiment the antenna was mounted on a bridge that crossed above a motorway, with the antenna pointing along the motorway. As the MS drove along the motorway the received signal level was recorded.

Figure 1.14 shows a set of received signal level versus distance curves. When driving near the BS the MS receiver was saturated as shown by the flat top curves. Because the radiated power from the back of the antenna was reduced by 25 dB compared to that from the front, the highway did not lose line-of-sight (LOS) with the antenna in the backward direction before the received signal level decreased to the noise floor. It was found that an inverse fourth power propagation law was always a good fit to the data. The radiation in the forward direction went farther than in the backward direction. Again the inverse fourth power law was appropriate in the forward direction, although the variation in the received signal level was as much as ±10 dB. Notice that the curve for the M25 motorway at junction 21A initially decreased rapidly. This was due to the motorway turning sharply into a cutting, losing LOS and causing the signal to experience a diffraction loss. However, having sustained this loss the propagation conformed to the inverse fourth power law.

The countryside in Southern England is essentially flat and the inverse fourth power propagation law observed in highway microcells can be explained by a two-path model consisting of a LOS path and a path reflected from the road surface. This simple model for highway microcells has been

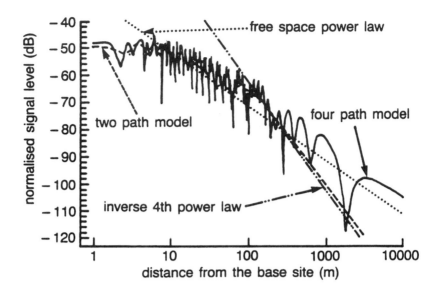

Figure 1.15: Signal level profiles for the two- and four-path propagation models. Also shown are the free space and inverse 4 th power laws (Green [10]).

investigated by both Rustako *et al.* [18] and Green [10]. The theoretical curves of normalised signal level versus distance of the MS from the BS are shown in Figure 1.15. The transmissions are at 900 MHz and the BS and MS antenna heights are 5 and 1.5 m, respectively. The received signal level for the two-path model is relatively constant for distances close to the BS. The signal level then decreases in a fluctuating manner until the MS is some 200 m from the BS when the curve becomes smoother and decreases according to an inverse fourth power law.

Vehicular MSs in cities travel significantly slower than on countryside highways. Indeed in some cities their average speed is comparable with 19 th century horse transportation speeds! The high density of vehicles on city streets means that vehicular microcells in cities should, in general, be shorter than highway microcells as the mobiles spend longer on a given length of road. A useful design criterion is to arrange for mobiles to spend similar times in microcells in order to harmonise the handover rates of communications between adjacent microcells. A vehicular MS driving through a city effectively proceeds along a canyon or trench. The BS antenna is at lamp-post elevation and the buildings restrain the radiation to be within the canyon. There is also penetration of electromagnetic energy into buildings which provides useful radio coverage there. Even ignoring the presence of other vehicles, there are a number of radiation paths between a BS and a MS, but the predominant paths are the LOS path, the ground reflected

path, and the two single reflected wall paths. When theoretical calculations are made based on these four most significant paths, and a perfectly
smooth canyon and road having homogeneous properties are assumed, we
obtain the curves shown in Figure 1.15.

The received signal level fluctuates significantly more than for the two-
path model as we are now vectorially summing two additional paths at the
MS antenna. Up to distances of 200 to 300 m the averages of the two-path
and the four-path curves are similar, but after these distances the curve for
the four-path model becomes closer to the free space characteristic than to
the curve for the inverse fourth power law.

Another way of considering the curves in Figure 1.15 is that they can be
modelled as a single pole circuit, rather like a resistance-capacitance integrator, having a break frequency. However, the 3 dB down break distance
is well within 100 m, and a break distance of more interest was proposed
by Green [10]. He divided his measured data into two parts and applied regression analysis to each part to obtain two-path loss straight line curves.
The distance b where the data divided was varied, curves obtained, and
the goodness of fit to the data found for both curves. The value of b that
gave the best fit was identified and the two-path loss laws established.
From measurements in Central London the double regression analysis at
900 MHz yielded variations from 1.7 to 2.14 for the exponent of the curve
for the data closest to the BS. For the data after the break distance b,
the exponent varied from 4.46 to 9.19, while b varied from 207 to 316 m.
Chia [19] making measurements at 1.7 GHz reported the first and second
exponents as 1.0 and 4.8, respectively, and b=115 m.

The above experimental results do not agree in detail with those predicted by the four-path model. This is not surprising as the buildings
are not homogeneous and large vehicles can cause multiple reflections and
diffractions. However, there are some general conclusions that can be made.
Close to the BS the path loss is less than in free space, then it decreases
with an exponent of two or less, and eventually it can plummet at a rapid
rate. By arranging for the microcell to be sufficiently small, eg., 100 to
300 m, the signal level is reasonably well maintained throughout the microcell. Outside the boundary of the microcell the signal level falls rapidly,
offering less interference to adjacent microcells.

Another important design parameter is the loss of signal level as a MS
turns a street corner. To find this parameter measurements were made
in the Harley Street area of Central London at 900 MHz [17]. This area
was characterised by a near rectilinear grid pattern of roads, as shown in
Figure 1.16. The power into the antenna was again 16 mW. A 10 dB
gain Yagi antenna with a front-to-back ratio of 15 dB, a 5 dB gain omni-
directional colinear array, and a 9 dB gain corner reflector antenna with a
25 dB front-to-back ratio were used.

With the BS antenna located at 6 or 9 m elevation in the position shown
in Figure 1.16, the signal level was measured by the MS as it drove around

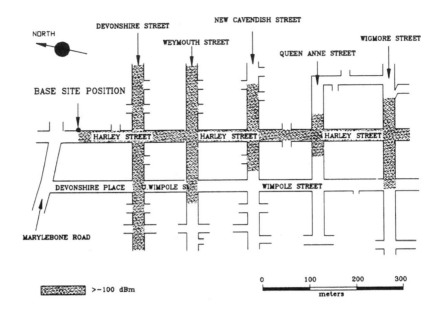

Figure 1.16: Street plan of the Harley Street area, Central London.

the area. Figure 1.17 shows the received signal level along Weymouth Street for different BS antenna heights. Whether the antenna was at 6 or 9 m made negligible difference, although the effect of antenna gains was evident. The slopes of the curves in the vicinity of the corners varied from 0.55 to 1.3 dB/m. This rapid loss of signal level as a MS turns a corner means that unless care is taken in siting microcellular BSs, calls may be forced to prematurely terminate if the received signal level goes below the receiver noise floor before the MS transmissions can be transferred to another BS that can provide a higher signal level.

The shaded area in Figure 1.16 shows the region where acceptable communications can probably be maintained. The two-dimensional microcell has a Christmas tree appearance with the BS near the foot of the tree. (A mirror image of the tree about its base is expected for an omni-directional antenna allowing for a concomitant increase in radiated power.) The height of the tree is about 620 m, and its width at the first branch level is 350 m. A propagation model by Chia based on diffraction theory offers path loss predictions which are in reasonable agreement with these measurements [17].

Figure 1.18 shows a microcell for vehicular MSs in Central London, while Figure 1.19 shows the received signal level along Piccadilly for this microcell [7]. Similar observations to those made in the Harley Street area apply here. We also show in Figure 1.18 a number of microcell BSs for pedestrian MSs. Typically we would arrange for the handover times be-

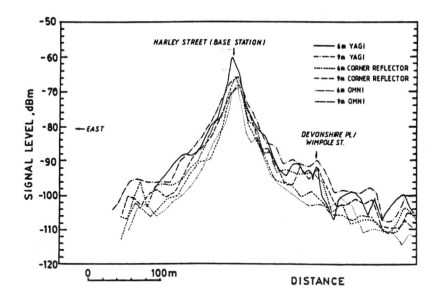

Figure 1.17: Received signal level profiles along Weymouth Street, Central London for different BS antennas (Chia *et al.* [17]).

tween pedestrian microcell BSs to be similar to the handover times for the larger vehicular microcell BSs. Consequently there may be many pedestrian microcells within a vehicular microcell. A key point to emphasise is that by deploying BS antennas on lamp posts or clamped to the sides of buildings at elevations of 5 to 12 m allows the buildings and the streets they frame to control the radio propagation. The building heights are unimportant, but not the ability of the buildings to absorb or duct electromagnetic radiation.

1.2.6.2 Fading in Street Microcells

Although anticipated by Steele and Prabhu [6], Green [10] demonstrated that microcells are not merely smaller, but have better propagation properties than large cells. The following experiment was performed along the A33 road in Southampton. A microcell antenna was located on a pedestrian cross-walk at 7.5 m above the road. Omni-directional transmissions at 1 mW and 900 MHz were radiated to the vehicular MS travelling along the A33 road. Another BS antenna was located at a height of 35 m on the tallest building on the Southampton University campus. This macrocellular BS was a mile from the microcellular BS and radiated 8 W at a similar frequency. The vehicular MS was also able to receive the macrocellular BS transmissions.

Figure 1.20 shows the received signal level and Rician K-factor distance

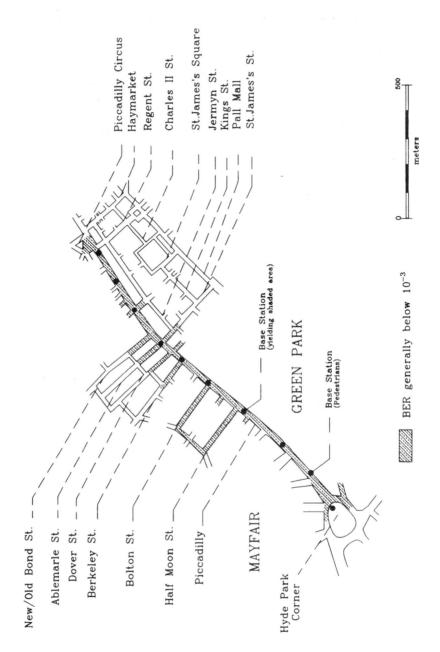

Figure 1.18: A microcell for vehicular MSs in Piccadilly, Central London (Steele [7]).

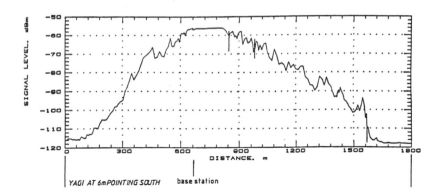

Figure 1.19: Received signal level profile along Piccadilly, London (Steele [7]).

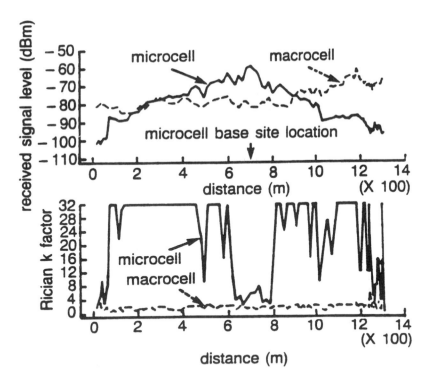

Figure 1.20: Received signal and Rician K-factor profiles for the microcellular BS and the oversailing macrocell BS (Green [10]).

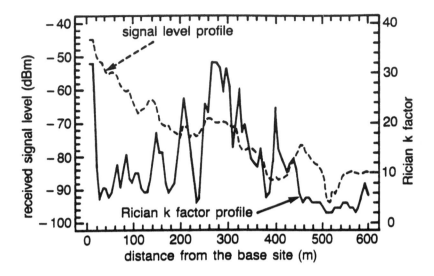

Figure 1.21: Received signal level and Rician K-factor profiles for Harley Street, Central London (Green [10]).

profiles at the MS for transmissions from the microcellular and macrocellular BSs. The received signal level for the macrocellular BS was relatively constant throughout the microcell, and K was 1 or 2, i.e., close to Rayleigh fading. By contrast, the LOS between the microcellular BS and the MS often produced high K-factors, virtually Gaussian channel conditions, due in part to the tree-line nature of the road. Near the transmitter the K values were low, and this was probably due to the antenna radiation pattern, and path loss nulls associated with the two- and four-path models. As expected, the received signal level from the microcellular BS decreased with distance along the road that formed the microcell, but because of the higher K factors the microcell will have a better performance for digital transmissions than can be achieved by the macrocellular BS.

The path loss is the important factor in determining the micro-cell boundary. The K-factors only become influential as the edge of the microcell is approached. If the K-factors are low in this region, the channel is poor with occasional deep fades causing bursts of digital errors. Should the K-factor be high, few or no errors will occur, even though the received signal level is becoming low. In this situation the signal level could descend below the noise floor level causing the quality of transmission to deteriorate rapidly and catastrophically. The same effect can happen when the MS experiences cochannel interference from another BS and the cochannel is operating with a low K. Thus being able to predict the value of K is desirable. It is also difficult. Further, how to arrange for K to be relatively

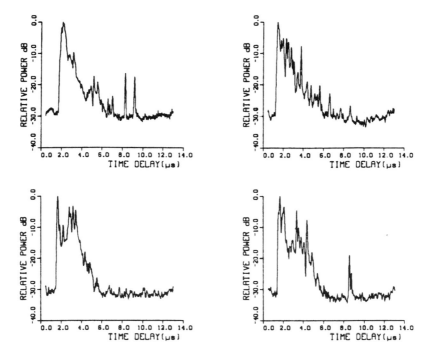

Figure 1.22: Some worst-case average impulse response envelopes for conventional size cells (Bultitude and Bedal [20]).

high at the micro-cell boundary is unknown at the present time.

Figure 1.21 shows the variations of signal level and K-factors for Harley Street made by Green at 900 MHz [10]. Chia [19] has measured K-distance profiles at 1.7 GHz showing high K values near the BS, but for distances from 20 to 300 m away from the BS the K values varied in a relatively random way between 2 and 20. He also found that the K factors differed significantly with BS location. Our own experiments confirm the widely differing K-profiles along similar streets, and until more is known about their generation the designer should work on worse case scenarios. We also note that the absolute value of K quoted by research workers depends on the windowing used, see Section 2.7.

Figures 1.22 and 1.23 show worst-case average impulse responses for small conventional size cells i.e., macrocells and for street micro-cells, obtained by Bultitude and Bedal [20] at 910 MHz in urban conditions. For the macrocells the transmitter antenna was mounted 78 m above the ground and the cell radius was approximately 1 km. The micro-cell had a length of three blocks, with the BS antenna at heights of 8.2 or 3.7 m. The MS antenna height was 3.5 m. The impulse responses for the macrocells are

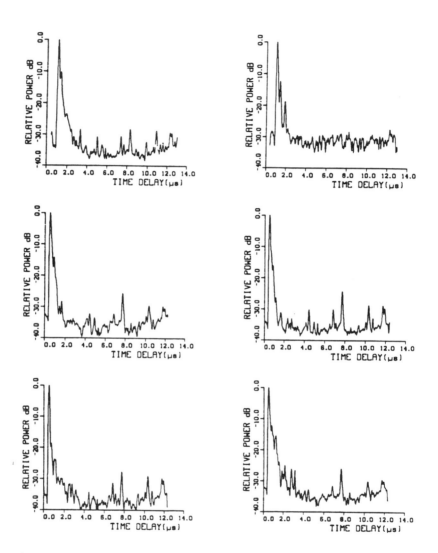

Figure 1.23: Some worst-case average impulse response envelopes for microcells (Bultitude and Bedal [20]).

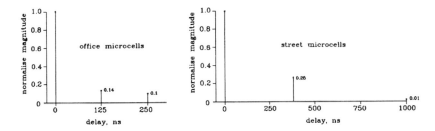

Figure 1.24: Discretised channel impulse responses.

shown in Figure 1.22 and relate to responses at the MS in the vicinity of different city blocks. The micro-cell channel impulse responses are shown in Figure 1.23. Each row corresponds to a specific location, while the first and second columns relate to BS antenna heights of 8.2 and 3.7 m, respectively. Although the maximum rms delay spreads for the micro-cells exceeded those of the macrocells the power of the long-delay components was not greater than –23 dB. However, the average rms delay spread was 3.7 times lower for micro-cells, and in general the micro-cell impulse responses indicated an ability to support higher bit rate transmissions without the need for equalisation compared to those for the macrocells. When typical average impulse responses were examined the micro-cell was clearly superior. From Bultitude's results we can extract suitable channel impulse responses for both street and for office micro-cells [21]. These are displayed in Figure 1.24.

1.2.7   Indoor Radio Propagation

Mobile communications within an office-type environment are often referred to as cordless telecommunications (CT). They can be achieved by introducing an optical or cable network within the building to which small fixed stations (FSs) are connected. These FSs communicate with portable stations (PSs) via radio. Notice that FS and PS correspond to base station (BS) and mobile station (MS) in cellular radio terminology. When cables are used it is possible to make them into leaky feeders, radiating electromagnetic energy directly to the PSs. Introducing an optical or cable local area network (LAN) into a building may be too expensive, and an alternative approach is to use a radio LAN, which, although easier to install, must be carefully designed to accommodate local movement of people and changes in the radio path due to resiting of office equipment.

Buildings exhibit great variation in size, shape and type of materials employed, making propagation prediction difficult. Compounding the problem is that radiation can exit from a building and return via buildings close by. Indoor propagation in a building as a function of its construction and its proximity to other buildings is very complex.

1.2.7.1   Path Loss

Keenan and Motley [22] made measurements in a modern multistory office block at 864 and 1728 MHz. The building was 500 m by 15 m, steel-framed with reinforced concrete and had plasterboard internal walls. The receiver was located at one end of the sixth floor and the transmitter moved on this and the next lower two floors. Adopting the traditional starting point, they opted for the straight-line representation of path loss, namely,

$$PL = L(\nu) + 10n\log_{10} d, (\text{dB}) \tag{1.22}$$

where d was the straight-line distance between the transmitter and receiver, and therefore passed through floors and walls. The propagation loss exponent was n. An additional loss was observed which was accommodated by a so-called clutter loss $L(\nu)$, which was easy to introduce but difficult to quantify its physical meaning. From Equation 1.22, $L(\nu)$ was the PL at $d = 1$ m. The distribution of $L(\nu)$ was lognormal with variance ν. For both propagation frequencies n was close to 4, but the regression lines provided a poor fit to the data. The PL equation was then modified to

$$PL = L(\nu) + 20\log_{10} d + n_f a_f + n_w a_w \tag{1.23}$$

i.e., n=2, where the attenuation in dBs of the floors and walls was a_f and a_w, and the number of floors and walls along the line d were n_f and n_w, respectively. The regression fit for this expression was much better. The values of $L(\nu)$ at 864 and 1728 MHz were 32 and 38 dB, with standard deviations of 3 and 4 dB, respectively. This law is elegant in its simplicity.

Owen and Pudney [23] found that the Keenan and Motley model provided a good fit to their data at short range, except they used the horizontal range rather than the three-dimensional straight line distance d. This was because for signals on floors adjacent to the one housing the transmitter there was only a small path loss in the vertical plane, as the lower signal levels on these floors was essentially due to the attenuation by the floors. Energy also arrived on other floors by means of stairwells and lift shafts. At 1650 MHz the floor loss factor was 14 dB, while the wall losses were 3 to 4 dB for the double plasterboard and 7 to 9 dB for breeze block or brick. The parameter $L(\nu)$ was 29 dB. When the propagation frequency was 900 MHz the floor loss factor was 12 dB and $L(\nu)$ was 23 dB. The higher $L(\nu)$ at 1650 MHz was due to the reduced antenna aperature at this frequency compared to that at 900 MHz. For a 100 dB path loss the BS to MS distance exceeded 70 m on the same floor, was 30 m for the floor above, and 20 m for the floor above that, when the top propagation frequency was 1650 MHz. The corresponding distances at 900 MHz were 70 m, 55 m and 30 m. The propagation decay exponent was approximately

3.5 for both frequencies. These results apply to a particular building, and they can be expected to vary significantly with the type and construction of the building, the furniture and equipment it houses, and the number and deployment of the people who populate it.

Bultitude *et al* [24] found that for line-of-sight (LOS) along a corridor, the exponent n was 1 for 1.7 GHz and 1 to 3 for 860 MHz transmissions. The average loss by a single partition into an office was 5 dB at both 860 MHz and 1.7 GHz. A large variation in loss between floors occurred which depended on the building structure. For example, at 1.7 GHz the losses to the adjacent floor varied from 26 to 41 dB, while over a distance of two floors the variation reduced from 45 to 52 dBs. For an open office, free space propagation plus a lumped attenuation that was both frequency and distance dependent occurred. At 1.7 GHz, this lumped attenuation was –1 dB for $d < 3$ m, and –9 dB for $d > 20$ m. The attenuation tended to be a couple of dBs less at 860 MHz. For FS and PS on the same floor, the furniture in an open office contributed 5 to 7 dB loss when LOS was achieved, and 12 to 17 dB when there was no LOS. The building penetration loss was found to be some 24 to 37 and 17 to 29 dBs depending on the angle of incidence for 1.7 GHz and 860 MHz, respectively.

Cox, a pioneer in mobile radio propagation, and his associates studied the problem of low level transmissions into and around suburban houses when the BS was located at a height of 27 ft and when the BS to MS distance varied up to 2500 ft [25]. This was equivalent to having a street micro-cell that also included the rooms in houses. However, the measurements were done within and in the immediate vicinity of the houses. The exponent n was found to vary from 3 to 6.2. The slow fading, e.g., between rooms, and between houses, was log normal.

1.2.7.2   Fading Properties

Based on measurements at 1.5 GHz in a narrow two-storey building, Saleh and Valenzuela [26] noted that the indoor channel impulse response changed slowly as it was related to people's movement, and in the absence of LOS it was independent of the polarisation of the transmitting and receiving antennas. The maximum delay spreads were up to 200 ns in rooms, and values in excess of 300 ns were found in hallways. The rms delay spread within rooms had a median value of 25 ns and a maximum value of 50 ns.

Of particular interest is their model of the channel impulse response. They observed that rays arrive in clusters, and the rays that trigger a cluster are due to the building superstructure. When a particular ray arrives in the vicinity of the transceiver multiple reflections occur in the local environment generating a sequence of received rays. The later the arrival of a ray the more reflections it has probably experienced and the smaller its magnitude is likely to be. While these rays are still arriving, albeit of negligible magnitude, another ray arrives to initiate the next cluster, and

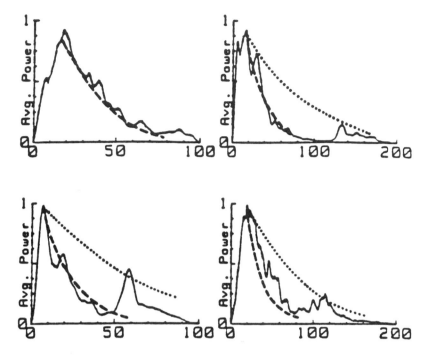

Figure 1.25: Spatially averaged power profiles within different rooms. The time axis is scaled in nanoseconds. (Saleh and Valenzuela [26]).

so on.

Figure 1.25 shows spatially averaged power profiles within four different rooms. The dashed lines are the responses generated by their proposed model, namely, that the baseband channel impulse response is

$$h(t) = \sum_{l=0}^{\infty} \left[\sum_{k=0}^{\infty} \beta_{k,l} e^{j\theta_{k,l}} \delta(t - T_l - \tau_{k,l}) \right]. \qquad (1.24)$$

In the square bracket we see the representation of an infinite number of rays in the l-th cluster. The arrival time of the ray that initiated the cluster is T_l, the arrival time of the k th ray measured from T_l is $\tau_{k,l}$. Both T_l and $\tau_{k,l}$ are independent of each other, and their inter-arrival times have exponential PDFs. Each ray is represented by a Dirac function $\delta(t - T_l - \tau_{k,l})$ whose weight is $\beta_{k,l}$ with phase $\theta_{k,l}$. The weights $\beta_{k,l}$ are independent Rayleigh variables whose variances decay exponentially for both the rays that initiate a cluster and for rays within a cluster. Their phase angles are independent uniformly distributed random variables over $(0, 2\pi]$. Typical parameter values are: cluster arrival rate = $1/300$ ns; ray arrival rate = $1/5$ ns, cluster delay time constant = 60 ns, ray delay time constant = 20 ns.

Point-to-point radio links, as distinct from BS to MS links, may be used in office environments to avoid the use of fixed fibre or cable links. The radio channel for these fixed radio links may be considered to be a nearly time invariant channel cascaded with a time variant channel, the latter due to the movement of people within the immediate environment of the terminals, while the former exists in the absence of movement. It appears that a notch may exist in the relatively static channel frequency response for long periods, regardless of the multi-path spread [24]. This means that even if the time variant channel does not require equalisation, the static channel does. For transmissions between fixed terminals, the average fading below and above the mean at 1.7 GHz was found to be 27 to 35 dB and 6 to 12 dB, respectively, for non-LOS transmission. The corresponding figures (only one set of measurements) for a LOS experiment was 25 and 4 dB. These large variations indicate that point-to-point communication channels within buildings will require complex channel conditioning circuits.

Devasirvatham [27] making measurements in a large building at 850 MHz observed a median rms time delay spread of 125 ns and a maximum delay spread of 250 ns. Bultitude et al. [28] compared the power delay profiles in two buildings at 910 MHz and 1.75 GHz. The average profiles differed significantly within the two buildings, although the standard deviation of the rms delay spreads was greater for 1.7 GHz. The coverage was less uniform at 1.75 GHz than at 910 MHz. Interestingly the median rms delay spread was about 30 ns, significantly less than the larger building measured in reference [27]. Even lower values of median rms delay spreads, 11 ns, were reported by Davies et al. [29] at 1.7 GHz.

1.2.7.3   60 GHz Propagation

Electromagnetic radiation is partially absorbed by oxygen molecules in the atmosphere. Resonant absorption lines occur in the band 50 to 70 GHz [30, 31]. They are resolvable at high altitudes where the molecular density is low, but broaden at the earth's surface inflicting significant attenuation on electromagnetic radiation. The spectrum from 51.4 to 66 GHz is the absorption band A_1. Another absorption band, A_2, stretches from 105 to 134 GHz. There are peaks in the attenuation of electromagnetic radiation due to water vapour absorption at 22 and 200 GHz. Figure 1.26 shows the oxygen and water vapour absorption curves. We notice that when the absorption in A_1 reaches a peak the attenuation due to water vapour is near a minimum [32]. The oxygen absorption is lower in A_2 than in A_1, while the water vapour attennuation is higher. These observations suggest that band A_1 is more suitable for communications than band A_2.

Alexander and Pulgliese [33] examined the prospects of using 60 GHz communications in buildings. Like them, Steele was attracted by the large bandwidths in band A_1, and the characteristic that the oxygen absorption

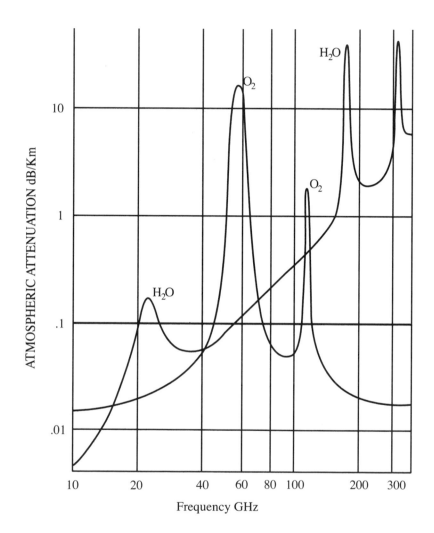

Figure 1.26: Oxygen and water vapour absorption curves.

would limit the cell size and hence improve frequency re-use [5]. The path loss at ground level due to oxygen absorption is some 14 dB/km, and there is an additional loss due to rain (that can in bad weather exceed that due to oxygen absorption). These losses are in addition to the usual path loss values discussed previously. The diffraction of 60 GHz radiation around a corner is only a metre or so, which means that a micro-cell is strictly defined by line-of-sight (LOS). This is both an advantage and a disadvantage. If micro-cells are to be sharply defined by building structures to limit co-channel interference; or to act as a communication node providing high bit rate LOS communications, then the use of band A_1 is desirable. If the signal is required to propagate beyond LOS, to provide coverage for an irregular shape micro-cell, then transmissions in band A_1 are unsuitable.

A study [34] of 60 GHz propagation for micro-cells, revealed that even along corridors in buildings free-space propagation applied as the wavelength is only 0.5 cm. Satisfactory communications were found to occur without the need to align the transmitter and receiver in offices and in lecture theatres of differing constructions. The transmissivity and reflectivity of aluminium or brass; wood; plasterboard; and glass; were measured in percentages as < 0.06, > 99; 6.3, 2; 63, 3; and 25, 16. The attenuation in signal strength was typically 8 dB for pedestrians, 10 to 14 dB for cars, 4 dB for bicycles and motorcycles, and 16 dB for buses, the measurements being made as they crossed the LOS path between the transmitter and receiver.

An attractive use for transmissions in band A_1 is as a micro-cellular point-to-point link, where the transmitters and receivers are above the height of pedestrians but below the urban sky-line. Because of the absorption properties of band A_1, the same frequencies can be repeatedly re-used across a city to give numerous point-to-point links [34]. These links can distribute 2B+D ISDN channels to each dwelling, or link micro-cellular BSs together. We note that these applications can also use lower frequencies, say 15-50 GHz, as the urban infrastructure will absorb the radiation and allow frequency re-use. 60 GHz transmissions are of particular value when buildings are widely spaced, e.g., in suburbia, as then the oxygen absorption will effectively truncate the path range.

This concludes our discourse on mobile radio propagation, and the reader is advised to consult Chapter 2 for an in-depth treatment. Our deliberations so far relate to propagation between a fixed and moving transceiver. Now we consider how multiple users can access the radio channel. After that we will describe in Section 1.4 the notion of a radio cell, clusters of cells, and the basic arrangement of a cellular network.

1.3   Principles of Multiple Access Communications

Mobile radio networks provide mobile radio communications for many users who must be able to access or receive calls from other mobile users, or from users connected to the fixed networks. Ideally, they should be able to communicate independently of their speed, location, or the time of day, subject to teletraffic demand. The way in which the multiple users access a communication system is typically by one of three methods based on either frequency, time or code allocation [35].

1.3.1   Frequency Division Multiple Access

In frequency division multiple access (FDMA) the spectrum provided by the regulatory bodies is sub-divided into contiguous frequency bands, and the bands assigned to the mobile users for their communications. For frequency division duplex (FDD) transmissions using FDMA there is a group of n contiguous sub-bands occupying a bandwidth W Hz for forward or down-link radio transmissions from a BS to its MSs, and a similar group of n sub-bands for the reverse or up-link transmissions from the MSs to their BS. A band of frequencies separates the two groups. The arrangement is shown in Figure 1.27. Interference in adjacent sub-bands due to non-perfect channel filtering is designed to be below an acceptable threshold.

Each MS is allocated a sub-band, or channel, in both of the FDD bands of W Hz for the duration of its call. All the first-generation analogue cellular mobile radio systems in service use FDMA/FDD with the speech conveyed by analogue frequency modulation (FM), while the control is performed digitally with the data transmitted via frequency shift keying (FSK). A serious disadvantage of FDMA is that a separate transceiver is required at the BS for each MS in its coverage area. Although each BS does not use all n channels, it may use many tens of them. Further, high power antenna combining networks are required to handle the simultaneous transmissions of many channels. A significant advantage of using FDMA in the first-generation systems is that as each user's transmissions are over a narrow channel of bandwidth W/n the fading is flat, and this is easier to handle than transmissions over dispersive channels.

FDMA can also be used with time division duplex (TDD). Here only one band is provided for mobile transmissions, so a time frame structure is used allowing transmissions to be done during one half of the frame while the other half of the frame is available to receive signals. For the same number of mobile users the bandwidth required is the same as for FDD, namely, $2W$ Hz. FDMA/TDD is used in the cordless telecommunications (CT), see Section 1.7.

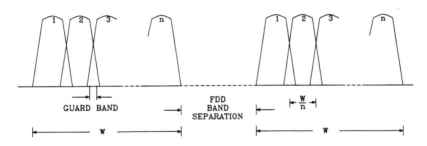

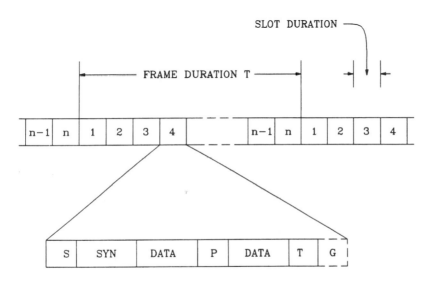

Figure 1.27: FDMA/FDD channel arrangement.

Figure 1.28: TDMA framing structure.

1.3.2   Time Division Multiple Access

Time division multiple access (TDMA) is a method to enable n users to access the assigned bandwidth W on a time basis. Each user accesses the full bandwidth W (not W/n as in FDMA) but for only a fraction of the time and on a periodic basis. Instead of requiring n radio carriers to convey the communications of n users as in FDMA, only one carrier is required in TDMA. Each user gains access to the carrier for $1/n$ of the time and generally in an ordered sequence.

In TDMA a framing structure is used as shown in Figure 1.28. Typically, a user is given a slot in a frame of duration T having n slots. If a user generates continuous data at a rate of R bits/s, it must be transmitted in a burst at a higher rate of at least nR during each frame when it is the user's

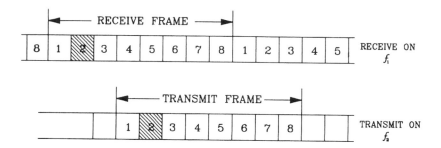

Figure 1.29: Transmitting and receiving slot assignments in TDMA.

turn to transmit. In practice the actual transmitted rate R_t of a user is considerably in excess of nR because the data is transmitted in each slot in a packetised format, as shown in Figure 1.28, with extra bits to aid the receiver. Starting bits (S) and tailing bits (T) are often added to the data, such as all zero sequences, to assist in data recovery. Sometimes the header includes bits (SYN) to assist the receiver in bit timing recovery and frame synchronisation. A sequence (P) may be inserted to estimate the impulse response of the radio channel. A guard space (G), which may be viewed as a number of guard bits, is located between packets to ease synchronisation at the receiver and to accommodate the different propagation delays between MSs close to the BS and those far away.

In general, the higher transmission rate of TDMA compared to FDMA means that the channel often becomes dispersive, adding considerable complexity in signal processing at the receiver. However, there are advantages in using TDMA. In TDMA/FDD there are two frequency bands, one for the transmission and the other for reception. When the BS transmits the MSs are switched to reception, and when the MSs transmit the BS receives their signals. This significantly simplifies transceiver design compared to FDMA where transmission and reception are performed simultaneously. Further, the BS needs only one transceiver to accommodate its n TDMA users, as it only processes them one at a time at instants according to their slot position.

Another feature of TDMA is the transceiver's ability to monitor other channels during slots when it is neither transmitting nor receiving signals. Figure 1.29 is drawn for a MS operating on channel 2, i.e., slot 2, with a frame of 8 slots, where the transmitting frame is three slot positions displaced from the receiving frame. The receiving and transmitting carriers are f_1 and f_2, respectively. Monitoring of another BS's signal can be done, say, in receiver time slot 7, but with the MS retuned to the frequency of the monitored BS. The MS transceiver will re-synthesise its oscillator as it changes from receiver to transmitter to monitor modes in this periodic sequence.

In mobile radio communications a number of TDMA carriers, each carrying n users, may be assigned unique carriers to produce a TDMA/FDMA multiple access arrangement. This approach where each TDMA/FDMA operates with FDD is employed by the pan-European GSM and the American IS-54 systems, see Section 1.6. The digital European cordless telecommunications (DECT) network described in Section 1.7.2 uses TDMA/TDD/FDMA.

1.3.3   Code Division Multiple Access

Spread spectrum multiple access (SSMA) is another method that allows multiple users to access the mobile radio communications network. Each user is allowed to use all the bandwidth, like TDMA, and for the complete duration of the call, like FDMA. The presence of all the mobile users having their signals occupying the entire bandwidth at the same time appears at first sight to be a nightmare, with each user interfering with every other user. So it is perhaps surprising to realise that spread spectrum communications was originally conceived to provide covert communications with an innate robustness to jamming.

There are two basic types of SSMA. One is direct sequence SSMA (DS/SSMA), which is more frequently referred to as code division multiple access (CDMA) [36–38], and the other is called frequency hopped SSMA (FH/SSMA). The latter scheme was proposed [39] as an alternative to FDMA in the first generation cellular radio systems. CDMA is a second-generation digital cellular scheme whose parameters are given in Section 1.6 [40, 41]. As CDMA is currently proving to be a successful multiple access method for cellular radio we will describe its salient points, and leave the reader to pursue FH/SSMA. CDMA is conceptually more complex than FDMA and TDMA, but not necessarily more difficult to implement because of the advances in microelectronics.

Let us consider a MS generating data at a rate of R bits/s. We "spread" each bit by representing it by a sequence of N_c pulses, known as chips, within the bit period T. Each chip has a duration T_c, and $T = 1/R = N_c \cdot T_c$. The bandwidth of the spread signal is much greater than the bandwidth of the data signal. Figure 1.30 shows the long term power spectra of the data and spread signals. Suppose there are M users, such that when a logical 1 bit is generated each MS uses its unique PN code of N_c chips, and when a logical 0 is formed the inverse of this PN code is transmitted. Suppose these M mobile users transmit to a BS using the same frequency band via binary phase shift keying (BPSK). This modulation means that the polarity of the chips controls the phase of the transmitted signal. For the k th user the transmitted CDMA signal is

$$s_k(t) = \sqrt{2P_s}.a_k(t).b_k(t)\cos(2\pi f_c t + \theta_k) \qquad (1.25)$$

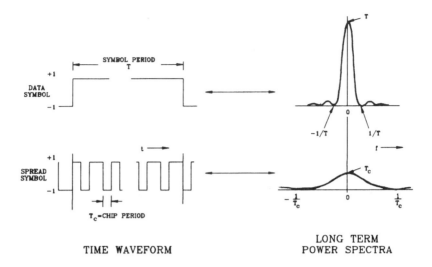

Figure 1.30: The long term spectra of the data and spread signals in CDMA.

where P_s is the transmitted power, $a_k(t)$ is the spreading code signal for the k th user, $b_k(t)$ is the k th user's data signal, f_c is the common carrier frequency used by all mobiles and θ_k is the phase of the k th user.

Figure 1.31 shows the arrangement of a BPSK modulator and demodulator. We may therefore view the CDMA signal $s_k(t)$ as the multiplication of the BPSK signal $b_k(t)\cos\omega_c t$ by the spreading signal $a_k(t)$. The recovery of the BPSK signal is achieved by multiplying the received signal $r(t)$ by $\hat{a}_k(t)$, where $\hat{a}_k(t)$ is $a_k(t)$ suitably delayed such that the $a_k(t)$ that arrives at the receiver is synchronised with $\hat{a}_k(t)$. Observe that the BPSK signal was spread at the modulator due to the multiplication by $a_k(t)$, and consequently all components of $r(t)$, other than the wanted component, will experience spreading due to the multiplication by $\hat{a}_k(t)$. For those unwanted components the BPSK/CDMA demodulator acts like a CDMA modulator! Only the wanted component is de-spread into a narrowband BPSK signal, as the product of $\hat{a}_k(t)\,a_k(t)$ is unity.

The power spectral density (PSD) of the received signal $r(t)$ is shown in Figure 1.32(a). The components of this PSD are the PSD of the M BPSK/CDMA users, i.e., the received k th BPSK/CDMA signal plus the $(M--1)$ other BPSK/CDMA users, and the PSD of the receiver noise. Included is the PSD of an arbitrary narrow-band interferer, such as a point-to-point radio signal whose carrier is f_I. After multiplication of $r(t)$ by $\hat{a}(k)$ the wanted component in $r(t)$, namely the BPSK signal is obtained. The PSD of the signal components applied to the BPSK demodulator in Figure 1.31 are displayed in Figure 1.32(b). The PSD of the wanted signal is a signal occupying a bandwidth of $2/T$ about the carrier. The narrowband

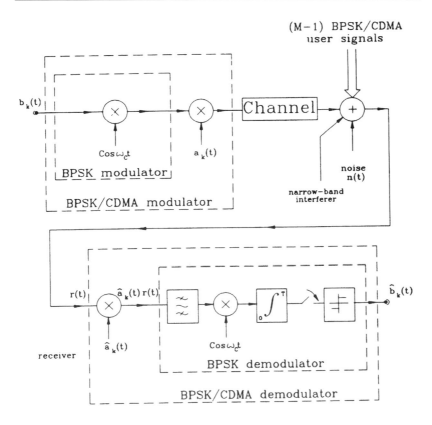

Figure 1.31: Basic BPSK/CDMA link

interference has been spread by $\hat{a}(t)$ into a wideband signal, and the $M--1$ other CDMA signals, and the receiver noise, remain wideband.

The signal $\hat{a}_k(t)r(t)$, whose PSD is shown in Figure 1.32(b), is applied to the band pass filter shown in Figure 1.31. This filter has a bandwidth from $f_c - (1/2T)$ to $f_c + (1/2T)$, which accepts the wanted component, and rejects most of the interference from the $(M--1)$ other CDMA users and from the narrow band interference, as well as from the receiver noise. In keeping with the spirit of this introductory chapter we will keep the mathematics to a minimum. Rather than proving the reduction in noise power, we can see that as the PSD of $r(t)\hat{a}_k(t)$ is composed of $sinc^2()$ functions, then if the filter has a bandwidth sufficient to pass the wanted signal, the noise and interference components are reduced by (T/T_c). This quotient is called the processing gain

$$G = \frac{T}{T_c} \tag{1.26}$$

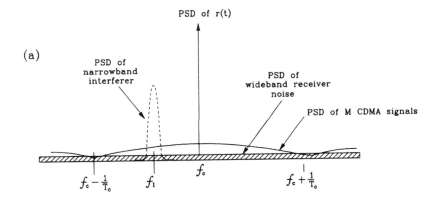

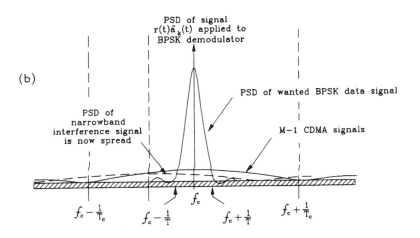

Figure 1.32: Power spectral density of, (a) received signal $r(t)$, and (b) after de-spreading in CDMA.

i.e., the ratio of the chip rate to the data rate. The greater the amount of spreading by having more chips per data bit, the larger G becomes, and the lower the noise power at the multiplier in the BPSK demodulator.

The signal at the output of the filter is multiplied by a coherent carrier, and the resulting signal integrated over a bit period T. The output of the integrator is sampled at the end of each integration period, and the polarity of the sample specifies the logical state of the recovered bit.

We notice that the receiver was faced with M BPSK/CDMA users and recovered the wanted signal as it knew $\hat{a}_k(t)$. A base station receiver must have $\hat{a}_k(t)$; $k = 1, 2, \ldots, M$, in order to recover the signals from all M mobile stations. The separation of the received signal into the M user signals is done on a code basis, and hence the term code division multiple access.

The description given has been simplified. For a mobile channel impulse response given by Equation 1.12, the received signal at a CDMA BS is

$$
r(t) = \sum_{k=1}^{M} \left[\sum_{i=1}^{L} \beta_i b_k(t - \tau_i - \Gamma_k) a_k(t - \tau_i - \Gamma_k) \right.
$$
$$
\left. \cdot \cos \{ \omega_c t - \omega_c (\tau_i + \Gamma_k) + \Psi_k + \phi_i \} \right] + n_k(t) \qquad (1.27)
$$

where for the k th MS, Γ_k is the delay when transmissions start, ψ_k is the phase of the carrier, and $n_k(t)$ is the receiver noise. If the BS is recovering the k th MS signal, it forms

$$
Z_k^j = \int_0^T r(t) a_k(t - \tau_j - \Gamma_k)
$$
$$
\cdot \cos \{ \omega_c t - \omega_c (\tau_j + \Gamma_k) + \psi_k + \phi_j \} \, dt \qquad (1.28)
$$

where the receiver is locked onto the j th path assuming β_j is the largest component. It is common for CDMA receivers to have the ability to repeat this process for other paths, say u and v, to give outputs Z_k^j, Z_k^u, Z_k^v. By combining these signals coherently we make our decision by sampling the combination and generating a logical one if the sample is positive, otherwise a logical zero is formulated. The procedure is known as correlation diversity or path diversity, or the RAKE process.

The ability to synchronise the locally generated version of the user's code prior to cross-correlation is essential if good performance is to be achieved [40]. Asynchronous CDMA has a poor performance unless different forms of diversity reception are added [37].

The ratio of the bandwidth $W (\simeq 1/T_c)$ of the spread signal to the bit rate $R = (1/T)$ of the data is the processing gain G. The more chips in the code the more unique it is and the higher G becomes. To maximise the number of users it is important that all MSs transmit at power levels such that the received power at the BS from each of them is, to a good approximation, the same. If one MS transmits at too high a power level the quality of the link for all MSs deteriorates. To ensure link quality is maintained in the presence of a rogue mobile it is necessary to reduce the number of mobile users. Consequently every effort is made to control the radiated power from each MS, no matter where it is in the cell, such that the received signal power at the BS is of the required value.

This signal power is the product of R (bits/s) and E_b (energy/bit), while the interference from the other MSs is the product of W (Hz) and $(J_o + N_o)$ (W/Hz), where J_0 and N_o are the interference and receiver noise PSDs. The signal-to-interference ratio is, for $J_0 >> N_o$,

$$SIR = \frac{RE_b}{WJ_o}. \tag{1.29}$$

If the wanted received power is P_R, and the received power of the other $(M{-}1)$ users is $P_R(M{-}1)$, we have an alternative expression for SIR of

$$SIR = \frac{P_R}{P_R(M-1)}. \tag{1.30}$$

Hence,

$$M \simeq \frac{W/R}{E_b/J_o} = \frac{G}{(E_b/J_o)} \tag{1.31}$$

and M increases with the processing gain. The ability to control the power P_R from each MS so that it is the same at the BS in the presence of Rayleigh fading is a major problem in CDMA, and essential if Equation (1.31) is to be valid.

CDMA transmissions in neighbouring cells using the same carrier frequency will cause interference which we can allow for by introducing a factor F. This factor reduces the number of users as the interference due to users in other cells is added to the interference caused by the other mobiles in the user's cell. When one speaker is listening in a conversation his bit rate can be significantly reduced to allow only the background noise to be transmitted. As on average people speak in conversations for only 40% of the time the interference they generate can be significantly decreased when they are not speaking. We designate the reduction in interference due to speaker inactivity by introducing a factor d. Notice that this inactivity results in the overall interference being reduced and this benefits all users. This is radically different to the situation in TDMA and FDMA where only cochannel users benefit. It is also commonplace in cellular radio to introduce sectorisation of cells where instead of using omnidirectional antennas at a base site, S antennas are used each radiating into a sector of $(360/S)^o$. These interference mitigating factors, lead to an increase in the number of

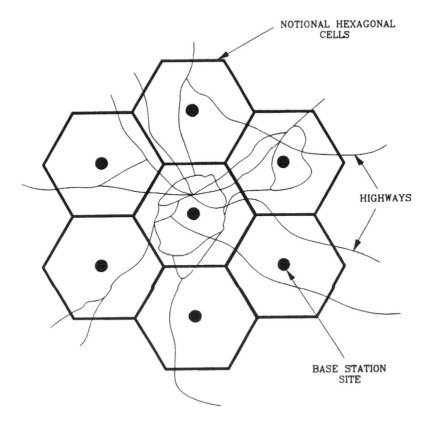

Figure 1.33: A cluster of hexagonal cells overlaying geographical area.

CDMA users to

$$M \simeq \frac{W/R}{E_b/J} \cdot \frac{1}{d} \cdot F.S. \qquad (1.32)$$

Imperfect power control can be allowed for by multiplying the right hand-side of the Equation 1.29 by another factor which is less than unity, and better than 0.5. Typically we may anticipate in a well run system that $d = 0.4$, $F = 0.6$, $S = 3$ to 6.

1.4 First-Generation Mobile Radio Systems

These systems are essentially concerned with the transmission of speech signals, although they are able to transmit data at relatively low bit rates. They are usually referred to as 'analogue' systems as the speech signals are not digitally encoded prior to transmission on a radio frequency (RF) carrier. However, all the command and control of the network is digital. The user accesses the systems by means of frequency division multiple access (FDMA). When a call is connected, the mobile user is assigned a frequency band exclusively for his or her use until the call is completed.

There are numerous analogue mobile radio networks [3, 4]. These include the Nordic Mobile Telephone (NMT) system, the American Advanced Mobile Phone Service (AMPS), the British Total Access Communications System (TACS), the German Netz C and D networks, and the Japanese Nippon Advanced Mobile Telephone System (NAMTS). There are only minor differences between these analogue mobile radio systems.

Communications between a roving mobile station (MS) and another MS, or between a MS and a fixed station, such as a telephone connected to the public switched telephone network (PSTN) or an integrated services digital network (ISDN), generally involves the use of cellular techniques. The basic concept in FDMA (and TDMA) cellular mobile radio is to divide the spectrum W assigned by the regulatory body into equal parts B_c, and to allocate B_c to each base station (BS) in a cluster of N BSs until all the bandwidth $W(= NB_c)$ is used. Conceptually, each BS may be viewed as being located at the centre of a hexagonal cell (which is not physically realised) and the hexagons are tessellated to provide a continuous mosaic over a geographical area as illustrated in Figure 1.33. The radiated power from a BS is sufficient to provide adequate radio coverage for all the MSs travelling in its domain, or cell.

As the assigned frequency band W is totally used by the cluster of N BSs, it must be repeatedly reused if contiguous radio coverage is to be nationally provided. This means that N-cell clusters must be tessellated as shown in Figure 1.34 where seven cell clusters are shown. The consequence of tessellating clusters is that a MS travelling in, say, cell 4 of a particular cluster will experience interference from BSs located in cell 4 of other clusters that are transmitting to MSs in their cells using the same frequency. For the seven-cell cluster there are six significant interfering BSs, although only four are shown in Figure 1.34. This interference is known as cochannel interference. The BSs that use the same radio frequencies must be sufficiently separated for the cochannel interference to be acceptably small. As a consequence mobile radio communications are often referred to as being interference limited in that the BSs are moved close together to increase the density of users, but not so close that the cochannel interference results in unacceptable speech quality to the mobile user. The relationship between the reuse distance D, the cell radius R and the number of cells N is $D = (3N)^{1/2}R$.

The same bandwidth B_c, but not the same frequency band, is assigned to each cell. If each user is allocated a bandwidth B_u, then the maximum number of channels per cell is B_c/B_u. As the cell covers an area A the density of channels is $B_c/(AB_u)$. By making A small this density increases. Consequently in city centres where the density of MSs is high the cell radius

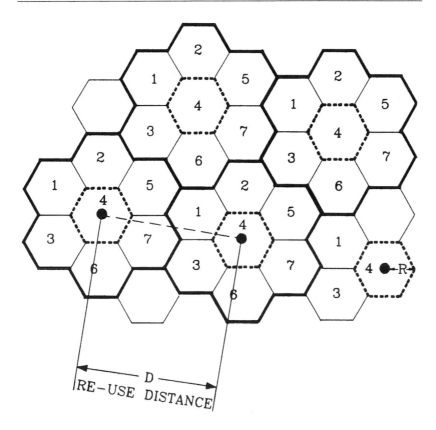

Figure 1.34: Tessellated clusters of cells; seven cells per cluster.

may be only 1 km, while in remote rural areas the radius may be some 35 km and yet still be sufficient to accommodate user demand.

The actual shapes of the cells are significantly different from the notional hexagonal ones. They are determined by the terrain, buildings, directivity of the antennas, radiated power level, and so forth. The BSs in the first-generation systems occupy a moderate size room in an office block as each radio channel requires its own transceiver. The BS antennas tend to be mounted on the tops of tall buildings or on the peaks of hills, and the radiated power may be substantial ($\simeq 100$W) when the cells are large.

As the MSs are generally beyond the line-of-sight (LOS) of the BS the received signal envelope in the first-generation systems experiences fast fading having a Rayleigh probability density function (PDF). As the band-width of the mobile radio channel B_u is either 25 or 30 kHz, the fading is said to be flat, i.e. all the frequencies across the band fade by the same amount. As a consequence the channel is devoid of dispersion and data transmissions do not experience intersymbol interference (ISI). The PDF of the slow fading signal level is log-normal. The path loss (PL) experienced by the MS decreases with an exponent of the order of 3.5 rather than the inverse square law and two-path plane earth exponents of 2 and 4, respectively.

The radio links in the first-generation systems are simple compared to those found in the second-generation system. Essentially the link is the transmission of frequency modulated (FM) speech, or frequency shift keying (FSK) data. The band used for TACS is 935-950 MHz for the forward band or downlink, and 890-950 MHz for the reverse band or uplink. The complexity of the first-generation system lies in their control procedure.

1.4.1   Network Aspects

We have discussed the behaviour of mobile radio propagation in Section 1.2, emphasising the fading nature of the channel, that it can be dispersive and noisy. When describing cellular clusters we introduced cochannel interference from other cells and there is adjacent channel interference from users of adjacent channels. Much of the emphasis of this book is devoted to the methods of establishing reliable communications of acceptable quality over these hostile channels. However, this is only part of the problem of enabling users to make and receive calls while on the move. We need to have a network that will establish and terminate calls with mobiles, track the mobiles as they travel, enable their calls to be switched between BSs to maintain call quality, and so forth. To achieve all these requirements we have the cellular network.

The control of the first-generation cellular network is digital. Although this control is simpler than the control procedures adopted for the second-generation digital networks, e.g., see those described in Chapter 8, it does have all the basic features found in these more complex networks. We will therefore describe the control aspects of the British TACS network.

In TACS the BSs are connected by permanent links to mobile switching centres (MSCs) which are computer controlled telephone exchanges specifically designed for handling cellular services. The MSCs in turn are connected to the PSTN and to other MSCs. The arrangement is shown in Figure 1.35. It enables MSs to communicate with other MSs and to non-mobile users. It also allows calls to be connected to MSs who are not in their home area.

The cellular network must keep track of all the MSs that are subscribers to the network. It does this by forming traffic areas which consist of groups of cells. The MSCs log the current locations of all MSs through a process known as registration. When the MS is switched on, but is not making a call, it constantly listens to one of the common signalling channels transmitted by the BSs to determine the traffic area in which it is currently

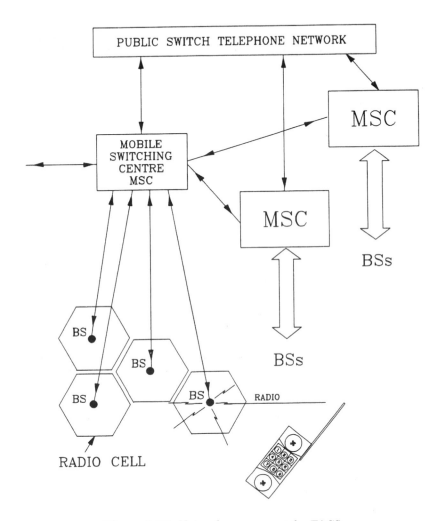

Figure 1.35: Network arrangement for TACS.

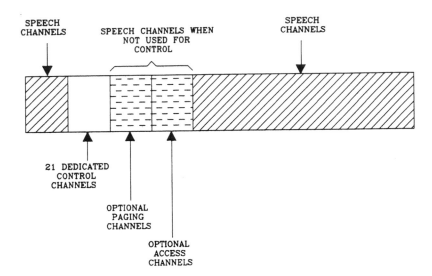

Figure 1.36: TACS control and speech channels.

residing. Should the quality of the current signalling channel deteriorate as the MS travels it scans the signalling channels until it finds one with an acceptable quality. If the traffic area has changed the MS automatically, and unknown to the user, calls the new BS, gives its identification number, and is thereby registered in the new traffic area, then the old BS de-registers the MS.

When a call has been established and a conversation is in progress the MS may travel near the edge of its cell. Should the communications quality become unsatisfactory the network is required to ensure that the MS changes its BS to one that can provide a better quality channel. The process of switching from one BS to another BS is called handover or handoff. The handover process is required to be automatic, with no interruption of the call.

To facilitate the handover procedure the BS constantly monitors the received signal levels from all mobiles with speech transmissions in progress. When a received signal level drops below a threshold it informs the MSC that a hand-off may be required. The MSC commands all the surrounding base stations to measure the signal strength of the mobile, and then chooses the BS with the strongest received signal to handle the call. Once the new base station has been informed of the handover request, and the radio channel allocation has been made, the original base station is commanded to send a control message to the mobile to have it re-tune to the newly assigned channel. This happens automatically within a few seconds, and the mobile user is aware of only a very brief break in transmission (about

400 ms) when the actual handover takes place.

The TACS system has both control and speech radio channels. There are 21 channels reserved for control which cannot be used as speech channels. Of the remaining channels, some may be used for control where traffic demands extra control channels; otherwise, they can all be used as speech channels. Figure 1.36 shows the TACS control and speech channels.

1.4.1.1 Control Channels

The TACS control channels are used for the co-ordination of mobiles and for all call set-up procedures. Functionally, there are three types of control channel: dedicated control channels, paging channels and access channels. All mobiles are permanently programmed with the channel numbers of the dedicated control channels and they scan these channels at switch-on. The dedicated control channels carry basic information about the network and inform the mobiles about the channel numbers of the paging channels.

The paging channels are used both to transmit messages to specific mobiles, for example, to alert them of incoming calls, and for general network information, such as traffic area identity, channel numbers of the access channels, access methods to be used by mobiles, etc. The access channels are used by mobiles for accessing the network to initiate out-going calls, to register their location and to respond to paging calls.

All three types of control channel carry status information in the sequences of data blocks called overhead messages. The type of overhead message depends on the function of the control channel. Where the functions of dedicated control channel, paging channel and access channel are combined, the overhead messages contain all the relevant information as one message train.

TACS has a total of four signalling channels as shown in Figure 1.37. The Forward Control Channel (FOCC) and the Reverse Control Channel (RECC) are used for setting-up calls, as well as maintaining contact between BS and MS when no calls are in progress. The Forward Voice Channel (FVC) and the Reverse Voice Channel (RVC) must be compatible with a standard voice channel. In order to maintain system flexibility, FOCC and RECC are also compatible with a standard voice channel. Data messages on the FVC and RVC are kept short because they are inserted periodically into the user's voice channel. The FOCC is transmitted at all times so that a mobile entering the system can acquire system specific information, e.g., which channels are allocated for voice traffic. The RECC is only active intermittently, e.g., when the MS moves from one BS to another (no conversation in progress) or the MS initiates a call.

Signalling between the mobile and the base station for call set-up, hand-off and other similar control functions is carried digitally by using frequency-shift keying (FSK) of the radio-frequency carrier with a deviation of 6.4 kHz. The basic data rate used is 8 kbit/s, but to facilitate clock

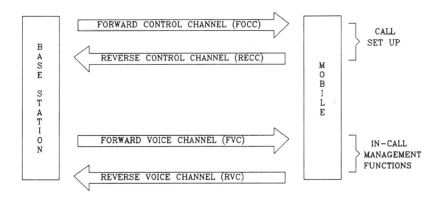

Figure 1.37: The TACS signalling channels.

recovery at the receiver the data is Manchester encoded before transmission to give a rate of 16 kbit/s.

It is necessary for all signalling information to be correctly received. Accordingly robust error protection is used having the combined techniques of repeated transmission with majority decision and forward error correction coding. The message to be sent is coded into a block and a parity word is generated from it by using Bose-Chaudhuri-Hocquenghem (BCH) coding. The parity word is appended to the message to form a signalling block. A complete signalling frame is then formed which starts with a bit- and word-synchronisation sequences, followed by the combined message and parity word repeated several times.

When the message is received, a bit-by-bit majority decision is carried out on the repeated blocks, and many of the errors that may have occurred are corrected. The parity word is then used to correct up to one remaining error, and to detect if more than one error is present. If there are two or more errors remaining after majority decision, they cannot be corrected and the message is rejected.

1.4.1.2   Supervision

Whilst a call is in progress, a supervisory audio tone (SAT) is transmitted by the base station and looped back by the mobile. Both base station and mobile require the presence of the SAT on the received signal to enable the audio path. Three different frequencies are used for the SAT, all around 6 kHz. During call set-up, the base station informs the mobile which SAT to expect on the speech channel. If the SAT is incorrect, the mobile does not enable the audio path, but starts a timer which, on expiry, returns the mobile to stand-by. Similarly, the base station expects to see its transmitted SAT returned and takes this as confirmation that the mobile is operating on the correct channel.

The three SAT frequencies are allocated to the BSs to give a three-cluster repeat pattern. Consequently a cell in the adjoining cluster using the same radio channel uses a different SAT. The effect of this arrangement is to reduce the probability of a co-channel interfering signal being stronger than the wanted signal at a mobile.

1.4.1.3 Call Origination

When a user makes an outgoing call, the number required is keyed into the mobile, or the required number is extracted from the short-code memory in the mobile. A SEND key is then pressed. This causes the mobile to perform a system access. To do this the mobile first scans the network's access channels. The channel numbers of the access channels are transmitted as part of the overhead information on the paging channels. The mobile chooses the two channels with the highest signal level and attempts to receive the overhead messages being transmitted. Should the mobile fail to receive these messages on the strongest channel, it tries again on the other channel. Contained in the overhead messages are parameters which inform the mobile of the access procedure. Once the mobile has received these parameters, it checks to see if the access channel is in use. If it is available the mobile transmits its message.

When the mobile has sent its message to the network, it turns off its transmitter and remains on the access channel awaiting a message from the base station. For MS call originations, the received message from the BS is normally a speech-channel allocation and contains the channel number and the SAT code. On receipt of the message, the mobile tunes to the required channel and starts to transpond the SAT. If the correct SAT is received at the BS, the audio paths are enabled and the user can hear the call being set-up.

If the access was as a result of a registration, the message received from the BS on the access channel is normally a registration confirmation. On receipt of this message, the mobile returns to the IDLE condition.

1.4.1.4 Call Receipt

When an incoming call for a MS is received, the MSC checks the current location of the mobile which has been obtained through the registration procedures. A paging call is then transmitted on the paging channel of all BSs in the mobile's current traffic area. When the MS receives a paging call, it accesses the network in the same way as for call originations, but the message sent to the BS informs the network that the access is as a result of receiving a page. The MS receives a speech-channel allocation from the BS, tunes to the new channel, and checks the SAT received. The BS then transmits an alert message to the MS, causing it to alert the user of the incoming call and to transmit a continuous 8 kHz signalling tone. When the user answers the call, the signalling tone is turned off, the audio paths are enabled and the call proceeds.

1.4.2   Power Levels and Power Control

The transmit power levels used at a BS are chosen to give the required coverage area. The maximum effective radiated power (ERP) per channel is limited to 100W. Mobile stations have a nominal power class, namely 1, 2, 3 and 4 with ERPs of 10W, 4W, 1.6W and 0.6W, respectively. However, Classes 2 and 4 are used by the majority of MSs. In addition, the BS instructs the MSs within its coverage area to adjust their transmitted power to the minimum level required for acceptable performance. The power control facility reduces interference within the system and improves spectral occupancy. A maximum of eight different power levels may be emitted by a MS. The minimum ERP is −22 dBW and the step changes of ERP are 4 dB. The ERP to be used is transmitted to the MS using a three-bit attenuation code. On receipt of the power control message from the BS, the mobile sends an acknowledgement to the BS and selects the appropriate power level.

1.4.2.1   Call Termination

When a mobile user completes a call and replaces the handset, the mobile transmits a long burst (1.8 s) of 8 kHz signalling tone to the BS, and then re-enters the control-channel scanning procedure. If the originator of a call on the PSTN clears down, a release message is sent to the mobile, which responds by sending a burst of 8 kHz signalling tone, after which it re-enters the control-channel scanning procedure.

We have shown how the TACS network operates to facilitate the ability of mobiles to make and receive calls while on the move. This is not the full story, but it does indicate that perhaps the major complexity of cellular radio lies in its network organisation. However, the quality of the communications is dependent on the radio links. We now examine the basic digital mobile radio link.

1.5   Digital Cellular Mobile Radio Systems

The second-generation cellular mobile radio systems are all digital, and once more the primary service is mobile telephony. FDMA is abandoned in favour of time division multiple access (TDMA), although one American system employs code division multiple access (CDMA). The consequence is that the transmitted bit rate per carrier is significantly increased compared to transmitting digital speech via FDMA, and this usually results in time variant, dispersive mobile radio channels.

1.5.1   Communication Sub-systems

Of paramount importance is the speech codec which encodes the speech into a digital format for transmission, and decodes the regenerated bits at the receiver output to provide the recovered speech signal. We have discussed the time varying nature of the mobile radio channel, emphasising the distortions it can impose on the transmitted signal. Consequently we condition the signals prior to transmission in order to assist the receiver in its difficult role of regenerating the data with an acceptably low bit error rate (BER). Conditioning sub-systems include forward error correcting (FEC) codecs that operate on the encoded speech signal, increasing the bit rate while allowing the receiver to perform bit error correction. We also need to introduce interleavers to scramble the FEC data. Scrambling the data at the transmitter enables the descrambling process at the receiver to convert burst errors into random errors, and this improves the performance of the FEC decoding. The RF modem is another important sub-system that determines spectral occupancy and battery power drain. We also need to consider equalisers to remove the effects of channel dispersion, and the use of diversity in order to receive a number of versions of the transmitted signal and to combine them in such a way that the BER is decreased. We will now briefly describe some of these important sub-systems although they are considered in depth in subsequent chapters.

1.5.1.1   Speech Codec

There are a range of speech codecs that enable us to decrease the bit rate for an increase in complexity, while maintaining speech quality [42–44]. Logarithmic pulse code modulation (log-PCM) has long been used by the wire networks and can be used in mobile radio when there is sufficient channel bandwidth [45–50]. Adaptive delta modulation (ADM) [51] has long been favoured by the military for mobile communications, and is well suited to cordless telecommunication (CT) applications, i.e., short range mobile communications as used in offices. The speech quality of ADM at 32 kb/s is not as good as that of adaptive differential pulse code modulation (AD-PCM) at the same bit rate when the transmissions are over ideal channels. However, over mobile radio channels where burst errors occur it generally provides better quality speech. In addition the battery consumption of ADM is significantly lower than that of ADPCM. Nevertheless, ADPCM is preferred in cordless communications, see Section 1.7, as it is an international standard, and ADM does not perform well when multiple tandem links are used in the network.

Sub-band coding (SBC) [52] at 16 kb/s is more appropriate for cellular radio where bandwidth is usually at a premium compared to CT applications. The lower bit rate of 16 kb/s is achieved with considerable extra complexity and delay compared to ADM and ADPCM. Analysis-by-synthesis (ABS) techniques are currently in vogue for cellular radio [53,54].

They operate effectively (after some channel coding) from 8 to 13 kb/s, and yield near toll quality speech. A popular ABS scheme is the code excited linear predictive codec (CELP) [53, 54]. The regular pulse excited linear predictive codec (RPE-LPC) had been adopted for the pan-European GSM network [55]. Low bit rate vocoders at 2.4 kb/s and below [56] that are used by the military are not currently employed in digital cellular mobile radio. Speech codecs are discussed in-depth in Chapter 3.

1.5.1.2   Channel Codec

The digitally encoded speech is channel coded [57] to enable the receiver to correct many of the symbol errors that occur during transmission [52]. As the symbol errors are associated with deep fades in the signal level they tend to occur in bursts. In order to correct symbol errors at the receiver we can either use a long block channel code that spans a number of error bursts, or use a short block code and interleave the data prior to transmission. The long codes work well because for most of the time there are few deep fades within the code words, and although error bursts occur during these fades the resulting erroneous symbols are only a small fraction of the total symbols in the code words. The consequence is that the channel decoder is able to correct the relatively few symbol errors. The penalty of using long codes is the delay, and more importantly, the complexity of the channel codec.

Short block codes do not have the high error correcting power of the long block codes. However, by interleaving the data prior to transmission, the errors in the bursts are randomised at the receiver during the de-interleaving process. Consequently during decoding there are relatively few errors in a block and satisfactory error correction is achieved. Short codes are usually preferred to long codes because of the lower complexity, but the overall delay is similar due to the delay incurred by the interleaving process. Important block codes are the Bose-Chaudhuri-Hocquenghem (BCH) and Reed-Solomon (RS) codes that use one bit per symbol or many bits per symbol, respectively. The RS codes are more complex than the BCH codes with a more powerful and reliable error detection and error correction capability. The abilities of these codes are known with mathematical certainty.

Another class of channel coding used in mobile radio is convolutional coding (CC). Interleaving must be used to combat error bursts, and the complexity of the codec increases exponentially with the error correcting power. CCs do not have the reliable error detecting capabilities of block codes, and the designer cannot guarantee the number of errors that will be corrected. The CCs are also subjected to error propagation effects.

However, CCs are often preferred to block codes in mobile radio applications. They can employ soft decoding whereby the signal applied for CC decoding is not a sequence of bits from the demodulator, but a multilevel signal. This approach yields an enhanced performance. Another feature of CCs is the use of 'puncturing', a process which provides a high rate code (i.e., the ratio of the information bits to the total bits) by periodically deleting some of the coded bits from the coder output. Puncturing reduces the complexity of the CC decoder compared to an identical rate non-punctured CC, while weakening its correcting power. Speech can be encoded, followed by convolutional coding. By puncturing the CC bits, control data can be inserted without affecting the speech encoding. The transmitted bit rate is unaltered, although the correcting power is marginally decreased.

In Chapter 4 we present the detailed operation of BCH and Reed Solomon block codes, convolutional codes, and the operation of different interleavers. The importance of channel coding and interleaving will be demonstrated.

1.5.1.3 Modulation

The modulation process converts the channel coded speech into a format suitable for transmission over the mobile radio channel [35]. It is desirable that the digital modulation technique employed has a high bandwidth efficiency, i.e., the bit rate per channel bandwidth is high for a given power expenditure and at a specific bit error rate (BER). The implementation costs should also be low. Constant-envelope modulated signals find favour in both digital cellular and cordless telecommunications (CT) radio links. They have acceptably good bandwidth efficiencies and relatively low battery power drain due to their use of class-C amplifiers.

Of particular importance in mobile radio is Gaussian minimum shift keying (GMSK), which is considered in detail in Chapter 6. The modulator is basically a Gaussian filter followed by a voltage controlled oscillator (VCO). The modulated signal has a spectrum that is narrow due to the deliberate introduction of ISI into the transmitted data stream. The ISI is generated by passing the input data through a digital filter having a Gaussian shaped impulse response which causes each bit to be spread over a number of bit intervals. The greater the bit spreading the narrower the spectrum of the GMSK signal and the better the adjacent channel interference performance. However, the clock recovery and symbol detection at the receiver becomes more difficult.

In fast frequency shift keying (FFSK) or minimum shift keying (MSK), logical 0 and logical 1 data bits are conveyed by assigning distinct carrier frequencies to them with a suitable frequency band between them to avoid detection ambiguities. To achieve a spectral occupancy that is less than that of MSK each data bit is filtered, to yield smooth transitions in the data signal before it modulates a VCO. We notice that filtering of the data is also done in GMSK, but an essential difference between MSK and GMSK is that in MSK the shaping of each data bit is done over only one bit period. When the shaping is done over one bit period it is referred to as a full response system, as distinct from a partial response system when it is implemented over more than one bit period. The consequence of MSK being a full response system is that the bandwidth of the modulated signal is larger than that for GMSK. However, the detection of MSK signals is simpler as its eye-pattern is more open.

The tighter spectrum of GMSK means that carrier spacings can be closer than in MSK for the same adjacent channel interference, producing a higher spectral efficiency. With the oscilloscope synchronised to symbol timing, the display of the demodulated signal has and eye-like appearance due to the data being filtered. The GMSK parameter BT is related to the number of bit periods over which each data bit is spread. Common values employed are $BT = 0.3$ and 0.5, e.g., $BT = 0.5$ means the spreading is over two bit periods. As this spreading does not occur in MSK its eye-pattern is wide open, and the bit regenerator samples the signal at the instants corresponding to the widest openings of the eye. For moderate amounts of signal noise, it is easy to identify the polarity of the signal and thereby regenerate the bits correctly. By contrast the GMSK eye-patterns are less open, making it more difficult in the presence of noise to identify the correct polarity of the received bit. By spreading each bit over a number of bit periods a multilevel signal and hence a multilevel eye-pattern is formed which depends on the variations in the patterns of logical ones and zeros in the data. Some patterns are easier to regenerate than others.

In contrast to these constant envelope modulation methods supporting one bit per symbol, we have multi-level modulators having n bits per symbol. For example, the modulated signal can have different discrete amplitudes, as in multi-level amplitude shift keying, known as m-ary ASK, where $m = 2^n$ signifies the number of levels. Sometimes the amplitude of the modulated phasor is constant, but its phase has one of m distinct values as in multi-level phase shift keying, called m-ary PSK. By combining the two types of modulation we obtain quadrature amplitude modulation (QAM) where the data are conveyed by both the magnitude and phase of the transmitted phasor [49, 50]. Many other types of multi-level modulation are possible. The twin phase modulation, known as star-QAM [58], has the desirable property that when used with differentially encoded data it effectively avoids the need for automatic gain control (AGC) and carrier recovery procedures at the receiver.

Figure 1.38 shows the constellations for these multi-level modulation signals. The dots or constellation points, correspond to the tips of phasors whose other extremities are at the centre of the constellation. Each constellation point is associated with a data word. For example, a 16-level QAM constellation supports 4-bit words on each constellation point or phasor. So during any symbol period the transmitted phasor, suitably filtered, will correspond to one point and convey 4-bits. By having 256-levels, 8 bits are conveyed by each transmitted symbol. Another multi-level modulation method is m-ary FSK where n bits of data are transmitted during each

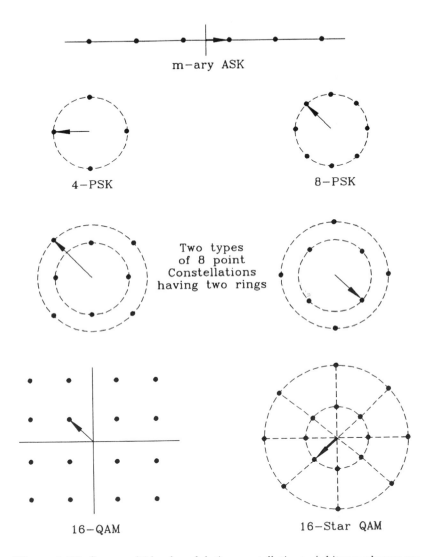

Figure 1.38: Some multi-level modulation constellations. Arbitrary phasors are drawn on the constellations.

symbol period as one of m unique carrier frequencies.

The desirable feature of multi-level modulation is that each symbol carries n-bits, $n \geq 2$, and as the symbol rate determines bandwidth occupancy, the signal bandwidth is lower compared to binary modulation methods where $n = 1$. However, the channel SNR is required to be higher for multi-level modulators as the regenerator does not make simple binary decisions. QAM modems also require linear ampliers which although having lower efficiencies can be used in micro-cellular mobile radio networks as the radiated power levels are much lower than those used with large cells.

1.5.2   FDMA Digital Link

Having briefly discussed some of the communication sub-systems, we will now describe how they are cascaded to form a basic FDMA digital mobile radio link. The transmitter contains a speech and channel codec, an interleaver to provide time diversity, and a modulator prior to amplification for transmission over the mobile radio channel. At the receiver demodulation ensues followed by clock recovery and bit regeneration. The resulting data stream is de-interleaved causing the burst errors to be partially randomised, and channel decoding and speech decoding executed to recover the speech signal. The block diagram of the basic FDMA digital link is shown in Figure 1.39. Also shown in the Figure is a switched diversity arrangement. The two radio channels are associated with each of the two receiver antennas. These antennas are spaced sufficiently far apart that their received signals are uncorrelated. The signal applied to the demodulator switches from one antenna to the other whenever the received signal level falls below a system threshold or above an acceptable BER. We emphasise that there are a variety of space diversity techniques [2]. The one shown in Figure 1.37 is the simplest and is known as second-order switched diversity. By using more antennas the order of the diversity increases pro rata. As the signal applied to the demodulator is frequently switched in order to provide a good signal level the statistics of the signal being demodulated changes from, say, Rayleigh when only a single antenna is continuously used, to near-Gaussian when the order of diversity is high. A Gaussian channel is the best we can achieve as all the fading is eliminated. The general case of switched diversity is known as selection diversity (SD), where each antenna has its own receiver and the one with the highest baseband SNR is selected to be the demodulated signal.

The weakness of selection diversity is that only one antenna is used at any instant, while all the others are disregarded. Maximal ratio combining diversity (MRCD) seeks to exploit the signals from each antenna by weighting each signal in proportion to their SNRs and then summing them. Accordingly in MRCD the individual signals in each diversity branch are cophased and combined, exploiting all the received signals, even those with poor SNRs. However, MRCD is more difficult to implement than SD.

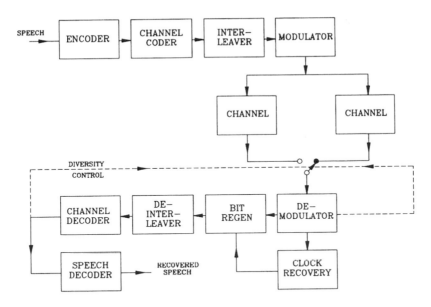

Figure 1.39: Basic FDMA digital link showing second-order switched diversity.

Space diversity, i.e., where the receiving antennas are spaced apart, is usually employed in flat fading environments as the deep fades can be effectively combatted at the expense of system complexity. However, space diversity is less useful with dispersive channels, although it can be effectively employed but not in the simple ways discussed above. In dispersive systems the diversity procedures must be integrated into the equalisation techniques, or into the adaptive correlation diversity methods used in spread spectrum receivers. The system shown in Figure 1.39 assumes that the symbol rate is sufficiently low that dispersion has not occurred. We now consider TDMA links where dispersion is usually prevalent.

1.5.3 TDMA Digital Link

In time division multiple access (TDMA) systems a number of user's signals are transmitted on a single RF carrier. Their transmissions are synchronised to occur in a particular time slot in each TDMA frame and hence the data are sent in packets where the packet duration is marginally shorter than the slot duration. If there are n slots in a TDMA frame, then it follows that if a user's coded data are generated at a rate R it will be transmitted at a rate in excess of nR, whereas in FDMA it is transmitted at R. The actual TDMA transmitted rate is above nR as described in connection with Figure 1.28.

Figure 1.40 shows the basic arrangement of a TDMA transmitter-

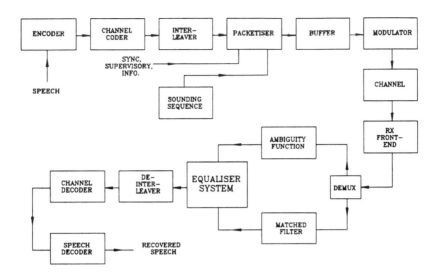

Figure 1.40: Basic TDMA radio link.

receiver link. At the transmitter a packetiser accepts the coded interleaved digital speech at a rate R, forms a packet over a period of a frame and releases it as a data burst to the modulator for transmission. The high speed data on the RF carrier suffer considerable distortion during their transmission over the mobile radio channel and this must be compensated for at the receiver. After the receiver has demodulated the RF signal, demultiplexing of the baseband signal ensues to yield the channel data and the propagation channel sounding data. Referring to Figure 1.40, the sounding channel data are applied to a matched filter whose impulse response is matched to the sounding sequence inserted into the packet prior to transmission. Should the sounding sequence at the transmitter be passed through the matched filter (i.e., an ideal channel) a sharp pulse of some two bits width is produced. If this sharp pulse were passed through the equivalent baseband channel we would obtain the impulse response of the channel, slightly degraded by the sharp pulse not being an ideal delta function. We see that the cascading of the sounding sequence with the matched filter and radio channel yields an estimate of the baseband channel impulse response $h'(t)$. Since the mobile radio channel is essentially linear, we obtain the same $h'(t)$ by transmitting the sounding sequence over the mobile channel, followed by the matched filtering, i.e. when using the arrangement shown in Figure 1.40.

Having obtained a measure of the complex baseband channel impulse response $h'(t)$ we are able to perform channel equalisation on the traffic data. Equalisation is a process which attempts to remove the ISI introduced by the channel [21,59-61]. In linear equalisers, or decision directed equalisers, the knowledge of $h'(t)$ enables the equaliser's coefficients to be found. In Viterbi equalisation, which is really a maximum likelihood sequence detection process, a baseband modulator is used at the receiver to generate a wide range of possible signals that could have been received over the real channel. The data signal and the locally generated estimates of the data signal are combined to form mean square error signals, called metrics, which are used in a Viterbi processor (VP) to identify the most probable sequence of data transmitted. The operation of the VP is explained in detail in Chapter 6. The ambiguity function in Figure 1.40 is a filter whose impulse response is the convolution of the sounding sequence with the matched filter impulse response. It is introduced to compensate for the channel being sounded by an impulse of this type and not a delta function, i.e., both the data and the locally generated estimates of the signal are subjected to the same distortion.

Having regenerated the data in the packet, de-interleaving is performed (this may be over a number of packets) followed by channel decoding and speech decoding.

1.6 Second-Generation Cellular Mobile Systems

In the USA the system to be introduced, known at the time of writing as IS-54, has the TDMA carriers spaced by 30 kHz to align with those in their analogue advanced mobile phone service (AMPS). Each carrier supports three users at the TDMA rate of 48.6 kb/s. The dual-mode transmissions are in the bands 824-849 MHz and 869-894 MHz. Vector sum excited linear prediction speech encoding (see Section 3.5) is used operating at 7.95 kb/s. After channel coding the rate becomes 13 kb/s, and allowing for control information the effective rate per user is 16.2 kb/s. The 48.6 kb/s TDMA rate is transmitted using $\pi/4$ shifted DQPSK modulation at 2 bits/symbol [62].

The Japanese are taking a similar approach. As their first-generation analogue system has carrier spacings of 25 kHz, they use this bandwidth to introduce a TDMA rate of 42 kb/s, again using $\pi/4$ shifted DQPSK. The speech channel coding rate is 11.2 kb/s.

The pan-European digital cellular mobile system is called GSM. These initials originally stood for, 'Groupe Speciale Mobile', after the name of the committee responsible for its specification, but it now represents, 'Global System for Mobile Communications'. The GSM network began to be deployed in Europe in July 1991. It is a much more ambitious network than IS-54. Operating in a TDMA mode, a regular pulse excited linear predictive coder (RPE-LPC) encodes the speech at 13 kb/s. This is followed by channel coding and bit interleaving to yield a voice rate of 22.8 kb/s. The data are assembled into packets with a propagation sounding sequence located in the centre of the packets, and transmitted via GMSK at a TDMA rate of 270.8 kb/s. The carrier spacing is 200 kHz. The frequency bands are 935-960 and 890-915 MHz. Because of the high transmission rate and the large cells that may be used (up to 35 km radius), the mobile radio channel is often dispersive and this requires receivers to employ channel equalisation as outlined in the previous section. Chapter 8 deals exclusively with the GSM system and we will refrain from discussing it further here.

In 1989 the British Government announced that service providers would be licensed to operate so-called personal communication networks (PCNs). This has now been done and these PCNs became operational around 1993. They use a modified form of GSM, referred to as DCS 1800, meaning a digital cellular system at 1800 MHz. Duplex bands of 75 MHz with a 20 MHz guard band will be used. Frequencies assigned are 1805-1880 MHz for the down-link (BS to MSs) and 1710-1785 MHz for the up-link (MSs to BS). Each service provider has a contiguous block of spectrum. The DCS 1800 specification is essentially that of GSM, with minor modifications as DCS 1800 operates in typically smaller cells than those used in GSM. This means that the radiated power levels are lower for DCS 1800, and hence the complexity of the channel equaliser could be significantly decreased. Both these features decrease the battery power drain in hand-held portables.

1.6.1   Qualcomm CDMA

Qualcomm Incorporated in the USA opted for CDMA as the multiple access method for mobile radio. Although CDMA has been well understood for a long time, its use in cellular radio had been avoided due mainly to the problems associated with power control. If the standard deviation of the received power from each mobile at the BS is not controlled to an accuracy of approximately ± 1 dB relative to the target received power the number of users supported by the system can be significantly curtailed. Other problems sited were whether there were sufficient codes available for a large number of mobile users, and difficulties of synchronisation. These major and many other minor problems have been successfully addressed by this CDMA system.

Qualcomm CDMA operates at the top of the AMPS band. The CDMA bandwidth required for each up- and down-link is 1.23 MHz, equivalent to 41 AMPS channels (41×30 KHz = 1.23 MHz). This CDMA network also operates in the 1.7 to 1.8 GHz band.

1.6.1.1   Qualcomm CDMA Down-link

There is one pilot channel, one synchronisation channel, and 62 other channels. All of the 62 channels can be used for traffic, but up to 7 can be used for paging. The 64 Walsh codes of length 64 are used for each of these channels. The first 64 Walsh codes are shown in Figure 1.41. Walsh code W_0, an all-one code is used for the pilot, the alternating polarity W_{32} is

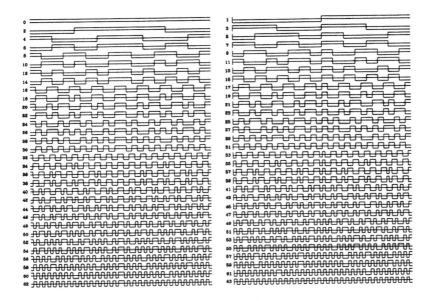

Figure 1.41: The first 64 Walsh basis functions.

used for the synchronisation channel, while the paging and traffic data use the other 62 Walsh codes. It is important to notice that the Walsh codes are used to identify the channel. Their modus operandi is very different on the reverse or up-link.

Figure 1.42 shows a block diagram of the BS transmitter. The pilot channel consists of a pair of pseudo random binary sequences (PRBS) at 1.2288 Mchip/s. The synchronisation channel data at 1200 b/s is convolutionally encoded to 2400 b/s, repeated to 4800 b/s and interleaved over the period of the pilot PRBS. Each of these interleaved symbols spans 4 Walsh symbols; so that when W_{32} generated at 1.2288 Mchip/s is exclusive-ORed with the sync data, a signal of 1.2288 Mchips/s is produced.

The speech is encoded by a variable rate vocoder that generates forward traffic channel data at rates of 1.2, 2.4, 4.8 or 9.6 kb/s, depending on speaker activity. As the frame duration is fixed at 20 ms, the number of bits per frame varies according to the traffic rate. Half rate convolutional encoding with a constraint length of 9 doubles the traffic rate to give rates from 2.4 to 19.2 ksymbols/s. To ensure the rate is always 19.2 ksymbols/s, data repetition is appropriately used at the lower speech rates. Interleaving is performed over 20 ms, and the higher the data repetition used, the lower is the transmission power of the symbols.

A long code of $2^{42}-1(=4.4 \times 10^{12})$ is generated containing the user's electronic serial number embedded in the mobile station's long code mask. This code, suitably processed, scrambles the output stream from the block

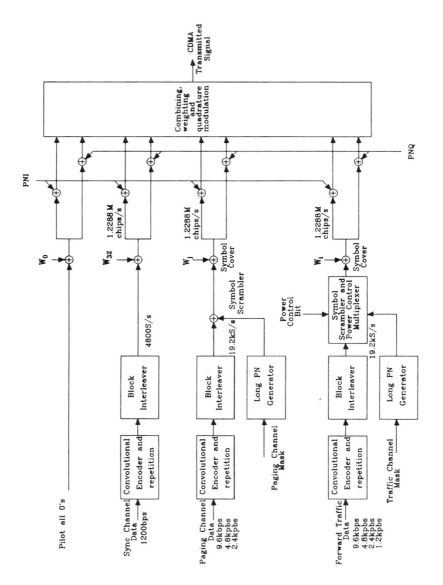

Figure 1.42: Block diagram of the Qualcomm CDMA BS transmitter.

interleaver. The scrambled data are multiplexed with the power control information which essentially steals bits from the scrambled data. The multiplex signal remains at 19.2 kb/s and is changed to 1.2288 Mchip/s by the Walsh code W_i assigned to the user's traffic channel. This signal is spread at 1.2288 Mchip/s by the pilot quadrature PRBS signals, and the resulting quadrature signals are then weighted.

The last set of channels are the paging channels. They provide the MSs with system information and instructions, in addition to acknowledgement messages following access requests made on the MSs' access channels. Essentially the paging channel data are processed in a similar way to the traffic channel data; but with the following exceptions. There is no variation in the power level on a per frame basis, and the 42 bit mask used to generate the long code contains different data.

All the 64 CDMA channels are combined to give single I and Q channels. These signals are applied to quadrature modulators and the resulting signals summed to form a CDMA/QPSK signal. The resulting CDMA signal is linearly amplified.

The pilot CDMA signal transmitted by a BS provides a coherent carrier reference for all MSs to use in their demodulation process. The transmitted pilot signal level for all BSs is some 4 to 6 dB higher than a traffic channel and is of constant value. The pilot signals are quadrature PRBS signals with a period of 32768 chips. As the chip rate is 1.2288 Mchip/s($= 128 \times 9600$, where 9600 is the maximum bit rate of the speech codec) the pilot PRBS corresponds to a period of 26.66 ms, equivalent to 75 pilot channel code repetitions every two seconds. The pilot signals from all BSs use the same PRBS, but each BS is characterised by a unique time offset of its PRBS. These offsets are in increments of 64 chips providing 511 unique offsets relative to the zero offset code. These large numbers of offsets ensure that unique BS identification can be performed, even in dense micro-cellular environments.

A MS processes the pilot channel and finds the strongest signal components. The processed pilot signal provides an accurate estimation of the time delay, phase and magnitude of three of the multipath components. These components are tracked in the presence of fast fading, and coherent reception with combining is used. The chip rate on the pilot channel, and on all channels is locked to precise system time, e.g., by using the Global Positioning System (GPS). Once the MS identifies the strongest pilot offset by processing the multipath components from the pilot channel correlator, it examines the signal on its synchronisation channel which is locked to the PRBS signal on the pilot channel.

All synchronisation channels use the same code W_{32} to spread their data. The information rate on the synchronisation channel is 1200 b/s (although its chip rate is 1.2288 Mchips/s). Because the synchronisation channel is time aligned with its BS's pilot channel, the MS finds on the synchronisation channel the information pertinent to this particular BS.

The synchronisation channel message contains time-of-day and long code synchronisation to ensure that the long code generators at the BS and MS are aligned and identical.

The MS now attempts to access the paging channel, and listens for system information. The MS enters the idle state when it has completed acquisition and synchronisation. It listens to the assigned paging channel and is able to receive and initiate calls. When told by the paging channel that voice traffic is available on a particular channel, the MS recovers the speech data by applying the inverse of the scrambling procedures shown in Figure 1.42.

1.6.1.2   Qualcomm CDMA Up-link

The up-link from MS to BS uses the same 32768 chip short code employed on the down-link, and the MS also uses its unique code embedded in the long $2^{42}-1$ PRBS. Speech is again convolutionally coded, this time using a rate 1/3 code of constraint length 9, and data are repeated depending on speech activity. Interleaving is then performed over the vocoder block length of 20 ms. However, the repeated symbols are not transmitted giving rise to a variable transmission duty cycle. Whereas the down-link uses one-bit symbols, the up-link groups the data into 6-bit symbols. Each symbol generates an appropriate 64-chip Walsh code which is then combined with both the long PRBS to bring the rate up to 1.2288 Mchip/s, and the short code to launch it onto the quadrature modulation channels. Notice that the Walsh codes are used in a totally different way in the two links. In the down-link the Walsh codes label the channels, while on the up-link they convey data to their only destination, namely the BS.

In the traffic operating mode, the Walsh coded signals at a MS are modulated by the long $2^{42}-1$ PRBS with a specific time offset that is unique to a particular MS, enabling the BS to distinguish signals arriving from different MSs. This long code is the same long PRBS that was used on the down-link. Further modulation by the quadrature 32768 chips PRBSs ensues, but with a fixed zero offset, followed by quadrature modulation. The reverse link does not use a pilot CDMA as to give each MS a pilot channel would be impracticable. The receiver at the BS has a tracking receiver and four receivers that each locks on to a significant path in the channel impulse response. The outputs of 64 correlators, one associated with each Walsh functions, are examined for each receiver. The outputs of the four correlation receivers are combined and the correlator number having the maximum output selected to identify the recovered 6-bit symbol. Progressive 6-bit symbols are serialised, de-interleaved and convolutionally decoded.

The reverse CDMA transmission can accommodate up to 62 traffic channels and up to 32 access channels per paging channel. The access channel enables the MS to communicate non-traffic information, such as originating calls and responding to paging. The access rate is fixed at 4.8 kb/s. The output duty cycle is 100%; and the access channel is identified by a long PN mask having an access number, a paging channel number (on the forward or down-link) associated with the access channel, and other system data.

Each BS transmits on the same frequency (initially there will only be one carrier frequency) a pilot signal of constant power. As mentioned above, this pilot signal is sent as a CDMA signal whose code identifies the BS to the MS. However, the received power level of the received pilot also enables the MS to estimate the BS to MS path loss (PL). Knowing the PL the MS adjusts its transmitted power such that the BS will receive the signal at the requisite power level. However, the MS transmits and receives in duplex frequency bands which although having similar average PL values, have different instantaneously received signal levels due to the independent fading in each transmit and receive band. To allow for this independent fading, the BS measures the MS received power and informs the MS to make the appropriate fine adjustment to its transmitter power. One command every 1.25 ms adjusts the transmitted power from the MS in steps of ± 0.5 dB. The dynamic range of the transmitted power is 85 dB.

The MS measures the SIR by comparing the desired signal power with the total interference and noise powers. If the SIR is below a threshold the MS requests the BS to increase its transmitter power, and vice versa when the SIR is above the threshold. The changes in MS transmitter power are small for this situation, $\simeq 0.5$ dB over a range of ± 6 dB, and are made at the vocoder frame rate.

An interesting feature of CDMA is that it can operate with single cell clusters. Neighbouring BSs transmit to their MSs using the same carrier frequency, although different codes are used for the pilot, set-up and traffic channels. As a MS moves to the edge of its single cell, i.e., cluster, the adjacent BS assigns a modem to the call, while the current BS continues to handle the call. The call is then handled by both BSs on a make-before-break basis. In effect, handover diversity occurs with both BSs handling the call until the MS moves sufficiently close to one of the BSs which then exclusively handles the call. This handover procedure is called a 'soft handover', as distinct to the more conventional break-before-make 'hard handover' method. Soft handovers are also made between sectors in a cell.

The Qualcomm CDMA system operates with a low E_b/N_o ratio, exploits voice activity, and uses sectorisation of cells. Each sector has 64 CDMA channels as previously described. It is a synchronised system, and there are three receivers to provide path diversity at the MS and four receivers are used at the cell site. The single cell cluster enables cells to be easily replicated along streets and into buildings. Complex frequency reassignment procedures required in TDMA and FDMA systems when small cells are introduced to alleviate teletraffic hot-spots are avoided. Nevertheless, CDMA/FDMA systems can be deployed to increase capacity if additional spectrum is available. Although conceived for the current range of conventional cell sizes, it is able to operate in micro-cells.

1.7   Cordless Telecommunications

Cordless telecommunications (CT) networks are designed for mobile radio coverage over relatively small distances, such as in office environments. Because of the small cells, or micro-cells, in which CTs operate the networks are basically much simpler than cellular ones.

The propagation environments in CT networks have less average delay spreads, but greater variability than cells in cellular networks due to the widely differing types of building construction. A discussion of the propagation in CT environments is provided in Section 1.2.7. Suffice to say that micro-cells and picocells are used in CT enabling very high bit rates per square kilometer or large values of Erlangs/MHz/km^2 to be achieved. This means that the complex speech codecs used in cellular radio can often be discarded in favour of simpler codecs that are inherently more robust to channel errors, have negligible encoding delays, and consume relatively little battery power. The higher bit rates that can be accommodated in CT also allow more bits to be used for synchronisation and control. The micro-cells yield other advantages. The excess path delay of the received radio signals is much lower than in cellular radio, and often much higher TDMA bit rates can be transmitted without the need for equalisation. Channel coding can often be avoided, a wider range of services accommodated, and so forth. We again emphasise that the most crucial factor determining system capacity in mobile radio system design is the cell size, and from a radio point of view, small is best!

1.7.1   CT2 System

The first operational digital CT is the British CT2 system [63]. CT2 has three principal applications; cordless PABX, cordless telephony, and telepoint in which the user can only call into the network. However, it must be stressed that in its non-telepoint mode it enables the cordless user either to be called, or to dial into the network.

CT2 operates in the band 864.1 to 868.1 MHz. The channel spacing is 100 kHz enabling 40 time division duplex (TDD or ping-pong) channels to be accommodated. In TDD one RF carrier supports transmissions on both the up-link and the down-link. To do this a frame structure shown in Figure 1.43 is used. ADPCM speech at 32 kb/s is sent and received as a B channel or bearer channel. The D channel or signalling channel is used to control the link. This arrangement, known as Multiplex One, is used to transfer signalling and data across an established link. The data are transmitted at rates 72 kb/s, with the first half of the frame for fixed

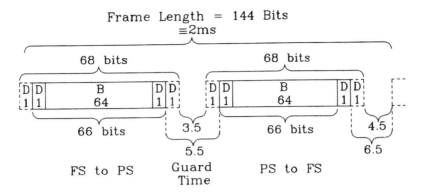

Figure 1.43: CT2 Multiplex One frame structure.

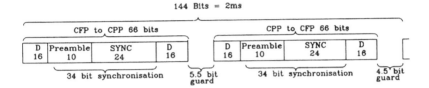

Figure 1.44: CT2 Multiplex Two frame structure.

station (FS) to portable station (PS) while the second half is for PS to FS transmissions. Notice that in CT terminology BS and MS becomes FS and PS, or cordless fixed part (CFP) and cordless portable part (CPP), respectively. Two or four bits on each link are assigned to the signalling channel at rates of 1 or 2 kb/s. All CT2 equipment must be able to operate using the 66-bit burst format, whereas the 68-bit burst is optional. When the shorter burst format is used the guard time is extended by two bits and hence the transmission rate remains at 72 kb/s, irrespective of the signalling rate. The frame length is 144 bits, corresponding to 2 ms.

Multiplex One is used once a link is established. However, in setting-up the link between the FS and the PS, Multiplex Two is used whose format is shown in Figure 1.44. The 34-bit synchronisation channel consists of a 10-bit alternate zero and one sequence, followed by a 24-bit sequence to facilitate burst synchronisation. The D channel has 32 bits resulting in a 16 kb/s control link. When the FS has a call for the PS it selects Multiplex Two, using its down-link part of the frame to call the PS with the information in the D channels, while the PS responds using the up-link part of the frame. The FS controls the timing of the PS transmissions.

When the PS initiates a call it uses Multiplex Three. The alternate up-link and down-link transmissions of Multiplexes One and Two are dis-

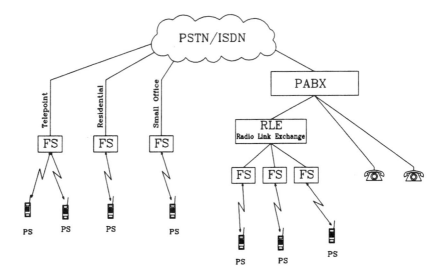

Figure 1.45: The DECT network.

carded. Instead the PS transmits five consecutive frames each having 144 bits. In the first frame, the first sub-multiplex group consists of a 6-bit preamble which is followed by a 10-bit D channel, by an 8-bit preamble, by a 10-bit D channel. The D channel is composed of 20 bits and contains the identity of the PS. Next come two sub-multiplex groups, each having an 8-bit preamble, followed by a 10-bit D channel, by another 8-bit preamble and 10-bit D channel. The last sub-multiplex group is the same as the two centre ones, except that it has a final 2-bit preamble. The same format of four sub-multiplex groups is repeated for the next three 144-bit frames. The fifth frame has a 12-bit preamble, followed by a 24-bit channel marker synchronisation word that informs the FS that a PS is calling. This arrangement is repeated three times to give a 144-bit fame. During the next two 144-bit frames the PS listens for a response from the FS. Multiplex Three continues to be used until a link is established, whence Multiplex One is used and speech transmissions commence.

1.7.2   Digital European Cordless Telecommunications System

In the Digital European Cordless Telecommunications (DECT) system, again, TDD is used, but with 12 channels per carrier and a carrier spacing of 1728 MHz in the 1880-1900 MHz band [63]. The TDMA rate is 1152 kb/s. DECT has a number of attractive features. It employs ADPCM at 32 kb/s; and GMSK modulation with the higher normalised bandwidth of 0.5 compared to the 0.3 used in GSM, thereby simplifying clock recovery procedures. No channel coding is used, neither is channel equalisation as the micro-cells are in general small and in office environments. Whereas GSM and CT2 have control channel rates of 967 and 2000 b/s, and control channel delays of 480 and 32 ms, respectively, the corresponding values for DECT are 6400 b/s and 5 ms. Thus DECT has a powerful control capability. A system similar to DECT is the Ericsson CT3 system. The DECT network arrangement is shown in Figure 1.45. The telepoint FSs are connected directly to the public switched telephone network (PSTN), as are the FSs installed in residential and office buildings. In larger offices, or complexes, e.g., airports, a number of FSs are connected to a radio link exchange (RLE) which distributes calls from neighbouring PABXs to the PSs they serve.

The DECT architecture consists of a concentrator which trunks N PSTN lines to L DECT lines. The link conversion unit (LCU) reformats the PSTN/DECT data to the new form. For example, incoming PSTN 64 kb/s A-law PCM speech is transcoded to 32 kb/s ADPCM for transmissions in the DECT network. The L lines carrying ADPCM speech are then conveyed to the RLE for distribution to the FSs and hence to the PSs. The transmission rate of the standard 2B+D ISDN channel is 144 kb/s, which is converted in the DECT system to $2 \times 32 + 6.4 = 70.4$ kb/s. It can also support the full ISDN rate of 144 kb/s.

The structure of the DECT TDMA frame is shown is Figure 1.46. The frame duration is 10 ms, and the TDMA slot length is 0.417 ms. In each slot a burst of 416 bits is transmitted at 1152 kb/s. At the commencement of a burst there is a 16-bit preamble followed by a 16-bit synchronisation code. Next comes 64 control bits conveyed on the C-channel, followed by 320 information bits on the I-channel. Notice that the information rate is 320 bits in 10 ms, i.e., 32 kb/s, the rate of the ADPCM speech encoder. The control rate is relatively high, 64 bits/10 ms = 6.4 kb/s. The control data are partitioned into control (C), paging (P) and broadcasting (Q), depending on the activity. A guard period of 64 bits equivalent to 52.1 μs is used. The preamble and sync code are present in every burst (the first 32 bits) to guarantee synchronisation of the time slot. The 16-bit synchronisation code on the down-link is inverted on the up-link transmissions.

A free channel is defined as one with a signal strength below a system threshold, or the channel with the lowest signal strength. This must be the situation on both links. When the FS has a call for a PS, it uses one or more free channels for signalling. On detecting its handshake code the PS responds. For calls initiated by the PS, one or more free channels are used to convey the PS handshake code to the FS for a period up to five seconds. Upon detection of the code the FS responds to the PS. The DECT system allows for handovers. The time between handovers must be at least three seconds, and one FS can use up to 75% of all available channels.

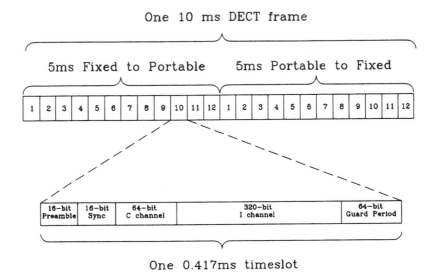

Figure 1.46: DECT TDMA frame structure.

Other features include multiple rate transmissions using more than one time slot per frame, and uni-directional transmissions, i.e., when the time slots in both halves of the frame can be used to receive or transmit data.

The data link layer is divided into two layers. Layer 2a comprises error protection, link quality assessment and handover control, while Layer 2b is concerned with the data link layer function. Layers 2a and 2b are called the Medium Access Control Layer and the Logical Link Control Layer, respectively. The 64-bit control field consists of an 8-bit header, followed by 40-bit content and finally a 16-bit cyclic redundancy check (CRC) code to combat transmission errors. The header defines the type of message, and whether the system is residential, business or telepoint. One header bit is used for paging and identifying the portable. The 40 bits of content are shared between layers 2a and 2b depending on the message type, and the service at the network being extended by DECT. The CRC code is a BCH (63,48,2) code.

1.7.3   Parameters of CTs and Cellular Systems

Both the digital cellular systems and the digital CT networks enable mobile users to access the national and international PSTNs and ISDNs. The first-generation cellular systems were conceived as mobile telephones. In the second-generation systems, however, a range of novel services, such as email, data and fax have been already incorporated. Nevertheless, second-generation cellular networks were designed for mobiles operating in large

System	TACS	GSM	DCS-1800	Qualcomm IS-95 CDMA	IS-54 DAMPS	JDC	CT2	DECT	PHS	PACS
Origin	UK	Europe	Europe	USA	USA	Japan	UK.	Europe	Japan	USA
Forward Band (MHz)	935-950	935-960	1805-1880	869-894	869-894	810-826 1477-1489 1501-1513 940-956 1429-1441 1453-1465	864-868	1880-1900	1895-1918	1930-1990
Reverse Band (MHz)	890-905	890-915	1710-1785	824-849	824-849		(TDD)	(TDD)	(TDD)	1850-1910
Multiple Access	FDMA	TDMA	TDMA	CDMA	TDMA	TDMA	FDMA	TDMA	TDMA	TDMA
Duplex	FDD	FDD	FDD	FDD	FDD	FDD	TDD	TDD	TDD	FDD
Carrier Spacing	25	200	200	1250	30	25	100	1728	300	300
Channels/carrier	1/pair	8/pair	8/pair	55-62	3	3	1	12	4	8/pair
Bandwidth/channel (kHz)	50	50	50	21	20	16.66	100	144	75	75
Modulation	FM	GMSK	GMSK	QPSK	$\pi/4$-DQPSK	$\pi/4$-DQPSK	FSK	GMSK	$\pi/4$-DQPSK	$\pi/4$-QPSK
Modulation Rate (kBd)	N/A	271	271	1228	48.6	42	72	1152	192	192
Voice+FEC Rate(kbps)	N/A	22.8	22.8	8/Var.	11.2	13	32	32	32	32
Speech codec	N/A	RPE-LTP	RPE-LTP	CELP	VSELP	VSELP	ADPCM	ADPCM	ADPCM	ADPCM
Unprotected Voice Rate(kbps)	N/A	13	13	1.2-9.6	7.95	6.7	32	32	32	32
Control Chan. Name		SACCH	SACCH	SACCH	SACCH	SACCH	D	C	SACCH	
Control Chan. Rate (bps)		967	967	800	600		2000	6400		4000
Control Message Size (bits)		184	184	1	65		64	64		10/2.5ms
Control Delay (ms)		480	480	1.25	240	0.3-3	32	10		
Peak Power (Mobile) (W)	0.6-10	2-20	0.25-2	0.6-3	0.6-3	0.3-3	10mW	250mW	80mW	200mW
Mean Power (Mobile) (W)	0.6-10	0.25-2.5	0.03-0.25	0.2-1	0.6-3	0.1-1	5mW	10mW	10mW	
Power Control	Yes	Yes	Yes	Yes	Opt.	Yes	No	No	Opt.	
Voice Activity Detection	Yes	Yes	Yes	Yes	Opt.	Yes	No	No	Yes	
Handover	Yes	Yes	Yes	Yes	Yes	Yes	No	Yes	Yes	Yes
Dynamic Channel Allocation	No	No	No	N/A	No	Opt.	Yes	Yes	Yes	autonom.
Min. Cluster Size	7	3	3	1	7	4	N/A	N/A	N/A	N/A
Capacity(Dpx ch/cell/MHz)	2.8	6.7	6.7	16.5 †	7		N/A	N/A	N/A	N/A
Frame duration(ms)	N/A	4.615	4.615	20	40	40	2	10	5	2.5
Speech FEC		Conv. (2,1,5)	Conv. (2,1,5)	Conv. Fwd:(2,1,9) Rev: R=1/3	Conv. (2,1,5)	Conv. R=9/17	No	No	CRC	CRC
Channel Eq.	N/A	Yes	Yes	Yes	Opt.	Opt.	No	No	No	No
Half-rate Codec (kbps)	N/A	5.6	5.6	No	No	3.2	No	No	16	No
Half-rate+FEC (kbps)	N/A	11.4	11.4			5.6				
Enhanced Full-rate (kbps)	N/A	12.2	12.2	Yes	7.4	No	No	No	No	No

DCS-1800: GSM-like European system in the 1800 MHz band
IS-95: American CDMA system
PHS: Japanese Personal Handyphone System
TACS: Total Access Communications System
CDMA: Code Division Multiple Access
FDMA: Frequency Division Multiple Access
TDMA: Time Division Multiple Access
FM: Frequency Modulation
DQPSK: Differential Quadrature Phase Shift Keying
RPE-LTP: Regular Pulse Excited - Long Term Predicted
VSELP: Vector Sum Excited Linear Predictive
SACCH: Slow Associated Control Channel

IS-54: American Digital Advanced Mobile Phone System (DAMPS)
CT2: British Cordless Telephone System
PACS: Personal Access Communications System
GSM: Global System of Mobile Communications
DECT: Digital European Cordless Telephone
FDD: Frequency Division Duplex
TDD: Time Division Duplex
GMSK: Gaussian Minimum Shift Keying
GFSK: Gaussian Phase Shift Keying
CELP: Code Excited Linear Predicted
ADPCM: Adaptive Differential Pulse Code Modulation
JDC: Japanese Digital Cellular

N/A means not applicable.

†Assumes $\frac{2}{3}$ frequency re-use

Table 1.1: Summary of second-generation mobile systems.

cells, although the arrival of the so-called DCS-1800 system - which is essentially a GSM-like system operating at 1800 MHz - heralded the commencement of the micro-cellular era.

CTs were designed for short-range communications in buildings and their immediate environments. CTs are essentially consumer products, being low cost, light-weight and inexpensive to buy and operate. Table 1.1 presents some parameters of the systems we have considered, spanning the range of first- and second-generation mobile systems as well as CTs operated across the globe. The exploration of their features is the subject of this book. Upon completing our portrayal of the mobile radio channel and the various system components in Chapter 2 - Chapter 7, Chapter 8 will provide an in-depth description of their interconnections in the context of the second-generation GSM system. This Chapter is then followed by the design of a variety of evolutionary wireless multimedia systems in Chapter 9 and by Chapter 10, which summarises the basic features of the forthcoming third-generation systems. Hence here we refrain from detailing the various system features in Table 1.1, noting that they will be detailed in the forthcoming chapters and will become tangible during our elaborations in the rest of the book. Let us now embark on a short discussion related to the various teletraffic features of wireless systems.

1.8   Teletraffic Considerations

Networks have far fewer mobile radio channels than users. This is because only a small percentage of subscribers make call attempts at any time. The network operator usually designs his network such that at the busy hour only a small percentage, say 2%, will have their call attempts blocked because all the channels are in use. The probability P_n of a request for a new call being denied must be significantly greater than the probability P_h of an existing call being forced to terminate due to a handover failure between two BSs.

There are many books on general traffic theory [64–66], but the literature dealing with the application of traffic theory to cellular radio is relatively sparse [67–69]. As this chapter attempts an overview of mobile radio we will only mention some of the salient points of the subject.

In conventional cellular radio where the cells are large, and handovers between BSs are relatively rare, we can assume that the number of MSs in a cell is so much greater than the number of BS channels N, that to a good approximation the number of users may be considered as infinite. In this situation the probability of a call attempt being blocked is given by

$$B = \frac{(\lambda/\mu)^N/N!}{\sum_{k=0}^{N}(\lambda/\mu)^k/k!} \tag{1.33}$$

where λ is the mean call arrival rate, and μ is the mean rate at which calls are terminated, i.e., cleared from the system. Equation (1.33) is known as the Erlang B formula or Erlang's first formula. It is derived on the basis that no queueing occurs so that if all N channels are busy the call is blocked and the user may try again. The offered traffic to the BS is

$$A = \lambda/\mu = \lambda T_c; \text{Erlangs} \qquad (1.34)$$

where T_c is the mean value of the call durations, or the mean channel holding time. The blocked traffic is AB, while the traffic carried by the BS is $A(1-B)$. The call requests arriving at the BS, and being cleared by the network, are independent events. It is interesting to notice that the arrival of call requests at the BS, and the number of motor vehicles and pedestrians (both potential MSs) passing a point on the road-side are Poisson distributed, while the separation between calls, motor vehicles and pedestrians are exponentially distributed. This observation is important when simulating mobile radio systems with a view to obtaining teletraffic results.

In micro-cells the number of MSs, M, may not be significantly different from N. If the total offered traffic to the micro-cellular BS is A, where A is small and the blocking probability is zero, then the average offered traffic per user is $a = (A/M) = (\alpha/\mu)$, where α is the reciprocal of the average inter-arrival time. When j of the N BS channels are in use the mean call arrival rate is $\lambda_j = (M-j)\alpha$, and $\mu_j = j\mu$. The probability of N channels being busy is

$$P_N = \frac{\binom{M}{N}a^N}{\sum_{k=0}^{N} \binom{M}{k}a^k} \qquad (1.35)$$

which is the fraction of time during which all channels are in use.

Another important probability is the probability that a call is attempted when all channels are in use. This probability is given by

$$P_B = \frac{\binom{M-1}{N}a^N}{\sum_{k=0}^{N} \binom{M-1}{k}a^k} \qquad (1.36)$$

and a can also be expressed as

$$a = \frac{A}{M - A(1-B)}. \qquad (1.37)$$

These equations are known as Engset formulae [64].

Determining the number of channels required at a BS in a conventional large cell system is straightforward. Having decided on the offered traffic per BS, i.e., the total offered traffic $A = \lambda/\mu$ in Erlangs, and the blocking probability B, the value of N is next determined using the Erlang B formula. This formula is tabulated [65] in terms of A, N and B, enabling N to be found quickly. It is a straightforward procedure to then calculate the effects on A, N and B of sectorisation of cells using directional antennas, and the splitting of a large cell into smaller cells.

Applying teletraffic theory to micro-cells is much more difficult due to MSs frequently making handovers [68, 69]. This means that the channel holding time in the micro-cell becomes a fraction of the call duration. The average channel holding time is

$$\overline{T}_H = \gamma_n \overline{T}_{Hn} + \gamma_h \overline{T}_{Hh} \qquad (1.38)$$

where $\overline{T}_{Hn}$ and $\overline{T}_{Hh}$ are the average holding times of the new call and the handover call, respectively, and

$$\gamma_n = (1 - \gamma_h) \qquad (1.39)$$

is the ratio of the carried new call rate to the total call rate, while γ_h is the ratio of carried handover rate to the new call rate. The total mean arrival rate at a BS is

$$\lambda_T = \lambda_N + \lambda_H \qquad (1.40)$$

where λ_N and λ_H are new call rate and the handover request rate, respectively.

The traffic carried by the micro-cell BS is

$$A_{cm} = A_T(1 - P_{bn}) \qquad (1.41)$$

where P_{bn} is the probability of a new call attempt being blocked. This traffic formula is applicable for the case where no priority assignment of channels is provided for handover requests. In this situation P_{bn} is the same as the probability of a call being forced to terminate when requesting a handover. To decrease the probability of handover failure we use a macrocell to oversail a cluster of micro-cells. The traffic carried by one macrocell BS is

$$A_{CM} = A_M(1 - P_{fhM}) \qquad (1.42)$$

where A_M is the traffic offered by the cluster of micro-cellular BSs to the macrocellular BS due to the former BSs not having sufficient channels to accommodate the demand for handovers, and P_{fhM} is the probability of handover failure in the macrocell. The total traffic carried by the network is

$$A_{CT} = C_m A_{cm} + C_M A_{CM} \qquad (1.43)$$

where C_m and C_M are the total number of micro-cells and macrocells, respectively.

The channel utilisation is defined as the carried traffic per channel. With each micro-cellular BS having N channels, and each macrocellular

BS having N_o channels, the average channel utilisation is

$$\rho = \frac{A_{cm} + A_{CM}}{N + N_o}.$$ (1.44)

If we define the spectral efficiency η in terms of Erlangs per Hz per m^2, and we assume that to a first approximation that the number of channels carried by the macrocells is very small compared to those carried by the micro-cells, then

$$\eta = \frac{A_{CT}}{S_T W} \simeq \frac{\rho}{S_m M_{mc} B_c}$$ (1.45)

where S_T is the total area covered by the network, S_m is the average area of each micro-cell, W is the total available bandwidth allocated to the network, B_c is the equivalent bandwidth per channel, and M_{mc} is the number of micro-cells per cluster. Notice that for high η: S_m should be small, i.e., deploy micro-cells, B_c should be small and depends on the modulation and multiple access method used; while the number of micro-cells per cluster should be small. One of the virtues of CDMA is that it meets these requirements, particularly $M_{mc} = 1$. It might appear that to increase ρ the number of channels should be decreased. However, if the grade-of-service (GOS) is to be maintained, the blocking probability must remain low. As a consequence it is more appropriate to increase N which will increase ρ, and therefore η, for the same blocking probability.

This opening chapter has attempted to introduce the reader to digital cellular radio using a bottom-up approach. The mathematics has been kept to a minimum, with the emphasis on concepts. We now embark on a series of chapters in which the subject treatment is more detailed and focused, commencing with a detailed discourse on mobile radio propagation in the next chapter.

Bibliography

[1] "Car radio telephones", *The Autocar*, pp.A22-A23, 30 October 1936.

[2] **W.C.Jakes**, "Microwave mobile communications", *John Wiley & Sons*, New York, 1974.

[3] **W.Y.C.Lee**, "Mobile cellular communications", *McGraw Hill*, New York, 1989.

[4] **J.D.Parsons** and **J.G.Gardiner**, "Mobile communication systems", *Blackie*, London 1989.

[5] **R.Steele**, "Towards a high capacity digital cellular mobile radio system", *Proc. of the IEE*, PtF, No.5, pp.405-415, August 1985.

[6] **R.Steele** and **V.K.Prabhu**, "Mobile radio cellular structures for high user density and large data rates", *Proc. of the IEE*, Pt F, No.5, pp.396-404, August 1985.

[7] **R.Steele**, "The cellular environment of lightweight hand-held portables", *IEEE Communications Magazine*, pp.20-29, July 1989.

[8] **V.H.MacDonald**, "The cellular concept", *Bell System Tech. J.*, Vol 58, No.1, pp.15-41, January 1979.

[9] **J.D.Parson** "The mobile radio propagation channel", Pentech Press, London, 1992.

[10] **E.Green**, "Radio link design for micro-cellular systems", *British Telecom Technology J*, Vol 8, No.1, pp.85-96 January 1990.

[11] GSM Recommendation 05.05, Annex 3, pp.13-16, November 1988.

[12] **K.Bullington**, "Radio propagation at frequencies above 30 Mc/s", *Proc. IRE 35*, pp.1122-1136, (1947).

[13] **R.Edwards** and **J.Durkin**, "Computer prediction of service area for VHF mobile radio networks", *Proc IRE 116 (9)* pp.1493-1500, 1969.

[14] **Y.Okumuma, E.Ohmori, T.Kawano** and **K.Fukuda**, "Field strength and its variability in VHF and UHF land-mobile radio service", *Review of the Elec. Comm. Lab.*, Vol 16, No.9 and 10, pp.825-873, 1968.

[15] **M.Hata**, 'Empirical formula for propagation loss in land mobile radio services", *IEEE Trans. Veh. Technol.*, Vol VT-29, No.3, pp.317-325, August 1980.

[16] **R.Steele**, "The importance of propagation phenomena in personal communication networks". *IEE 7-th International Conference on Antennas and Propagation*, ICAP, York, No.333, Pt I, pp.1-5, April 1991.

[17] **S.T.S.Chia, R.Steele, E.Green** and **A.Baran**, "Propagation and bit error ratio measurements for a micro-cellular system", *J IERE*, Vol 57, No.6 (Supplement), pp.S255-S266, November/December 1987.

[18] **A.J.Rustako, N.Amitay, G.J.Owens** and **R.S.Roman**, "Radio propagation measurements at microwave frequencies for micro-cellular mobile and personal communications", *IEEE ICC'89*, pp.482-488, Boston, USA, 1989.

[19] **S.T.S.Chia**, "Radiowave propagation and handover criteria for microcells", *British Telecom Tech J*, Vol 8, No.4, pp.50-61, October 1990.

[20] **R.Bultitude** and **G.Bedal**, "Propagation characteristics on micro-cellular urban mobile radio channels at 910 MHz", *IEEE J-SAC*, Vol 7, No.1, pp.31-39, January 1989.

[21] **W.T.Webb** and **R.Steele**, "Equaliser techniques for QAM transmission over dispersive mobile radio channels", *IEE Proc.-I*, Vol 138, No.6, pp.566-576, December 1991.

[22] **J.M.Keenan** and **A.J.Motley**, "Radio coverage in buildings", *British Telecom Technol J*, Vol 8, No.1, pp.19-24, January 1990.

[23] **F.C.Owen** and **C.D.Pudney**, "In-building propagation at 900 MHz and 1650 MHz for digital cordless telephones", *6th Int Conf on Antennas and Propagation*, ICAP'89, Pt 2: Propagation, Conf Pub No.301, pp.276-280, 1989.

[24] **R.J.C.Bultitude, P Melancon** and **J.LeBel**, "Data regarding indoor radio propagation", *Wireless'90*, Calgary, Canada, July 1990.

[25] **D.C.Cox**, "Universal digital portable radio communications", *Proc IEEE*, Vol 75, No.4, pp.436-477, April 1987.

[26] **A.A.M.Saleh** and **R.A.Valenzuela**, "A statistical model for indoor multipath propagation", *IEEE J-SAC*, pp.128-137, February 1987.

[27] **D.M.J.Devasirvatham**, "Time delay spread measurements of wideband radio signals within a building", *Electronics Letters*, Vol 20, No.23, pp.951-952, 8 November 1984.

[28] **R.J.C.Bultitude, S.A.Mahmoud** and **W.A.Sullivan**, "A comparison of indoor radio propagation characteristics at 910 MHz and 1.75 GHz", *IEEE JSAC*, Vol 7, No.1, pp.20-30, January 1989.

[29] **R.Davies, A.Simpson** and **J.P.McGeehan**, "Propagation measurements at 1.7 GHz for micro-cellular urban communications", *Electronic Letters*, Vol 26, No.14, pp.1053-1054, 5 July 1990.

[30] **E.Damosso, L.Stola** and **G.Brussaard**, "Characterisation of the 50-70 GHZ band for space communications", *ESA J*, 7, pp.25-43, 1983.

[31] **R.H.Ott** and **M.C.Thompson**, "Atmospheric amplitude spectra in an absorption region", *Proceedings of IEEE AP-S symposium*, Amherst, MA, USA, pp.594-597, 1976.

[32] **O.E.De Lange, A.F.Dietrich** and **D.C.Hogg**, "An experiment on propagation of 60 GHz waves through rain", *Bell Syst. Tech.J.*, 54, pp.165-176, 1975.

[33] **S.E.Alexander** and **G.Pulgliese**, "Cordless communication within buildings: results of measurements at 900 MHz and 60 GHz", *Br Telecom Tech.,J*, 1, pp.99-105, 1983.

[34] **S.Chia, D.Greenwood, D.Rickard, C.R.Shephard** and **R.Steele**, "Propagation studies for a point-to-point 60 GHz micro-cellular system for urban environments", *IEE Communications '86*, Birmingham, pp.28-32, 13-15 May 1986.

[35] **J.G.Proakis**, "Digital communications", *McGraw Hill*, 1989.

[36] **W.C.Y.Lee**, "Overview of cellular CDMA", *IEEE Trans. on Veh. Technol*, Vol 40, No.2, pp.291-302, May 1991.

[37] **W.H.Lam** and **R.Steele**, "Performance of direct-sequence spread spectrum multiple access systems in mobile radio", *IEE Proc-I*, Vol 138 No.1, pp.1-14, February 1991.

[38] **J.K.Holmes**, "Coherent spread spectrum systems", *John Wiley*, New York, 1981.

[39] **G.R.Cooper** and **R.W.Nettleton**, "A spread-spectrum technique for high capacity mobile communications", *IEEE Trans. Veh. Technol*, Vol VT-27, pp.264-275, November 1978.

[40] **K.S.Gilhousen, I.M.Jacobs, R.Padovani, A.J.Viterbi, L A Weaver** and **C E Wheatley**, "On the capacity of a cellular CDMA system", *IEEE Trans. Veh. Technol*, Vol 40, No.2, pp.303-312, May 1991.

[41] **A.J.Viterbi**, "Wireless digital communication: a view based on three lessons learned", *IEEE Communications Magazine*, pp.33-36, September 1991.

[42] **B.G.Haskell** and **R.Steele**, "Audio and video bit rate reduction", *Proc. IEEE*, Vol 69, No.2, pp.252-262, February 1981.

[43] **N.S.Jayant** and **P.Noll**, "Digital coding of waveforms", *Prentice-Hall*, 1984.

[44] **T.Aoyama, W.R.Daumer** and **G.Modena**, "Voice coding for communications", *Special Issue of IEEE JSAC*, Vol 6, No.2, pp.225-452, February 1988.

[45] **W.C.Wong, R.Steele, B.Glance** and **D.Horn**, "Time diversity with adaptive error detection to combat Rayleigh fading in digital mobile radio", *IEEE Trans*, COM-31, pp.378-387, March 1983.

[46] **C-E.Sundberg, W.C.Wong** and **R.Steele**, "Weighting strategies for companded PCM transmitted over Rayleigh fading and Gaussian channels", *Bell Syst. Tech. J.*, Vol 63, No.4, pp.587-626, April 1984.

[47] **R.Steele, C-E.Sundberg** and **W.C.Wong**, "Transmission errors in companded PCM over Gaussian and Rayleigh fading channels", *Bell Syst. Tech. J.*, Vol 63, No.6, pp.955-990, July/ August 1984.

[48] **W.C.Wong, R.Steele** and **C-E.W.Sundberg**, "Soft decision demodulation to reduce the effect of transmission errors in logarithmic PCM transmitted over Rayleigh fading channels", *Bell Syst. Tech. J.*, Vol 63, No.10, pp.2193-2213, December 1984.

[49] **R.Steele, C-E.W.Sundberg** and **W.C.Wong**, "Transmission of log-PCM via QAM over Gaussian and Rayleigh fading channels", *IEE Proc.* Vol 134, Pt.F, No.6, pp.539-556, October 1987.

[50] **C-E.W.Sundberg, W.C.Wong** and **R.Steele**, "Logarithmic PCM weighted QAM transmission over Gaussian and Rayleigh fading channels", *IEE proc.*, Vol 134, Pt.F, No.6, pp.557-570, October 1987.

[51] **R.Steele**, "Delta modulation systems", *Pentech Press*, London, 1975.

[52] **L.Hanzo, R.Steele** and **P-M.Fortune**, "A subband coding, BCH coding and 16-QAM system for mobile radio speech applications", *IEEE Trans. Veh. Technol.*, Vol 39, No.4, pp.327-339, November 1990.

[53] **R.C.Cox** "Robust CELP coders for noisy background and noisy channels", *IEEE Proc ICASSP'89*, Glasgow, pp.739-742, 23-26 May 1989.

[54] **L.Hanzo, R.Salami** and **R.Steele**, " A 2.1 kBd speech transmission system for Rayleigh fading channels", *IEE Colloquium on Speech Coding*, London, Digest No.1989/112, pp.10/1-10/5, 9 October 1989.

[55] **P.Vary**, "GSM speech codec", *Digital Cellular Radio Conference*, Hagen, Germany, paper 2a, October 1988.

[56] **J.L.Flanagan**, "Speech analysis, synthesis and perception", *Springer-Verlag*, Berlin, 1972.

[57] **R.E.Blahut**, "Theory and practice of error control codes", *Addison-Wesley*, 1983.

[58] **W.T.Webb, L.Hanzo** and **R.Steele**, "Bandwidth efficient QAM schemes for Rayleigh fading channels", *IEE Proc.-I*, Vol 138, No.3, pp.169-175, June 1991.

[59] **A.P.Clark**, "Advanced data-transmission systems", *Pentech Press*, London, 1977.

[60] **A.Duel-Hallen** and **C.Heegard**, "Delayed decision-feedback sequence estimation," *IEEE Trans. on Comms.*, Vol COM-37, pp.428-436, May 1989.

[61] **J.C.S.Cheung**, "Adaptive equalisers for wideband time division multiple access mobile radio", *PhD Thesis*, University of Southampton, England, 1992.

[62] **D.J.Goodman**, "Second-generation wireless information networks", *IEEE Trans. Veh. Technol.*, No.2, pp.336-374, May 1991.

[63] **W.H.W.Tuttlebee** (Ed), "Cordless telecommunication in Europe", *Springer-Verlag*, 1990.

[64] **M.Schwartz**, "Telecommunication Networks: protocols, modeling and analysis", *Addison-Wesley*, 1987.

[65] **J.R.Boucher**, "Voice teletraffic systems engineering", *Artech House*, 1988.

[66] **L.Kleinrock**, " Queueing systems volume 1: theory", *John Wiley & Sons, New York*, 1975

[67] **D.Hong** and **S.S.Rappaport**, "Traffic model and performance analysis for cellular mobile radio telephone systems with prioritized and nonprioritized hand-off procedures", *IEEE Trans. Veh. Technol.*, Vol VT-3, pp.77-92, August 1986.

[68] **S.A.Dolil, W.C.Wong** and **R.Steele**, "Teletraffic performance of highway micro-cells with overlay macrocell", *IEEE JSAC*, Vol 7, No.1., pp.71-78, January 1989.

[69] **R.Steele** and **M.Nofal**, "Teletraffic performance of micro-cellular performance of micro-cellular personal communications networks", *IEE Proc.-I*, Vol. 139, No.4, August 1992.

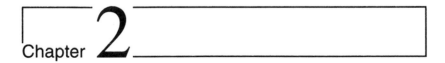

Chapter 2

Characterisation of Mobile Radio Channels

D. Greenwood[1] and L. Hanzo[2]

Modern society is continually demanding more and better communications services [1]. There is a very real market for a global network allowing voice and data communications between any two points on the earth's surface, no matter how remote. A key factor in the realisation of such a network is the ability to provide multimedia services via cellular mobile radio systems. Unfortunately, mobile radio channels are extremely harsh media for information transmission. It is the intention of this chapter to introduce the reader to the behaviour of mobile radio channels and to their characterisation.

In Section 2.2 several mobile radio channel types are defined, and their usage is discussed. The characteristics of each of these channels are examined in terms of their physical structure in Section 2.3. By considering the bandwidth and duration of the information signals they carry, channels can be classified according to their invariance properties. This concept is discussed in Section 2.4 along with the resulting simplifications.

Section 2.5 introduces the Bello system functions—a set of functions which describe general linear time-variant channels with a powerful mathematical elegance. These functions are commonly used to describe mobile radio channels not only for their simplicity and ease of manipulation, but also because they assist in the intuitive understanding of channel behaviour. Simplifications to the general theory are discussed as statistical constraints are placed on the channel, leading to the development of the

[1]University of Southampton
[2]University of Southampton and Multiple Access Communications Ltd

QWSSUS (Quasi Wide Sense Stationary Uncorrelated Scattering) channel. Section 2.6 discusses the application of the QWSSUS model to the characterisation of mobile radio channels, while Section 2.7 portrays a practical approach to their description for the system-designer.

The chapter commences with a review of the complex notation often used in the study of time-variant linear channels. This notation will be used extensively throughout this chapter.

2.1   Complex Baseband Representation of Bandpass Signals and Systems

A useful tool for studying bandpass communication systems is the complex lowpass equivalent notation. It provides a mathematical shorthand which bypasses the tedious trigonometry that generally accompanies the mathematical manipulation of signals modulated onto a sinusoidal carrier. The following description pursues the approach of Stein and Jones [2].

2.1.1   Bandpass Signals

The general form of a bandpass signal, $x(t)$, having a carrier frequency f_c is

$$x(t) = A(t) \cos[2\pi f_c t + \phi(t)], \tag{2.1}$$

where either, or both, of the amplitude, $A(t)$, and phase, $\phi(t)$, are used to carry the message information, such as digitally encoded speech. Trigonometric expansion of this equation yields

$$x(t) = u_I(t) \cos 2\pi f_c t - u_Q(t) \sin 2\pi f_c t, \tag{2.2}$$

where

$$u_I(t) = A(t) \cos \phi(t) \tag{2.3}$$

and

$$u_Q(t) = A(t) \sin \phi(t) \tag{2.4}$$

are the envelopes of the two quadratic carrier frequency components. Using complex notation, we may write

$$x(t) = \Re \left\{ x_+(t) \right\}, \tag{2.5}$$

where $x_+(t)$ is referred to as the pre-envelope of $x(t)$ [3] or the analytic signal. The pre-envelope of a bandpass signal can be represented as a phasor, with the bandpass signal given by the image of the pre-envelope along the real axis. Figure 2.1 shows a phasor representation of $x_+(t)$.

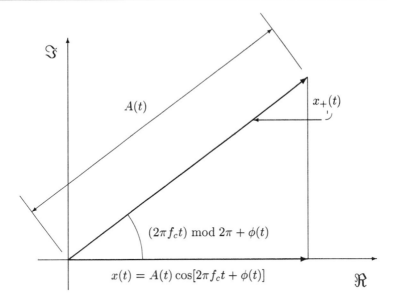

Figure 2.1: Phasor representation of the pre-envelope of a bandpass signal.

It is seen from Equations 2.2 and 2.5 above that

$$x_+(t) = u(t) \exp(j2\pi f_c t), \qquad (2.6)$$

where

$$u(t) = u_I(t) + j u_Q(t). \qquad (2.7)$$

The waveform $u(t)$ is referred to as the complex envelope, or complex low-pass equivalent, of $x(t)$. It is also a phasor, as shown in Figure 2.2. The pre-envelope, $x_+(t)$, is obtained by rotating the phasor $u(t)$ with an angular velocity $2\pi f_c$. In many modulators, the modulation waveform is generated from its complex envelope. Figure 2.3 shows how this is accomplished.

Equations 2.5 and 2.7 reveal that

$$x(t) = \Re \left\{ u(t) \exp(j2\pi f_c t) \right\}. \qquad (2.8)$$

This equation shows that knowledge of $u(t)$ and f_c completely describes the signal $x(t)$. As all the message information is represented by $u(t)$, it is common to describe $x(t)$ by its complex lowpass equivalent alone—the presence of a carrier frequency being implied.

To establish how the frequency spectra of $x(t)$ and $u(t)$ [$X(f)$ and $U(f)$ respectively] are related, consider the spectrum of the bandpass signal $x(t)$. That is,

$$X(f) = \int_{-\infty}^{\infty} x(t) \exp(-j2\pi f t)\, dt. \qquad (2.9)$$

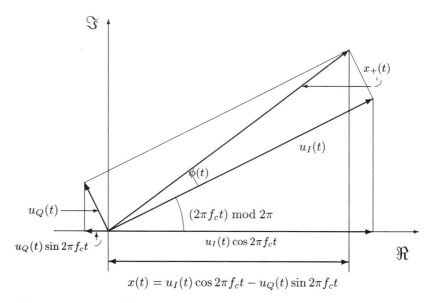

Figure 2.2: Phasor representation of the complex envelope of a bandpass signal.

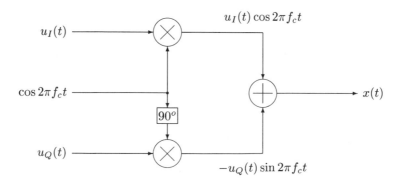

Figure 2.3: Generation of a bandpass signal from its complex envelope.

Substituting for $x(t)$ in the above equation from Equation 2.8 gives

$$X(f) = \int_{-\infty}^{\infty} \Re\{u(t)\exp(j2\pi f_c t)\}\exp(-j2\pi ft)\,dt. \qquad (2.10)$$

It is easily shown that the real part of a complex variable, z, can be written as

$$\Re\{z\} = \frac{1}{2}\{z + z^*\}, \qquad (2.11)$$

where z^* is its complex conjugate. Hence

$$X(f) = \frac{1}{2}\int_{-\infty}^{\infty}[u(t)\exp(j2\pi f_c t) + u^*(t)\exp(-j2\pi f_c t)]\exp(-j2\pi ft)\,dt. \qquad (2.12)$$

Defining the spectrum $U(f)$ as the Fourier transform of $u(t)$,

$$U(f) = \int_{-\infty}^{\infty} u(t)\exp(-j2\pi ft)\,dt, \qquad (2.13)$$

enables us to express Equation 2.12 as

$$X(f) = \frac{1}{2}[U(f - f_c) + U^*(-f - f_c)]. \qquad (2.14)$$

Figure 2.4 shows an example amplitude spectrum for a bandpass signal $X(f)$ and the corresponding lowpass spectrum of $U(f)$.

In the cellular mobile radio environment the bandwidth of information carrying signals is always very much less than the carrier frequency. Thus, as shown in Figure 2.4, the two components of $X(f)$ in Equation 2.14 do not overlap in frequency. If they did overlap then the complex notation could still be employed, but would require the explicit use of Hilbert transforms [3]- [5].

2.1.2   Linear Bandpass Systems

Linear bandpass systems, exemplified here by the mobile radio channel, can also be described using complex notation. The impulse response of such a bandpass system, $g(t)$, can be written as the inverse Fourier transform of its frequency response, $G(f)$,

$$g(t) = \int_{-\infty}^{\infty} G(f)\exp(j2\pi ft)\,df. \qquad (2.15)$$

A representative frequency response for $G(f)$ is shown in Figure 2.5. It is seen to comprise identical components centred at $f = \pm f_c$. In order to derive a complex lowpass equivalent representation for a bandpass system we must first establish $g(t)$ in terms of just one of these components, say

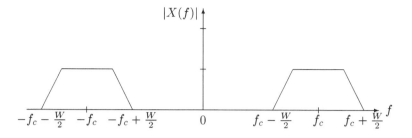

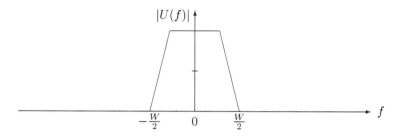

Figure 2.4: Amplitude spectra relationships.

that centred at $f = f_c$. This component can then be frequency translated to baseband, resulting in the frequency response for the complex lowpass equivalent system.

This is accomplished in the following manner. The range of integration of the integral in Equation 2.15 is split as,

$$g(t) = \int_0^\infty G(f) \exp(j2\pi ft)df + \int_{-\infty}^0 G(f) \exp(j2\pi ft)\,df, \qquad (2.16)$$

and the variable $f' = -f$ is substituted into the second integral to give

$$g(t) = \int_0^\infty G(f) \exp(j2\pi ft)df + \int_0^\infty G(-f') \exp(-j2\pi f't)\,df. \qquad (2.17)$$

It is noted that a physical system must possess a real impulse response, and that for a real $g(t)$ the following equality holds,

$$G(-f) = G^*(f). \qquad (2.18)$$

Equation 2.17 then becomes,

$$g(t) = \int_0^\infty G(f) \exp(j2\pi ft)df + \int_0^\infty G^*(f') \exp(-j2\pi f't)\,df, \qquad (2.19)$$

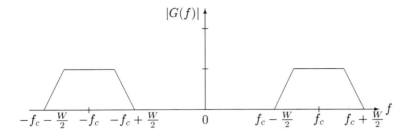

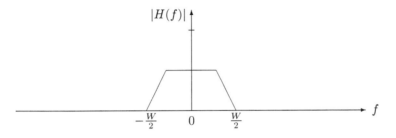

Figure 2.5: Frequency response of a bandpass system and its lowpass complex representation.

which when compared with Equation 2.11 reveals that

$$g(t) = 2\Re \left\{ \int_0^\infty G(f) \exp(j2\pi ft) \, df \right\}. \qquad (2.20)$$

Having obtained $g(t)$ in terms of the positive frequency component of $G(f)$, the bandpass to complex lowpass equivalent can be effected. The spectrum of a lowpass equivalent system is defined such that it is equal to the positive frequency component of $G(f)$ centred on zero frequency. That is,

$$H(f - f_c) = \begin{cases} G(f) & f > 0 \\ 0 & f < 0. \end{cases} \qquad (2.21)$$

Equation 2.20 is then equivalent to

$$g(t) = 2\Re \left\{ \int_{-\infty}^\infty H(f - f_c) \exp(j2\pi ft) \, df \right\}, \qquad (2.22)$$

or

$$g(t) = 2\Re \left\{ h(t) \exp(j2\pi f_c t) \right\}, \qquad (2.23)$$

where

$$h(t) = \int_{-\infty}^\infty H(f) \exp(j2\pi ft) \, dt \qquad (2.24)$$

is the complex lowpass equivalent impulse response of the system. As with $u(t)$, $h(t)$ can be written in terms of its inphase and quadrature components

$$h(t) = h_I(t) + jh_Q(t). \tag{2.25}$$

It has thus been shown that in much the same way as bandpass signals, bandpass systems may be fully described by knowledge of their complex lowpass equivalent, and their centre frequency.

It is easily shown from Equations 2.18 and 2.21 that the frequency response of the system is described by

$$G(f) = H(f - f_c) + H^\star(-f - f_c). \tag{2.26}$$

This equation differs in form from that of Equation 2.14 by a factor of a half (compare Figures 2.4 and 2.5). The reason for this difference will become apparent after reading the next section.

2.1.3   Response of a Linear Bandpass System

The response of a linear bandpass system, $y(t)$, must be a bandpass signal, even if the input to the system is not bandlimited. Referring to Section 2.1.1 we see that this response can therefore be represented in the manner of Equation 2.8, i.e.,

$$y(t) = \Re\left\{z(t)\exp(j2\pi f_c t)\right\}, \tag{2.27}$$

where $z(t)$ is the complex lowpass equivalent signal, namely, the complex envelope of $y(t)$. Furthermore, it is seen from Equation 2.14 that the signal's spectrum is given by,

$$Y(f) = \frac{1}{2}[Z(f - f_c) + Z^\star(-f - f_c)]. \tag{2.28}$$

A bandpass system, described in the frequency domain by the transfer function $G(f)$, with an input signal spectrum of $X(f)$, has an output

$$Y(f) = G(f)X(f). \tag{2.29}$$

Using Equations 2.14, and 2.26, Equation 2.29 is expanded to give,

$$Y(f) = \frac{1}{2}[H(f - f_c) + H^\star(-f - f_c)][U(f - f_c) + U^\star(-f - f_c)]. \tag{2.30}$$

As indicated on page 95, the bandwidth of signals encountered in the mobile radio environment are small compared with the carrier frequency. Hence the terms in the product $H(f - f_c)U^\star(-f - f_c)$ do not overlap in frequency, and the product equates to zero. Similarly, the product $H^\star(-f - f_c)U(f - f_c)$ is equal to zero.

The spectrum of the received RF signal is therefore

$$Y(f) = \frac{1}{2}[H(f - f_c)U(f - f_c) + H^*(-f - f_c)U^*(-f - f_c)]. \quad (2.31)$$

This Equation may be compared with Equation 2.28 to obtain the complex lowpass equivalent relationship,

$$Z(f) = H(f)U(f), \quad (2.32)$$

from which the complex envelope of $y(t)$ may be deduced as,

$$z(t) = \int_{-\infty}^{\infty} h(\xi)u(t - \xi) \, d\xi. \quad (2.33)$$

This may also be written as

$$z(t) = h(t) \star u(t), \quad (2.34)$$

where $\star$ represents convolution.

Substituting for $h(t)$ and $u(t)$ into Equation 2.34 from Equations 2.25 and 2.7, respectively, gives the complex envelope of the output signal as the sum of four convolutions

$$z(t) = h_I(t) \star u_I(t) - h_Q(t) \star u_Q(t) + j[h_I(t) \star u_Q(t) + h_Q(t) \star u_I(t)], \quad (2.35)$$

which in terms of the inphase and quadrature components of $z(t)$ yield

$$z_I(t) = h_I(t) \star u_I(t) - h_Q(t) \star u_Q(t) \quad (2.36)$$

$$z_Q(t) = h_I(t) \star u_Q(t) + h_Q(t) \star u_I(t). \quad (2.37)$$

This reveals that the structure of an equivalent complex lowpass system has the form shown in Figure 2.6.

Let us now return to the question of why Equation 2.26 differs from Equation 2.14 by a factor of two. The answer is that, it is simply to ensure that Equation 2.32 has the same form as Equation 2.29. Had Equation 2.21 been defined such that the frequency domain relationship for bandpass systems was analogous to that of a bandpass signal, i.e.,

$$G(f) = \frac{1}{2}[H(f - f_c) + H^*(-f - f_c)],$$

then Equation 2.32 would have read

$$Z(f) = \frac{1}{2}H(f)U(f).$$

If this were the case, complex lowpass equivalent signals and systems theory

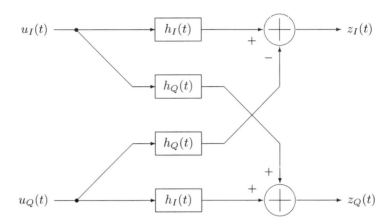

Figure 2.6: Complex lowpass equivalent of a bandpass system.

would not mirror conventional theory. That is, the response of a complex lowpass system would not be the convolution of its impulse response and the complex lowpass input signal.

Conceptually, the bandpass to complex lowpass transformation of signals can be viewed as mapping both positive and negative frequency components of the signals' spectra to baseband and summing them. Systems, however, do not possess frequency spectra, they respond to the spectra of signals (i.e., they are described by frequency responses). Hence, the reason that a linear bandpass system has a positive and a negative component to its frequency response is so that it will shape both components of the input signal.

After a bandpass to complex lowpass transformation, the two components of a signal's spectrum are then centred on zero frequency. Only one component of the system's frequency response needs to be mapped to baseband, because this will then shape both of the signal's components at once.

The results derived in this section illustrate that we can reduce problems involving high frequency bandpass-type radio signals and systems to baseband schemes using their complex lowpass equivalents. This is essential if computer simulations of mobile radio channels and equipment are to be carried out, as it is impractical to simulate a high frequency radio carrier.

Nevertheless, in order to be able to apply the complex lowpass equivalent representation to the simulation of practical bandpass communication systems, there is one further equivalence that must be derived. That is, the baseband representation of the ubiquitous additive white Gaussian noise

(AWGN).

2.1.4 Noise in Bandpass Systems

In communications systems noise from all sources are referenced to the receiver input and are represented by a single noise source added directly to the received signal. This is illustrated in Figure 2.7. The dominant noise source is the Gaussian distributed thermal noise generated within the receiver. As the spectrum of thermal noise extends to frequencies of the order of 10^{13} Hz, the additive noise source is assumed to be white (possessing all frequencies).

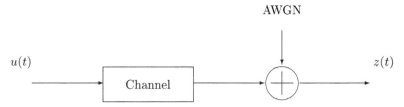

Figure 2.7: The arrangement assumed when analysing the noise properties of a bandpass system.

AWGN has an infinite spectrum. It therefore follows from the earlier discussion (see page 95) that the use of Hilbert transforms is required in order to derive a complex lowpass equivalent representation. Not only is this difficult, it is also unnecessary.

A simplified approach is adopted based on the assumption that bandpass systems are unaware of signals that lie outside their frequency band. The AWGN can therefore be regarded as having been passed through an ideal block filter prior to its addition to the received signal. Figure 2.8 illustrates this modified model. The block filter has the ideal frequency response,

$$H(f) = \left\{ \begin{array}{ll} 1 & f_c - \frac{B}{2} \leq f \leq f_c + \frac{B}{2} \\ 0 & \text{otherwise,} \end{array} \right. \tag{2.38}$$

where B is greater than the bandwidth of the system, but not large when compared to the centre frequency.

To noise falling in band, the filter is transparent. Noise power outside the system bandwidth receives infinite attenuation. The resulting bandpass Gaussian noise process, $n_B(t)$, is then added to the received signal.

The bandpass noise process, $n_B(t)$, can be represented in the form of Equation 2.8. That is,

$$n_B(t) = \Re \left\{ n(t) \exp(j2\pi f_c t) \right\}. \tag{2.39}$$

The complex lowpass equivalent noise signal, $n(t)$, remains Gaussian distri-

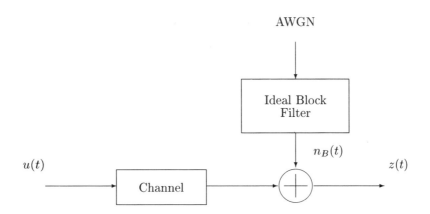

Figure 2.8: The noise model used when analysing a bandpass system.

buted, because the signal statistics are unaffected by the frequency translation.

It is possible to represent $n(t)$ as the sum of two quadrature Gaussian noise processes,

$$n(t) = n_I(t) + jn_Q(t). \tag{2.40}$$

The quadrature components, $n_I(t)$ and $n_Q(t)$, are independent Gaussian variables both with the same mean and variance as $n(t)$.

2.2 Mobile Radio Channel Types

The meaning of a communications channel is not universally agreed, and it is often used in an imprecise way. We may view a particular channel as the link between two points along a path of communications. When defining a specific channel we shall indicate under what conditions the channel exhibits either or both of the properties of linearity and reciprocity. Linearity is often described as follows.

If signals x_1 and x_2 applied to a channel give rise to the output signals y_1 and y_2, respectively, then the channel is said to be linear if an input signal $x = x_1 + x_2$ produces an output signal $y = y_1 + y_2$.

Often a channel behaves in a linear fashion only over certain regions of input voltage, temperature, supply voltage, etc. When this is the case, we refer to the regions of linear operation of the channel. The linearity of a channel is important when amplitude sensitive modulation schemes, such as quadrature amplitude modulation (QAM) are employed.

A channel is called a reciprocal channel if its behaviour is identical

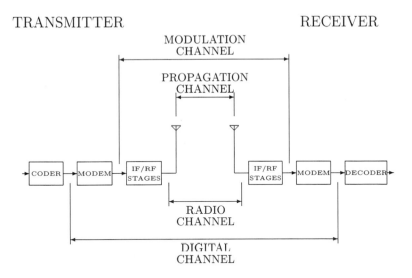

Figure 2.9: Channel types arising in radio communications.

regardless of the direction of information flow. It follows that a reciprocal channel need only be investigated in one direction. Figure 2.9 shows the channels we consider to be of use to the systems engineer.

2.2.1 The Propagation Channel

The propagation channel is the physical medium that supports electromagnetic wave propagation between a transmit and a receive antenna. In other words, it consists of everything that influences propagation between two antennas.

It is assumed that in the mobile radio environment propagating waves will only encounter media which are both bilateral and linear (an example of when this is not the case is in an ionised plasma). This assumption implies that mobile radio propagation channels are both linear and reciprocal. The channel is also time-variant due to the movement of the mobile.

2.2.2 The Radio Channel

The transmitter antenna, propagation channel and receiver antenna viewed collectively constitute the radio channel. As the propagation channel is reciprocal, so reciprocity of the radio channel depends on the antennas used. It can be shown [6] that antennas exhibit the same transmit and receive radiation patterns in free space if they are bilateral, linear and passive. Under these circumstances the antennas are reciprocal, and therefore so is the radio channel.

Non-linearities can occur in antenna systems due to rust, ice, and mounting structures, but they are usually small and we shall assume they can be neglected.

2.2.3   The Modulation Channel

The modulation channel extends from the output of the modulator to the input of the demodulator, and is composed of the transmitter front-end, receiver front-end, and the radio channel. It is of particular interest to designers of modulation schemes, and trellis coding systems.

Assuming a linear radio channel, the linearity of the modulation channel is determined by the transfer characteristics of the front-ends of the transmitter and receiver. Modulation systems that employ multilevel amplitude modulation, such as quadrature amplitude modulation (QAM), require the modulation channel to be approximately linear.

To achieve linearity, amplifiers are biased to operate in their linear regions, low distortion mixers are used, and linear phase filters employed. Because linear phase filters (Bessel or Gaussian) have a slow attenuation roll-off, more stages are required to obtain the same selectivity as that of steeper, non-linear-phase filter families. Linear amplifiers are more expensive than non-linear versions having the same output power.

Although on a one off basis the cost of having a linear front-end may not present a problem, in a commercial cellular system where every base station and mobile has to be equipped, it becomes a major consideration.

Power efficiency is an additional problem. Amplifiers operated in their linear region (Class A) are inefficient, compared with non-linear (for example Class C) amplifiers. In a mobile environment power efficiency is of paramount concern, because the size and weight of a hand-held portable is governed largely by the batteries it uses.

Understandably, system designers avoid using linear front-ends unless it is justified by the need for high bit rate transmission in microcellular environments where the radiated power levels are relatively low.

The modulation channel is non-reciprocal, since amplifiers and other front-end components are non-reciprocal. This is not generally a problem, because a transceiver uses separate front-end equipment for the transmit and receive operations. The two radio sections are then connected to the antenna via a duplexer. Hence, in a cellular radio system, the modulation channel from base station to mobile is different than that from the mobile to base station.

2.2.4   The Digital Channel

A further channel has been proposed by Aulin [7] for the case of digital transmissions. Called the digital channel, it consists of all the system components (including the radio channel) linking the unmodulated digital sequence at the transmitter to the regenerated sequence at the receiver. The digital channel is of value to source coding and channel coding engineers.

The digital channel is non-linear because the output can only take on certain fixed values. Reciprocity does not hold for the same reasons as described in Section 2.2.3.

2.2.5 A Channel Naming Convention

The propagation channel in the mobile radio environment is called the mobile radio propagation channel. Formally the radio channel is called the mobile radio radio channel. Similarly, the modulation and digital channels are referred to as the mobile radio modulation channel and the mobile radio digital channel, respectively.

Nevertheless, we will often refer to just the mobile radio channel. Unless otherwise stated, whenever the mobile radio channel is refered to, the mobile radio radio channel is implied.

2.3 Physical Description of the Channels

A prerequisite to combating the impairments experienced by a radio signal when it is transmitted over a mobile radio channel is to understand how these impairments originate. In the following section we examine the modifications to the transmitted signal by each of the above channels.

2.3.1 The Propagation Channel

The transmitted signal follows many different paths before arriving at the receiving antenna, and it is the aggregate of these paths that constitutes the mobile radio propagation channel. Figure 2.10 shows (in two dimensions only) two simplified propagation scenarios.

Each path may support a unique combination of propagation phenomena, nevertheless the effect of an individual path when viewed by the receiver is to attenuate, delay and phase shift the transmitted signal. The receiver antenna has a voltage induced in it that is the superposition of many scaled and phased echoes of the transmitted waveform.

Motion of the mobile and nearby scatterers, such as trucks and buses, may cause Doppler frequency shifts in each received signal component.

Maxwell's equations tell us that wherever there exists a time varying electric field there must also be a time varying magnetic field and vice versa. In the following theory the signal $x_p(t)$ is a generic symbol which can be used to represent either an electric or a magnetic field component of the transmitted signal. Similarly $y_p(t)$ is a generic symbol representing a field component of the received signal.

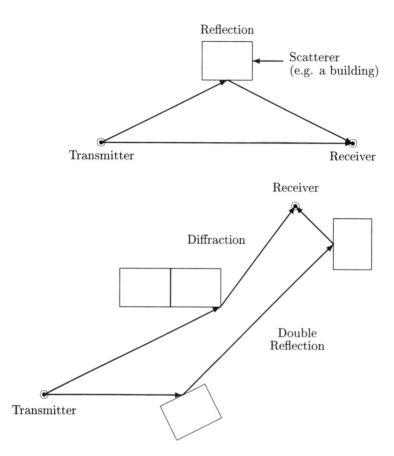

Figure 2.10: Two possible propagation channel scenarios.

2.3.1.1 The Received Signal

Consider a signal $x_p(t)$, of bandwidth B_x and centre frequency f_c radiated from a perfect isotropic radiator into a mobile radio propagation channel (the subscript p indicates an association of the variable with the propagation channel).

If B_x is small, relative to the centre frequency, the characteristics of each propagation path may be regarded as being independent of frequency (even though the propagation channel itself may be frequency dispersive). However if B_x is large compared to f_c, then the propagation phenomena (e.g. reflection and diffraction) can no longer be regarded as frequency independent over the band, and may cause appreciable signal distortion over individual paths.

It is assumed that signals of practical interest have bandwidths sufficiently narrow for the channel to be non-dispersive.

The component, $y_{pi}(t)$, of the received signal due to the i th path will then be a replica of the transmitted waveform, delayed by $\tau_i(t)$ seconds, attenuated by a factor $a_i(t)$, and phase retarded (due to reflections and diffractions) by $\theta_i(t)$ radians. That is,

$$y_{pi}(t) = \Re\{a_i(t)x_p[t - \tau_i(t)] \exp j[\omega_c(t - \tau_i(t)) - \theta_i(t)]\}. \qquad (2.41)$$

Summing the received components over all the propagation paths supported by the channel yields the total received signal,

$$y_p(t) = \Re\left\{\sum_{i=0}^{I-1} a_i(t)x_p[t - \tau_i(t)] \exp j[\omega_c(t - \tau_i(t)) - \theta_i(t)]\right\} \qquad (2.42)$$

where I is the number of paths comprising the channel. Using the complex notation of Section 2.1 we can write,

$$z_p(t) = \sum_{i=0}^{I-1} a_i(t)u_p[t - \tau_i(t)] \exp -j[\omega_c\tau_i(t) + \theta_i(t)], \qquad (2.43)$$

where $u_p(t)$ and $z_p(t)$ are the complex envelopes of $x_p(t)$ and $y_p(t)$, respectively.

2.3.1.2 The Impulse Response of the Channel

In the mobile radio environment it is normally impossible to establish the exact value of I. The summation over the number of paths is therefore replaced by the integral over all the possible delays. This allows the complex envelope of $y_p(t)$ to be written as,

$$z_p(t) = \int_0^\infty a_{p\tau}(t)u_p(t - \tau) \exp -j[\omega_c\tau + \theta_{p\tau}(t)]\, d\tau. \qquad (2.44)$$

The variables $a_{p\tau}(t)$ and $\theta_{p\tau}(t)$ are given by,

$$a_{p\tau}(t) = \left| \sum_{\tau_i(t)=\tau} a_i(t) \exp{-j\theta_i(t)} \right| \tag{2.45}$$

and

$$\theta_{p\tau}(t) = \arg\left(\sum_{\tau_i(t)=\tau} a_i(t) \exp{-j\theta_i(t)} \right), \tag{2.46}$$

respectively.

Both Equation 2.43 and Equation 2.44 provide accurate descriptions of the received signal. However, Equation 2.44 is in a mathematically much more convenient form.

The function, $h_p(t,\tau)$ is defined as,

$$h_p(t,\tau) \triangleq a_{p\tau}(t) \exp{-j[\omega_c\tau + \theta_{p\tau}(t)]} \tag{2.47}$$

so that Equation 2.44 can be written as,

$$z_p(t) = \int_0^\infty h_p(t,\tau) u_p(t-\tau)\,d\tau. \tag{2.48}$$

Comparing Equation 2.48 with Equation Equation 2.33 in Section 2.1.3 $h_p(t,\tau)$ is identified as the time varying impulse response of the propagation channel. Specifically, $h_p(t,\tau)$ is the response of the lowpass equivalent channel at time t to a unit impulse τ seconds in the past. It is known as the input delay-spread function, and is one of eight system functions described by Bello [8] which can be used to fully characterise linear time-variant channels. These functions are discussed in Section 2.5.

2.3.1.3   The Effect of Time Variations on the Channel

To examine the effects of the time dependence of the channel, we return to Equation 2.43 and express all the channel parameters at instant t_0 as linear functions of time, that is,

$$a_i(t_0 + \delta t) = \dot{a}_i(t_0)\delta t + a_i, \tag{2.49}$$

$$\tau_i(t_0 + \delta t) = \dot{\tau}_i(t_0)\delta t + \tau_i, \tag{2.50}$$

and

$$\theta_i(t_0 + \delta t) = \dot{\theta}_i(t_0)\delta t + \theta_i, \tag{2.51}$$

where a_i, τ_i and θ_i are values taken at time $t = t_0$, δt is measured from time t_0, and a dot above a symbol signifies differentiation with respect to time. Without loss of generality we can choose t_0 to equal the time origin

$t = 0$. Substituting Equations 2.49 through 2.51 into Equation 2.43 gives,

$$z_p(t_0 + \delta t) = \sum_{i=0}^{I-1} [\dot{a}_i(t_0)\delta t + a_i]u_p[t - \dot{\tau}_i(t_0)\delta t - \tau_i] \cdot$$

$$\exp - j\{\omega_c[\dot{\tau}_i(t_0)\delta t + \tau_i] + \dot{\theta}_i(t_0)\delta t + \theta_i\}, \quad (2.52)$$

an equation illustrating the complexity of the time dependence of the mobile radio propagation channel.

For very small δt, we can write

$$\dot{a}_i(t_0)\delta t \approx 0, \quad \dot{\tau}_i(t_0)\delta t \approx 0, \quad \dot{\theta}_i(t_0)\delta t \approx 0. \quad (2.53)$$

In the cellular mobile radio environment, however, $\omega_c \gg 10^6$ Hz. Therefore, in spite of the approximations given above, the product $\omega_c\dot{\tau}_i(t_0)\delta t$ in Equation 2.52 is not negligible.

Equation 2.52 can be simplified for small time periods as,

$$z_p(t_0 + \delta t) = \sum_{i=0}^{I-1} a_i u_p(t - \tau_i) \exp j(2\pi\nu_i(t_0)\delta t - \phi_i), \quad (2.54)$$

where

$$2\pi\nu_i(t_0) = -\omega_c\dot{\tau}_i(t_0) \quad (2.55)$$

and

$$\phi_i = \omega_c\tau_i + \theta_i. \quad (2.56)$$

The variable $\nu_i(t_0)$ represents the Doppler frequency shift due to changes in the electrical length, $l_i(t_0)$, of the i th path at time t_0. Figure 2.11 illustrates (in plan) the geometry associated with a small movement of the mobile. Referring to this figure it is seen that if the difference between $\alpha_i(t_0 + \delta t)$ and $\alpha_i(t_0)$ is small, then the two arrival paths are approximately parallel and the change in path length in time δt is

$$\delta l_i = l_i(t_0 + \delta t) - l_i(t_0) = -\delta s \cos\alpha_i(t_0). \quad (2.57)$$

The rate of change of the delay associated with the path is then

$$\dot{\tau}_i(t_0) = \lim_{\delta t \to 0} -\frac{\delta s}{\delta t}\frac{1}{c}\cos\alpha_i(t_0), \quad (2.58)$$

where c is the velocity of electromagnetic waves in free space.

In the limit Equation 2.58 becomes

$$\dot{\tau}_i(t_0) = -\frac{v}{c}\cos\alpha_i(t_0), \quad (2.59)$$

where v is the velocity of the mobile. Combining this result with Equa-

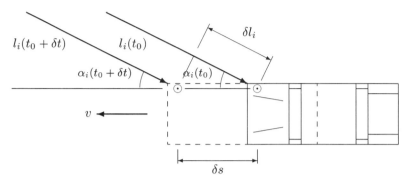

Figure 2.11: Signal geometry for small movements of the mobile.

tion 2.55 gives the Doppler frequency shift

$$\nu_i(t_0) = \frac{\omega_c v}{2\pi c} \cos \alpha_i(t_0) = \frac{v}{\lambda} \cos \alpha_i(t_0), \tag{2.60}$$

where $\lambda = c/f_c$ is the propagation wavelength. Movement of the scatterers effecting the i th path (we assume that the base station is stationary) will also cause Doppler shifting.

Once more (cf. Section 2.3.1.2) it is noted that the number of paths comprising the propagation channel is generally indeterminate. Delay and Doppler shifts are therefore represented as continuous domains. It is then deduced from Equations 2.54 to 2.56 that,

$$z_p(t) = \int_0^\infty \int_{-\infty}^\infty S_p(\tau, \nu) u_p(t - \tau) \exp j2\pi\nu t \, d\nu \, d\tau, \tag{2.61}$$

where

$$S_p(\tau, \nu) = \sum_{\substack{\tau_i(t)=\tau \\ \nu_i(t)=\nu}} a_i(t) \exp -j\theta_i(t). \tag{2.62}$$

$S_p(\tau, \nu)$ is another of the Bello system functions, called the delay-Doppler-spread function. Equations 2.45 through 2.48 with Equation 2.61 and Equation 2.62 reveal the relationship

$$h_p(t, \tau) = \int_{-\infty}^\infty S_p(\tau, \nu) \exp j2\pi\nu t \, d\nu. \tag{2.63}$$

That is, the delay-Doppler-spread and input delay-spread functions form a Fourier transform pair over the time and Doppler shift variables, while τ is a fixed parameter. This relationship exemplifies the elegance with which the Bello functions are related. The delay-Doppler-spread function

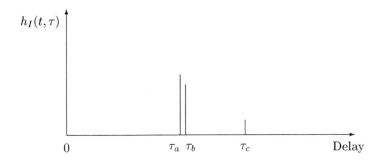

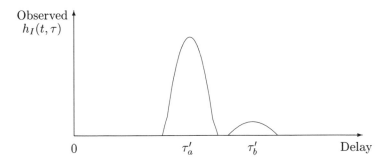

Figure 2.12: Inphase component of the channel impulse response observed by a system of low delay resolution.

is interesting in that it explicitly illustrates both the time and frequency dispersion of a channel.

2.3.1.4   Channel Effects on Systems of Finite Delay Resolution

All radio communications systems have a finite delay resolution related to the reciprocal of their transmission bandwidths. Two propagation paths separated by less than the system's delay resolution will appear to the receiver as one path. This is illustrated in Figure 2.12. The actual channel impulse response shown comprises three impulses at delays τ_a, τ_b and τ_c, nevertheless the system of low delay resolution only sees two signals, with the apparent delays τ_a' and τ_b'.

To investigate the effect of finite delay resolution on the received signal we represent all the paths arriving with delays in the range

$$\tau_n - \frac{\Delta\tau}{2} \leq \tau < \tau_n + \frac{\Delta\tau}{2} \tag{2.64}$$

as a single path of delay τ_n. The delay range can then be partitioned such that,

$$\tau_{n+1} = \tau_n + \Delta\tau, \tag{2.65}$$

for $n \in 0, 1, \ldots$ and arbitrarily,

$$\tau_0 = \frac{\Delta\tau}{2}. \tag{2.66}$$

The delay range defined in Equation 2.64 is called the n th delay bin. The width, $\Delta\tau$, of the delay bin is chosen to be less than or equal to the delay resolution of the system using the channel.

From Equation 2.48 the received signal due to the n th delay bin is then

$$z_{pn}(t) \approx \int_{\tau_n - \frac{\Delta\tau}{2}}^{\tau_n + \frac{\Delta\tau}{2} - \varepsilon} u_p(t - \tau)h_p(t, \tau)\, d\tau, \tag{2.67}$$

where ε is vanishingly small but not equal to zero. Adopting a bin width at least as narrow as the delay resolution of the system, ensures that the receiver cannot register the difference between a signal arriving with delay $\tau_n + \Delta\tau/2 - \varepsilon$ and one arriving with delay $\tau_n - \Delta\tau/2$. The approximation

$$u_p(t - \tau_n + \xi) \approx u_p(t - \tau_n), \tag{2.68}$$

where

$$|\xi| \le \frac{\Delta\tau}{2}, \tag{2.69}$$

can therefore be applied to Equation 2.67 to give

$$z_{pn}(t) \approx u_p(t - \tau_n) \int_{\tau_n - \frac{\Delta\tau}{2}}^{\tau_n + \frac{\Delta\tau}{2} - \varepsilon} h_p(t, \tau)\, d\tau. \tag{2.70}$$

Defining the channel response for the n th delay bin as

$$h_{p\Delta\tau}(t, \tau_n) \triangleq \int_{\tau_n - \frac{\Delta\tau}{2}}^{\tau_n + \frac{\Delta\tau}{2} - \varepsilon} h_p(t, \tau)\, d\tau \tag{2.71}$$

provides the approximation to the complex envelope of the received signal as

$$z_p(t) \approx \int_0^\infty h_{p\Delta\tau}(t, \tau)u_p(t - \tau)\, d\tau, \tag{2.72}$$

where the function

$$h_{p\Delta\tau}(t, \tau) = \sum_{n=0}^\infty h_{p\Delta\tau}(t, \tau_n)\delta(\tau - \tau_n) \tag{2.73}$$

may be regarded as the band-limited input delay-spread function of the

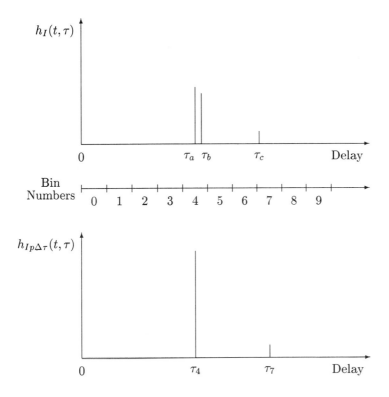

Figure 2.13: The inphase component of the band-limited impulse response of a channel.

propagation channel, with the factor $\Delta\tau$ determining the response bandwidth.

Figure 2.13 shows how the channel impulse response considered in Figure 2.12 maps into its band-limited version.

If $\Delta\tau$ is allowed to tend towards zero, the bandwidth over which the channel is observed increases until in the limit, $h_{p\Delta\tau}(t,\tau)$ is identical to the theoretical impulse response of the channel, $h_p(t,\tau)$.

As a vector sum of several phasors, $h_{p\Delta\tau}(t,\tau_n)$ is itself a phasor, and can be represented as,

$$h_{p\Delta\tau}(t,\tau_n) = a_{p\Delta\tau}(t,\tau)\exp{-j[\omega_c\tau_n + \theta_{p\Delta\tau}(t,\tau)]}, \qquad (2.74)$$

where $a_{p\Delta\tau}(t,\tau)$ and $\theta_{p\Delta\tau}(t,\tau)$ are random variables. The distribution of $\theta_{p\Delta\tau}(t,\tau)$ will be over $[0,2\pi]$ if $\Delta\tau \geq 1/f_c$ because the phase error in approximating a particular signal from the n th delay bin as having delay τ_n may be as great as $\pm\pi$. Furthermore, since the path delay $\tau \gg \Delta\tau$,

$\theta_{p\Delta\tau}(t,\tau)$ is uniformly distributed.

Fitting an accurate distribution to the random variable $a_{p\Delta\tau}(t,\tau)$ is not straightforward, and problems of this nature have taxed many minds over the last century. The most successful distribution, in terms of popularity, was proposed by Lord Rayleigh in 1880. His distribution, known as the Rayleigh distribution, refers to a specific situation, and may not always accurately describe observed distributions of $a_{p\Delta\tau}(t,\tau)$. Nevertheless, it has maintained its position in mobile radio theory by virtue of representing a worst case scenario.

2.3.1.5   Channel Effects on Systems of Finite Doppler Resolution

In a manner similar to the delay domain, the Doppler domain may be partitioned into Doppler bins, such that,

$$\nu_{m+1} = \nu_m + \Delta\nu, \tag{2.75}$$

for $m \in \ldots, -1, 0, 1, \ldots$ and

$$\nu_0 = 0. \tag{2.76}$$

The width of the Doppler delay bins, $\Delta\nu$, is chosen to be less than the frequency resolution of the system. Therefore all signals subject to a Doppler shift falling in the m th Doppler bin can be considered to be shifted by ν_m Hz.

The function $S_{p\Delta\tau\Delta\nu}(\tau_n, \nu_m)$ can then be defined as,

$$S_{p\Delta\tau\Delta\nu}(\tau_n, \nu_m) \triangleq \int_{\tau_n - \frac{\Delta\tau}{2}}^{\tau_n + \frac{\Delta\tau}{2} - \varepsilon} \int_{\nu_m - \frac{\Delta\nu}{2}}^{\nu_m + \frac{\Delta\nu}{2} - \epsilon} S(\tau, \nu)\, d\nu\, d\tau, \tag{2.77}$$

where both ε and ϵ are vanishingly small but not equal to zero. We therefore obtain the approximation,

$$z_p(t) \approx \int_0^\infty \int_{-\infty}^\infty S_{p\Delta\tau\Delta\nu}(\tau, \nu) u_p(t - \tau) \exp j2\pi\nu t\, d\nu\, d\tau, \tag{2.78}$$

where

$$S_{p\Delta\tau\Delta\nu}(\tau, \nu) = \sum_{n=0}^\infty \sum_{m=-\infty}^\infty S_{p\Delta\tau\Delta\nu}(\tau_n, \nu_m)\delta(\tau - \tau_n)\delta(\nu - \nu_m). \tag{2.79}$$

2.3.2   The Radio Channel

Antennas provide the means for interfacing communication equipment with the propagation channel. Currents in the transmitter antenna generate electromagnetic radiation. This radiation travels via the propagation channel to the receiver where electromagnetic coupling generates currents in the receiver antenna.

It is impossible to build an antenna which radiates equally in all directions (an isotropic radiator). Real antennas radiate more strongly in certain directions than in others. The radiation pattern of an antenna is the gain of that antenna as a function of the direction of radiation. When the antenna is reciprocal, the radiation pattern for transmission is identical to that of reception. In spherical coordinates the direction of radiation is defined by the zenith angle, θ, and the azimuthal angle, ϕ, as shown in Figure 2.14.

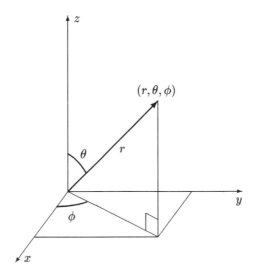

Figure 2.14: The spherical polar coordinates system.

The directive gain function of an antenna is often defined relative to that of an isotropic radiator as

$$G(\theta, \phi) \triangleq \frac{K(\theta, \phi)}{W/4\pi},\qquad (2.80)$$

where W is the total radiated power, and 4π is the total number of steradians in a sphere. The radiation intensity $K(\theta, \phi)$ is defined as the power radiated in a given direction per unit solid angle such that,

$$W = \int_0^\pi \int_0^{2\pi} K(\theta, \phi) \sin\theta \, d\theta \, d\phi.\qquad (2.81)$$

$G(\theta, \phi)$ may also be defined relative to other standard antennas, such as the quarter-wave monopole or the short dipole [6].

Real antennas may also produce a phase shift which is once more a function of the direction of transmission. Thus a continuous equiphase surface (a wavefront) propagating in all directions from the antenna need not be equal to the surface of a sphere.

Denoting the complex envelope of a signal transmitted across a mobile radio radio channel by $u_r(t)$, we can write the complex envelope of the input to the i th path of the propagation channel as,

$$u_{pi}(t) = \sqrt{G_t(\theta_{ti}, \phi_{ti})}\, u_r(t) \exp j\psi_t(\theta_{ti}, \phi_{ti}), \qquad (2.82)$$

where θ_{ti} and ϕ_{ti} define the direction of transmission of the radiation following the i th path relative to the zenith and the azimuthal angles respectively, $G_t(\theta, \phi)$ is the directive gain of the transmitter antenna, and $\psi_t(\theta, \phi)$ is its directive phase shift.

The complex envelope of the signal at the output of the radio channel due to the i th propagation path is

$$z_{ri}(t) = \sqrt{G_r(\theta_{ri}, \phi_{ri})}\, z_{pi}(t) \exp j\psi_r(\theta_{ri}, \phi_{ri}), \qquad (2.83)$$

where θ_{ri} and ϕ_{ri} identify the angle of arrival of the i th path, $G_r(\theta, \phi)$ is the directive gain of the receiver antenna and $\psi_r(\theta, \phi)$ is its directive phase shift.

Replacing the real signals in Equation 2.41 by their complex lowpass equivalents gives the complex envelope of the received signal for the i th path of the *propagation* channel as:

$$z_{pi}(t) = a_i(t)u_{pi}[t - \tau_i(t)] \exp -j[\omega_c \tau_i(t) + \theta_i(t)]. \qquad (2.84)$$

In the above equation $u_{pi}(t)$ is used, because the inputs to each propagation path are not necessarily identical. Substituting for $z_{pi}(t)$ into Equation 2.83 from Equation 2.84 yields,

$$
\begin{aligned}
z_{ri}(t) \;=\;& \sqrt{G_r(\theta_{ri}, \phi_{ri})} \\
& \cdot a_i(t)u_{pi}[t - \tau_i(t)] \exp -j[\omega_c \tau_i(t) + \theta_i(t)] \exp j\psi_r(\theta_{ri}, \phi_{ri}).
\end{aligned}
\qquad (2.85)
$$

The combination of Equations 2.82 and 2.85 yields,

$$
\begin{aligned}
z_{ri}(t) \;=\;& \sqrt{G_r(\theta_{ri}, \phi_{ri})G_t(\theta_{ti}, \phi_{ti})}\, a_i(t)u_r[t - \tau_i(t)] \\
& \cdot \exp -j[\omega_c \tau_i(t) + \theta_i(t)] \exp j[\psi_r(\theta_{ri}, \phi_{ri}) + \psi_t(\theta_{ti}, \phi_{ti})],
\end{aligned}
\qquad (2.86)
$$

which can then be summed over all the paths to give the output of the radio channel as

$$z_r(t) = \sum_{i=0}^{I-1} \sqrt{G_r(\theta_{ri}, \phi_{ri})G_t(\theta_{ti}, \phi_{ti})} \; a_i(t)u_r[t - \tau_i(t)]$$
$$\cdot \exp -j[\omega_c\tau_i(t) + \theta_i(t)] \exp j[\psi_r(\theta_{ri}, \phi_{ri}) + \psi_t(\theta_{ti}, \phi_{ti})].$$

$$(2.87)$$

Integrating over the delay domain instead of summing over all the propagation paths allows $z_r(t)$ to be written in the form

$$z_r(t) = \int_0^\infty h_r(t, \tau)u_r(t - \tau)\,d\tau, \qquad (2.88)$$

where

$$h_r(t, \tau) = \sum_{\tau_i(t)=\tau} w_i a_i(t) \exp -j\theta_i(t) \qquad (2.89)$$

is the input delay-spread function of the radio channel, and the antenna weighting function for the i th path is

$$w_i = \sqrt{G_r(\theta_{ri}, \phi_{ri})G_t(\theta_{ti}, \phi_{ti})} \; \exp j[\psi_r(\theta_{ri}, \phi_{ri}) + \psi_t(\theta_{ti}, \phi_{ti})]. \qquad (2.90)$$

The antenna weighting function can be measured for a given pair of transmit and receive angles. However, it is difficult, if not impossible to measure all the angles of transmission and reception associated with each propagation path [9]. A measurement system can seldom measure the true response of a propagation channel. It must measure a radio channel, which can at best only be an approximation to the propagation channel. Further, unless $a_i(t)$ and $\theta_i(t)$ appearing in Equation 2.89 can be established for each path comprising the propagation channel, $h_r(t, \tau)$ cannot be accurately deduced for any other radio channel (that is, for any other combination of transmitter and receiver antennas).

2.3.3   The Modulation Channel

The modulation channel combines the front-end of the radio equipment and the radio channel as shown in Figure 2.9. It represents the complete signal path between the output of the modulator and the input to the demodulator. If both front-ends are linear, then the complex lowpass impulse response of the modulation channel is

$$h_m(t, \tau) = h_T(\tau) \star h_r(t, \tau) \star h_R(\tau), \qquad (2.91)$$

where $h_T(\tau)$, $h_r(t, \tau)$ and $h_R(\tau)$ represent the complex lowpass equivalent impulse responses of the transmitter front-end, the radio channel and the receiver front-end, respectively, and $\star$ represents convolution. As the transmitter and receiver front-ends are assumed to be time-invariant, they are not considered functions of t. If either or both of the front-ends are non-linear then the modulation channel cannot be described fully by an impulse response.

2.3.4 The Digital Channel

The digital channel was introduced by Aulin [7]. It consists of all the system components (including the radio channel) linking the unmodulated digital sequence at the transmitter to the regenerated sequence at the receiver. It is characterised by bit error patterns.

This channel is of great use to engineers working at baseband. For example a knowledge of the digital channel will enable a speech codec to be designed for a particular set of bit error statistics, without the designer needing to know the complexities of the propagation channel, modem and transceiver behaviour.

The digital channel is non-linear, so the relationship between it and the modulation channel is not simply described. It is also non-reciprocal, because modems are non-reciprocal.

2.4 Classification of Channels

The impulse response of a mobile radio channel generally exhibits both delay and Doppler spreading. Delay spreading results in two effects, time dispersion and frequency-selective fading. Doppler spreading leads to frequency dispersion and time-selective fading.

Although all four effects are displayed by mobile radio channels, whether they are apparent to systems operating over a channel is dependent on the nature of the transmitted signal. The channel *as perceived by the system* can therefore be classified according to which effects are dominant.

In order to develop a system of classification we shall first examine each of the effects mentioned above.

2.4.1 Time Dispersion and Frequency-Selective Fading

Time dispersion and frequency-selective fading are both manifestations of multipath propagation with delay spread. The presence of one effect perforce implies the presence of the other.

Time dispersion stretches a signal in time so that the duration of the received signal is greater than that of the transmitted signal. Frequency-selective fading filters the transmitted signal, attenuating certain frequencies more than others. Two frequency components closely spaced receive approximately the same attenuation; however, if they are far apart they often receive vastly different attenuations.

Time dispersion is a result of the signals taking different times to cross the channel by different propagation paths. Frequency-selective fading occurs because the electrical length of each propagation path can be expressed as a function of frequency.

If the bandwidth, B_x, of the transmitted signal is sufficiently narrow, then all the transmitted frequency components will receive about the same amount of attenuation, and the signal will be passed undistorted, without frequency-selective fading.

As the transmission bandwidth is increased, the frequency components at the extremes of the transmitted spectrum will start to be attenuated by different amounts. Thus the channel is having a filtering effect, and is distorting the transmitted waveform, that is frequency-selective fading is experienced. The distortion increases as B_x is increased.

For very large transmission bandwidths the receiver may be able to observe distinct echoes of the transmitted waveform. At this point the system is able to recognize time dispersion, since the delay spread of the channel is greater than the delay resolution of the receiver. In digital systems this results in intersymbol interference.

The minimum transmission bandwidth at which time dispersion is observable is inversely proportional to the maximum excess delay of the channel, τ_m, where the excess delay is the actual delay minus the delay of the first arrival path. The constant of proportionality is system dependent, but shall be taken here to be $\frac{1}{4}$.

There are thus two observable effects of delay spread; distortion and dispersion.

A measure of the transmission bandwidth at which distortion becomes appreciable is often based on the channel's coherence bandwidth. The coherence bandwidth $B_c(t)$ indicates the frequency separation at which the attenuation of the amplitudes of two frequency components becomes decorrelated such that the envelope correlation coefficient, $\rho(\Delta f, \Delta t)$, reaches a predesignated value. This value has in the past been taken as 0.9 [10], 0.5 [11, 12] and $1/e$ (0.37) [13]- [15]. The amount of signal distortion required before a specific system's performance is effected is heavily dependent on the modulation and demodulation techniques employed. A particular system may start to have problems when the transmission bandwidth corresponds to a value of 0.9 for the envelope correlation coefficient, whereas a more robust system may perform perfectly satisfactorily up to a transmission bandwidth corresponding to an envelope correlation coefficient of 0.37.

In line with the two major works on mobile radio communications Lee [11] and Jakes [12] the coherence bandwidth is taken to correspond to an envelope correlation coefficient of 0.5. That is

$$\rho(B_c(t), 0) = 0.5, \tag{2.92}$$

where [12]

$$\rho(\Delta f, \Delta t) \triangleq \frac{\langle a_1 a_2 \rangle - \langle a_1 \rangle \langle a_2 \rangle}{\sqrt{[\langle a_1^2 \rangle - \langle a_1 \rangle^2][\langle a_2^2 \rangle - \langle a_2 \rangle^2]}}. \tag{2.93}$$

In the above definition $\langle \rangle$ denotes the ensemble average. Variables a_1 and a_2 represent the amplitudes of signals at frequencies f_1 and f_2, respectively, and at times t_1 and t_2, respectively, where $|f_2 - f_1| = \Delta f$ and $|t_2 - t_1| = \Delta t$.

To derive a value for $B_c(t)$, we shall employ the approximation that in the mobile radio environment the amplitude of each received signal is unity and that the probability of receiving a signal with delay τ is given by

$$p(\tau) = \frac{1}{2\pi\sigma(t)} \exp \frac{-\tau}{\sigma(t)}, \tag{2.94}$$

where $\sigma(t)$ is the delay spread of the channel. This may at first appear a widely inaccurate approximation for the mobile radio environment, since we realise that signals of different delays rarely arrive with equal amplitudes. It should, however, be remembered that this is a mathematical model. The same results would be produced from a model that uses signals arriving at fixed delays with exponentially distributed amplitudes—intuitively a more acceptable scenario.

The delay spread, $\sigma(t)$ is equal to the square root of the second central moment of the channel's power-delay profile, $P_h(\tau)$. That is

$$\sigma(t) = \frac{\int_0^\infty (\tau - d(t))^2 P_h(\tau) \, d\tau}{\int_0^\infty P_h(\tau) \, d\tau}, \tag{2.95}$$

where $d(t)$ is the mean propagation delay, given by

$$d(t) = \frac{\int_0^\infty \tau P_h(\tau) \, d\tau}{\int_0^\infty P_h(\tau) \, d\tau}. \tag{2.96}$$

The power-delay profile of the channel, to be described in Section 2.6.3, is given as

$$\begin{aligned} P_h(\tau) &= |h(t, \tau)|^2 \\ &= h_I(t, \tau)^2 + h_Q(t, \tau)^2. \end{aligned} \tag{2.97}$$

Approximating $p(\tau)$ as shown in Equation 2.94 does not of course describe all mobile radio channels, since the specific environment of each system varies. However, results indicate that it is not an unreasonable assumption [16, 17]. It can then be shown [11, 12] that the envelope correlation coefficient for two signals separated by Δf Hz and Δt seconds is

$\rho(\Delta f, 0)$

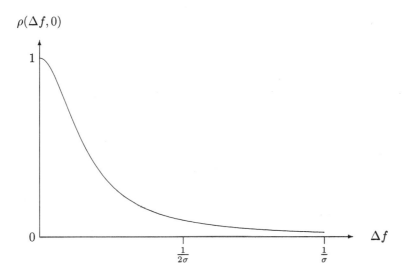

Figure 2.15: Envelope correlation with frequency separation.

equal to

$$\rho(\Delta f, \Delta t) = \frac{J_0^2(2\pi f_m \Delta t)}{1 + (2\pi \Delta f)^2 \sigma^2}, \qquad (2.98)$$

where $J_0()$ is the zero order Bessel function of the first kind and $f_m = v/c$ is the maximum Doppler shift for a vehicular velocity of v, with c representing the velocity of light. To observe the decorrelation of two signals as their frequency separation is increased, Δt is set equal to zero in Equation 2.98. This gives the frequency correlation function as

$$\rho(\Delta f, 0) = \frac{1}{1 + (2\pi \Delta f)^2 \sigma^2}. \qquad (2.99)$$

The graphical representation of Equation 2.99 is shown in Figure 2.15, where the envelope correlation decreases with the frequency separation of the signals. The correlation bandwidth is obtained from Equations 2.92 and 2.99 as

$$B_c(t) = \frac{1}{2\pi \sigma(t)}. \qquad (2.100)$$

A typical delay spread value of $2\mu s$ for conventional size cells in an urban environment [18] results in a coherence bandwidth of about 80kHz. In practice, the correlation function $\rho(\Delta f, 0)$ does not decrease monotonically with increasing frequency separation [19] and the presence of a strong echo with excess delay τ_{ex}, will lead to an oscillatory component in the correlation coefficient of frequency $1/\tau_{ex}$ [20]. In this case, as Δf is increased the first occurrence of the envelope correlation coefficient dropping below 0.5 is taken as the channel coherence bandwidth.

2.4.2   Frequency Dispersion and Time-Selective Fading

When a channel is time variant it is referred to as possessing time-selective fading. Time-selective fading can cause signal distortion, because the channel may change its characteristics whilst the transmitted signal is in flight. The channel seen by the leading edge of the signal is not the same as that seen by the trailing edge.

In Section 2.3.1.3 it was shown that when the response of a channel is time-variant, Doppler spreading (frequency dispersion) occurs. Frequency dispersion results in the signal bandwidth being stretched so that the received signal's bandwidth is different (greater or less) from that of the transmitted signal.

If a signal has a short duration then it is passed through the channel before any significant change in the channel characteristics can take place. As the signal's duration is increased, the channel is able to change whilst the signal is still in flight, thereby causing distortion. The distortion increases as the signal duration is increased. At the same time, Doppler spreading of the signal increases relative to the transmission bandwidth until it is possible to observe significant widening of the received spectrum. That is, when the maximum Doppler frequency is larger than the Doppler resolution of the receiver,

$$f_m > \Delta\nu. \tag{2.101}$$

The minimum signal duration at which frequency dispersion becomes noticeable is inversely proportional to the magnitude of the maximum Doppler shift experienced by the signal, f_m. The constant of proportionality is again somewhat arbitrary, but will be taken as being equal to $\frac{1}{4}$.

In a manner similar to that of Section 2.4.1, we can estimate at what transmitted signal duration distortion becomes noticeable by referring to the channel's coherence time, $T_c(t)$. Analogous to the channel's coherence bandwidth (Equation 2.92) the coherence time is defined as,

$$\rho(0, T_c(t)) = 0.5. \tag{2.102}$$

Setting $\Delta f = 0$ in Equation 2.98 gives

$$\rho(0, \Delta t) = J_0{}^2(2\pi f_m \Delta t), \tag{2.103}$$

which is plotted in Figure 2.16.   From the previous two equations

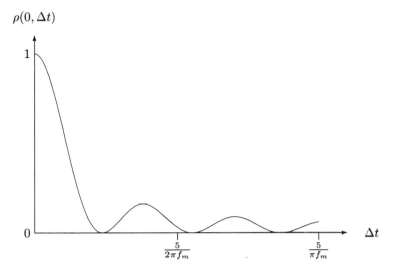

$\rho(0, \Delta t)$

Figure 2.16: Envelope correlation with time separation.

$$T_c(t) = \frac{{J_0}^2(2\pi f_m \Delta t)}{2} \approx \frac{9}{16\pi f_m}. \qquad (2.104)$$

For example, for a vehicular speed of 30 ms^{-1} (67.5 mph), a channel centered on 1.7 GHz exhibits a coherence time of approximately 1 ms. This corresponds to a transmitted bit rate of 1 kb/s. Signals with bit rates in excess of 1 kb/s can therefore assume the channel to be non-distorting in time.

2.4.3   Channel Classifications

The coherence bandwidth and coherence time are properties of a channel which may be used to assess how it will appear to transmitted signals. If the bandwidth of a transmission is less than the coherence bandwidth of the channel, the frequency-selective fading, and therefore the time dispersion of the channel appear to be transparent to the signal. The channel is viewed by the system as having a flat response across the transmission band, and is therefore referred to as being frequency-flat.

Similarly, if the duration of the received waveform is less than the coherence time, the channel will appear to the signal to be time-invariant. Notice that we have specified the *received* waveform duration, since this is the time for which the signal is in flight. It is generally taken as the trans-

mitted signal duration (the symbol period for digital transmissions) plus
the channel delay spread, $\sigma(t)$ (defined in Equation 2.95). As the channel
response appears to be constant for the duration of the signal's flight, the
channel is referred to as time-flat.

When a channel is flat in both frequency and time, it is called a flat-
flat channel. When a channel is flat neither in frequency nor in time, it
is often referred to as a doubly dispersive channel. This nomenclature,
however, is somewhat misleading because such a channel need only cause
signal distortion not dispersion. Hence we shall refer to a channel that is
neither time-flat nor frequency-flat as a non-flat channel. Figure 2.17 shows
the classification of channels following the above approach. The shaded
region of the figure indicates the physical restriction that it is impossible
for the time bandwidth product of a signal to be less than 1/2 [21]. A
more rigorous system of classification, emphasising the differences between
distorting and dispersive channels, is shown in Figure 2.18.

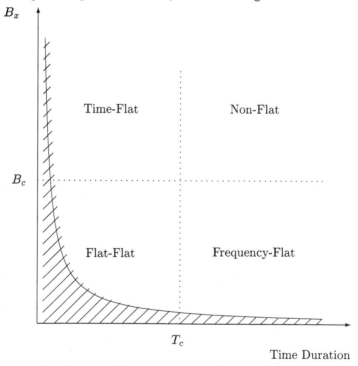

Figure 2.17: Channel classifications.

The flat-flat, or doubly-flat channel does not fade with either time or
frequency. Using the approximate values derived above, it is seen that a
signal of bandwidth less than $B_c(t)$ Hz and duration less than $T_c(t)$ seconds
will observe a flat-flat channel at time t. For example, in this category re-

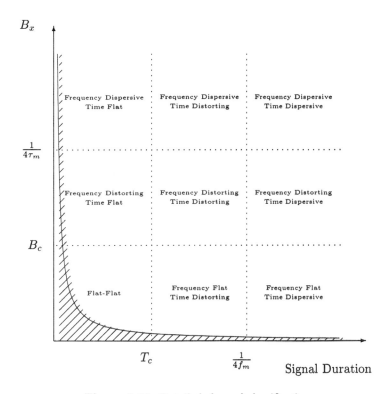

Figure 2.18: Detailed channel classifications.

sides the uplink from the mobile stations (MS) to the base station (BS) of the MATS-D system [22]. MATS-D is a hybrid mobile radio system that was put forward as a contender for the pan-European cellular radio system. It employs a narrowband frequency division multiple access (FDMA) scheme for the uplink. The transmissions employ generalised tamed frequency modulation (GTFM) to give a bandwidth of approximately 25 kHz for a bit rate of 19.5 kb/s.

The frequency-flat fading channel is observed by narrowband channel sounders. These sounders transmit a monochromatic (single tone) signal continuously, and so approximate to a signal of infinitesimal bandwidth and infinite duration. The envelope of the received process therefore varies in sympathy with the channel.

For a delay spread of $2\mu s$, the time-flat fading channel applies to all mobile radio systems using digital transmissions with bit rates in excess of 80 kb/s (see page 121). Many of the systems put forward for the

Variable	Notation	Units
Time	t	Seconds
Frequency	f	Hertz
Delay	τ	Seconds
Doppler Shift	ν	Hertz

Table 2.1: Variables used to describe linear time-variant channels.

pan-European cellular system (e.g., S900D [23], DMS90 [24], MATS-D (downlink) [22], and CD900 [25]) including the GSM system that has been adopted [26], fall into this category.   The situation where the channel is flat with neither time nor frequency does not occur with mobile radio channels. This is because of the short delay spreads and low mobile speeds. Such a channel may be encountered in satellite to aircraft communications where greater mobile (aircraft) speeds are combined with large excess delays due to ground reflections.   Although $2\mu s$ is a typical delay spread value for example in New York, it may not be appropriate for all mobile radio environments. Suburban environments tend to show less delay spread [19, 27], whilst some measurements in cities have yielded delay spreads of $5\mu s$ and greater [28, 29]. In extreme environments, e.g., hilly terrain, delay spreads of up to $17\mu s$ have been recorded [30]. In order to measure the characteristics of a time-flat channel, wideband sounding techniques must be employed.

2.5   A Systems Approach to Linear Time-Variant Channels

Bello [8] proposed a set of eight system functions to describe linear time-variant channels. Each function embodies a complete description of the channel, and full knowledge of one function allows calculation of any of the others. Each one uses two of the four variables shown in Table 2.1.

2.5.1   The Variables Used For System Characterisation

The familiar time and frequency variables, t and f, are by definition [31] dual network variables, whilst the delay and Doppler shift variables, τ and ν, are dual operators describing time and frequency translation. The concept of duality has been discussed at length by Bello [31]; however, for the purposes of this discussion, it is sufficient to understand that two operators (functions, elements, or systems) are dual when the behaviour of one with reference to a time-related domain (the time or the delay domain) is identical to the behaviour of the other referenced to the corresponding frequency-related domain (i.e., the frequency or the Doppler shift domain, respectively).

A common mistake when first encountering the delay variable is to assume that since τ is measured in seconds it is linearly dependent on the time variable t. This is not the case (i.e., $\tau \not\propto t$). The delay variable is orthogonal to the time variable, and as such can be drawn on a set of Cartesian coordinates. It is easiest to appreciate that the variables t and τ are independent, if the electrical lengths of propagation paths are considered.

A channel's electrical length is one of its physical properties, and is related to τ by

$$l_e = \frac{\tau}{c}. \tag{2.105}$$

Even if it were possible to freeze time, a path would still possess an electrical length of l_e, and therefore an associated delay. Although the electrical length of a particular path may vary with time, perhaps due to the motion of the mobile, in general there may exist a path at any instant in time possessing any positive electrical length. That is, the two variables are independent, and from Equation 2.105 we deduce that τ is also independent of t.

It is perhaps more difficult to understand that ν is orthogonal to f, because the Doppler shift associated with a particular path is a function of the frequency of transmission and physically it is caused by a change in the delay or electrical length of a path, as evidenced by Equation 2.55. The rate of change of l_e with respect to time is expressed in ms^{-1}. Equations 2.55 and 2.105 show that the Doppler shift is actually the rate of change of the physical length of the path, dl/dt, scaled by the signal's frequency. The frequency scaling occurs because a signal perceives length in terms of wavelengths, not absolute measures. As dl/dt is independent of frequency, it is theoretically possible for a path to exist possessing any Doppler shift value at any particular frequency.

2.5.2 The Bello System Functions

Two of the Bello system functions were encountered in Section 2.3.1 when the propagation channel was analysed in the time domain. The first one is the **input delay-spread function**, $h(t, \tau)$, defined by

$$z(t) = \int_{-\infty}^{\infty} h(t, \tau)u(t - \tau)\, d\tau \tag{2.106}$$

and interpreted as the response of the channel at time t to a unit impulse input τ seconds in the past. This function describes the channel in terms of the t–τ domain.

It is helpful to visualise Equation 2.106, as the output from a densely tapped delay line, where $h(t, \tau)d\tau$ is the tap weighting for delay τ, as seen in Figure 2.19. Notice that $h(t, \tau)$ is called the *input* delay-spread function because the delay is associated with the *input* port of the channel. A further

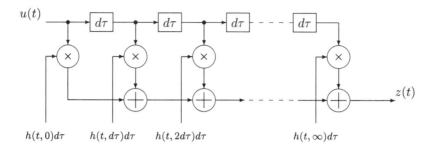

$h(t,0)d\tau \qquad h(t,d\tau)d\tau \qquad h(t,2d\tau)d\tau \qquad\qquad h(t,\infty)d\tau$

Figure 2.19: Tapped delay line representation of the input delay-spread function.

Bello function, $g(t,\tau)$, will be introduced which has the delay associated with the output port of the channel. This function is then called the output delay-spread function.

Computer simulations generally employ the tapped delay line approach to model mobile radio channels, using the complex baseband equivalent system of Figure 2.6.

The second Bello function, which has already been presented is the **delay-Doppler-spread function**, $S(\tau,\nu)$. This function was defined by Equation 2.61 as:

$$z(t) \triangleq \int_{-\infty}^{\infty} \int_{-\infty}^{\infty} S(\tau,\nu)u(t-\tau)\exp j2\pi\nu t\,d\nu\,d\tau \qquad (2.107)$$

and is interpreted as the gain experienced by signals suffering first delay in the range $[\tau,\tau+d\tau]$ then Doppler shift in the range $[\nu,\nu+d\nu]$.

$S(\tau,\nu)$ uses the τ–ν domain to describe the channel. It is seldom used in the simulation of channels because the double integral in the above equation requires more computation than the single integral of Equation 2.106. Nevertheless, the delay-Doppler-spread function has found favour as a means of displaying the dispersive characteristics of a channel, because it explicitly shows both time and frequency dispersion.

Equation 2.63 shows that $h(t,\tau)$ and $S(\tau,\nu)$ form a complex Fourier pair over the variables ν and t, with the common variable τ. It may come as somewhat of a surprise that the time domain Fourier transformation transforms into the Doppler shift domain and not into the frequency domain. This is because it is a change in the channel's behaviour as a function of time that causes a Doppler shift, whilst it is the frequency response of the channel at a specific time as a function of the delay variable that determines the channel's spectrum. As ν is the dual of τ, and t is the dual of f, we can deduce from the above relationship that the delay domain Fourier transforms with the frequency domain.

From these two Fourier relationships it is possible to deduce why Bello functions require two variables, and also why there are only eight Bello functions.

Fourier transform pairs contain the same information, presented in different forms. The time and Doppler shift domains contain information describing frequency dispersion, whilst the frequency and delay domains contain time dispersion information. To *fully* describe a channel, a function must use at least two variables, one from each Fourier pair. Should a third variable be used, its information would be redundant, while if only one variable were to be used, the description would be incomplete.

The number of ways of permuting two variables from four is 12. However, Bello has only defined eight functions. This is because four of these permutations are due to variables from the same Fourier pair (i.e., t, ν; ν, t; τ, f; and f, τ;). Functions defined for these permutations would contain information about only one type of dispersion.

The six Bello functions that have yet to be presented will be discussed below.

The time-variant transfer function, $T(f, t)$ is defined by the equation

$$z(t) = \int_{-\infty}^{\infty} T(f, t) U(f) \exp j2\pi f t \, df. \qquad (2.108)$$

It is interpreted as the complex envelope of the received signal for a cissoidal input at the carrier frequency. As the name implies, $T(f, t)$ is the time-variant equivalent of the conventional (time-invariant) system transfer function. Equations 2.106 and 2.108 may be manipulated as follows to reveal that $T(f, t)$ and $h(t, \tau)$ form a Fourier pair with the common variable t.

Replacing $u(t - \tau)$ in Equation 2.106 with its Fourier transform gives

$$z(t) = \int_{-\infty}^{\infty} h(t, \tau) \int_{-\infty}^{\infty} U(f) \exp j2\pi f(t - \tau) \, df \, d\tau, \qquad (2.109)$$

which on rearranging yields

$$z(t) = \int_{-\infty}^{\infty} \left\{ \int_{-\infty}^{\infty} h(t, \tau) \exp -j2\pi f\tau \, d\tau \right\} U(f) \exp j2\pi f t \, df. \qquad (2.110)$$

Equating the above equation to Equation 2.108 shows the required Fourier relationship, that is:

$$h(t, \tau) = \int_{-\infty}^{\infty} T(f, t) \exp j2\pi f\tau \, df. \qquad (2.111)$$

The output Doppler-spread function, $H(f, \nu)$ describes the channel in the $f-\nu$ domain, and is defined by

$$Z(f) = \int_{-\infty}^{\infty} U(f - \nu)H(f - \nu, \nu)\, d\nu. \tag{2.112}$$

The interpretation of the function $H(f, \nu)$ is as the spectral response of the channel at a frequency ν Hz above a cissoidal input at f Hz.

A differential circuit representation of Equation 2.112 is given in Figure 2.20. The circuit is a densely tapped frequency conversion chain.

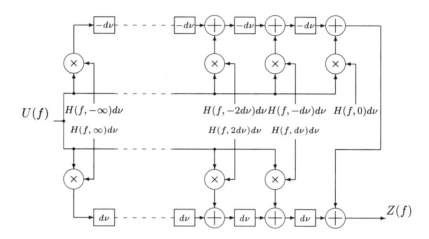

Figure 2.20: Tapped frequency conversion chain representation of the output Doppler-spread function.

The Doppler shift $d\nu$ is associated with the output of the channel, hence the name *output* Doppler-spread function. $H(f, \nu)$ offers an alternative approach to computer simulation instead of using $h(t, \tau)$. The two models (Figures 2.19 and 2.20) are equally easy to translate into program code, allowing a worker to deal with channel inputs in either the time or frequency domains.

It is possible to show that $H(f, \nu)$ and $T(f, t)$ are a Fourier pair by examining the received spectrum when a cissoidal input at $f = f'$ is present. From the previous equation we then have:

$$Z(f) = \int_{-\infty}^{\infty} \delta(f - f' - \nu)H(f - \nu, \nu)\, d\nu. \tag{2.113}$$

Taking the Fourier transform of both sides of this equation with respect to

f, we get:

$$z(t) = \int_{-\infty}^{\infty}\int_{-\infty}^{\infty} \delta(f - f' - \nu)H(f - \nu, \nu) \exp j2\pi ft \, d\nu \, df, \qquad (2.114)$$

which reduces to

$$z(t) = \int_{-\infty}^{\infty} H(f', \nu) \exp j2\pi(f' + \nu)t \, d\nu. \qquad (2.115)$$

Now from Equation 2.108, the received signal for a cissoidal input at $f = f'$ may also be written as,

$$z(t) = T(f', t) \exp j2\pi f't. \qquad (2.116)$$

Equating the above two equations gives the required relationship,

$$T(f, t) = \int_{-\infty}^{\infty} H(f, \nu) \exp j2\pi\nu t \, d\nu. \qquad (2.117)$$

$H(f, \nu)$ also forms a Fourier transform pair with $S(\tau, \nu)$, this time with the common variable ν. This may be derived as follows. Fourier transformation of both sides of Equation 2.107 with respect to t gives

$$Z(f) = \int_{-\infty}^{\infty}\int_{-\infty}^{\infty}\int_{-\infty}^{\infty} S(\tau, \nu)u(t - \tau) \exp j[2\pi(\nu - f)t] \, d\nu \, d\tau \, dt, \quad (2.118)$$

which can be rearranged as

$$Z(f) = \int_{-\infty}^{\infty}\int_{-\infty}^{\infty} S(\tau, \nu) \left\{ \int_{-\infty}^{\infty} u(t - \tau) \exp j[2\pi(\nu - f)t] \, dt \right\} d\nu \, d\tau. \qquad (2.119)$$

The term in braces is evaluated using the shifting property of Fourier transform, so that

$$Z(f) = \int_{-\infty}^{\infty}\int_{-\infty}^{\infty} S(\tau, \nu) \exp j[2\pi(f - \nu)\tau]U(f - \nu) \, d\nu \, d\tau. \qquad (2.120)$$

Comparing this equation with Equation 2.106 reveals the Fourier relationship

$$S(\tau, \nu) = \int_{-\infty}^{\infty} H(f, \nu) \exp j2\pi\tau f \, d\tau. \qquad (2.121)$$

Figure 2.21 shows how the four Fourier relationships derived so far (Equations 2.63, 2.111, 2.117 and 2.121) allow the functions to be arranged in a symmetric pattern.

Just as the delay process is associated with the input of the channel for $h(t, \tau)$, we can derive a further Bello function by associating the Doppler

LEGEND : F Fourier Transform
           F^{-1} Inverse Fourier Transform

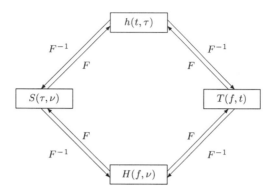

Figure 2.21: Fourier relationships amongst the first set of Bello functions.

shift with the input of the channel. The resulting function is called the **input Doppler-spread function**, $G(f,\nu)$. The model describing this situation is shown in Figure 2.22, from which we see that

$$Z(f) = \int_{-\infty}^{\infty} G(f,\nu)U(f - \nu)\,d\nu. \tag{2.122}$$

Comparison of this equation and Equation 2.106 shows that they are identical in form and both represent a convolution. Equation 2.122 is in fact the dual relationship to that of Equation 2.106, and $G(f,\nu)$ is the dual function of $h(t,\tau)$. Duality was introduced at the start of this section. To reiterate, dual systems, or operators, for example, behave in an identical manner, however one exists in a dual domain to the other.

Bello [31] has presented the techniques used to manipulate time-frequency duality relationships. In brief, to obtain a dual relationship replace each function by its dual and make the substitutions shown in Table 2.2.

By applying duality relations to the defining equations of the functions in Figure 2.21, the definitions of the remaining system functions are found. The input Doppler-spread function, $G(f,\nu)$, has already been introduced, as seen in Equation 2.122 above. The **output delay-spread function**, $g(t,\tau)$, is the dual of the output Doppler-spread function, $H(f,\nu)$ and its defining equation is obtained from Equation 2.112 applying duality as:

$$z(t) = \int_{-\infty}^{\infty} u(t - \tau)g(t - \tau, \tau)\,d\tau. \tag{2.123}$$

The tapped delay line representation of $g(t,\tau)$ is shown in Figure 2.23. The

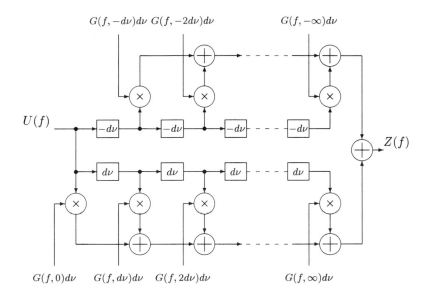

Figure 2.22: Tapped frequency conversion chain representation of the input Doppler-spread function.

Original notation	Dual notation
t	f
f	t
τ	ν
ν	τ
$\exp(\cdot)$	$\exp -(\cdot)$
$\exp -(\cdot)$	$\exp(\cdot)$
$x(t)$	$X(f)$
$X(f)$	$x(t)$

Table 2.2: Notational changes to establish dual relations.

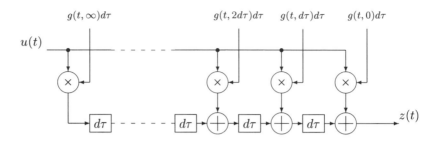

Figure 2.23: Tapped delay line representation of the output delay-spread function.

Doppler-delay-spread function, $V(\nu, \tau)$, is the dual of $S(\tau, \nu)$, and from Equation 2.107 is given by

$$Z(f) = \int_{-\infty}^{\infty}\int_{-\infty}^{\infty} V(\nu, \tau) U(f - \nu) \exp{-j2\pi\tau f}\, d\tau\, d\nu. \qquad (2.124)$$

Finally, **the frequency-dependent modulation function,** $M(t, f)$, is the dual of the time-variant transfer function, $T(f, t)$. From Equation 2.108

$$Z(f) = \int_{-\infty}^{\infty} M(t, f) u(t) \exp{-j2\pi ft}\, dt. \qquad (2.125)$$

Figure 2.24 shows that this second set of Bello functions can also be arranged symmetrically with respect to complex Fourier transforms.

A summary of the definitions and interpretations of all the eight Bello functions are presented in Table 2.3.

In order to move between the two sets of functions, relationships between sister functions are exploited. Sister functions are functions that describe the channel in terms of the same domain. For example, the input and output delay-spread functions, $h(t, \tau)$ and $g(t, \tau)$, are sister functions because they both describe the channel in the t-τ domain. From Equations 2.106 and 2.123 it is seen that

$$h(t, \tau) = g(t - \tau, \tau). \qquad (2.126)$$

The relationships between the other three pairs of sister functions are easily derived from their defining equations. They are summarised in Table 2.4. As sister functions describe the channel in the same domain, it is usual to employ just one set of functions. The most commonly used set [32]- [34] is the one illustrated in Figure 2.21. The functions in this set generally occur as shown in Table 2.5.

Name	Notation	Definition	Interpretation	Dual
Input delay-spread function	$h(t,\tau)$	$z(t) = \int h(t,\tau)u(t-\tau)d\tau$	The channel response at time t to a unit impulse input τ seconds in the past	$G(f,\nu)$
Time-variant Transfer function	$T(f,t)$	$z(t) = \int T(f,t)U(f)\exp j2\pi ft df$	The complex envelope of the received signal due to a cissoidal input at the carrier frequency	$M(t,f)$
Output Doppler-spread function	$H(f,\nu)$	$Z(f) = \int H(f-\nu,\nu)u(f-\nu)d\nu$	The spectral response of the channel at a frequency ν Hz above the cissoidal input at frequency f Hz	$g(t,\tau)$
Delay-Doppler-spread function	$S(\tau,\nu)$	$z(t) = \int\int S(\tau,\nu)u(t-\tau)\exp j2\pi\nu\tau d\nu d\tau$	The gain afforded signals suffering first delay in $[\tau,\tau+d\tau]$ then Doppler shift in $[\nu,\nu+d\nu]$	$V(\nu,\tau)$
Ouput delay-spread function	$g(t,\tau)$	$z(t) = \int g(t-\tau,\tau)u(t-\tau)d\tau$	The channel response τ seconds in the future to a unit impulse input at time t	$H(f,\nu)$
Frequency-dependent modulation function	$M(t,f)$	$Z(f) = \int M(t,f)u(t)\exp -j2\pi tf dt$	The complex amplitude spectrum of the received signal for a unit impulse input at time $t=0$	$T(f,t)$
Input Doppler-spread function	$G(f,\nu)$	$Z(f) = \int G(f,\nu)u(f-\nu)d\nu$	The spectral response of the channel at a frequency f Hz due to a cissoidal input ν Hz below f	$h(t,\tau)$
Doppler-delay-spread function	$V(\nu,\tau)$	$Z(f) = \int\int V(\nu,\tau)U(f-\nu)\exp -j2\pi\tau\nu d\tau d\nu$	The gain afforded signals suffering first Doppler shift in $[\nu,\nu+d\nu]$ then delay in $[\tau,\tau+d\tau]$	$S(\tau,\nu)$

Table 2.3: Bello's system functions.

LEGEND : F Fourier Transform
            F^{-1} Inverse Fourier Transform

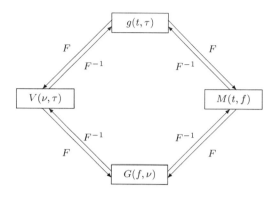

Figure 2.24: Fourier relationships amongst the second set of Bello functions.

Function	Sister funct.	Relationship
$h(t,\tau)$	$g(t,\tau)$	$h(t,\tau) = g(t-\tau,\tau)$
$T(f,t)$	$M(t,f)$	$\displaystyle\int\int M(t',f')\exp-j2\pi(f-f')(t-t')\,df'\,dt'$
$H(f,\nu)$	$G(f,\nu)$	$H(f,\nu) = G(f+\nu,\nu)$
$S(\tau,\nu)$	$V(\nu,\tau)$	$S(\tau,\nu) = V(\nu,\tau)\exp-j2\pi\nu\tau$

Table 2.4: Sister functions.

Function	Normal Occurrence
$h(t,\tau)$	Measured directly by time domain wideband sounders. Used in computer simulations.
$T(f,t)$	Measured directly by narrowband (single tone) sounders.
$H(f,\nu)$	Measured by frequency domain wideband sounders. Used in computer simulations.
$S(\tau,\nu)$	Used to display the time and frequency dispersion of the channel simultaneously.

Table 2.5: Occurrences of the first set of Bello functions.

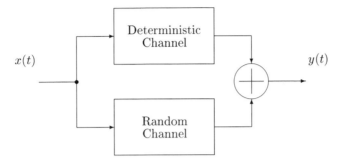

Figure 2.25: The decomposition of a channel into its deterministic and random
components.

So far, we have introduced a family of system functions which fully describe a time-variant channel, and have provided techniques for manipulating these functions. Attention will now be given to randomly time-variant
channels. Such channels are particularly useful to study, because, as will
be seen in Section 2.6, mobile radio channels can be regarded as being
randomly time-variant.

2.5.3 Description of Randomly Time-Variant Channels

A general linear time-variant channel can be viewed as the superposition
of a deterministic channel and a purely random, zero ensemble average
channel, shown in Figure 2.25.

The deterministic channel may be fully characterised by applying directly the system functions described above. However, the functions become stochastic processes when they are used to describe the randomly
varying component of the channel.

Unfortunately, a full statistical description of the system functions requires the determination of multidimensional pdf's for the functions, and
this is not a trivial task. A less stringent, but practical approach [8, 35]
to characterising purely random channels involves the determination of the
correlation functions for any one of the Bello system functions.

2.5.3.1 Autocorrelation of a Bandpass Stochastic Process

The autocorrelation $R_y(t_1, t_2)$ of a stochastic process, $y(t)$, is the ensemble
average, or expected value, of the product $y(t_1)y(t_2)$ and may be written

as [36]

$$R_y(t_1, t_2) = \int_{-\infty}^{\infty}\int_{-\infty}^{\infty} y_1 y_2 f(y_1, y_2; t_1, t_2)\, dy_1\, dy_2, \qquad (2.127)$$

where $f(y_1, y_2; t_1, t_2)$ is the second order density of the process $y(t)$, given by

$$f(y_1, y_2; t_1, t_2) \triangleq \frac{\partial^2 F(y_1, y_2; t_1, t_2)}{\partial y_1 \partial y_2}. \qquad (2.128)$$

The cumulative density function $F(y_1, y_2; t_1, t_2)$ is defined as the joint probability,

$$F(y_1, y_2; t_1, t_2) \triangleq \Pr\{y(t_1) < y_1, y(t_2) < y_2)\}. \qquad (2.129)$$

The ensemble average of a quantity, $(\cdot)$, is denoted by angle brackets, like $\langle(\cdot)\rangle$. The autocorrelation of $y(t)$ can therefore be written as,

$$R_y(t_1, t_2) = \langle y(t_1) y(t_2)\rangle. \qquad (2.130)$$

This autocorrelation may be expressed in terms of the signal's complex envelope, $z(t)$, namely

$$R_y(t_1, t_2) = \langle \Re\{z(t_1)\exp j2\pi f_c t_1\} \Re\{z(t_2)\exp j2\pi f_c t_2\}\rangle, \qquad (2.131)$$

or, applying Equation 2.11 we have:

$$\begin{aligned} R_y(t_1, t_2) &= \frac{1}{2}\Re\{\langle z(t_1) z(t_2)\rangle \exp j2\pi f_c(t_2 + t_1)\} \\ &\quad + \frac{1}{2}\Re\{\langle z^*(t_1) z(t_2)\rangle \exp j2\pi f_c(t_2 - t_1)\}, \qquad (2.132) \end{aligned}$$

where $\star$ identifies a complex conjugate, and the exponentials have been taken outside the averaging process, as they are deterministic. The auto-correlation of a real bandpass stochastic process is seen to be the sum of two autocorrelation functions that depend upon the complex envelope of the process $y(t)$ at instants t_1 and t_2. Bello has reported [8] that most narrowband processes are constituted such that

$$\langle z(t_1) z(t_2)\rangle = 0. \qquad (2.133)$$

If the channel exhibits wide-sense stationarity (WSS) in the time variable, which will be discussed later in Section 2.5.3.3, the above equation must be true. This is because the WSS criterion implies that the process's time-domain characteristics, such as the autocorrelation, cannot be dependent on absolute time, only on the time difference, $t_2 - t_1$. It is assumed that Equation 2.133 is applicable for all mobile radio channels, enabling

Equation 2.132 to be simplified to

$$R_y(t_1, t_2) = \frac{1}{2} \Re \left\{ R_z(t_1, t_2) \exp j2\pi f_c(t_2 - t_1) \right\},$$ (2.134)

where

$$R_z(t_1, t_2) = \langle z^*(t_1) z(t_2) \rangle.$$ (2.135)

2.5.3.2 General Randomly Time-Variant Channels

The Bello functions are defined in terms of the complex lowpass equivalent notation. Hence, in line with Equation 2.135 above, the correlation function for the input delay-spread function is given by

$$R_h(t_1, t_2; \tau_1, \tau_2) = \langle h^*(t_1, \tau_1) h(t_2, \tau_2) \rangle.$$ (2.136)

Expressions relating the autocorrelation functions of the channel output to the correlation functions of the Bello system functions are easily derived from the functions' defining equations given in Table 2.3. For example, from Equation 2.106

$$
\begin{aligned}
z^*(t_1) z(t_2) &= \int_{-\infty}^{\infty} h^*(t_1, \tau_1) u^*(t_1 - \tau_1) \, d\tau_1 \int_{-\infty}^{\infty} h(t_2, \tau_2) u(t_2 - \tau_2) \, d\tau_2 \\
&= \int_{-\infty}^{\infty} \int_{-\infty}^{\infty} h^*(t_1, \tau_1) h(t_2, \tau_2) u^*(t_1 - \tau_1) u(t_2 - \tau_2) \, d\tau_1 d\tau_2.
\end{aligned}
$$ (2.137)

Taking the ensemble average of both sides of the above equation gives

$$\langle z^*(t_1) z(t_2) \rangle = \int_{-\infty}^{\infty} \int_{-\infty}^{\infty} \langle h^*(t_1, \tau_1) h(t_2, \tau_2) u^*(t_1 - \tau_1) u(t_2 - \tau_2) \rangle \, d\tau_1 d\tau_2.$$ (2.138)

If the channel input is deterministic, it can be removed from the ensemble average to yield

$$R_z(t_1, t_2) = \int_{-\infty}^{\infty} \int_{-\infty}^{\infty} R_h(t_1, t_2; \tau_1, \tau_2) u^*(t_1 - \tau_1) u(t_2 - \tau_2) \, d\tau_1 d\tau_2.$$ (2.139)

However if $u(t)$ is a random process which is assumed to be independent of the channel characteristics, Equation 2.137 becomes

$$R_z(t_1, t_2) = \int_{-\infty}^{\infty} \int_{-\infty}^{\infty} R_h(t_1, t_2; \tau_1, \tau_2) R_u(t_1 - \tau_1, t_2 - \tau_2) \, d\tau_1 d\tau_2,$$ (2.140)

where

$$R_u(t_1, t_2) = \langle u^*(t_1) u(t_2) \rangle.$$ (2.141)

Autocorrelation function	Input-output correlation function relationships
$R_h(t_1, t_2; \tau_1, \tau_2) = \langle h^*(t_1, \tau_1) h(t_2, \tau_2) \rangle$	$R_z(t_1, t_2) = \iint \mathcal{F}_u(t_1 - \tau_1, t_2 - \tau_2) R_h(t_1, t_2; \tau_1, \tau_2) d\tau_1 d\tau_2$
$R_T(f_1, f_2; t_1, t_2) = \langle T^*(f_1, t_1) T(f_2, t_2) \rangle$	$R_z(t_1, t_2) = \iint \mathcal{F}_U(f_1, f_2) R_T(f_1, f_2; t_1, t_2) \exp j2\pi (f_2 t_2 - f_1 t_1) df_1 df_2$
$R_H(f_1, f_2; \nu_1, \nu_2) = \langle H^*(f_1, \nu_1) H(f_2, \nu_2) \rangle$	$R_Z(f_1, f_2) = \iint \mathcal{F}_U(f_1 - \nu_1, f_2 - \nu_2) R_H(f_1 - \nu_1, f_2 - \nu_2; \nu_1, \nu_2) d\nu_1 d\nu_2$
$R_S(\tau_1, \tau_2; \nu_1, \nu_2) = \langle S^*(\tau_1, \nu_1) S(\tau_2, \nu_2) \rangle$	$R_z(t_1, t_2) = \iint\!\!\iint \mathcal{F}_u(t_1 - \tau_1, t_2 - \tau_2) R_S(\tau_1, \tau_2; \nu_1, \nu_2) \exp j2\pi (\nu_2 t_2 - \nu_1 t_1) d\nu_1 d\tau_1 d\nu_2 d\tau_2$
$R_g(t_1, t_2; \tau_1, \tau_2) = \langle g^*(t_1, \tau_1) g(t_2, \tau_2) \rangle$	$R_z(t_1, t_2) = \iint \mathcal{F}_u(t_1 - \tau_1, t_2 - \tau_2) R_g(t_1 - \tau_1, t_2 - \tau_2; \tau_1, \tau_2) d\tau_1 d\tau_2$
$R_M(t_1, t_2; f_1, f_2) = \langle M^*(t_1, f_1) M(t_2, f_2) \rangle$	$R_Z(f_1, f_2) = \iint \mathcal{F}_u(t_1, t_2) R_M(t_1, t_2; f_1, f_2) \exp -j2\pi (t_2 f_2 - t_1 f_1) dt_1 dt_2$
$R_G(f_1, f_2; \nu_1, \nu_2) = \langle G^*(f_1, \nu_1) G(f_2, \nu_2) \rangle$	$R_Z(f_1, f_2) = \iint \mathcal{F}_U(f_1 - \nu_1, f_2 - \nu_2) R_G(f_1, f_2; \nu_1, \nu_2) d\nu_1 d\nu_2$
$R_V(\nu_1, \nu_2; \tau_1, \tau_2) = \langle V^*(\nu_1, \tau_1) V(\nu_2, \tau_2) \rangle$	$R_Z(f_1, f_2) = \iint\!\!\iint \mathcal{F}_U(f_1 - \nu_1, f_2 - \nu_2) R_V(\nu_1, \nu_2; \tau_1, \tau_2) \exp -j2\pi (\tau_2 f_2 - \tau_1 f_1) d\tau_1 d\nu_1 d\tau_2 d\nu_2$

Table 2.6: Correlation functions of the Bello system functions.

Defining

$$F_u(t_1, t_2) = \begin{cases} u^\star(t_1)u(t_2) & u(t) \text{ deterministic} \\ \langle u^\star(t_1)u(t_2) \rangle & u(t) \text{ random,} \end{cases} \quad (2.142)$$

Equations 2.139 and 2.140 can be combined as

$$R_z(t_1, t_2) = \int_{-\infty}^{\infty} \int_{-\infty}^{\infty} R_h(t_1, t_2; \tau_1, \tau_2) F_u(t_1 - \tau_1, t_2 - \tau_2) \, d\tau_1 \, d\tau_2. \quad (2.143)$$

Table 2.6 details the relations between the channel output autocorrelation functions and the correlation functions of the remaining Bello functions. Notice that the dual function of F_U is employed in the table, that is,

$$F_U(f_1, f_2) = \begin{cases} U^\star(f_1)U(f_2) & U(f) \text{ deterministic} \\ \langle U^\star(f_1)U(f_2) \rangle & U(f) \text{ random.} \end{cases} \quad (2.144)$$

As one would expect, since the system functions can be arranged symmetrically with respect to their Fourier relationships, the correlation functions can also be arranged symmetrically, this time with respect to their double Fourier relationships. As an example consider the relationship between the functions $T(f, t)$ and $H(f, \nu)$. From Equation 2.117

$$T^\star(f_1, t_1)T(f_2, t_2) = \int_{-\infty}^{\infty} \int_{-\infty}^{\infty} H^\star(f_1, \nu_1)H(f_2, \nu_2)$$
$$\cdot \exp j2\pi(\nu_2 t_2 - \nu_1 t_1) \, d\nu_1 \, d\nu_2. \quad (2.145)$$

Taking ensemble averages of both sides

$$R_T(f_1, f_2; t_1, t_2) = \int_{-\infty}^{\infty} \int_{-\infty}^{\infty} R_H(f_1, f_2; \nu_1, \nu_2) \exp j2\pi(\nu_2 t_2 - \nu_1 t_1) \, d\nu_1 \, d\nu_2.$$
$$(2.146)$$

The above equation shows that $R_T(f_1, f_2; t_1, t_2)$ is the two-dimensional Fourier transform of the correlation function $R_H(f_1, f_2; \nu_1, \nu_2)$ with the convention that when transforming from a pair of time variables to a pair of frequency variables a positive exponential connects the first variable in each pair and a negative the second.

Figure 2.26 illustrates the symmetric double Fourier transform relationships of the correlation functions of the first set of Bello functions. If required, a table showing the relationships between the correlation functions of sister Bello functions can easily be derived from Tables 2.4 and 2.6.

The relationships derived above for the correlation functions of Bello's system functions may be applied to any time-variant linear channel. However, if the statistical behaviour of a channel obeys certain constraints, then it is possible to simplify them. Specifically, if a channel is wide-sense-stationary in the time domain and/or the frequency domain then its

LEGEND : DF Double Fourier Transform
           DF^{-1} Inverse Double Fourier Transform

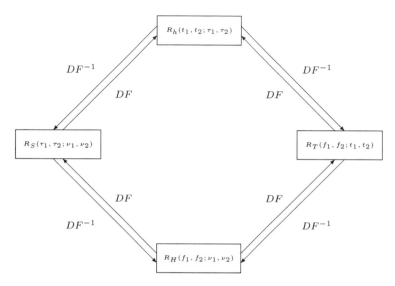

Figure 2.26: Fourier relations of the correlations for the first set of Bello functions.

correlation functions can be simplified. The following sections investigate how.

2.5.3.3   Wide-Sense Stationary Channels

Firstly, we shall examine a channel that exhibits wide-sense (second-order) stationarity. A process is called wide-sense stationary with respect to time if its first two moments (mean and autocorrelation) are independent of absolute time. That is, the correlation function for a wide-sense stationary (WSS) channel depends on time difference, and not absolute time. For example, the correlation of the input delay-spread function for a WSS channel becomes

$$R_h(t_1, t_2; \tau_1, \tau_2)\Big|_{\text{WSS}} \equiv R_h(\Delta t; \tau_1, \tau_2), \qquad (2.147)$$

where

$$\Delta t = t_2 - t_1. \qquad (2.148)$$

Substituting for $R_h(t_1, t_2; \tau_1, \tau_2)$ from Equation 2.147 into Equation 2.143 gives

$$R_z(t_1, t_2) = \int_{-\infty}^{\infty} \int_{-\infty}^{\infty} R_h(\Delta t; \tau_1, \tau_2)\mathcal{F}_u(t_1 - \tau_1, t_2 - \tau_2)\, d\tau_1 d\tau_2. \qquad (2.149)$$

The autocorrelation function of the channel output is still a function of absolute time even though the channel correlation function is expressed in terms of differential time. This is because the channel output is of course dependent on its input, which need not have any statistical constraints placed upon it.

Earlier discussion indicated that the Doppler shift domain contains the same information as the time domain. This implies that wide-sense stationarity in the time variable, t, must also manifest itself in the Doppler domain. To illustrate the way in which this occurs we shall look at the correlation function of the delay-Doppler-spread function $R_S(\tau_1, \tau_2; \nu_1, \nu_2)$. Applying the information contained in Figure 2.26, we can write

$$R_S(\tau_1, \tau_2; \nu_1, \nu_2) = \int_{-\infty}^{\infty}\int_{-\infty}^{\infty} R_h(t_1, t_2; \tau_1, \tau_2) \exp j2\pi(\nu_1 t_1 - \nu_2 t_2)\, dt_1\, dt_2.$$

$$(2.150)$$

Substituting Equations 2.147 and 2.148 for the WSS channel gives

$$R_S(\tau_1, \tau_2; \nu_1, \nu_2)\Big|_{\text{WSS}} = \int_{-\infty}^{\infty} \exp j2\pi t_1 (\nu_1 - \nu_2)\, dt_1$$
$$\cdot \int_{-\infty}^{\infty} R_h(\Delta t; \tau_1, \tau_2) \exp -j2\pi\nu_2 \Delta t\, d(\Delta t).$$

$$(2.151)$$

The second integral may be recognised as the Fourier transform of an autocorrelation function, which from the Wiener-Khinchine theorem results in a power spectral density function. The other integral is zero except for the case $\nu_1 = \nu_2$, when it is infinite. This is recognised as the definition of a unit impulse at $\nu_2 = \nu_1$. We can thus write,

$$R_S(\tau_1, \tau_2; \nu_1, \nu_2)\Big|_{\text{WSS}} \equiv P_S(\tau_1, \tau_2; \nu_2)\delta(\nu_2 - \nu_1), \qquad (2.152)$$

where $P_S(\tau_1, \tau_2; \nu_2)$ is the cross power spectral density of $h(t, \tau_1)$ and $h(t, \tau_2)$. The impulse function in Equation 2.152 implies that, for a WSS channel, signals arriving with different Doppler shift values are uncorrelated.

As ν_1 does not feature in the function $P_S(\tau_1, \tau_2; \nu_2)$ we can drop the suffix on ν_2 to get,

$$R_S(\tau_1, \tau_2; \nu_1, \nu_2)\Big|_{\text{WSS}} \equiv P_S(\tau_1, \tau_2; \nu)\delta(\nu_2 - \nu_1). \qquad (2.153)$$

Column 2 in Table 2.7 lists the correlation functions of the Bello system functions for the WSS channel. The dual functions for a general time-variant linear channel are no longer duals for a WSS channel. This is because we have applied statistical constraints to variables t and ν, but not to their dual variables, f and τ respectively. Figure 2.27 shows, how

LEGEND : $(D)F$ (Double) Fourier Transform
                   $(D)F^{-1}$ Inverse (Double) Fourier Transform

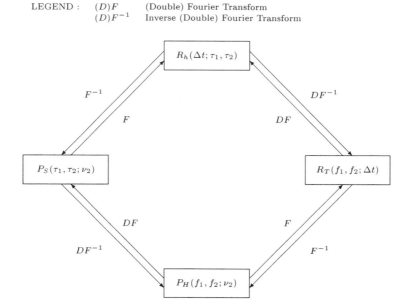

Figure 2.27: Fourier relationships amongst the correlation functions of the first set of Bello functions for WSS channels.

the correlation functions for the first set of Bello functions are related on a WSS channel.

In order to fully characterise a WSS channel, the correlation functions must be established for all frequencies.

2.5.3.4   Uncorrelated Scattering Channels

The dual of the WSS channel is the uncorrelated scattering (US) channel. For this channel, the statistics describing signals arriving with different delays are uncorrelated.

The WSS channel was seen to be wide-sense stationary in the time domain and to possess uncorrelated scattering in the Doppler shift domain. By applying duality we can state immediately that a channel possessing uncorrelated scattering in the delay domain will be wide-sense stationary in the frequency domain. Hence, for the output Doppler-spread function, we can write,

$$R_H(f_1, f_2; \nu_1, \nu_2)\Big|_{US} \equiv R_H(\Delta f; \nu_1, \nu_2), \qquad (2.154)$$

where

$$\Delta f = f_2 - f_1. \qquad (2.155)$$

The correlation function of the delay-Doppler-spread function is derived from $R_H(f_1, f_2; \nu_1, \nu_2)$ by double Fourier transform. (See Figure 2.26.)

General Channel	WSS Channel	US Channel	WSSUS Channel
$R_h(t_1, t_2; \tau_1, \tau_2)$	$R_h(\Delta t; \tau_1, \tau_2)$	$P_h(t_1, t_2; \tau_2)\delta(\tau_2 - \tau_1)$	$P_h(\Delta t; \tau_2)\delta(\tau_2 - \tau_1)$
$R_T(f_1, f_2; t_1, t_2)$	$R_T(f_1, f_2; \Delta t)$	$R_T(\Delta f; t_1, t_2)$	$R_T(\Delta f; \Delta t)$
$R_H(f_1, f_2; \nu_1, \nu_2)$	$R_H(f_1, f_2; \nu_2)\delta(\nu_2 - \nu_1)$	$R_H(\Delta f; \nu_1, \nu_2)$	$R_H(\Delta f; \nu_2)\delta(\nu_2 - \nu_1)$
$R_S(\tau_1, \tau_2; \nu_1, \nu_2)$	$R_S(\tau_1, \tau_2; \nu_2)\delta(\nu_2 - \nu_1)$	$R_S(\tau_2; \nu_1, \nu_2)\delta(\tau_2 - \tau_1)$	$R_S(\tau_2; \nu_2)\delta(\nu_2 - \nu_1)\delta(\tau_2 - \tau_1)$
$R_g(t_1, t_2; \tau_1, \tau_2)$	$R_g(\Delta t; \tau_1, \tau_2)$	$P_g(t_1, t_2; \tau_2)\delta(\tau_2 - \tau_1)$	$P_g(\Delta t; \tau_2)\delta(\tau_2 - \tau_1)$
$R_M(t_1, t_2; f_1, f_2)$	$R_M(\Delta t; f_1, f_2)$	$R_M(t_1, t_2; \Delta f)$	$R_M(\Delta t; \Delta f)$
$R_G(f_1, f_2; \nu_1, \nu_2)$	$R_G(f_1, f_2; \nu_2)\delta(\nu_2 - \nu_1)$	$R_G(\Delta f; \nu_1, \nu_2)$	$R_G(\Delta f; \nu_2)\delta(\nu_2 - \nu_1)$
hline $R_V(\nu_1, \nu_2; \tau_1, \tau_2)$	$R_V(\nu_2; \tau_1, \tau_2)\delta(\nu_2 - \nu_1)$	$R_V(\nu_1, \nu_2; \tau_2)\delta(\tau_2 - \tau_1)$	$R_V(\nu_2; \tau_2)\delta(\tau_2 - \tau_1)\delta(\nu_2 - \nu_1)$

Table 2.7: Correlation functions for different channel types.

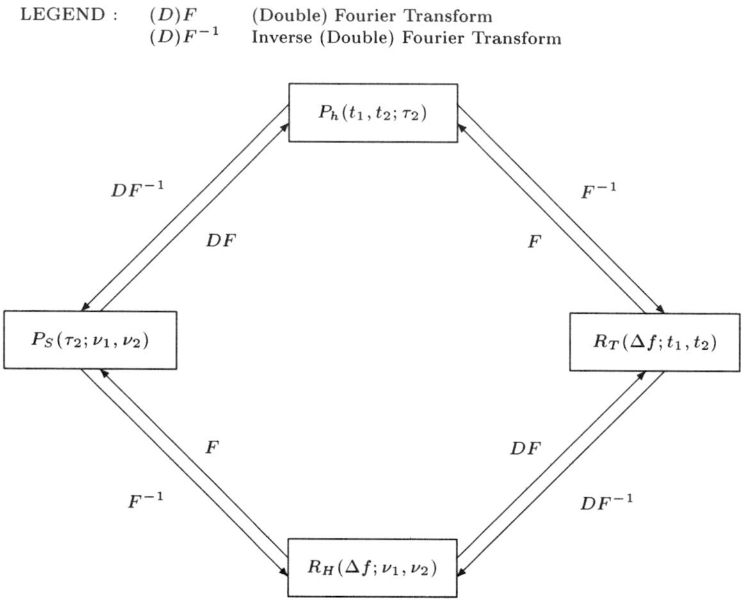

LEGEND : $(D)F$ (Double) Fourier Transform
             $(D)F^{-1}$ Inverse (Double) Fourier Transform

Figure 2.28: Fourier relationships amongst correlation functions of the first set of Bello functions for US channels.

Applying Equation 2.154 we have:

$$R_S(\tau_1, \tau_2; \nu_1, \nu_2)\Big|_{US} = \int_{-\infty}^{\infty} \exp j2\pi f_1(\tau_1 - \tau_2)\, df_1$$
$$\cdot \int_{-\infty}^{\infty} R_H(\Delta f; \nu_1, \nu_2) \exp -j2\pi\tau_2\Delta f\, d(\Delta f).$$

$$(2.156)$$

Then expressing $R_S(\tau_1, \tau_2; \nu_1, \nu_2)$ in terms of the cross-power spectral density of $H(f, \nu_1)$ and $H(f, \nu_2)$ reveals that

$$R_S(\tau_1, \tau_2; \nu_1, \nu_2)\Big|_{US} \equiv P_S(\tau; \nu_1, \nu_2)\delta(\tau_2 - \tau_1), \qquad (2.157)$$

which illustrates the uncorrelated scattering in the delay domain. The remaining correlation functions are listed in Column 3 of Table 2.7, and the relationships between the functions for the first set of Bello functions are displayed in Figure 2.28. As with a WSS channel, dual functions for a general time-variant linear channel are not dual for a US channel.

A full description of the US channel requires the correlation functions to be established for all time.

2.5.3.5   Wide-Sense Stationary Uncorrelated Scattering Channels

The most useful channel as far as the mobile radio engineer is concerned is a hybridisation of the above channels. Referred to as the wide-sense stationary uncorrelated scattering (WSSUS) channel, its first and second order statistics are invariant under translation in time and frequency. This means that the correlation functions for a WSSUS channel need only be worked out once, since they apply for all time and all frequency.

The correlation functions are easily deduced by applying first the wide-sense stationary criteria and then the uncorrelated scattering criteria, or vice versa.

Consider $R_h(t_1, t_2; \tau_1, \tau_2)$. Under wide sense stationary conditions,

$$R_h(t_1, t_2; \tau_1, \tau_2)\Big|_{\text{WSS}} \equiv R_h(\Delta t; \tau_1, \tau_2)\delta(\nu_2 - \nu_1), \qquad (2.158)$$

then adding the uncorrelated scattering restriction,

$$R_h(t_1, t_2; \tau_1, \tau_2)\Big|_{\text{WSSUS}} \equiv P_h(\Delta t; \tau)\delta(\tau_2 - \tau_1), \qquad (2.159)$$

where $P_h(\Delta t; \tau)$ is the cross-power spectral density of $T(f, t_1)$ and $T(f, t_1 + \Delta t)$.

Table 2.7 lists the correlation functions for a WSSUS channel in Column 4, and Figure 2.29 illustrates their inter-relations for the first set of Bello functions. Notice from this table that dual functions under the general time-variant linear channel are still duals under a WSSUS channel. This is because although we have applied certain statistical constraints to t and ν, we have also applied the dual constraints to f and τ.

WSSUS channels are of particular significance because they are the simplest channels to analyse, that exhibit both time and frequency fading. Workers are therefore disposed to approximate real channels by WSSUS channels.

2.5.3.6   Quasi-Wide-Sense Stationary Uncorrelated Scattering Channels

In order to utilise the benefits of a WSSUS channel in the characterisation of real channels, the Quasi-WSSUS (QWSSUS) channel was introduced [8]. A QWSSUS channel behaves as a WSSUS channel for a restricted interval of time T and a band of frequencies B. Outside this region, the channel correlation functions can no longer be assumed invariant with time, frequency or both.

Bello suggested in [8] that a useful method of describing real channels is to work out the correlation functions over time and frequency intervals small enough for the channel to be described by a hypothetical WSSUS channel. That is, successively apply a QWSSUS model. Then determine

LEGEND : F Fourier Transform
           F^{-1} Inverse Fourier Transform

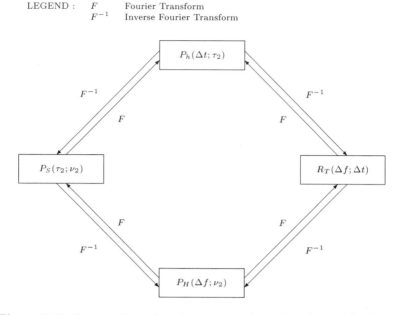

Figure 2.29: Fourier relationships between correlation functions of the first set of Bello functions for WSSUS channels.

the statistics of these correlation functions over longer time periods to fully characterise the channel.

2.6 Channel Description by Bello Functions

This section discusses how the Bello functions introduced previously are applied to the characterisation of mobile radio channels. It explains that mobile radio channels are purely random, and describes how a practical approach to their characterisation has evolved based on the QWSSUS concept.

2.6.1 Space-variance

Consider a mobile station (MS) roaming through an area illuminated by a fixed base station (BS) transmitting a constant single tone. As it moves the MS will see random variations in the amplitude and phase of the signal it receives. Assuming that all scatterers comprising the channel are stationary, then whenever the MS stops, the amplitude and phase of the received signal both remain constant. That is, the channel appears to be time-invariant. When the MS starts to move again the channel once more appears time-variant.

The channel characteristics are therefore dependent on the position of the MS. In the case of single tone transmission, the MS is seen to move through a standing electromagnetic field of random amplitude and phase.

When the MS transmits to the BS, a different physical scenario exists. The receiver is stationary and the field at the BS is changing due to movement of the transmitter. Nevertheless, consideration of the reciprocity of the channel reveals that the effect can be modelled in the same way as before. That is, the BS can be regarded as being a mobile terminal, 'moving through' the same standing electromagnetic field that the MS would experience if the BS were the transmitting terminal.

Hence, regardless of the direction of transmission, the characteristics of the mobile radio channel can be regarded as being dependent on the spatial position of the MS, rather than on absolute time.

2.6.2 Statistical Characteristics

The earlier discussion of the physical structure of mobile radio channels (see Section 2.3) outlined the statistical nature of the amplitude and phase of the received signal component corresponding to a particular delay bin. The phase of the component was reasoned to be a random variable uniformly distributed over $[0, 2\pi]$. This being the case, the ensemble average of any one of the Bello functions when used to describe the mobile radio channel must be zero. Put another way, the mobile radio channel is a purely random channel.

In Section 2.5.3.5 it was explained that it is advantageous to be able to describe a random channel in terms of a WSSUS channel, and that to achieve this with practical channels the QWSSUS concept was proposed.

For the case of the mobile radio channel an approach based upon the QWSSUS method has evolved, where instead of restricting the frequency and time intervals over which a hypothetical WSSUS channel will adequately describe the real channel, the bounds are defined in terms of a frequency interval and a physical area.

The original QWSSUS approach described in Section 2.5.3.6 partitions a channel in frequency and time such that the second-order channel statistics are invariant to frequency translations within a bandwidth B, and time translations within an interval of duration T. That is, the channel must be WSS in both frequency[3] and time.

That mobile radio channels are WSS in the frequency variable was stated earlier in Section 2.3.1.1. This stationarity is a result of the physical properties of the propagation media.

As described above, for the mobile radio channel, wide-sense stationarity in the time variable may be commuted to wide-sense stationarity with respect to the position of the MS. In line with the original QWSSUS ap-

---

[3]Wide-sense stationarity in the frequency variable is generally referred to as uncorrelated scattering (US).

proach we want to partition the area through which the MS will roam into small areas within each of which the channel characteristics are WSS with position.

Small changes in the position of the MS can cause dramatic changes in the amplitude and phase of the received signal corresponding to a particular delay bin. This effect is referred to as fast fading. Although the characteristics of each path (i.e., $\{a_i(t), \tau_i(t), \theta_i(t)\}$) will have changed almost imperceptibly, the ω_c multiplicative factor described on page 109 may produce a relatively large phase shift in the received signal component due to a particular propagation path. The interference of the received signal components from all the propagation paths comprising the channel can therefore change from say, predominantly constructive to predominantly destructive over a very short distance.

For the fast fading statistics to be WSS there should be no change in the mean and variance of the fading process. This implies that the characteristics of each path, such as amplitude, delay and phase retardation, remain unchanged. Hence for small areas, the channel can be considered WSS with respect to position.

Furthermore, combining the two stationarity criteria shows that mobile radio channels, partioned as small areas, are QWSSUS and can therefore be described in terms of the functions shown in Figure 2.29, and listed in Column 4 of Table 2.7.

Motion of the MS over large areas results in a second fading effect, called slow fading. This is the result of significant changes in any of the three propagation path characteristics. It could be due to the obscuring of one building by another, or by a change in the position of a scatterer relative to the MS.

A mobile roaming over a large area will experience both types of fading, the fast fading being superimposed on the slow fading.

It should be understood that this method of applying QWSSUS channels to mobile radio channels derives from the assumption that changes in the channel characteristics are due essentially to movement of the mobile, and that variations due to moving scatterers are a second order effect. In support of this premise, Cox has reported that in New York City cars and trucks generally produce only minor multipath effects [18].

2.6.3   Small-Area Characterisation

A small-area is generally taken to have a radius approximately equal to a few tens of wavelengths [18, 35]. In general such areas are arbitrarily chosen, however care must be exercised as features of the local topology may cause significant changes over relatively short distances. This may be the case close to road junctions in urban environments.

In Section 2.5.3 it was noted that a practical approach to channel characterisation is adopted that involves taking the mean and correlation of one of the Bello functions. However, as the mobile radio channel is purely random it possesses zero ensemble average. Thus, the approach reduces to the establishment of the correlation function for any one of the eight Bello functions.

Consider the input delay-spread function, $h(t, \tau)$. Commuting time to position, this function can be rewritten as $h(p, \tau)$, where p denotes the position of the MS. The correlation function of $h(p, \tau)$ for a small-area is $P_h(\Delta s, \tau)$, where Δs is the distance between the points at which samples of $h(p, \tau)$ are taken. If Δs is set to zero, $P_h(0, \tau)$ (or just $P_h(\tau)$) is obtained. $P_h(\tau)$ is referred to as the power-delay profile of the channel, and is the power spectral density of the channel as a function of delay.

Each measurement of $P_{hp}(\tau)$ is a sample value of the product $h^\star(p, \tau)h(p, \tau)$ for a specific position of the MS. To establish the channel correlation function applicable to the small-area, the ensemble average, $< h^\star(p, \tau)h(p, \tau) >$, taken across the whole area, must be evaluated.

Hence the correlation function is the average power-delay profile, given by

$$P_h(\tau) = \frac{1}{K} \sum_{k=1}^{K} P_{hp}(\tau), \qquad (2.160)$$

where K is the number of samples of the power-delay profile taken over the small-area.

The statistics of each delay bin comprising the mobile radio channel are often assumed to be Gaussian in nature. This is because the local scatterers around the terminals give rise to many propagation paths of virtually identical delay. The central limit theorem [11] can then be applied to reach the assumption of Gaussian statistics.

$P_h(\tau)$ provides a complete description of the channel over the small-area if the channel statistics are Gaussian, because in this case WSS implies strict-sense stationarity (SSS) [11].

Analysis of the statistical distribution of $P_{hp}(\tau)$ across small-areas has in general supported the Gaussian assumption [11,12], although this is not always the case [37]. The amplitude distribution, $\sqrt{P_{hp}(\tau)}$, for a given delay has often been found to fit either a Rayleigh or a Ricean distribution, thereby implying Gaussian statistics.

In narrowband propagation studies, it is often $R_{Tp}(0,0)$ which is measured. From Figure 2.29, it is seen that,

$$R_T(\Delta f; \Delta t) = \int_{-\infty}^{\infty} P_h(\Delta t; \tau) \exp{-j2\pi f \tau d\tau}. \qquad (2.161)$$

From which it is seen that evaluation of $R_T(0,0)$ for the narrowband channel is identical to that of $P_h(\tau)$.

2.6.4   Large-Area Characterisation

By analysing the results obtained within small-areas over large-areas, a description of the slow fading process results. Large-areas are taken as covering geographical districts of similar constitution, such as suburban, or rural. Large-area characterisation takes one of two forms.

In the first form, characterisation is by means of statistical analysis of the variation of channel descriptors derived from the small-area results. For example, for wideband channels, large-areas may be described by the distribution of the delay-spread and mean delay of the average power-delay profile of the channel, see Equations 2.95 and 2.96, respectively. Another descriptor often used is the coherence bandwidth. Narrowband channel descriptors include level crossing rates and fade durations.

The second approach is to analyse the variation in the Bello functions over large-areas. In the case of the average power-delay profile this involves determining the probability of occupancy of a delay bin, and the distribution of the amplitude and phase retardation associated with each bin [38–40]. For narrowband channels, this reduces to measuring the variation in the mean signal strength.

Further analysis of large-areas over grossly dissimilar propagation areas results in gross channel descriptors useful for the prediction of radio coverage during cellular system planning. Gross descriptors are used to modify the basic free space path loss equation, in order to be able to predict the gross channel characteristics for a given propagation environment. 'Rule of thumb' descriptors have evolved this way. For instance, a loss factor representing the degree of urbanisation. Fast fading and slow fading effects are still superimposed upon the predicted channel response.

2.7   Practical Description of Mobile Channels

Having portrayed the mobile radio environment in theoretical terms in time, frequency, delay and Doppler-shift domains by means of Bello functions, we now embark upon the practical characterisation of mobile channels for the practising engineer. Our main goal in this section is to describe the wave-propagation environment as simply as possible, while deriving a set of relevant parameters for power budget and system designers. In harmony with this ambition we characterise the channel by the help of Figure 2.30 in terms of:

1. Propagation pathloss law,

2. Slow fading statistics,

3. Fast fading statistics,

which in general will vary as a function of the propagation frequency, surrounding natural and man-made objects, vehicular speed, etc. Clearly, a

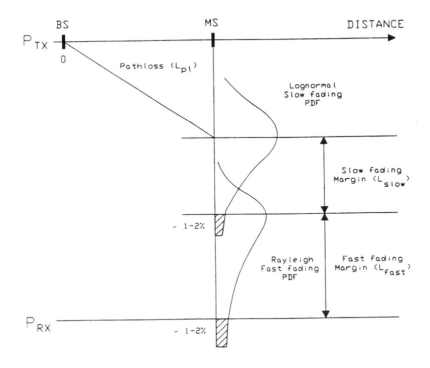

Figure 2.30: Power-budget design.

deterministic treatment is not possible due to the unpredictable variation of channel features, hence we resort to statistical methods. The general approach is to develop theoretical models and check their validity by statistical methods against various real propagation environments.

In this chapter we cannot elaborate on the propagation channels of each existing and perspective mobile radio system, spanning maritime mobile satellite systems, public land mobile radio (PLMR) services, private mobile radio (PMR) schemes, high capacity personal communications networks (PCN) penetrating buildings, halls, etc. The channel and terrain features, multiple access, modulation and signal detection methods, bandwidths, vehicular speeds, etc. are so vastly different that a generic model would be extremely complicated, yet inaccurate. For an all-encompassing up-to-date reference Parson's excellent book published in this series is recommended [42]. This book is concerned mainly with PLMR whilst offering outlook to the emerging PCN. We therefore restrict our discourse to channels in the 900-1800 MHz band, at the same time recognising that the methodology used can be applied to all mobile channels. These frequencies fall in the so-called Ultra High Frequency (UHF) band where convenient antenna sizes are associated with power efficient transmitters and compact receivers. Also the wave propagation is coveniently curtailed by the horizon thereby limiting cochannel interference, when the frequencies are reused in neighbouring cell clusters. At these frequencies, even if there is no line-of-sight path between transmitter and receiver, by means of wave scattering, reflection and diffraction generally sufficient signal power is received to ensure communications.

The prediction of the expected mean or median received signal power plays a crucial role in determining the coverage area of a specific base station and for known interference tolerance also determines the closest acceptable reuse of the propagation frequency deployed. For high antenna elevations and large rural cells a more slowly decaying power exponent is expected than for low elevations and densely built-up urban areas. As suggested by Figure 2.30, the received signal is also subjected to slow or shadow fading which is mainly governed by the characteristic terrain features in the vicinity of the mobile receiver. When designing the system's power budget and coverage area pattern, the slow fading phenomenon is taken into account by including a shadow fading margin as demonstrated by Figure 2.30.

Statistically speaking this requires increasing the transmitted power P_{tx} by the shadow fading margin L_{slow}, which is usually chosen to be the 1-2% quantile of the slow fading probability density function (PDF) to minimise the probability of unsatisfactorily low received signal power P_{rx}. Additionally, the short term fast signal fading due to multipath propagation is taken into account by deploying the so-called fast fading margin L_{fast}, which is typically chosen to be also a few percent quantile of the fast fading distribution. In the worst-case scenario both of these fading margins are simultaneously exceeded by the superimposed slow and fast envelope fading. This situation is often referred to as 'fading margin overload', resulting in a very low-level received signal almost entirely covered in noise. The probability of these cases can be taken to be the sum of the individual margin overload probabilities, when the error probability is close to 0.5, since the received signal is essentially noise. Clearly, the system's error correction codec must be designed to be able to combat this worst-case average bit error probability. This reveals an important trade-off in terms of designed fading margin overload probability, transmitted signal power and error correction coding 'power', which will be made more explicit in the concluding part of this section.

2.7.1   Propagation Pathloss Law

In our probabilistic approach it is difficult to give a worst-case pathloss exponent for any mobile channel. However, it is possible to specify the most optimistic scenario. That is propagation in free space. The free-space pathloss, L_{pl} is given by [42]:

$$L_{pl} = -10\log_{10} G_T - 10\log_{10} G_R + 20\log_{10} f^{Hz} + 20\log_{10} d^m - 147.6 \text{ dB},$$
$$(2.162)$$

where G_T and G_R are the transmitter and receiver antenna gains, f^{Hz} is the propagation frequency in Hz and d^m is the distance from the BS antenna in m. Observe that the free-space pathloss is increased by 6 dB every time, the propagation frequency is doubled or the distance from the mobile is doubled. This corresponds to a 20 dB/decade decay and at d=1 km, f=1 GHz and $G_T = G_R = 1$ a pathloss of $L_F = 92.4$ dB is encountered. Clearly, not only technological difficulties, but also propagation losses discourage the deployment of higher frequencies. Nevertheless, spectrum is usually only available in these higher frequency bands.

In practice, for UHF mobile radio propagation channels of interest to us, the free-space conditions do not apply. There are however a number of useful pathloss prediction models that can be adopted to derive other prediction bounds. One such case is the 'plane earth' model. This is a two-path model constituted by a direct line of sight path and a ground-reflected one, as discussed in Section 1.2.6 and depicted in Figure 1.15, which ignores the curvature of the earth's surface. Assuming transmitter base station (BS) and receiver mobile station (MS) antenna heights of $h^m_{BS}, h^m_{MS} \ll d$, respectively, the plane earth pathloss formula [42] can be derived:

$$L_{pl} = -10\log_{10} G_T - 10\log_{10} G_R - 20\log_{10} h^m_{BS} - 20\log_{10} h^m_{MS} + 40\log_{10} d^m,$$
$$(2.163)$$

where the dependence on propagation frequency is removed. Observe that a 6 dB pathloss reduction is resulted, when doubling the transmitter or receiver antenna elevations, and there is an inverse fourth power law decay with increasing the BS-MS distance d. In the close vicinity of the transmitter antenna, where h_{BS} or $h_{MS} \ll d$ does not hold, Equation 2.163 is no longer valid. Instead, distance-dependent periodic received signal level maxima and minima are experienced, as suggested by Figure 1.15 [67].

The urban microcellular channels are more realistically described by a four-path model including two more reflected waves from building walls along the streets [43], [44], [67]. In this scenario it is assumed that the transmitter antenna is below the characteristic urban skyline. The four-path model of Figure 1.15 used by Green [67] assumed smooth reflecting surfaces yielding specular reflections with no scattering, finite permittivity and conductivity, vertically polarised waves and half-wave dipole antennas. The resultant pathloss profile vs. distance becomes rather erratic with received signal level variations in excess of 20 dB, which renders pathloss modelling by a simple power exponent rather inaccurate, however attractive it would appear due to its simplicity.

There exists a wide variety of further refined models with different strengths, weaknesses and applicability, which take into account other channel imperfections neglected so far. The most widely used of these in the mobile radio environment is the Hata pathloss model [49], which will be discussed in the following section. Parsons [42] gives a detailed comparative study of how various multiple diffractions can be taken into account in illuminating shadowed or obstructed areas, highlighting a number of published pathloss prediction models. Further pathloss model comparisons are readily found in [45]- [47]. Here we use the comprehensively tabulated summary of [46] to provide a quick overview in Table 2.8.

It is quite plausible that more sophisticated models guarantee generally better predictions but are more difficult to evaluate. Irrespective of the prediction model deployed, estimated values always have to be verified by measurements other than those utilised to derive the empirical model invoked. If necessary, correction factors have to be derived and introduced in further predictions. This is the approach we will adopt in our further discussions.

For the sake of illustration here we attempt to verify the applicability of the Hata model [49] to the 1.8 GHz microcellular environment. In doing so we fit minimum mean squared error regression lines to our measurement data, compare this model to Hata's predictions and derive appropriate correction factors to generate further pathloss estimates.

2.7.1.1 The Hata Pathloss Models

Hata developed three pathloss models described below. These were developed from an extensive data base derived by Okumura *et al.* [48] from measurements in and around Tokyo. *The typical urban Hata model* is defined as:

$$
\begin{aligned}
L_{Hu} \;=\; & 69.55 + 26.16\log_{10}f - 13.82\log_{10}h_{BS} - a(h_{MS}) \\
& + (44.9 - 6.55\log_{10}h_{BS})\log_{10}d \text{ dB},
\end{aligned}
\tag{2.164}
$$

where f is the propagation frequency in MHz, h_{BS} and h_{MS} are the BS and MS antenna elevations in terms of metres, respectively, $a(h_{MS})$ is a terrain dependent correction factor, while d is the BS-MS distance in km. The correction factor $a(h_{MS})$ for small and medium sized cities was found to be

$$
a(h_{MS}) = (1.1\log_{10}f - 0.7)h_{MS} - (1.56\log_{10}f - 0.8),
\tag{2.165}
$$

while for large cities is frequency-parameterised:

$$
a(h_{MS}) = \begin{cases} 8.29[\log_{10}(1.54h_{MS})]^2 - 1.1 & \text{if } f \le 200 \text{ MHz} \\ 3.2[\log_{10}(11.75h_{MS})]^2 - 4.97 & \text{if } f \ge 400 \text{ MHz} \end{cases}.
\tag{2.166}
$$

Name	Ref	Date	Frequency	Environment	Remarks
Allesbrook & Parsons	[56]	1977	85-441MHz	urban	Correction factor for VHF
BBC	[52]	1974	UHF(VHF)	urban/rural	Correction factor for VHF
Blomquist & Ladell	[63]	1974	30-900MHz	urban	Deterministic model for pathloss
Bullington	[60]	1947	>30MHz	rural	Equivalent knife-edge for diffraction
Deygout	[62]	1966	VHF	hilly	Main knife-edge diffraction model
Egli	[54]	1957	40-900MHz	rural	Correction factor of irregular terrain
Edwards & Durkin	[51]	1969	30-300MHz	urban/suburban	Plane earth model
Epstein & Peterson	[61]	1953	850MHz	hilly	Multiple knife-edge diffraction
Hata	[49]	1980	100MHz-3GHz	urban/suburban	Computerised Okumura's graphical model
Ibrahim & Parsons	[57]	1981	150-450MHz	urban	Empirical & Semi-empirical models
Japanese Atlas	[64]	1957	30MHz-10GHz	hilly	Multiple knife-edge diffraction
Kessler & Wiggins	[58]	1977	VHF/UHF	urban/rural	Semi-empirical model
Lee	[11]	1982	UHF	urban/suburban	Based on inverse-square law, simple
Leubbers	[65]	1984	VHF	rural	Based on Uniform theory of diffraction
Longley & Rice	[50]	1968	VHF/UHF	rural	Two-ray model + correction factors
Lustgarten & Madison	[59]	1977	VHF/UHF	rural	For quick computation of pathloss
Murphy	[55]	1970	VHF/UHF	rural	Statistical model for irregular terrain
Okumura	[48]	1968	100MHz-3GHz	urban/suburban	Graphical results, mainly for Japan
Palmer	[53]	1979	UHF	urban/rural	Terrain data-base computer model

Table 2.8: Comparison of various pathloss models [46].

The typical suburban Hata model applies a correction factor to the urban model yielding:

$$L_{Hsuburban} = L_{Hu} - 2[\log_{10}(f/28)]^2 - 5.4 \text{ dB.} \qquad (2.167)$$

The rural Hata model modifies the urban formula differently, as seen below:

$$L_{Hrural} = L_{Hu} - 4.78(\log_{10} f)^2 + 18.33 \log_{10} f - 40.94 \text{ dB.} \qquad (2.168)$$

Before we try to interpret these formulae in terms of power-loss exponents, the fundamental limitations of its parameters have to be listed:

$$
\begin{aligned}
f &: \quad 150 - 1500 \text{ MHz} \\
h_{BS} &: \quad 30 - 200 \text{ m} \\
h_{MS} &: \quad 1 - 10 \text{ m} \\
d &: \quad 1 - 20 \text{ km.}
\end{aligned}
$$

For a 900 MHz PLMR system these conditions can be usually satisfied but for a 1.8 GHz typical PCN urban microcell all these limits have to be slightly stretched.

In what follows we now evaluate the Hata prediction for a specific set of values used in our experiments to check its applicability and accuracy in urban microcells. The measured and predicted values are then compared for a large set of measurements to derive relevant correction factors to Hata's model allowing its deployment in microcellular environments. The measurements were carried out in typical urban environments in Southern England at a propagation frequency of 1.8 GHz, BS antenna heights (AH) of 6.4 m, 8.9 m, 11.4 m, 14 m, 17.1 m and 22.1 m, and MS antenna height of 2 m. Using the urban Hata model and the above mentioned parameters the predicted pathlosses are plotted as a function of logarithmic distance in Figure 2.31, where, for example, abscissa values 2 and 3 correspond to $10^2 = 100$ and $10^3 = 1000$ m, respectively. As expected, the higher the antenna elevation, the less steep the pathloss prediction. The power-loss exponents for these parameters vary between 3.962 (39.62 dB/decade) and 3.61 (36.1 dB/decade) for AH=6.4 m and AH=22.1 m, respectively. These exponent values are reasonably close to the inverse fourth power law of the two-path 'plane earth' model. However, they provide a better approximation of the expected measured pathloss in various propagation environments.

The measurement results were collected by sampling and logging the received signal strength at the MS at distances of 3.22 mm and averaging these samples over 2000-sample long windows to remove the effects of fast fading. This delivered a received signal value every 6.44 m. We then fitted a minimum mean squared error regression line to the averaged measured data points and compared it to the appropriate Hata model, as seen in Figures 2.32-2.35 for the antenna heights of AH=6.4 m, 8.9 m, 11.4 m and

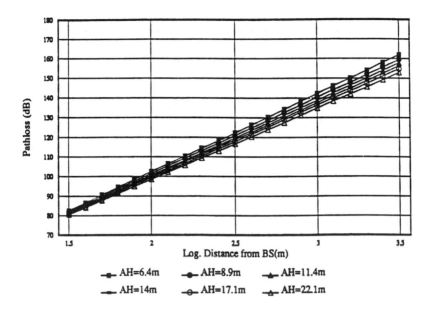

Figure 2.31: Hata pathloss in urban environment at various antenna heights.

AH(m)	Regression gradient (dB/decade)	Hata gradient (dB/decade)	Gradient difference (dB/decade)
14	24.6	37.4	−12.8
11.4	29.8	38	−8.2
8.9	34.5	38.7	−4.2
6.4	36.3	39.6	−3.6

Table 2.9: Measured and predicted pathloss gradients.

AH=14 m.

Observe that as the antenna height is increased from AH=6.4m to AH=14m, the regression lines fitted to the measured data become increasingly more optimistic than the corresponding Hata estimates, which is attributable to the fact that the antenna is gradually elevated beyond the urban skyline. Naturally, local building and terrain features do influence these findings, but the larger the measured data-base, the more consistent the predictions become. The pathloss regression lines for the antenna elevations AH=6.4 m, 8.9 m, 11.4 m and AH=14 m are summarised in Figure 2.36 along with the two extreme Hata models corresponding to AH=6.4 m and AH=14 m. The regression and Hata pathloss law gradients for our experiments are summarised in Table 2.9, which show reasonable

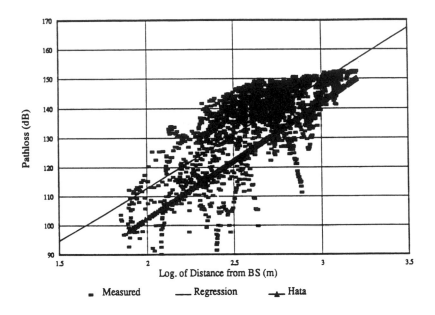

Figure 2.32: Fitting regression line to measured data for AH=6.4m.

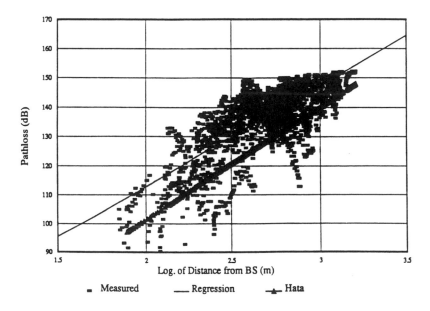

Figure 2.33: Fitting regression line to measured data for AH=8.9m.

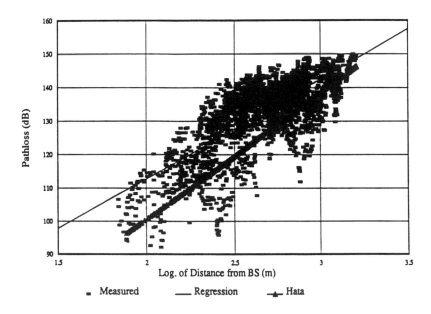

Figure 2.34: Fitting regression line to measured data for AH=11.4m.

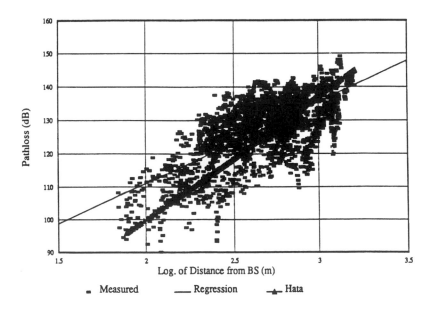

Figure 2.35: Fitting regression line to measured data for AH=14m.

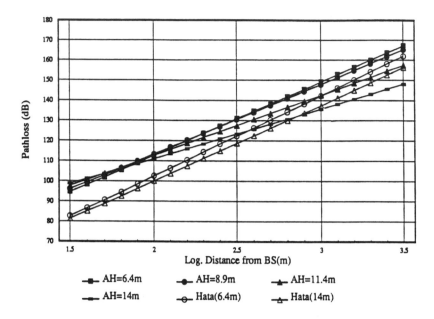

Figure 2.36: Comparison of pathloss regression lines with corresponding Hata models.

agreement only for the lower antenna elevations, below the urban skyline. For the higher elevations clearly different propagation phenomena dominate. We have to point out that the regression line fitting is inevitably biased towards measurements between 500 m and 1000 m, since there are more streets to be measured farther away from the BS than in its immediate vicinity, as inferred from the clustering of measurement points in Figures 2.32-2.35. Nevertheless, from Figure 2.36 we see that in the most important d=100-1000 m region the Hata model gives approximately 10 dB more pessimistic estimates for the parameters considered than the fitted regression lines, if this extremely simplistic model is acceptable.

2.7.2   Slow Fading Statistics

Having derived the propagation pathloss law from our measurements we briefly focus our attention on the characterisation of the slow fading phenomena, which constitutes the second component of the overall power budget design of mobile radio links, as portrayed in Figure 2.30. In slow fading analysis the effects of fast fading and pathloss have to be ignored. The fast fading fluctuations have already been removed for pathloss modelling by averaging over 6.4 m distances. The slow fading fluctuations are simply separated by subtracting the best-fit pathloss regression estimate from

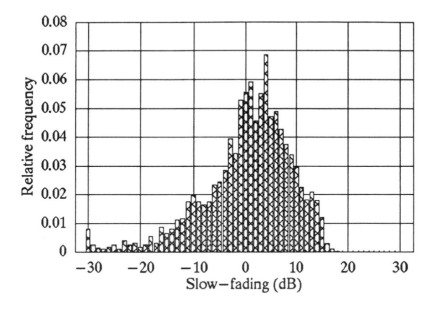

Figure 2.37: Typical microcellular slow fading histogram for AH=6.4m.

each individual 6.4 m-spaced averaged received signal value. The slow fading histograms derived this way from Figures 2.32-2.35 for the previously
used four antenna heights of AH=6.4 m, 8.9 m, 11.4 m and 14 m are depicted in Figures 2.37-2.40. As expected, these figures suggest a lognormal
distribution in terms of dBs due to normally distributed random shadowing effects. Indeed, when subjected to rigorous Kolmogorov-Smirnov and
χ^2 (Chi-square) distribution fitting techniques (see later in Section 2.7.3.4)
using the lognormal hypothesis, the hypothesis is confirmed at a high confidence level. The associated standard deviations are 6.5 dB, 6.8 dB, 7.3 dB
and 7.8 dB for AH=6.4 m, 8.9 m, 11.4 m and 14 m, respectively. When
amalgamating all four slow fading histograms, Figure 2.41 is derived, which
has an even smoother lognormal distribution due to the higher number of
measured points.

2.7.3   Fast Fading Evaluation

2.7.3.1   Analysis of Fast Fading Statistics

Irrespective of the distribution of the numerous individual constituent propagation paths of both quadrature components (a_i, a_q) of the received signal,
their distribution is normal due to the central limit theorem. Then the complex baseband equivalent signal's amplitude and phase characteristics are

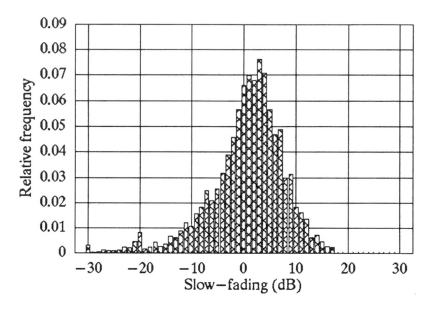

Figure 2.38: Typical microcellular slow fading histogram for AH=8.9m.

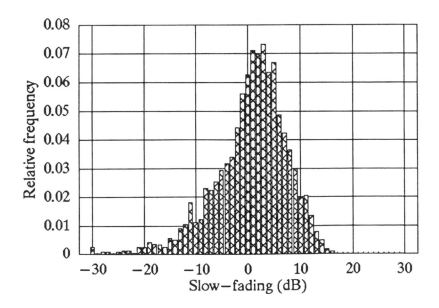

Figure 2.39: Typical microcellular slow fading histogram for AH=11.4m.

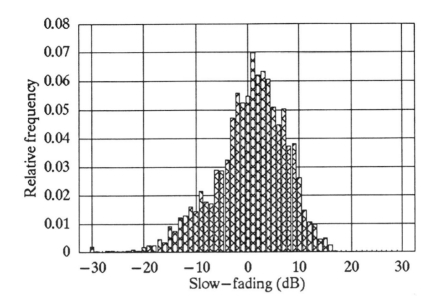

Figure 2.40: Typical microcellular slow fading histogram for AH=14m.

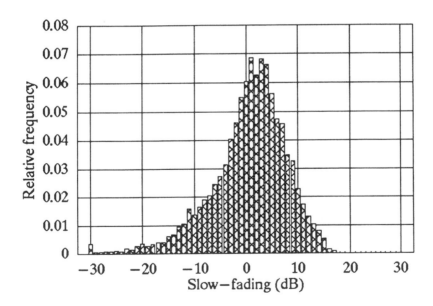

Figure 2.41: Typical microcellular slow fading histogram for various antenna elevations.

given by:

$$a(k) = \sqrt{a_i^2(k) + a_q^2(k)} \qquad (2.169)$$

$$\phi(k) = \arctan[a_q(k)/a_i(k)]. \qquad (2.170)$$

Our aim is now to determine the distribution of the amplitude $a(k)$, if $a_i(k)$ and $a_q(k)$ are known to have a normal distribution. In general, for n normally distributed random constituent processes with means $\overline{a_i}$ and identical variances σ^2, the resultant process $y = \sum_{i=1}^{n} a_i^2$ has a so-called χ^2 distribution with a PDF given below [66]:

$$p(y) = \frac{1}{2\sigma^2} \left(\frac{y}{s^2}\right)^{(n-2)/4} \cdot e^{-(s^2+y)/2\sigma^2} \cdot I_{(n/2)-1}\left(\sqrt{y}\frac{s}{\sigma^2}\right) \qquad (2.171)$$

where

$$y \geq 0 \qquad (2.172)$$

and

$$s^2 = \sum_{i=1}^{n} (\overline{a_i})^2 \qquad (2.173)$$

is the so-called non-centrality parameter computed from the first moments of the component processes $a_1 \cdots a_n$. If the constituent processes have zero means, the χ^2 distribution is central, otherwise non-central. Each of these processes has a variance of σ^2 and $I_k(x)$ is the modified k th order Bessel-function of the first kind, given by

$$I_k(x) = \sum_{j=0}^{\infty} \frac{(x/2)^{k+2j}}{j!\Gamma(k+j+1)} , \quad x \geq 0. \qquad (2.174)$$

The Γ function is defined as

$$\Gamma(p) = \int_0^{\infty} t^{p-1}e^{-t}dt \quad \text{if } p > 0$$

$$\Gamma(p) = (p-1)! \quad \text{if } p > 0 \text{ integer} \qquad (2.175)$$

$$\Gamma(\tfrac{1}{2}) = \sqrt{\pi} , \quad \Gamma(\tfrac{3}{2}) = \frac{\sqrt{\pi}}{2}.$$

In our case we have two quadrature components, i.e. $n = 2$, $s^2 = (\overline{a_i})^2 + (\overline{a_q})^2$, the envelope is computed as $a = \sqrt{y} = \sqrt{a_i^2 + a_q^2}, a^2 = y, p(a)da = p(y)dy$, and hence $p(a) = p(y)dy/da = 2ap(y)$ yielding the Rician PDF

$$p_{\text{Rice}}(a) = \frac{a}{\sigma^2}e^{-(a^2+s^2)/2\sigma^2}I_o\left(\frac{as}{\sigma^2}\right) \quad a \geq 0. \qquad (2.176)$$

Formally introducing the Rician K-factor as

$$K = s^2/2\sigma^2 \qquad (2.177)$$

renders the Rician distribution's PDF to depend on one parameter only:

$$p_{\text{Rice}}(a) = \frac{a}{\sigma^2} \cdot e^{-\frac{a^2}{2\sigma^2}} \cdot e^{-K} \cdot I_o \left(\frac{a}{\sigma} \cdot \sqrt{2K} \right), \qquad (2.178)$$

where K physically represents the ratio of the power received in the direct line-of-sight path, to the total power received via indirect scattered paths. Therefore, if there is no dominant propagation path, $K = 0$, $e^{-K} = 1$ and $I_0(0) = 1$ yielding the worst-case Rayleigh PDF:

$$p_{\text{Rayleigh}}(a) = \frac{a}{\sigma^2} e^{-\frac{a^2}{2\sigma^2}}. \qquad (2.179)$$

Conversely, in the clear direct line-of-sight situation with no scattered power, $K = \infty$, yielding a 'Dirac-delta shaped' PDF, representing a step-function-like CDF. The signal at the receive antenna then has a constant amplitude with a probability of one. Such a channel is referred to as a Gaussian channel. This is because although there is no fading present, the receiver will still see the additive white Gaussian noise (AWGN) referenced to its input, as seen in section 2.1.4. Clearly, if the K-factor is known, the fast fading envelope's distribution is described perfectly.

The Rician CDF takes the shape of [66]

$$
\begin{aligned}
C_{\text{Rice}}(a) &= 1 - e^{-\left(K + \frac{a^2}{2\sigma^2}\right)} \sum_{m=0}^{\infty} \left(\frac{s}{a}\right)^m \cdot I_m \left(\frac{a\,s}{\sigma^2}\right) \\
&= 1 - e^{-\left(K + \frac{a^2}{2\sigma^2}\right)} \sum_{m=0}^{\infty} \left(\frac{\sigma\sqrt{2K}}{a}\right)^m \cdot I_m \left(\frac{a\sqrt{2K}}{\sigma}\right).
\end{aligned}
$$
$$(2.180)$$

Clearly, this formula is more difficult to evaluate than the PDF of Equation 2.178 due to the summation of an infinite number of terms, requiring double or quadruple precision and it is avoided in numerical evaluations, if possible. However, in practical terms it is sufficient to increase m to a value, where the last term's contribution becomes less than 0.1%.

A range of Rician CDFs evaluated from Equation 2.180 are plotted on a linear scale in Figure 2.42 for $K = 0, 1, 2, 4, 8$ and 15. Figure 2.43 shows the same Rician CDFs plotted on a more convenient logarithmic scale, which reveals the enormous difference in terms of deep fades for the K values considered. When choosing the fading margin overload probability, Figure 2.43 expands the high-attenuation tails of the CDFs, where for example for a Rician CDF with $K = 1$ the 15 dB fading margin overload probability is seen to be approximately 10^{-2}. Lastly, a set of Rician PDFs computed

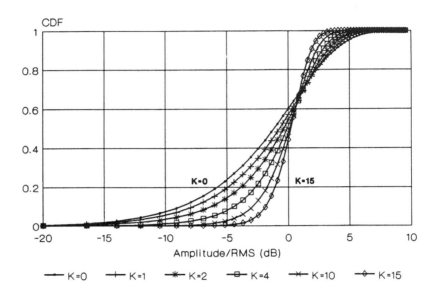

Figure 2.42: Rician CDFs on linear scale.

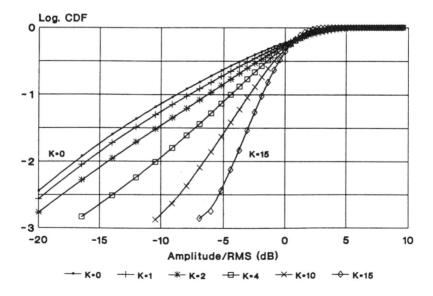

Figure 2.43: Rician CDFs on logarithmic scale.

from Equation 2.178 are seen in Figure 1.6, while the corresponding fading envelopes are depicted in Figure 1.5.

2.7.3.2   The Relation of Rician and Gaussian PDFs

In this subsection we show that the Rician PDF tends to the Gaussian one, as the K factor tends to infinity. To do so we simplify Equation 2.178 by introducing the transformation [41]

$$\alpha = \frac{a}{\sigma\sqrt{2K}}\,, \tag{2.181}$$

which yields

$$p(\alpha) = p(a)\, da/d\alpha = p(a)\, \sigma\sqrt{2K}\,. \tag{2.182}$$

Substituting $p_{\text{Rice}}(a)$ from Equation 2.178 in Equation 2.182 we have:

$$p(\alpha) = 2K\alpha \cdot e^{-K(\alpha^2+1)} \cdot I_0(2K\alpha). \tag{2.183}$$

For large x values

$$I_0(x) \approx \frac{e^x}{\sqrt{2\pi x}} \left[1 + \frac{1^2}{1!\,8x} + \frac{1^2 \cdot 3^2}{2!\,(8x)^2} + \dots \right] \approx \frac{e^x}{\sqrt{2\pi x}}. \tag{2.184}$$

Hence from Equation 2.183 we have:

$$
\begin{aligned}
\lim_{K\to\infty} p(\alpha) &\approx \lim_{K\to\infty} 2K\alpha \cdot e^{-K(\alpha^2+1)} \cdot \frac{e^{2K\alpha}}{\sqrt{2\pi}\sqrt{2K\alpha}} \\
&\approx \lim_{K\to\infty} \sqrt{\alpha}\, \frac{1}{\sqrt{2\pi} \cdot 1/\sqrt{2K}} \cdot e^{-\frac{(\alpha-1)^2}{2\cdot 1/2K}}\,,
\end{aligned}
\tag{2.185}
$$

which tends to a Gaussian PDF with a mean of one and a variance of $1/2K \approx 0$, yielding a Dirac delta function when K tends to infinity.

2.7.3.3   Extracting Fast Fading Characteristics

To determine the fast fading statistics of the received signal envelope one has to remove the effects of the path loss as well as that of the slow fading. The standard recognised method to extract the fading envelope is to normalise the received signal to its local RMS value, as proposed by Clarke [17] and used by other authors since then [67]. For the received sample $r(x_i)$ the local RMS is given by

$$\text{RMS} = \left(\frac{1}{W} \sum_{i-W/2}^{i+W/2} [r(x_i)]^2 \right)^{\frac{1}{2}} \tag{2.186}$$

where W represents the window-length for the computation. This local RMS estimate is computed for each individual received sample in a sliding window and the normalised samples $r(x_i)$/RMS are subjected to distribution fitting algorithms. The adequacy of this normalisation depends on the appropriate selection of W. Lee [11] suggested a window of 40 wavelengths (λ) for conventional cell sizes, but in agreement with other authors [67], we found that in microcells the local RMS received signal level undergoes quite large fluctuations in such wide windows. This could affect the local statistics. Our experiments with a range of W values suggested that for any signal envelope sampling rate the window size must 'cover' a computation interval of about $4\lambda - 10\lambda$. This gives a sound RMS estimate and does not distort the fast fading statistics, hence we opted for $W = 200$ samples. After this smoothing-normalisation the fast fading envelope is stored in the computer ready for distribution fitting.

Knowledge of the expected Rician K-factor is important in system design, as it allows estimation of the fast fading margin required in the link budget calculation. A long term average of the K-factor gives an estimation of the average performance that can be expected. However, more efficient system design requires knowledge of the variation in K-factor as the mobile moves. In extreme, an individual K-factor could be calculated for each fade. This would require a very short computational window but as the computational window becomes shorter, the K-factor variations become more erratic. That is, since the distribution gradually changes, the confidence in the goodness-of-fit reduces. Furthermore, such very short windows would result in such a vast amount of information that computation and analysis become impractical.

In this paragraph we set out to determine the optimum window size of the fast fading distribution fitting. To achieve this we synthesised a file of 80,000 fading envelope values with an overall $K=3$ dB or $K \approx 2$ using Figure 1.9 in Chapter 1. This represents moderately severe fading, as can be seen from the CDFs in Figures 2.42 and 2.43. Observe that localised K-factors will have increased variation, as the window size is reduced, revealing the true fast fading profile. Accordingly, the K-factors were then evaluated by the distribution fitting algorithms to be highlighted in the next subsection using computation windows of length $D=1000$, 2000, 4000, 8000 and 80,000 samples. As expected, using $D=80,000$, the overall K-factor was measured to be $K=3$ dB or $K \approx 2$, with a very high degree of confidence, i.e. the significance level in the goodness-of-fit test was almost unity. As the computation window size, D, was decreased, the variation in the K-factors increased. The derived K-factor profiles and their associated significance levels are shown for window sizes of $D=2000$, 4000 and 8000 in Figures 2.44–2.46. where the whole file of 80,000 points is represented in each case and a significance level in excess of 0.1 implies a high degree of confidence in the fit. Analysis of these figures and a variety of similar profiles using different window sizes show that the variation in the

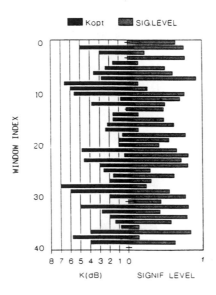

Figure 2.44: K-factor and significance-level profiles for $D = 2000$.

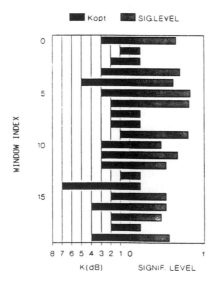

Figure 2.45: K-factor and significance-level profiles for $D = 4000$.

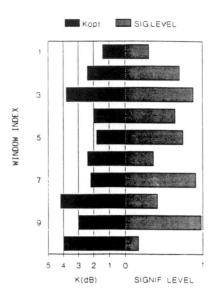

Figure 2.46: K-factor and significance-level profiles for $D = 8000$.

K-factors does not increase significantly for window sizes below D=1000-2000, suggesting that the window is sufficiently short to track the change in the K-factor. However, reducing the window size further significantly lowers the confidence measures associated with the K-factor, implying a poor fit. Observe that although the file of 80,000 samples gives a K-factor of 2, the average of the K values over the windows in Figures 2.44–2.46 is generally a different value. Analysis of other data files revealed similar results confirming our choice of D=1000-2000 for distribution fitting when tracking K-profiles.

The erratic K-factor variations seen in Figure 2.44 are fairly typical, making their efficient use in power budget design rather difficult. Clearly, the K-variations must be statistically characterized in terms of their distributions, which were found to have 'near-normal' PDFs for a variety of scenarios, but the expected value and variance of the K-factors depended on the local paraphernalia. The interested reader is referred to Figures 1.20 and 1.21 for some measured K-profiles. With the analysis parameters settled, one can proceed to analyse the measured data.

2.7.3.4 Goodness-of-fit Techniques

K-profiles of measured propagation data are conveniently evaluated by computing the PDFs and CDFs of the measured samples for windows of, for example, $D = 1000 \dots 2000$ smoothed values and then comparing them with a set of hypothesis PDFs and/or CDFs. Each comparison yields a probability, confidence measure or significance level proportional to the likelihood that the measured distribution is really a representative of the assumed hypothesis distribution.

The specific comparison giving the highest significance level is checked against the minimum acceptance level allowing rejection or acceptance of the initial hypothesis.

There is an abundance of goodness-of-fit methods for testing the statistical relevance of a match between a measured and a hypothesis distribution, all of which have different strengths and weaknesses [68,69]. Here we briefly consider the Chi-square (χ^2) and the Kolmogorov-Smirnov (KS) methods.

2.7.3.4.1 Chi-square Goodness-of-fit Test The method devised in 1900 by Pearson is based on a normalised quadratic sum of the deviations of the observed occurrences (n_i) from their hypothesised expected values ($n \cdot p_i$), where n_i and p_i represent the observed number of occurrences in bin i and their expected probabilities, respectively. There are k bins and a total of n samples in the experiment. Then the degree of freedom is $(k-1)$, since the only linear restriction present is that

$$\sum_{j=1}^{k} n_j = n. \tag{2.187}$$

Other restrictions can be introduced for distributions where some parameters, such as mean or variance, have to be estimated. Each maximum likelihood parameter estimate reduces the degree of freedom by one. The measure of deviation is then computed as:

$$\chi^2 = \sum_{i=1}^{k} \frac{(n_i - n \cdot p_i)^2}{n \cdot p_i}, \tag{2.188}$$

which can be shown to have a Chi-square distribution, if the differences between n_i and $n \cdot p_i$ are non-deterministic. Therefore the confidence measure associated with a specific χ^2 value in Equation 2.188 gives the probability that such a χ^2 value could have come from the χ^2 distribution of Equation 2.171, which is also given in the Tables of [68] and [69].

2.7.3.4.2 Kolmogorov-Smirnov (KS) Goodness-of-fit Test In contrast to the Chi-square fitting, where the binned measured PDF was used, the KS-test uses the CDFs. According to Kolmogorov and Smirnov the limiting distribution of the maximum CDF deviation D_n between the measured distribution $C_n(x)$ and hypothesis $C(x)$

$$\sqrt{n} \cdot D_n = \sqrt{n} \cdot \text{Max}|C_n(x) - C(x)| \tag{2.189}$$

is characterised by [69]

$$H(x) \quad = \quad \lim_{n \to \infty} \ [CDF_{D_n}(x)] = \lim_{n \to \infty} \ P[\sqrt{n}D_n \leq x]$$

$$= \quad \left[1 - 2\sum_{j=1}^{\infty} (-1)^{(j-1)} e^{-2j^2 x^2} \right] \cdot I_{0,\infty}(x) \qquad (2.190)$$

where the indicator function $I_{(0,\infty)}(x)$ is

$$I_{(0,\infty)}(x) = \left\{ \begin{array}{ll} 1 & \text{if } 0 < x < \infty \\ 0 & \text{otherwise} \end{array} \right. . \qquad (2.191)$$

Observe that the CDF $H(x)$ in Equation 2.190 does not depend on the distribution $C_n(x)$, which explains the versatility of the method. On the other hand, the KS method does not allow testing of composite hypothesies, where some parameters have to be estimated, as this would reduce the degree of freedom.

In practical hypothesis testing the measured data of each window of D samples has to be sorted in ascending order, which actually gives a fine but non-uniformly spaced representation of its CDF. Then for each individual measured and sorted sample the corresponding CDF value has to be found from the hypothesis CDF and the maximum deviation over the window must be remembered for every hypothesis CDF. The hypothesis CDF with the smallest maximum deviation D_n has the highest confidence, the value of which is computed from Equation 2.190 or from tables given in [70]. The KS test is computationally more demanding than the χ^2 test, because the computation of the Rician CDFs from Equation 2.180 implies the evaluation of a high number of summation terms, necessitating double or quadruple precision computations. Furthermore, as opposed to the χ^2 test, where the hypothesis PDFs can be prestored, here the hypothesis CDF of Equation 2.180 must be evaluated 'on-line' for every single, non-uniformly spaced abscissa value, constituted by the measured fading samples. Hence for large-scale measurement programmes the χ^2 test is preferred.

2.7.3.4.3   Goodness-of-fit of the Hypothesis Distribution   In distribution fitting the higher the number of samples used to compute the measured CDF and PDF, the smaller is the tolerable discrepancy for a specific significance level between the best-fit theoretical PDF or CDF and its experimental counterpart. This can be inferred from Equations 2.188 and 2.190 for both the χ^2 and the KS test, since the error terms in these equations increase as the number of samples n in a fitting window is increased. The reason for this is plausible, since for larger number of samples the statistical relevance of the experimental data is enhanced, allowing di-

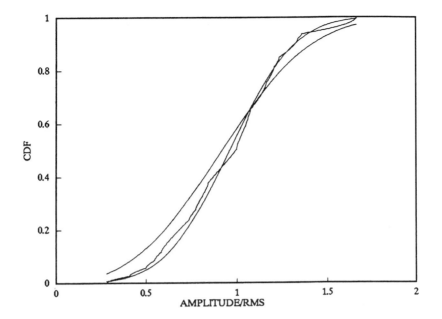

Figure 2.47: KS CDF-fitting using a window size of $D = 1000$.

minishing random differences only between the measured and hypothesis statistics. This is clearly illustrated by Figures 2.47–2.49, where the measured CDFs of a $K=3$ dB Rician channel generated by theory are depicted using window sizes of $D=1000$, 2000 and 4000 samples, together with their best matching theoretical counterparts. Specifically, Figure 2.47 displays the goodness-of-fit to be expected, if a window of D=1000 samples is used. As expected, the KS distribution fitting procedure favours the CDF, representing $K=3$ dB from the preselected set of 80 hypothesis CDFs, to yield the best fitting theoretical CDF. For comparison we also depicted the $K=7$ dB CDF, which exhibits a consistently different shape, while the differences with respect to the $K=3$ dB CDF appear to be random. When the window size is increased to D=2000 and 4000 samples, as seen in Figures 2.48 and 2.49, the differences become less significant, as the higher number of samples encountered gives a more adequate representation of the random process.

Figure 2.50 displays the significance level as a function of the Chi-square distribution parameter χ^2 computed from Equation 2.171. Here we utilised a window-size of 1000 samples and 80 'bins' for categorising the data between 0 and $4 \cdot RMS$. Using the χ^2 error term computed from Equation 2.171 the significance level, referred to also as confidence measure (CM), is read from Figure 2.50a as the intercept of the CM-curve with the appropriate vertical grid line at the specific χ^2-value. For CMs lower than

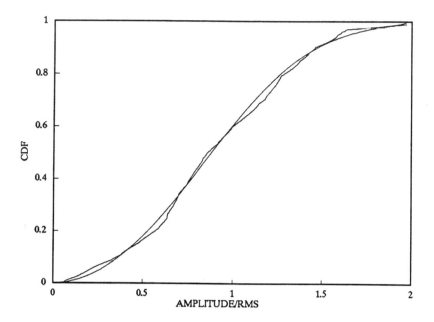

Figure 2.48: KS CDF-fitting using a window size of $D = 2000$.

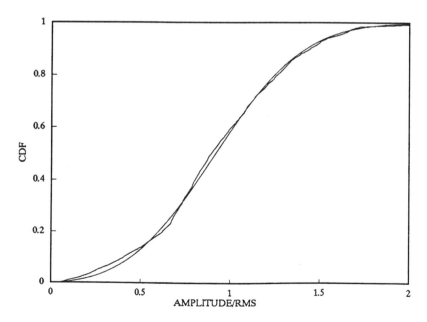

Figure 2.49: KS CDF-fitting using a window size of $D = 4000$.

10^{-2} the logarithmically scaled curve of Figure 2.50b is more preferable. Observe that for a χ^2-value of 60 the confidence level is near-unity, while a value of 120 only achieves a CM of approximately 10^{-3}. In practice any hypothesis PDF having $\chi^2 > 120$, i.e. CM $< 10^{-3}$ has to be rejected.

Similarly, the linear and logarithmic Kolmogorov-Smirnov (KS) significance levels are depicted in Figure 2.51, where the abscissa values are scaled by the square-root of the sample number (SAMPNO). Again, the sharply decaying curves ensure a well-defined acceptance or rejection of the hypothesis distribution and the logarithmically scaled curve in Figure 2.51b conveniently expands the lower end of the CM-scale.

2.7.4 Summary

Below we summarise the practical characterisation of microcellular mobile radio channels by way of an example using our three-step approach portrayed in Figure 2.30 and the results from our previous deliberations.

1) We estimate the pathloss using the Hata-model and deploy a correction factor corresponding to the antenna elevation, deduced from measurements: $L_{pl} = L_{Hata} + L_{corr}$.

2) Using the characteristic slow fading variance of, say $\sigma=7$ dB, assuming lognormal slow fading PDF and allowing for a 1.4% slow 'fading margin overload' probability we introduce a 'slow fading margin' of $L_{slow} = 2 \cdot \sigma = 14$ dB.

3) Assuming a typical Rician fading with $K = 10$ and a fast 'fading margin overload' probability of 1% a 'fast fading margin' of $L_{fast}=7$ dB is inferred from Figure 2.42.

Summing the pathloss and the two fading margin components from above yields a total pathloss of: $L_{total} = L_{pl} + L_{slow} + L_{fast}$. With the knowledge of the receiver sensitivity P_{rec} this then allows us to compute the required minimum transmitted power as: $P_{tx} = P_{rec} + L_{total}$.

For the sake of illustration let us compare the transmitted power requirements for an urban microcellular environment with cell radii of 300 m and 100 m, using the above mentioned $L_{slow}=14$ dB and $L_{fast}=7$ dB fading margins. From Figure 2.36 we find that our extensive measurement programme in this environment suggests a pathloss of $L_{pl}=130$ dB at a distance of 300 m from the BS, when the antenna elevation is AH=6.4 m, while at 100 m $L_{pl}=110$ dB. Then we have pathlosses of $L_{300}=130$ dB$+14$ dB$+7$ dB$=151$ dB and $L_{100}=131$ dB. Assuming a receiver sensitivity of -104 dBm, as in the Pan-European digital mobile radio system, the corresponding transmitted power requirements are: $P_{300}=47$ dBm ≈ 50 W and $P_{100}=27$ dBm ≈ 0.5 W. Clearly, for this low antenna elevation microcellular scenario a cell radius of larger than 100 m is becoming unrealistic in terms of transmitted power. Increasing the antenna height

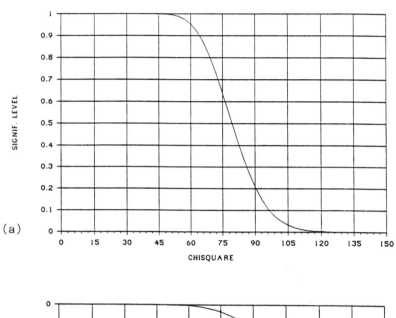

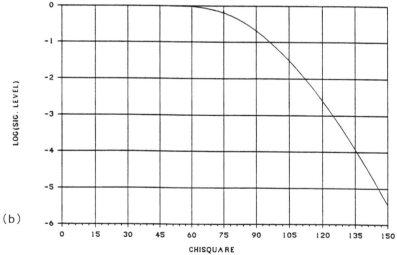

Figure 2.50: Chi-square significance level variation (a) on linear scale, or (b) on logarithmic scale.

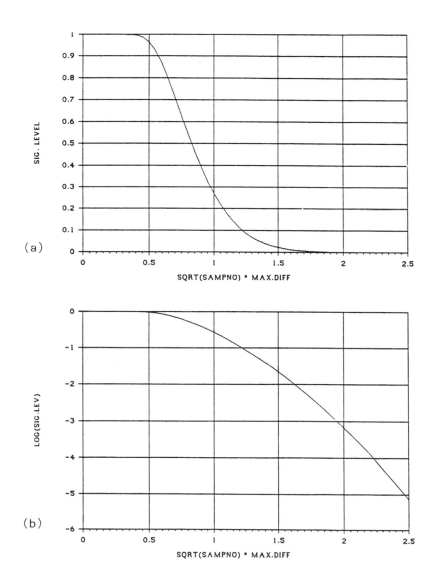

Figure 2.51: Kolmogorov-Smirnov significance level variation (a) on linear scale
(b) on logarithmic scale.

substantially reduces the transmitted power requirement or extends the cell radius. When the BS antenna is elevated beyond the urban sky-line, substantially higher coverage area can be achieved at the same transmitted power. However, the fading gradually becomes Rayleigh, requiring a higher fading margin and thereby reducing the gains achieved.

Observe that the power budget design highlighted outlines the requirements imposed on the error correction codec as well. Namely, in the event when the slow and fast fading exceed the total fading margin of $L_{fading} = L_{slow} + L_{fast}$=21 dB, the channel codec is faced with an error probability of 50 %, hence it must be designed to correct an error rate of about half of the fading margin overload probability.

This chapter aimed at introducing the reader to the intricaces of mobile radio propagation. After introducing the baseband representation of systems and channels the so-called Bello-functions were defined. Issues of time- and frequency-dispersive channels were discussed, leading to a practically motivated discussion on the statistical characterisation of wireless channels in terms of pathloss, slow- and fast-fading. The chapter was concluded with the power-budget design of mobile systems. Let us now embark on considering speech codecs in the next chapter, which constitute an important component of wireless communications systems.

Bibliography

[1] **D. C. Cox**. "Universal digital portable radio communications". *Proc. IEEE, Vol. 75, No. 4*, pages pp.436–477, April 1987.

[2] **S. Stein** and **J. J. Jones**. *Modern Communication Principles*. McGraw-Hill, 1967.

[3] **J. Dugundji**. "Envelopes and pre-envelopes of real waveforms". *IRE Trans., Vol. IT-4, No. 1*, pages pp.53–57, March 1958.

[4] **M. Schwartz**, **W. R. Bennett**, and **S. Stein**. *Communications Systems and Techniques*. McGraw-Hill, 1966.

[5] **D. J. Sakrison**. *Communications Theory: Transmission of Waveforms and Digital Information*. John Wiley & Sons, 1968.

[6] **S. Y. Liao**. *Engineering Applications of Electromagnetic Theory*. West Publishing Company, 1988.

[7] **T. Aulin**. "Characteristics of a digital mobile radio channel". *IEEE Trans., Vol. VT-30, No. 2*, pages pp.45–53, May 1981.

[8] **P. A. Bello**. "Characterization of randomly time-variant linear channels". *IEEE Trans., Vol. CS-11, No. 4*, pages pp.360–393, December 1963.

[9] **W. C. Y. Lee**. "Finding the approximate angular probability density function of wave arrival by using a directional antenna". *IEEE Trans., Vol. AP-21, No. 3*, pages pp.328–334, May 1973.

[10] **D. C. Cox** and **R. P. Leck**. "Correlation bandwidth and delay spread multipath propagation statistics for 910–MHz urban mobile radio channels". *IEEE Trans., Vol. COM-23, No. 11*, pages pp.1271–1280, November 1975.

[11] **W. Y. C. Lee**. *Mobile communications engineering*. McGraw Hill Book Co., 1982.

[12] **W. C. Jakes, ed.** *Microwave mobile communications*. Wiley, New York, 1974.

[13] **P. A. Bello** and **B. D. Nelin**. "The effect of frequency selective fading on the binary error probabilities of incoherent and differentially coherent matched filter receivers". *IEEE Trans., Vol. CS–11, No. 2*, pages pp.170–186, June 1963.

[14] **H. F. Schmid**. "A prediction model for multipath propagation of pulse signals at VHF and UHF over irregular terrain". *IEEE Trans., Vol. AP–18, No. 2*, pages pp.253–258, March 1970.

[15] **P. Melançon** and **J. Le Bel**. "A characterization of the frequency selective fading of the mobile radio channel". *IEEE Trans., Vol. VT–35, No. 4*, pages pp.153–161, November 1986.

[16] **W. R. Young, Jr.** and **L. Y. Lacy**. "Echoes in transmission at 450 megacycles from land-to-car radio units". *Proc. IRE, Vol. 38*, pages pp.255–258, March 1950.

[17] **R. H. Clarke**. "A statistical theory of mobile-radio reception". *B. S. T. J., Vol. 47*, pages pp. 957–1000, July–August 1968.

[18] **D. C. Cox**. "910 MHz urban mobile radio propagation: Multipath characteristics in New York city". *IEEE Trans., Vol. VT–22, No. 4*, pages pp.104–110, November 1973.

[19] **A. S. Bajwa** and **J. D. Parsons**. "Small-area characterisation of UHF urban and suburban mobile radio propagation". *IEE Proc., Vol. 129, Pt. F, No. 2*, pages pp.102–109, April 1982.

[20] **J. Zander**. "A stochastical model of the urban UHF radio channel". *IEEE Trans., Vol. VT–30, No. 4*, pages pp.145–155, November 1981.

[21] **D. Gabor**. "Theory of communications". *Journal of the IEE, Part III, Vol. 93*, pages pp. 429–457, November 1946.

[22] **D. G. Appleby, P. M. Fortune, Y. F. Ko, W. H. Lam, R. Steele, I. J. Wassell**, and **K. H. H. Wong**. "The propose multiple access methods for the pan-European mobile radio systems". *IEE Colloquium on "Multiple Access Techniques in Radio Systems"*, Savoy Place, London, pages pp.1.1–1.26, October 1986.

[23] GSM Doc. No. 85/85. *"Description of the Experimental S–900D for Digital Radiotelephone"*.

[24] **J. Udenfeldt**. *"DMS90—An Experimental TDMA Digital Mobile Telephone System"*. Ericsson Radio System AB, Stockholm, Sweden.

[25] **K. D. Eckert** and **G. Höfgen**. "The fully digital cellular radio telephone system CD900". *Nordic Seminar on Digital Land Mobile Radio Communication, Espoo, Finland*, pages pp.249–259, February 1985.

[26] *GSM Specifications*, 1988.

[27] **D. C. Cox**. "Delay doppler characteristics of multipath propagation at 910 MHz in a suburban mobile radio environment". *IEEE Trans., Vol. AP-20, No. 5*, pages pp.625–635, September 1972.

[28] **T. Takeuchi, F. Ikegami, S. Yoshida**, and **N. Kikuma**. "Comparison of multipath delay spread characteristics with BER performance of high speed digital mobile transmission". *38th IEEE Vehicular Technology Conference, Philadelphia, Pennsylvania*, pages pp. 119–126, 15–17 June 1988.

[29] **M. Nilson**. "Measurements of some characteristics of multipath propagation at 900 MHz in the city of Stockholm". *COST 207 TD (85)14*, pages pp. 117–132, 11 February 1985.

[30] **P. Lo Muzio, G. Guidotti, N. Benvenuto**, and **S. Pupolin**. "Experimental characterization of VHF propagation in a mountain environment". *Alta Frequenza, Vol. 56, No. 6*, pages pp.273–282, August 1987.

[31] **P. A. Bello**. "Time-frequency duality". *IEEE Trans., Vol. IT-10, No. 1*, pages pp.18–33, January 1964.

[32] **J. D. Parsons** and **A. S. Bajwa**. "Wideband characterisation of fading mobile radio channels". *IEE Proc., Vol. 129, Pt. F, No. 2*, pages pp.95–101, April 1982.

[33] **P. W. Huish** and **E. Gurdenli**. "Propagation measurements and planning requirements for digital cellular systems". *Second Nordic Seminar on Digital Land Mobile Radio Communications, Stockholm*, pages pp.199–204, 14–16 October 1986.

[34] **P. Olivier** and **J. Tiffon**. "Transfer function measurement as a characterisation of the urban mobile radio channel". *IEE Fifth Int. Conf. on Antennas and Propagation, ICAP 87, IEE Conf. Pub. No. 274, Pt. 2*, pages pp.95–98, 30 March–2 April 1987.

[35] **J. D. Parsons** and **A. S. Bajwa**. "Wideband characterisation of fading mobile radio channels". *IEE Proc., Vol. 129, Pt. F, No. 2*, pages pp.95–101, April 1982.

[36] **A. Papoulis**. *Probability Random Variables and Stochastic Processes*. McGraw-Hill, 1981.

[37] IEEE Vehicular Technology Society Committee on Radio Propagation. "Coverage prediction for mobile radio systems operating in the 800/900 MHz frequency range". *IEEE Trans., Vol. VT-37, No. 1*, pages pp.3–72, February 1988.

[38] **G. L. Turin, F. D. Clapp, T. L. Johnston, S. B. Fine**, and **D. Lavry**. "A statistical model of urban multipath propagation". *IEEE Trans., Vol. VT-21, No. 1*, pages pp.1–9, February 1972.

[39] **H. Suzuki**. "A statistical model for urban radio propagation". *IEEE Trans., Vol. COM-25, No. 7*, pages pp.673–680, July 1977.

[40] **H. Hashemi**. "Simulation of the urban radio propagation channel". *IEEE Trans., Vol. VT–28, No. 3*, pages pp.213–225, August 1979.

[41] **L. Pap.** *Private Communication*, 1992

[42] **J. D. Parsons.** "The Mobile Radio Propagation Channel". *Pentech Press*, 1991

[43] **A. Rustako, N. Amitay, G.J. Owens, R.S. Roman.** "Propagation Measurments at Microwave Frequencies for Microcellular Mobile and Personal Communications". *Proc. of 39th IEEE VTC*, pp. 316-320, 1989

[44] **A.J. Rustako, N. Amitay, G.J. Owens, R.S. Roman.** "Radio Propagation Measurements at Microwave Frequencies for Microcellular Mobile and Personal Communications". *IEEE Int. Conf. on Comm's (ICC'89)*, pp. 482-488, Boston, U.S.A.

[45] **J.F. Aurand, R.E. Post.** "A Comparison of Prediction Methods for 800 MHz Mobile Radio Propagation". *IEEE Tr. VT-34, No 4*, pp. 149-153, Nov. 1985

[46] **S.T.S. Chia.** "Propagation Studies for Microcellular Mobile Radio". *PhD. Thesis*, University of Southampton, U.K., 1987

[47] **G.Y. Delisle, J. Lefevre, M. Lecours, J. Chouinard.** "Propagation Loss Prediction: A Comparative Study with Application to Mobile Radio Channel". *IEEE Trans. VT-34, No 2*, pp. 86-96, May 1985

[48] **Y. Okumura, E. Ohmori, T. Kawano, K. Fukuda.** "Field Strength and its Variability in VHF and UHF Land Mobile Service". *Review of the Electrical Communication Laboratory, Vol 16*, pp. 825-873, Sept.-Oct. 1968

[49] **M. Hata.** "Empirical Formula for Propagation Loss in Land Mobile Radio". *IEEE Trans. VT-29*, pp. 317-325, August 1980

[50] **A.G. Longley, P.L. Rice.** "Predictions of Tropospheric Radio Transmission Loss over Irregular Terrain - A Computer Method". *ESSA Tech. Rep. ERL79-ITS67*, 1968

[51] **R. Edwards, J. Durkin.** "Computer Prediction of Service Areas for VHF Mobile Radio Networks". *Proc. IEE 116, 9*, pp. 1493-1500, 1969

[52] **J.H. Causebrook.** "Computer Prediction of UHF Broadcast Service Areas". *BBC Research Report RD 1974-4*

[53] **F.H. Palmer.** "VHF/UHF Path-Loss Calculations Using Terrain Profiles Deduced from a Digital Topographic Data Base". *AGARD Conf. Proc., No 269*, pp. 26-1-26-11, 1979

[54] **J.J. Egli.** "Radio Propagation above 40 Mc over Irregular Terrain". *Proc. IRE*, pp. 1383-1391, 1957

[55] **J.D. Murphy.** "Statistical Propagation Model for Irregular Terrain Paths between Transportable and Mobile Antennas". *AGARD Conf. Proc., No 70*, pp. 49-1-49-20, 1970

[56] **K. Allesbrook, J.D. Parsons.** "Mobile Radio Propagation in British Cities at Frequencies in the VHF and UHF Bands". *IEEE Trans VT-26, No 4*, pp. 313-323, 1977

[57] **M.F. Ibrahim, J.D. Parsons.** "Signal Strength Prediction in Built Up Area". *Part I, Proc IEE, Vol 130, pt F*, pp. 377-384, August 1983

[58] **W.J. Kessler, M.J. Wiggins.** "A Simplified Method for Calculating UHF Base-to-Mobile Statistical Coverage Contours over Irregular Terrain". *27th IEEE Veh. Techn. Conf.*, pp. 227-236, 1977

[59] **M.N. Lustgarten, J.A. Madison.** "An Empirical Propagation Model". *IEEE Trans. EMC-19, No 3*, pp. 301-309, 1977

[60] **K. Bullington.** "Radio Propagation at Frequencies above 30 Mc". *Proc. IRE, 35*, pp. 1122-1136, 1947

[61] **J. Epstein, D.W. Peterson.** "An Experimental Study of Wave Propagation at 850 Mc". *Proc IRE, 41, 5*, pp. 595-611, 1953

[62] **J. Deygout.** "Multiple Kinfe-Edge Diffraction of Microwaves". *IEEE Trans. AP-14*, pp. 480-489, 1966

[63] **A. Blomquist, L. Ladell.** "Prediction and Calculation of Transmission Loss in Different Types of Terrain". *NATO-AGARD Conf. Publ. CP-144*, 1974

[64] "Atlas of radio Waves Propagation Curves for Frequencies between 30 and 10000 Mc/s". *Radio Research Laboratory, Ministry of Postal Services, Tokyo, Japan*, pp. 172-179, 1957

[65] **R.J. Leubbers.** "Finite Conductivity Uniform GTD versus Knife Edge Diffraction in Prediction of Propagation Path Loss". *IEEE Trans. AP-32, No 1*, pp. 7-76, January 1984

[66] **J.G. Proakis.** "Digital Communications". *McGraw-Hill*, 1983

[67] **E.Green**, "Radio link design for microcellular systems", *British Telecom Technology J*, Vol 8, No.1, pp.85-96 January 1990.

[68] **R.B. D'Agostino** and **M.A. Stephens (Ed.)** *Goodness-of-fit Techniques* Marcel Dekker Inc., 1986

[69] **A.M. Mood, F.A. Garybill, D.C. Boes** *Introduction to the Theory of Statistics*, McGraw-Hill, 1985

[70] **F.J. Massye Jr.** *The Kolmogorov-Smirnov Test for Goodness-of-fit*, J. Amer. Statist. Ass., pp 46-70, 1951

Chapter 3

Speech Coding

R.A. Salami[1], L. Hanzo[2], F.C.A Brooks[3], and R. Steele[4]

3.1 Introduction

A recurrent theme in digital cellular mobile radio is spectral efficiency, which is generally taken to mean the user density for the allotted spectrum. We have argued in Chapter 1 that the most influential factor in determining the spectral efficiency is the cell size. Microcells reuse the allotted spectrum over a smaller geographical area and thereby produce a more spectrally efficient system. For a given cell size and bandwidth allocation there are a set of sub-systems, such as the speech encoders, channel coders, interleavers, modulators, and so forth, that are influential in determining the number of mobile users that can be accommodated. By reducing the bit rate of the speech encoders, or the amount of channel coding required, the number of users and therefore the system's spectral efficiency can be increased.

A classic overview of various early speech coding techniques can be found for example in [1]. The family of so-called **waveform coders** aims for the best possible waveform representation of the speech signal. Well-known representatives of these codecs are the 64 kb/s (kbps) ITU Pulse Code Modulated (PCM) scheme, sampling the speech signal at a sampling rate of 8 kbps and transmitting 8-bit logarithmically companded

[1]University of Southampton
[2]University of Southampton and Multiple Access Communications Ltd
[3]University of Southampton
[4]University of Southampton and Multiple Access Communications Ltd

samples [2, 3]. The coding rate can be reduced to 32 kbps, when using the 4 bit/sample ITU Adaptive Differential Pulse Code Modulated (AD-PCM) scheme [3]. Due to its low complexity, delay and good speech quality, this scheme was adopted by a number of cordless telephone systems, such as the Pan-European DECT system, the British CT2 scheme, the Japanese PHS system and the Pan-American PACS, which were breifly summarized in Chapter 1. A bit-rate between 16 and 32 kbps can be maintained also by Delta Modulation (DM) schemes [4].   Although these waveform codecs typically exhibit a low complexity, their bit-rate is excessive for cellular applications. In contrast, the so-called source coders, such as *Linear Predictive Coding (LPC) vocoders* [5, 6], operate at bit rates as low as 2 kb/s, however, the synthetic speech quality of vocoded speech is not broadly appropriate for commercial telephone applications. Linear predictive coding [9] in its basic form has been mainly used in secure military communications, where speech must be carried at very low bit rates. At the time of writing some of the coding principles borrowed from vocoders have however reached a maturity, where their application in commercial systems is becoming a reality in conjunction with a range of so-called hybrid techniques.

The need to produce toll-quality speech at bit rates below 10 kb/s for mobile radio applications over band-limited channels has drawn the interest of researchers to look at more efficient algorithms for LPC speech coding. The main limitation of LPC vocoding is the assumption that speech signals are either voiced or unvoiced [12, 13]; hence, the source of excitation of the so-called all-pole synthesis filter [3] is either a train of pulses (for voiced speech) or random noise (for unvoiced speech). In fact, there are more than two modes, in which the vocal tract is excited and often these modes can occur near-simultaneously. Even when the speech waveform is voiced, it is a gross simplification to assume that there is only one point of excitation in the entire pitch period. In 1982, Atal and Remde [7] proposed a new model for the excitation which is known as *multi-pulse excitation*. In this model, no prior knowledge of a voiced/unvoiced decision or pitch period is needed. The excitation is modelled by a number of pulses (usually 4 per 5 ms) whose amplitudes and positions are determined by minimizing the perceptually weighted error between the original and synthesized speech.

The introduction of this model has generated a great deal of interest, and it was the first of a new generation of analysis-by-synthesis speech coders capable of producing high quality speech at bit rates around 10 kb/s and down to 4.8 kb/s. This new generation of coders use the same all-pole synthesis filter (source model of speech production) as used by LPC vocoders. However, the excitation signal is carefully optimized and efficiently coded using waveform coding techniques. All analysis-by-synthesis coders share the same basic structure in which the excitation is determined by minimizing the perceptually weighted error between the original and synthesized speech. They differ in the way the excitation is modelled. The original multi-pulse approach assumes that both the pulse positions and amplitudes are initially unknown, then they are determined inside the minimization loop one pulse at a time. The *Regular Pulse Excitation* (RPE) approach [8] assumes that the pulses are regularly spaced and the amplitudes are then computed by solving a set of $M \times M$ equations where M is the number of pulses. In the *Code-Excited Linear Prediction* (CELP) [10, 11], the excitation signal is an entry of a very large stochastically populated codebook. The complexity of these coders increases as the bit rate is reduced. For example, CELP can produce good quality speech at bit rates as low as 4.8 kb/s at the expense of high computational demands due to the exhaustive search of the large excitation codebook (usually 1024 entries) for determining the optimum innovation sequence.

The state-of-art of speech compression was documented in a range of excellent monographs by O'Shaughnessy [14], Furui [15], Kondoz [16], Kleijn and Paliwal [17] and in a tutorial review by Gersho [19]. More recently the 5.6 kbits/s half-rate GSM quadruple-mode Vector Sum Excited Linear Predictive (VSELP) speech codec standard developed by Gerson *et al.* [20] was approved, which will be detailed in Chapter 8 portraying the Global System of Mobile Communications. Similarly, the 12.2 kbps so-called enhanced full-rate GSM speech codec will be the topic of Chapter 8. In Japan the 3.45 kbits/s half-rate JDC speech codec invented by Ohya, Suda and Miki [21] using the Pitch Synchronous Innovation (PSI) CELP principle was standardized.

This book is concerned with mobile radio communications and speech compression is treated here as one of the important components of these systems, mainly concentrating on the important subclass of the so-called analysis-by-synthesis (ABS) codecs, since they guarantee good speech quality at rates below 16 kbps. However, we refrain from a full discussion on all the existing standard and propriatary speech codecs. In the next section, a general model for high quality analysis-by-synthesis speech coding is described. The different parts of the model will be discussed in detail. These parts include: The LPC synthesis filter (for modelling the short-term spectral envelope of speech), the pitch predictor (for modelling the spectral fine structure), the error weighting filter and the error minimization procedure. The definition of the excitation sequence plays a dominant role in determining the coder performance and complexity. Different excitation models will be described in detail including the multi-pulse, regular-pulse, code-excited, self-excited and binary-pulse excited schemes. The associated speech quality, bit-rate, delay, error resilience and complexity of these excitation models will be addressed. Let us commence by considering the general schematic of ABS codecs in the next section.

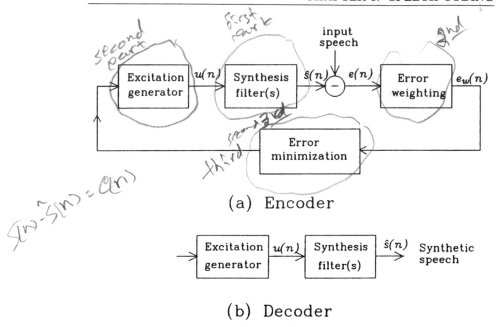

(a) Encoder

(b) Decoder

Figure 3.1: General model for analysis-by-synthesis LPC coding.

3.2   General Model for Analysis-by-Synthesis Speech Coding

The basic structure of the general model for analysis-by-synthesis predictive coding of speech is shown in Figure 3.1. The model consists of three main parts. The first part is the synthesis filter which is an all-pole time-varying filter for modelling the short-time spectral envelope of the speech waveform. It is often called *short-term correlation* filter because its coefficients are computed by predicting a speech sample from few previous samples (usually previous 8-16 samples, hence the name short term). The synthesis filter could also include a long-term correlation filter cascaded to the short-term correlation filter. The long-term predictor models the fine structure of the speech spectrum. The second part of the model is the excitation generator. This generator produces the excitation sequence which is to be fed to the synthesis filter to produce the reconstructed speech at the receiver. The excitation is optimized by minimizing the perceptually weighted error between the original and synthesized speech. As it is shown in Figure 3.1, a local decoder is present inside the encoder, and the analysis method for optimizing the excitation uses the difference between the original and synthesized speech as an error criterion, and it chooses the sequence of excitation which minimizes the weighted error.

The efficiency of this analysis-by-synthesis method comes from the closed loop optimization procedure, which allows the representation of the prediction residual using a very low bit rate, while maintaining high speech quality. This explains the superiority of analysis-by-synthesis predictive coding over other predictive coders which have open-loop structures such as *Residual Excited Linear Prediction* (RELP) coders [23]. The key point in the closed-loop structure is that the prediction residual is quantized by minimizing the perceptually weighted error between the original and re-constructed speech rather than minimizing the error between the residual and its quantized version as in open-loop structures. The third part of this model is the criterion used in the error minimization. The most common error minimization criterion is the mean squared error (mse). In this model, a subjectively meaningful error minimization criterion is used, where the error $e(n)$ is passed through a perceptual weighting filter which shapes the noise spectrum in a way to make the power concentrated at the formant frequencies of the speech spectrum so that the noise is masked by the speech signal.

The encoding procedure includes two steps: firstly, the synthesis filter parameters are determined from the speech samples (10–30 ms of speech) outside the optimization loop. Secondly, the optimum excitation sequence for this filter is determined by minimizing the weighted error criterion. The excitation optimization interval is usually in the range of 4–7.5 ms which is less than the LPC parameter update frame. The speech frame is therefore divided into sub-blocks, or subframes, where the excitation is determined individually for each subframe. The quantized filter parameters and the quantized excitation are sent to the receiver. The decoding procedure is performed by passing the decoded excitation signal through the synthesis filters to produce the reconstructed speech.

In the following subsections, we will discuss the LPC synthesis and pitch synthesis filters and the computation of their parameters, as well as the error weighting filter and the selection of the error criterion. The definition of every excitation method will be discussed in separate sections.

3.2.1   The Short-Term Predictor

The short-term predictor models the short-time spectral envelope of the speech. The spectral envelope of a speech segment of length L samples can be approximated by the transmission function of an all-pole digital filter of the form

$$H(z) = \frac{1}{1 - P_s(z)} = \frac{1}{1 - \sum_{k=1}^{p} a_k z^{-k}}, \tag{3.1}$$

where

$$P_s(z) = \sum_{k=1}^{p} a_k z^{-k} \tag{3.2}$$

is the short-term predictor. The coefficients $\{a_k\}$ are computed using the method of Linear Prediction (LP). The set of coefficients $\{a_k\}$ is called the *LPC parameters* or the *predictor coefficients*. The number of coefficients p is called the *predictor order*. The basic idea behind linear predictive analysis is that a speech sample can be approximated as a linear combination of past speech samples (8-16 samples), i.e.

$$\tilde{s}(n) = \sum_{k=1}^{p} a_k s(n - k), \tag{3.3}$$

where $s(n)$ is the speech sample and $\tilde{s}(n)$ is the predicted speech sample at sampling instant n. The prediction error, $e(n)$, is defined as

$$e(n) \;=\; s(n) - \tilde{s}(n)$$

$$\;=\; s(n) - \sum_{k=1}^{p} a_k s(n - k). \tag{3.4}$$

Taking the z-transform of Equation (3.4), we get

$$E(z) = S(z)A(z), \tag{3.5}$$

where

$$A(z) = 1 - \sum_{k=1}^{p} a_k z^{-k}. \tag{3.6}$$

$A(z)$ is the inverse of $H(z)$ in Equation (3.1), hence, $A(z)$ is called the *inverse filter*.

Because of the time-varying nature of the speech, the prediction coefficients should be estimated from short segments of speech signal (10-20 ms). The basic approach is to find a set of predictor coefficients that will minimize the *mean-squared prediction error* over a short segment of speech waveform. The resulting parameters are then assumed to be the parameters of the system function $H(z)$ in the model of speech production in Equation (3.1). The short-time average prediction error is defined as

$$E \;=\; \sum_{n} e^2(n)$$

$$\;=\; \sum_{n} \left[s(n) - \sum_{k=1}^{p} a_k s(n - k) \right]^2 . \tag{3.7}$$

To find the values of $\{a_k\}$ that minimize E, we set $\partial E / \partial a_i = 0$ for $i =$

$1, \ldots, p$. Then

$$\frac{\partial E}{\partial a_i} = -2 \sum_n \left\{ \left[s(n) - \sum_{k=1}^{p} a_k s(n-k) \right] s(n-i) \right\} = 0, \qquad (3.8)$$

which gives

$$\sum_n s(n)s(n-i) = \sum_n \sum_{k=1}^{p} a_k s(n-k)s(n-i). \qquad (3.9)$$

Changing the order of the summation in the right-hand side of Equation (3.9)

$$\sum_n s(n)s(n-i) = \sum_{k=1}^{p} a_k \sum_n s(n-k)s(n-i), \qquad i = 1, \ldots, p. \qquad (3.10)$$

If we define

$$\phi(i,k) = \sum_n s(n-i)s(n-k), \qquad (3.11)$$

then Equation (3.10) can be written as

$$\sum_{k=1}^{p} a_k \phi(i,k) = \phi(i,0), \qquad i = 1, \ldots, p. \qquad (3.12)$$

This set of p equations in p unknowns can be solved efficiently for the unknown predictor coefficients $\{a_k\}$. First, we must compute $\phi(i,k)$ for $i = 1, \ldots, p$ and $k = 0, \ldots, p$. To compute $\phi(i,k)$ from Equation (3.11), the limits of the summation must be specified. For short-time analysis, the limits must be over a finite interval. Two known methods for linear prediction analysis emerge out of a consideration of the limits of the summation, the *autocorrelation method* and the *covariance method* [24,25]. There is a third method which computes the reflection coefficients, that are an equivalent representation of the filter parameters, directly from the speech samples, by-passing an estimate of the autocorrelation coefficients. This approach is termed the *covariance lattice method* [26] or the *Burg algorithm* [27]. Although the Burg algorithm has different applications, commonly, the detection of sinusoidal signals in additive noise, Gray and Wong [27] showed that in speech analysis applications, Burg's method does not appear any more useful than the other techniques. In the following two subsections, we will discuss the more commonly used autocorrelation and covariance methods.

3.2.1.1 The Autocorrelation Method

In this approach, we assume that the error in Equation (3.7) is computed over the infinite duration $-\infty < n < \infty$. Since this cannot be done in practice, it is assumed that the waveform segment is identically zero outside the interval $0 \leq n \leq L_a - 1$, where L_a is the LPC analysis frame length. This is equivalent to multiplying the input speech by a finite length window $w(n)$ that is identically zero outside the interval $0 \leq n \leq L_a - 1$. Considering Equation (3.7), $e(n)$ is nonzero only in the interval $0 \leq n \leq L_a + p - 1$. Thus

$$\phi(i,k) = \sum_{n=0}^{L_a+p-1} s(n-i)s(n-k), \qquad \begin{array}{l} i = 1, \ldots, p, \\ k = 0, \ldots, p. \end{array} \qquad (3.13)$$

Setting $m = n - i$, Equation (3.13) can be expressed as

$$\phi(i,k) = \sum_{m=0}^{L_a-1-(i-k)} s(m)s(m+i-k). \qquad (3.14)$$

Therefore, $\phi(i,k)$ is the short-time autocorrelation of $s(m)$ evaluated for $(i-k)$. That is

$$\phi(i,k) = R(i-k), \qquad (3.15)$$

where

$$R(j) = \sum_{n=0}^{L_a-1-j} s(n)s(n+j) = \sum_{n=j}^{L_a-1} s(n)s(n-j). \qquad (3.16)$$

Therefore, the set of p equations in (3.12) can be expressed as

$$\sum_{k=1}^{p} a_k R(|i-k|) = R(i), \qquad i = 1, \ldots, p. \qquad (3.17)$$

Equation (3.17) is expressed in matrix form as

$$\begin{pmatrix} R(0) & R(1) & R(2) & \cdots & R(p-1) \\ R(1) & R(0) & R(1) & \cdots & R(p-2) \\ R(2) & R(1) & R(0) & \cdots & R(p-3) \\ \vdots & \vdots & \vdots & \ddots & \vdots \\ R(p-1) & R(p-2) & R(p-3) & \cdots & R(0) \end{pmatrix} \begin{pmatrix} a_1 \\ a_2 \\ a_3 \\ \vdots \\ a_p \end{pmatrix} = \begin{pmatrix} R(1) \\ R(2) \\ R(3) \\ \vdots \\ R(p) \end{pmatrix}$$
$$(3.18)$$

The $p \times p$ matrix of autocorrelation values is a symmetric Toeplitz matrix, i.e., all the elements along a given diagonal are equal. This special property can be exploited to obtain an efficient algorithm for the solution of Equation (3.18). The most efficient solution is a recursive procedure known as

Durbin's algorithm, which can be stated as follows [24]:

$$E(0) = R(0)$$

For $i = 1$ to p do

$$k_i = \left[R(i) - \sum_{j=1}^{i-1} a_j^{(i-1)} R(i-j) \right] / E(i-1) \qquad (3.19)$$

$$a_i^{(i)} = k_i$$

For $j = 1$ to $i - 1$ do

$$a_j^{(i)} = a_j^{(i-1)} - k_i a_{i-j}^{(i-1)} \qquad (3.20)$$

$$E(i) = (1 - k_i^2) E(i-1). \qquad (3.21)$$

The final solution is given as

$$a_j = a_j^{(p)} \qquad j = 1, \ldots, p. \qquad (3.22)$$

The quantity $E(i)$ in Equation (3.21) is the prediction error of a predictor of order i. The intermediate quantities k_i are known as the *reflection coefficients*. They are the same coefficients which appear in the lossless tube model of the vocal tract [24]. The value of k_i is in the range

$$-1 \leq k_i \leq 1. \qquad (3.23)$$

This condition imposed on the parameter k_i is necessary and sufficient for all the roots of the polynomial $A(z)$ to be inside the unit circle, thereby guaranteeing the stability of the system $H(z)$. It is found that the auto-correlation method always leads to a stable filter $H(z)$.

As mentioned earlier in this section, the speech samples, $s(n)$, are identically set to zero outside the interval $0 \leq n \leq L_a - 1$. However, a sharp truncation of the speech segment is likely to create a large increase in the prediction error at the beginning and the end of the segment being analysed. This problem is avoided by using tapered windows such as a Hamming window whose amplitude falls to zero in a gradual manner. The Hamming window is given by

$$w(n) = 0.54 - 0.46 \cos(2\pi n/(L_a - 1)), \qquad 0 \leq n \leq L_a - 1, \quad (3.24)$$

where L_a is the LPC analysis frame length. The length L_a of the Hamming window (i.e. the length of the LPC analysis frame) is usually chosen longer than the length of the speech update frame L. The overlapped window-ing gives a smoothing effect in LPC analysis, that is, it alleviates abrupt changes in LPC coefficients between analysis blocks.

3.2.1.2   The Covariance Method

In contrast to the autocorrelation method, here we assume that the error E in Equation (3.7) is minimized over a finite interval, $0 \le n \le L - 1$. Therefore $\phi(i, k)$ of Equation (3.11) is given by

$$\phi(i, k) = \sum_{n=0}^{L-1} s(n - i)s(n - k), \qquad \begin{matrix} i = 1, \dots, p \\ k = 0, \dots, p. \end{matrix} \qquad (3.25)$$

The set of equations in (3.12) can be written in matrix form as

$$\begin{pmatrix} \phi(1,1) & \phi(1,2) & \phi(1,3) & \dots & \phi(1,p) \\ \phi(2,1) & \phi(2,2) & \phi(2,3) & \dots & \phi(2,p) \\ \phi(3,1) & \phi(3,2) & \phi(3,3) & \dots & \phi(3,p) \\ \vdots & \vdots & \vdots & \ddots & \vdots \\ \phi(p,1) & \phi(p,2) & \phi(p,3) & \dots & \phi(p,p) \end{pmatrix} \begin{pmatrix} a_1 \\ a_2 \\ a_3 \\ \vdots \\ a_p \end{pmatrix} = \begin{pmatrix} \phi(1,0) \\ \phi(2,0) \\ \phi(3,0) \\ \vdots \\ \phi(p,0) \end{pmatrix}.$$

$$(3.26)$$

Since $\phi(i, k) = \phi(k, i)$, the $p \times p$ matrix is symmetric but it is not Toeplitz. Cholesky decomposition [28] can be used to solve Equation (3.26) where the $p \times p$ matrix $\mathbf{\Phi}$ is decomposed into

$$\mathbf{\Phi} = \mathbf{V}\mathbf{D}\mathbf{V}^T, \qquad (3.27)$$

where $\mathbf{V}$ is a lower triangular matrix with diagonal elements equal to 1, $\mathbf{D}$ is a diagonal matrix, and T denotes transpose.

Another form of Cholesky decomposition is

$$\mathbf{\Phi} = \mathbf{U}\mathbf{U}^T, \qquad (3.28)$$

where $\mathbf{U}$ has a lower triangular structure. $\mathbf{U}$ in Equation (3.28) and $\mathbf{V}$ in Equation (3.27) are related by

$$\mathbf{U} = \mathbf{V}\mathbf{D}^{1/2}. \qquad (3.29)$$

The square root in Equation (3.29) requires the matrix $\mathbf{\Phi}$ to be positive definite if the decomposition form in Equation (3.28) is to be used.

Unlike the autocorrelation method, the covariance method does not always guarantee the stability of the all-pole filter $H(z)$. To guarantee the stability of $H(z)$, the stabilized covariance method can be used [29]. In the stabilized covariance method, the matrix of covariances $\mathbf{\Phi}$ is decomposed according to Equation (3.28). Therefore Equation (3.26) becomes

$$\mathbf{U}\mathbf{U}^T\mathbf{a} = \mathbf{c} \qquad (3.30)$$

where $\mathbf{a} = [a_1 \dots a_p]^T$ and $\mathbf{c} = [\phi(1,0) \dots \phi(p,0)]^T$. The elements of $\mathbf{U}$ are computed from the elements of $\boldsymbol{\Phi}$ by the following recursion:

$$\text{For } j = 1 \text{ to } p \text{ do}$$

$$u_{jj} = \sqrt{\phi_{jj} - \sum_{k=1}^{j-1} u_{jk}}$$

$$\text{For } i = j+1 \text{ to } p \text{ do}$$

$$u_{ij} = \frac{1}{u_{jj}} \left(\phi_{ij} - \sum_{k=1}^{j-1} u_{ik} u_{jk} \right). \tag{3.31}$$

Let

$$\mathbf{g} = \mathbf{U}^T \mathbf{a}, \tag{3.32}$$

then Equation (3.30) is reduced to

$$\mathbf{U}\mathbf{g} = \mathbf{c}. \tag{3.33}$$

Since the matrix $\mathbf{U}$ has a lower triangular structure, Equation (3.33) is easily solved for $\mathbf{g}$ by

$$g_i = \frac{1}{u_{ii}} \left(c_i - \sum_{k=1}^{i-1} u_{ik} g_k \right), \qquad i = 1, \ldots, p. \tag{3.34}$$

The reflection coefficients are found from the elements of the vector $\mathbf{g}$ by

$$k_i = \frac{g_i}{\sqrt{\phi_{00} - \sum_{k=1}^{i-1} g_k^2}} \qquad i = 1, \ldots, p. \tag{3.35}$$

The predictor coefficients a_i can be now computed from the reflection coefficients k_i using the same recursive relation as in Durbin's algorithm, namely Equation (3.20), i.e.

$$\text{For } i = 1 \text{ to } p \text{ do}$$

$$a_i^{(i)} = k_i$$

$$\text{For } j = 1 \text{ to } i-1 \text{ do}$$

$$a_j^{(i)} = a_j^{(i-1)} - k_i a_{i-j}^{(i-1)}. \tag{3.36}$$

Atal has further improved his originally proposed stabilized covariance method by introducing frequency compensation [22] and error weighting [73].

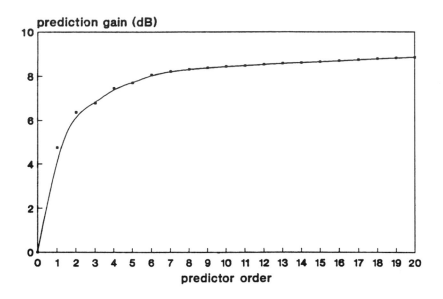

Figure 3.2: Prediction gain versus predictor order in LPC analysis.

3.2.1.3   Considerations in the Choice of LPC Analysis Conditions

The variables in the LPC analysis are: the analysis method, the predictor order and the update frame length. We used the multi-pulse LPC approach (to be discussed in a later section) in order to gain some insight into the effect of the above mentioned analysis conditions on the speech quality. Regarding the analysis method, we found that both the autocorrelation method and the stabilized covariance method discussed earlier lead to very similar results. It is difficult to distinguish between both methods under the same analysis conditions. The autocorrelation method is used throughout our investigations. The second variable in the LPC analysis is the number of prediction coefficients p. To reduce the number of bits needed to encode the LPC parameters, it is desirable to use the minimum number of parameters necessary to accurately model the short-term spectral envelope of the speech. It was shown in [25], Chapter 4, that to adequately represent the vocal tract under ideal circumstances, the memory of the model $A(z)$ must be equal to twice the time required for sound waves to travel from the glottis to the lips, that is, $2\ell/c$, where ℓ is the length of the vocal tract and c is the speed of sound. For example, the representative values $c = 34\,\text{cm/ms}$ and $\ell = 17\,\text{cm}$ result in a necessary memory of $1\,\text{ms}$. When the sampling frequency is f_s samples/s, the period of $1\,\text{ms}$ corresponds to $f_s/1000$ samples. At $8\,\text{kHz}$ sampling rate, the predictor order p must be at least 8 for this ideal model. It is generally necessary to add several more

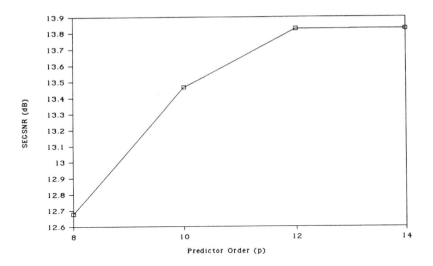

Figure 3.3: SEGSNR versus predictor order for MPE-LPC.

coefficients (4 or 5) to accommodate other factors excluded from the ideal acoustic tube model (the glottal and lip radiation and the fact that the digitized speech waveforms are not exactly all-pole waveforms). Figure 3.2 shows the average prediction gain against the predictor order p for 20 s of speech uttered by two males and two females. The prediction gain for a speech frame is given by

$$G = \frac{\sum_n s^2(n)}{\sum_n r^2(n)}, \tag{3.37}$$

where $r(n)$ is the prediction residual. In the case of the autocorrelation method the prediction gain can be written as

$$G = \frac{R(0)}{R(0) - \sum_{i=1}^{p} a_i R(i)} = \frac{1}{\prod_{i=1}^{p}(1 - k_i^2)}. \tag{3.38}$$

The average prediction gain in dB is given by

$$G_{av} = \frac{1}{N_f} \sum_{i=0}^{N_f-1} 10 \log G_i \qquad \text{dB} \tag{3.39}$$

where N_f is the number of frames. It is clear from Figure 3.2 that the prediction gain starts to saturate for predictor orders larger than 8. Figure 3.3 shows the SEGSNR of the MPE-LPC by varying the predictor order from 8 to 14. The degradation in speech quality becomes noticeable when p is reduced to 8 and the quality saturates when p is above 12. We found that the choice of $p = 10$ is very reasonable to adequately represent the vocal tract.

The third variable in LPC analysis is the updating interval L. Like the predictor order p, the choice of the updating interval is a trade-off between quality and bit rate. It is usually desirable to perform spectral analysis within an interval where the vocal tract movement is negligible. For most vowels, this interval is on the order of 15-20 ms, and it is usually shorter for unvoiced sounds. In fact, the burst associated with the release of an unvoiced stop consonant in the initial position such as $/t/$ may exist for only a few ms. Asynchronous analysis (arbitrary placement of the time interval regardless of the pitch period) will often extend the averaging into the voiced portion following the $/t/$ or the silence preceding the $/t/$ release. Therefore, for accurate analysis of transient sounds, an interval on the order of 10 ms is desirable, while for quasi-periodic sounds like most of the vowels, a 15-20 ms interval is adequate. When 10 predictor coefficients are used, they are usually quantized with 40 bits using scalar quantization of the so-called Log Area Ratios. If the updating interval is 20 ms, 2 kb/s are needed to quantise the LPC parameters. On the other hand, if the coefficients are updated every 10 ms, their bit rate rises to 4 kb/s. This means an increase of 2 kb/s, which is very significant in low bit rate coding of speech. Figure 3.4 shows the decrease in the segmental SNR of MPE with increasing the updating interval. We can conclude that for low bit rate speech coding, a 20 ms interval is sufficient to maintain good speech quality, although this would introduce a little degradation in some sounds which have fast changing spectral characteristics, like transient sounds.

3.2.1.4   Quantization of the LPC parameters

The spectral envelope represented by the set of LPC parameters a_i can be quantized using either scalar or vector quantization methods. In scalar quantization, the LPC parameters are quantized individually using either uniform or nonuniform quantization. In vector quantization (VQ) [32,33] the set of LPC parameters is considered as one entity, and they are quantized using a large trained codebook by minimizing a specified spectral distortion measure [34]. Generally speaking, vector quantisers yield smaller quantization error than scalar quantisers at the same bit rate, however, the high complexity associated with VQ algorithms has hindered their use in real time implementations. Conventional vector quantisers use algorithms to design a codebook of LPC parameters based on a long training sequence of LPC vectors [35,36]. These VQs usually lack robustness when speakers outside the training sequence are tested. Using 10 bit codebooks, conventional vector quantization results into consistently noticeable spectral distortions. Increasing the number of bits used to encode the LPC parameters causes the codebook size to grow exponentially. Accordingly, the storage needs and processing time make such large codebooks impractical in real time applications. Therefore vector quantization of LPC parameters

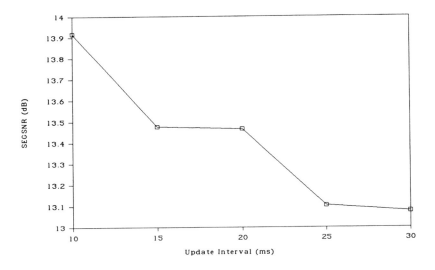

Figure 3.4: SEGSNR versus updating interval of LPC parameters for MPE-LPC.

has been limited to applications where coarse quantization of the spectral envelope is sufficient [37]. VQ methods have not proved to be useful in high quality speech coding due to their excessive complexity and poor performance with practically small book sizes. VQ of LPC parameters was mostly used with very low bit rate vocoders which inherently limit the achievable speech quality. Another disadvantage of VQs is their vulnerability to transmission errors, especially when vector prediction is used. Therefore the attention in the rest of this section will be focused on scalar quantization of LPC parameters since it is the most useful method in high quality speech coding.   The set of LPC parameters $\{a_k\}$ represents the coefficients of the LPC synthesis filter $H(z) = 1/A(z)$. When quantizing the set of prediction coefficients, one has to insure the stability of the synthesis filter. In other words, the poles of the quantized synthesis filter should lie inside the unit circle, a task which is hard to achieve if the set of parameters $\{a_k\}$ is to be quantized directly. $\{a_k\}$ has to be quantized with 10 bits per parameter to insure the synthesis filter stability. Therefore, it is necessary to transform the LPC coefficients into another set of parameters which are related in a one-to-one manner to the coefficients of a stable synthesis filter. The new set of parameters should possess well behaved statistical properties, and there should be a criterion for guaranteeing the stability of the quantized synthesis filter.

3.2.1.4.1 Reflection Coefficients Such a transformation is the set of reflection coefficients or partial correlation (PARCOR) coefficients. For a stable LPC synthesis filter, the reflection coefficients have the property

$$|k_i| < 1. \tag{3.40}$$

The reflection coefficients are computed as a byproduct of solving the set of $p \times p$ equations in (3.12) using either Durbin's algorithm or the stabilized covariance method. Because for values $|k_i|$ approaching 1, the poles of $H(z)$ approach the unit circle, small changes in k_i can result in large changes in the spectrum. Previous studies have shown that the spectral sensitivity function of the reflection coefficients is approximately U-shaped, with increasing sensitivity as the magnitude of the reflection coefficient approaches unity [38]. Therefore uniform quantization of the reflection coefficients is not efficient. For quantization purposes, the reflection coefficients are transformed to another set of coefficients that exhibit lower spectral sensitivity as k_i approaches 1. Two popular transformations of the reflection coefficients are the inverse sine transformation

$$S_i = \sin^{-1}(k_i) \tag{3.41}$$

and the log-area ratios

$$LAR_i = \log \frac{1 - k_i}{1 + k_i}. \tag{3.42}$$

The terminology *log-area ratio* as the term *reflection coefficient* are obtained from the acoustic tube analogy of the vocal tract [24].

The probability density functions (PDF) of the LARs of an eighth-order filter are shown in Figure 3.5. It is clear that the dynamic range of the parameters LAR_i decreases as the index i increases. Therefore more bits are allocated for quantizing the first LARs. The 8 LARs are usually quantized with 6, 6, 5, 5, 4, 4, 3 and 3, respectively. A total of 36 bits are used in this case. For 20 ms LPC parameter updating frames, 1.8 kb/s are needed to quantise the filter coefficients, and the bit rate is reduced to 1.2 kb/s in the case of 30 ms updating frames.

To measure the efficiency of a certain LPC parameter quantiser, the log spectral distortion measure is usually used. The log spectral distortion of a speech frame is given by [31]

$$SD = \frac{1}{2\pi} \int_{-\pi}^{\pi} \left(10 \log |H(\omega)|^2 - 10 \log \left| \hat{H}(\omega) \right|^2 \right)^2 d\omega \qquad (\text{dB})^2$$

$$= \frac{1}{2\pi} \int_{-\pi}^{\pi} \left(10 \log \frac{\left| \hat{A}(\omega) \right|^2}{|A(\omega)|^2} \right)^2 d\omega \qquad (\text{dB})^2, \tag{3.43}$$

where $\hat{H}(z)$ and $\hat{A}(z)$ are the quantized synthesis filter and inverse filter, respectively. The log spectral distortion is then averaged over a large num-

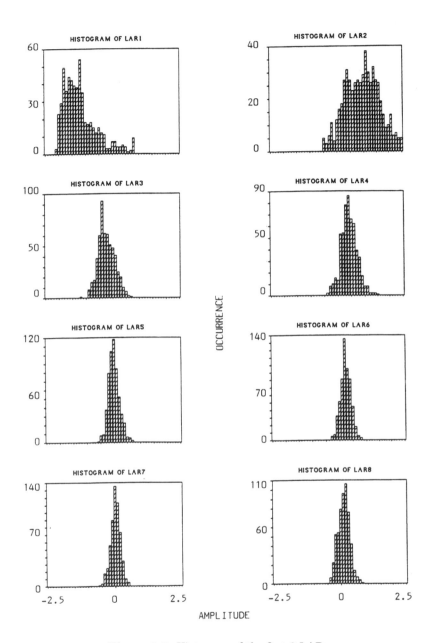

Figure 3.5: Histogram of the first 8 LARs.

ber of speech frames. A spectral deviation of $1\,dB$ is considered as the perceptual difference limen for coding the LPC parameters [39]. Utilizing nonuniform quantization reduces the number of bits needed to quantise the LARs to about 36 bits per frame with maintaining a spectral distortion less than 1 dB. Decreasing the number of bits below 30 bits per LPC frame results in noticeable spectral distortion (larger than 1 dB).

3.2.1.4.2   Line Spectrum Pairs   Besides the LARs, another important transformation of LPC parameters is the set of *line spectrum pairs* (LSP) [40] or *line spectrum frequencies* [41]. The inverse filter $A(z)$ associated with nth order LPC analysis satisfies the following recursive relation [42]

$$A_n(z) = A_{n-1}(z) - k_n z^{-n} A_{n-1}(z^{-1}), \quad n = 1, \dots, p, \tag{3.44}$$

with $A_0(z) = 1$ and k_n being the n th reflection coefficient. Extending the filter order to $n = p + 1$ Equation (3.44) becomes

$$A_{p+1}(z) = A_p(z) - k_{p+1} z^{-(p+1)} A_p(z^{-1}). \tag{3.45}$$

Consider the two extreme artificial boundary conditions $k_{p+1} = 1$ and $k_{p+1} = -1$. These two conditions correspond, respectively, to a complete closure and complete opening at the glottis in the acoustic tube model. Under these conditions, we obtain the two following polynomials:

$$\begin{aligned} P(z) &= A(z) - z^{-(p+1)} A(z^{-1}) \\ &= 1 + p_1 z^{-1} + p_2 z^{-2} + \cdots - p_2 z^{-(p-1)} - p_1 z^{-p} - z^{-(p+1)} \end{aligned}$$

$$\tag{3.46}$$

for $k_{p+1} = 1$, and

$$\begin{aligned} Q(z) &= A(z) + z^{-(p+1)} A(z^{-1}) \\ &= 1 + q_1 z^{-1} + q_2 z^{-2} + \cdots + q_2 z^{-(p-1)} + q_1 z^{-p} + z^{-(p+1)} \end{aligned}$$

$$\tag{3.47}$$

for $k_{p+1} = -1$. Notice that the polynomials $P(z)$ and $Q(z)$ are, respectively, antisymmetric and symmetric. It can be shown that the polynomials $P(z)$ and $Q(z)$ possess the following important properties [43]

1. All roots of $P(z)$ and $Q(z)$ are on the unit circle.

2. The roots of $P(z)$ and $Q(z)$ alternate each other on the unit circle.

3. Minimum phase property of $A(z)$ (or the stability of $H(z)$) is easily preserved after quantizing the roots of $P(z)$ and $Q(z)$.

Since the roots of $P(z)$ and $Q(z)$ are on the unit circle, they are given by $e^{j\omega_i}$, and it is easily shown that for the case of even predictor order p, $P(z)$ and $Q(z)$ can be expressed as

$$P(z) = (1 - z^{-1}) \prod_{i=2,4,\ldots,p} (1 - 2\cos(2\pi f_i)z^{-1} + z^{-2}), \qquad (3.48)$$

and

$$Q(z) = (1 + z^{-1}) \prod_{i=1,3,\ldots,p-1} (1 - 2\cos(2\pi f_i)z^{-1} + z^{-2}). \qquad (3.49)$$

The frequencies f_i in Equations (3.48) and (3.49) are normalized by the sampling frequency f_s. The parameters f_i, $i = 1, \ldots, p$, are defined as the *line spectrum frequencies* or the *line spectrum pairs*. It is important to note that $f_0 = 0$ and $f_{p+1} = 0.5$ are always fixed corresponding to the fixed roots $z = 1$ and $z = -1$ of $P(z)$ and $Q(z)$ respectively. Therefore these two fixed roots are excluded from the LSP parameter set. The LSFs can be interpreted as the resonant frequencies of the vocal tract under the two extreme boundary conditions at the glottis (complete closure and complete opening). The second property of the LSFs can be stated as

$$f_0 < f_1 < f_2 < \cdots < f_{p-1} < f_p < f_{p+1}, \qquad (3.50)$$

where $f_0 = 0$ and $f_{p+1} = 0.5$. This relation is known as the *ordering property* of the LSFs. As long as the ordering property is preserved while quantizing the LSFs, the stability of the synthesis filter $H(z)$ is insured.

Several methods for efficient computation of the LSFs can be found in the literature [41,43–46]. It has been reported that quantizing the LSP parameter set gives 25% reduction in bit rate compared with the LARs when straight uniform quantization of the LSFs is used [47], and 30% reduction has been obtained when quantizing the LSF differences [43]. However, Atal *et al.* [48] concluded that quantizing the LSFs does not offer any advantage over quantizing the LARs or the inverse sines. In general, the LSP parameters possess the following properties which allow them to be quantized more efficiently than the LARs

1. The LSFs have well-behaved statistical properties, and the stability of the quantized synthesis filter is easily insured by preserving the ordering property of Equation (3.50). Further, the ordering property can be efficiently used to detect transmission errors in the LSP parameters, and accordingly, substitution algorithms can be utilized to overcome channel errors in LSFs with zero-redundancy.

2. There is evidence of the existence of a direct relation between the LPC spectrum and LSFs. Past studies have shown that a concentration of line spectrum frequencies in a certain frequency band approximately

corresponds to a resonance in that band [49]. Further investigation showed that it is not trivial to derive the resonance frequencies from the LSFs [50, 51].

3. The LSP parameters between adjacent frames are highly correlated. This can be exploited to reduce the bit rate by employing predictive quantization. It also leads to efficient interpolation procedures in which the LSFs are transmitted every odd frame, which dramatically reduces their bit rate.

Figure 3.6 shows the histograms of the LSFs of a 10 th order LPC filter.

Since their introduction by Itakura, the LSP parameters have been intensively studied, and various approaches have been proposed for efficient quantization of the LSFs. These approaches attempted to exploit the previously mentioned properties, namely, the ordering property, the relation between the LSFs and the LPC spectrum, and the correlation between adjacent LSP frames. Recently, Soong and Juang have described a dynamic programming procedure for globally optimizing the allocation of bits as well as the distribution of levels for nonuniform quantization of LSF differences [52]. Starting from zero bits per frame, they successively assigned increasing number of bits to parameters which provided the minimum marginal improvement in quantization distortion. LSP-based LPC vocoders have been formally tested by Kang and Fransen [53]. In 800 b/s vocoding, they used vector quantization of the LSFs, and in their 4800 b/s vocoder, they quantized each pair scalarly in terms of its centre frequency $(f_i + f_{i+1})/2$ and offset frequency $(f_{i+1} - f_i)/2$. Crosmer and Barnwell [54] tried to exploit the relation between the LSFs and the formant frequencies in quantizing the LSP parameters taking into account spectral features known to be important in perceiving speech signals. In their approach they assumed that the odd LSFs approximately correspond to the locations of formant centre frequencies, and the closer two LSFs are together, the narrower the bandwidth of the corresponding pole of the vocal tract. Based on this interpretation they described an approach for quantizing the LSFs. Sugamura and Farvardin [39] described and compared several schemes to quantise the LSFs utilizing the ordering property. They obtained the 1 dB difference limen of spectral distortion with 32 bits/frame. Wong and Boyd [55] obtained similar results (32 bits/frame with 1 dB spectral distortion) by quantizing the LSF differences between adjacent frames.

3.2.1.4.3 Interpolation of LPC parameters   As we mentioned earlier, the excitation frame length is usually smaller than the LPC frame. The LPC frame is, therefore, divided into several subframes, and the excitation parameters are updated every subframe. Figure 3.7 demonstrates the relationship between the frame, subframe, and Hamming window used to derive the LPC parameters. In Figure 3.7 the speech frame is of length

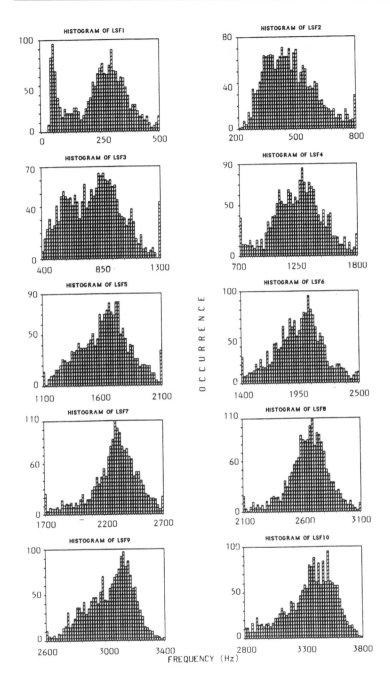

Figure 3.6: Histograms of the LSFs of a 10th order filter.

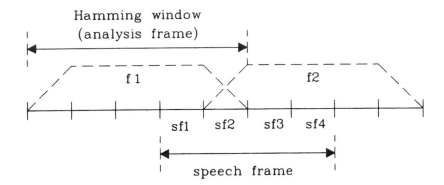

Figure 3.7: Relationship between the frame, subframe, and Hamming window.

160 samples (20 ms), the subframe 40 samples (5 ms), and the Hamming
window 200 samples (25 ms). In this example a new set of LPC parameters is transmitted every 20 ms. Interpolation of LPC parameters between
adjacent frames can be used to obtain a different set of parameters for every subframe. In our example, interpolation enables the updating of filter
parameters every 5 ms while transmitting them every 20 ms, i.e. without
needing the higher bit rate associated with shorter updating frames.

The set of predictor coefficients $\{a_i\}$ cannot be used for interpolation,
because the interpolated parameters in this case do not guarantee a stable
synthesis filter. The interpolation is, therefore, performed using a transformed set of parameters where the filter stability can be easily guaranteed,
e.g. using the LARs or the LSFs. If $\mathbf{f}_n$ is the quantized LPC vector in the
present frame and $\mathbf{f}_{n-1}$ is the quantized LPC vector from the past frame,
then the interpolated LPC vector $\mathbf{sf}_k$ in a subframe k is given by

$$\mathbf{sf}_k = \delta_k \mathbf{f}_{n-1} + (1 - \delta_k)\mathbf{f}_n, \tag{3.51}$$

where δ_k is a fraction between 0 and 1. δ_k is gradually decreased with the
subframe index. For our specific example, a good choice of the values of δ_k
is 0.75, 0.5, 0.25, and 0 for $k = 1, \ldots, 4$, respectively. Using these values,
the interpolated LPC vectors in the 4 subframes are given by

$$\begin{aligned}
\mathbf{sf}_1 &= 0.75\mathbf{f}_{n-1} + 0.25\mathbf{f}_n \\
\mathbf{sf}_2 &= 0.5\mathbf{f}_{n-1} + 0.5\mathbf{f}_n \\
\mathbf{sf}_3 &= 0.25\mathbf{f}_{n-1} + 0.75\mathbf{f}_n \\
\mathbf{sf}_4 &= \mathbf{f}_n.
\end{aligned}$$

Interpolation has been recently used to reduce the bit rate associated with
the spectral parameters by transmitting them every odd frame [56] or even
every third frame [57]. For the 20 ms speech frame in our example, if the

LSFs are quantized with 36 bits, their corresponding bit rate is 1.8 kb/s. If the LSFs are transmitted every odd frame their bit rate is reduced to 0.9 kb/s. Similarly, if the LSFs are transmitted every third frame their bit rate is further reduced to 0.6 kb/s. In the frames where the LSFs are not transmitted, they are interpolated using the available LSFs similar to Equation (3.51) with a proper weighting fraction δ. This reduction in bit rate offered by interpolation is crucial for encoding at bit rates below 4.8 kb/s. The disadvantage of interpolation is increasing the codec delay. By using interpolation in every odd frame the speech frame size is effectively doubled, and for our example, the end-to-end delay is increased from about 40 ms to 80 ms.

3.2.2   The Long-Term Predictor

While the short-term predictor models the spectral envelope of the speech segment being analysed, the long-term predictor (LTP), or the pitch predictor, is used to model the fine structure of that envelope. Inverse filtering of the speech input removes some of the redundancy in the speech by subtracting from the speech sample its predicted value using the past p samples. This is called short-term prediction since only the previous p samples (usually 10) are used to predict the present sample of speech. The short-term prediction residual, however, still exhibits, to a lesser extent, some periodicity (or redundancy) related to the pitch period of the original speech when it is voiced. This periodicity is on the order of 20–160 samples (50–400 Hz pitch frequencies). Adding the pitch predictor to the inverse filter further removes the redundancy in the residual signal and turns it into a noise-like process. It is called *pitch predictor* since it removes the pitch periodicity, or *long-term predictor* since the predictor delay is between 20 and 160 samples. The LTP is not an essential part in medium bit rate LPC coding (e.g. MPE-LPC and RPE-LPC), although including a pitch predictor in these coders improves their performance. However, the long-term predictor is very essential in low bit rate speech coders, as in the CELP, where the excitation signal is modelled by a Gaussian process, therefore long-term prediction is necessary to insure that the prediction residual is very close to random Gaussian noise process.

The general form of a long-term correlation filter is

$$\frac{1}{P(z)} = \frac{1}{1 - P_l(z)} = \frac{1}{1 - \sum_{k=-m_1}^{m_2} G_k z^{-(\alpha+k)}} \tag{3.52}$$

where

$$P_l(z) = \sum_{k=-m_1}^{m_2} G_k z^{-(\alpha+k)} \tag{3.53}$$

is the long-term predictor. For $m_1 = m_2 = 0$, we have a one-tap predictor, and for $m_1 = m_2 = 1$, we have a three-tap predictor. The delay α usually represents the pitch period (or a multiple of it).

The parameters α and $\{G_m\}$ are determined by minimizing the mean-squared residual error after short-term and long-term prediction over a period of N samples. For a one-tap predictor, the long-term prediction residual $e(n)$ is given by

$$e(n) = r(n) - Gr(n - \alpha) \qquad (3.54)$$

where $r(n)$ is the residual signal after short-term prediction. The mean-squared residual E is given by

$$E = \sum_{n=0}^{N-1} e^2(n) = \sum_{n=0}^{N-1} [r(n) - Gr(n - \alpha)]^2 . \qquad (3.55)$$

Setting $\partial E/\partial G = 0$ yields

$$G = \frac{\sum_{n=0}^{N-1} r(n)r(n - \alpha)}{\sum_{n=0}^{N-1} [r(n - \alpha)]^2} \qquad (3.56)$$

and substituting G into Equation (3.55) gives

$$E = \sum_{n=0}^{N-1} r^2(n) - \frac{\left[\sum_{n=0}^{N-1} r(n)r(n - \alpha)\right]^2}{\sum_{n=0}^{N-1} [r(n - \alpha)]^2} . \qquad (3.57)$$

Minimizing E means maximizing the second term in the right-hand side of Equation (3.57), which represents the normalized correlation between the residual $r(n)$ and its delayed version. This term is computed for all possible values of α over its specified range, and the value of α which maximizes this term is chosen. The energy $\mathcal{E}$ in the denominator can be easily updated from delay $(\alpha - 1)$ to α instead of recomputing it afresh by

$$\mathcal{E}_\alpha = \mathcal{E}_{\alpha-1} + r^2(-\alpha) - r^2(-\alpha + N) \qquad (3.58)$$

which requires 2 instructions (addition plus multiplication). The updating of the term to be maximized requires $N + 4$ instructions for each new delay.

If one-tap LTP is used, then Equation (3.56) is used to compute the gain G. For K-tap LTP, the LTP delay α is first determined by maximizing the second term of Equation (3.57) and then a set of $K \times K$ equations is solved to compute the K predictor gains. For example, if 3-tap LTP is used, the gains are computed by solving the matrix equation

$$\begin{pmatrix} \psi(\alpha - 1, \alpha - 1) & \psi(\alpha - 1, \alpha) & \psi(\alpha - 1, \alpha + 1) \\ \psi(\alpha, \alpha - 1) & \psi(\alpha, \alpha) & \psi(\alpha, \alpha + 1) \\ \psi(\alpha + 1, \alpha - 1) & \psi(\alpha + 1, \alpha) & \psi(\alpha + 1, \alpha + 1) \end{pmatrix}$$

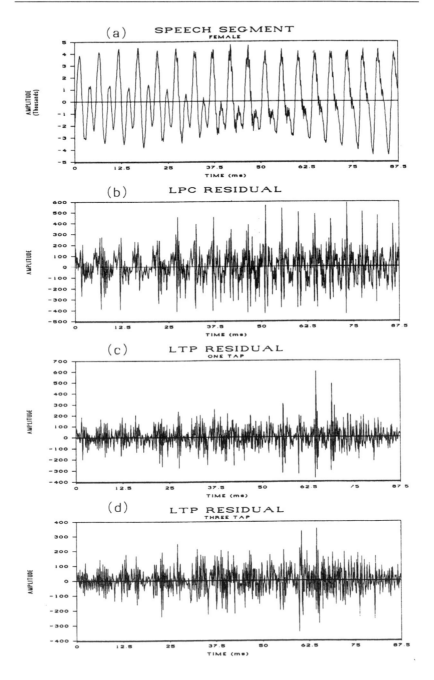

Figure 3.8: (a) 87.5 ms of speech; (b) LPC residual; (c) 1-tap LTP residual; (d) 3-tap LTP residual.

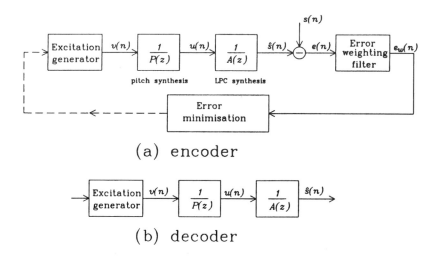

(a) encoder

(b) decoder

Figure 3.9: General analysis-by-synthesis LPC model with LTP.

$$\cdot \left(\begin{array}{c} G_{-1} \\ G_0 \\ G_1 \end{array} \right) = \left(\begin{array}{c} \psi(0, \alpha - 1) \\ \psi(0, \alpha) \\ \psi(0, \alpha + 1) \end{array} \right) \qquad (3.59)$$

where

$$\psi(i, j) = \sum_{n=0}^{N-1} r(n - i)r(n - j). \qquad (3.60)$$

The stability of the pitch synthesis filter $1/P(z)$ is not always guaranteed. For one-tap predictor, the stability condition is $|G| \leq 1$. Therefore, the stabilization can be easily carried out by setting $|G| = 1$ whenever $|G| > 1$. For 3-tap predictor, another stabilization procedure can be used [59]. However, the instability of the pitch synthesis filter is not that harmful to the quality of the reconstructed speech. The unstable filter will persist for a few frames (increasing the energy), but eventually, periods of stable filters are encountered, so that the output does not continue to increase with time. Figure 3.8 shows an example of a segment of voiced speech, the residual signal after short-term prediction, and the remaining signal after 1-tap and 3-tap long-term prediction. From Figure 3.8, it is seen that it is enough to use 3-tap filtering to remove the quasi-periodicity in the short-term prediction residual. Usually, to reduce the number of bits needed to encode the predictor gains, it is sufficient to use 1-tap LTP. When using LTP, the general codec schematic seen in Figure 3.1 is modified, as portrayed in Figure 3.9.

3.2.2.1 Computing the LTP parameters inside the loop: the adaptive codebook approach

In the block diagram of Figure 3.1, the LTP parameters can be determined outside the error minimization loop (directly from the LPC residual signal) as in equations (3.57) and (3.59). However, a significant improvement is achieved when the LTP parameters are optimized inside the analysis-by-synthesis loop [30]. In this case, the computation of the parameters contributes directly to the weighted error minimization procedure. Assuming one-tap long-term prediction, the output of the pitch synthesis filter is given by

$$u(n) = v(n) + Gu(n - \alpha). \tag{3.61}$$

We first assume that no excitation has been determined, so that Equation (3.61) reduces to

$$u(n) = Gu(n - \alpha). \tag{3.62}$$

The weighted synthesized speech is given by

$$\hat{s}_w(n) = \sum_{i=0}^{n} u(i)h(n - i) + \hat{s}_0(n), \tag{3.63}$$

where $h(n)$ is the impulse response of the weighted synthesis filter $1/A(z/\gamma)$ (see Section 3.2.3) and $\hat{s}_0(n)$ is the zero-input response of the weighted synthesis filter, that is, the output of the filter due to its initial states. The weighted error between the original and synthesized speech is given by

$$e_w(n) = x'(n) - \sum_{i=0}^{n} u(i)h(n - i), \tag{3.64}$$

where

$$x'(n) = s_w(n) - \hat{s}_0(n) \tag{3.65}$$

and $s_w(n)$ is the weighted input speech. Substituting Equation (3.62) into Equation (3.64) gives

$$e_w(n) = x'(n) - Gy_\alpha(n), \tag{3.66}$$

where

$$y_j(n) = u(n - j) * h(n) = \sum_{i=0}^{n} u(i - j)h(n - i). \tag{3.67}$$

The mean squared weighted error is given by

$$E_w = \sum_{n=0}^{N-1} [x'(n) - Gy_\alpha(n)]^2. \tag{3.68}$$

Setting $\partial E_w / \partial G = 0$ leads to

$$G = \frac{\sum_{n=0}^{N-1} x'(n) y_\alpha(n)}{\sum_{n=0}^{N-1} [y_\alpha(n)]^2}. \tag{3.69}$$

Substituting Equation (3.69) into Equation(3.68) gives

$$E_w = \sum_{n=0}^{N-1} [x'(n)]^2 - \frac{\left[\sum_{n=0}^{N-1} x'(n) y_\alpha(n) \right]^2}{\sum_{n=0}^{N-1} [y_\alpha(n)]^2}. \tag{3.70}$$

The pitch delay α is selected as the delay which maximizes the second term in Equation (3.70), and G is then computed from Equation (3.69). Significant speech quality improvement over the open loop solution is achieved when the long-term predictor parameters are computed inside the optimization loop. The disadvantage of the closed loop solution is the extra computational load needed to compute the convolution in Equation (3.67) over the range of the delay α. A fast procedure to compute this convolution $y_\alpha(n)$ for all the possible delays is to compute it for the first value in the range and then update it by

$$\begin{aligned} y_j(0) &= u(-j) h(0) \\ y_j(n) &= u(-j) h(n) + y_{j-1}(n-1), \qquad n = 1, \ldots, N-1. \end{aligned} \tag{3.71}$$

Equation (3.71) requires N operations to determine the convolution $y_j(n)$, while $N(N+1)/2$ operations are needed when Equation (3.67) is used. The term to be maximized requires $3N + 2$ instructions for each new delay. Another approach (the autocorrelation approach) can be used to update the energy in the denominator in Equation (3.70) with fewer instructions, than in the case of the convolution approach [92] (especially for large frame sizes). This approach will be described in a later section while discussing the CELP overlapping codebook approach.

The past synthetic excitation $u(n)$ is stored in an adaptive shift-storage register from $-L_p$ to -1, where L_p is the register or buffer length (usually 147). The contents of this buffer are updated every subframe (excitation frame) by introducing N new samples and dropping the last N samples, that is

$$u(n) \leftarrow u(n+N), \qquad n = -L_p, \ldots, -1. \tag{3.72}$$

The shift-storage register can be represented by an adaptive codebook, where each codeword is obtained by shifting the previous codeword to the left by one sample. The codewords are given by

$$c_j(n) = u(-j + n), \qquad \begin{array}{l} n = 0, \ldots, N-1, \\ j = N, \ldots, L_p. \end{array} \qquad (3.73)$$

For pitch delays less than the excitation frame length N, only the first j values of the codeword $c_j(n)$ are available. In natural speech the pitch delay varies from 20 to 160 samples (from 50 to 400 Hz) and the subframe length N is usually larger than 20 ($N = 60$ is commonly used in 4.8 kb/s coding). For these delays which are less than N the codewords $c_j(n)$ are constructed by repeating the available values until the codeword is completed. That is, for $j < N$

$$c_j(n) = \begin{cases} u(-j + n), & n = 0, \ldots, j-1, \\ u(-2j + n) & n = j, \ldots, 2j-1, \end{cases} \qquad (3.74)$$

and so on until the codeword is completed. The delay range 20–147 is commonly used (7 bits). For delays in the range 20 to $N-1$, the relation in Equation (3.71) has to be modified to accommodate for these delays. For $j < N$ the codeword $c_j(n)$ can be expressed by (assuming $j \geq N/2$)

$$c_j(n) = c_j^{(1)}(n) + c_j^{(2)}(n), \qquad (3.75)$$

where

$$c_j^{(1)}(n) = \begin{cases} u(-j + n), & n = 0, \ldots, j-1, \\ 0 & n = j, \ldots, N-1, \end{cases} \qquad (3.76)$$

and

$$c_j^{(2)}(n) = \begin{cases} 0 & n = 0, \ldots, j-1, \\ u(-2j + n), & n = j, \ldots, N-1 \end{cases} . \qquad (3.77)$$

Accordingly, the filtered codeword is given by

$$\begin{aligned} y_j(n) &= \left(c_j^{(1)}(n) + c_j^{(2)}(n) \right) * h(n) \\ &= y_j^{(1)}(n) + y_j^{(2)}(n). \end{aligned} \qquad (3.78)$$

From Equations (3.76) and (3.77)

$$c_j^{(2)}(n) = c_j^{(1)}(n - j), \qquad n = j, \ldots, N-1,$$

which yields

$$y_j^{(2)}(n) = y_j^{(1)}(n - j), \qquad n = j, \ldots, N-1. \qquad (3.79)$$

$y_j^{(1)}(n)$ can be updated using the relation in (3.71) from $j = 21$ to 147. For delays $j < N$, $y_j(n)$ is computed from $y_j^{(1)}(n)$ by

$$
\begin{aligned}
y_j(n) &= y_j^{(1)}(n), & \text{for } n = 0, \ldots, j-1, & \quad (3.80)\\
&= y_j^{(1)}(n) + y_j^{(1)}(n-j), & \text{for } n = j, \ldots, 2j-1, & \quad (3.81)\\
&= y_j^{(1)}(n) + y_j^{(1)}(n-j) + y_j^{(1)}(n-2j), & n = 2j, \ldots, N-1. &
\end{aligned}
$$
$$(3.82)$$

Notice that Equation (3.82) is only applied when $j < N/2$. A simpler approach for accommodating delays less than the frame length N is to extend the excitation buffer by the short-term prediction residual. That is

$$
u(n) = r(n), \qquad n = 0, \ldots, N - \alpha_{\min} - 1, \qquad (3.83)
$$

where $\alpha_{\min}$ is the minimum value in the range of the pitch delay. In this case, the delays $\alpha < N$ are not treated separately.

The pitch predictor performance can be improved by utilizing noninteger pitch delays. It often happens that the pitch delay does not coincide with the sampling instant. In this case the integer delay nearest to the real pitch delay, or a multiple of it, will be chosen. To find a delay closer to the real one, higher sampling resolution is needed [61–63]. This is done by up-sampling the synthetic excitation signal $u(n)$ (the content of the adaptive buffer) to obtain interpolated codewords in the adaptive codebook. The upsampling factor is determined by the required resolution. Upsampling by a factor m is accomplished by inserting $m-1$ zeros between each two samples and then low-pass filtering using a filter with cut-off frequency at π/m. A Hamming windowed truncated sinc function is commonly used for FIR low-pass filtering. The standard 4.8 kb/s DoD (Department of Defence, U.S.A.) CELP coder [64] uses 128 integer delays and 128 noninteger delays with the noninteger delays nonuniformly distributed among the integer delays (higher resolution at smaller pitch delays to obtain improvement in speech quality for typical female speakers). The use of noninteger delays introduces a substantial amount of complexity to the pitch search. In fact, the size of the adaptive codebook is doubled, and interpolation is used to find the codewords corresponding to noninteger delays. To avoid this increase in complexity, the following search procedure can be used [65] :

- The match score function is determined for the integer delays only. The match score is the square root of the second term on the right-hand side of Equation (3.70), and it is given by

$$
\chi(\alpha) = \frac{\left| \sum_{n=0}^{N-1} x'(n) y_\alpha(n) \right|}{\sqrt{\sum_{n=0}^{N-1} [y_\alpha(n)]^2}}, \qquad \alpha = \alpha_{\min}, \ldots, \alpha_{\max}. \qquad (3.84)
$$

LTP method	SEGSNR (dB)	LTP bit rate (kb/s)
IN1	13.3181	2.00
IN2	12.7172	1.80
OUT1	10.3038	2.00
OUT2	9.1432	0.50
OUTIN1	11.0102	2.00
OUTIN2	11.3637	2.00
OUTIN3	10.5513	0.95
OUTIN4	11.3780	1.35

Table 3.1: The effect of long-term prediction on the SEGSNR of CELP.

- The interpolation is performed on the match score function and its maximum values are searched. This avoids the need to filter the interpolated codewords and compute their match score according to Equation (3.84). In fact, the interpolated codewords are not determined in the first place. Only if the maximum of the interpolated match score function corresponded to a noninteger delay, would the excitation buffer be interpolated to determine the required codeword.

- Only a few interpolated points of the match score need to be computed. The interpolated points are determined around the integer delay which maximizes the score function and its submultiples. The delay (or the fractional delay) which maximizes the match score is selected. If the match score at a submultiple of chosen delay is larger than 0.95 of the maximum match score, then the submultiple delay is favoured. This avoids the selection of the multiples of the actual pitch value and results in a smooth pitch contour [66] which is useful in delta coding of the delay. It is also useful for detecting transmission errors in the delay.

To take advantage of both the simplicity of the open loop and the high performance of the closed loop, the pitch delay can be determined using the open loop solution by maximizing the second term in Equation (3.57) and then the gain can be computed inside the loop using Equation (3.69) [60]. In this case the convolution $y_\alpha(n)$ is computed only once for the value of α determined outside the loop. Nonexhaustive search of the adaptive codebook can be also utilized to reduce the search complexity. This can be done by subsampling, delta coding, and/or hierarchical search.

Table 3.1 shows the SEGSNRs of a CELP coder obtained using the different approaches for determining the LTP parameters. A CELP coder is used as the LTP is an essential part to the coder. A 20 ms speech frame is used with 5 ms excitation frames (subframes). A 9-bit ternary codebook is used (to be explained in a later section). The LTP gain is quantized with 3 bits and the delay with 7 (only noninteger delays). IN1 denotes computing the LTP parameters inside the loop (the adaptive codebook approach). IN2 denotes computing the parameters inside the loop with delta coding the delay in even subframes, where the delay in every even subframe is encoded with 5 bits. This reduces the bit rate by 0.2 kb/s at the expense of reducing the SNR by 0.5 dB. The important advantage of delta coding, however, is reducing the search complexity as the adaptive codebook size in even subframes is reduced to 32. OUT1 denotes computing the LTP parameters outside the analysis-by-synthesis loop with 5 ms updating interval (every subframe). This resulted in significant degradation in the CELP performance (3 dB drop in SNR). OUT2 denotes computing the parameters outside the loop with 20 ms updating interval (every speech frame). Only 0.5 kb/s transmission rate is needed to quantise the parameters but the speech quality is reduced dramatically (4 dB drop in SNR). OUTIN1 denotes computing the delay outside the loop and the gain inside the loop and OUTIN2 is the same except that the delay range $\alpha \pm 2$ is searched inside the loop, where α is the delay determined outside the loop. This approach resulted in 1 dB improvement over OUT1 with keeping the low complexity of the open loop search. OUTIN3 and OUTIN4 are similar to OUTIN1 and OUTIN2, respectively, except that the delay is updated every speech frame. In OUTIN4 a delay is determined every speech frame outside the loop and then in every subframe, the delay range $\alpha - 1, \ldots, \alpha + 2$ is searched inside the loop (the delay offset is encoded with 2 bits every subframe). Using OUTIN4 the bit rate is reduced by 0.6 kb/s and the search complexity is significantly reduced at the expense of 2 dB reduction in SNR.

3.2.2.2 Quantization of LTP parameters

The LTP parameters are the delay α and gain G (or the adaptive codebook index and gain). The delay is quantized with 6–8 bits depending on the range used. Most commonly a 7 bits range is utilized where 128 possible values are used in the range 20–147. To reduce the number of bits the LTP delays can be delta coded in even subframes with 5 bits. Another way to reduce the delay as well as the encoding rate (and also encoder complexity) is to determine a delay outside the loop and then to search for a few codewords around that delay in every subframe. When noninteger delays are used, 8 bits are usually needed (256 entry adaptive codebook). The histogram of the LTP gain is shown in Figure 3.10. It is sufficient to quantise the gain with 3 or 4 bits (the GSM full-rate coder uses only 2 bits as the gain is restricted to be positive). Due to the nonuniform distribution of the gain, nonuniform quantization has to be used. The quantization levels are determined from a large data base using a Lloyd-Max quantiser. The absolute value of the gain sometimes exceeds 10, so when designing the quantiser the gain value can be truncated. In the DoD

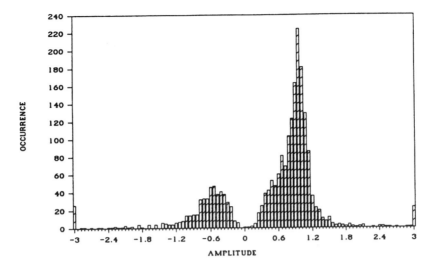

Figure 3.10: Histogram of LTP gain.

4.8 kb/s CELP coder [64], the range from -1 to 2 is used. In voiced speech segments, the gain value is very close to 1. Negative gains are usually obtained in unvoiced speech frames. Restricting the gain to the range 0–1.2 was found satisfactory. Notice that the gain values larger than 1 or less than −1 correspond to an unstable pitch synthesis filter (poles outside the unit circle). However, the speech quality is not affected by these short unstable periods as the LTP parameters are usually updated every 5 ms. These large gain values are obtained in the transient periods from silence to speech. A detailed treatment of long-term predictors can be found in a series of papers by Kabal and Ramachandran [59,67,68].

3.2.3   The Error Weighting Filter

In this section, we address the selection of a suitable error criterion in the general model of speech coding of Figure 3.1. Traditionally, speech coding algorithms have attempted to minimize the rms difference between the original and coded speech waveforms. However, it is now well recognized that the subjective perception of the signal distortion is not based on the rms error alone. The theory of auditory masking suggests that noise in the formant regions would be partially or totally masked by the speech signal. Thus, a large part of the perceived noise in a coder comes from the frequency regions where the signal level is low. Therefore, to reduce the perceived noise, its flat spectrum is shaped so that the frequency component in the noise around the formant regions is allowed to have higher energy relative to the components in the inter-formant regions. Now comes the question of how to choose this error shaping (or weighting) filter which appears in Figure 3.1. Incorporating noise shaping in APC [58], Atal and Schroeder showed that the quantization noise appearing in the reconstructed speech signal is given by

$$|N(f)|^2 = |\hat{S}(f) - S(f)|^2 = |\Delta(f)|^2 \left|\frac{1 - F(f)}{1 - P_s(f)}\right|, \qquad (3.85)$$

where $|\Delta(f)|^2$ is the power spectrum of the noise at the output of the quantiser, $P_s(z)$ is the short-term predictor, and $F(z)$ is a feedback filter. In [69], Atal and Schroeder described an efficient method for determining the weighting filter by minimizing the subjective loudness of the quantiza-tion noise. In the model of Figure 3.1, the weighting filter $W'(z)$ can be expressed as

$$W'(z) = \frac{1 - P_s(z)}{1 - F(z)} = \frac{A(z)}{B(z)}. \qquad (3.86)$$

Equation (3.86) is derived from Equation (3.85) where

$$\begin{aligned}
\Delta(f) &= |\hat{S}(f) - S(f)|\frac{1 - P_s(f)}{1 - F(f)} \\
&= N(f)W(f). \qquad (3.87)
\end{aligned}$$

Details about the selection of $B(z)$ are found in [58]. An appropriate choice was found to be $B(z) = A(z/\gamma)$ which gives

$$W'(z) = \frac{A(z)}{A(z/\gamma)} = \frac{1 - \sum_{k=1}^{p} a_k z^{-k}}{1 - \sum_{k=1}^{p} a_k \gamma^k z^{-k}} \qquad (3.88)$$

where γ is a fraction between 0 and 1. The value of γ is determined by the degree to which one wishes to deemphasize the formant regions in the error spectrum. Note that decreasing γ increases the bandwidth of the poles of $W'(z)$. The increase in the bandwidth ω is given by

$$\omega = -\frac{f_s}{\pi} \ln(\gamma), \qquad (3.89)$$

where f_s is the sampling frequency. The choice $\gamma = 0$ gives $W'(z) = A(z)$. In this case, the coder output noise has the same envelope as the original speech. On the other hand, the choice $\gamma = 1$ gives $W'(z) = 1$ which is equivalent to no weighting. A good choice is to use a value of γ between 0.8 and 0.9, which corresponds to an increase in the bandwidth of the poles of $W'(z)$ by 570 down to 270 Hz approximately. Figure 3.11 shows an example of the spectrum of $1/A(z)$ and $A(z)/A(z/\gamma)$ for different values of γ. Makhoul and Berouti [70] discussed other choices of $W'(z)$, where they assumed $W'(z)$ as an all-pole filter or an all-zero filter, but this choice was

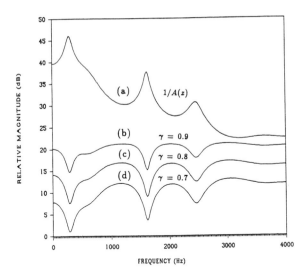

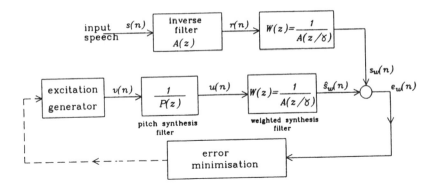

Figure 3.11: Spectrum of $1/A(z)$ and $A(z)/A(z/\gamma)$ for different values of γ.

Figure 3.12: Basic structure for analysis-by-synthesis predictive coding using $W'(z) = A(z)/A(z/\gamma)$.

found to be inferior to the pole-zero filter of Equation (3.88).

Using the error weighting filter given in Equation (3.88), and weighting the original speech and the synthesized one separately before their subtraction, the configuration of Figure 3.1 is reduced to the form shown in Figure 3.12. In this new configuration, the synthesis filter is combined with the error weighting filter to produce the filter

$$W(z) = \frac{1}{A(z/\gamma)} = \frac{1}{1 - \sum_{k=1}^{p} a_k \gamma^k z^{-k}}. \qquad (3.90)$$

We will refer to $W(z)$ as the *weighted synthesis filter*. From now on, the configuration of Figure 3.12 will be used as the basic structure for high-quality analysis by synthesis predictive coding.

3.3 Multi-pulse and Regular-pulse Excitation

In the previous sections we described an efficient basic structure of a model of speech coding which is capable of producing near toll quality speech in the range 4.8–16 kb/s. We showed how the model removes the redundancy in the speech signal by employing short-term and long-term predictors. We discussed also the utilization of an efficient perceptually weighted error minimization criterion. The crucial part of this model which has not been discussed yet is determining an appropriate excitation signal to drive the synthesis filters to produce the synthesized speech. The excitation signal should be defined in a clever way to make the number of bits needed to encode it as small as possible. In this section we discuss the multi-pulse [71] and regular-pulse approaches, and in a later sections we will discuss the code-excited approach. The multi-pulse excited (MPE) and regular-pulse excited (RPE) approaches assume that the excitation signal is modelled by a definite number of pulses in a short time period (5–15 ms). We will denote this time period the excitation frame of length N. The LPC parameters update frame L is usually divided into subframes of length N (L is a multiple of N), and the excitation is determined for every subframe. The difference between MPE and RPE is the way the positions of the pulses are determined. In the next subsection, a general mathematical formulation for determining the pulse positions and amplitudes will be given. In the later subsections, the MPE and RPE algorithms will be discussed and evaluated individually.

3.3.1 Formulation of the Pulse Amplitudes and Positions Computation

In multi-pulse excitation, an excitation frame of length N contains M pulses with amplitudes β_i at positions m_i. Therefore, the excitation signal $v(n)$ is defined as

$$v(n) = \sum_{k=0}^{M-1} \beta_k \delta(n - m_k), \qquad\qquad n = 0, \ldots, N-1, \qquad (3.91)$$

where β_k are the pulse amplitudes, m_k are the pulse positions and M is the number of pulses modelling an excitation sequence of length N samples. The pulse amplitudes and positions are determined by minimizing the mean squared weighted error between the original and synthesized speech.

The residual signal after short-term prediction, $r(n)$, is found by filtering the original speech through the inverse filter $A(z)$. The residual $r(n)$ is given by

$$r(n) = s(n) - \sum_{k=1}^{p} a_k s(n-k). \qquad (3.92)$$

The weighted input speech is found by recursive filtering the residual signal $r(n)$ through the weighted synthesis filter $W(z)$ of Equation (3.90). This can be expressed as

$$s_w(n) = r(n) + \sum_{k=1}^{p} a_k \gamma^k s_w(n-k). \qquad (3.93)$$

The weighted input speech, $s_w(n)$, can also be found by convolving the residual signal $r(n)$ with $h(n)$, the impulse response of the weighted synthesis filter $W(z)$, that is

$$s_w(n) = \sum_{i=-\infty}^{n} r(i)h(n-i) = \sum_{i=0}^{n} r(i)h(n-i) + s_0(n), \qquad (3.94)$$

where $s_0(n)$ is the zero-input response of the filter $W(z)$ in the upper branch of Figure (3.1), i.e. the output of $W(z)$ due to its initial states. The memoryless convolution

$$r(n) * h(n) = \sum_{i=0}^{n} r(i)h(n-i) \qquad (3.95)$$

is referred to as the zero-state response of $W(z)$ to the input $r(n)$. It is preferred to use the recursive relation in Equation (3.93) to compute $s_w(n)$ since it requires only p multiply/add operations per speech sample.

The impulse response, $h(n)$, of the weighted synthesis filter is deduced directly from Equation (3.90) and is given by

$$h(n) = \delta(n) + \sum_{k=1}^{p} a_k \gamma^k h(n-k), \qquad n = 0, \ldots, N-1, \qquad (3.96)$$

where $h(0) = 1$ and $h(n) = 0$ for $n < 0$.

The weighted synthesized speech $\hat{s}_w(n)$ is computed by convolving the excitation signal $v(n)$ with the impulse response of the combination of the pitch synthesis filter $1/P(z)$ and the weighted synthesis filter $W(z)$. The combined filter $C(z)$ is given by

$$C(z) = \frac{1}{P(z)} \cdot W(z) = \frac{1}{(1 - Gz^{-\alpha})(1 - \sum_{k=1}^{p} a_k \gamma^k z^{-k})}. \qquad (3.97)$$

The impulse response $h_c(n)$ of the combined filter $C(z)$ is given by

$$h_c(n) = f(n) * h(n), \qquad (3.98)$$

where $f(n)$ is the impulse response of the pitch synthesis filter $1/P(z)$ given by

$$
\begin{aligned}
f(n) &= \delta(n) + Gf(n-\alpha) \\
&= \sum_{i=0}^{n_p} G^i \delta(n - i \cdot \alpha), \qquad n = 0, \ldots, N-1, \qquad (3.99)
\end{aligned}
$$

where n_p is the number of pitch periods in the excitation frame of length N and it usually varies from 0 to 3. Substituting Equation (3.99) into Equation (3.98) gives

$$
\begin{aligned}
h_c(n) &= \sum_{i=0}^{n_p} G^i h(n - i \cdot \alpha) \\
&= h(n) + Gh(n-\alpha) + \cdots + G^{n_p} h(n - n_p \cdot \alpha). \qquad (3.100)
\end{aligned}
$$

It is plausible from Equation (3.100) that for pitch periods α larger than the excitation frame length N, $h_c(n)$ is equal to $h(n)$ for the values $n < N$. Therefore if the LTP delay α is restricted to be larger than the excitation frame length N, the pitch synthesis filter will not contribute to the impulse response of the combined filter $C(z)$ and it will only be considered when the zero-input response of the combined filter is computed. Also when the adaptive codebook concept is used $h_c(n)$ becomes equal to $h(n)$ as the pitch synthesis filter is replaced by a codebook. From now on we will assume that the pitch synthesis filter is replaced by an adaptive codebook. In this case the excitation at the input of the LPC synthesis filter is given by

$$u(n) = v(n) + Gc_\alpha(n), \qquad (3.101)$$

where G is the adaptive codebook gain (or the LTP gain) and $c_\alpha(n)$ is the codeword selected from the adaptive codebook (or α is the LTP delay).

At the beginning the parameters α and G are determined as was described in Section 3.2.2. The weighted synthesized speech can now be expressed by

$$\hat{s}_w(n) = v(n) * h(n) + Gc_\alpha(n) * h(n) + \hat{s}_0(n), \qquad (3.102)$$

where the convolution is a memoryless process (as in Equation (3.95)) and $\hat{s}_0(n)$ is the zero-input response of the weighted synthesis filter in the lower branch of Figure 3.12. If the adaptive codebook concept is not used $c_\alpha(n)$ is replaced by $u(n-\alpha)$ and $h(n)$ is the impulse response defined in Equation (3.96). The zero-input response $\hat{s}_0(n)$ is given by

$$\hat{s}_0(n) = \sum_{k=1}^{p} a_k \gamma^k \hat{s}_0(n-k), \qquad n = 0, \ldots, N-1, \qquad (3.103)$$

where initially,

$$\hat{s}_0(n) = \hat{s}_w(N+n) \qquad \text{for} \qquad n = -p, \ldots, -1, \qquad (3.104)$$

which involves buffering the states in the taps of the filter $W(z)$ at the end of the previous excitation frame. Note that $\hat{s}_0(n)$ has already been determined while computing the adaptive codebook parameters (see Equation (3.63)). Substituting Equation (3.91) into Equation (3.102) leads to

$$
\begin{aligned}
\hat{s}_w(n) &= \sum_{i=0}^{n} \left(\sum_{k=0}^{M-1} \beta_k \delta(n - m_k) \right) h(n-i) + Gc_\alpha(n) * h(n) + \hat{s}_0(n), \\
&= \sum_{k=0}^{M-1} \beta_k h(n - m_k) + Gy_\alpha(n) + \hat{s}_0(n), \qquad (3.105)
\end{aligned}
$$

where

$$y_\alpha(n) = c_\alpha(n) * h(n)$$

is the zero-state response of the weighted synthesis filter to the codeword c_α chosen from the adaptive codebook. Now, the weighted error between the original speech and the synthesized speech is given by

$$
\begin{aligned}
e_w(n) &= s_w(n) - \hat{s}_w(n), \\
&= s_w(n) - Gy_\alpha(n) - \hat{s}_0(n) - \sum_{k=0}^{M-1} \beta_k h(n - m_k), \\
&= x(n) - \sum_{k=1}^{M} \beta_k h(n - m_k), \qquad (3.106)
\end{aligned}
$$

where

$$x(n) = s_w(n) - Gy_\alpha(n) - \hat{s}_0(n). \qquad (3.107)$$

The signal $x(n)$ is computed by updating $x'(n)$ of Equation (3.65), that is

$$x(n) = x'(n) - Gy_\alpha(n). \qquad (3.108)$$

The mean squared weighted error is given by using Equation (3.106) as

$$E_w = \sum_{n=0}^{N-1} e_w^2(n),$$

$$= \sum_{n=0}^{N-1} \left[x(n) - \sum_{k=0}^{M-1} \beta_k h(n - m_k) \right]^2 . \tag{3.109}$$

The task now is to find the pulse amplitudes β_k and the pulse positions m_k which minimize the mean squared weighted error in Equation (3.109). By setting $\partial E_w / \partial \beta_i = 0$ for $i = 0, \ldots, M-1$ we get

$$\frac{\partial E_w}{\partial \beta_i} = -2 \sum_{n=0}^{N-1} \left[x(n) - \sum_{k=0}^{M-1} \beta_k h(n - m_k) \right] h(n - m_i) = 0. \tag{3.110}$$

Therefore

$$\sum_{n=0}^{N-1} x(n) h(n - m_i) = \sum_{n=0}^{N-1} \left[\sum_{k=0}^{M-1} \beta_k h(n - m_k) \right] h(n - m_i). \tag{3.111}$$

Reordering the summations in the right-hand side of Equation (3.111) yields

$$\sum_{k=0}^{M-1} \beta_k \sum_{n=0}^{N-1} h(n - m_k) h(n - m_i) = \sum_{n=0}^{N-1} x(n) h(n - m_i), \qquad i = 0, \ldots, M-1. \tag{3.112}$$

Define

$$\phi(i, j) = \sum_{n=0}^{N-1} h(n - i) h(n - j) \tag{3.113}$$

to be the autocorrelation of the impulse response $h(n)$, and

$$\psi(i) = \sum_{n=0}^{N-1} x(n) h(n - i) \tag{3.114}$$

to be the cross-correlation between $x(n)$ and $h(n)$, then the set of M equations in (3.112) can be written as

$$\sum_{k=0}^{M-1} \beta_k \phi(m_i, m_k) = \psi(m_i), \qquad i = 0, \ldots, M-1. \tag{3.115}$$

The set of M equations in (3.115) can be written in matrix form as

$$\begin{pmatrix} \phi(m_0, m_0) & \phi(m_0, m_1) & \cdots & \phi(m_0, m_{M-1}) \\ \phi(m_1, m_0) & \phi(m_1, m_1) & \cdots & \phi(m_1, m_{M-1}) \\ \vdots & \vdots & \ddots & \vdots \\ \phi(m_{M-1}, m_0) & \phi(m_{M-1}, m_1) & \cdots & \phi(m_{M-1}, m_{M-1}) \end{pmatrix}$$

$$
\cdot \begin{pmatrix} \beta_0 \\ \beta_1 \\ \vdots \\ \beta_{M-1} \end{pmatrix} = \begin{pmatrix} \psi(m_0) \\ \psi(m_1) \\ \vdots \\ \psi(m_{M-1}) \end{pmatrix} \tag{3.116}
$$

We will now proceed in deriving an expression for the weighted mean-squared error using the optimal pulse positions and amplitudes from Equation (3.115). Using Equation (3.109)

$$
\begin{aligned}
E_w &= \sum_{n=0}^{N-1} x^2(n) - 2 \sum_{n=0}^{N-1} x(n) \sum_{k=0}^{M-1} \beta_k h(n - m_k) \\
&+ \sum_{n=0}^{N-1} \left[\sum_{k=0}^{M-1} \beta_k h(n - m_k) \right]^2, \\
&= \sum_{n=0}^{N-1} x^2(n) - 2 \sum_{k=0}^{M-1} \beta_k \psi(m_k) + \sum_{i=0}^{M-1} \sum_{k=0}^{M-1} \beta_i \beta_k \phi(m_i, m_k).
\end{aligned}
\tag{3.117}
$$

Using the optimum solution for the pulses in Equation (3.115) we have

$$
\begin{aligned}
\sum_{i=0}^{M-1} \beta_i \psi(m_i) &= \sum_{i=0}^{M-1} \beta_i \sum_{k=0}^{M-1} \beta_k \phi(m_i, m_k), \\
&= \sum_{i=0}^{M-1} \sum_{k=0}^{M-1} \beta_i \beta_k \phi(m_i, m_k).
\end{aligned}
\tag{3.118}
$$

Substituting the relation of Equation (3.118) in Equation (3.117), the minimum mean squared weighted error between the original and the synthesized speech is given by

$$
E_{\min} = \sum_{n=0}^{N-1} x^2(n) - \sum_{k=0}^{M-1} \beta_k \psi(m_k). \tag{3.119}
$$

To find the optimum pulse positions and amplitudes, Equation (3.116) has to be solved. In Equation (3.116), we have a set of M equations with $2M$ unknowns, expressed in a matrix form. The unknowns are M pulse positions and M pulse amplitudes. Therefore, it is very difficult to find an optimal solution for the pulse positions and amplitudes. In fact, the optimal solution is to solve Equation (3.116) for all the possible combinations of pulse positions, and select the pulse amplitudes which minimize the error in Equation (3.119). The number of pulse position combinations is $^N C_M = N!/((N - M)!M!)$, and for typical values of $N = 40$ and $M = 4$ the number of combinations is 91390. This shows the complexity of an optimal solution. In the following sections, we will discuss two sub-optimal approaches for solving Equation (3.116) using the minimum error expression in Equation (3.119). The so-called multi-pulse excited approach (MPE) determines one pulse at a time and the regular pulse excited method (RPE) assumes predefined regularly spaced positions. We will start with the MPE approach.

3.3.2 The Multi-pulse Approach

The multi-pulse algorithm is a sub-optimal approach for solving Equation (3.116). The algorithm determines one pulse at a time in an M-stage process. At every stage j, a new pulse amplitude β_j and position m_j are computed by using the previously determined pulse positions and amplitudes at stages less than j. At the beginning, we assume that there is only one pulse with amplitude β_0 at position m_0. Now, Equation (3.116) is reduced to

$$\beta_0 = \frac{\psi(m_0)}{\phi(m_0, m_0)}. \tag{3.120}$$

With only one pulse, the minimum error expression in Equation (3.119) is reduced to

$$E_{\min} = \sum_{n=0}^{N-1} x^2(n) - \beta_0 \psi(m_0). \tag{3.121}$$

Substituting Equation (3.120) in (3.121)

$$E_{\min} = \sum_{n=0}^{N-1} x^2(n) - \frac{\psi^2(m_0)}{\phi(m_0, m_0)}. \tag{3.122}$$

To find the first pulse, we search for the value m_0 which minimizes $E_{\min}$ or, equivalently, the value which maximizes the second term in the right-hand side of Equation (3.122). Having determined the first pulse position m_0, the first pulse amplitude β_0 is computed from Equation (3.120). Introducing a second pulse, the mean-squared weighted error is now given by

$$
\begin{aligned}
E_w^{(1)} &= \sum_{n=0}^{N-1} [x(n) - \beta_0 h(n - m_0) - \beta_1 h(n - m_1)]^2 \\
&= \sum_{n=0}^{N-1} \left[x^{(1)}(n) - \beta_1 h(n - m_1) \right]^2
\end{aligned}
\tag{3.123}
$$

where

$$x^{(1)}(n) = x(n) - \beta_0 h(n - m_0). \tag{3.124}$$

Setting $\partial E_w^{(1)}/\partial \beta_1$ to zero leads, similar to Equations (3.120) and (3.122), to the relations

$$\beta_1 = \frac{\psi^{(1)}(m_1)}{\phi(m_1, m_1)}, \tag{3.125}$$

and

$$E_{\min}^{(1)} = \sum_{n=0}^{N-1} \left[x^{(1)}(n) \right]^2 - \frac{\left[\psi^{(1)}(m_1) \right]^2}{\phi(m_1, m_1)} \tag{3.126}$$

where

$$\begin{aligned}
\psi^{(1)}(n) &= \sum_{i=0}^{N-1} x^{(1)}(i)h(i-n) \\
&= \sum_{i=0}^{N-1} \left[x(i) - \beta_0 h(i - m_0) \right] h(i-n) \\
&= \psi(n) - \beta_0 \phi(m_0, n). \tag{3.127}
\end{aligned}$$

Therefore, the value of $\psi(n)$ is updated, as in Equation (3.127), by removing the effect of the first pulse, and the second pulse position is determined by the value which maximizes the second term in Equation (3.126). This process is continued until all the pulses are determined. We first initialize $\psi^{(0)}(n) = \psi(n)$. At every stage j, $j = 1, \ldots, M - 1$, the value $\psi^{(j)}(i)$ is found from

$$\psi^{(j)}(i) = \psi^{(j-1)}(i) - \beta_{j-1}\phi(m_{j-1}, i), \qquad i = 0, \ldots, N - 1, \tag{3.128}$$

and the position of the jth pulse, m_j, is determined by the value i which maximizes the normalized correlation, given by

$$T(i) = \frac{\left[\psi^{(j)}(i) \right]^2}{\phi(i, i)}. \tag{3.129}$$

The pulse amplitude is then computed by

$$\beta_j = \frac{\psi^{(j)}(m_j)}{\phi(m_j, m_j)}. \tag{3.130}$$

When the speech segment is voiced, the multi-pulse algorithm tends to locate more than one pulse at the same position, which virtually reduces the number of pulses in the entire period. This can be avoided by setting $\psi(m_k) = 0$ for $k = 0, \ldots, i - 1$, where i is the index of the pulse being searched, so that no more than one pulse is located at the same position.

Figure 3.13 (a) shows the signal power with time for the sentence 'to reach the end he needs much courage' uttered by a female speaker. Figure 3.13 (b) shows the variation of SEGSNR with time for this speech

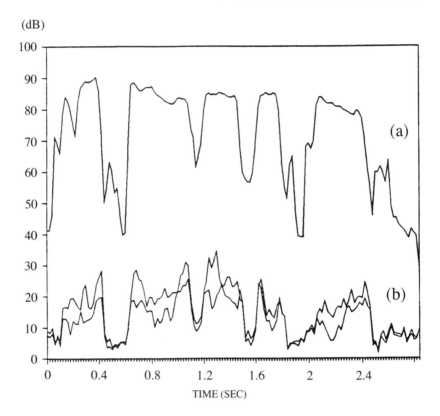

Figure 3.13: (a) Power variation of the sentence *'to reach the end he needs much courage'*; (b) Variation of SEGSNR vs. time for MPE using four pulses/5 ms with and without LTP (see upper and lower curves, respectively).

sequence when the multi-pulse algorithm is utilized. Four pulses are placed in an excitation search frame of 40 samples (5 ms) and the LPC parameters update frame is 20 ms. The upper curve is obtained when long-term prediction is utilized and the lower curve without LTP. It is observed that the SNR is high when the speech power is high. The SNR is high in the periods where the speech is quasi-periodic (voiced) compared with the unvoiced or transient periods.

Figure 3.14 shows how the multi-pulse algorithm models the LPC residual signal in the absence of a pitch predictor. Figure 3.14 (a) shows a 87.5 ms segment of female speech, Figure 3.14 (b) displays the LPC residual for this speech segment, and the multi-pulse excitation is shown in Figure 3.14 (c) for the case where no LTP is used. The reconstructed speech is depicted in Figure 3.14 (d). It can be clearly seen how the quasi-

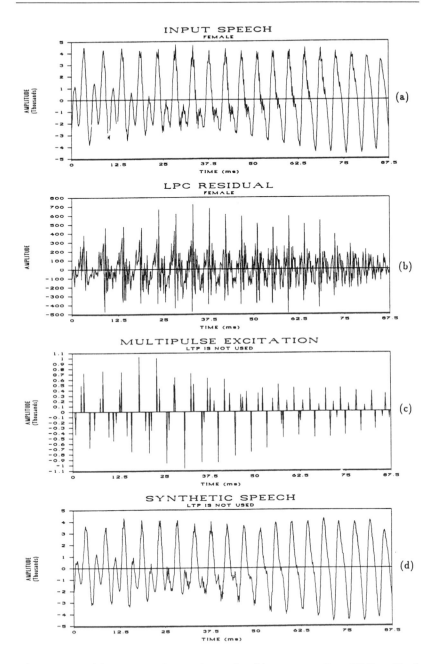

Figure 3.14: (a) 87.5 ms of voiced speech; (b) corresponding LPC residual; (c) multipulse excitation without LTP (4 pulses/5 ms); (d) reconstructed speech.

periodicity in the LPC residual is preserved by the multi-pulse algorithm without having any prior knowledge of the pitch period or whether the speech is voiced or unvoiced. Figure 3.15 shows the excitation and reconstructed speech for the same speech segment when LTP is utilized. The excitation pulses are interpolated inside the taps of the pitch synthesis filter, and the excitation of the LPC synthesis filter does not contain zeros as in the multi-pulse without LTP. The excitation signal is much closer to the LPC residual than in the case where the LTP is not utilized. It is also clear that the fine details in the speech waveform are preserved when LTP is used.

3.3.3   Modification of the MPE Algorithm

The multi-pulse algorithm described earlier is suboptimal because the algorithm implicitly assumes that the amplitudes of the past pulses remain constant during the search for the location of the present pulse. Thus the determined pulse amplitudes and positions do not satisfy the set of equations in (3.116). This results in speech quality degradation, especially when the pulses are closely spaced [72]. The pulse amplitudes can be reoptimized, after the last pulse position has been determined, by solving the set of equations in (3.116) using the determined pulse positions. However, the pulse positions remain suboptimal as they have been determined using the nonoptimal amplitudes. A better solution is obtained if the pulse amplitudes are reoptimized at every stage of the search. Thus after determining the second pulse position m_1, the amplitudes of the first two pulses β_0 and β_1 are recomputed using Equation (3.116) (2×2 matrix equation), and the correlation ψ is updated using these reoptimized pulses, that is

$$\psi^{(2)}(n) = \psi(n) - \beta_0^{(1)}\phi(m_0, n) - \beta_1^{(1)}\phi(m_1, n).$$

$\psi^{(2)}(n)$ is used to search for the position of the third pulse m_2 and then another set of reoptimized pulses is determined using Equation (3.116) (3×3 matrix equation), and the correlation is updated by

$$\psi^{(3)}(n) = \psi(n) - \beta_0^{(2)}\phi(m_0, n) - \beta_1^{(2)}\phi(m_1, n) - \beta_2^{(2)}\phi(m_2, n).$$

This process is continued until the last pulse position is found, then the pulse amplitudes are finally recomputed using Equation (3.116). For M pulses, this approach requires solving 2×2, 3×3, ..., and $M \times M$ matrix equations in order to reoptimize the pulse amplitudes at each stage of the search. The method becomes rather complex as the number of pulses is increased. Singhal [73] developed a computationally efficient algorithm for reoptimizing the pulse amplitudes without needing to solve Equation (3.116) at every stage. His algorithm is based on the Cholesky decomposition of the matrix of autocorrelations and it is detailed in [72, 73].

The MPE algorithm is simplified by utilizing the autocorrelation formu-

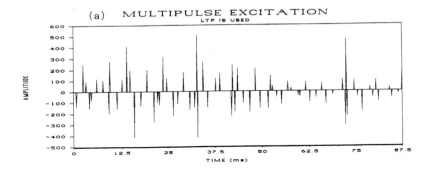

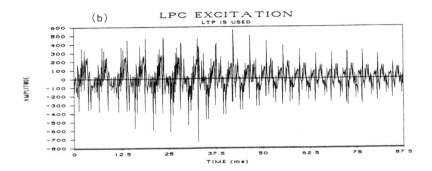

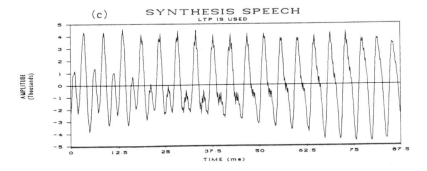

Figure 3.15: (a) Multi-Pulse excitation utilizing LTP; (b) Excitation of LPC synthesis filter; (c) reconstructed speech.

lation [74,75]. This is done by extending the summation limits in the error minimization to $-\infty$ and ∞, and windowing with a window having zero values outside the range 0 to $N - 1$. Doing this will reduce the expression in Equation (3.113) to

$$\phi(i, j) = \phi(|i - j|) = \sum_{n=|i-j|}^{N-1} h(n)h(n - |i - j|). \qquad (3.131)$$

The autocorrelation approach will be explained in more detail later while describing the CELP. Using the autocorrelation approach, the term to be maximized in Equation (3.129) is reduced to

$$\mathcal{T}(i) = \frac{\left[\psi^{(j)}(i)\right]^2}{\phi(0)}. \qquad (3.132)$$

This is maximized by maximizing the absolute value of $\psi^{(j)}(i)$, which reduces the number of multiplications (or divisions) to search M pulses by $2MN$.

3.3.4 Evaluation of the Multi-pulse Algorithm

In this section, we study the effect of the different MPE encoder parameters on the quality of the synthesized speech. For the results in this section, the LPC parameters are quantized with 36 bits using LSPs. The adaptive codebook gain is quantized with 4 bits and the index with 7 bits in odd subframes and 5 bits in even subframes. The pulse amplitudes are not quantized. The parameters which we are going to take into consideration are

1. The number of pulses per excitation frame.

2. The length of the excitation frame.

We have already shown, in Section 3.2.1.3, the effect of changing the predictor order p, and the considerations in choosing the LPC analysis method and the updating frame length. Table 3.2 shows the default analysis conditions used in this chapter. This choice of frame lengths is suitable for the bit rate of 9.6 kb/s. At bit rates below 8 kb/s, larger LPC and excitation update frames will be necessary. The selection of predictor order $p = 10$ is satisfactory as we discussed earlier.

3.3.4.1 Number of Pulses per Excitation Frame

As we are aiming to achieve low bit rate while maintaining a high synthesized speech quality, it is desired to use as few pulses as possible for modelling the excitation signal. Using the analysis conditions stated in

sampling frequency	8000 Hz
LPC analysis frame	200 samples (25 ms)
LPC parameter update frame	160 samples (20 ms)
predictor order	10
analysis method	autocorrelation
LPC quantization	LSPs with 36 bits
excitation frame	40 samples (5 ms)
LTP predictor taps	1
LTP parameter update	40 samples (5 ms)
LTP analysis	adaptive codebook integer delays (20-147) delta coding (7,5,7,5)

Table 3.2: Default analysis conditions used in this chapter.

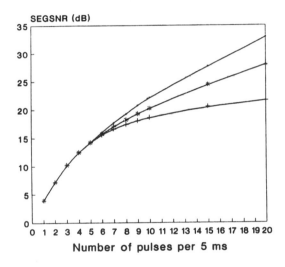

Figure 3.16: SEGSNR versus number of pulses per 5 ms with and without pulse amplitudes reoptimization (without LTP).

Table 3.2, and excluding the pitch predictor from the coder, the SEGSNR has been computed with number of pulses varying from 2 to 20 pulses per 5 ms excitation frame, and the results are shown Figure 3.16. Figure 3.17 shows the SEGSNR against number of pulses when long-term prediction is utilized. It is noticed from the lower curves in Figures 3.16 and 3.17 that after few pulses have been placed, the SEGSNR tends to saturate with the increased number of pulses. This is due to the suboptimal solution of the algorithm. The upper curves in the figures show the SEGSNR with the

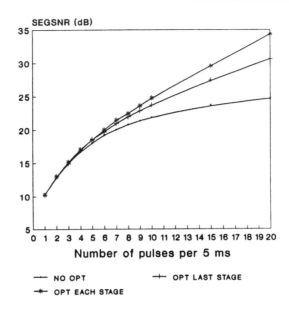

Figure 3.17: SEGSNR versus number of pulses per 5 ms subsegment with and without pulse amplitude reoptimization utilizing LTP.

pulses recomputed at every stage, and the middle curve with the pulses recomputed at the last stage. The improvement becomes more significant when the number of pulses is increased. For 9.6 kb/s MPE-LPC coding, at most four pulses per 5 ms are used, and deploying amplitude reoptimization does not introduce any significant improvement in speech quality in this case. It is sufficient to reoptimize the pulse amplitudes at the last stage.

Figure 3.18 shows the SEGSNR against number of pulses with and without long-term prediction using last stage reoptimization. A gap of 4–6 dB is noticed between the two curves. When 4 pulses are used, the SNR is increased by 4.5 dB if LTP is deployed. Quantizing the LTP parameters is almost equivalent to quantizing one pulse (amplitude and position) in terms of number of bits used. However, with LTP fewer pulses are needed to model the excitation signal. Using 2 pulses per 5 ms frame with LTP gives similar quality to using 4 pulses without LTP. Therefore deploying LTP results in improved quality at lower bit rates.

Figure 3.19 shows a comparison between the covariance method and autocorrelation method in determining the pulses with and without LTP. The covariance approach gives better SNRs as the number of pulses is increased. For practical values of 3 or 4 pulses per 5 ms the degradation due to using the autocorrelation approach is negligible (about 0.1 dB). Bearing in mind that the autocorrelation approach is computationally more efficient than the covariance approach, it is preferred to be used in practice where

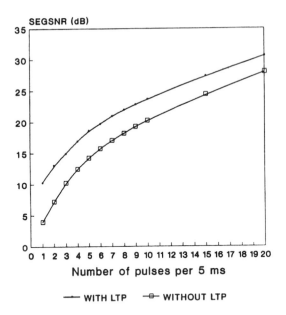

Figure 3.18: SEGSNR versus number of pulses per 5 ms with and without long-term prediction (pulse amplitudes are reoptimized at last stage).

the number of pulses is less than 5.

3.3.4.2   The Length of the Excitation Frame

In the multi-pulse approach, M pulses are located in a frame of N samples. The choice of the frame size is a trade-off between quality and complexity. It can be shown that the number of operations (multiplication/addition) needed per speech sample is proportional to NM. Therefore, to reduce the complexity, one wishes to reduce the frame size N. On the other hand, the excitation frame size N can not be chosen too small ($< 5\,\text{ms}$) to avoid nonoptimal pulse allocation. For example, for voiced speech the multi-pulse algorithm tends to locate the pulses around the major pitch pulse. If the pitch period is greater than the frame size N (as in low pitch frequency voiced speech), we are modelling a less important part of the pitch period with more pulses than it is necessary. Figure 3.20 shows the SEGSNR against the frame size for 10% pulse rate using multi-pulse excitation. We can see the increase in SNR for larger frame sizes. To reduce the complexity the choice of 5 ms excitation search frame is reasonable.

Another concern with choosing the search frame size is related to *block edge effects* [76]. The elements $\phi(i,j)$ of the correlation matrix become small for values of i or j close to N (due to the existence of few terms in the summation in Equation (3.113)). Therefore for pulses located towards the

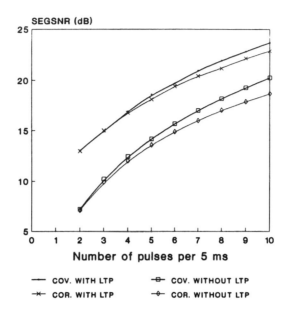

Figure 3.19: SEGSNR versus number of pulses per 5 ms with and without LTP for the covariance and autocorrelation approaches (pulses are re-optimized at last stage).

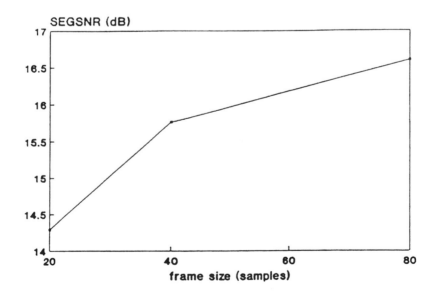

Figure 3.20: SEGSNR vs search frame size.

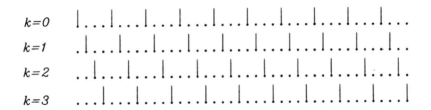

Figure 3.21: Candidate excitation patterns in RPE for $N = 40$ and $D = 4$.

end of the frame the solution in (3.115) becomes ill-conditioned resulting in artificially high pulse amplitudes [72]. To avoid this effect the search frame is made larger than the excitation frame by overlapping with the beginning of the next frame, and the pulses falling in the overlap region are recomputed in the next frame. An overlap region of 2.5 ms was found adequate to prevent ill-conditioned solutions [72].

3.3.5   Regular-Pulse Excitation Approach

Instead of determining one pulse (β_j, m_j) at every stage j assuming that the pulses up to stage $j - 1$ have been determined, Kroon et. al [8,77] suggested another suboptimal approach for the solution of Equation (3.116) by assuming predefined pulse positions, regularly spaced by distance D. The same approach was also proposed, at a similar time, by Adoul et al. [78], and was called generalized decimation.

According to the Regular-Pulse Excitation (RPE) approach, the excitation sequence for a frame of length N consists of M pulses, regularly spaced by a distance D, where $M = N$ DIV D and DIV denotes integer division (N is not necessarily a multiple of D). Depending on where the first pulse is positioned, D different excitation patterns are obtained. The pulse positions are given by

$$m_i^{(k)} = k + iD, \qquad \begin{array}{l} k = 0, \ldots, D - 1, \\ i = 0, \ldots, M - 1, \end{array} \qquad (3.133)$$

where k is the position of the first pulse or the *initial phase*. As an example, Figure 3.21 shows the possible excitation patterns for $N = 40$ and $D = 4$. The RPE algorithm consists of solving Equation (3.116) D times (for every possible excitation pattern) to obtain D sets of amplitudes $\{\beta_i^{(k)}\}$ at initial phase k. The mean squared weighted error of Equation (3.119) is then evaluated for every set of computed amplitudes and the set which minimizes the error is chosen. The RPE algorithm requires solving a set of M simultaneous linear equations D times. Typically, $M = 10$ and $D = 4$. The solution can be performed using Cholesky decomposition. This is the main computational load in the RPE algorithm.

Figure 3.22 shows the regular-pulse excitation signal and reconstructed speech for the speech segment shown in Figure 3.14-a without LTP, and Figure 3.23 shows the RPE excitation, the LPC excitation, and the reconstructed speech when LTP is deployed.

It should be noted that the multi-pulse algorithm needs fewer pulses than the RPE algorithm to achieve the same speech quality. This is because the pulse positions in the MPE algorithm are optimized, unlike the RPE where the positions are predefined. However, although fewer pulses are used in the MPE case, the pulse positions have to be quantized, while in the RPE case, only the position of the first pulse is quantized with 2 bits usually. Therefore, both the MPE and RPE approaches lead to similar bit rates for the same speech quality. The complexity of the RPE algorithm is higher than that of the MPE. This is because the RPE approach requires the solution of D $(M \times M)$ matrix equations (typically, $M = 10$ and $D = 4$). In later sections, we will look at some methods to reduce the complexity of the RPE algorithm. In the next section, we will examine the effect of different analysis parameters on the quality of the synthesized speech.

3.3.6   Evaluation of the RPE Algorithm

The effect of changing the coder parameters has already been studied with the multi-pulse approach. The same conclusions can be drawn in the RPE case. The only parameters which we will consider in this section are the pulse spacing D and the excitation search frame length N.

3.3.6.1   Pulse Spacing

Increasing the pulse spacing reduces the number of excitation pulses. Similar to the multi-pulse approach, the choice of the number of excitation pulses is a trade-off between quality and bit rate. Figure 3.24 shows the SEGSNR for different pulse rates from 1600 to 4000 pulses/s corresponding to pulse spacing from 5 down to 2. For speech coding at 9.6 kb/s, a good choice of the pulse spacing is $D = 4$ or 5. At this bit rate, the multi-pulse algorithm usually needs 3–4 pulses per 5 ms. The RPE approach normally gives slightly better quality than the MPE at the same bit rate, at the cost of more complexity. For 5 ms excitation frame and pulse spacing of $D = 4$, we have 10 excitation pulses every 40 samples, and to compute the optimum pulse amplitudes, the RPE algorithm requires solving a 10×10 matrix equation 4 times.

Using LTP gives 2 dB improvement in SNR when $D = 4$. At pulse spacing $D = 2$ (20 pulses per 40 samples) the LTP does not give any improvement as the number of pulses is large enough to model the LPC excitation.

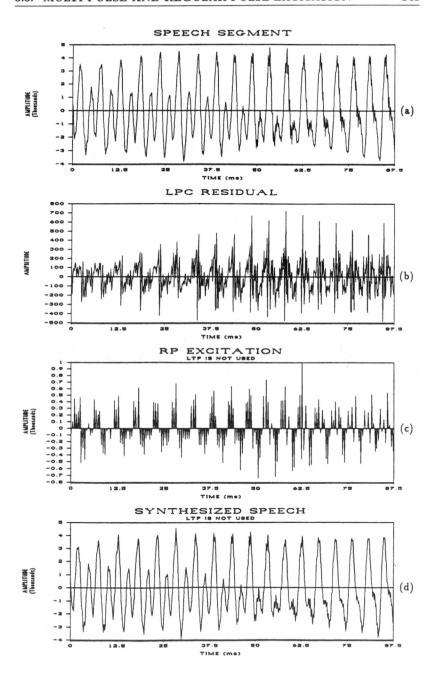

Figure 3.22: (a) 87.5 ms of voiced speech; (b) corresponding short-term prediction residual; (c) regular-pulse excitation ($D = 4$); (d) reconstructed speech.

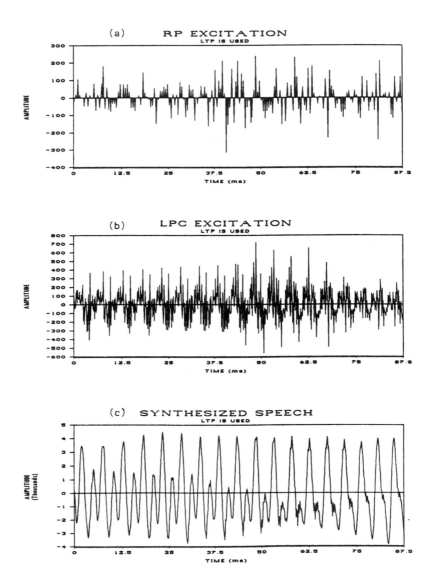

Figure 3.23: (a) regular-pulse excitation ($D = 4$); (b) LPC excitation; (c) reconstructed speech.

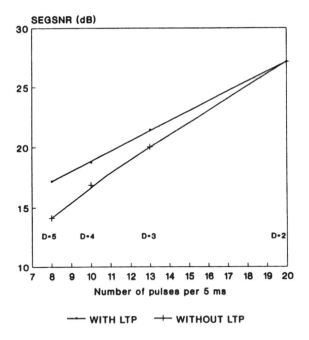

Figure 3.24: Variation of SNR with pulse rate ($D = 2, \ldots, 5$) for RPE without and with LTP inside the optimization loop.

3.3.6.2   Excitation Search Frame Length

In the MPE algorithm, we have seen that longer search frames give better results because the pulse positions would be optimized more efficiently. In the RPE algorithm, however, the opposite is true. Since the pulses are regularly spaced, the only factor which controls the positions is the initial phase, or the position of the first pulse, and this changes in every search frame. Therefore, shorter search frames give more flexibility in selecting the pulse positions (higher position updating rate). Figure 3.25 shows the SNR of the RPE with pulse spacing $D = 4$ for the search frame sizes of 20, 40, 52 and 80 (2.5 to 10 ms). The higher SNR is obtained at $N = 20$, and it is not considerably better than the other frame sizes. Another consideration in choosing the excitation frame size is the coder complexity. The longer the frame is, the more complex the coder becomes. For example, at LPC parameters frame of 160 and $D = 4$, if $N = 20$, we have 8 subframes, and we have to solve a 5×5 matrix equation 4 times in each frame. For $N = 80$, we have 2 subframes, and we have to solve a 20×20 matrix equation 4 times in each frame. Bearing in mind that solving an $M \times M$ matrix equation is proportional to M^3, then for $N = 20$, the complexity is proportional to 5^3 while for $N = 80$, it is proportional to 20^3. Therefore, shorter excitation

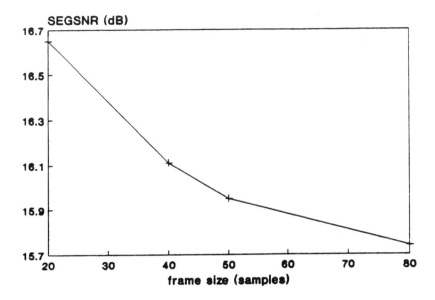

Figure 3.25: SEGSNR variation of the RPE codec with different excitation frame lengths.

frames means less complexity, but on the other hand, we need more bits to encode the positions of the first pulse.

3.3.7   Simplification of the RPE Algorithm

From the evaluation of the RPE algorithm, we have shown that it delivers slightly better quality than the MPE at the same bit rate. The disadvantage of the RPE is the high computational load arising from the necessity to solve 10 simultaneous linear equations every 5 ms (when the pulse spacing $D = 4$). A gross simplification can be achieved by using the autocorrelation method for defining the limits of the summation in computing $\phi(i, j)$. This will have a crucial role in simplifying the matrix in Equation (3.116) due to the regularity of the pulse spacing. We will discuss the autocorrelation approach in the next subsection. Further simplification is obtained by eliminating the matrix inversion in Equation (3.116) and by employing a fixed error weighting filter.

3.3.7.1   The Autocorrelation Approach

Recalling Equation (3.131) and using the autocorrelation approach, $\phi(i,j)$ can be substituted by

$$\phi(i,j) = \phi(|i-j|) = \sum_{n=|i-j|}^{N-1} h(n)h(n-|i-j|). \qquad (3.134)$$

Now, the elements of the matrix in Equation (3.116) are reduced to

$$\phi(m_i^{(k)}, m_j^{(k)}) = \phi(|m_i^{(k)} - m_j^{(k)}|). \qquad (3.135)$$

Using the relation in the definition of the pulse positions in Equation (3.133)

$$
\begin{aligned}
m_i^{(k)} - m_j^{(k)} &= k + iD - (k + jD), \\
&= (i-j)D, \qquad i,j = 0, \ldots, M-1. \qquad (3.136)
\end{aligned}
$$

Exploiting the result in Equation (3.136), Equation (3.116) becomes

$$
\begin{pmatrix}
\phi(0) & \phi(D) & \cdots & \phi([M-1]D) \\
\phi(D) & \phi(0) & \cdots & \phi([M-2]D) \\
\vdots & \vdots & \ddots & \vdots \\
\phi([M-1]D) & \phi([M-2]D) & \cdots & \phi(0)
\end{pmatrix}
\cdot
\begin{pmatrix}
\beta_0^{(k)} \\
\beta_1^{(k)} \\
\vdots \\
\beta_{M-1}^{(k)}
\end{pmatrix}
=
\begin{pmatrix}
\psi(m_0^{(k)}) \\
\psi(m_1^{(k)}) \\
\vdots \\
\psi(m_{M-1}^{(k)})
\end{pmatrix}.
\qquad (3.137)
$$

In Equation (3.137) two complexity reductions can be observed. Firstly, only M values of $\phi(i)$, $i = 0, D, 2D, \ldots, (M-1)D$, are computed. The total number of operations needed is $M(N-D)/2$. The second simplification is that the matrix in Equation (3.137) is independent of the initial phase, thus it is inverted only once every LPC frame rather than D times. Further, the matrix is Toeplitz and it can be solved more efficiently than the symmetric matrix in Equation (3.116) by the use of Levinson's algorithm.

3.3.7.2   Eliminating the Matrix Inversion

A closer look at the matrix of correlations in Equation (3.137) suggests that the matrix is strongly diagonal, where the off-diagonal elements become smaller farther away from the diagonal, i.e. $\phi(0) > |\phi(D)| > \cdots > |\phi([M-1]D)|$. If $g(n)$ is the impulse response of the synthesis filter $1/A(z)$ then it

Autocorrelation

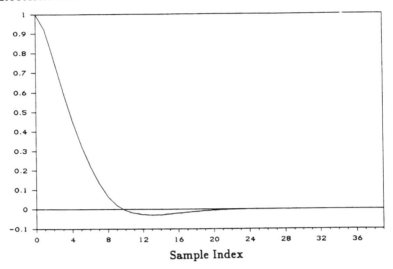

Sample Index

Figure 3.26: Autocorrelation of the typical weighted synthesis filter's impulse response.

is related to $h(n)$, the impulse response of $1/A(z/\gamma)$, by

$$h(n) = \gamma^n g(n). \qquad (3.138)$$

The impulse response $g(n)$ is already a decaying function, and the presence of the factor γ^n in Equation (3.138) causes $h(n)$ to decay even faster as γ is less than 1. Figure 3.26 shows an example of the autocorrelation of the impulse response $h(n)$. For a spacing $D = 4$, the diagonals of the autocorrelation matrix are equal to $\phi(0)$, $\phi(4)$, $\phi(8)$, ..., $\phi(4[M-1])$.

If all the off-diagonal elements of the autocorrelation matrix are set to zero, the matrix is reduced to $\phi(0)\mathbf{I}$, where $\mathbf{I}$ is the identity matrix. In this case Equation (3.137) becomes

$$\beta_i^{(k)} = \frac{1}{\phi(0)} \psi(m_i^{(k)}), \qquad i = 0, \ldots, M - 1. \qquad (3.139)$$

Recall Equation (3.114) for computing $\psi(n)$

$$\psi(n) = \sum_{i=n}^{N-1} x(i)h(i - n) = x(n) * h(-n), \qquad (3.140)$$

where, from Equation (3.107), and assuming that $\alpha \geq N$

$$x(n) = s_w(n) - Gu(n - \alpha) * h(n) - \hat{s}_0(n), \qquad (3.141)$$

and $\hat{s}_0(n)$ is the zero-input response of the weighted synthesis filter $1/A(\frac{z}{\gamma})$ in the lower branch of Figure 3.12. From Equations (3.94) and (3.141), $x(n)$ can be expressed as:

$$x(n) = r(n) * h(n) - Gu(n - \alpha) * h(n) + s_0(n) - \hat{s}_0(n). \qquad (3.142)$$

If we assume that the zero-input responses of the weighted synthesis filters $W(z)$ in both branches of Figure 3.12 are equal, then Equation (3.142) is reduced to

$$
\begin{aligned}
x(n) &= [r(n) - Gu(n - \alpha)] * h(n) \\
&= d(n) * h(n).
\end{aligned}
\qquad (3.143)
$$

where

$$d(n) = r(n) - Gu(n - \alpha). \qquad (3.144)$$

The signal $d(n)$ can be viewed as the residual after both short-term and long-term prediction. Using the result of Equation (3.143), $\psi(n)$ in (3.140) can be now written as

$$\psi(n) = d(n) * h(n) * h(-n). \qquad (3.145)$$

Note that

$$\phi(n) = \sum_{i=n}^{N-1} h(i)h(n - i) = h(n) * h(-n). \qquad (3.146)$$

Therefore

$$
\begin{aligned}
\psi(n) &= d(n) * \phi(n) \qquad &(3.147) \\
&= \sum_{i=0}^{N-1} d(i)\phi(|n - i|). \qquad &(3.148)
\end{aligned}
$$

Note that $\phi(n) = \phi(-n)$, $n = -(N - 1), \ldots, N - 1$. Let us define

$$z(n) = \frac{\phi(n)}{\phi(0)}, \qquad n = -(N - 1), \ldots, N - 1, \qquad (3.149)$$

to be the normalized autocorrelation of the impulse response $h(n)$. $z(n)$ is a double sided symmetric function where $z(n) = z(-n)$. The pulse amplitudes in Equation (3.139) are now given by

$$\beta_i^{(k)} = \frac{\psi(m_i^{(k)})}{\phi(0)} = d(n) * z(n), \qquad \text{at} \quad n = m_i^{(k)} \qquad (3.150)$$

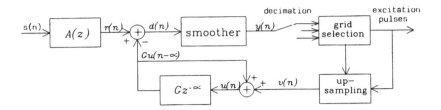

Figure 3.27: Schematic diagram of the simplified RPE structure.

and the mean squared weighted error of Equation (3.119) can be now written as

$$
\begin{aligned}
E^{(k)} &= \sum_{n=0}^{N-1} x^2(n) - \sum_{i=0}^{M-1} \beta_i^{(k)} \psi(m_i^{(k)}) \\
&= \sum_{n=0}^{N-1} x^2(n) - \phi(0) \sum_{i=0}^{M-1} [\beta_i^{(k)}]^2. \quad k = 0, \dots, D-1.
\end{aligned}
$$

$$(3.151)$$

Equations (3.150) and (3.151) are the key equations of an efficient and simple method for RPE coding without having solve the set of M linear equations in (3.137). The coder structure of this simplified RPE method is shown in the schematic diagram of Figure 3.27. The method can be described as follows: the short-term prediction residual $r(n)$ is obtained by inverse filtering the original speech through $A(z)$. The residual after long-term prediction, $d(n)$, is formed by subtracting from $r(n)$ its estimated value $Gu(n - \alpha)$ (the previously quantized excitation) as in (3.144). The LTP residual $d(n)$ is convolved with the smoothing function $z(n)$. The smoothed LTP residual is given by

$$
\begin{aligned}
y(n) &= d(n) * z(n) \\
&= \sum_{i=0}^{N-1} d(i)z(|n-i|).
\end{aligned}
$$

$$(3.152)$$

Now, the smoothed LTP residual $y(n)$ is decomposed into D sets of M amplitudes given by

$$
\{\beta_i^{(k)}\} = \{y(m_i^{(k)})\}, \qquad k = 0, \dots, D-1. \tag{3.153}
$$

The energy $T^{(k)}$ is then computed for every set by

$$
T^{(k)} = \sum_{i=0}^{M-1} [\beta_i^{(k)}]^2. \tag{3.154}
$$

According to Equation (3.151), the set $\{\beta_i^{(k)}\}$ which has maximum energy minimizes the error, thus is chosen to be the excitation signal, where the first pulse β_0 starts at position k and the pulses are separated by a distance D.

As we discussed earlier in this section, the autocorrelation value $\phi(n)$ drops very significantly as n increases. Therefore $z(n)$ can be truncated at $|n| = Q$, where $Q \ll N$, to further reduce the number of terms in the summation of Equation (3.152). This simplified RPE method is detailed in Algorithm 3.1. The closed analysis-by-synthesis loop is broken in this simplified RPE approach. The LTP parameters are determined by minimizing the mean-squared error

$$E = \sum_{n=0}^{N-1} (r(n) - Gu(n - \alpha))^2, \qquad (3.155)$$

and α is limited to be larger than $N - 1$. From the definition of the smoothing function $z(n)$ in Equation (3.149), it is the time-varying normalized autocorrelation of the impulse response of the weighted synthesis filter. Further simplification is obtained when the smoothing function is made fixed. From Figure 3.27 the similarity between the simplified RPE structure and the RELP structure is evident. It is natural therefore to choose the fixed smoothing function as a low-pass filter with cut-off frequency $f_s/(2D)$ where D is the decimation factor. Notice that the smoothing function $z(n)$, where $z(n) = z(-n)$, $n = -Q, \ldots, Q$, is noncausal. A proper approach for designing the low-pass filter is to use a windowed sinc function. For a cut-off frequency $f_s/(2D)$ and a Hamming windowed sinc function the coefficients of the fixed smoothing function are given by [79]

$$z(n) = \frac{D}{n\pi} \sin\left(\frac{n\pi}{D}\right) (0.54 + 0.46 \cos\left(\frac{n\pi}{Q}\right)), \qquad -Q \le n \le Q. \quad (3.156)$$

Usually Q is less than 10. If $z(n)$, $-Q \le n \le Q$, is shifted to the right by Q positions, an FIR filter $f(n)$, $0 \le n \le 2Q$, is obtained. The resulting FIR filter $f(n)$ is a linear phase filter where $f(n) = f(2Q - n)$. The relation between the double-sided smoother $z(n)$ and the FIR filter $f(n)$ is given by

$$f(n) = z(n - Q), \qquad n = 0, \ldots, 2Q. \qquad (3.157)$$

The smoothed residual is given by

$$y(n) = d(n) * z(n) = \sum_{i=-Q}^{Q} z(i)d(n - i), \qquad (n - i) \ge 0, \qquad (3.158)$$

Algorithm 3.1 (simplified RPE) *This algorithm determines M RPE pulses with a simplified method which does not need solving a set of M linear equations. The M pulses β_i are regularly spaced by a distance D in an excitation frame of length N, with the first pulse positioned at k_0.*

1) Compute the short-term prediction residual
$$r(n) = s(n) - \sum_{i=1}^{p} a_i s(n-i).$$

2) Determine the LTP parameters.

3) Compute the LTP residual.
$$d(n) = r(n) - Gu(n-\alpha).$$

4) Compute the smoothed LTP residual $y(n)$
$$y(n) = \sum_{i=0}^{N-1} d(i)z(|n-i|), \quad |n-i| \le Q.$$

5) Decompose $y(n)$ into D sets and compute the energy of every set
$$\text{for } k = 0 \text{ to } D-1 \text{ do}$$
$$\beta_i^{(k)} = y(k+iD), \ i = 0, \ldots, M-1$$
$$E^{(k)} = \sum_{i=0}^{M-1} [\beta_i^{(k)}]^2$$

6) Choose the set of pulses with maximum energy
$$\text{ref} = 0$$
$$\text{for } k = 0 \text{ to } D-1 \text{ do}$$
$$\text{if } E^{(k)} > \text{ref then}$$
$$\text{ref} = E^{(k)}$$
$$\beta_i = \beta_i^{(k)}, \ i = 0, \ldots, M-1$$
$$k_0 = k$$

or, equivalently,

$$
\begin{aligned}
y(n) &= d(n) * f(n+Q), \\
&= \sum_{i=-Q}^{Q} f(i+Q)d(n-i), \qquad n = 0, \ldots, N-1, \qquad n \ge i, \\
&= \sum_{i=0}^{2Q} f(i)d(n+Q-i), \quad n = 0, \ldots, N-1. \quad (n+Q) \ge i,
\end{aligned}
$$

$$(3.159)$$

Note that when convolving the segment $d(n)$ of length N with the FIR filter $f(n)$ of length $2Q+1$, the number of resulting output samples is $N+2Q$. The N samples resulting from the convolution in (3.159) are the central samples of the convolution $d(n) * f(n)$. Equation (3.159) suggests, therefore, that when the FIR filter is used as a smoother, block filtering is deployed where the first Q samples of the output are discarded.   An FIR low-pass filter with 11 taps is given in [80] for a pulse spacing of $D = 3$. The smoother is a low-pass filter with a cut-off frequency at $1333\,\text{Hz}$ ($D = 3$) and the taps are given by

$$z(\pm 5) = f(0) = f(10) = -0.016356$$
$$z(\pm 4) = f(1) = f(9) = -0.045649$$
$$z(\pm 3) = f(2) = f(8) = 0$$
$$z(\pm 2) = f(3) = f(7) = 0.250793$$
$$z(\pm 1) = f(4) = f(6) = 0.70079$$
$$z(0) = f(5) = 1.$$

With decimation $D = 4$ the filters cut-off frequency is $1000\,\text{Hz}$. The coefficients of a Hamming windowed low-pass filter at this cut-off frequency, and with $Q = 7$, are given by

$$z(0) = 1$$
$$z(\pm 1) = 0.859303$$
$$z(\pm 2) = 0.5263605$$
$$z(\pm 3) = 0.1927755$$
$$z(\pm 4) = 0$$
$$z(\pm 5) = -0.045591$$
$$z(\pm 6) = -0.0319721$$
$$z(\pm 7) = -0.0102893.$$

Using the fixed low-pass filter as a smoothing function has given better results than using the changing smoothing function which is equal to the normalized autocorrelation of the weighted synthesis filter as in Equation (3.149). This is due to the gross simplification in deriving the structure with changing smoother where the off-diagonal elements of the autocorrelation matrix were set to zero. Using a low-pass filter at cut-off frequency $f_s/(2D)$ is more sensible as the simplified structure bears a close similarity with the baseband coder, or residual excited linear predictive coder (RELP) [23]. In RELP coders the LPC residual is low-pass filtered, decimated, and the extracted baseband residual is quantized and used to excite the LPC synthesis filter after using interpolation to recover the full band residual (regenerating the residual high frequencies is usually accomplished by spectral folding where zeros are inserted between the baseband samples). The main advantage of the RPE over the RELP is its flexibility in choosing the position of the first excitation pulse, which produces a more appropriate excitation signal.   Another difference is the presence of the pitch predictor in the RPE (although pitch prediction has been suggested in RELP coders to reduce the effect of tonal noise [81]). Finally, when the residual signal is smoothed with an FIR low-pass filter of length $2Q + 1$ the residual subframes are smoothed individually (there is no continuous filtering of the residual) using block filtering where only the central N samples of the resulting $N+2Q$ samples are considered. With these advantages, the simplified RPE delivers better speech quality than the RELP, but it is still inferior to the original RPE because the analysis-by-synthesis optimization loop is broken in the simplified structure. The number of operations (add/multiply) needed to determine the optimum excitation is $2QN$ for the convolution (smoothing) and DM for the energy computation. For the GSM coder [82], $N = 40$, $Q = 5$ (11-tap), $D = 3$ and $M = 13$. The total number of operations needed in this case is 11 operations per speech sample. This illustrates the simplicity of the RPE coder with fixed smoothing filter. In fact, the coder simplicity was a decisive factor in choosing this coder for the pan-European digital mobile radio system. Figures 3.28 (a) and (b) show the SEGSNR obtained by the covariance, correlation and simplified RPE approaches described earlier for pulse spacings between $D = 2$ and $D = 5$ with and without LTP, respectively.   At $D = 4$ the SNR using the simplified RPE structure is 2.5 dB less than the covariance (original) approach, 2 dB less than the autocorrelation approach, and it is similar to that of the MPE with 4 pulses. For decimation values of 4 or 5, using the autocorrelation approach is a good choice. Some degradation in speech quality results when the simplified structure is used with the great advantage of significantly reducing the coder complexity. Another conclusion is that using a decimation factor of $D = 3$ does not give any improvement over using $D = 4$. Therefore it is neither necessary nor desirable to deploy this lower decimation factor, as it would inevitably increase the transmission bit rate.

3.3.8   Quantization of the Excitation in MPE and RPE Coders

In MPE and RPE coders, the excitation is described by the pulse amplitudes and pulse positions. Specifically, in RPE codecs the pulse positions are defined by the initial phase, or the position of the first pulse, as the pulses are regularly spaced. A decimation factor of 3 or 4 is usually used, and the initial phase is quantized with 2 bits in this case. The RPE pulse amplitudes are quantized using adaptive block quantization. The M pulses in a subframe are scaled by their rms value, or maximum value. The histograms of the maximum pulse and rms value of the pulses are shown in Figure 3.29 (a) and (b), respectively. In this case $D = 4$, $N = 40$, and $M = 10$. It is clear from the histograms that the scaling value cannot be efficiently quantized using a uniform quantiser. The scaling value is quantized either logarithmically or by using nonuniform quantisers. The histograms of the logarithms of the maximum pulse and the pulses' rms value are shown in Figure 3.30 (a) and (b), respectively.   It is clear that uniform quantization of the logarithm is adequate. When nonuniform quantization is used the quantization and decision levels are usually designed from a training data set using a Lloyd-Max quantiser [83, 84]. The scaling

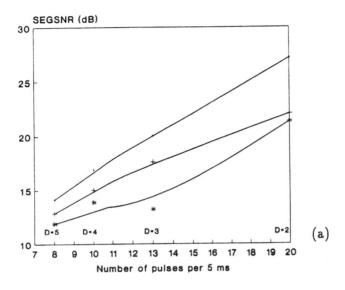

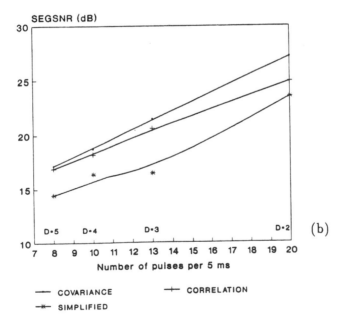

Figure 3.28: SEGSNR for different RPE approaches (a) with LTP; (b) without LTP.

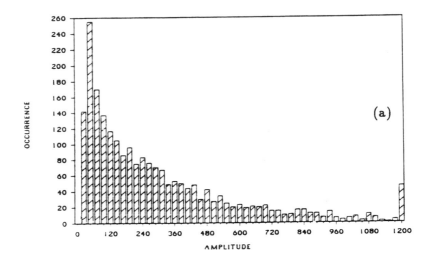

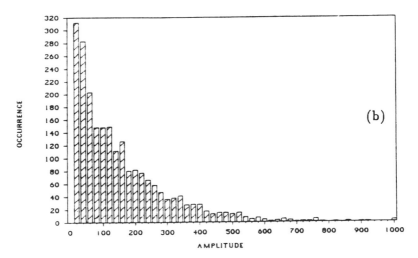

Figure 3.29: (a) Histogram of the maximum RPE pulse; (b) Histogram of the
rms value of the RPE pulses.

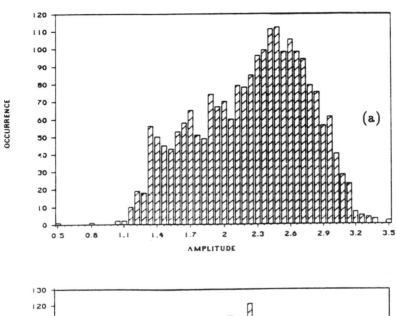

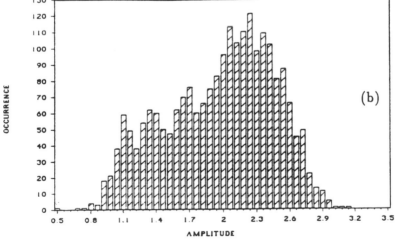

Figure 3.30: (a) Histogram of the logarithm of the maximum RPE pulse; (b) Histogram of the logarithm of the RPE pulses' rms.

value is quantized with 5 or 6 bits. Using fewer bits results in noticeable
degradation in speech quality. The histograms of the RPE pulses scaled by
their maximum value and by their rms value are shown in Figure 3.31 (a)
and (b), respectively. The normalized pulses are adequately quantized with
3 bits using nonuniform quantization. For a subframe of length 40 (5 ms)
and 10 RPE pulses, the number of bits needed to quantise the excitation
is: 2 for the initial phase, 6 for the scaling value, and 30 for the normalized
pulses. The bit rate associated with the excitation in this case is 7.6 kb/s.
Reducing the bit rate can be achieved by reducing the number of pulses
(increasing the decimation factor) and/or reducing the bits needed to quantise the pulses. Increasing the subframe length also reduces the bit rate as
the scaling factor is updated less frequently. Using a subframe of $N = 60$
(7.5 ms) and $D = 5$ (12 pulses), the bits needed are: 2 for the initial phase
(by limiting it to 4 positions rather than 5), 6 for the maximum pulse, and
36 for the normalized pulses. The excitation bit rate is reduced to 6 kb/s
in this case. Taking into account the bit rate needed for LTP and LPC
parameters, it is difficult to maintain high quality speech below 9.6 kb/s
using the RPE.

In the case of MPE a lower number of pulses is needed to obtain the
same speech quality as compared to the RPE, since the pulse positions are
also optimized. However, the major bit rate contribution in quantizing the
excitation is allocated to the pulse positions. If M pulses are allocated in an
excitation frame of length N then the total number of position combinations
is

$$L_p = \left(\begin{array}{c} N \\ M \end{array} \right) = \frac{N!}{(N - M)!M!}. \tag{3.160}$$

The minimum number of bits needed to quantise M pulse positions is

$$N_{\text{bit}} = \log_2 L_p. \tag{3.161}$$

For the typical values $N = 40$ and $M = 4$ at least 17 bits are needed to
quantise the positions of the 4 pulses. In this example if the positions are
separately quantized 6 bits are needed for each pulse which results in a
total of 24 bits. Using differential encoding the pulses can be quantized
each with 5 bits for this specific example. The pulses are reordered such
that $m_0 < m_1 < m_2 < m_3$ and the quantized quantities are m_0, $m_1 - m_0$,
$m_2 - m_1$, and $m_3 - m_2$. In this case the distance between adjacent pulses is
restricted to be less than 33, and the positions of the four pulses are encoded
with 20 bits.   The most efficient method, which needs the minimum
number of bits (N_{bit}) to quantise the pulse positions, is a combinatorial
coding scheme [75]. The scheme is given in [85] and is referred to as an
enumerative source coding technique. The total number of pulse positions
L_p in Equation (3.160) can be represented by an imaginary list from 0
to $L_p - 1$, and the value of the index corresponds to the required pulse
positions. The encoding is done by scanning the excitation sequence and

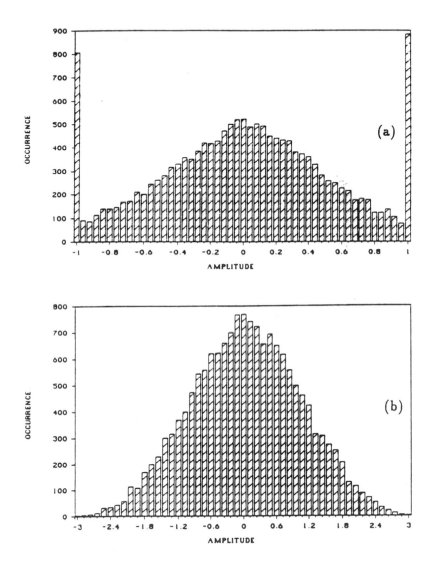

Figure 3.31: Histogram of the normalized RPE pulses (a) scaled by the maximum pulse; (b) by the rms value of the pulses.

incrementing the index every time a pulse is encountered, where the final value of the index is sent. The increment is given by

$$
\begin{aligned}
\mathcal{I} &= \binom{n}{m} = \frac{n!}{(n-m)!m!} \quad \text{for } n \geq m \\
&= 0 \qquad\qquad\qquad\quad \text{for } n < m,
\end{aligned}
\tag{3.162}
$$

where n is the remaining number of samples and m is the number of pulses yet to be encoded including the current pulse. The encoding and decoding procedures are described in [85].

Similar to the RPE the pulse amplitudes in MPE are quantized using adaptive block quantization. The pulses are scaled by their rms value, or maximum value, which is quantized with 5–6 bits, and the normalized pulses are quantized with 3–4 bits each. Figure 3.32 (a) shows the histogram of the rms value of the pulses while Figure 3.32 (b) shows the histogram of its logarithm. The histogram of the MPE pulses, normalized by their rms value is shown in Figure 3.33. The normalized pulses are adequately quantized with 3 bits using nonuniform quantization, while the scaling gain is quantized with 5 or 6 bits. Using 4 pulses in an excitation frame of 40 samples, the bits needed are: 6 for the scaling value, 12 bits for the normalized pulses, and 17 bits for the positions using the combinatorial coding scheme. The bit rate needed to quantise the excitation in this case is 7 kb/s. When pitch prediction is used, fewer pulses are needed to represent the excitation signal. Using 3 pulses in a 60 samples subframe reduces the excitation bit rate to 4 kb/s. More efficient approaches can also be used to quantise the amplitudes of the excitation pulses, however, the overall bit rate is not significantly reduced as the quantization of the pulse positions reserves a substantial proportion of the total bit rate. Soheili et al. [86] described several adaptation schemes in quantizing the pulse magnitudes which do not need side information. They found that the most promising adaptation technique is to scale the pulses by the pitch filter memory energy. In this way the scaling gain, which was previously quantized with 5 or 6 bits, is not transmitted and this reduces the bit rate by about 1 kb/s.

3.4   Code-Excited Linear Prediction

There is currently a high demand for speech coding techniques which are able to produce high quality speech at bit rates below 8 kb/s. Since the full rate GSM speech codec recommendation has been finalized, there has been an intensive research activity devoted to half-rate codecs around 6.5 kb/s encoding rates. The MPE and RPE coders discussed in the previous section can be used to produce good quality speech at bit rates as low as 9.6 kb/s. When the bit rate is reduced below 9.6 kb/s, the MPE and RPE fail in maintaining good speech quality. This is due to the large number

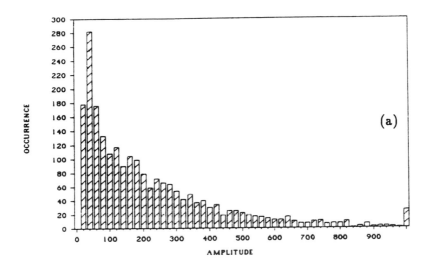

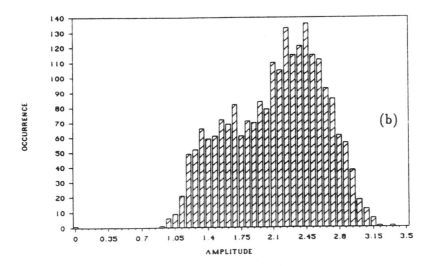

Figure 3.32: (a) Histogram of the rms value of the MPE pulses; (b) Histogram of the logarithm of the MPE pulses' rms value.

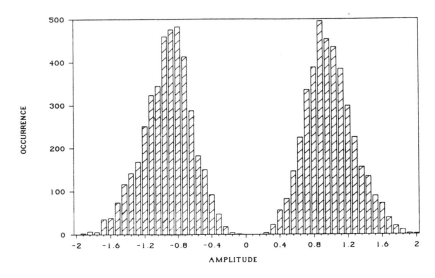

Figure 3.33: Histogram of normalized MPE pulses scaled by the rms value.

of bits needed to encode the excitation pulses, and the quality is deteriorated when these pulses are coarsely quantized, or when their number is reduced, to reduce the bit rate. Therefore, if the analysis-by-synthesis structure is to be used for producing good quality at bit rates below 8 kb/s, more subtle approaches have to be used for defining the excitation signal. The implementation of a long-term predictor in the analysis-by-synthesis loop becomes of prominent importance to remove the redundancy of the speech as much as possible. The residual signal after short-term and long-term prediction becomes noise-like, and it is assumed that the residual can be modelled by a zero-mean Gaussian process with slowly varying power spectrum. This is the key point in implementing stochastic coders, where the excitation frame is vector quantized using a large stochastic codebook. Stochastic coding, or code-excited linear prediction (CELP) coding was first introduced by Atal and Schroeder in 1984 [10, 11]. A similar approach was proposed by Copperi and Sereno in 1985 [87]. In the CELP approach, a 5 ms (40 samples) excitation frame is modelled by a Gaussian vector chosen from a large Gaussian codebook by minimizing the perceptually weighted error between the original and synthesized speech. Usually, a codebook of 1024 entries is needed, and the optimum innovation sequence is chosen by the exhaustive search of the codebook. Using the CELP approach, an excitation frame (5–7.5 ms) can be encoded with 15 bits only (10 bits for the book address and 5 bits for the scaling gain). This illustrates the dramatic reduction in bit rate compared, for example, to the GSM RPE-LTP codec where 47 bits are needed to encode the same excitation. However, until recently, the high complexity of the CELP algorithm hindered its real time implementation. The complexity comes from the exhaustive search of the excitation codebook, where the weighted synthesized speech has to be computed for all possible codebook entries and compared with the weighted original speech.   In the last few years, research activity has been focused on reducing the complexity of the originally proposed CELP coder and achieving its real time implementation using the current DSP technology. Significant simplification of the CELP encoder has been achieved by using sparse excitation codebooks [88,89], or centre-clipped codebooks [90], in which most of the excitation samples in the Gaussian random vector are set to zero. Another significant simplification in the codebook structure is to use ternary excitation codebooks where the elements of the excitation vector are set to -1, 0, or 1 [90,91]. Overlapping sparse or ternary codebooks [90] have also been used to reduce the computational complexity and storage requirements in CELP systems. Binary pulse codebooks [93,94] are ternary codebooks whereby the excitation pulses are regularly spaced. This specific structure allows for efficient nonexhaustive search procedures to be used. Another efficient way of defining the excitation signal is using the vector sum excitation (VSELP) [95], where the excitation vector is a linear combination of a number of basis vectors weighted with -1 or 1. This special excitation structure yields a significant reduction in the computation needed to identify the optimum excitation vector. An 8 kb/s VSELP coder was recently selected for the future American digital mobile radio system. Algebraic codebooks have also been utilized to reduce the CELP complexity [97] where the codebook is generated using special binary error-correcting codes. Another simplified approach has been proposed [98] in which a CELP system operates on the baseband of the LPC residual signal. The structure of this coder is similar to the GSM RPE coder, and the bit rate reduction is achieved by vector quantizing the smoothed residual with a CELP codebook. The self-excitation concept [99] (or backward excitation recovery [100]) has also been used for producing high quality speech at bit rates below 6.4 kb/s. In this approach, the excitation is obtained by searching through the past excitation signal, and the segment which minimizes the perceptually weighted error between the original and synthesized speech is chosen. The self-excited LPC can be seen as another variant of the CELP in which the codebook is changing, and it has the advantage of less computational demand and less storage requirement with the disadvantage of lacking robustness over noisy transmission environments.

In this section we describe the CELP coder and the different approaches used for generating the excitation codebook. In the next section we give a detailed description to the CELP encoding algorithm, we then discuss methods to reduce the complexity of the codebook search procedure and we discuss the use of sparse excitation, ternary excitation, overlapping codebooks, algebraic codebooks, and binary pulse excitation.

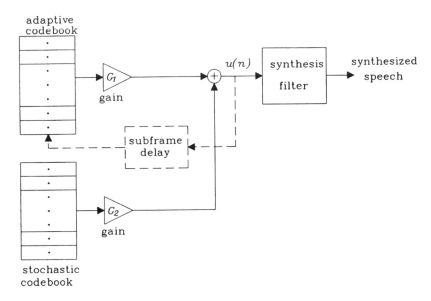

Figure 3.34: Schematic diagram of the CELP synthesis model.

3.4.1   CELP Principle

After short-term prediction and long-term prediction of the speech signal, the redundancies in the speech signal are almost removed, and the residual signal has very little correlation. A Gaussian process with slowly varying power spectrum can be used to represent the residual signal, and the speech waveform is generated by filtering white Gaussian innovation sequences through the time-varying linear long-term and short-term synthesis filters. The optimum innovation sequence is selected from a codebook of random white Gaussian sequences by minimizing the subjectively weighted error between the original and the synthesized speech. The schematic diagram of the CELP synthesis model is shown in Figure 3.34. The pitch correlation filter is replaced here by an adaptive overlapping codebook as was discussed in Section 3.2.2.1. The address selected from the adaptive codebook and the corresponding gain (the pitch delay and gain) along with the address selected from the stochastic codebook and the corresponding scaling gain are sent to the decoder, which uses the same codebooks (in the absence of channel errors) to determine the excitation signal at the input of the LPC synthesis filter to produce the synthesized speech.

The excitation codebook contains L codewords (stochastic vectors) of length N samples (typically $L = 1024$ and $N = 40$ corresponding to a 5 ms excitation frame). The excitation signal of a speech frame of length N is chosen by the exhaustive search of the codebook after scaling the Gaussian vectors by a gain factor β.

The filter $W(z)$ is the weighted synthesis filter given by

$$W(z) = \frac{1}{A(z/\gamma)} = \frac{1}{1 - \sum_{k=1}^{P} a_k \gamma^k z^{-k}}. \tag{3.163}$$

Having determined the adaptive codebook parameters (pitch delay and gain) as was described in Section 3.2.2.1, the weighted synthesized speech can be written as

$$\hat{s}_w(n) = \beta c_k(n) * h(n) + G y_\alpha(n) + \hat{s}_0(n), \tag{3.164}$$

where the convolution is memoryless, $c_k(n)$ is the excitation codeword at index k, β is a scaling factor, $h(n)$ is the impulse response of the weighted synthesis filter $W(z)$, $\hat{s}_0(n)$ is the zero input response of the weighted synthesis filter, G is the adaptive codebook gain and $y_\alpha(n) = c'_\alpha(n) * h(n)$ is the zero-state response of the weighted synthesis filter to the codeword $c'_\alpha(n)$ selected from the adaptive codebook.

The weighted error between the original and synthesized speech is given by

$$\begin{aligned} e_w(n) &= s_w(n) - \hat{s}_w(n) \\ &= x(n) - \beta c_k(n) * h(n), \end{aligned} \tag{3.165}$$

where

$$x(n) = s_w(n) - G y_\alpha(n) - \hat{s}_0(n). \tag{3.166}$$

The signal $x(n)$ is computed by updating $x'(n)$ of Equation (3.65), that is

$$x(n) = x'(n) - G y_\alpha(n), \tag{3.167}$$

as $x'(n)$ has already been determined while searching the adaptive codebook.

The mean squared weighted error is given by

$$E = \sum_{n=0}^{N-1} [e_w(n)]^2 = \sum_{n=0}^{N-1} [x(n) - \beta c_k(n) * h(n)]^2. \tag{3.168}$$

Setting $\partial E / \partial \beta = 0$, we get

$$\beta = \frac{\sum_{n=0}^{N-1} x(n)[c_k(n) * h(n)]}{\sum_{n=0}^{N-1} [c_k(n) * h(n)]^2}, \tag{3.169}$$

and substituting β in Equation (3.168) gives

$$E = \sum_{n=0}^{N-1} x^2(n) - \frac{\left[\sum_{n=0}^{N-1} x(n)[c_k(n) * h(n)]\right]^2}{\sum_{n=0}^{N-1} [c_k(n) * h(n)]^2} \tag{3.170}$$

Equations (3.169) and (3.170) can be written in matrix form as

$$\beta = \frac{\mathbf{x}^T \mathbf{H} \mathbf{c}_k}{\mathbf{c}_k^T \mathbf{H}^T \mathbf{H} \mathbf{c}_k} \tag{3.171}$$

and

$$
\begin{aligned}
E &= \|\mathbf{x} - \beta \mathbf{H} \mathbf{c}_k\|^2 \\
&= \mathbf{x}^T \mathbf{x} - \frac{\left(\mathbf{x}^T \mathbf{H} \mathbf{c}_k\right)^2}{\mathbf{c}_k^T \mathbf{H}^T \mathbf{H} \mathbf{c}_k},
\end{aligned}
\tag{3.172}
$$

where $\mathbf{x}$ and $\mathbf{c}_k$ are N-dimensional vectors given by

$$\mathbf{x}^T = \begin{pmatrix} x_0 & x_1 & \ldots & x_{N-1} \end{pmatrix} \tag{3.173}$$

$$\mathbf{c}^T = \begin{pmatrix} c_0 & c_1 & \ldots & c_{N-1} \end{pmatrix} \tag{3.174}$$

and $\mathbf{H}$ is a lower triangular convolution matrix of the impulse response $h(n)$ given by

$$
\mathbf{H} = \begin{pmatrix}
h_0 & 0 & 0 & \ldots & 0 \\
h_1 & h_0 & 0 & \ldots & 0 \\
h_2 & h_1 & h_0 & \ldots & 0 \\
\vdots & \vdots & \vdots & \ddots & \vdots \\
h_{N-1} & h_{N-2} & h_{N-3} & \ldots & h_0
\end{pmatrix}.
\tag{3.175}
$$

Let

$$\mathbf{\Phi} = \mathbf{H}^T \mathbf{H} \tag{3.176}$$

then $\mathbf{\Phi}$ is a symmetric matrix containing the correlations of the impulse response $h(n)$ given by

$$\phi(i,j) = \sum_{n=\max(i,j)}^{N-1} h(n-i)h(n-j), \qquad i,j = 0, \ldots, N-1. \tag{3.177}$$

and let

$$\mathbf{\Psi}^T = \mathbf{x}^T \mathbf{H} \tag{3.178}$$

be a vector with elements

$$\psi(i) = x(i) * h(-i) = \sum_{n=i}^{N-1} x(n)h(n-i), \qquad i = 0, \ldots, N-1. \tag{3.179}$$

The mean squared weighted error can now be minimized by maximizing the second term of Equation (3.172), which is given by

$$\mathcal{T}_k = \frac{(\mathcal{C}_k)^2}{\mathcal{E}_k} = \frac{\left(\mathbf{x}^T \mathbf{H} \mathbf{c}_k\right)^2}{\mathbf{c}_k^T \mathbf{H}^T \mathbf{H} \mathbf{c}_k} = \frac{\left(\mathbf{\Psi}^T \mathbf{c}_k\right)^2}{\mathbf{c}_k^T \mathbf{\Phi} \mathbf{c}_k} \tag{3.180}$$

where $\mathcal{C}_k$ is the cross-correlation between $\mathbf{x}$ and the filtered codeword $\mathbf{H} \mathbf{c}_k$ and it is given by

$$\mathcal{C}_k = \sum_{n=0}^{N-1} x(n) \left[c_k(n) * h(n) \right] = \sum_{n=0}^{N-1} \psi(n) c_k(n) \tag{3.181}$$

and $\mathcal{E}_k$ is the energy of the filtered codeword c_k and it is given by

$$\begin{aligned} \mathcal{E}_k &= \sum_{n=0}^{N-1} [c_k(n) * h(n)]^2 \\ &= \sum_{i=0}^{N-1} c_k^2(i) \phi(i,i) + 2 \sum_{i=0}^{N-2} \sum_{j=i+1}^{N-1} c_k(i) c_k(j) \phi(i,j). \end{aligned} \tag{3.182}$$

$\psi(i)$ and $\phi(i,j)$ are computed outside the optimization loop, and the term $\mathcal{T}_k$ in Equation (3.180) is evaluated for $k = 0$ to $L - 1$, where L is the codebook size. The codeword with index k which maximizes this term is chosen, and the scalar gain β is then computed from Equation (3.171). In this approach the codeword $c_k(n)$ and the gain β are not jointly optimized since the gain has to be quantized, and the term in Equation (3.180) has been derived using the value of the unquantized gain. The gain and the excitation vector can be jointly optimized as follows: for the codeword with index k the cross-correlation $\mathcal{C}_k$ and the energy $\mathcal{E}_k$ are computed from Equations (3.181) and (3.182), respectively. The gain is computed as in Equation (3.171) by

$$\beta_k = \frac{\mathcal{C}_k}{\mathcal{E}_k}. \tag{3.183}$$

The gain is then quantized to obtain the value $\hat{\beta}_k$, and this quantized value is substituted in Equation (3.168) to obtain the minimum error

$$\begin{aligned} E &= \sum_{n=0}^{N-1} [x(n) - \hat{\beta}_k c_k(n) * h(n)]^2 \\ &= \mathbf{x}^T \mathbf{x} - 2\hat{\beta}_k \mathbf{x}^T \mathbf{H} \mathbf{c}_k + \hat{\beta}_k^2 \mathbf{c}_k^T \mathbf{H}^T \mathbf{H} \mathbf{c}_k \\ &= \mathbf{x}^T \mathbf{x} - 2\hat{\beta}_k \mathcal{C}_k + \hat{\beta}_k^2 \mathcal{E}_k \end{aligned} \tag{3.184}$$

Thus the term to be maximized is now given by

$$\mathcal{T}_k = \hat{\beta}_k (2\mathcal{C}_k - \hat{\beta}_k \mathcal{E}_k). \tag{3.185}$$

This term is computed for every codeword and the one which maximizes the term is chosen along with the quantized gain. This joint optimization approach does not introduce any considerable complexity as the correlation C and the energy $\mathcal{E}$ are computed once for every codeword similar to the case when Equation (3.180) is used. The extra computational load is that the the gain has to be quantized for every possible codeword.

The number of instructions needed to evaluate the expression in Equation (3.185) is approximately N^2 (when Equations (3.181) and (3.182) are used to compute C and $\mathcal{E}$). For a codebook with 1024 entries and an excitation frame of length 40 samples, around 40000 multiplications per speech sample are needed to search the codebook. When the convolution is computed by recursive filtering, the codewords $c_k(n)$ are filtered through the zero-state filter $1/A(z/\gamma)$, where the convolution needs Np instructions, the energy computation in $\mathcal{E}_k$ requires N, and the cross-correlation evaluation in C_k also N instructions, yielding a total of $N(p+2)$ operations. For a 1024 size codebook and a predictor of order 10, around 12000 multiplications per speech sample are required to search the codebook.

It can be seen from the previous discussion that the exhaustive search of the excitation codebook is a computationally demanding procedure, which is difficult to implement in real time. We will now look at some methods which simplify the codebook search procedure without affecting the quality of the output speech.

3.4.2   Simplification of the CELP Search Procedure Using the Autocorrelation Approach

Different approaches have been introduced to simplify the codebook search procedure. The frequency domain can be used [101] so that the convolution $c_k(n) * h(n)$ which appears in Equation (3.170) is reduced to the multiplication $C(i)H(i)$, where $C(i)$ is a Gaussian vector and $H(i)$ is the DFT of the impulse response $h(n)$. The number of operations is reduced this way, but we need to compute the DFTs of $h(n)$ and $x(n)$. Another method similar to the frequency domain approach is also proposed in [101]. In this method, the singular value decomposition (SVD) is used to reduce the matrix $\mathbf{H}$ which appears in Equation (3.172) to a diagonal form by expressing it as $\mathbf{H} = \mathbf{UDV}^T$, where $\mathbf{D}$ is diagonal, while $\mathbf{U}$ and $\mathbf{V}$ are orthogonal matrices. The properties of orthogonal matrices can be used to reduce the mean squared weighted error in Equation (3.172) to a form where only $4N$ multiplications are needed to evaluate the term to be maximized. This reduces the multiplications needed to search a 1024 entries codebook to 4000 multiplications/sample. However, this approach requires the extra burden of computing the SVD of the matrix $\mathbf{H}$ for every new set of filter parameters, which is proportional to N^3 operations, and for the typical value $N = 40$, more than 1600 operations per speech sample are introduced, which cannot be neglected.

A common approach to simplify the search procedure is to use the autocorrelation method [101]. In this approach, the matrix of covariances $\mathbf{\Phi} = \mathbf{H}^T\mathbf{H}$ is reduced to a Toeplitz form by modifying the summation limits in Equation (3.177) so that

$$\phi(i,j) = \phi(|i-j|) = \sum_{n=|i-j|}^{N-1} h(n)h(n-|i-j|). \tag{3.186}$$

The autocorrelation approach results from modifying the $N \times N$ convolution matrix of Equation (3.175) into a $(2N-1) \times N$ matrix of the form

$$\mathbf{H} = \begin{pmatrix} h_0 & 0 & 0 & \cdots & 0 \\ h_1 & h_0 & 0 & \cdots & 0 \\ h_2 & h_1 & h_0 & \cdots & 0 \\ \vdots & \vdots & \vdots & \ddots & \vdots \\ h_{N-1} & h_{N-2} & h_{N-3} & \cdots & h_0 \\ 0 & h_{N-1} & h_{N-2} & \cdots & h_0 \\ 0 & 0 & h_{N-1} & \cdots & h_0 \\ \vdots & \vdots & \vdots & \ddots & \vdots \\ 0 & 0 & 0 & \cdots & h_{N-1} \end{pmatrix}. \tag{3.187}$$

The convolution $\mathbf{H}c_k$ using this matrix results in a $2N-1$ length vector, obtained when convolving two segments each of length N. Notice that in the covariance approach only the first N samples of the obtained convolution are considered and any samples beyond the subframe limit are not taken into consideration. Remembering that the impulse response $h(n)$ is a sharply decaying function (see Figure 3.26), it can be truncated at a value $R-1 < N$ (say $R = 25$) without introducing any perceptually noticeable error. In this case the dimensions of the matrix in Equation (3.187) become $(2R-1) \times N$ and the matrix of autocorrelations $\mathbf{\Phi}$ becomes a band matrix (when $R-1 < N$) with $\phi(i) = 0$ for $i \geq R$. Henceforth, we will assume that the impulse response $h(n)$ is truncated at $R-1$.

Using the autocorrelation approach the energy of the filtered codeword $c_k(n)$ in Equation (3.182) can be written as

$$\mathcal{E}_k = \sum_{i=0}^{N-1} c_k^2(i)\phi(0) + 2\sum_{i=0}^{N-2}\sum_{j=i+1}^{N-1} c_k(i)c_k(j)\phi(j-i). \tag{3.188}$$

Defining $\mu_k(i)$ to be the autocorrelation of the codeword $c_k(i)$ given by

$$\mu_k(i) = \sum_{n=i}^{N-1} c_k(n)c_k(n-i), \tag{3.189}$$

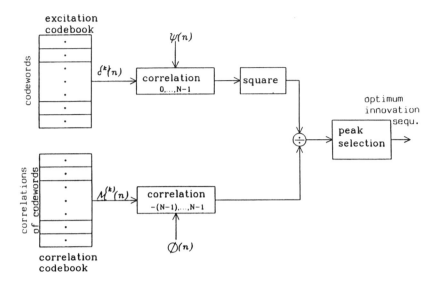

Figure 3.35: Searching for the optimum CELP innovation sequence using the autocorrelation approach.

Equation (3.188) can be written as

$$\mathcal{E}_k = \mu_k(0)\phi(0) + 2\sum_{i=1}^{R-1} \mu_k(i)\phi(i). \qquad (3.190)$$

Evaluating the energy now requires R instructions, and the term $\mathcal{T}_k$ in Equation (3.185) requires $N + R + 3$ instructions. Figure 3.35 shows the optimum innovation sequence search procedure using the autocorrelation approach. For $N = 40$ and $R = 25$ less than 2000 multiplications per synthesized speech sample are needed when a 1024 sized codebook is used. The autocorrelations of the codewords $\mu_k(i)$ are precomputed and stored in another codebook. The autocorrelation approach has the disadvantage of needing a second codebook at the encoder to store the autocorrelations of the excitation codebook.

3.4.2.1 Using Structured Codebooks

The autocorrelation approach discussed in the previous section simplifies the representation of the mean squared weighted error in order to reduce the excessive computational load needed to search for the optimum innovation sequence. In this section, we look at methods to simplify the CELP system by utilizing structured codebooks where the codebook structure enables fast search procedures. We will discuss sparse, ternary, overlapping, and algebraic codebooks.

3.4.2.2 Sparse Excitation Codebooks

In the sparse excitation code-book, most of the excitation pulses in an excitation vector are set to zero. This is done by using centre-clipping, where a zero-mean unit-variance Gaussian random process is used to populate the codebook, and the random variables are set to zero whenever their absolute value is below a specified threshold. The threshold value controls the codebook sparsity. Threshold values of 1.2 and 1.65 result in 77% and 90% sparsity, respectively. Sparse excitation (or centre-clipping) was first proposed by Atal in his pioneering work on Adaptive Predictive Coding (APC) [22], where he found that it is sufficient to quantise the high-amplitude portions of the prediction residual for achieving low perceptual distortion in the decoded speech. Using centre-clipping does not necessarily result in equal number of nonzero pulses in every excitation vector. To obtain w number of nonzero pulses per excitation vector, the largest w samples in the vector are retained and the remaining samples are set to zero. The use of sparse excitation vector codebooks was introduced by different authors independently [88, 90]. It was proposed by Davidson and Gersho [88] motivated by the multi-pulse LPC. It was shown that in MPE-LPC, about 8 pulses per pitch period are required to synthesize natural-sounding speech [73]. We have seen in Section 3.3.4.1 that it is sufficient to use 4 pulses in a 5 ms excitation frame (40 samples). In fact, when the speech is voiced, a few pulses are sufficient to represent the excitation signal, and setting most of the samples in the residual signal to zero does not affect the perceived speech quality. As the pulses in the excitation vector are not individually optimized, it is preferred to use sparse excitation vectors in the case of voiced speech segments. On the other hand, in the case of unvoiced speech segments, using nonsparse stochastic excitation vectors is more sensible. We found in our simulation that using 4 nonzero pulses in an excitation vector of 40 samples gives similar results to the original CELP where the whole excitation vector is populated from a Gaussian random process.

Using sparse excitation codebooks reduces the complexity of the CELP system by a factor around 10 when 4 nonzero pulses are used in an excitation vector of length 40. Since most of the excitation vector samples are equal to zero, most of the autocorrelations of the excitation vectors are also zero. The number of nonzero autocorrelations is not necessarily equal to the number of nonzero pulses. Using the autocorrelation approach with w nonzero pulses in the excitation vector, the cross-correlation term in Equation (3.181) can be expressed as

$$\mathcal{C}_k = \sum_{i=0}^{w-1} \psi(m_i) g_k(i), \qquad (3.191)$$

where w is the number of nonzero pulses, $g(i)$ are the pulse amplitudes and m_i are the nonzero pulse positions. The energy term in Equation (3.182) is now given by

$$\mathcal{E}_k = \mu_k(0)\phi(0) + 2 \sum_{i=1}^{Q-1} \mu_k(n_i)\phi(n_i), \qquad (3.192)$$

where Q is the number of nonzero autocorrelations for the excitation vector at index k, and n_i are the indices of the nonzero autocorrelations μ. As the number of nonzero pulses w is much less than N, the computational effort needed to evaluate the term in Equation (3.185) is significantly reduced. Using sparse excitation vectors does not only simplify the search procedure but it also reduces the storage requirements of the codebooks. The excitation codebook will contain w pulses (usually 4) and their positions, and the autocorrelation codebook will contain the nonzero autocorrelations and their positions.

3.4.2.3 Ternary Codebooks

The sparse excitation vector codebook search can be further simplified by using the ternary excitation approach [90, 91]. A ternary excitation vector is a sparse excitation vector in which the nonzero pulses are set to either -1 or 1. Similar to sparse excitation codebooks, ternary codebooks can be populated using a Gaussian random process with centre-clipping where the random variables are set to zero if their values fall below a certain threshold, otherwise the variable is set equal to its sign (-1 or 1). This populating procedure, however, results in a different number of nonzero pulses in every excitation vector. It is better to fix the number of nonzero pulses in every excitation vector in order to simplify the storage and search procedure. In this case the positions of the w pulses are chosen to be uniformly distributed between 0 and $N-1$ and their amplitudes are randomly chosen to be either -1 or 1.

The ternary approach is simpler than using sparse excitation vectors, since $g(i)$ in Equation (3.191) is either -1 or 1 which means that multiplications in the numerator of the term are reduced to summations. Another advantage is the reduction in codebook storage requirement, since for every nonzero pulse, its position and sign can be stored in one byte with the most significant bit in the byte reserved for the pulse sign [91]. If the excitation vector contains 4 nonzero pulses, 4 bytes are needed to store each codeword. Simulation results have shown that ternary codebooks perform as well as Gaussian or sparse codebooks. A geometric representation of the CELP excitation codebook was utilized in [91, 102] to show that sparse excitation vector codebooks and ternary codebooks are equivalent, in terms of coding performance, to the initially proposed Gaussian random codebook. We mentioned earlier that the CELP approach is based on representing the residual signal after short-term and long-term prediction by a slowly-varying power spectrum Gaussian random process. Due to the existence of the gain factor in the analysis-by-synthesis loop, all the codewords in the excitation codebook can be assumed to have unit energy, and the gain factor introduces the flexibility in changing the power spectrum of the excitation. Therefore, the excitation codebook of size L can be represented by L points on the surface of a unit sphere in N-dimensional space centred at the origin. In fact, the Gaussian process by which the excitation signal is modelled can be considered as all the points on the surface of the unit sphere. Now since the excitation codewords are chosen at random from this Gaussian process, and due to the spherical symmetry of the multi-dimensional Gaussian distribution, the L points representing the codewords are uniformly distributed over the surface of the sphere. In the case of sparse excitation or ternary excitation vectors, there are w nonzero pulses whose positions are chosen at random. The total number of position combinations is $^{N}C_w$. Provided that the codebook size L is much less than the number of position combinations, the sparse excitation or ternary excitation codebooks can still be considered as L points uniformly distributed over the surface of the unit sphere. To explain this, let us take the ternary case with the typical values $N = 40$, $w = 4$ and $L = 1024$. The number of possible position combinations is 91390, and since we have 4 pulses with amplitudes -1 or 1, the total number of the possible codewords from which the codebook is chosen is 91390×16. Since these 1.5 million points are distributed all over the surface of the unit sphere, and since the excitation codebook is randomly chosen from this huge number of possibilities, the resulting ternary codebook can be considered as 1024 points uniformly distributed over the surface of the unit sphere in the 40-dimensional space. This illustrates the equivalence of the ternary codebook to the originally proposed one where all the N pulses in the codeword are Gaussian random variables. Ternary codebooks can be properly structured to result into fast codebook search algorithms. Spherical lattice codebooks were proposed by Ireton and Xydeas [104] where large excitation codebooks can be used without requiring the CELP complexity. Regular pulse ternary codebooks were also proposed [105, 106] where the codebook can be efficiently exhaustively searched using Gray codes [95]. Adoul *et al.* [97, 102, 103] used algebraic codes for populating the ternary codebook where the special structure of the codebook leads to fast search algorithms. A special structure of algebraic codes will be discussed in the next subsection.

3.4.2.4 Algebraic codebooks

Algebraic codes can be used to populate the excitation codebooks. Efficient codebook search algorithms can be obtained using the highly structured algebraic codes. Initially, algebraic codebooks were obtained using binary error-correction codes [102]. We describe here an algebraic code whereby

Amp.	Potential positions
1	0, 8, 16, 24, 32, 40, 48, 56
-1	2, 10, 18, 26, 34, 42, 50, 58
1	4, 12, 20, 28, 36, 44, 52, (60)
-1	6, 14, 22, 30, 38, 46, 54, (62)

Table 3.3: Amplitudes and possible positions of the excitation pulses in the 12 bit algebraic code.

the excitation vectors are derived using interlaced permutation codes (IPC).

In the interlaced permutation codes, an excitation vector contains a few number of nonzero pulses with predefined interlaced sets of positions. The pulses have their amplitudes fixed to 1 or -1, and each pulse has a set of possible positions distinct from the positions of the other pulses. The sets of positions are interlaced. The excitation code is identified by the positions of its nonzero pulses. Thus, searching the codebook is in essence searching the optimum positions of the nonzero pulses. To further explain the codebook structure, we describe a 12 bit codebook used to encode 60 samples excitation vectors (utilized in 4.8 kb/s speech coding). The excitation vector contains 4 nonzero pulses having amplitudes of 1, -1, 1, and -1, respectively. Each pulse can take one of 8 possible positions, and each position is encoded with 3 bits resulting in a 12 bit codebook. If the sets of positions are denoted by $m_i^{(j)}$, $j = 0, \ldots, 3$ and $i = 0, \ldots, 7$ then

$$m_i^{(j)} = 2j + 8i, \qquad \begin{matrix} j = 0, \ldots, 3, \\ i = 0, \ldots, 7. \end{matrix} \qquad (3.193)$$

The pulse amplitudes and sets of positions are given in Table 3.3.

This codebook structure has several advantages. Firstly, it does not require any storage. Secondly, it has inherent robustness against channel errors as the pulse positions are transmitted and one channel error will alter only the position of one pulse. The most important advantage, however, is that the codebook can be very efficiently searched. Denoting the pulse positions by m_i, $i = 0, \ldots, 3$, then the cross-correlation term of Equation (3.181) is given by

$$\mathcal{C} = \psi(m_0) - \psi(m_1) + \psi(m_2) - \psi(m_3), \qquad (3.194)$$

and the energy term of Equation (3.182) is given by

$$\begin{aligned} \mathcal{E} \quad = \quad & \phi(m_0, m_0) \\ & + \phi(m_1, m_1) - 2\phi(m_1, m_0) \\ & + \phi(m_2, m_2) + 2\phi(m_2, m_0) - 2\phi(m_2, m_1) \\ & + \phi(m_3, m_3) - 2\phi(m_3, m_0) + 2\phi(m_3, m_1) - 2\phi(m_3, m_2) \end{aligned}$$

By changing only one pulse position at a time, the correlation and energy terms can be very easily updated. The search is accomplished in 4 nested loops where in the inner-most loop, the correlation is updated with one addition and the energy with 4 additions and one multiplication. Despite the efficiency of the search procedure, the exhaustive search becomes rapidly complicated as the codebook size exceeds 2^{12}. For searching huge excitation codebooks, a focused search strategy has been developed [111]. In this approach, a very small subset of the codebook is searched while guaranteeing a performance very close to that of full search.

3.4.2.5   Overlapping Codebooks

Another efficient codebook structure is represented by overlapped excitation codebooks [90]. This overlapping concept can be combined with sparse or ternary excitation concepts yielding very efficient codebook search algorithms and minimal storage requirements. In an overlapping shift by k codebook, each codeword is obtained by shifting the previous codeword by k samples and adding k new samples. Therefore two adjacent codewords share all but k samples. The first advantage of overlapping codebooks is reducing the codebook storage requirement. For a codebook of size L with N dimension vectors, $N + k(L - 1)$ samples need to be stored. Using stochastically derived ternary excitation most of these samples (about 80%) are zero and the rest are -1 or 1. A shift by 2 was found very efficient and it gave identical results to those obtained using non-overlapping codebooks [92]. Besides reducing the codebook storage, the other advantage of overlapping codebooks is reducing the computational load needed to exhaustively search the codebook.

Consider the stochastic sequence $q(n)$, $n = 0, \ldots, k(L-1) + N - 1$. For a shift by k codebook the codewords are given by

$$c_j(n) = q(kj + n), \qquad \begin{aligned} &n = 0, \ldots, N - 1, \\ &j = 0, \ldots, L - 1, \end{aligned} \qquad (3.195)$$

where N is the excitation vector length and L is the codebook size. The first observation is the reduction in the codebook storage. For a shift by 2 codebook ($k = 2$), $2L + N - 2$ samples are needed to be stored instead of NL samples. The main advantage is the reduction in the search complexity, when evaluating Equation (3.170). The convolution

$$\rho_j(n) = c_j(n) * h(n)$$

requires $N(N + 1)/2$ instructions and it has to be determined for every codeword. For a shift by k codebook

$$\rho_j(n) \;\;=\;\; \sum_{i=0}^{n} c_j(i)h(n - i) = \sum_{i=0}^{n} q(kj + i)h(n - i),$$

$$= \sum_{i=0}^{k-1} q(kj+i)h(n-i) + \sum_{i=k}^{n} q(kj+i)h(n-i) \qquad (3.196)$$

$$= \sum_{i=0}^{k-1} q(kj+i)h(n-i) + \sum_{m=0}^{n-k} q(k(j+1)+m)h(n-k-m).$$

From Equation (3.196)

$$\rho_j(n) = \sum_{i=0}^{k-1} q(kj+i)h(n-i) + \rho_{j+1}(n-k). \qquad (3.197)$$

For a shift by 1 codebook, as for the adaptive codebook, the following relation is obtained:

$$
\begin{aligned}
\rho_j(0) &= q(j)h(0) \\
\rho_j(n) &= q(j)h(n) + \rho_{j+1}(n-1), \qquad n = 1,\dots,N-1. \ (3.198)
\end{aligned}
$$

The convolution of the last codeword is first determined by

$$\rho_{L-1}(n) = \sum_{i=0}^{n} q(L-1+i)h(n-i), \qquad (3.199)$$

and then the relation in (3.198) is used to update the convolution from $j = L - 2$ down to 0. Updating the convolution $\rho_j(n)$ requires N instructions. The impulse response, $h(n)$, of the weighted synthesis filter can be truncated at $R - 1$ where R is usually 25 without any loss in accuracy [92] (see Figure 3.26). In this case R instructions are needed to update $\rho_j(n)$. In the case of shift by 2 codebooks Equation (3.197) is reduced to

$$
\begin{aligned}
\rho_j(0) &= q(2j)h(0) \\
\rho_j(1) &= q(2j)h(1) + q(2j+1)h(0) \\
\rho_j(n) &= q(2j)h(n) + q(2j+1)h(n-1) + \rho_{j+1}(n-2), \quad n = 2\dots N-1.
\end{aligned}
$$

$$(3.200)$$

The value of $\rho_{L-1}(n)$ is initially computed as in Equation (3.199) then the relation in (3.200) is used to update $\rho_j(n)$ from $j = L - 2$ down to 0. If $h(n)$ is truncated at $R - 1$, then $2R - 1$ instructions are needed to update $\rho_j(n)$.

In the case of the adaptive LTP codebook, the registration buffer $q(n)$ contains the excitation history at the input of the weighted synthesis filter $1/A(z/\gamma)$, that is $u(n)$ from $n = -L_a$ to -1 where L_a is the buffer length and its contents are updated in every new subframe by shifting the buffer contents to the left by N positions and introducing new N values. The term $\mathcal{T}_k$ to be maximized in this case requires about $2N + R$ instructions

(R for the convolution ρ_j, N for the energy $\mathcal{E}_j$, and N for the correlation $\mathcal{C}_j$).

In the case of the stochastic codebook, sparse or ternary sequences are usually used, and the ternary approach is preferred as it reduces the storage and complexity. In the ternary case the sequence $q(n)$ contains values -1, 1, or 0. The number of zeros (the sparsity) is usually 80-90%. The ternary sequence is derived by centre-clipping a unit variance Gaussian sequence at a certain threshold (which determines the sparsity). In the DoD coder [66] a threshold of 1.2 is used which results in 77% sparsity. A shift by 2 sparse stochastic codebook has been found equivalent in performance to a non-overlapping codebook. Using sparse codebooks reduces the complexity dramatically. For a shift by 2 codebook, when either $c_j(0)$ or $c_j(1)$ is zero (that is $q(2j)$ and $q(2j + 1)$) the number of instructions in the relation of (3.200) is reduced to R. Further, if both $c_j(0)$ and $c_j(1)$ are zero, no more instructions are needed to update $\rho_j(n)$.

The autocorrelation approach can also be used to update the energy term [92] which is given by

$$\mathcal{E}_j = \mu_j(0)\phi(0) + 2 \sum_{n=1}^{N-1} \mu_j(n)\phi(n), \qquad (3.201)$$

where $\mu_j(n)$ is the autocorrelation of the codeword $c_j(n)$ and for a shift by k codebook it is given by

$$\mu_j(n) = \sum_{i=n}^{N-1} c_j(i)c_j(i - n) = \sum_{i=n}^{N-1} q(kj + i)q(kj + i - n). \qquad (3.202)$$

It can be easily shown that the correlations $\mu_j(n)$ are updated by

$$\mu_{j+1}(n) = \mu_j(n) + \sum_{i=N}^{N+k-1} q(kj+i)q(kj+i-n) - \sum_{i=n}^{n+k-1} q(kj+i)q(kj+i-n).$$
$$(3.203)$$

Based on Equation (3.203) it can be easily shown that for a shift by 1 codebook the filtered codeword energy, using the autocorrelation approach, is updated by

$$\mathcal{E}_{j+1} = \mathcal{E}_j - q(j)\left[\phi(0) + 2 \sum_{n=1}^{N-1} q(j + n)\phi(n)\right]$$
$$+ q(j + N)\left[\phi(0) + 2 \sum_{n=1}^{N-1} q(j + N - n)\phi(n)\right]. (3.204)$$

When $\phi(n)$ is truncated at $R - 1$, where R is typically 25, $2R + 2$ instructions are needed to update the energy. Similarly, in the case of shift by 2 codebooks the energy is updated by

$$
\begin{aligned}
\mathcal{E}_{j+1} = \mathcal{E}_j \;&-\; q(2j)\left[\phi(0) + 2\sum_{n=1}^{N-1} q(2j+n)\phi(n)\right] \\
&-\; q(2j+1)\left[\phi(0) + 2\sum_{n=1}^{N-1} q(2j+1+n)\phi(n)\right] \\
&+\; q(2j+N)\left[\phi(0) + 2\sum_{n=1}^{N-1} q(2j+N-n)\phi(n)\right] \\
&+\; q(2j+N+1)\left[\phi(0) + 2\sum_{n=1}^{N-1} q(2j+N+1-n)\phi(n)\right].
\end{aligned}
$$

$$(3.205)$$

In this case $4R + 4$ instructions are required to update the energy. Using the convolution approach $2R + N - 1$ operations are required. For $4.8\,\text{kb/s}$ coding where $N = 60$ both approaches are similar in computational sense. For smaller frame sizes the convolution approach is preferred. Another great advantage of the convolution approach is its efficiency with sparse codebooks. In the convolution approach, when both $q(2j)$ and $q(2j+1)$ are zero, no operations are needed to update the convolution, and the energy is updated by

$$\mathcal{E}_{j+1} = \mathcal{E}_j - [\rho_{j+1}(N-2)]^2 - [\rho_{j+1}(N-1)]^2 \qquad (3.206)$$

which requires 2 instructions. In the case of the autocorrelation approach $q(2j)$, $q(2j+1)$, $q(2j+N)$ and $q(2j+N+1)$ have to be zero in order to update the energy without requiring any instructions, and this is less likely to happen. Thus for shift by 2 stochastic sparse codebooks the convolution approach is more attractive than the autocorrelation approach.

3.4.2.6   Self-Excitation

Another approach to define the excitation signal is the self-excitation concept. The self-excited LPC [99] can be seen as another variant of CELP in which the codebook is changing. The self-excitation structure offers some simplicity since the excitation can be viewed as a shift by 1 overlapping adaptive codebook (similar to the pitch codebook), but it has a serious disadvantage of propagating channel errors. In self-excitation systems, the excitation sequence is determined by searching through a buffer which contains the previous history of the excitation (or the past decoded speech [100]). Both the encoder and decoder use the same excitation buffer, and the buffers at the encoder and decoder are initially filled with the same Gaussian random sequence. The excitation sequence in the present frame

Parameter	Number of Bits
LSF's	36 (3,3,4,4,4,4,4,4,3,3)
LTP delays	24 (7,5,7,5)
LTP gains	16 (4 × 4)
book indices	36 (4 × 9)
excitation gains	20 (4 × 5)
Total	132 bits per 30 ms

Table 3.4: Bit allocation for 4.4 kb/s CELP coding.

is determined by searching through the excitation buffer for the sequence which minimizes the weighted error between the original and synthesized speech. The excitation is determined by a delay and a corresponding gain factor, and the excitation buffer is updated in every new frame using the excitation determined in the previous frame. The excitation buffers at both the encoder and decoder should have the same content in order to generate identical synthesized speech. This is true in the absence of channel errors. However, in practical applications, the encoded speech parameters could be perturbed due to the noise in the transmission channel, and a single bit error occurring to one of the excitation parameters will cause a mismatch between the excitation buffers at the encoder and decoder, and this will persist for the forthcoming frames.

The same algorithms used in overlapping codebooks and described in the previous section can be deployed in the self-excited (SE) coder as the excitation can be represented by an overlapping shift by 1 adaptive codebook. However, overlapping fixed stochastic codebooks give similar performance with much less complexity (due to their sparsity), and they are more robust against channel errors than the self-excited approach.

3.4.3   CELP Performance

The CELP coder has been evaluated at the bit rates from 4.8 to 8 kb/s. The resulting speech quality ranged from communications quality at 4.8 kb/s to near-toll quality at 8 kb/s. The bit allocations at 4.4 and 8 kb/s are shown in Tables 3.4 and 3.5, respectively. In 4.4 kb/s coding the speech frame is 30 ms long divided into 4 subframes of 7.5 ms (60 samples) while in 8 kb/s coding a 16 ms speech frame is used and is divided in 4 subframes of 4 ms. The histogram of the magnitude of the excitation gain and its logarithm are shown in Figure 3.36 (a) and (b), respectively.

The sign of the gain is quantized with one bit and the magnitude can be efficiently quantized with 4 bits using either logarithmic or non-uniform quantization.

Figures 3.37 and 3.38 show a speech segment, the CELP excitation, the synthesis filter excitation, and the reconstructed speech in cases of gaussian

Parameter	Number of Bits
LSF's	36 (3,3,4,4,4,4,4,4,3,3)
LTP delays	24 (7,5,7,5)
LTP gains	12 (4 × 3)
book indices	36 (4 × 9)
excitation gains	20 (4 × 5)
Total	128 bits per 16 ms

Table 3.5: Bit allocation for 8 kb/s CELP coding.

Codebook Population	SEGSNR (dB)
Gaussian	14.03
Sparse	14.06
Ternary	13.81
Overlapping sparse	14.09

Table 3.6: SEGSNR for different CELP approaches at 7.8 kb/s coding.

codebook and ternary codebook, respectively. A 5 ms excitation frame and a 512 sized codebook are used (6.6 kb/s coding). The variation of speech power and SEGSNR vs. time for the sentence *"to reach the end he needs much courage"* uttered by a female speaker is seen in Figure 3.39.

The different CELP approaches described earlier were compared and they all showed similar performances. Table 3.6 shows the SEGSNR for the different approaches with 4 ms excitation vectors (32 samples) and 512 sized codebooks (at a bit rate of 7.8 kb/s). The equivalence of the different approaches for populating the excitation codebook becomes clear from the dB figures quoted. To our satisfaction, informal subjective listening tests did not show any perceivable difference among them either.

Figure 3.40 shows the SEGSNR against the number of codebook address bits using a ternary excitation codebook with 5 ms excitation vectors. We found that at least a 9-bit codebook is needed to maintain high speech quality. Larger than 10-bit codebooks become impractical due to the exponential increase in the coder complexity.

3.5 Binary Pulse Excitation

We discussed in the previous section code-excited linear prediction coding and described several methods which reduce the coder complexity. These computationally efficient methods have reduced the excessive complexity of the original algorithm but the exhaustive search of the excitation codebook has still to be performed. In this section, a novel approach for representing

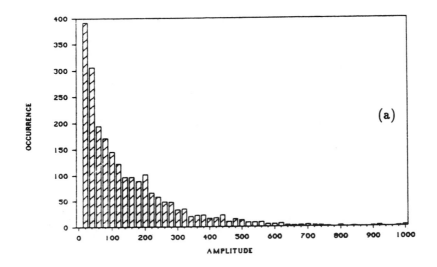

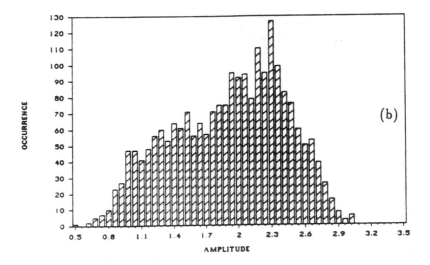

Figure 3.36: (a) Histogram of the codebook gain magnitude; (b) Histogram of the logarithmic codebook gain magnitude.

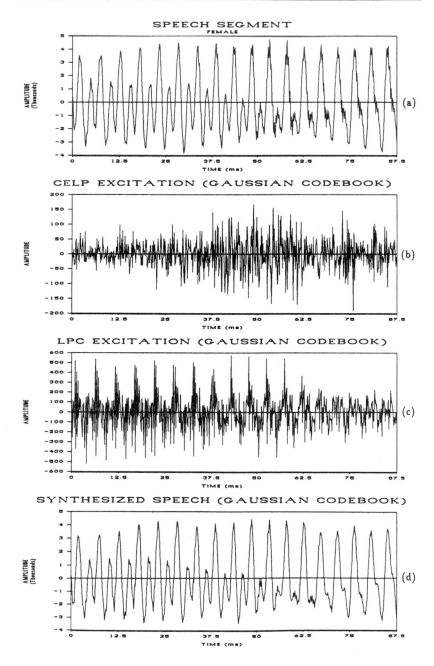

Figure 3.37: (a) A 87.5 ms speech segment; (b) CELP excitation; (c) LPC excitation; (d) reconstructed speech, in the case of a Gaussian codebook.

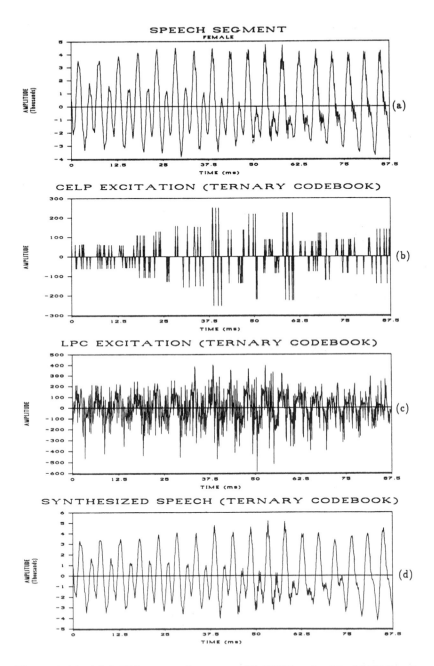

Figure 3.38: (a) A 87.5 ms speech segment; (b) CELP excitation; (c) LPC excitation; (d) reconstructed speech, in the case of a ternary codebook.

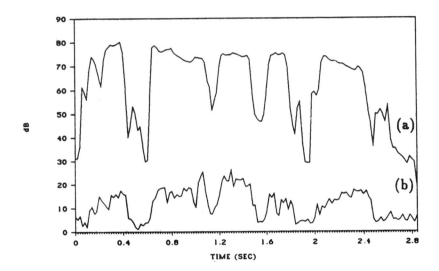

Figure 3.39: (a) Speech power and (b) SEGSNR variation versus time for the
CELP codec using a ternary codebook.

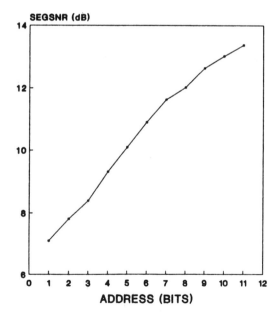

Figure 3.40: SEGSNR against codebook size in terms of the number of address
bits using ternary codebook with 5 ms excitation vectors.

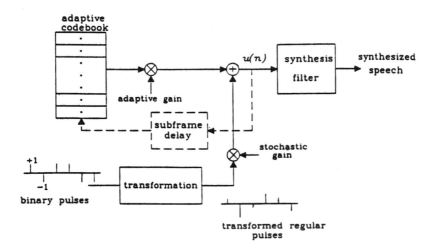

Figure 3.41: Block diagram of the transformed BPE coder.

the excitation signal called *transformed binary pulse excitation (TBPE)* is described. In this approach the excitation signal consists of regularly spaced stochastically derived pulses, where very efficient algorithms for determining the excitation sequence can be obtained. We will describe the excitation definition and derive efficient algorithms for exhaustive and non-exhaustive search of the excitation sequence. A performance comparison between the TBPE and CELP will be given.

3.5.1 Transformed Binary Pulse Excitation

The block diagram of the TBPE coder is shown in Figure 3.41. The coder has common features with both the RPE and CELP. The excitation signal consists of a number of pseudo-stochastic pulses with predefined pulse positions. In an excitation frame of length N, we suppose that there are M nonzero pulses separated by $D - 1$ zeros, where $M = N \operatorname{DIV} D$, and DIV denotes integer division. The excitation vector is given by

$$v(n) = \beta \sum_{i=0}^{M-1} g_i \delta(n - m_i), \qquad n = 0, \ldots, N - 1, \qquad (3.207)$$

where $\delta(n)$ is the Kronecker delta, g_i are the pulse amplitudes, m_i are the pulse positions, and β is a scalar gain similar to that which appears in the CELP. As in the RPE approach [8], there are D sets of pulse positions given by

$$m_i^{(k)} = k + iD, \qquad \begin{aligned} i &= 0, \ldots, M - 1, \\ k &= 0, \ldots, D - 1, \end{aligned} \qquad (3.208)$$

where D is the pulse spacing, and k is the position of the first pulse. In RPE coders, the optimum pulse amplitudes and first pulse position are determined by minimizing the mean-squared weighted error between the original and synthesized speech, and this requires solving a set of $M \times M$ equations D times. Further, the pulse amplitudes in RPE are each quantized with 3 bits after scaling by the maximum pulse, or the rms value of the pulses, which is quantized with 5 or 6 bits. The large number of bits needed to quantise the pulse amplitudes in RPE makes it difficult to achieve high quality speech below 9.6 kb/s. In our TBPE approach, the pulses are pseudo-stochastic random variables, similar to the CELP concept, and they are quantized only with one bit per pulse, in addition to the scaling gain β.

Instead of obtaining the pulse amplitudes g_i, $i = 0, \ldots, M-1$, from a large stochastic codebook as in the CELP approach, the pulses are determined by the transformation of a binary vector. That is

$$\mathbf{g} = \mathbf{Ab}, \tag{3.209}$$

where $\mathbf{b}$ is an $M \times 1$ binary vector with elements -1 or 1, $\mathbf{A}$ is an $M \times M$ transformation matrix, and $\mathbf{g}$ is the excitation vector containing the pulse amplitudes. The vector $\mathbf{b}$ could be one of 2^M possible binary patterns, which means 2^M different excitation vectors can be obtained using the transformation in Equation (3.209). Thus, this transformation is equivalent to a 2^M sized codebook with the need to only store an $M \times M$ matrix. The equivalent of smaller codebook sizes can be obtained by setting some of the binary pulses to fixed values, or by omitting some of the columns of the matrix $\mathbf{A}$. If the hypothetical codebook size is to be reduced by a factor m, either m pulses in the binary vector are made fixed (say -1), or m columns are omitted from the matrix $\mathbf{A}$ resulting in an $M \times Q$ transformation matrix and $Q \times 1$ binary vector, where $Q = M - m$. On the other hand, the equivalent of larger codebooks is obtained by utilizing several transformation matrices. Using m different transformation matrices is equivalent to a book of size 2^{M+m}. When the transformation matrix is of dimension $N \times M$ the resulting excitation vector is of dimension $N \times 1$. The excitation in this case is not sparse, and this transformation becomes equivalent to the VSELP approach [95,96]. The M columns of the $N \times M$ transformation matrix are equivalent to the M basis vectors of the VSELP. If the columns of the $N \times M$ matrix are given by $\mathbf{c}_i$, $i = 0, \ldots, M-1$, then the excitation vector $\mathbf{v}$ can be expressed as

$$\mathbf{v} = \sum_{i=0}^{M-1} b_i \mathbf{c}_i, \tag{3.210}$$

where b_i are the elements of the binary vector with values -1 or 1. The equivalence of an $N \times M$ transformation to the VSELP approach becomes plausible in Equation (3.210). However, using an $M \times M$ transformation reduces the complexity (as the excitation vectors become sparse with regularly spaced pulses) without affecting the speech quality.   For the special case where the transformation matrix is equal to the identity matrix $\mathbf{I}$, the excitation pulses are binary with values -1 or 1. This regular pulse binary codebook has been suggested by many authors [105, 106, 108]. In fact, binary codebooks were first proposed by Le Guyader *et al.* [109] where the excitation is non-sparse and the pulses are -1 or 1. Sparse codebooks yield better performance, and setting most of the binary pulses to zero results in ternary excitation vectors [90, 91]. Regular pulse binary vectors are ternary vectors where the nonzero pulse positions are predefined to be equally spaced. This eliminates the codebook storage and yields very efficient search algorithms as we will see later in this chapter.

The regular binary pulse excitation vectors can be viewed as 2^M points regularly distributed over the surface of a sphere in N-dimensional space. When the transformation matrix $\mathbf{A}$ is orthogonal (i.e. $\mathbf{A}^T \mathbf{A} = \mathbf{I}$), the transformation results in a vector containing Gaussian random variables. Generating binary pulses at random and examining the distribution of the pulses g_i resulting from the orthogonal transformation reveals that the variables g_i follow a Gaussian distribution with zero mean and unit variance. Applying an orthogonal transformation to the binary vectors rotates the vectors without changing their distribution in the N-dimensional space. Both identity and orthogonal transformations exhibited similar objective performances; however, using an orthogonal transformation resulted in slight improvement in the subjective speech quality.

A further speech quality improvement was achieved when the transformation matrix was derived from a training set of RPE vectors as the RPE approach gives the optimum amplitudes of the excitation pulses. Any other iterative algorithm which minimizes the expectation of the perceptually weighted error between the original and synthesized speech can be used to derive the transformation matrix. For example, the iterative method used to optimize the VSELP basis vectors [95] can be used here. In fact, the advantage of incorporating the transformation is that it introduces a general framework for defining the excitation characteristics, where the transformation matrix can be chosen in a way to obtain some desired codebook properties. Laflamme *et al.* [110] have recently proposed an elegant approach for defining the transformation, or the shaping matrix where the matrix is a function of the LPC filter $A(z)$, resulting in a codebook which is dynamically frequency-shaped. In the their implementation [111], the transformation is a lower triangular matrix containing the impulse response of the filter

$$F(z) = (1 - \mu z^{-1}) \frac{A(z/\gamma_1)}{A(z/\gamma_2)}. \tag{3.211}$$

This filter has a similar role to that of postfiltering (see Section 3.6) with the advantage of being implemented inside the analysis-by-synthesis loop.

3.5.2   Excitation Determination

The weighted error between the original and synthesized speech is given by

$$e_w(n) = s_w(n) - \hat{s}_w(n), \tag{3.212}$$

where $s_w(n)$ is the weighted input speech and $\hat{s}_w(n)$ is the weighted synthesized speech. The weighted synthesized speech can be written as

$$\hat{s}_w(n) = \sum_{i=0}^{n} v(i)h(n-i) + Gc_\alpha(n) * h(n) + \hat{s}_0(n), \tag{3.213}$$

where $h(n)$ is the impulse response of the weighted synthesis filter $1/A(z/\gamma)$, G is the adaptive codebook gain (or the LTP gain), $c_\alpha(n)$ is the codeword chosen from the adaptive codebook (or α is the LTP delay), and $\hat{s}_0(n)$ is the zero-input response of the weighted synthesis filter.

From Equations (3.212) and (3.213), the weighted error can be written as

$$e_w(n) = x(n) - \sum_{i=0}^{n} v(i)h(n-i), \tag{3.214}$$

where

$$x(n) = s_w(n) - Gc_\alpha(n) * h(n) - \hat{s}_0(n). \tag{3.215}$$

Now, substituting the excitation signal $v(n)$ from Equation (3.207) into Equation (3.214) gives

$$
\begin{aligned}
e_w(n) &= x(n) - \sum_{i=0}^{n} \beta \sum_{k=0}^{M-1} g_k \delta(i - m_k) h(n-i), \\
&= x(n) - \beta \sum_{k=0}^{M-1} g_k h(n - m_k), \quad n = 0, \ldots, N-1,
\end{aligned}
\tag{3.216}
$$

where $h(n-m_k) = 0$ for $n < m_k$. The excitation parameters are determined by minimizing the mean square of the weighted error $e_w(n)$ which is given by

$$E = \sum_{n=0}^{N-1} \left[x(n) - \beta \sum_{i=0}^{M-1} g_i h(n - m_i) \right]^2. \tag{3.217}$$

Setting $\partial E/\partial \beta$ to zero leads to

$$\beta = \frac{\sum_{i=0}^{M-1} g_i \psi(m_i)}{\sum_{i=0}^{M-1} \sum_{j=0}^{M-1} g_i g_j \phi(m_i, m_j)}, \tag{3.218}$$

where ψ is the correlation between $x(n)$ and the impulse response $h(n)$, given by

$$\psi(i) = \sum_{n=i}^{N-1} x(n)h(n-i), \tag{3.219}$$

and ϕ is the autocorrelation of the impulse response $h(n)$ given by

$$\phi(i,j) = \sum_{n=\max(i,j)}^{N-1} h(n-i)h(n-j). \tag{3.220}$$

By substituting Equation (3.218) into Equation (3.217), the minimum mean squared weighted error between the original and synthesized speech can be written as

$$
\begin{aligned}
E &= \sum_{n=0}^{N-1} x^2(n) - \beta \sum_{i=0}^{M-1} g_i \psi(m_i), \\
&= \sum_{n=0}^{N-1} x^2(n) - \frac{\left[\sum_{i=0}^{M-1} g_i \psi(m_i)\right]^2}{\sum_{i=0}^{M-1}\sum_{j=0}^{M-1} g_i g_j \phi(m_i, m_j)}.
\end{aligned} \tag{3.221}
$$

Equation (3.221) can be written in matrix form as

$$E = \mathbf{x}^T\mathbf{x} - \frac{(\mathbf{\Psi}^T\mathbf{g})^2}{\mathbf{g}^T\mathbf{\Phi}\mathbf{g}} = \mathbf{x}^T\mathbf{x} - \frac{(\mathbf{\Psi}^T\mathbf{Ab})^2}{\mathbf{b}^T\mathbf{A}^T\mathbf{\Phi}\mathbf{Ab}}, \tag{3.222}$$

where $\mathbf{x}$ is an $N \times 1$ vector, $\mathbf{\Psi}$ and $\mathbf{b}$ are $M \times 1$ vectors, and $\mathbf{\Phi}$ is an $M \times M$ symmetric matrix with elements $\phi(m_i, m_j)$, $i, j = 0, \ldots, M-1$. The autocorrelation approach can be used to express $\phi(m_i, m_j) = \phi(|m_i - m_j|)$. In this case, the matrix $\mathbf{\Phi}$ is reduced to a Toeplitz symmetric matrix with diagonal $\phi(0)$ and off-diagonals $\phi(D), \phi(2D), \ldots, \phi([M-1]D)$, respectively (see Equation (3.137)). Defining

$$\mathbf{z} = \mathbf{A}^T\mathbf{\Psi}, \tag{3.223}$$

and

$$\mathbf{\Theta} = \mathbf{A}^T\mathbf{\Phi}\mathbf{A}, \tag{3.224}$$

Equation (3.222) becomes

$$E = \mathbf{x}^T\mathbf{x} - \frac{(\mathbf{z}^T\mathbf{b})^2}{\mathbf{b}^T\mathbf{\Theta}\mathbf{b}}. \tag{3.225}$$

The excitation gain and binary code can be jointly optimized similar to the CELP case, and the mean squared weighted error becomes (see Equation (3.184))

$$E = \mathbf{x}^T\mathbf{x} - \hat{\beta}(2\mathcal{C} - \hat{\beta}\mathcal{E}), \tag{3.226}$$

where $\hat{\beta}$ is the quantized value of the gain $\beta = \mathcal{C}/\mathcal{E}$, $\mathcal{C}$ is the cross-correlation between $x(n)$ and the filtered excitation (see Equation (3.181)) given by

$$\mathcal{C} = \mathbf{z}^T \mathbf{b}, \tag{3.227}$$

and $\mathcal{E}$ is the energy of the filtered excitation (see Equation (3.182)) given by

$$\mathcal{E} = \mathbf{b}^T \Theta \mathbf{b}. \tag{3.228}$$

The optimum excitation vector is the one which maximizes the second term in Equation (3.226) given by

$$\mathcal{T} = \hat{\beta}(2\mathcal{C} - \hat{\beta}\mathcal{E}). \tag{3.229}$$

3.5.2.1   Efficient Exhaustive Search: The Gray Code Approach

To determine the optimum innovation sequence, one could exhaustively search through all possible binary patterns and select the pattern which maximizes the term in Equation (3.229). This can be easily done using a Gray code counter [95,107], where the Hamming distance between adjacent binary patterns is 1. As for every new pattern only one pulse is changed, $\mathcal{C}$ and $\mathcal{E}$ can be simply updated taking into account the pulse which has been toggled. Using a Gray code counter, the cross-correlation $\mathcal{C}$ is updated by

$$\mathcal{C}_k = \mathcal{C}_{k-1} + 2z_j b_j^{(k)}, \tag{3.230}$$

where k is the index of the Gray code and j is the index of the pulse which has been toggled. The energy of the filtered excitation $\mathcal{E}$ can be similarly updated by

$$\mathcal{E}_k = \mathcal{E}_{k-1} + 4b_j^{(k)} \sum_{\substack{i=0 \\ i\neq j}}^{M-1} b_i^{(k)} \theta(i,j). \tag{3.231}$$

To get rid of the multiplications by 2 and 4 in Equations (3.230) and (3.231) we define [95]

$$\mathcal{C}' = \mathcal{C}/2 \qquad \text{and} \qquad \mathcal{E}' = \mathcal{E}/4.$$

The term in (3.229) now becomes

$$\mathcal{T} = 4\hat{\beta}(\mathcal{C}' - \hat{\beta}\mathcal{E}'). \tag{3.232}$$

Equations (3.230) and (3.231) are reduced to

$$\mathcal{C}'_k = \mathcal{C}'_{k-1} + z_j b_j^{(k)}, \tag{3.233}$$

and

$$\mathcal{E}'_k = \mathcal{E}'_{k-1} + b_j^{(k)} \sum_{\substack{i=0 \\ i \neq j}}^{M-1} b_i^{(k)} \theta(i,j). \tag{3.234}$$

Equations (3.233) and (3.234) offer a very efficient method to exhaustively search for the best binary excitation pattern. For every new pattern, $M+1$ operations are needed to update both $\mathcal{C}'$ and $\mathcal{E}'$. This is much more efficient than the overlapping codebook approach described in Chapter 4. The Gray code approach is also used in the VSELP coder [95] which has been selected for the future American digital mobile radio system. The VSELP differs from the TBPE in defining the transformation matrix where an $N \times M$ dimension matrix is used (M basis vectors of length N), which results in nonsparse excitation vectors.

3.5.2.2 Non-exhaustive Search

Although the exhaustive search based Gray code approach is very efficient, the regular structure of the excitation pulses results into a much simpler excitation determination procedure in which the exhaustive search is ruled out [105, 107].

A closer look at the autocorrelation matrix $\mathbf{\Phi}$ in Equation (3.137) suggests that it is strongly diagonal, because the magnitude of $\phi(nD)$ (D is usually 4) is much less than $\phi(0)$ (see Figure 3.26). As $\mathbf{A}^T\mathbf{A}$ is equal to the identity matrix $\mathbf{I}$ (when $\mathbf{A}$ is orthogonal), $\mathbf{\Theta}$ of Equation (3.224) is also strongly diagonal. Therefore, as $\mathbf{b}^T\mathbf{b}$ is constant ($= M$), the denominator in Equation (3.225) (the energy of the filtered excitation $\mathcal{E}$) can be approximated by a constant equal to $M\phi(0)$. Figure 3.42 shows an example of the variation of the magnitude of both the numerator $\mathcal{C}^2$ and the denominator $\mathcal{E}$ with varying the binary pattern ($M = 10$). It is clear that the change in the term to be maximized $\mathcal{C}^2/\mathcal{E}$ is dominated by the value of the numerator. Thus, minimization of the error in Equation (3.225) can be performed by maximizing the numerator, i.e. maximizing the absolute value of the cross-correlation $\mathcal{C} = \mathbf{z}^T\mathbf{b}$, and this can be simply done by choosing the pulses to be equal to the signs of $\mathbf{z}$, i.e.

$$b_i = \text{sign}\{z_i\}, \qquad i = 0, \ldots, M-1. \tag{3.235}$$

Equation (3.235) offers an extremely simple excitation determination procedure in which no exhaustive search is needed.

Figure 3.43 shows the histogram of the Hamming distance between the binary vector $\mathbf{b}_0$ determined using the simple relation of Equation (3.235) and the optimum binary vector $\mathbf{b}_{opt}$ determined by the exhaustive search through all the possible binary vectors for the one which minimizes the mean squared weighted error. The exhaustive search is performed with the joint optimization of the binary vector and excitation gain. We notice

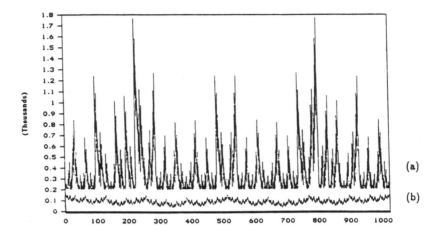

Figure 3.42: Variation of (a) the numerator $\mathcal{C}^2$, and (b) denominator $\mathcal{E}$, with varying the binary pattern from 0 to 1023 ($M = 10$).

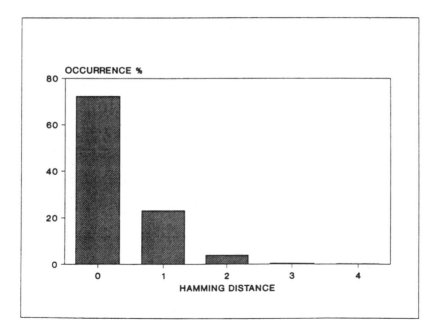

Figure 3.43: Frequency of occurrence vs. Hamming distance between the Binary code determined by Equation (3.235) and the one determined by exhaustive search.

Parameter	Number of Bits
LSFs	36 (3,3,4,4,4,4,4,4,3,3)
LTP delays	24 (7,5,7,5)
LTP gains	12 (4 × 3)
Binary pulses	48 (4 × 12)
Pulse positions	8 (4 × 2)
Excitation gains	20 (4 × 5)
Total	144 bits per 30 ms

Table 3.7: Bit allocation for 4.8 kb/s BPE coding.

that the optimum vector is properly computed 72% of the time since the Hamming distance between $\mathbf{b}_0$ and $\mathbf{b}_{opt}$ over that time is zero. For about 23% of the time the Hamming distance is one, which means that whenever Equation (3.235) fails to determine the optimum binary vector the computed vector differs from the optimum one by only one sign.

This observation has led us to the following efficient search procedure. An initial binary vector is first determined using Equation (3.235), then the second term of Equation (3.226) is evaluated using the initial vector and the other M vectors which have a Hamming distance of one from the initial vector. In this efficient procedure the search of a book of size 2^M is reduced to searching a local book of size $M + 1$, yet guaranteeing that 95% of the time the optimum binary vector is identified. Notice that for the $M + 1$ sized local codebook the efficient Gray code procedure is used to update $\mathcal{C}_k$ and $\mathcal{E}_k$ as in Equations (3.230) and (3.231).

3.5.3 Evaluation of the BPE Coder

The BPE coder was evaluated at different bit rates in the range from 4.8 to 8 kb/s. The subjective and objective speech quality was indistinguishable from that of the CELP coder at similar bit rates. Tables 3.7 and 3.8 show the bit allocation for BPE at 4.8 and 7.5 kb/s, respectively.   In 4.8 kb/s coding a 30 ms speech frame is used and it is divided into 4 excitation frames of 7.5 ms (60 samples). In 7.5 kb/s coding a 24 ms speech frame is used and it is divided into 6 excitation frames of 4 ms (32 samples). Figure 3.44 shows SEGSNR against bit rate from 4.8 to 9.6 kb/s.

Figure 3.45 shows a speech segment, the binary pulse excitation, the synthesis filter excitation, and the reconstructed speech, where 8 pulses in a 5 ms excitation frame are used (6.6 kb/s coding). The variation of the speech energy and the SEGSNR with time for the sentence *"to reach the end he needs much courage"* uttered by a female speaker is shown in Figure 3.46.

The SNRs of the different search procedures described earlier are shown

Parameter	Number of Bits
LSFs	36 (3,3,4,4,4,4,4,3,3)
LTP delays	36 (7,5,7,5,7,5)
LTP gains	18 (6 × 3)
Excitation pulses	48 (6 × 8)
Pulse positions	12 (6 × 2)
Matrix identifier	6 (6 × 1)
Excitation gains	24 (6 × 4)
Total	180 bits per 24 ms

Table 3.8: Bit allocation for 7.5 kb/s BPE coding.

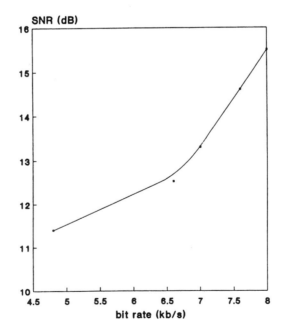

Figure 3.44: BPE codec's segmental SNR versus bit rate.

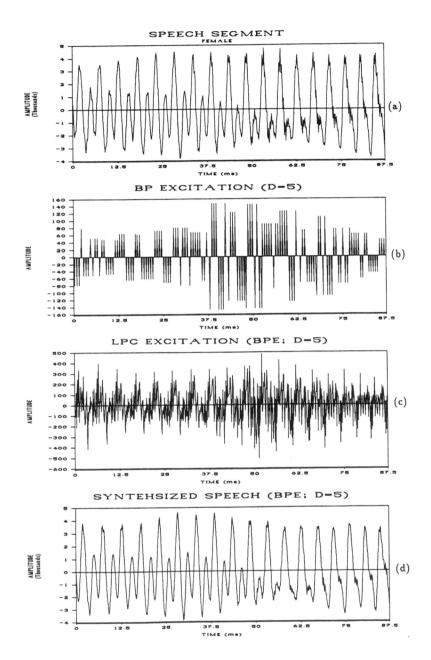

Figure 3.45: (a) A 87.5 ms speech segment; (b) binary excitation; (c) LPC excitation; (d) reconstructed speech.

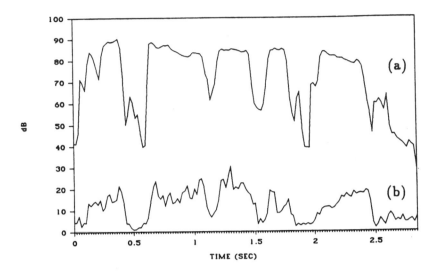

Figure 3.46: Variation of (a) speech power; (b) SEGSNR versus time for the sentence *"to reach the end he needs much courage"* uttered by a female speaker using BPE.

Search procedure	BP1	BP2	BP3
SEGSNR (dB)	13.130	13.350	13.373

Table 3.9: SNRs of different search procedures in BPE.

in Table 3.9. In a 5 ms excitation frame, 10 pulses are used. BP1 represents the simple approach using Equation (3.235), BP2 denotes the two stage approach in which an $M + 1$ sized local codebook is searched (M is the number of pulses), and BP3 denotes the exhaustive search. Using the simple search reduces the SNR by 0.24 dB as compared to the exhaustive search approach which is insignificant. Using the nonexhaustive two stage search brings the SNR closer to the exhaustive search case.

Table 3.10 shows the segmental SNRs of the BPE and ternary CELP coding using 4 ms and 5 ms excitation vectors. In the case of the ternary codebook a 9-bit stochastic codebook was utilised with the gain quantized using 5 bits (4 bits for the magnitude and 1 for the sign). In the case of BPE, 8 binary pulses were used with the first pulse position quantized using 2 bits and the gain using 4 bits (the BPE gain is always positive as the sign information is carried by the pulses themselves). It is clear from the segmental SNR figures in Table 3.10 that the objective quality of BPE is very close to that of the CELP. In fact, subjective listening tests did not show any difference in speech quality in either case.

.	Binary regular pulses	Ternary excitation
5 ms excitation vectors	12.5208 dB	12.6356 dB
4 ms excitation vectors	13.85 dB	13.81 dB

Table 3.10: SEGSNR for the BPE and ternary CELP with 4 ms and 5 ms excitation vectors.

Number of matrices	SEGSNR (dB)
0	13.2734
1	13.3800
2	13.7100
4	14.3353
8	14.6198

Table 3.11: SNR improvement with increasing the number of transformation matrices.

When the excitation vectors were not transformed, implying the utilisation of binary regular pulses, there was a slight degradation in speech quality compared to the transformed case. Table 3.11 shows the improvement in SEGSNR with increasing the number of transformation matrices, where 10 pulses are used in a 5 ms excitation vector. Zero number of matrices refers to untransformed binary pulses. Using 8 matrices gives 1.4 dB improvement in SNR at the expense of increasing the bit rate by 0.6 kb/s. Using untransformed binary pulses reduces the complexity since computing $\mathbf{z}$ and Θ in Equations (3.223) and (3.224) is not needed. In 8 kb/s coding, better performances were obtained using two transformation matrices, where one of the matrices was set to the identity matrix (no transformation) to reduce the complexity and storage requirement.

The TBPE coder has several advantages over the CELP. The main advantage is the significant reduction in the computational complexity. As shown in Section 3.5.2.2, the search of an excitation codebook of size 2^M is reduced to searching a local book of size $M+1$. In the case of untransformed vectors, computing the second term of Equation (3.225) requires about $M^2 + M$ instructions and for the next M vectors in the local codebook, about $M(M + 3)$ instructions are needed to update the term. This is repeated D times (for all possible first pulse positions) which results in a total of $2M + 4$ instructions per speech sample. Using the transformation requires the computation of $\mathbf{z}$ and Θ in Equations (3.223) and (3.224), where $\mathbf{z}$ needs M^2 instructions and it is computed D times, which results in M instructions per speech sample, while Θ is computed only once since it is independent of the first pulse position, and it requires about $(3M^2 + M)/(2D)$ instructions per speech sample. For $M = 8$ and $D = 4$, and with no transformation, about 20 instructions per speech sample are required to jointly optimize the binary vector and the first pulse position, and this number rises to 50 when a transformation is used.

The second advantage of the TBPE is the ability to improve the speech quality by utilizing several transformation matrices, which is equivalent to using a very large excitation codebook; a task which becomes impractical with the CELP when the codebook address exceeds 10 bits. Another advantage is the reduction in the storage requirements of the excitation codebook. The equivalent of a 2^M sized codebook is obtained by storing an $M \times M$ matrix, and the storage is eliminated in the case of untransformed pulses. Finally, the TBPE possesses an inherent robustness against transmission errors. As the excitation pulses are directly derived from the transmitted binary vector, a transmission error in the binary vector will cause little change in the transformed excitation vector, while in CELP coders a transmission error in the codebook address will cause the receiver to use an entirely different excitation vector.

3.5.4 Complexity Comparison Between BPE and CELP Codecs

Here we will attempt here to give a comparison between BPE codecs and the different CELP approaches in terms of complexity and storage requirement. Consider the case where 40 samples excitation frames (5 ms) are used. In the case of CELP, a 9-bit codebook is used with the gain quantized using 5 bits. In case BPE, we use 8 pulses spaced by a distance of $D = 5$ with 2 bits reserved for the first pulse position and 4 bits for the gain. We will assume that the impulse response $h(n)$ is truncated at $R - 1 = 24$, and the computation of $\psi(n)$ and $\phi(n)$ is not considered in the complexity assessment. Only the arithmetical operations are considered here (fetching stored data from memory is not considered).

Using the original CELP with the autocorrelation approach, $N + R = 65$ instructions (one instruction is one addition plus one multiplication) are needed to compute $\mathcal{C}$ and $\mathcal{E}$ in Equations (3.181) and (3.190), respectively. This results in a total of about 850 instructions per speech sample to search the 512 sized codebook. Assuming that every real-valued sample requires 4 bytes, then 160 kbytes are needed to store both the excitation and correlation codebooks.

In the case of a ternary excitation codebook, with 4 nonzero pulses, and using the autocorrelation approach, 4 instructions are needed to compute $\mathcal{C}$ and about 6 to compute $\mathcal{E}$. The term $\mathcal{C}^2/\mathcal{E}$ requires about 12 instructions, and the codebook search requires about 150 instructions per speech sample. Concerning the book storage, every excitation pulse position and sign are stored in one byte and the excitation codebook requires 2 kbytes. The problem here is the storage of the correlations of the codewords. For 4 nonzero excitation pulses, there are at most 6 nonzero correlations for every excitation vector [91]. The correlation $\mu(0)$ is always equal to 4 and need not to be stored. The other correlations have integer values and can be stored with 2 bytes each (amplitude and position). A maximum of 5 kbytes is then required to store the 512 sized correlations codebook. Storage of the correlation codebook can be eliminated if the energies of the filtered codewords are computed on-line. The energy is expressed by

$$\mathcal{E} = 4\phi(0) + 2\sum_{i=0}^{2} \sum_{j=i+1}^{3} b_i b_j \phi(|m_i - m_j|),$$

where b_i is the amplitude $(-1,1)$ of the ith nonzero pulse and m_i is its position. Computing $\mathcal{E}/2$ requires 13 instructions and the term to be maximized now requires 19 instructions. A total of about 250 instructions are now needed with only 2 kbytes needed to store the excitation codebook.

In the case of a ternary shift by 1 overlapping codebook, the excitation buffer contains $40+511 = 551$ samples with values $(-1,0,1)$. Assuming that every sample is stored in 2 bits, about 140 bytes are needed to store the excitation. The convolution $c(n) * h(n)$ is computed for the first codeword and it requires $N(N + 1)/2$ instructions. The correlation $\mathcal{C}$ needs N and the energy $\mathcal{E}$ needs N instructions. For the next codeword, the convolution is updated by $R = 25$ instructions, the correlation by N and the energy by N instructions. Remember that if the new pulse in the next codeword is zero then the convolution and energy need not to be updated. For a sparsity factor $1 - S$ (S is the ratio of nonzero pulses) and codebook size L, the convolution requires $N(N + 1)/2 + S(L - 1)R$, the energy $N + S(L - 1)N$, and the correlation NL instructions. For the typical values of $N = 40$, $L = 512$, $R = 25$, and $1 - S = 0.9$, over 600 instructions per speech sample are needed to search the codebook. Note that the shift by 2 codebook used by the DoD codec requires even more computational load in order to search the codebook. The only obvious advantage of overlapping codebooks is their storage efficiency. It is clear from the previous discussion that Xydeas' ternary approach [91] is far more computationally efficient than the DoD codec's overlapping codebook approach [64], although both approaches utilize similar excitation definition (stochastic ternary).

In the VSELP approach [95] with 512 sized codebook, there are 9 basis vectors of dimension 40. The correlation term $\mathcal{C}$ is expressed by

$$\mathcal{C} = \mathbf{\Psi}^T \mathbf{A} \mathbf{b}$$

where $\mathbf{\Psi}$ is a 40×1 vector, $\mathbf{A}$ is a 40×9 matrix whose columns are the 9 basis vectors, and $\mathbf{b}$ is a 9×1 binary vector with elements -1 or 1. The filtered codeword energy is given by

$$\mathcal{E} = \mathbf{b}^T \mathbf{A}^T \mathbf{\Phi} \mathbf{A} \mathbf{b}$$

, where $\mathbf{\Phi}$ is the 40×40 matrix of autocorrelations. For the first binary code, $\mathcal{C}$ requires $40 \times 9 + 9$ instructions and $\mathcal{E}$ requires about $40 \times 9 + 40 \times 40$ instructions. When the Gray code search is used, $\mathcal{C}$ is updated with 1 instruction and $\mathcal{E}$ with 9 instructions (see Equations (3.233) and (3.234)). Note that only half the codebook is searched since the complement of a binary code $\mathbf{b}$ yields the same value of $\mathcal{C}^2/\mathcal{E}$, and only the sign of the gain $\mathcal{C}/\mathcal{E}$ is changed. The total number of operations needed to search the codebook is about 130 instructions per speech sample. The codebook storage requires 40×9 real values which is about 1440 bytes.

In the BPE approach, when 8 binary pulses are used with no transformation, no codebook storage is required. The correlation $\mathcal{C}$ in Equation (3.227) requires 8 instructions and the energy $\mathcal{E}$ in Equation (3.228) requires about 8^2 instructions. This is repeated 4 times for all the possible positions. When the simple search approach is used, less than 10 instructions per speech sample are needed. When an 8×8 transformation matrix is used, the storage requirement is 64 real values which is about 0.25 kbytes. The extra computational load is computing $\mathbf{z}$ in Equation (3.223) 4 times which requires $8^2 \times 4$ instructions, and computing $\mathbf{\Theta}$ one time (independent of the pulse positions) which requires 800 instructions. The total is now less than 40 instructions per speech sample. Using the local codebook approach requires to search another M binary codes. As the Gray code approach is used, which needs $M + 1$ operations to update $\mathcal{C}$ and $\mathcal{E}$, the extra load is $8(8 + 1)/40$ which is less than two instructions per speech sample and this can be neglected. Table 3.12 shows the complexity and storage requirement figures described earlier. Objective and subjective performances of the approaches described in Table 3.12 were very close. However, the superiority of the BPE in both computational and storage efficiency is evident.

3.6   Postfiltering

Below $8\,\mathrm{kb/s}$ the quality of the reconstructed speech starts to degrade and speech enhancement techniques can be deployed to improve the speech quality at lower bit rates. Postfiltering has been efficiently used with AD-PCM [112] and the same concept can be utilized to enhance the quality of analysis-by-synthesis predictive coders. The postfilter emphasizes the speech formants and it has a spectral shape which lies between an all-pass filter and the synthesis filter. A commonly used postfilter is given by [113]

$$F(z) = \frac{A(z/\beta)}{A(z/\alpha)} = \frac{1 - \sum_{k=1}^{p} a_k \beta^k z^{-k}}{1 - \sum_{k=1}^{p} a_k \alpha^k z^{-k}}, \qquad (3.236)$$

where $0 \leq \alpha, \beta \leq 1$. The filter spectrum is controlled by the values of α and β. The parameters α and β increase the bandwidth of the spectrum resonances. Figure 3.47 shows an example of the spectra of the synthesis filter $1/A(z)$ and the filters $1/A(z/0.8)$ ($\alpha = 0.8$ and $\beta = 0$) and

Approach	Complexity	Storage (kbytes)
Original CELP	850	160
Ternary with Correlation codebook	250	7
Ternary without Correlation codebook	250	2
Overlapping shift by 1 ternary	600	0.13
VSELP	130	1.4
binary (no transformation)	10	0
binary transformed pulses	40	0.25

Table 3.12: Comparison between the BPE and different CELP approaches in terms of codebook search complexity and storage requirements. The complexity is given by the number of instructions per speech sample. A 512 sized codebook is used with an excitation frame of 40 samples.

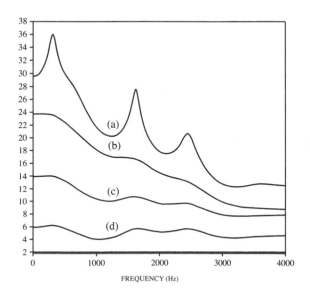

Figure 3.47: Spectra of (a) $1/A(z)$; (b) $1/A(z/0.8)$; (c) $A(z/0.5)/A(z/0.8)$; (d) $(1 - 0.5z^{-1})A(z/0.5)/A(z/0.8)$.

$A(z/0.5)/A(z/0.8)$ ($\alpha = 0.8$ and $\beta = 0.5$). When $\beta = 0$ the filter $1/A(z/\alpha)$ is a bandwidth expanded synthesis filter with bandwidth expansion given by Equation (3.89). Due to the spectral tilt of the filter it acts as a low-pass filter and the postfiltered speech is somewhat muffled. Using the numerator reduces the spectral tilt while keeping the enhancement of the spectral resonances and this improves the postfilter performance. Choosing the parameters α and β depends on the bit rate used, and at 4.8 kb/s the values $\alpha = 0.8$ and $\beta = 0.5$ have been suggested [113]. To further remove the spectral tilt the following form can be used:

$$F(z) = (1 - \mu z^{-1})\frac{A(z/\beta)}{A(z/\alpha)}. \tag{3.237}$$

The spectrum of this filter is shown in Figure 3.47 (d) for a value of $\mu = 0.5$. A value of $\mu = 0.5$ was found adequate [113]; however, it is better to use adaptive postfiltering in which μ is changed according to the speech characteristics. The value of μ can be derived using the first reflection coefficient.

Pitch postfiltering has also been proposed to enhance the pitch periodicity in the reconstructed speech. The pitch postfilter has the form

$$\frac{1}{P'(z)} = \frac{1}{1 - \epsilon G z^{-M}}, \tag{3.238}$$

where G is the LTP gain, M is the LTP delay, and ϵ is a fraction which lies around 0.3 [85]. The pitch and spectral postfilters are cascaded where pitch postfiltering is applied first. Another enhancement configuration was proposed in [95] where the pitch enhancing filter is applied to the decoded LPC excitation rather than the reconstructed speech (the filter is called prefilter in this case). The pitch prefilter in this case enhances the periodicity in the excitation signal. The speech is then reconstructed using the prefiltered excitation and spectral postfiltering is applied afterwards. It is argued that this configuration reduces the artifacts in the reconstructed speech due to waveform discontinuities which pitch postfiltering sometimes introduces.

In general, the postfilter causes an amplification of the speech signal and gain control techniques have to be used to compensate for the gain differences. This can be done either on a sample-by-sample basis [113] or on a block basis.

In previous sections of this chapter we have considered a range of analysis-by-synthesis speech codecs, such as multi-pulse excited, regular-pulse excited and code-excited linear predictive schemes, which were capable of delivering toll-quality speech between 13 and 4.8 kbps. Below 4.8 kbps, however, even CELP codecs fail to guarantee toll- or near-toll quality. In the forthcoming section we will focus our attention on a variety of techniques, which can be invoked for bit rates below 4.8 kbps.

3.7   Coding at Rates Below 2.4 kbps [114]

3.7.1   Overview and Background

A range of recently investigated low-rate speech coding schemes are Prototype Waveform Interpolation (PWI) proposed by Kleijn [115], Multi-Band
Excitation (MBE) suggested by Griffin and Lim [116] and Interpolated Zinc
Function Prototype Excitation (IZFPE) codecs advocated by Hiotakakos
and Xydeas [117]. Lastly, the standardization of the 2.4 kbps US Department of Defence (DoD) codec led to intensive research in this very low-rate
range.

The seven 2.4 kbps DoD candidate coders fell, disproportionately, into
two categories. Harmonic coders constitute the first family, which can be
further sub-divided into Multiband Excitation (MBE) [116, 118] and sinusoidal coders [119, 120]. Four of the candidate speech coders fell into
the harmonic coder category, while the remaining candidate coders were
highlighted in references [18, 121, 122]. Following a set of rigorous comparative tests, the Mixed Excitation Linear Predictive (MELP) codec by Texas
Instruments was selected for standardization [123].

Against this background, our elaborations in this section are centred
around the popular low bit rate speech compression technique of waveform
interpolation (WI), pioneered by Kleijn [115]. In waveform interpolation a
characteristic waveform - which is also referred to as the prototype waveform - is periodically located in the original speech signal. Between these
selected prototype segments smoothly evolving interpolation is employed
in order to reproduce the continuous synthesized speech signal. The interpolation can be performed in either the frequency or time domain, dis-
tinguishing two basic interpolation sub-classes. Since only these prototype
segments have to be encoded, the required bit rate is low, while maintaining
good perceptual speech quality.   Most WI architectures rely on Kleijn's
frequency domain approach [18, 115], although there are schemes, such as
the proposed one, which employ time-domain based coding [117, 125]. A
complication with any WI scheme is the need for interpolation between
two prototype segments, which have different lengths. In this section we
proposed a parametric excitation, which permits simple time-domain interpolation, as it will become explicit during our further discourse.

In traditional vocoders the decision as regards to the extent and nature
of voicing in a speech segment is critical. Pitch detection is an arduous task
due to a number of factors, such as the non-stationary nature of speech, the
effect of the vocal tract on the pitch frequency and the presence of noise.
Incorrect voicing decisions cause distortion in the reconstructed speech, and
distortion is also apparent if the common phenomenon of pitch doubling
occurs. Thus for any low bit rate speech codec the pitch detector chosen is vital in determining the resulting synthesized speech quality. Many
different methods exist for pitch detection of speech signals [124], giving

an indication of the difficulty involved in producing a robust pitch detector. Perhaps the most commonly used approaches are the autocorrelation based methods, where following the determination of the Autocorrelation Function (ACF) for a segment of speech, the time-offset corresponding to the normalized correlation maximum is deemed to be the pitch duration. The normalizing parameter is the autocorrelation at zero delay, namely the signal's energy. If the maximum value exceeds a certain threshold, the segment of speech is considered voiced; beneath this threshold an unvoiced segment is indicated.   Recently the wavelet transform has been applied to the task of pitch detection [126,127]. The wavelet approach to pitch detection is event-based, implying that both the pitch period and the instant of glottal closure are determined. In the proposed speech codec we employed a wavelet-based pitch detector.

Apart from the wavelet-based pitch-detection, this section additionally investigates the low bit rate technique of multiband and mixed excitation, which is used in conjunction with the WI scheme. Multiband and mixed excitation both attempt to reduce the artifact termed 'buzziness', by eliminating the binary classification into entirely voiced or unvoiced segments. This 'buzzy' quality is particularly apparent in portions of speech, which have dominant voicing in some frequency regions, but dominant noise in other frequency bands of the spectrum.   We commence our discussions in Section 3.7.2 by a brief overview of wavelets, specifically the polynomial spline wavelets introduced by Mallat and Zhong [128]. Following the introduction to wavelets we then proceed by applying the theory of wavelets to pitch detection, which was proposed before for example in references [126,127,132]. However, **we refined these wavelet-based pitch-detection techniques and incorporated them in a correlation-based pitch-detector, which allowed us to achieve a substantial pitch-detection complexity reduction, as it will be demonstrated in Figure 3.55.** Subsequently we introduced the wavelet-based pitch detector into a waveform interpolation speech codec. In Section 3.7.5 we highlight the concept of the pitch prototype segment selection process of Figure 3.56 and the features of the Zinc basis function excitation of Figure 3.53, which was originally proposed by Sukkar, LoCicero and Picon [134] and dramatically refined by Hiotakakos and Xydeas [117]. **Our contribution in the Zinc-function excitation (ZFE) optimization is a further refinement of the technique, leading to the realization that the position of these ZFE pulses does not have to be explicitly signalled to the decoder, which reduces the bit-rate.** We then refine the processing of voiced-unvoiced transitions and discuss the process of smooth excitation interpolation between prototype segments, as it will be highlighted in Figure 3.56 and characterize the Zinc-excited codec performance. Finally, **we propose combining the well-known technique of mixed voiced/unvoiced multiband excitation [116] (MBE) with our ZFE-based codec in Figure 3.58 of Section 3.7.13, in order to**

reduce the binary voiced/unvoiced decision induced 'buzziness' and hence to further improve the reconstructed speech quality. We summarize our findings in Sections 3.7.15 and 3.7.16. Let us now focus our attention on the proposed wavelet-based pitch detector.

3.7.2 Wavelet-Based Pitch Detection

In recent years wavelets have stimulated significant research interests in a variety of applications. Historically, the theory of wavelets was recognized as a distinct discipline in the early 1980s. Daubechies [129] and Mallat [130] have substantially advanced this field in various signal processing applications. In this section wavelets are harnessed to reduce the computational complexity and improve the accuracy of an autocorrelation function (ACF) based pitch detector. We commence with a brief introduction to wavelet theory.

A wavelet is an arbitrary function, which obeys certain conditions [129] that allow it to represent a signal $f(x)$ by a series of basis functions, which is described by:

$$f(x) = \sum_{jk} d_{jk} \psi_{jk}(x), \qquad (3.239)$$

where d_{jk} are the coefficients of the decomposition and ψ_{jk} are the basis functions.

The wavelet transform can be used to analyse a signal $f(x)$, but unlike the short-time Fourier transform, its localization varies over the time-frequency space. This flexibility in resolution makes it particularly useful for analysing discontinuities, where during a short time period an extensive range of frequencies is present. It can be noted that the instance of glottal closure is represented by a discontinuity in the speech waveform.

The characterization of wavelets centres around the so-called mother wavelet ψ, from which a class of wavelets can be derived as follows [131]:

$$\psi_{a,b}(t) = \frac{1}{\sqrt{a}} \psi\left(\frac{t-b}{a}\right), \qquad (3.240)$$

where a is the frequency, or dilation variable and b is the position, or time-domain translation, parameter. Thus, wavelets exist for every combination of a and b. Koornwinder's book [131] is suggested for further augmenting these concepts.

In this section the polynomial splines suggested by Mallat and Zhong [128] are adopted for the wavelets. These polynomial spline wavelets can be implemented effectively using a pyramidal algorithm similar to subband filtering, as shown in Figure 3.48, where the high-pass filtered signals are D_j, while the low-pass filtered signals are A_j.

The discrete dyadic wavelet transform $D_Y WT$ of a 20 ms segment of speech is exemplified in Figure 3.49. As the wavelet scales increase towards

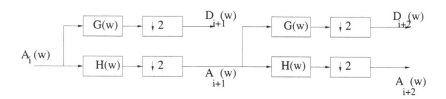

Figure 3.48: Pyramidal algorithm for multiresolution analysis, ©IEEE, Brooks and Hanzo 1998 [114].

the bottom of the Figure, the periodicity of the speech signal becomes more evident from both the time- and frequency domain plots shown in Figure 3.49 for the $D_j(\omega)$ signals. We note here that although the time-domain waveforms of Figure 3.49 are plotted on the finest scale, corresponding to a sampling frequency of 8 kHz, they are waveforms that can be sub-sampled by a factor of 2^j, thus have lower effective time-domain resolution, as evidenced by their frequency-domain plots. Hence, while the higher wavelet scales give a clear indication of the pitch frequency, the lower scales give the most accurate description of the time-domain position of discontinuities, which are typically associated with the glottal closure instants (GCI).

The reduction in computational complexity of the ACF is achieved by reducing the number of pitch periods which require their autocorrelation determined, thus the wavelet analysis determines a set of candidate pitch periods, which are then passed to the ACF process. The selection of these candidate pitch periods is described next.

Observing Figure 3.49 we concluded that some form of preprocessing must be performed, in order to determine the instants of glottal closure, and hence the fundamental or pitch frequency of the speech waveform. In Figure 3.49 the maxima and minima during each scale of the $D_Y WT$ provide the most pertinent information about the speech waveform's pitch period. Hence, the left-hand trace of Figure 3.50 illustrates the initial preprocessing, whereby positive impulses are placed at the time-domain waveform maxima and negative impulses at the minima. Each of these impulses is assumed to represent possible instants of glottal closure.

The highest permitted fundamental frequency is 400 Hz, corresponding to a pitch period of 2.5 ms, thus the impulses of Figure 3.50 (a) placed at the maxima must be at least 2.5 ms apart, and similarly, the impulses placed at the minima are also at least 2.5 ms apart. Additionally, only impulses which occur in every wavelet scale are considered as potential glottal pulse locations. Finally, the glottal closure instants are normalized and combined, where the impulse magnitudes indicate our confidence in the assumed position of the glottal closure instant. This process is described

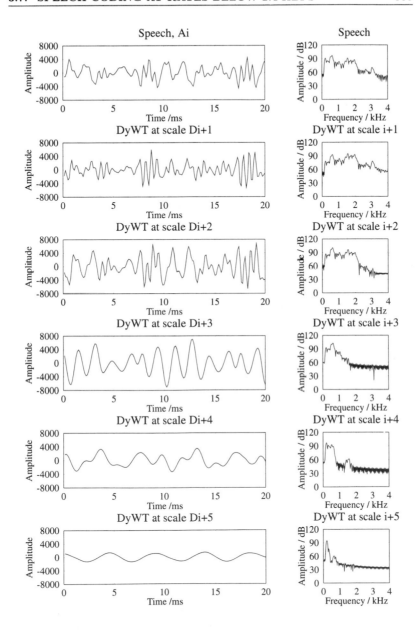

Figure 3.49: The $D_Y WT$ of 20ms of speech for a male speaker uttering "live". For each scale of the $D_Y WT$ the time and frequency domain response are portrayed, enabling the process of the $D_Y WT$ to be clearly interpreted. The $D_Y WT$ scales are 2000-4000 Hz, 1000-2000 Hz, 500-1000 Hz, 250-500 Hz, and 125-250 Hz, respectively; ©IEEE, Brooks and Hanzo 1998 [114].

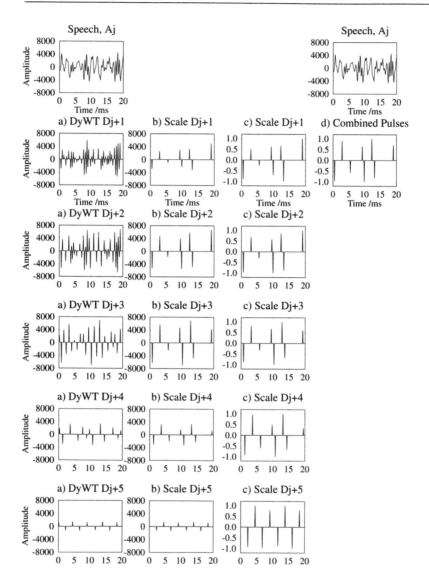

Figure 3.50: The $D_Y WT$ of 20 ms of speech for a male speaker uttering "live".
In column a) the corresponding impulses have been placed at
the locations of the maxima and minima of the detail signals,
$D_{j+1} \cdots D_{j+5}$ of Figure 3.49. In column b) only those maxima
and minima which persist in every scale are kept. In column c) all
the maxima and minima are normalized. Finally, in column d) the
scales are amalgamated to produce combined maxima and minima
representing all scales; ©IEEE, Brooks and Hanzo 1998 [114].

in Figure 3.50 (c) and (d). Assuming that the largest positive and negative pulses are true glottal pulse locations, a range of possible pitch periods can be calculated. Namely, the candidate pitch periods are classified on the basis of the time durations between the largest positive pulse and all other positive pulses, or the largest negative pulse and all other negative pulses.

The candidate pitch periods from the $D_Y WT$ can now be used to reduce the computational complexity of the ACF. However, first a brief description of how wavelet analysis can be used for voiced-unvoiced decisions is given.

3.7.3   Voiced-Unvoiced Decisions

The ability of the $D_Y WT$ to categorize speech as voiced or unvoiced has been shown previously [126, 132], and hence similar to these two methods, we briefly describe a voiced-unvoiced decision method based on the energy of the speech signal. The process of the $D_Y WT$ across the scales gradually removes the higher frequency components present in the speech waveform, as it was shown in Figure 3.49. For unvoiced speech most energy is present in the higher frequencies, while voiced speech has more of a low-pass nature. A suitable parameter for evaluating voiced/unvoiced decisions was found to be the ratio of the RMS energy in the frequency range 2kHz→4kHz, to that in the frequency band 0kHz→2kHz.

3.7.4   Pitch Detection

Figure 3.51 displays the potential pitch periods for each speech frame in a speech file. The resultant graph is fairly complex; however, it can be observed that the candidate pitch periods are commonly placed at the true pitch period and its harmonics. Typically the true pitch period and two or three harmonics are present. There can be at most 15 candidate pitch periods. Namely, in each 20 ms speech frame there can be a maximum of seven pitch intervals that are spaced at least 20 samples apart, for both positive and negative pitch-related pulses. Additionally, the previous pitch period may be reintroduced, yielding a total of 15 potential GCI locations, for which the autocorrelation has to be computed.

The performance of this wavelet-based ACF pitch detector was compared to the performance of the ACF pitch detector without the $D_Y WT$ pre-processing invoked to determine candidate pitch periods. The performance of both pitch detectors was evaluated using 20 s of mixed male and female speech, which had been manually pitch-tracked. A pitch-detection error was recorded, when either the voiced-unvoiced detector operated incorrectly or if a gross pitch error occurred. The results are shown in Table 3.13, where it can be seen that the wavelet-based ACF pitch period detector has the lowest overall error rate of $W_U + W_V + P_G$ of 3.9%, which was defined in the caption of Table 3.13.

The performance of the wavelet-based ACF pitch detector and that

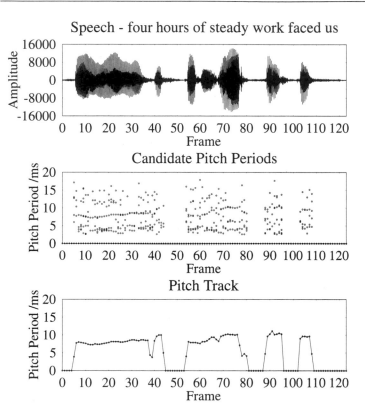

Figure 3.51: The original speech signal (top trace), the candidate pitch periods (middle trace) for a British male speaker and the decided pitch periods (bottom trace). The potential pitch periods tend to consist of the true pitch and its harmonics. The detector suffers from pitch halving around frame 40 and 80; ©IEEE, Brooks and Hanzo 1998 [114].

Pitch Detector	$W_U\%$	$W_V\%$	$P_G\%$	Total %
wavelets-based ACF	1.3	0.3	2.3	3.9
ACF	1.6	5.3	5.8	12.7

Table 3.13: A comparison between the performance of ACF pitch detection with and without incorporating wavelet analysis. W_U represents the percentage of frames that are labelled voiced when they should have been identified as unvoiced. W_V indicates the number of frames that have been labelled as unvoiced when they are actually voiced. P_G represents the number of frames where a gross pitch error has occurred. The total number of incorrect frames is given as $W_U + W_V + P_G$. ©IEEE, Brooks and Hanzo 1998 [114].

of the stand-alone ACF pitch detector can also be compared in terms of computational complexity. For the ACF pitch detector the computational complexity is based on the autocorrelation value determined for the legitimate pitch periods of $20 \rightarrow 147$ samples at a sampling frequency of 8 kHz, or 127 values, producing a computational complexity of 3.35MFLOPS. The computational complexity of the wavelet-based ACF pitch detector is dependent on two factors, namely the wavelet analysis and the autocorrelation calculation for the 15 candidate pitch periods. These produced complexities of 2.23MFLOPS and 0.62MFLOPS, respectively. Thus, the wavelet-based ACF pitch detector has a lower overall computational complexity of 2.85MFLOPS.

Following the above employment of the wavelet transform to reduce the complexity of an autocorrelation based pitch detector, this pitch detector was included in a waveform interpolation low bit rate speech codec, which is the topic of the next section.

3.7.5 Basic Zinc-excited Coding Algorithm

Our WI codec of Figure 3.52 operates on 20 ms speech frames, for which LPC analysis is performed. The LPC coefficients are transformed to line spectrum frequencies (LSFs) and vector quantized to 18 bits/frame using an LSF coding scheme similar to that of the G.729 ITU codec [133]. Following LPC analysis, pitch detection and a voiced-unvoiced (V/U) decision are performed, where the pitch-detection algorithm is based on the novel technique of employing the wavelet transform described above in Section 3.7.2. For this pitch detector the pitch period is the distance between two adjacent located glottal closure instants (GCIs). For an unvoiced frame the Root-Mean-Square (RMS) value of the LPC residual is determined, allowing random Gaussian noise to be scaled appropriately and used as unvoiced excitation. The speech waveform is perceptually weighted in the frequency regions of the speech formants and hence masks the effects of the coding noise.

Since voiced speech segments typically exhibit a higher perceptual importance than unvoiced frames, these segments are more comprehensively defined in our codec, as it will be detailed below. For a voiced perceptually weighted speech frame a prototype segment is selected, in order to represent a full cycle of the pitch period, which is then passed to an analysis-by-synthesis loop, as portrayed in Figure 3.52 and detailed throughout the rest of the section, for the selection of the best voiced excitation. Explicitly, we opted for using the orthogonal Zinc basis functions of Figure 3.53, in order to model the voiced prototype segments, which, owing to their specific shapes were shown by Sukkar, Cicero and Picone [134] to out-perform the Fourier series based representation of the prediction residual in analysis-by-synthesis coding of speech. Accordingly, analysis-by-synthesis excitation optimization is invoked, in order to determine the best Zinc function ex-

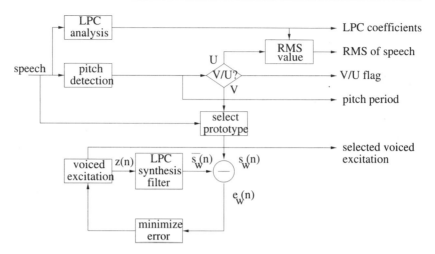

Figure 3.52: Schematic of a time domain prototype WI arrangement; ©IEEE, Brooks and Hanzo 1998 [114].

citation (ZFE) for each prototype segment of voiced speech, a technique proposed by Hiotakakos and Xydeas [117]. The proposed speech codec is termed a prototype waveform interpolation zinc function excitation (PWI-ZFE) scheme. These issues will be elaborated on in Section 3.7.7 and will also be detailed with reference to the characteristic waveforms of Figure 3.56. The ZFE is then quantized and the corresponding parameters detailed in Section 3.7.7 are passed to the decoder. At the decoder the excitation is determined by interpolating between the adjacent excitation prototype segments, an issue to be treated in more depth in Section 3.7.11 in the context of Figure 3.56. Subsequently the excitation generated by interpolation is passed through the LPC synthesis filter of Figure 3.52 in order to reproduce the synthesized speech signal.

Following the above rudimentary overview of the speech codec, the next section offers a more detailed discussion on the different sections of the speech codec shown in Figure 3.52. Particular emphasis will be placed on the ZFE optimization process. Let us continue by considering the pitch prototype segment's identification.

3.7.6   Pitch Prototype Segment

The determination of the prototype segment in a sequence of voiced speech frames commences by selecting the pitch prototype segment for the first frame in a voiced frame sequence [117]. If P is the pitch of the voiced frame, which was determined by our wavelet-based pitch-detector of Section 3.7.2, then the prototype segment will also be P samples in length. The prototype segment is deemed to begin at a zero crossing immediately to the left of a speech waveform maximum near the centre of the speech frame, an approach suggested and detailed by Hiotakakos and Xydeas in [117].

For the subsequent prototype segments within the voiced sequence of frames the current pitch prototype segment is found by employing Kleijn's cross-correlation based technique [115]. Explicitly, the prototype segment is located by finding the position of maximum cross correlation between the current speech frame and the previous prototype segment, ensuring as much similarity between prototype segments as possible. Let us now highlight the features of the Zinc basis function.

3.7.7 Zinc Function Excitation

As mentioned above, the voiced excitations of our codec were derived from the orthogonal Zinc basis functions [134], which have previously been advocated by Hiotakakos and Xydeas [117] for a sophisticated higher bit rate interpolation scheme. The Zinc function $z(t)$ was defined by Sukkar *et al.* [134] as:

$$z(t) = A \cdot \mathrm{sinc}(t - \lambda) + B \cdot \mathrm{cosc}(t - \lambda) \tag{3.241}$$

where

$$\mathrm{sinc}(t) = \frac{\sin(2\pi f_c(t - \lambda))}{2\pi f_c(t - \lambda)} \tag{3.242}$$

and

$$\mathrm{cosc}(t) = \frac{1 - \cos(2\pi f_c(t - \lambda))}{2\pi f_c(t - \lambda)}. \tag{3.243}$$

For discrete time processing with a speech bandwidth of f_c=4 kHz and a sampling frequency of f_s=8 kHz we have [117]:

$$z(n) = A \cdot \mathrm{sinc}(n - \lambda) + B \cdot \mathrm{cosc}(n - \lambda) \tag{3.244}$$

$$= \begin{cases} A & n - \lambda = 0 \\ \frac{2B}{(n-\lambda)\pi} & n - \lambda = \text{odd} \\ 0 & n - \lambda = \text{even} \end{cases} .$$

The ZFE model's typical shape is shown by Figure 3.53, where the coefficients A and B describe the function's amplitude and λ defines its position.

Once the A and B parameters have been determined, they are Max-Lloyd scalar quantized with 6-bits for each A and B parameter. This requires knowledge of their Probability Density Function (PDF), which is portrayed for a given training sequence in Figure 3.54. The Max-Lloyd quantizer was used to create 4, 5 and 6-bit scalar quantizers for both the A and B parameters. Table 3.14 show the SNR values for the quantized A and B parameters for the various quantization schemes. On the basis of our subjective and objective investigations we concluded that the 6-bit quantization constitutes the best compromise in terms of bit rate and

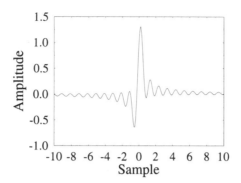

Figure 3.53: Typical shape of a Zinc basis function, using the expression $z(n) = A \cdot \mathrm{sinc}(n - \lambda) + B \cdot \mathrm{cosc}(n - \lambda)$; ©IEEE, Brooks and Hanzo 1998 [114].

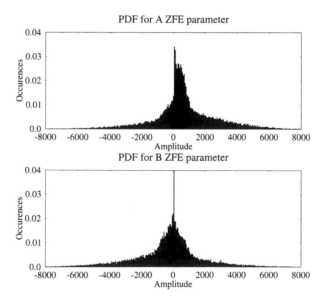

Figure 3.54: Typical PDF of the A and B ZFE parameters, created from 8 mins of BBC Radio 4's Book at Bedtime; ©IEEE, Brooks and Hanzo 1998 [114].

speech quality. We now continue our discourse by considering the excitation optimization process.

Quant. Scheme	SNR /dB for A	SNR /dB for B
4-bit	10.45	10.67
5-bit	18.02	19.77
6-bit	26.47	27.07

Table 3.14: SNR values for scalar quantization of the A and B ZFE Parameters.

3.7.8 Excitation Optimization

From Figure 3.52 the perceptually weighted error signal $e_w(n)$ can be described by [117]:

$$e_w(n) \quad = \quad s_w(n) - \bar{s}_w(n) \tag{3.245}$$
$$= \quad s_w(n) - m(n) - (z(n) * h(n)), \tag{3.246}$$

where $m(n)$ is the memory of the LPC synthesis filter due to previous excitation segments, while $h(n)$ is the impulse response of the synthesis filter. Thus, the optimization of the excitation signal involves comparing the perceptually weighted error signal $e_w(n)$ for all legitimate values of λ in the range of $[1 \rightarrow \text{pitch period}]$, and calculating the corresponding optimum A and B values, which minimize the weighted error for the given λ. The filter memory $m(n)$ was the same as used by Hiotakakos and Xydeas [117].

There are four possible phases of the ZFE, produced by four combinations of positive or negative valued A and B parameters. If the ZFE phase defined in this way is not maintained throughout a voiced frame sequence, simply because the optimum A or B value has changed sign, the smooth ZFE interpolation process will introduce a sign change for A or B. This results in some small valued interpolated ZFEs, as the values of A or B pass through zero. For each legitimate Zinc pulse position of λ, the sign of A and B are initially checked during the excitation optimization process, and only if the phase restriction of the voiced frame sequence is maintained is the excitation deemed valid. It is feasible that a suitably phased ZFE will not be found. If this occurs, then the previous ZFE is scaled using the RMS value of the LPC residual and repeated for the current voiced frame [117].

3.7.9 Complexity Reduction

The complexity of the error minimization process described in Section 3.7.8 is critical in terms of determining the practicality of the codec. The associated complexity for the optimization is evaluated as follows. The ZFE optimization has a computational complexity dominated by the convolution between the sinc and cosc functions and the impulse response $h(n)$, which is necessary, according to the schematic of Figure 3.52, for the optimization loop. This complexity is dependent on the pitch period, or length of the

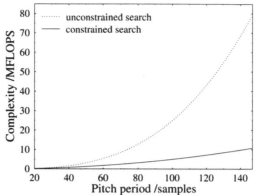

Figure 3.55: Computational complexity for the permitted pitch period range of 20 to 147 sample duration, for both an unrestricted and constrained search; ©IEEE, Brooks and Hanzo 1998 [114].

prototype segment, where the complexity dependence on the pitch period is created by the prototype segment length, over which the convolution is performed that may vary from 20 to 147 samples or 50 Hz to 400 Hz fundamental frequency. The dashed line curve of Figure 3.55 demonstrates the relationship between the ZFE optimization complexity and pitch period, when no restrictions are imposed on this optimization process.

This curve indicates that if every location λ within the prototype segment were examined, the complexity of ZFE optimization would be prohibitive for real-time implementations. The complexity increase is exponential, as shown by Figure 3.55, where it can be seen that any pitch period greater than 90 samples in duration will exceed a complexity of 20 MFLOPS in terms of the ZFE optimization search.

The complexity of the ZFE minimization procedure can be reduced by considering the glottal closure instants (GCI) introduced in Section 3.7.2. In Section 3.7.2 wavelet analysis was harnessed to produce a pitch detector, where the pitch period was determined as the distance between two GCIs. These GCIs indicate the snapping shut, or closure, of the vocal folds, which provides the impetus for the following pitch period. The energy peak caused by the GCI will typically be in close proximity to the position of the ZFE placed by the ZFE optimization process. This allows the complexity reduction of the analysis-by-synthesis process. As noted before, Figure 3.55 shows that as the number of possible ZFE positions increases linearly, the computational complexity increases exponentially. Hence, constraining the number of ZFE positions will ensure that the computational complexity remains at a realistic level. This constraining process is described next.

The first frame in a voiced frame sequence has no minimization procedure; simply a single ZFE pulse is situated at the glottal pulse location

	Unconstrained search	Constrained search
No phase restrictions	3.36 dB	2.68 dB
Phase restrictions	2.49 dB	1.36 dB

Table 3.15: SEGSNR results for the ZFE optimization process in voiced segments, where in contrast to unvoiced segments the SEGSNR can be computed, with and without phase restrictions, or a constrained search; ©IEEE, Brooks and Hanzo 1998 [114].

within the prototype segment. For the other voiced frames, in order to maintain a moderate computational complexity, the number of possible ZFE positions is restricted as if the pitch periods were always 20 samples. A suitable constraint is to have the ZFE located within ±10 samples of the instant of glottal closure situated in the pitch prototype segment. In Figure 3.55 the solid line represents the computational complexity of a restricted search procedure in locating the ZFE. The maximum complexity for a 147 sample pitch period is 11MFLOPS.

We note, however that constraining the location of the ZFE pulse to within ±10 positions with respect to the GCIs reduces the SEGSNR value, as shown in Table 3.15. The major drawback of the constrained search is the possibility that the optimization process is degraded through the limited range of ZFE locations searched. Additionally, it is possible to observe a slight degradation to the Mean Squared Error (MSE) optimization, caused by the phase restrictions imposed on the ZFEs, necessary to permit smooth interpolation. Table 3.15 displays the SEGSNR values of the concatenated purely voiced prototype speech segments, for which the SegSNR values can be computed. By contrast, the unvoiced segments are ignored, since these speech spurts are represented by noise, thus a SEGSNR value would be meaningless.

Observing Table 3.15 for a totally unconstrained search, the SEGSNR achieved by the ZFE optimization loop is 3.36 dB. The process of either implementing the above-mentioned ZFE phase restriction or constraining the permitted ZFE locations to the vicinity of the GCIs reduces the voiced segments' SEGSNR after ZFE optimization by 0.87 dB and 0.68 dB, respectively. Restricting both the phase and the ZFE locations reduces the SEGSNR by 2 dB. However, in perceptual terms the ZFE interpolation procedure, described in Section 3.7.11, actually improves the subjective quality of the decoded speech due to the smooth speech waveform evolution facilitated, despite the SEGSNR degradation of about 0.87 dB caused by imposing phase restrictions. To assess the impact of constraining the location of the ZFEs, listening tests were conducted, where eight listeners were asked to express a preference between the output speech from constraining the ZFE locations and not constraining the ZFE locations. Three sentences were played to each listener. It was found that 45.8% of listeners preferred the output speech where the ZFE locations had been constrained, while 54.2% of listeners preferred the output speech, where the ZFE locations had not been constrained. The preference values allowed us to justify constraining the ZFE locations, yielding the corresponding computational complexity reductions seen in Figure 3.55.

We now proceed by devoting some attention to improving the representation quality of voiced-unvoiced transitions.

3.7.10 Voiced-Unvoiced Transition

In low bit rate speech codecs typically the worst represented portion of speech is the rapidly evolving on-set of voiced speech. Previous speech codecs have been found to produce better quality speech by locating the emergence of voicing as precisely as possible [117, 135]. Once again, the GCIs inferred from our wavelet transform based pitch detector of Section 3.7.2 are used to determine the onset of voicing. Specifically, if frame N is voiced and frame $N - 1$ is unvoiced, then the end of frame $N - 1$ is examined for the evidence of an emerging voiced segment. If GCIs exist at or near the locations, which would maintain the periodicity of voiced speech in frame $N - 1$, then the voiced speech region is extended to cover the end of the predominantly unvoiced frame $N - 1$, otherwise the region of speech belonging to frame $N - 1$ is confirmed as purely unvoiced. A similar procedure is implemented at the end of a string of voiced frames. We marked the location of the voiced-unvoiced transition by the parameter b_s, which encodes the number of voiced pitch-duration speech cycles within unvoiced frames. Following this description of the speech encoder the interpolation process harnessed in the decoder is examined.

3.7.11 Excitation Interpolation

The adopted ZFE parametric representation of the voiced excitation permits simple linear interpolation at the decoder to reinsert the Zinc-pulse locations, which were not transmitted. These issues are detailed below with reference to Figure 3.56. Figure 3.56 follows the spirit of the work by Hiotakakos and Xydeas [117], and shows an example of the ZFE excitation based reconstruction of a 60 ms speech segment for a female speaker. Specifically, the top trace shows a 60 ms segment of the original speech signal, the second trace displays the prototype segments identified, while the third one shows the corresponding ZFE. In the fourth trace the ZFE amplitude parameters A and B are linearly interpolated between the corresponding A and B values of the prototype segments. The bottom trace shows the reproduced speech waveform which exhibits a close waveform similarity.

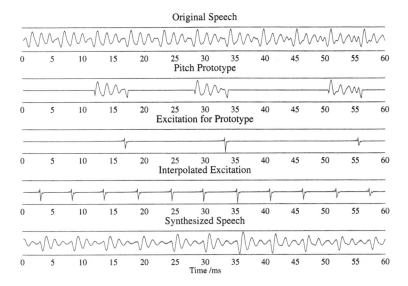

Figure 3.56: Example of 60 ms segments of the original speech (top to bottom), the pitch prototype and its Zinc-model as well as the interpolated excitation and the synthesized speech for a voiced utterance by a female speaker uttering /ɔ/ in 'dog'; ©IEEE, Brooks and Hanzo 1998 [114].

Interpolating the position of the ZFEs in a similar manner to their amplitudes does not produce a smoothly evolving excitation signal. Instead, the pulse position within each prototype segment is kept stationary throughout a voiced frame sequence. This introduces time misalignment between the original and synthesized waveforms, but maintains a smooth excitation signal. In order to compensate for changes in the length of prototype segments the normalized location of the initial ZFE position is calculated according to:

$$\lambda_r = \frac{\lambda_1(N)}{P(N)}, \tag{3.247}$$

where $P(N)$ is the pitch period of the first frame in the voiced frame sequence. For all subsequent frames in the voiced frame sequence the position of the ZFE is calculated by:

$$\lambda_1(N) = \text{rint}\{\lambda_r * P(N)\} \tag{3.248}$$

where $\text{rint}\{\cdot\}$ represents rounding to the nearest integer.

In the λ_1 transmission procedure, although λ_1 is transmitted every frame, only the first λ_1 in every voiced sequence is used in the interpolation process, thus, λ_1 is predictable and hence it contains redundancy. Furthermore, when constructing the excitation waveform at the decoder, every ZFE is permitted to extend over three interpolation regions, namely its allotted region together with the previous and the next region. This allows ZFEs near the interpolation region boundaries to be fully represented in the excitation waveform, while ensuring that every ZFE will have a tapered low energy value when it is curtailed. Hence, as an improvement of the scheme in [117] it is suggested that the true position of the ZFE pulse, λ_1, is arbitrary and need not be transmitted. Following this hypothesis, our experience shows that we can set $\lambda_1 = 0$ at the decoder, which has no degrading effect on the speech quality.

In order to perform the interpolation procedure described above the zero-crossing parameter of the prototype segments, which was intorduced in [117] must be transmitted to the decoder. However, it can be observed that the zero-crossing values of the prototype segments are approximately a frame length apart, thus following the principle of interpolating between prototype segments near the centre of the frame, which are also transmitted every frame. Hence, instead of explicitly transmitting the zero-crossing parameter, it can be assumed that the start of the consecutive prototype segments is a frame length apart. An arbitrary starting point for the prototype segments could be $FL/2$, where FL is the speech frame length.

3.7.12   1.9 kbps ZFE-WI Codec Performance

The speech segment displayed in Figure 3.57(a) was recorded for a female speaker. As seen by comparing Figures 3.57(a) and 3.57(c) in the time domain, the basic shape of the original waveform is more or less preserved. Additionally, by observing the Figures 3.57(a) and 3.57(c) we note from the frequency-domain plots that the formant location is preserved; however, it is noticeable that the inclusion of unvoiced speech above 1800 Hz is not modelled well by the distinct voiced-unvoiced nature of the PWI-ZFE scheme. Observing the ZFE waveform of Figure 3.57(b), a flat excitation frequency domain envelope is produced, while its spectral fine-structure reflects the pitch-dependent needle-like behaviour. Informal listening tests showed that the reproduced speech contained slight 'buzziness', and hence it was somewhat less transparent, than the original speech.

The bit allocation for the ZFE coder is summarized in Table 3.16, where 18 bits are reserved for LSF vector-quantization [133], while a one-bit flag is used for the V/U classifier. For unvoiced speech the RMS parameter is scalar quantized with 5-bits. The b_s offset requires a maximum of 3-bits to encode the voiced-unvoiced transition point in terms of the number of voiced speech cycles within unvoiced frames, since assuming a minimum pitch duration of 20 samples, a maximum of eight pitch cycles can fit in a 160-sample speech frame. For voiced speech the pitch can vary from $20 \rightarrow 147$ samples, thus requiring 7 bits for transmission. The ZFE ampli-

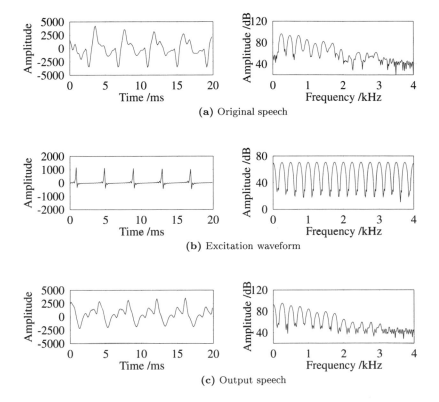

Figure 3.57: Time and frequency domain comparison of the (a) original speech,
(b) ZFE waveform and (c) output speech after the pulse dispersion
filter. The 20 ms speech frame is the liquid /r/ in the utterance
'*rice*' for a female speaker; ©IEEE, Brooks and Hanzo 1998 [114].

Parameter	Unvoiced	Voiced
LSFs	18	18
V/U flag	1	1
RMS value	5	-
b_s	3	-
Pitch	-	7
A	-	6
B	-	6
total/20 ms	27	38
bit rate	1.35 kbps	1.90 kbps

Table 3.16: Bit allocation of the 1.9 kbps PWI-ZFE speech codec;©IEEE,
Brooks and Hanzo 1998 [114].

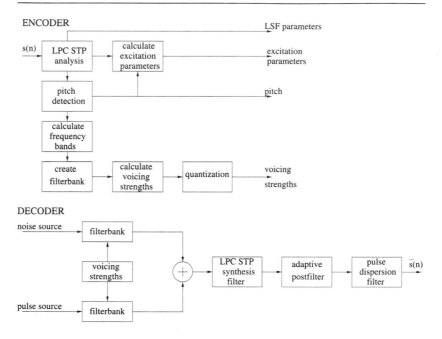

Figure 3.58: Schematic of the MMBE encoder and decoder; ©IEEE, Brooks
and Hanzo 1998 [114].

tude parameters A and B are scalar quantized with 6-bits. Following the
above investigations of the proposed PWI-ZFE speech codec, we now invoke mixed multiband excitation (MMBE) in order to improve the quality
of the speech codec, while increasing the bit rate from 1.9 kbps to 2.4 kbps.

3.7.13 Multiband Excited Codec

Speech typically contains a mixture of voiced and unvoiced excitation across
its frequency bandwidth. Thus, the division of speech into voiced-unvoiced
frames, which has been performed so far, does not follow the true nature
of the speech signal. The well-known speech coding technique of multiband excitation (MBE) [116], which is briefly explained next, is capable of
allowing a mixture of voiced and unvoiced excitation in each speech frame.

3.7.14 The MMBE Coding Algorithm

The encoder and decoder schematics of a MMBE architecture are shown
in Figure 3.58. Following the encoder schematic after LP analysis has been
performed on the 20 ms speech frame, the pitch detection is invoked in
order to locate any evidence of voicing. A frame deemed unvoiced has the

RMS value of its LPC residual quantized and sent to the decoder.

Speech frames labelled as voiced are split into M frequency bands, with M constrained to be a time-invariant constant value. In our scheme $M = 3$ bands were used. Each of the $M = 3$ frequency bands is examined for evidence of voicing [136], and has a voicing strength assigned that was scalar quantized to eight different strengths, allowing us to assign a total of $3 \times 3 = 9$ bits per 20 ms to the three bands. Hence a total rate of 0.45 kbps was required for voicing strength quantization. The 4 kHz frequency spectrum was divided into three frequency bands depending on the pitch period of the speech frame [137]. The chosen method for voiced excitation must also be determined and its parameters sent to the decoder. In our proposed scheme the previously described PWI-ZFE speech codec model was employed. At the decoder, following Figure 3.58, both unvoiced and voiced speech frames have a pair of filter banks created [136]. A consequence of the time-variant pitch period is the need for the filterbank to be reconstructed every frame, in both the encoder and decoder, as shown in Figure 3.58, thus increasing the computational costs. However, for unvoiced frames, the filterbank excitation is declared fully unvoiced and hence no voiced excitation is created. Following Figure 3.58, both the voiced and unvoiced decoder filter banks are created using the knowledge of the pitch period and the number of frequency bands, M. Specifically, the filter bandwidths have to be an integer multiple of the pitch period [116]. For the voiced filter banks the filter coefficients are scaled by the quantized voicing strengths determined at the encoder. A value of 1 represents full voicing, while a value 0 signifies a frequency band of noise. Intermediate values represent a mixed excitation source. For the unvoiced filter bank the voicing strengths are adjusted, ensuring that the voicing strengths of each voiced and unvoiced frequency band combine to unity. This constraint ensures that the combined signal from the filter banks is spectrally flat over the entire frequency range. The mixed excitation speech is then synthesized, as shown in Figure 3.58, where the LPC filter determines the spectral envelope of the speech signal.

3.7.15   2.35 kbps ZFE-MMBE-WI Codec Performance

The combined 2.35 kbps 3-band PWI-ZFE-MMBE scheme was studied in terms of speech quality. The speech frame examined in Figure 3.59 is the same utterance as that characterized in Figure 3.57. Observing Figure 3.59(b) above 2 kHz a mixture of voiced and unvoiced excitation is harnessed. From Figure 3.59(c) it can be seen that the presence of noise above 2 kHz produces a better representation of the frequency spectrum than Figure 3.57(c).

Listening tests were conducted to assess the performance of the developed speech coders. Pairwise-comparison tests were performed where eight listeners were played three different sentences. For each different sentence the listeners were played two versions, version A and version B. Having

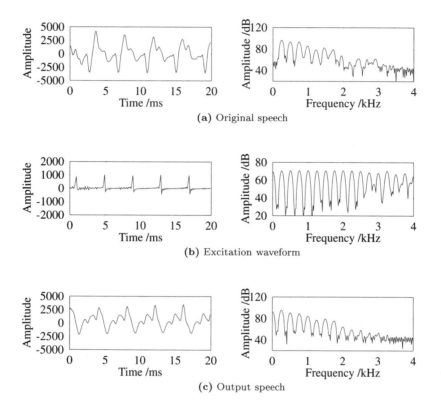

Figure 3.59: Time and frequency domain comparison of the (a) original speech,
(b) **3-band MMBE ZFE** waveform and (c) output speech after
the so-called pulse dispersion filter. The 20 ms speech frame is
the liquid /r/ in the utterance 'rice' . For comparison with the
full-band process refer to Figure 3.57; ©IEEE, Brooks and Hanzo
1998 [114].

Speech Coder A	Speech Coder B	Prefer A%	Prefer B%
1.9 kbps PWI-ZFE	2.35 kbps 3-band MMBE with PWI-ZFE	4.2	95.8
2.35 kbps 3-band MMBE with PWI-ZFE	2.35 kbps 5-band MMBE with a single pulse	79.2	20.8

Table 3.17: Listening tests.

played each version twice, the listeners were asked to express a preference for version A or version B. The 1.9 kbps PWI-ZFE speech coder was compared with the 2.35 kbps speech coders where MMBE has been added for three frequency bands. From Table 3.17 it can be seen that 95.8% of listeners preferred the 2.35 kbps speech coder; note that this speech quality improvement came at the cost of a higher bit rate. The 2.35 kbps 3-band MMBE speech coder with PWI-ZFE was also compared with a 2.35 kbps 5-band MMBE speech coder where a single pulse was used to represent the voiced excitation. From Table 3.17 it can be seen that 79.2% of listeners preferred the PWI-ZFE for representing the voiced excitation.

3.7.16   Summary and Conclusions

The proposed wavelet-based techniques substantially reduced the pitch-search complexity of our codec. The refined PWI-ZFE codec reduced the bit-rate of the scheme proposed in [117], but due to the binary voiced/unvoiced classification it exhibited some 'buzziness', which was mitigated by introducing an 8-level voicing strength in each of the three sub-bands of the MBE-ZFE codec, resulting in a 2.35 kbps arrangement. Our future work is targetted at creating sinusoidally excited benchmarkers.

In this chapter we reviewed a range of analysis-by-synthesis speech codecs, which have found applications in various mobile radio systems over the past decade. Multi-pulse excited and regular-pulse excited schemes were considered, paving the way for introducing code-excited linear predictive schemes, which constitute the most successful class of speech codecs at the time of writing. The chapter was concluded with the portrayal of novel research trends in the field of multi-band excited codec operating around 2.4 kbps. In the forthcoming chapter we will consider the range of error correction codecs often employed in wireless systems.

Bibliography

[1] **J.L. Flanagan, M.R. Schroeder, B.S. Atal, R.E. Crochiere, N.S. Jayant**and **J.M. Tribolet.** "Speech coding". *IEEE Trans. on Commun., vol.27, no.4,* pp. 710-737, April 1979.

[2] **K.W. Cattermole.** "Principles of pulse code modulation". *Iliffe,* London, 1969.

[3] **N.S. Jayant** and **P. Noll.** "Digital coding of waveforms". *Prentice-Hall,* 1984.

[4] **R. Steele.** "Delta modulation systems". *Pentech Press,* London, 1975.

[5] **F.I. Itakura** and **S.I. Saito.** "Analysis-synthesis theory based on the maximum likelihood method". *Proc. of 6th Int. Congress on Acoustics, Tokyo,* pp. c17-20, 1968.

[6] **B.S. Atal.** "Speech analysis and synthesis by linear prediction of speech wave". *J. Acoust. Soc. Am., vol.47,* 1970.

[7] **B.S. Atal** and **J.R. Remde.** "A new model of LPC excitation for producing natural-sounding speech at low bit rates". *Proc. ICASSP'82,* pp. 614–617, 1982.

[8] **P. Kroon, E.F. Deprettere** and **R.J. Sluyter.** "Regular-pulse excitation _ A novel approach to efficient multi-pulse coding of speech". *IEEE Trans. ASSP, Vol. 34, No. 5,* pp. 1054–1063, October 1986.

[9] **B.S. Atal** and **M.R. Schroeder.** "Predictive coding of speech signals". *Bell System Tech. J.,* pp. 1973-1986, October 1970.

[10] **B.S. Atal** and **M.R. Schroeder.** "Stochastic coding of speech signals at very low bit rates". *IEEE Int. Conf. Commun.,* May 1984.

[11] **M.R. Schroeder** and **B.S. Atal.** "Code-Excited Linear Prediction (CELP): High quality speech at very low bit rates". *Proc. ICASSP'85, Tampa, Florida, USA* pp. 937–940, 26-29 March, 1985.

[12] **J.L. Flanagan.** "Speech analysis synthesis and perception". *Springer-Verlag* Berlin, 1972.

[13] **C.R. Rabiner** and **R.W. Schafer.** "Digital Processing of Speech Signals". *Prentice Hall, 1978.*

[14] **D. O'Shaughnessy** *Speech communication: human and machine.* Addison-Wesley, 1987.

[15] **S. Furui.** *Digital Speech Processing, Synthesis and Recognition,* Marcel Dekker Inc., 1989

[16] **A.M. Kondoz.** *Digital Speech: Coding for low bit rate communications systems.* Wiley, 1994.

[17] **W.B.Keijn** and **K.K.Paliwal, Ed.** *Speech Coding and Synthesis,* Elsevier Science, 1995.

[18] **W.B. Kleijn** and **J. Haagen.** "A speech coder based on decomposition of characteristic waveforms," in *Proceedings of ICASSP 95,* pp. 508–511, 1995.

[19] **A. Gersho.** "Advances in Speech and Audio Compression", *Proceedings of the IEEE,* pp. 900–918, June 1994

[20] **I.A. Gerson, M.A. Jasiuk, J-M. Muller, J.M. Nowack** and **E.H. Winter,** "Speech and channel coding for the half-rate GSM channel," *Proceedings ITG-Fachbericht,* vol. 130, pp. 225–233, November 1994.

[21] **T. Ohya, H. Suda** and **T. Miki,** "3.45 kbits/s PSI-CELP of the half-rate PDC speech coding standard ", *Proceeding of the IEEE Conference on Vehicular Technology,* pp. 1680–1684, June 1994

[22] **B.S. Atal.** "Predictive coding of speech at low bit rates". *IEEE Trans. on Commun., vol.30,* pp. 600-614, April 1982.

[23] **C.K. Un** and **D.T. Magill.** "The residual-excited linear prediction vocoder with transmission rate below 9.6 Kbits/s". *IEEE Trans. on Commun., vol.23, no.12,* pp. 1466-1474, December 1975.

[24] **L.R. Rabiner** and **R.W. Shafer.** "Digital processing of speech signals". *Prentice-Hall Int.,* 1978.

[25] **J.D. Markel** and **A.H. Gray, Jr.** "Linear prediction of speech". *Springer-Verlag, New York,* 1976.

[26] **J. Makhoul.** "Stable and efficient lattice methods for linear prediction". *IEEE Trans. on ASSP, vol.25,* pp. 423-428, October 1977.

[27] **A.H. Gray, Jr.** and **D.Y. Wong.** "The Burg algorithm for LPC speech analysis/synthesis". *IEEE Trans. on ASSP, vol.28, no.6,* pp. 609-615, Dec. 1980.

[28] **A. Jennings.** "Matrix computation for engineers and scientists". *Wiley and Sons Ltd.*, 1977.

[29] **B.S. Atal** and **M.R. Schroeder.** "Predictive speech signal coding with reduced noise effects". *U.S. Patent, no. 4,133,976*, January 1979.

[30] **S. Singhal** and **B.S. Atal.** "Improving performance of multi-pulse LPC coders at low bit rates". *Proc. ICASSP'84*, pp. 1.3.1-1.3.4, 1984.

[31] **A.H. Gray, Jr.** and **J.D. Markel.** "Distance measures for speech processing". *IEEE Trans. ASSP, vol.24, no.5*, pp. 380-391, October 1976.

[32] **J. Makhoul, S. Roucos** and **H. Gish.** "Vector quantization in speech coding". *Proc. of IEEE, vol.73, no.11*, pp. 1551-1588, November 1985.

[33] **R.M. Gray.** "Vector quantization". *IEEE ASSP Magazine*, pp. 4-29, April 1984.

[34] **R.M. Gray, A. Buzo, A.H. Gray** and **Y. Matsuyama.** "Distortion measures for speech processing". *IEEE Trans. ASSP, vol.28, no.4*, pp. 367-376, August 1980.

[35] **Y. Linde, A. Buzo** and **R.M. Gray.** "An algorithm for vector quantiser design". *IEEE Trans. on Commun., vol.28, no.1*, pp. 84-95, January 1980.

[36] **L.R. Rabiner, M.M. Sondhi** and **S.E. Levinson.** "Note on the properties of a vector quantiser for LPC coefficients". *Bell Sys. Tech. J., vol.26, no.8*, pp. 2603-2616, October 1983.

[37] **D.Y. Wong, B.H. Juang** and **A.H. Gray, Jr.** "An 800 bit/s vector quantization LPC vocoder". *IEEE Trans. ASSP, vol.30, no.5*, pp.770-780, October 1982.

[38] **R. Viswanathan** and **J. Makhoul.** "Quantization properties of transmission parameters in linear predictive systems". *IEEE Trans. ASSP, vol.23*, pp. 309-321, June 1975.

[39] **N. Sugamura** and **N. Farvardin.** "Quantizer design in LSP analysis-synthesis". *IEEE J. on Selec. Areas in Commun., vol.6, no.2*, pp. 432-440, February 1988.

[40] **F. Itakura** "Line spectral representation of linear predictive coefficients of speech signals". *J. Acoust. Soc. Amer., vol.57, Supplement no.1, S35*, 1975.

[41] **G.S. Kang** and **L.J. Fransen.** "Low-bit rate speech encoders based on line-spectrum frequencies (LSFs)". *NRL Report 8857*, November 1984.

[42] **J. Makhoul.** "Linear prediction: a tutorial review". *Proc. of IEEE, vol.63, no.4*, pp. 561-580, April 1975.

[43] **F.K. Soong** and **B.H. Juang.** "Line spectrum pair (LSP) and speech data compression". *Proc. ICASSP'84*, pp. 1.10.1-1.10.4, 1984.

[44] **P. Kabal** and **R.P. Ramachandran.** "The computation of line spectral frequencies using Chebyshev polynomials". *IEEE Trans. ASSP, vol.34, no.6*, pp. 1419-1426, December 1986.

[45] **M. Omologo.** "The computation and some spectral considerations on line spectrum pairs (LSP)". *Proc. EUROSPEECH'89*, pp. 352-355, 1989.

[46] **B.M.G. Cheetham.** "Adaptive LSP filter". *Electronics Letters, vol.23, no.2*, pp. 89-90, 16th January 1987.

[47] **F. Itakura** and **N. Sugamura.** "LSP speech synthesizer". *Tech. Rept. 5, Speech Group, Acoustical Soc. of Japan*, November 1979.

[48] **B.S. Atal, R.V. Cox** and **P. Kroon.** "Spectral quantization and interpolation for CELP coders". *Proc. ICASSP'89, Glasgow, UK*, pp. 69-72, 23-26 May, 1989.

[49] **N. Sugamura** and **F. Itakura.** "Speech analysis and synthesis methods developed at ECL in NTT–From LPC to LSP". *Speech Commun., vol.5*, pp. 199-215, June 1986.

[50] **A. Lepschy, G.A. Mian** and **U. Viaro.** "A note on line spectral frequencies". *IEEE Trans. ASSP, vol.36, no.8*, pp. 1355-1357, August 1988.

[51] **B.M.G. Cheetham** and **P.M. Huges.** "Formant estimation from LSP coefficients". *Proc. IERE 5th Int. Conf. on Digital Processing of Signals in Commun., Univ. of Loughborough*, pp. 183-189, September 20-23, 1988.

[52] **F.K. Soong** and **B.H. Juang.** "Optimal quantization of LSP parameters". *Proc. ICASSP'88*, pp. 394-397, 1988.

[53] **G.S. Kang** and **L.J. Fransen.** "Application of line-spectrum pairs to low-bit-rate speech encoders". *Proc. ICASSP'85, Tampa, Florida, USA* pp. 244-247, 26-29 March, 1985.

[54] **J.R. Crosmer** and **T.P. Barnwell, III.** "A low bit rate segment vocoder based on line spectrum pairs". *Proc. ICASSP'85, Tampa, Florida, USA* pp. 240-243, 26-29 March, 1985.

[55] **W.T.K. Wong** and **I. Boyd.** "Optimal quantization performance of LPC parameters for speech coding". *Proc. EUROSPEECH'89*, pp. 344-347, 1989.

[56] **C. Laflamme, J-P. Adoul** and **S. Morissette.** "A real time 4.8 kb/s CELP on a single DSP chip (TMS320C25)". *IEEE Workshop on Speech Coding for Telecom., Vancouver, Canada*, September 5-8, 1989.

[57] **D. Lin** and **B.M. McCarthy.** "Efficient quantization and interpolation of LPC spectral parameters". *IEEE Workshop on Speech Coding for Telecom., Vancouver, Canada*, September 5-8, 1989.

[58] **B.S. Atal** and **M.R. Schroeder.** "Predictive coding of speech and subjective error criteria". *IEEE Trans. ASSP, vol.27, no.3*, pp. 247-254, June 1979.

[59] **R.P. Ramachandran** and **P. Kabal.** "Stability and performance analysis of pitch filters in speech coders". *IEEE Trans. ASSP, vol.35, no.7,* pp. 937-946, July 1987.

[60] **W.P. LeBlanc, S. Hanna** and **S.A. Mahmoud.** "Performance of a low complexity CELP speech coder under mobile channel fading conditions". *Proc. IEEE Vehicular Tech. Conf.,* pp. 647-651, May 1989.

[61] **P. Kroon** and **B.S. Atal.** "On improving the performance of pitch predictors in speech coding systems". *IEEE Workshop on Speech Coding for Telecom., Vancouver, Canada,* September 5-8, 1989.

[62] **P. Kroon** and **B.S. Atal.** "Pitch predictors with high temporal resolution". *Proc. ICASSP'90, Albuquerque, New Mexico, USA,* pp. 661-664, 3-6 Apr., 1990.

[63] **J.S. Marques, I.M. Trancoso, J.M. Tribolet** and **L.B. Almeida.** "Improved pitch prediction with fractional delays in CELP coding". *Proc. ICASSP'90, Albuquerque, New Mexico, USA,* pp. 665-668, 3-6 Apr., 1990.

[64] **Proposed Federal Standard 1016.** "Telecommunications: Analog to digital conversion of radio voice by 4,800 bit/second code excited linear prediction (CELP)". *First draft,* September 1, 1989.

[65] **C. Laflamme.** Unpublished work.

[66] **J.P. Campbell, Jr., T.E. Tremain** and **V.C. Welch.** "The DoD 4.8 kbps standard (proposed federal standard 1016)". in *Advances in Speech Coding,* Kluwer Academic Publishers, pp. 121-133, 1990.

[67] **R.P. Ramachandran** and **P. Kabal.** "Pitch prediction filters in speech coding". *IEEE Trans. ASSP, vol.37, no.4,* pp. 467-478, April 1989.

[68] **P. Kabal** and **R.P. Ramachandran.** "Joint optimization of linear predictors in speech coding". *IEEE Trans. ASSP, vol.37, no.5,* pp. 642-650, May 1989.

[69] **B.S. Atal** and **M.R. Schroeder.** "Optimizing predictive coders for minimum audible noise". *Proc. ICASSP'79,* pp. 453-455, 1979.

[70] **J. Makhoul** and **M. Berouti.** "Adaptive noise spectral shaping and entropy coding in predictive coding of speech". *IEEE Trans. ASSP, vol.27, no.1,* pp. 63-73, February 1979.

[71] **P. Kroon** and **E.F. Deprettere.** "Experimental evaluation of different approaches to the multi-pulse coder". *Proc. ICASSP'84,* pp. 10.4.1-10.4.4, 1984.

[72] **S. Singhal** and **B.S. Atal.** "Amplitude optimization and pitch prediction in multi-pulse coders". *IEEE Trans. ASSP, vol.37, no.3,* pp. 317-327, March 1989.

[73] **S. Singhal.** "Reducing computation in optimal amplitude multi-pulse coders". *Proc. ICASSP'86,* pp. 2364-2367, 1986.

[74] **T. Aresaki, K. Ozawa, S. Ono** and **K. Ochiai.** "Multi-pulse excited speech coder based on maximum cross correlation search algorithm". *Globecom 83*, pp. 794-798, 1983.

[75] **M. Berouti, H. Garten, P. Kabal** and **P. Mermelstein.** "Efficient computation and coding of the multipulse excitation for LPC". *Proc. ICASSP'84*, pp. 10.1.1-10.1.4, 1984.

[76] **J-P. Lefevre** and **O. Passien.** "Efficient algorithms for obtaining multipulse excitation for LPC coders". *Proc. ICASSP'85, Tampa, Florida, USA* pp. 957-960, 26-29 March, 1985.

[77] **E.F. Deprettere** and **P. Kroon.** "Regular excitation reduction for effective and efficient LP-coding of speech". *Proc. ICASSP'85, Tampa, Florida, USA* pp. 965-968, 26-29 March, 1985.

[78] **J-P. Adoul, F. Didelot, P. Mabilleau** and **S. Morissette.** "Generalization of the multipulse coding for low bit rate coding purposes: The generalized decimation". *Proc. ICASSP'85, Tampa, Florida, USA* pp. 256-259, 23-26 March, 1985.

[79] **R.D. Strum** and **D.E. Kirk.** "First principles of discrete systems and digital signal processing". *Addison-Wesley Pub. Comp.*, 1988.

[80] **K. Helwig, R. Hofman, P. Vary** and **R.J. Sluyter.** "MATS-D speech codec: Regular pulse excitation LPC". *Proc. 2nd seminar on Land Mobile Digital Radio Communication, Stockholm*, 14-16 October 1986.

[81] **R.J. Sluyter, G.J. Bosscha** and **H.M.P.T. Schmitz.** "A 9.6 kb/s speech coder for mobile radio applications". *Proc. ICC'84*, pp. 1159-1162, 1984.

[82] **P. Vary** *et al.* "Speech codec for the European mobile radio system". *Proc. ICASSP'88*, pp. 227-230, 1988.

[83] **S.P. Lloyd.** "Least squares quantization in PCM". *IEEE Trans. Inf. Th., vol.28, no.2*, pp. 129-137, March 1982.

[84] **M.J. Noah.** "Optimal Lloyd-Max quantization of LPC speech parameters". *Proc. ICASSP'84*, pp. 1.8.1-1.8.4, 1984.

[85] **P. Kroon** and **E.F. Deprettere.** "A class of analysis-by-synthesis predictive coders for high quality speech coding at rates between 4.8 and 16 kb/s". *IEEE J. on Selected. Areas in Commun., vol.6, no.2*, pp. 353-363, February 1988.

[86] **R. Soheili, A.M. Kondoz** and **B.G. Evans.** "New innovations in multipulse speech coding for bit rates below 8 kb/s". *Proc. EUROSPEECH*, pp. 298-301, 1989.

[87] **M. Copperi** and **D. Sereno.** "Vector quantization and perceptual criteria for low bit rate coding of speech". *Proc. ICASSP'85, Tampa, Florida, USA* pp. 252-255, 23-26 March, 1985.

[88] **G. Davidson** and **A. Gersho.** "Complexity reduction methods for vector excitation coding". *Proc. ICASSP'86,* pp. 3055-3058, 1986.

[89] **G. Davidson, M. Yong** and **A. Gersho.** "Real-time vector excitation coding of speech at 4800 bps". *Proc. ICASSP'87,* pp. 2189-2192, 1987.

[90] **D. Lin.** "New approaches to stochastic coding of speech sources at very low bit rates". *Signal Processing III: Theories and Applications (Proc. of EUSIPCO-86),* pp. 445-448, 1986.

[91] **C.S. Xydeas, M.A. Ireton** and **D.K. Baghbadrani.** "Theory and real time implementation of a CELP coder at 4.8 and 6.0 Kbits/sec using ternary code excitation". *Proc. of IERE 5th Int. Conf. on Digital Processing of Signals in Commun., Univ. of Loughborough,* pp. 167-174, 20-23 September 1988.

[92] **W.B. Kleijn, D.J. Krasinsky** and **R.H. Ketchum.** "An efficient stochastically excited linear predictive coding algorithm for high quality low bit rate transmission of speech". *Speech Commun., vol.7, no.3,* pp. 305-316, October 1988.

[93] **R.A. Salami.** "Binary pulse excitation: a novel approach to low complexity CELP coding" in *Advances in speech coding,* Kluwer Academic Publishers, pp. 145-156, 1991.

[94] **M. Delprat, M. Lever** and **C. Gruet.** "A 6 kbps regular pulse CELP coder for mobile radio communications". in *Advances in speech coding,* Kluwer Academic Publishers, pp. 179-188, 1991.

[95] **I.A. Gerson** and **M.A. Jasiuk.** "Vector sum excitation linear prediction (VSELP) speech coding at 8 kbps". *Proc. ICASSP'90, Albuquerque, New Mexico, USA,* pp. 461-464, 3-6 Apr., 1990.

[96] **I.A. Gerson** and **M.A. Jasiuk.** "Vector Sum Excited Linear Prediction (VSELP)". *IEEE Workshop on Speech Coding for Telecom., Vancouver, Canada,* September 5-8, 1989.

[97] **J-P. Adoul** *et al.* "Fast CELP coding based on algebraic codes". *Proc. ICASSP'87,* pp. 1957-1960, 1987.

[98] **A.M. Kondoz** and **B.G. Evans.** "CELP base-band coder for high quality speech coding at 9.6 to 2.4 kbps". *Proc. ICASSP'88,* pp. 159-162, 1988.

[99] **R.C. Rose** and **T.P. Barnwell III.** "Quality comparison of low complexity 4800 bps self excited and code excited vocoders". *Proc. ICASSP' 87,* pp. 1637-1640, 1987.

[100] **N. Gouvianakis** and **C. Xydeas.** "Advances in analysis by synthesis LPC speech coders". *J. IERE, vol.57, no.6 (supplement),* pp. S272-S286, November/December 1987.

[101] **I.M. Trancoso** and **B.S. Atal.** "Efficient procedures for finding the optimum innovation in stochastic coders". *Proc. ICASSP'86,* pp. 2375-2378, 1986.

[102] **J-P. Adoul** and **C. Lamblin.** "A comparison of some algebraic structures for CELP coding of speech". *Proc. ICASSP'87*, pp. 1953-1956, 1987.

[103] **C. Lamblin, J-P. Adoul, D. Massaloux** and **S. Morissette.** "Fast CELP coding based on the Barnes-Wall lattice in 16 dimensions". *Proc. ICASSP'89, Glasgow, UK*, pp. 61-64, 23-26 May, 1989.

[104] **M.A. Ireton** and **C.S. Xydeas.** "On improving vector excitation coders through the use of spherical lattice codebooks (SLC's)". *Proc. ICASSP'89, Glasgow, UK*, pp. 57-60, 23-26 May, 1989.

[105] **M. Lever** and **M. Delprat.** "RPCELP: A high quality and low complexity scheme for narrow band coding of speech". *Proc. EUROCON*, pp. 24-27, June 1988.

[106] **R.A. Salami.** "Binary code excited linear prediction (BCELP): new approach to CELP coding of speech without codebooks". *Electronics Letters, vol.25, no.6*, pp. 401-403, 16 March 1989.

[107] **R.A. Salami** and **D.G. Appleby.** "A new approach to low bit rate speech coding with low complexity using binary pulse excitation (BPE)". *IEEE Workshop on Speech Coding for Telecomm., Vancouver, Canada*, September 5-8, 1989.

[108] **J.-M. Müller, H. Scheuermann** and **B. Wächter.** "GSM half rate codec: a possible candidate". *IEEE Workshop on Speech Coding for Telecomm., Vancouver, Canada*, September 5-8, 1989.

[109] **A. Le Guyader, D. Massaloux** and **F. Zurcher.** "A robust and fast CELP coder at 16 kb/s". *Speech Communication, vol.7*, pp. 217-226, 1988.

[110] **C. Laflamme, J-P. Adoul, H.Y. Su** and **Morissette.** "On reducing computational complexity of codebook search in CELP coder through the use of algebraic codes". *Proc. ICASSP'90, Albuquerque, New Mexico, U.S.A.*, April 3-6, 1990.

[111] **C. Laflamme, J-P. Adoul, R. Salami, S. Morissette** and **P. Mabilleau.** "16 kbps wideband speech coding technique based on algebraic CELP". *Proc. ICASSP'91, Toronto, Canada* pp. 13-16, 14-17 May, 1991.

[112] **V. Ramamoorthy** and **N.S. Jayant.** "Enhancement of ADPCM speech by adaptive postfiltering". *Bell Sys. Tech. J., vol.63*, pp. 1465-1475, October 1984.

[113] **J. Chen** and **A. Gersho.** "Real-time vector APC speech coding at 4800 bps with adaptive postfiltering". *Proc. ICASSP'87*, pp. 2185-2188, 1987.

[114] **F.C.A.Brooks** and **L. Hanzo.** "A Multiband Excited Waveform Interpolated 2.35 kbps Speech Codec for Bandlimited Channels", submitted to IEEE Tr. on Veh. Techn, 1998

[115] **W.B. Kleijn,** "Encoding speech using prototype waveforms," *IEEE Transactions on Speech and Audio Processing*, vol. 1, pp. 386–399, October 1993.

[116] **D.W. Griffin** and **J.S. Lim**, "Multiband excitation vocoder," *IEEE Transactions on Acoustics, Speech and Signal Processing*, vol. 36, no. 8, pp. 1223–1235, 1988.

[117] **D.J. Hiotakakos** and **C.S. Xydeas**, "Low bit rate coding using an interpolated zinc excitation model," in *Proceedings of the ICCS 94*, pp. 865–869, 1994.

[118] **K.A. Teague, B. Leach** and **W. Andrews**, "Development of a high-quality MBE based vocoder for implementation at 2400bps," in *Proceedings of the IEEE Wichita Conference on Communications, Networking and Signal Processing*, pp. 129–133, April 1994.

[119] **H. Hassanein, A. Brind'Amour, S. Déry** and **K. Bryden**, "Frequency selective harmonic coding at 2400bps," in *Proceedings of the 37th Midwest Symposium on Circuits and Systems*, vol. 2, pp. 1436–1439, 1995.

[120] **R.J. McAulay** and **T.F. Quatieri**, "The application of subband coding to improve quality and robustness of the sinusoidal transform coder," in *Proceedings of ICASSP 93*, vol. 2, pp. 439–442, 1993.

[121] **A.V. McCree** and **T.P. Barnwell**, "A mixed excitation LPC vocoder model for low bit rate speech coding," *IEEE Transactions on Speech and audio Processing*, vol. 3, no. 4, pp. 242–250, 1995.

[122] **P.A. Laurent** and **P. de La Noue**, "A robust 2400bps subband LPC vocoder," in *Proceedings of ICASSP 95*, pp. 500–503, 1995.

[123] **A.V. McCree, Kwan Truong, E.B. George, T.P. Barnwell** and **V. Viswanathan**, "A 2.4kb/s coder candidate for the new U.S. Federal standard," in *Proceedings of ICASSP 96*, pp. 200–203, 1996.

[124] **W.Hess.** "Pitch determination of speech signals: algorithms and devices", *Berlin: Springer Verlag*, 1983

[125] **Y. Hiwasaki** and **K. Mano**, "A new 2-kb/s speech coder based on normalized pitch waveform," in *Proceedings of ICASSP 97*, pp. 1583–1586, 1997.

[126] **S. Kadambe** and **G.F. Boudreaux-Bartels**, "Application of the wavelet transform for pitch detection of speech signals," *IEEE Transactions on Information Theory*, vol. 38, pp. 917–924, March 1992.

[127] **N. Gonzalez** and **D. Docampo**, "Application of singularity detection with wavelets for pitch estimation of speech signals," in *SIGNAL PROCESSING VII: Theories and Applications*, pp. 1657–1660, 1994.

[128] **S. Mallat** and **S. Zhong**, "Characterization of signals from multiscale edges," *IEEE Transactions on Pattern Analysis and Machine Intelligence*, vol. 14, pp. 710–732, July 1992.

[129] **I. Daubechies**, "The wavelet transform, time-frequency localization and signal analysis," *IEEE Transactions on Information Theory*, vol. 36, pp. 961–1005, September 1990.

[130] **S. Mallat**, "A theory for multiresolution signal decomposition: the wavelet representation," *IEEE Transactions on Pattern Analysis and Machine Intelligence*, vol. 11, pp. 674–693, July 1989.

[131] **T.H. Koornwinder**, *Wavelets: An Elementary Treatment of Theory and Applications.* World Scientific, 1993.

[132] **J. Stegmann, G. Schröder** and **K.A. Fischer**, "Robust classification of speech based on the dyadic wavelet transform with application to CELP coding," in *Proceedings of ICASSP 96*, pp. 546–549, 1996.

[133] **CCITT**, *Coding of speech at 8 kb/s using Conjugate-Structure Algebraic CELP, G.729*, December 1995.

[134] **R.A. Sukkar, J.L. LoCicero** and **J.W. Picone**, "Decomposition of the LPC excitation using the zinc basis functions," *IEEE Transactions on Acoustics, Speech and Signal Processing*, vol. 37, no. 9, pp. 1329–1341, 1989.

[135] **K. Yaghmaie** and **A.M. Kondoz**, "Multiband prototype waveform analysis synthesis for very low bit rate speech coding," in *Proceedings of ICASSP 97*, pp. 1571–1574, 1997.

[136] **R.J. McAulay** and **T.F. Quatieri**, "Sinusoidal coding," in *Speech Coding and Synthesis* (W.B.Keijn and **K.K.Paliwal**, ed.), ch. 4, Elsevier Science, 1995.

[137] **S. Yeldner, A.M. Kondoz** and **B.G. Evans**, "Multiband linear predictive speech coding at very low bit rates," *IEE Proceedings in Vision, Image and Signal Processing*, vol. 141, pp. 284–296, October 1994.

Chapter 4

Channel Coding

K.H.H. Wong[1] and L. Hanzo[2]

4.1 Introduction

Sparked off by Shannon's pioneering discoveries back in 1948, the 1950s witnessed feverish activities in the field of forward error correction coding (FEC), which led to the introduction of the single error correcting Hamming block code [1] in 1950 and to the birth of the first convolutional code in 1955 [2]. The theory of FEC is lavishly published in classic references [3]-[11], however real-time implementations constituted serious limitations to their deployment in the past. Constrained by the error statistics of existing communications links, such as satellite channels or cables, almost exclusively random error statistics, i.e. memoryless channels, have been studied. With the evolution of digital mobile radio communication it is important to provide a comprehensive overview of FEC techniques tailored for these hostile bursty channels.

The key to efficient FEC via bursty fading channels is the deployment of appropriately matched binary or non-binary channel interleavers to randomise the bursty error statistics and hence render the channel memoryless. If this condition is met, most of the memoryless theory applies. Whence we commence our discourse with a rudimentary description of a number of interleaving schemes, followed by the encoding and decoding algorithms as well as theoretical and simulated performance of various convolutional codes using both soft and hard decisions. Block codes, in particular Bose-Chaudhuri-Hocquenghem (BCH) codes and Reed-Solomon (RS) codes are

---

[1]University of Southampton
[2]University of Southampton and Multiple Access Communications Ltd

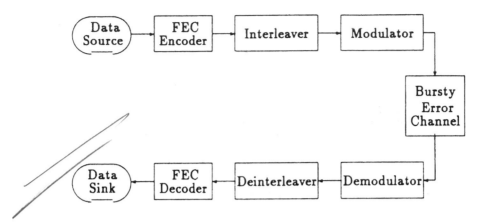

Figure 4.1: Typical communications system with FEC coding and interleaving.

portrayed in contrast to convolutional codes in terms of encoding and decoding algorithms, complexity as well as theoretical and measured performance. A particularly powerful combination, concatenated codes, are also highlighted and compared to convolutional and block codes in the context of the interplay of complexity, coding gain, as well as the ability to detect and correct various channel error distributions for the tranmission of speech and data signals.

4.2 Interleaving Techniques

Interleaving is a process of rearranging the ordering of a sequence of binary or non-binary symbols in some unique one-to-one deterministic manner. The reverse of this process is the deinterleaving which is to restore the sequence to its original ordering.

In many applications in communication technology, interleaving is used in conjunction with forward error correction (FEC) codes to enhance error correction performance. The interleaver is inserted between the channel encoder and the modulator as shown in Figure 4.1. Most block or convolutional codes are designed to combat random independent errors which usually occur in a memoryless channel. For channels having memory, burst errors are observed that are due to the mutually dependent signal transmission impairments. An example of such a channel is a fading channel. The fading arises because the signals arriving at the antenna have transversed different paths having various attenuations and delays. The effective received signal is the vector sum of these multipath signals and exhibits fades which depend on numerous factors, such as mobile speed, propagation delay spread and frequency. The received signal fading causes burst errors during the signal fades.

Interleaving is deployed to disperse the burst errors when the received signal level fades, and to reduce the concentration of the errors that must be corrected by the channel code. Before a sequence of symbols is transmitted the symbols from several codewords are interleaved. Then when an error burst occurs the errors will be shared among the interleaved codewords and a less powerful code is required to correct them. Thus interleaving effectively makes the channel appear like a random error channel to the decoder. The idea behind interleaving is to separate the codeword symbols in time, thereby reducing the memory of the channel. As the interleaving period increases, the error performance can be expected to improve in the sense that noise bursts are more dispersed. On the other hand, the delays due to interleaving and deinterleaving increase. Consequently, there is always a tradeoff between error performance and interleaving delay.

The interleaver shuffles the code symbols, each consisting of m bits, over a span of several codewords. If m is equal to one, bit interleaving is applied. The span required is often determined by the burst duration. Four types of interleavers are described in the Sections 4.2.1 to 4.2.4. The effects of the interleaving on the bit and symbol error distributions are presented in Sections 4.2.5-4.2.7.

4.2.1 Diagonal Interleaving

A diagonal interleaver [12] accepts coded symbols in blocks from the encoder, permutes these symbols, and feeds them to the modulator. The ith input block of coded symbols is interleaved in a diagonal way with the previous $(i-1)$th block and the following $(i+1)$th blocks as shown in Figure 4.2. A block of n coded symbols, each symbol consisting of m bits, is divided into s subblocks where $B_j^i, j = 1, 2, \ldots, s$ is the jth subblock of the ith block. The interleaved output is formulated by reading vertically two subblocks belonging to the $(i-1)$th and ith blocks or to the ith and $(i+1)$th blocks.

If the number of subblocks is equal to the number of symbols in a block, i.e., $s = n$, the subblock has only a single symbol and the interleaver shuffles the block on a symbol-by-symbol basis such that the output is a sequence of alternate symbols from two successive blocks. This is apparent from Figure 4.3, where the interleaved output is $\ldots z_1 y_0 z_2 y_1 z_3 x_0 y_2 x_1 y_3 x_2 \ldots$, and x, y and z are the symbols of the $(i-1)$th, ith and $(i+1)$th block, respectively. If burst errors occurred such that the error symbol e in the received sequence is $\ldots z_1 e e y_1 z_3 x_0 y_2 e e x_2 \ldots$, then the deinterleaved sequence becomes $\ldots z_1 e z_3 e y_1 y_2 e x_0 e x_2 \ldots$, where the errors are distributed over two successive blocks.

The interleaver and deinterleaver end-to-end delay is three times the block length, i.e., $3 \, nm$ bits. This is because the interleaving results in a delay of one and a half blocks, see Figure 4.2, and the deinterleaving results in a similar delay. This short delay is particularly important in some

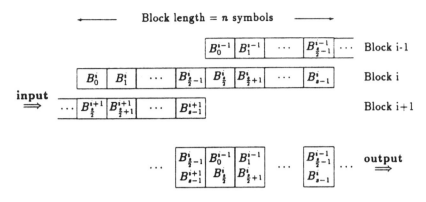

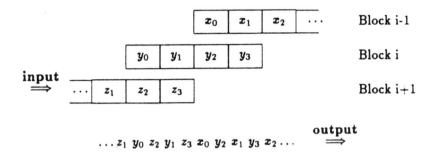

Figure 4.2: Diagonal interleaver.

Figure 4.3: Example of diagonal interleaving with single symbol subblock.

applications such as digital speech transmission. However, the dispersion of the burst errors is limited to the adjacent blocks which implies that the burst errors are halved and shared by only two successive blocks.

4.2.2   Block Interleaving

A block interleaver takes a codeword of n symbols and writes them by rows into a matrix with a depth of D rows and width of W columns as shown in Figure 4.4. Suppose W is equal to n, i.e. a row of symbols in the interleaver corresponds to a codeword. Each codeword is composed of k information symbols and $n-k$ parity symbols. After the matrix is completely filled, the symbols are fed to the modulator one column at a time and transmitted over the channel. At the receiver, the deinterleaver performs the inverse permutation of the matrix by feeding in one column at a time until the matrix is filled, and removing the symbols a row at a time. The important characteristics of this interleaving approach are that any burst of errors of length $b \leq D$ results in single errors in a codeword. Also any burst of

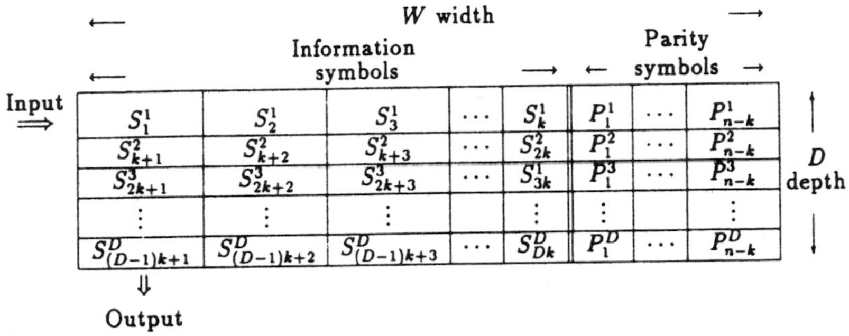

Figure 4.4: Block interleaver.

length $b = rD(r > 1)$ results in bursts of no more than r symbol errors in a codeword. However, a periodic sequence of single errors spaced by D symbols results in a single codeword error, where every symbol in the codeword is incorrect. The interleaver and deinterleaver end-to-end delay is $2WD$ symbols. To be precise, only $W(D-1)+1$ storage needs to be filled before transmission can begin, i.e. as soon as the first symbol of the last row of the $D * W$ matrix is filled. The same delay applies to the deinterleaver. Therefore, the end-to-end delay is $(2WD - 2W + 2)$ symbols. The memory requirement is WD symbols in both the interleaver and deinterleaver. Another block interleaver [13], shown in Figure 4.5 is a derivation from the arrangement shown in Figure 4.4. Instead of writing rows of symbols into the matrix and appending the parity symbols to the k successive information symbols, the symbols are now written a column at a time. The parity symbols are encoded from a row of k information symbols each separated by D symbols in their natural order. As symbols $S_1^1, S_2^2, S_3^3, \ldots, S_{kD}^D$ occur they are placed into the matrix and also transmitted at the same time. Thus by the time S_{D+1}^1 arrives, symbols $S_1^1, S_2^2, \ldots, S_D^D$ have been transmitted. Armed with a knowledge of all the information symbols, the parity symbols P_1^1 to P_{n-k}^D are calculated immediately after symbol S_{kD}^D has entered the matrix. The advantage of this interleaving scheme is that information symbols are transmitted in their natural order. Hence, the interleaver delay is negligible; the end-to-end delay is WD symbols which is due to the deinterleaver.

The interleaver parameters D and W must be selected so that all expected burst lengths are less than D. However, this type of interleaver lacks robustness when a periodic sequence of single errors spaced by D symbols occurs. In this situation all the symbols in a row are erroneous and this overloads the channel codec. The interleaving scheme to be described in the next section exhibits the ability of dispersing bursty noise as well as periodic noise.

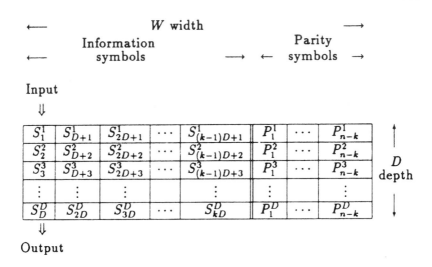

Figure 4.5: Modified block interleaver.

4.2.3   Inter-Block Interleaving

The inter-block interleaver [12] takes an input block of NB symbols and disperses N symbols to each of the next B output blocks. Consider a coded symbol x from the encoder and an output symbol y from the interleaver. The mapping from the mth symbol of the ith coded input block to the $(j + Bt)$th interleaved symbol of the $(i + j)$th output block is given by

$$y(i + j, j + Bt) = x(i, m), \quad \text{for all } i \tag{4.1}$$
$$\text{with } j = m \bmod B$$
$$\text{and } t = m \bmod N.$$

An example of inter-block interleaving with $B = 3$ and $N = 2$ is illustrated in Figure 4.6. The symbols of the three successive ith, $(i + 1)$th and $(i + 2)$th coded input blocks are denoted as a, b and c respectively. Here, $y(i + j, j + 3t) = x(i, m)$, for all i, with $j = m \bmod 3$, and $t = m \bmod 2$. For $m = 0, y(i, 0) = x(i, 0)$ and $m = 1, y(i + 1, 4) = x(i, 1)$ and so on. It is noted that the successive symbols of the ith input block are mapped to the next B output blocks consecutively, but with the irregular offset position $(j + Bt)$ in the block. This irregular offset has the advantage of randomising the periodic noise. In order to make sure that the mapping is one-to-one, B and N cannot have a common multiple. Usually this places a constraint on the block size of BN symbols. Another disadvantage of this interleaving scheme is the dispersive nature in that the output sequence is expanded by $(B - 1)$ blocks, the interleaving delay is $B^2 N$ symbols composed of the delay BN due to buffering the input block, plus the extra delay $(B-1)BN$

Input blocks:

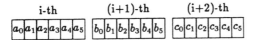

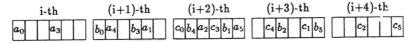

Output block:

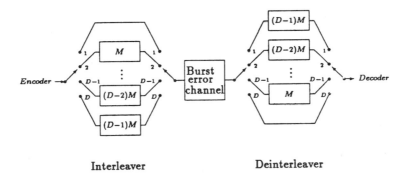

Figure 4.6: Example of inter-block interleaving with $B = 3$ and $N = 2$.

Figure 4.7: Convolutional interleaver and deinterleaver.

due to the dispersion of the symbols.

4.2.4   Convolutional Interleaving

Convolutional interleavers have been proposed by Ramsey [14] and Forney [11]. The structure proposed by Forney is shown in Figure 4.7. The
code symbols are sequentially shifted into the bank of D registers, where
each successive register provides M symbols more memory than the preceding one. With the switch in position 1 the data are passed directly to
the channel. With each new code symbol the commutator switches to the
next register, and the oldest symbol in that register is shifted out. After the
switch has reached position D, it returns to position 1 and the cycle of the
switching continues. The deinterleaver performs the inverse operation, and
the input and output commutators for both interleaver and deinterleaver
must be synchronised.

The performance of a convolutional interleaver is very similar to that of

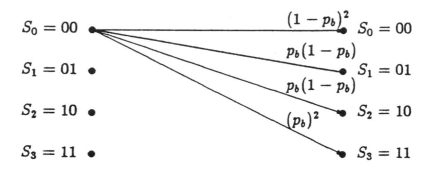

Figure 4.8: Non-binary and non-symmetric channel.

a block interleaver. The important advantage of convolutional over block interleaving is that with convolutional interleaving the end-to-end delay is $W(D-1)$ symbols, where $W = DM$, and the memory required is $W(D-1)/2$ in both interleaver and deinterleaver. Therefore, there is a reduction of one-half in the delay and the memory of a convolutional interleaver compared to the block interleaver.

4.2.5 Discrete Memoryless Channel

The FEC encoded symbols consisting of m-bits are serially transmitted. The receiver performs symbol regeneration prior to FEC decoding. If the channel is Gaussian, or if sufficient interleaving is employed in the case of a fading channel, the probability of any bit being in error is p_b. The probability of the regenerated symbol being erroneous depends on the value of m. For example, Figure 4.8 shows the probabilities of a transmitted two bits symbol 00 being regenerated as 00, 01, 10 and 11. The probability of receiving an error symbol with i number of error bits can be expressed as

$$p_{s,i} = \left(\begin{array}{c} m \\ i \end{array} \right) p_b^i (1 - p_b)^{m-i}. \tag{4.2}$$

The symbol error probability p_s is the sum of all the error symbols with i bit errors and is given by

$$p_s = \sum_{i=1}^{m} \left(\begin{array}{c} m \\ i \end{array} \right) p_b^i (1 - p_b)^{m-i}. \tag{4.3}$$

This equation of symbol error probability is based on the assumption of the memoryless channel. However, the mobile channel has considerable memory and therefore Equation 4.3 is only valid when the interleaving period is sufficiently long. Unfortunately, the tolerable delay of some trans-

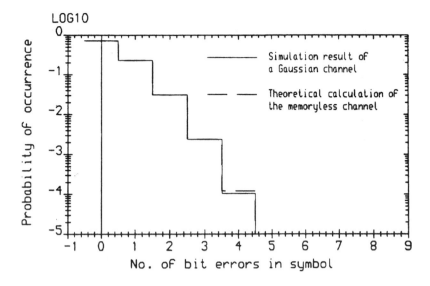

Figure 4.9: Histogram of the number of bit errors in 8-bit symbols via Gaussian channel with SNR=2 dB and BER=3.74×10^{-2}.

missions, such as digital speech communications restrict the interleaving period and prevent the establishment of a true memoryless channel. The next section illustrates the memoryless condition on the channel by using different interleaving methods with various delays.

4.2.6 The Effect of Interleaving on Symbol Error Distribution

Under certain conditions, for example in some microcells [15], the mobile radio channel has minimal memory, i.e., it behaves as a Gaussian channel. In this channel the bit errors are random and there is no point in interleaving, as demonstrated by Figure 4.9, where the bit error distribution within eight bit symbols has been determined using computer simulation when the transmissions are over a Gaussian channel using MSK modulation. The bit error probability for a channel SNR of 2 dB is 3.74×10^{-2}. Inserting this p_b value into Equation 4.2 enabled the theoretical curves to be plotted. The simulation and theoretical results are in very close agreement, demonstrating that the Gaussian channel is indeed memoryless.

By constrast, the Rayleigh fading channel has memory as error bursts occur due to the presence of deep fades resulting from the movement of the mobile station. We conducted an experiment where data at 16 kbit/s formulated into blocks, subjected to interleaving, and conveyed via MSK over a mobile radio channel that exhibited Rayleigh fading [16]. The mobile

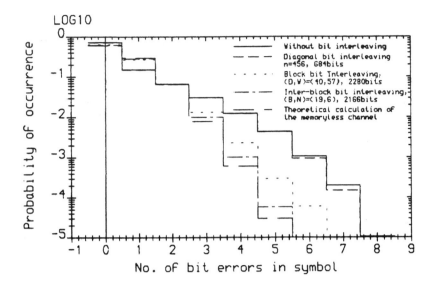

Figure 4.10: Histogram of the number of bit errors in 8-bit symbols via Rayleigh channel with SNR=2 dB and BER= 5.79×10^{-2}.

station travelled at 30 mph, and the propagation frequency was 900 MHz. The channel SNR at the mobile station was 5 dB. In Figure 4.10 we display our findings. As bench markers we show the histogram of the number of bit errors in a symbol in the absence of interleaving, and also for the memoryless channel computed from Equation 4.2. The performance in the absence of interleaving indicates the bursty nature of the channel as there was a relatively high probability of 10^{-5} that all 8-bits of a symbol are erroneous. The randomising performance of the diagonal interleaver (see Figure 4.2) is poor because the block size was relatively small at 456 bits, and the interleaving only involves reallocating the bits over two adjacent blocks. This method of interleaving is not efficient at dispersing the errors, although it does have the advantage of a relatively small delay penalty. Clearly, when the diagonal bit interleaving is deployed, the histogram is similar to that of the no interleaving scenario and it is very different from that of the ideal memoryless channel.

When the block interleaver of Figure 4.4 was used having a long inter-leaving delay of 2280 bits, with $D = 40$ and $W = 57$, the probability of many bit errors per symbol significantly decreased compared to the diag-onal interleaver case. The inter-block interleaver shown in Figure 4.6 was deployed with $B = 19$ and $N = 6$, numbers which produce an interleaving delay of 2166 bits, i.e., a number close to the 2280 bits used for the block interleaver. The inter-block interleaver had the best performance due to its ability to disperse the bits in the interleaving process in a near random

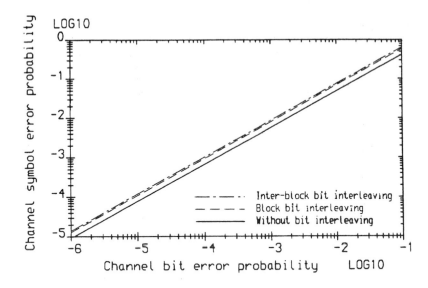

Figure 4.11: The effect of bit interleaving on 8-bit symbols via Rayleigh fading channels.

fashion, rather than the periodic way of the block interleaver. Its histogram suggests a near-random in-symbol bit error distribution, when compared to that of the memoryless channel. The dispersion of the burst errors by bit interleaving results in the random distribution of bit errors. As we will show in the forthcoming sections, this is desirable for the Viterbi decoding of convolutional codes and the trellis decoding of block codes since both decoding methods operate on a bit-by-bit basis. However, the random distribution of the bit errors increases the symbol error probability as shown in Figure 4.11. This figure implies that dispersion of the burst errors results in higher symbol error probability. This phenomenon is not desirable to those decoding methods operating on symbol basis such as the Berlekamp-Massey decoding of Reed-Solomon codes. In this case, symbol interleaving is applied in order to constrain the bit errors in the symbol to itself without spreading out to other symbols such that the symbol error probability after deinterleaving remains the same as on the channel. The channel symbol error probability of the fading channel is less than that of the memoryless channel. Hence, the symbol interleaving on the fading channel has a lower symbol error probability than the theoretical value calculated for the bit interleaved memoryless channel.

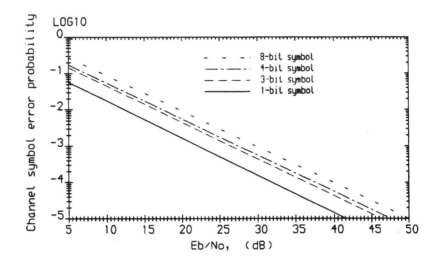

Figure 4.12: The effect of symbol size on symbol error probability via Rayleigh fading channels.

4.2.7 Effect of Symbol Size on Symbol Error Probability

From Equation 4.3 in Section 4.2.5, we notice that the symbol error probability p_s increases with the symbol size m. This increase is due to the summation of all the i, $i = 1, 2, \ldots, m$, bit errors in a symbol. Figure 4.12 shows the channel symbol error probability for the case of 1, 3, 4 and 8 bit symbols. This implies that large symbol sizes suffer from higher symbol t.o. probability which then discourages the use of long codewords in block coding.

4.3   Convolutional Codes

Convolutional codes were first suggested by Elias [2] in 1955. Shortly afterwards, a sequential decoding algorithm for these codes was proposed by Wozencraft [17, 18], and its implementation was independently described by Fano [19] and by Massey [3] in 1963. Called threshold decoding, it has been successfully implemented in numerous communications systems. In 1967 Viterbi [20] proposed a maximum likelihood decoding algorithm that provided optimum error rate performance and achieved a shorter decoding delay compared to sequential decoding. The implementation of the Viterbi decoder became feasible with the advent of integrated circuit technology and was used in the deep-space and satellite communications of the early 1970s. It was also adopted by the GSM committee in the late 1980s for the

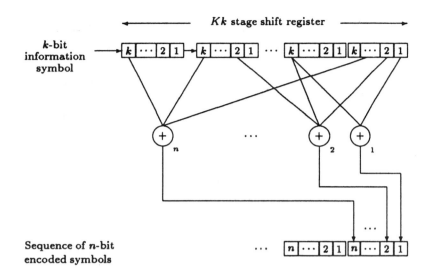

Figure 4.13: Convolutional encoder for the $CC(n, k, K)$ code.

Pan-European digital cellular mobile radio system.

4.3.1 Convolutional Encoding

A convolutional code (CC) is a sequence of encoded symbols which is generated by passing the information sequence through a binary shift register as shown in Figure 4.13. At each symbol instant, a k-bit information symbol is inserted into the input stages of the shift register. The register consists of Kk binary stages that constitute the present k-bit information symbol and the $(K-1)$ previous k-bit input symbols. The parameter K is known as the constraint length and it determines the memory length of the shift register. The n linear algebraic function generators, $g_1, g_2, \ldots, g_n$, defined by their connections to various register stages yield an n-bit convolutional coded symbol via the set of modulo-2 adders. At each symbol instant, a new information symbol enters the register, the other symbols move to the next symbol location and one symbol leaves the register. A new encoded symbol is then generated as previously described. The process continues, encoding the sequence of information symbols into a new sequence of output symbols.

The convolutional code is basically described by three parameters n, k and K and is denoted by $CC(n, k, K)$. The coding rate, defined by

$$R = k/n \tag{4.4}$$

represents the amount of information per encoded bit. If k is equal to one,

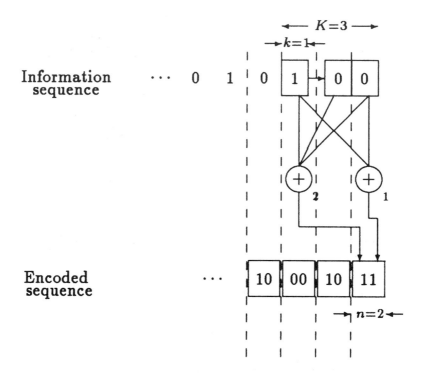

Figure 4.14: Convolutional encoder for the $CC(2, 1, 3)$ code.

namely one bit per symbol, a binary CC is generated, whereas for k greater than one, the CC code is referred to as, not surprisingly, a non-binary code.

In order to generate the $CC(n, k, K)$ code, a set of n generators, $g_1, g_2, \ldots, g_n$, is used to produce an n-bit symbol. Each generator is described as a vector with a dimension of full register length of Kk bits. The ith element of the vector specifies the existence or otherwise of the connection of the ith bit position of the register to the modulo-2 adder. The value of the vector is defined by assigning a logical 1 to where the connections are made, and all other positions in the register are assigned to be logical zeros. We see in Figure 4.13 that various bit positions in the register are connected to modulo-2 adders. For example, we observe that the inputs to the modulo-2 adder defined by g_1 are in the last two symbol stages of the register. So generator g_1 is $[00 \ldots 0010000001]$ for $k = 4$, i.e., a sequence having only two logical ones. Notice that a generator defines the positions of the connections from the register to the modulo-2 adder. The output of the adder depends, of course, on the logical values of the information bits applied to its inputs.

For the sake of simplicity, we demonstrate the encoding process by considering the binary $(k = 1)$, half-rate $(R = k/n = 1/2)$, convolutional code, $CC(2, 1, 3)$ with constraint length of three binary stages $(K = 3)$. Once the basic concepts are established by this example, the same principle can be generalised to any convolutional code. The generators in Figure 4.14 are described by the vectors,

$$g_1 = [1\ 0\ 1] \quad \text{and} \quad g_2 = [1\ 1\ 1] \tag{4.5}$$

or are written in equivalent form of polynomials as

$$g_1(z) = 1 + z^2 \quad \text{and} \quad g_2(z) = 1 + z + z^2 \ . \tag{4.6}$$

The shift register has three binary stages, the first stage is for the present bit and the latter stages are for the two previous input bits. The status of the register is defined by the logical values of its bits, while the *state* of the code is represented by the two previous input bits. Initially, the shift register is set to the all-zero state and then an input bit of logical value 1 is applied. Generators g_1, associated with the first and third stages of the register, and g_2, associated with all three stages of the register, cause both the outputs from the two modulo-2 adders to produce output bits of logical 1 that formulate the encoded symbol 11. At the next instant, data move into the register from the right which is equivalent to the register being shifted to the left by $k = 1$ bit. The new input bit is a logical 0 and the state of the register is changed from 00 to 10. The new encoded symbol becomes 10. As the encoding process continues, the next input bits are ...,0,1, and the state transition becomes ..., 01, 10, 01. The string of the subsequent encoded symbols ..., 10, 00, is generated, and passed to the modulator for transmission. Convolutional codes, as implied by the name, can be viewed as the convolution of the impulse response of the encoder with the information sequence. In our example, if an information sequence with a single logical 1 followed by all zeros is shifted into the register, the output sequence is referred to as the impulse response of the encoder and is given by $11, 10, 11, 00, 00, 00, \ldots$. For any arbitrary input sequence, the encoder output is the modulo-2 addition of the impulse responses arising from each logical 1 at the input, where each response is positioned in accordance with the location of the ones in the input sequence. The output sequence O corresponding to any arbitrary input sequence I can be found by multiplying the input vector by a generator matrix G, namely

$$\mathbf{O} = \mathbf{I}^\mathrm{T}\mathbf{G}, \tag{4.7}$$

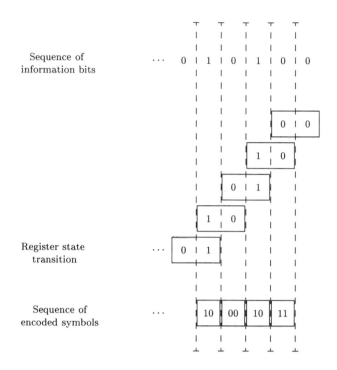

Figure 4.15: The action of the encoding process.

where G can be constructed as

$$G = \begin{bmatrix} 11\ 10\ 11\ 00\ 00\ 00\ \dots \\ 00\ 11\ 10\ 11\ 00\ 00\ 00\ \dots \\ 00\ 00\ 11\ 10\ 11\ 00\ 00\ 00\ \dots \\ \vdots \end{bmatrix}. \tag{4.8}$$

As convolutional codes do not have a defined length, the generator matrix G is a semi-infinite matrix. For the example in Figure 4.14, $\mathbf{I}^T = 101000\dots$, the output sequence is obtained by adding the first and the third rows of G to produce $\mathbf{O}^T = 11, 10, 00, 10, 11, 00, 00, \dots$.

4.3.2   State and Trellis Diagrams

Figure 4.15 shows the action of the encoding process from an information sequence to a chain of state transitions and finally to a sequence of encoded symbols. As the information sequence is shifted into the register by

Present State   Next State

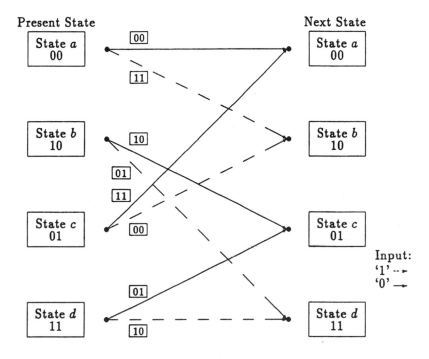

Figure 4.16: State transition diagram.

$k = 1$ bit at a time, the encoder state obtains its value by observing the information sequence with a window size of $(K - 1)k = 2$ bits. A chain of state transitions of ..., 01, 10, 01, 10, 00, is obtained from an information sequence of ...010100. New encoded symbols are formed in response to these state transitions and the sequence of the encoded symbols becomes ..., 10, 00, 10, 11.

The encoder state is given by the last $(K - 1)k$ register stages, namely 2 in this example. The number of possible states is $2^{(K-1)k}$ ($= 4$ here). The state at each instant can be either 00, 10, 01 or 11, which we label as state a, b, c or d, respectively. Figure 4.16 tabulates the possible state transitions at any symbol instant. The branch emanating from the present state to the next state indicates the state transition. The broken line branch is the transition initiated by input bit of logical 1, whereas the solid branch is due to the input bit being a logical 0. The symbol attached to each branch is an $n = 2$ bit word and is the encoded symbol. The number of outgoing branches emanating from the present state is 2^k ($=2$ here), which corresponds to the number of possible input patterns. Figure 4.16 represents the fundamental structure that can be used to construct the state diagram and trellis diagram of the convolutional code.

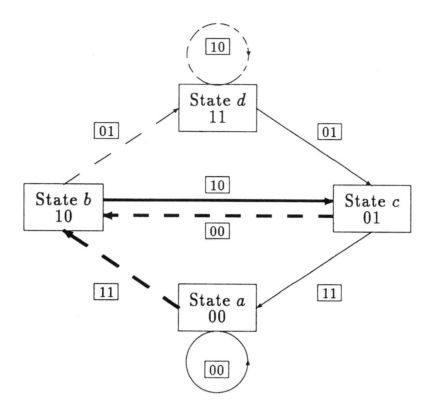

Figure 4.17: The state diagram.

The state diagram corresponding to the state transition diagram is shown in Figure 4.17. It consists of a total number of $2^{(K-1)k} = 4$ states connected by all the possible transitions shown in the state transition diagram. In the example being considered, the shift register is initially cleared setting the encoder to state a. The present input is a logical 1, the state changes from a to b, as illustrated by the dotted branch emanating from state a in Figure 4.17. The encoder output is the symbol 11 that is assigned to the transition branch. At the next instant, the present state becomes state b and the encoder receives a new input of logical 0. This causes the state transition from b to c, and the encoded symbol becomes 10. The encoding cycle is repeated for subsequent inputs, which change the states and then generate the encoded symbols. By following the change of states throughout the encoding process, a particular path associated with states $a \rightarrow b \rightarrow c \rightarrow b \rightarrow c \rightarrow \dots$, can be observed. This path is unique to the input sequence.

Another representation of the encoding process is the trellis diagram

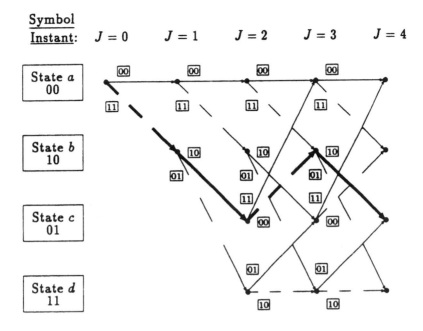

Figure 4.18: The trellis diagram.

shown in Figure 4.18. This is formed by concatenating the consecutive instants of the state transition diagram of Figure 4.16 starting from the reset all-zero condition. The diagram illustrates all the possible paths, namely 16, that can occur for 4-bit information sequences. We observe in Figure 4.18 that after the initial transient, i.e. after instant $J = 1$, the trellis contains four nodes at every symbol instant. The trellis is extended, for example to the $J = 5$ symbol instant, by appending the state transition diagram of Figure 4.16 every node, i.e., state. In our example, the path corresponding to the input sequence is drawn by the thick line in Figure 4.18 corresponding to the encoded sequence of 11, 10, 00, 10, ...

4.3.3   Maximum Likelihood Decoding

We have seen in Section 4.3.1 that an information sequence changes the encoder states which in turn generates a sequence of encoded symbols. The encoded sequence is interleaved to combat the effects of signal fading in the channel, modulates a carrier and is transmitted over the mobile radio channel. The channel impairments will distort the received signal. After demodulation and deinterleaving, the convolutional decoder inverts the encoding process. The convolutional decoder operates on the input sequence by estimating the most likely path of state transitions in the trellis.

Once identified, the corresponding information sequence is delivered as the decoded sequence. If the decoder employs the Viterbi algorithm [20, 21], *all* the possible paths in the trellis are searched, and their distances to the sequence at the decoder input are compared. The path with the smallest distance is then selected, and the information sequence regenerated. This method is known as maximum likelihood decoding in the sense that the most likely sequence from *all* the paths in the trellis is selected. It therefore results in the minimum bit error rate.

We have seen in Section 4.3.1 that a convolutional code is a long sequence of encoded symbols without a well-defined block length. In practice, convolutional codes can be truncated to a fixed length and concatenated, i.e., as if in a sequence of packets. The code is terminated by appending $(K-1)k$ logical zeros to the last information bit for the purpose of clearing the shift register of all information bits. By this means, the encoder returns to a known all-zero state. As the stuffing of logical zeros carries no information, the actual coding rate is now below R. In order to maintain the coding rate close to R, the encoded sequence (period of truncation) needs to be very long.

4.3.3.1   Hard-decision Decoding

The Viterbi algorithm is best explained using the trellis diagram of the simple binary $CC(2,1,3)$ code that was previously used as an example. We will assume that the demodulator provides only hard-decisions when regenerating the information sequence (soft-decision decoding is considered later). In this case the Hamming distances between the received symbols and the estimated transmitted symbols in the trellis are used as a metric, i.e., a confidence measure. Figure 4.19 records the history of the paths selected by the Viterbi decoder. Suppose there are no channel errors, and the input sequence to the decoder is the same as the encoded sequence 11, 10, 00, 10, ... as illustrated in Figure 4.14. At the first instant $J = 1$, the received symbol is 11 which is compared with the possible transmitted symbols 00 and 11 of the branches from node a to a and from node a to b, respectively. The metrics of these two branches are their Hamming distances, namely the differences between the possible transmitted symbols 00 or 11 and between the received symbol 11. Their distances are 2 and 0, respectively.

Now we define that the branch metric is the Hamming distance of an individual branch, and the path metric at the Jth instant is the sum of the branch metrics at all its branches from $J = 0$ to the Jth instant. Hence the path metrics, printed on top of each node in Figure 4.19, at instant $J = 1$ are 2 and 0 for the paths $a \rightarrow a$ and $a \rightarrow b$, respectively. At the second instant $J = 2$, the received symbol is 10 and the branch metrics are 1, 1, 0 and 2 for the branches $a \rightarrow a$, $a \rightarrow b$, $b \rightarrow c$ and $b \rightarrow d$, respectively. The path metrics, or accumulated metrics, are 3, 3, 0 and

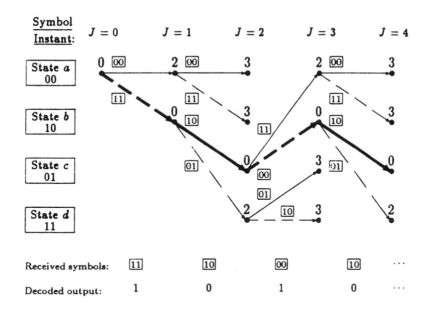

Figure 4.19: Example of Viterbi decoding

2 for the corresponding paths $a \to a \to a$, $a \to a \to b$, $a \to b \to c$ and $a \to b \to d$. At the third instant, the received symbol is 00. There are eight possible branches, as seen also in Figure 4.17 and their branch metrics are 0, 2, 2, 0, 1, 1, 1 and 1 for the branches $a \to a$, $c \to a$, $a \to b$, $c \to b$, $b \to c$, $d \to c$, $b \to d$ and $d \to d$, respectively. Let α_1 and α_2 denote the corresponding paths $a \to a \to a \to a$ and $a \to b \to c \to a$ that begin at the initial node a and remerges at node a at $J = 3$. Their respective path metrics are 3 and 2. Any further branches with $J > 3$ stemming from the node a at $J = 3$ will add identical branch metrics to the path metric of both paths α_1 and α_2, and this means that the path metric of α_1 is larger at $J = 3$ and will continue to be larger at $J > 3$. The Viterbi decoder selects the path with the smallest metric and therefore decides to discard the α_1 path and retain the α_2 path. The α_2 path is called the *survivor*. This procedure is also applied at the other nodes b, c and d at $J = 3$. Notice that paths $a \to a \to a$ and $a \to a \to b$ cannot survive as their path metrics are larger than those of their counterparts of the merging pairs and they are therefore eliminated from the decoder memory. Thus there are four paths that survive at each instant and there are four path metrics, one for each of the surviving paths. Similarly, at the instant $J = 4$, there are again four survivors and the survivor with the smallest metric is the path $a \to b \to c \to b \to c$. This path corresponds to the transmitted sequence and hence the correct information sequence 1010... is delivered

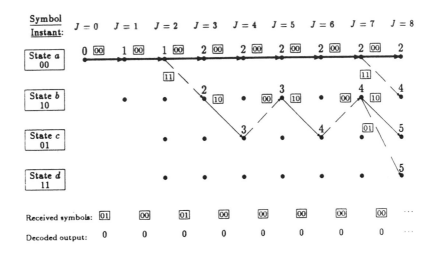

Figure 4.20: Example of correct decoding.

to the decoder output. This decoding method is optimum in the sense that it minimises the probability that the entire sequence is in error.

4.3.3.1.1   Correct Decoding   We will now introduce channel errors, and demonstrate how the Viterbi decoding corrects them. Figure 4.20 shows four survivor paths in the trellis at the instant $J = 8$. Suppose that the all-zero sequence is transmitted, and the received sequence is 01 00 01 00 00 ... , where the logical ones are the channel error bits. The first received symbol is 01 and the metric at $J = 1$ is 1 which reflects the number of error bits in the first received symbol. Similarly, the next error bit in the third received symbol 01 increments the metric of the all-zero path as well. At instant $J = 8$, four survivors are the estimated transmitted sequences which have Hamming distances of 2, 4, 5 and 5 for the received sequence shown in Figure 4.20. The Viterbi decoder favours the all-zero path, and its metric shows the number of channel error bits.

4.3.3.1.2   Incorrect Decoding   When the number of channel errors exceeds the correcting capability of the code, incorrect decoding will occur as illustrated in Figure 4.21. Again, an all-zero sequence is transmitted and the received sequence due to the channel impairments contains three channel error bits. The incorrect decoding occurs in the three initial branches which results in a single decoded information bit error determined at $J = 8$. The path metric of the selected path, drawn as a thick line in Figure 4.21, is 2 and no longer corresponds to the actual number of the channel error bits, namely 3.

The last two examples of correct and incorrect decoding, are related to

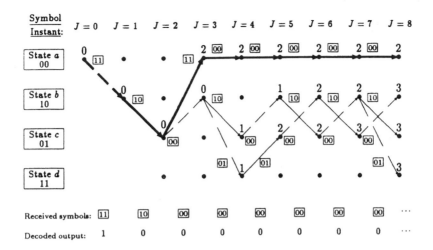

Figure 4.21: Example of incorrect decoding.

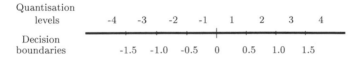

Figure 4.22: 8-level received signal quantisation.

decisions that depend on whether the Hamming distance of the received sequence to the correct path is smaller than the distances to other paths in the trellis. If it is closer, we obtain correct decoding. We observe that the Hamming distance between the paths in the trellis plays an important role in the correcting capability of the code, and this will be discussed in Section 4.3.4.

4.3.3.2 Soft-decision Decoding

So far we have limited our discussion to hard-decision decoding. That is, the demodulated signal at the demodulator output is sampled and hard-limited to regenerate the binary signal for channel decoding. We now explore the techniques of soft-decision decoding. In this approach, the signal variations at the output of the demodulator are sampled and quantised. For an additive white Gaussian noise (AWGN) channel, hard quantisation of the received signal results in a loss of about $2\,\mathrm{dB}$ in E_b/N_0 compared with infinitely fine quantisation [22], while an 8-level quantisation reduces the loss compared to infinite fine quantisation to less than $0.25\,\mathrm{dB}$. This indicates that quantisation with 8-levels is adequate for our purposes.

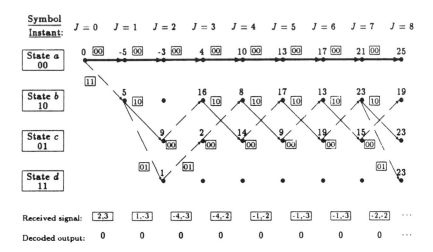

Figure 4.23: Example of Viterbi decoding with soft-decision.

Figure 4.22 shows the range of the sampled output from a binary de-modulator with decision boundaries spaced by 0.5 and eight quantisation levels, where the transmitted signal level is unity. The magnitude of the quantised level {1, 2, 3, 4} represents the confidence we have in the received sample, where the higher the number the greater the confidence. For example, a level ±4 gives a high confidence that the bit is a logical ±1, whereas a level of ±2 gives us less confidence as to the polarity of the bit.

Figure 4.23 applies the Viterbi decoding with soft-decision to the same example as in Figure 4.21, which produced an erroneous output when hard-decision decoding was used. By using soft-decision, the demodulator passes a sequence of quantised levels to the FEC decoder instead of a sequence of data bits. Assuming that an all-zero sequence is transmitted, in the absence of channel noise and fading the received signal at the output of the demodulator is quantised to a large negative level. Should the quantised output be positive, i.e., a transmission error occurs, the channel impairments have caused a stronger, but opposite polarity voltage against the transmitted signal. In Figure 4.23 we see that the first three quantised levels are positive, namely 2,3,1 are channel error bits in the absence of channel decoding. The first received symbol has confidence levels 2,3, and is compared with the transmitted symbol of the possible transitions of 00 and 11 in the trellis. We observe that levels 2 and 3 give us a measure of confidence that the transition should be 11. Comparing transition 00 with 2,3, the branch metric for path $a \rightarrow a$ at $J = 1$ is $-2 + (-3) = -5$. However, when we compare the possible transition 11 with levels 2,3, the branch metric is $2 + 3 = 5$, when a transition $a \rightarrow b$ occurred. Similar arguments at $J = 2$ result in accumulated confidences of $-3, 9, 1$ for paths

$a \to a \to a$; $a \to b \to c$ and $a \to b \to d$. Let us denote by α_1 and α_2 the paths $a \to a \to a \to a$ and $a \to b \to c \to a$ (not shown in Figure 4.23) respectively. The path metrics at $J = 3$ for paths α_1 and α_2 are 4 and 2 ($= 9 - 7$), respectively. The Viterbi decoder selects the path with the largest metric because of its stronger accumulated confidence. Hence, path α_1 is selected. At instant $J = 8$, the all-zero path has the largest metric among the four survivors. This example shows that the incorrect decoding, which occurred when hard-decision decoding was used can be rectified by using soft-decision decoding.

4.3.3.3   The Viterbi Algorithm

From the examples in Subsections 4.3.3.1 and 4.3.3.2, we found that the complexity of the trellis diagram directly reflects the computation and memory requirement of the Viterbi decoder. In general, for a $CC(n, k, K)$ code, there are $2^{(K-1)k}$ possible states in the encoder. In the Viterbi decoder all states are represented by a single column of nodes in the trellis at every symbol instant. At each node in the trellis, there are 2^k merging paths and the one with the minimum distance is selected as the survivor. As a consequence there are 2^k path comparisons at each merging node, and the same number of comparisons is repeated for all $2^{(K-1)k}$ nodes at a sampling instant. The computation increases exponentially with K and k, and this restricts them to relatively small values.

The Viterbi decoder selects and updates $2^{(K-1)k}$ surviving sequences and stores them in a memory at each instant. At the end of the encoded sequence or packet, the decoder selects the survivor with the minimum distance, i.e., the minimum Hamming distance with hard-decision decoding, or the maximum confidence measure with soft-decision decoding. In practice, the encoded packets are usually very long, and it is impractical to store the entire length of the surviving sequences before making a decision as to the information sequence because of the long decoding delay that would accrue. Instead, only the most recent L information bits in each of the surviving sequences are stored. Once the survivor with the minimum distance is identified the symbol associated with this path L periods ago is conveyed to the output as a decoded information symbol. In Section 4.3.1 we described how the present encoded symbol depends on the $(K - 1)$ previous symbols. The ramification of this is that the joint probabilities between symbols tends to be inversely proportional to the time separation between them. As a consequence we arrange for the parameter L to be made sufficiently large, normally $L \geq 5K$, for the present symbol of the surviving sequences to have a minimum effect on the decoding of the L previous symbols.

We have described the principles of hard-decision and soft-decision Viterbi decoding. The path metric in hard-decision decoding is a measure of the Hamming distance, and the path with the minimum metric is identified by the Viterbi decoder to yield the recovered sequence. With soft-decision decoding the path metric is a confidence measure and the path with the maximum metric provides the highest confidence and is therefore selected by the Viterbi decoder. Let us now summarise the essential features of the Viterbi algorithm that apply for hard-decision and for soft-decision decoding. For a $\text{CC}(n, k, K)$ code, there are $2^{(K-1)k}$ number of states and each state can change to 2^k possible states at the next instant in the trellis. The state transition in the trellis is represented by the branch, and there are 2^K possible branches at each instant. For a particular information sequence, a unique path is observed in the trellis. A path consists of a chain of branches and each branch in the trellis is associated with an encoded symbol. The symbol attached to the path forms the sequence of encoded symbols that are transmitted. At the receiver, the Viterbi decoder matches the received symbol to every possible encoded symbol in the trellis. Suppose c_{ji} denotes the bit at the instant J in the trellis, where j indicates the jth branch and i indicates the ith bit in the symbol. The value of c_{ji} in the trellis is an estimate of the corresponding bit r_{ji} at the output of the demodulator. The Viterbi decoding process can be described by the following steps:

1. **Branch metric calculation**

   The branch metric $m_j^{(\alpha)}$ at the Jth instant of the α path through the trellis is defined as the logarithm of the joint probability of the received n-bit symbol $r_{j1}r_{j2}\ldots r_{jn}$ conditioned on the estimated transmitted n-bit symbol $c_{j1}^{(\alpha)}c_{j2}^{(\alpha)}\ldots c_{jn}^{(\alpha)}$ for the α path. That is,

$$m_j^{(\alpha)} = \ln \prod_{i=1}^{n} P(r_{ji} \mid c_{ji}^{(\alpha)})$$

$$= \sum_{i=1}^{n} \ln P(r_{ji} \mid c_{ji}^{(\alpha)}) . \qquad (4.9)$$

2. **Path metric calculation**

   The path metric $M^{(\alpha)}$ for the α path at the Jth instant is the sum of the branch metrics belonging to the α path from the first instant to the Jth instant. Therefore,

$$M^{(\alpha)} = \sum_{j=1}^{J} m_j^{(\alpha)} . \qquad (4.10)$$

3. **Information sequence update**

   There are 2^k merging paths at each node in the trellis and the decoder selects from the paths $\alpha_1, \alpha_2, \ldots, \alpha_{2^k}$, the one having the largest metric, namely,

$$\max \left(M^{(\alpha_1)}, M^{(\alpha_2)}, \dots, M^{(\alpha_{2^k})} \right) \qquad (4.11)$$

and this path is known as the survivor.

4. **Decoder output**

When all of the $2^{(K-1)k}$ survivors have been determined at the Jth instant, the decoder outputs the $(J - L)$th information symbol from its memory of the survivor with the largest metric.

Let us apply the path metric calculation, defined in the above steps, for hard-decision decoding, where r_{ji} is either 0 or 1. We again use our previous example of Viterbi decoding, see Figure 4.19. The two paths α_1 and α_2 that begin at the initial node a at $J = 0$ and merge to node a after three branches in the trellis have Hamming distances of 2 and 3 with the received sequence 10 00 01 ..., respectively. According to Equations 4.9 and 4.10, the path metrics for α_1 and α_2 are

$$M^{(\alpha_1)} = \sum_{j=1}^{3} \sum_{i=1}^{2} \ln P(r_{ji} \mid c_{ji}^{\alpha^1}) = 4\ln(1 - p_b) + 2\ln(p_b)$$

$$M^{(\alpha_2)} = \sum_{j=1}^{3} \sum_{i=1}^{2} \ln P(r_{ji} \mid c_{ji}^{\alpha^2}) = 3\ln(1 - p_b) + 3\ln(p_b) \qquad (4.12)$$

where p_b is the probability of a channel bit error. Assuming that $p_b < \frac{1}{2}$, then $M^{(\alpha_1)}$ is greater than $M^{(\alpha_2)}$ and therefore the α_1 path is selected as the survivor. We also note that, as expected, the path α_1 has the smallest Hamming distance. If soft-decision is employed, each quantised sample at the output of the demodulator indicates a confidence measure of its associated data bit. This measure is the Euclidean distance of the received signal vector from the signal boundary in the constellation. For an M-ary modulation scheme, the signal points in the constellation are surrounded by more than one boundary. The smaller the Euclidean distance of the received vector to a particular signal point, the stronger is our confidence in the value of the vector. However, for binary modulation, the two signal points are separated by a single boundary. The larger the Euclidean distance of the received signal from the boundary, the more confidence we have that the received signal has the correct polarity. As an example, if the modulation is minimum shift keying (MSK) [23], there is an in-phase (I) and quadrature phase (Q) signalling channel. The signal boundaries of the I and Q channels are spaced by a phase angle of 90° and they are therefore orthogonal. The confidence measure on one channel does not affect the measure on the other channel. Both I and Q can be then treated as independent binary signalling channels and our confidence is proportional to the Euclidean distance of the received vector from the signal boundary.

Suppose the MSK transmitted signal is coherently detected to yield

$$r_{ji} = \frac{A_s T}{2} a_{ji}^{(\alpha)} + N(T),\tag{4.13}$$

where A_s is the transmitted signal amplitude, T is the bit duration, $a_{ji}^{(\alpha)}$ is -1 for $c_{ji}^{(\alpha)} = 0$; and 1 for $c_{ji}^{(\alpha)} = 1$, i.e., $a_{ji}^{(\alpha)} = 2c_{ji}^{(\alpha)} - 1$, and $N(T)$ is a Gaussian random noise signal with zero mean and variance $\sigma_N^2 = N_0 T/8$, and $N_0/2$ is the double-sided power spectral density of the receiver thermal noise. For the AWGN channel, the demodulator output signal has a probability density function (PDF) of

$$f(r_{ji} \mid c_{ji}^{(\alpha)}) = \frac{1}{\sqrt{2\pi}\sigma_N} \exp\left(-\frac{(r_{ji} - \frac{A_s T}{2}a_{ji}^{(\alpha)})^2}{2\sigma_N^2}\right).\tag{4.14}$$

By substituting Equation 4.14 into Equation 4.9, we obtain the branch metric which is then inserted into Equation 4.10 to obtain the path metric,

$$
\begin{aligned}
M^{(\alpha)} &= \sum_{j=1}^{J}\sum_{i=1}^{n} \ln P\left(r_{ji} \mid c_{ji}^{(\alpha)}\right) \\
&= \sum_{j=1}^{J}\sum_{i=1}^{n} \left[\ln\left(\frac{1}{\sqrt{2\pi}\sigma_N}\right) - \frac{(r_{ji} - \frac{A_s T}{2}a_{ji}^{(\alpha)})^2}{2\sigma_N^2}\right] \\
&= \sum_{j=1}^{J}\sum_{i=1}^{n} \left[\ln\left(\frac{1}{\sqrt{2\pi}\sigma_N}\right) - \frac{r_{ji}^2}{2\sigma_N^2} + \frac{\frac{A_s T}{2}r_{ji}a_{ji}^{(\alpha)}}{2\sigma_N^2} - \frac{(\frac{A_s T}{2}a_{ji}^{(\alpha)})^2}{2\sigma_N^2}\right].
\end{aligned}
\tag{4.15}
$$

The first and second terms of Equation 4.15 are common to all paths. Also $(a_{ji}^{(\alpha)})^2$ is always 1, and the fourth term is a constant and again is common to all paths. As the terms common to all path metrics do not change the path selection, they can be ignored from the calculation of path metrics. Futhermore, $(A_s T/4\sigma_N^2)$ in the third term can also be neglected. Therefore the path metric difference becomes

$$\Delta M^{(\alpha)} = \sum_{j=1}^{J}\sum_{i=1}^{n} r_{ji}a_{ji}^{(\alpha)}.\tag{4.16}$$

Equation 4.16 implies that the path metric at the Jth instant of the α path is the accumulated bit confidence measure. The path with the higher confidence reflects the larger metric and thus the Viterbi decoder selects the path with the larger metric as the survivor. This decision method has been demonstrated in our previous soft-decision decoding example in Figure 4.23.

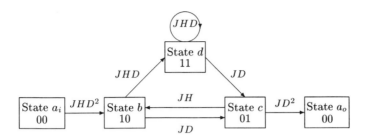

Figure 4.24: State diagram for the CC$(2, 1, 3)$ code.

4.3.4 Distance Properties of Convolutional Codes

From the examples in Figure 4.20 and Figure 4.21, we observe that the Hamming distance of the paths in the trellis determines the correcting capability of the code. The bit error rate (BER) performance of a convolutional code thus depends on its distance property. As convolutional encoding involves modulo-2 linear operations on the information sequence, the convolutional code is linear and therefore the distance separations of all the paths from a code sequence in the trellis is independent of which particular code sequence is considered.

For the sake of simplicity, we assume that the all-zero sequence is transmitted. Consequently if any erroneous decoding occurs, the non-zero path in the trellis is favoured by the Viterbi decoder. For example, in Figure 4.21, we see that the trace of the incorrect path leaves the all-zero sequence before eventually merging back to the all-zero path. The trace can also be observed from the state diagram of the code shown in Figure 4.17. The initial state is at node a, and if there is no channel error, the received symbol is 00 and the next state is again at node a. Self-looping at node a occurs. Suppose the received sequence is no longer an all-zero sequence because of the channel noise and that the decoder fails to correct the errors. The state at node a changes to the other nodes for some instants before merging back to node a again. By splitting the node a of the state diagram in Figure 4.17 into an input stage a_i and an output stage a_o, the traces of all the incorrect paths are revealed by the possible connections from the input to the output as shown in Figure 4.24.

The transition from node a_i to b represents the state transition of leaving the correct path, whereas that from node c to a_o is the transition of merging back to the all-zero path. The branches of the state diagram are labelled as either D^0, D^1 and D^2, where the exponent of D denotes the Hamming distance of the received symbol to the all-zero symbol in bits. The factor H is introduced into those branches activated by the information bit 1. Also, a factor of J is introduced into each branch such that the exponent of J

will serve as a counter to indicate the number of instants in any given path before merging back to the correct sequence.

Let us illustrate the situation by considering an example of the factors J, H and D associated with the branch $a_i \to b$ in Figure 4.24. This branch consists of a single transition and therefore the exponent of the factor J is unity. The state transition from 00 to 10 is due to a single information bit 1 to the input of the encoder, the exponent of H is thus 1. As the transition produces an encoded symbol of 11 having a distance of 2 bits from the all-zero sequence, the factor D^2 is attached to this branch. Consequently, the branch $a_i \to b$ is labelled by JHD^2. Let X_s be a variable representing the accumulated weight of each path that enters state s. The transfer function associated with all transitions from state a_i to a_o then provides the required enumeration of path weights. That is, we consider,

$$T(D, H, J) = X_{a_o}/X_{a_i} \qquad (4.17)$$

such that all the possible incorrect paths originating from the input stage and terminating at the output stage are illustrated. From Figure 4.24, the state equations provide the following recursive relationships:

$$
\begin{aligned}
X_b &= JHD^2 X_{a_i} + JHX_c \\
X_c &= JDX_b + JDX_d \\
X_d &= JHDX_b + JHDX_d \\
X_{a_o} &= JD^2 X_c .
\end{aligned}
\qquad (4.18)
$$

The transfer function of the $CC(2,1,3)$ code is obtained by determining X_{a_o}/X_{a_i} from Equation 4.18 and substituting the result into Equation 4.17, namely

$$T_{\mathrm{CC}213}(D, H, J) = \frac{J^3 H D^5}{1 - JHD(1 + J)} \qquad (4.19)$$

and on dividing out,

$$
\begin{aligned}
T_{\mathrm{CC}213}(D, H, J) &= J^3 H D^5 + J^4 H^2 D^6 + J^5 H^2 D^6 + J^5 H^3 D^7 \\
&\quad + 2J^6 H^3 D^7 + J^7 H^3 D^7 + \ldots
\end{aligned}
\qquad (4.20)
$$

The first term of the transfer function $T_{\mathrm{CC}213}(D, H, J)$ indicates that there is an incorrect path having a Hamming distance of five bits (exponent of D) from the all-zero sequence that merges back to node a after three instants (exponent of J), and there is an erroneous information bit 1 (exponent of H). From the trellis diagram shown in Figure 4.25, the first term of $T_{\mathrm{CC}213}(D, H, J)$ is the trace of the path which is observed from node a, through b, c and then merging back to node a again. Similarly, the second term of $T_{\mathrm{CC}213}(D, H, J)$ is shown up as the path $a \to b \to d \to c \to a$ in

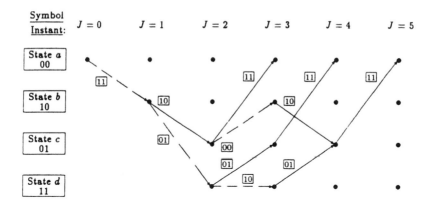

Figure 4.25: Pictorial description of $T_{\mathrm{CC}(2,1,3)}(D, H, J)$.

Figure 4.25. This path of four branches, leaving at $J = 0$ and returning at $J = 4$, has a Hamming distance of six bits with the all-zero sequence and induces two erroneous information bits at the decoder output.

Each term of the transfer function represents an incorrect trellis path. The total number of incorrect paths increases exponentially with J and therefore the transfer function has infinite terms. An important property of the transfer function is that it provides the distance properties of all the paths of the convolutional code. The minimum distance between two paths of the code is called the *minimum free distance* and is denoted as d_{free}. The d_{free} of CC$(2, 1, 3)$ is equal to five, the exponent of D in the first summation term in Equation 4.20.

The factor J of the transfer function is to determine the number of branches spanned by the paths. If the convolutional code sequence is truncated after q instants, then the transfer function for the truncated code is obtained by truncating $T(D, H, J)$ at the term J^q. However, for a very long code sequence, the transfer function tends to have an infinite number of terms and therefore J is no longer important in determining the truncation. The factor J can be suppressed by setting $J = 1$ in Equation 4.20 to yield,

$$T_{\mathrm{CC}(2,1,3)}(D, H, 1) = HD^5 + 2H^2 D^6 + 4H^3 D^7 + \ldots \qquad (4.21)$$

The transfer function in Equation 4.21 does not depend on the path length J. For instance, the second and third terms in Equation 4.20 indicate that either term has two non-zero, i.e. erroneous, information bits and a Hamming distance of six bits with the all-zero path, but both terms are different in J such that their path lengths span over four and five instants, respectively. These two terms are combined together to form the second term in Equation 4.21 regardless of their path lengths. Furthermore, H is set to unity in Equation 4.21 in order to ignore the number of erroneous information bits associated with the path. The transfer function now depends only on D and the Hamming distance of all the incorrect paths to the all-zero sequence becomes:

$$T_{\text{CC}(2,1,3)}(D,1,1) = D^5 + 2D^6 + 4D^7 + \ldots \; . \tag{4.22}$$

As the correct path is assumed to be an all-zero sequence, the Hamming distance between an incorrect and correct path is the weight (number of logical ones) of the incorrect path. In general, if d denotes the Hamming distance of a weight-d path, Equation 4.22 can be expressed as

$$T(D) = \sum_{d=d_{min}}^{\infty} A_d D^d, \tag{4.23}$$

where the coefficient A_d is the number of incorrect paths of weight-d regardless of the information bits on the path.

An important property of the code is the weight distribution, which determines the number of information bit errors at the decoder output if the weight-d path is incorrectly selected. The weight distribution $W_{\text{CC}213}(d)$ of the CC$(2,1,3)$ code is characterised by the total number of erroneous information bits of all weight-d trellis paths and is formally obtained by differentiating $T_{\text{CC}213}(D,H)$ with respect to H and setting $H = 1$:

$$W_{\text{CC}213}(d) \;\; = \;\; \left. \frac{d(T_{\text{CC}213}(D,H,1))}{dH} \right|_{H=1.} \tag{4.24}$$

This is true, since differentiating $T(D,H)$ effectively yields the multiplication of the total number of incoming paths with the number of incorrect decoded bits per such path, giving the total number of incorrect decoded bits in all weigth-d paths. Substituting Equation 4.21 into 4.24, we have

$$W_{\text{CC}213}(d) = D^5 + 4D^6 + 12D^7 + \ldots \; . \tag{4.25}$$

In general, the weight distribution of a code can be described by

$$W(d) = \sum_{d=d_{min}}^{\infty} W_d D^d, \tag{4.26}$$

where the coefficient W_d is defined as the total number of erroneous information bits for all weight-d paths. The minimum distance among all these paths is d_{free} and therefore $W_d = 0$ for $d < d_{min}$.

The CC$(2,1,3)$ code of our example is a *non-systematic* code because neither of the polynomial generators produces output bits that are identical

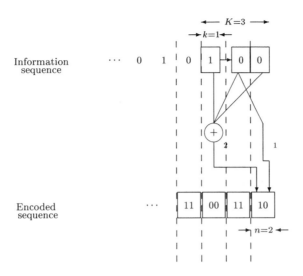

Figure 4.26: Convolutional systematic encoder for the $CC(2,1,3)$ code.

to the information bits. Conversely, if either of the generator polynomials $g_1(z)$ or $g_2(z)$ has only a single connection to the register, the information sequence appears directly at the encoder output and the codes are described as *systematic* codes. Non-systematic convolutional codes are usually preferred because of their higher error correcting capability compared to systematic codes, while for block codes the converse is true. The reasons become apparent if we compare the distance property of the systematic $CC(2,1,3)$ with our example of a non-systematic $CC(2,1,3)$. The generators $g_1(z)$ and $g_2(z)$ of the systematic code are, as shown in Figure 4.26:

$$g_1(z) = z^2 \quad \text{and} \quad g_2(z) = 1 + z + z^2 . \tag{4.27}$$

The generator $g_1(z)$ is a direct hard-wire connection and therefore copies the information sequence to its output. The optimum performance code is formulated by computer search of optimum connections for the generator polynomials. As the $g_1(z)$ connection is fixed to be z^2, it reduces the degree of freedom for the search of the optimum code. The code cannot therefore achieve the maximum distances between encoded sequences as it can for non-systematic codes. By eliminating one of the adders, there is a reduction in the minimum free distance.

If we calculate the transfer function of the systematic $CC(2,1,3)$ code with the generator polynomials given in Equation 4.27, then the weight distribution of this code is obtained with the aid of Equation 4.24 as

$$W_{CC213}(d) = 3D^4 + 15D^6 + 58D^8 + \dots . \tag{4.28}$$

The d_{free} of this code is 4, which is less than that of its non-systematic counterpart ($d_{free} = 5$). We note that the d_{free} is the minimum separable distance of the code, i.e., any two encoded sequences must be separated by a Hamming distance of at least d_{free}. The performance of the code is proportional to d_{free}. The smaller d_{free} of the systematic code compared with the non-systematic reflects the reduction of the relative distances between encoded sequences. In fact, Bucher and Heller [24] have shown that for large K, the performance of a systematic code of constraint length K is approximately the same as that of a non-systematic code of constraint length $K(1-R)$. Thus for $R = 1/2$ and very large K, systematic codes have the performance of non-systematic codes of half the constraint length, while demanding the same decoder complexity.

So far, we have used the $CC(2,1,3)$ code as our simple example to illustrate the principle of convolutional coding. This code is weak because its contraint length $K = 3$ is short. The code advocated by the GSM recommendation [12] for speech communications over the Pan-European cellular mobile radio system is $CC(2,1,5)$. The generator polynomials of the full-rate speech channel [25] for this code are described by

$$g_1(z) = 1 + z^3 + z^4 \quad \text{and} \quad g_2(z) = 1 + z + z^3 + z^4 . \tag{4.29}$$

This is a binary half rate code with a constraint length of 5 bits. The state diagram has $2^{(K-1)k} = 16$ states and the transfer function of the code is found by solving 16 state equations. For the binary code where $k = 1$, the complexity of the computation grows exponentially with K. An alternative way of obtaining the transfer function for large values of K is to trace through every possible non-zero path in the trellis by an exhaustive computer search and record their path distances. The weight distribution of the code $CC(2,1,5)$ is obtained by recording the total weight of all information sequences which produce paths of distance d from the all-zero path. From our computer search we found the weight distribution to be

$$W_{CC215}(d) = 4D^7 + 12D^8 + 26D^9 + \dots . \tag{4.30}$$

The coefficient W_7 of the first term in Equation 4.30 indicates that a total of 4 information bit errors are associated with all weight-7 paths. Similarly, the coefficients W_8 and W_9 of the second and third terms have 12 and 20 information bit errors associated with weight-8 and weight-9 paths, respectively. The minimum distance d_{free} among all of these paths is 7, the exponent of D in the first term in Equation 4.30.

The code used for satellite communications [25] is $CC(2,1,7)$, a binary half rate code with a longer constraint length of 7 bits. The generator polynomials [26] used in the encoder are given by

$$g_1(z) = 1 + z^2 + z^3 + z^5 + z^6 \quad \text{and} \quad g_2(z) = 1 + z + z^2 + z^3 + z^6 \ . \quad (4.31)$$

The weight distribution of this code is obtained by computer search in order to avoid the complexity of solving its $2^{(K-1)k} = 64$ state equations, and is given by

$$W_{\text{CC}217}(d) = 36D^{10} + 211D^{12} + 1404D^{14} + \dots \ . \quad (4.32)$$

The minimum free distance of this code is $d_{free} = 10$. Although all of the $CC(2,1,3)$, $CC(2,1,5)$ and $CC(2,1,7)$ are half rate binary codes, the effect of increasing their constraint length is to increase the number of states in the code and their minimum free distance and thereby enhance their error correcting capability.

4.3.5 Punctured Convolutional Codes

For a convolutional code of rate k/n there are 2^k merging paths at each node in the trellis. The decoding of this code by the Viterbi algorithm selects the path with the highest metric out of the 2^k possibilities at each node. The number of calculations per selection at each node grows exponentially with k, rendering the implementation of the codec for operation at high speed a difficult task, particularly in the case of high-rate codes. Yamada et al. propose a syndrome-former trellis [27] to decode high-rate codes. The method achieves the same performance [28] as the Viterbi decoding algorithm with reduced number of computations. However, the Viterbi decoding of high rate codes, where $k > 1$ can also be significantly simplified by employing punctured convolutional codes [29,30]. Puncturing allows us to obtain a high-rate code by periodically deleting some of the coded bits from a low-rate encoder output. In addition, puncturing of the low-rate $1/n$ code results in the decoding trellis operating with k equal to unity.

Let us consider a high rate $R = 2/3$ code, which can be achieved by either the $CC(3,2,2)$ code or the punctured $PCC(2,1,3)$ code. Both examples are chosen with the same number of states, i.e., $2^{(K-1)k} = 4$. The generator polynomials [31] of the $CC(3,2,2)$ code are

$$\begin{aligned} g_1(z) &= 1 + z + z^2 + z^3 \\ g_2(z) &= z + z^2 \\ g_3(z) &= 1 + z + z^3 \ . \end{aligned} \quad (4.33)$$

At each instant, a 2-bit information symbol is inserted into the encoder and the three generators produce a 3-bit output symbol. The coding rate R is therefore equal to $2/3$. Figure 4.27 illustrates all the possible state transitions at a symbol instant. The number of possible states of this code is $2^{(K-1)k} = 4$ and the symbol attached to each state transition is an $n = 3$

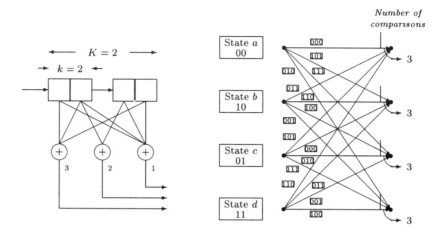

Figure 4.27: Encoder and trellis diagram for the CC(3, 2, 2), $R = 2/3$ code.

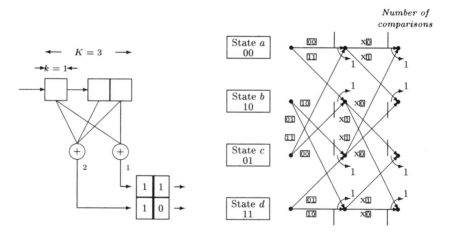

Figure 4.28: Encoder and trellis diagram for the PCC(2, 1, 3), $R = 2/3$ code.

bit encoded symbol. Each state transition is activated by a new $k = 2$ bit information symbol at the encoder input. A 2-bit shift of the encoder state at each instant induces one of four possible state transitions. For each node in the trellis shown in Figure 4.27 there are four merging branches and thus three pairwise comparisons are required to select the survivor. The comparisons are repeated for the other nodes and the total number of comparisons for decoding every three received bits is 12.

The same coding rate can also be achieved by periodically deleting bits

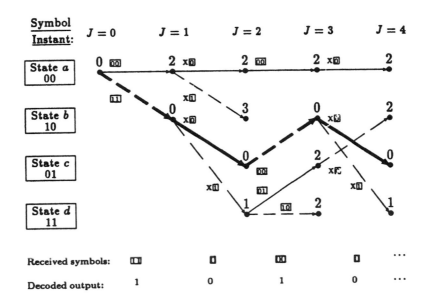

Figure 4.29: Example for the Viterbi decoding of a punctured code.

from the $CC(2,1,3)$ half rate code as demonstrated in Figure 4.28. A puncturing matrix is assigned at the output of a half rate encoder, where an element of 1 in the matrix allows its input bit to appear at the output, whereas an element of 0 in the matrix deletes the incoming bit at its input. The top row of the matrix consists of 11 which does not delete any bits from the output of adder 1. The second row of the matrix is 10 which deletes every alternate bit from the output of adder 2. The encoded sequence is formulated by sampling the adders output alternately. In this case, every fourth encoded bit is deleted, the encoder will produce three output bits for every two input bits resulting in a $R = 2/3$ code. The trellis of the punctured code shown in Figure 4.28 is basically constructed from Figure 4.16. It is equivalent to the trellis of the half rate code except that the X indicates the position of the deleted bits. There are only two merging paths to each node and therefore only a pairwise comparison is required at each node. For every three bits received over two symbol instants, the number of comparisons is 8, which is less than the 12 required by the $CC(3,2,2)$ code. Puncturing the code reduces its minimum free distance, in this case from 5 to 3. However, this is the largest minimum free distance [32] for any $CC(3,2,2)$ code, and therefore in this case there is no reduction in the minimum free distance due to puncturing.

Let us now illustrate an example of punctured coding by using the encoder described in Figure 4.28. It consists of a half-rate $CC(2,1,3)$ encoder as used in Figure 4.14, followed by a puncturing matrix at its output. Suppose we use the same input sequence to the encoder as in Figure 4.14, i.e., $\ldots, 0, 1, 0, 1$. The output sequence is punctured every fourth bit and becomes $\ldots, 0, 00, 0, 11$. On the basis of no channel errors, the input sequence to the decoder is the same as the encoded sequence. The received sequence can be either hard-decision or soft-decision decoded. Let us apply hard-decision Viterbi decoding as shown in Figure 4.29. At the first instant, $J = 1$, the received symbol is 11 which is compared with the possible transmitted symbols 00 and 11 of the branches $a \rightarrow a$ and $a \rightarrow b$. Their Hamming distances are 2 and 0, respectively. At the second instant, $J = 2$, the received symbol consists of only a single bit 0, as the other bit of the symbol was punctured at the encoder. The corresponding bit being punctured in the trellis is marked by X in Figure 4.29. The received bit is then compared with the possible transmitted bits 0, 1, 0 or 1 for the corresponding branches $a \rightarrow a$, $a \rightarrow b$, $b \rightarrow c$ and $b \rightarrow d$. The path metrics are 2, 3, 0 and 1 for the branches $a \rightarrow a \rightarrow a$, $a \rightarrow a \rightarrow b$, $a \rightarrow b \rightarrow c$ and $a \rightarrow b \rightarrow d$, respectively. Similarly, the Viterbi decoding algorithm is applied to the third and fourth instants. At the instant $J = 4$, there are four survivors and the survivor with the smallest metric is the path $a \rightarrow b \rightarrow c \rightarrow b \rightarrow c$. This path corresponds to the transmitted sequence and hence the correct information sequence $1010 \ldots$ is delivered to the decoder output.

Let us consider a punctured $R = 2/3$ code with a constraint length of 5 bits. This may be produced by puncturing the $CC(3, 1, 5)$ code where the generator polynomials are given by [30]

$$
\begin{aligned}
g_1(z) &= 1 + z + z^4 \\
g_2(z) &= 1 + z^2 + z^3 + z^4 \qquad\qquad (4.34) \\
g_3(z) &= 1 + z^2 + z^4 \,.
\end{aligned}
$$

This punctured code is designed to produce the optimum performance at BER$= 10^{-5}$ in the presence of random bit errors. The puncturing pattern of the encoded bit sequence is to delete every alternate output bit, starting from the first bit generated by $g_1(z)$ and $g_2(z)$, and also every alternate output bit, starting from the second bit produced by $g_3(z)$. Hence, three output bits remain for every two input bits that results in a 2/3 rate code.

4.3.6   Hard-decision Decoding Theory

As the convolutional code is linear, we assume that the all-zero path is transmitted, knowing that our findings can be generalised for other non-zero paths. Convolutional codes, unlike block codes, do not necessarily have a fixed length. The Viterbi decoder selects the survivors in the trellis at every instant while the sequence is being received. Bit errors at the decoder output are due to selecting the incorrect path in the trellis. In order to derive the post-decoding bit error rate performance, we define the first-event error probability, P_{FE}, as the probability when for the first time an incorrect path merges to the correct path at a node with a metric that is smaller than the one for the correct path. We recall that for hard-decision decoding, the metrics in the Viterbi algorithm are the Hamming distances between the received sequence and the paths in the trellis. Let us assume that the decoder receives a sequence of demodulated bits having an average bit error rate of p_b over a memoryless channel. Suppose that the non-zero path merging at node a is separated by a Hamming distance of d from the all-zero path. If the number of error bits in the received sequence exceeds $d/2$, the received sequence is compared with both non-zero and all-zero paths and their Hamming distances are $< d$ and $> d$, respectively. The non-zero path has a smaller metric and is therefore favoured by the Viterbi decoder. Consequently an erroneous decoding occurs. If the Hamming distance between the non-zero and the all-zero paths is odd, i.e., d is odd, the probability of an incorrect decoding is the probability that the number of channel errors is $\geq (d+1)/2$,

$$P_{ICD}(d) = \sum_{i=(d+1)/2}^{d} \binom{d}{i} p_b^i (1 - p_b)^{d-i} \; ; \qquad d \text{ is odd.} \qquad (4.35)$$

However, when the Hamming distance between the non-zero and the all-zero paths is even, and the number of channel errors equals $d/2$, the received sequence has equal distances from the non-zero and the all-zero paths. The metrics of both paths are therefore equal. In this case, the Viterbi decoder will randomly select one of the paths and therefore an erroneous decoding occurs for half of the time. Hence the probability of the incorrect decoding for even values of d is

$$P_{ICD}(d) = \sum_{i=d/2+1}^{d} \binom{d}{i} p_b^i (1 - p_b)^{d-i} + \frac{1}{2} \binom{d}{d/2}$$
$$p_b^{d/2}(1 - p_b)^{d/2} \; ; \quad d \text{ is even.} \qquad (4.36)$$

From Equation 4.35, we may obtain an upper bound by noting that

$$\begin{aligned}
P_{ICD}(d) &< \sum_{i=(d+1)/2}^{d} \binom{d}{i} p_b^{\frac{d}{2}} (1 - p_b)^{\frac{d}{2}} \\
&= p_b^{\frac{d}{2}} (1 - p_b)^{\frac{d}{2}} \sum_{i=(d+1)/2}^{d} \binom{d}{i} \\
&< p_b^{\frac{d}{2}} (1 - p_b)^{\frac{d}{2}} \sum_{i=0}^{d} \binom{d}{i} \\
&= 2^d p_b^{\frac{d}{2}} (1 - p_b)^{\frac{d}{2}} \; . \qquad (4.37)
\end{aligned}$$

Similarly, it can be shown that Equation 4.37 is an upper bound on P_{ICD} for d even.

The transfer function of Equation 4.20 describes all the possible non-zero paths in the trellis that initially leave the all-zero path and eventually merge at node a after some symbol instants. The distance from the non-zero path increases with the number of instants of separation. Any non-zero paths terminating at node a with different distances have the possibility of being selected by the Viterbi decoder as the most likely path and will result in a first-event error. The union bound of the first-event error probability is obtained by summing the probability of incorrect decoding over all the possible non-zero paths merging at node a,

$$P_{FE} < \sum_{d=d_{min}}^{\infty} A_d P_{ICD}(d) \qquad (4.38)$$

where the coefficient of A_d is the number of non-zero paths with weight d. Substituting the upper bound of P_{ICD} from Equation 4.37 into Equation 4.38 yields the upper bound on the first-event error probability,

$$P_{FE} < \sum_{d=d_{min}}^{\infty} A_d 2^d p_b^{\frac{d}{2}} (1-p_b)^{\frac{d}{2}} . \qquad (4.39)$$

Furthermore, by substituting $D = 2\sqrt{p_b(1-p_b)}$ into Equation 4.39 and using Equation 4.23, enables the upper bound of P_{FE} to be expressed as

$$P_{FE} < \sum_{d=d_{min}}^{\infty} A_d D^d = T(D) \Bigg|_{D = 2\sqrt{p_b(1-p_b)}} . \qquad (4.40)$$

If a first-event error occurs, the non-zero path is selected and the decoder outputs the corresponding erroneous information sequence. As an all-zero sequence was transmitted, the number of post-decoding bit errors is equal to the number of non-zero bits in the information sequence. Hence, if each event error probability term is weighted by the total number of non-zero information bits on all the weight-d paths, W_d namely, the first-event error probability bound can be modified to provide a bound on the post-decoding bit error, p_{bp}. In addition, for an $R = k/n$ code, there is a k-bit symbol decoded at each instant. Thus, the hard-decision post-decoding bit error probability p_{bp} is union bounded by

$$p_{bp} < \frac{1}{k} \sum_{d=d_{min}}^{\infty} W_d P_{ICD}(d) \qquad (4.41)$$

where the coefficient W_d is the total number of information bit errors for all weight-d paths and P_{ICD} is obtained from Equations 4.35 and 4.36.

However, if the upper bound of P_{ICD} is substituted into Equation 4.41, the upper bound of p_{bp} is expressed as

$$p_{bp} < \frac{1}{k} \sum_{d=d_{min}}^{\infty} W_d \left[4p_b(1-p_b)\right]^{d/2} \qquad (4.42)$$

and from Equation 4.24 and Equation 4.26, we can write,

$$p_{bp} < \frac{1}{k} \frac{d\,T(D,H)}{dH} \bigg|_{H=1, D=2\sqrt{p_b(1-p_b)}} \cdot \qquad (4.43)$$

In our example of the $CC(2,1,3)$ code, the weight distribution is obtained by differentiating $T_{CC_{213}}(D,H)$ with respect to H, see Equation 4.25. On substituting our result into Equation 4.42, the upper bound of p_{bp} is given by

$$p_{bp} < D^5 + 2D^6 + 4D^7 + \cdots \bigg|_{D=2\sqrt{p_b(1-p_b)}} \cdot \qquad (4.44)$$

However, if the expressions of P_{ICD} in Equations 4.35 and 4.36 are used to substitute into Equation 4.41, a tighter bound of p_{bp} can be obtained.

4.3.7 Soft-decision Decoding Theory

In Section 4.3.6 we evaluated the performance of convolutional codes for hard-decision decoding. With this type of decoding the signal at the output of the demodulator is sampled and binary quantised and the resulting bits are passed to the input of the decoder. However, with soft-decision decoding the demodulated signal is quantised and conveyed to the decoder. In deriving the probability of bit error for convolutional codes we assume that the voltage levels at the demodulator output are statistically independent from each other. To simplify the analysis we transmit the all-zero sequence, knowing that our results are applicable to other transmitted sequences.

Suppose that the transmitted signal is MSK modulated, and that coherent demodulation is used at the receiver. The sampled demodulated signal is applied directly to the decoder without the samples being quantised. The path metric of the α path at J instant is given by Equation 4.16 and the Viterbi decoder selects the survivor with the largest metric among all the competing paths merging at node a. Let us denote α_0 and α_1 as the all-zero path and the non-zero path, respectively, merging to node a. If $M^{(\alpha_1)} > M^{(\alpha_0)}$, the non-zero path is selected that results in incorrect decoding, and the probability of an incorrect decoding is

$$\begin{aligned} P_{ICD}(d) &= P\left(M^{(\alpha_1)} > M^{(\alpha_0)}\right) \\ &= P\left(M^{(\alpha_1)} - M^{(\alpha_0)} \geq 0\right) \end{aligned}$$

$$= P\left(\sum_{j=1}^{J}\sum_{i=1}^{n} r_{ji}(a_{ji}^{(\alpha_1)} - a_{ji}^{(\alpha_0)}) \geq 0\right). \qquad (4.45)$$

Suppose the non-zero path α_1 that merges with the all-zero path α_0 differs in d number of encoded bits, i.e., there are d logical 1 bits in the non-zero sequence. We can therefore simplify Equation 4.45 by evaluating $P_{ICD}(d)$ only at the d positions to yield

$$P_{ICD}(d) = P\left(\sum_{p=1}^{d} r_p > 0\right) \qquad (4.46)$$

where the index p is the position of those d bits in which the two paths differ, and r_p represents the demodulator output corresponding to one of these d bits. For an AWGN channel, the signal r_p is a Gaussian random variable with a PDF described by Equation 4.14. A new random variable $r = \sum_{p=1}^{d} r_p = r_p d$ can be derived, where r is also Gaussian with a mean and variance equal to μd and $\sigma_N^2 d$, respectively, and $\mu = A_s T/2$. The probability $P_{ICD}(d)$ of selecting the non-zero path is the probability of r (i.e., summation of r_p) being positive in Equation 4.46,

$$P_{ICD}(d) = \frac{1}{2}\mathrm{erfc}\left(\sqrt{\Gamma d}\right)$$
$$= \frac{1}{2}\mathrm{erfc}\left(\sqrt{E_b R d/N_0}\right) \qquad (4.47)$$

where $\Gamma = A_s^2 T/\eta_0$ is the channel signal-to-noise ratio (SNR) and E_b is the energy per information bit.

Equation 4.47 gives the probability of incorrectly decoding from the pairwise comparisons into a path of distance d from the all-zero path. There are, of course, many other possible paths with different distances that merge with the all-zero path at a given node. However, we can upper bound the error probability by summing the incorrect decoding probabilities of the pairwise comparisons over all possible paths that merge with the all-zero path at a given node. By doing so we form the union bound of the first-event error probability as

$$P_{FE} < \sum_{d=d_{min}}^{\infty} A_d P_{ICD}(d)$$
$$= \frac{1}{2}\sum_{d=d_{min}}^{\infty} A_d \mathrm{erfc}\left(\sqrt{E_b R d/N_0}\right) \qquad (4.48)$$

where A_d is the coefficient of the transfer function $T(D, H)$, representing the number of non-zero paths with weight d. Furthermore, the expression of the union bound in Equation 4.47 can be simplified to obtain the upper bound by replacing the complementary error function by the exponential function because

$$\text{erfc}\left(\sqrt{E_b Rd/N_0}\right) \leq \exp\left(-E_b Rd/N_0\right) \qquad (4.49)$$

and therefore Equation 4.47 becomes

$$P_{ICD}(d) \leq \frac{1}{2}\exp\left(-E_b Rd/N_0\right). \qquad (4.50)$$

The probability of post-decoding bit errors p_{bp} is obtained by weighting the first-event error probability term with the weight (i.e., number of bit 1 in the non-zero information sequence), of the incorrect path. Hence, similarly to hard decision, the union bound of the post-decoding bit error probability is given by

$$
\begin{aligned}
p_{bp} \quad &< \quad \frac{1}{k}\sum_{d=d_{min}}^{\infty} W_d P_{ICD}(d) \\
&< \quad \frac{1}{2k}\sum_{d=d_{min}}^{\infty} W_d \text{erfc}\left(\sqrt{E_b Rd/N_0}\right). \qquad (4.51)
\end{aligned}
$$

By substituting Equation 4.50 into 4.51, we obtain the upper bound of the post-decoding bit error probability,

$$
\begin{aligned}
p_{bp} \quad &< \quad \frac{1}{2k}\sum_{d=d_{min}}^{\infty} W_d \exp\left(-E_b Rd/N_0\right) \\
&< \quad \frac{1}{2k}\frac{d\,T(D,H)}{d\,H}\bigg|_{H=1,\,D=\exp(-E_b Rd/N_0)} \cdot \qquad (4.52)
\end{aligned}
$$

4.3.8 Convolutional Code Performance

The Viterbi decoding of convolutional codes is analysed in references [20, 33], and their performance over additive white Gaussian noise (AWGN) channels is widely known [22,30,31] for various coding rates and constraint lengths. The AWGN channel results in every bit having an equal probability of being erroneous. Convolutional codes rely on adjacent bits to correct an error. As burst errors are infrequent on a random channel, convolutional codes are appropriate. However, in the mobile radio environment where the narrowband transmission link is modelled by a Rayleigh fading channel, burst errors occur due to the deep fades. The result is that convolutional codes become occasionally overloaded and the BER performance deteriorates. In order to decrease the BER, interleaving techniques are introduced. Unfortunately interleaving introduces delay which may be unacceptable for digital speech transmissions. This situation has motivated us to examine various interleaving methods having minimum delay, while still providing an acceptable performance. We studied the effect of code parameters, such as the constraint length, the coding rate, and the performance of hard and soft decisions on the received signals. Our results emphasised the gain in performance achieved by using the Viterbi decoding with soft-decisions when the transmissions were via Rayleigh fading channels.   The system block diagram used in our simulations is shown in Figure 4.1. The source data were protected by convolutional codes and scrambled by diagonal interleaving, block interleaving or inter-block interleaving. The interleaved data were MSK modulated and transmitted over AWGN or Rayleigh fading channels. Previous experimental results [34] have showed that the mobile radio channel in highway microcells is Rician, although they can approach Gaussian or Rayleigh channels on occasions. In our experiments the received signals were demodulated into symbols if hard-decision decoding was used, or the demodulated signal was sampled and quantised into values representing the confidence of the received signals if soft-decision decoding was applied. The demodulated data were deinterleaved and convolutionally decoded to give the recovered data.

4.3.8.1   Convolutional Code Performance via Gaussian Channels

Figure 4.30 displays a set of theoretical and simulation results using Minimum Shift Keying (MSK) for the the half-rate $R = 1/2$ convolutional code $CC(2,1,5)$ decoded by hard-decision Viterbi decoding [VD-HD]. The post-decoding BER measured at the decoder output is shown as a function of signal-to-noise ratio (E_b/N_0) for the AWGN channel. The union bound and the upper bound in the Figure are obtained by substituting the weight distribution $W_{CC_{215}}(d)$ of the $CC(2,1,5)$ code, given by Equation 4.30, into Equations 4.41 and 4.42, respectively. The union bound and the simulation results were in good agreement, especially at the high E_b/N_0 values. At a BER of 10^{-6}, the difference between the upper bound and the simulation result was < 1 dB, whereas the union bound and the simulation agreed within a 0.1 dB. We concluded that the simulation results for the hard-decision decoding agreed with our theoretical calculations.

The minimum free distance d_{min} of the two-third rate punctured $PCC(3,2,5)$ code generated by using Equation 4.34 was five, a value which was the same as for the half rate $CC(2,1,3)$ code. The codes were recovered by hard-decision decoding and their BER performances against E_b/N_0 are displayed in Figure 4.31. It is interesting to note that the $CC(2,1,3)$ code had 1 dB loss over the $CC(2,1,5)$ code as shown in Figure 4.30 at a BER of 10^{-6}. Comparing the simulation results of the $PCC(3,2,5)$ code with the $CC(2,1,5)$ code again in Figure 4.30, we observe that it is only 0.1 dB and 0.6 dB inferior to the $CC(2,1,5)$ code at BERs of 10^{-6} and 10^{-2}, respectively. The performance was therefore improved by having codes with

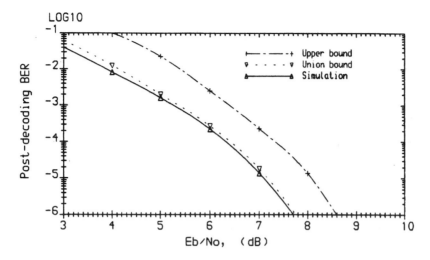

Figure 4.30: Post-decoding BER of the $R=1/2$, CC$(2,1,5)$ code with [VD-HD] via Gaussian channel.

longer constraint length K, rather than by reducing the coding rate R. The constraint length of the code determined the coder complexity as the number of states of the binary code were 4 and 16, for $K = 3$ and $K = 5$, respectively. Although the PCC$(3,2,5)$ is more complex to implement than the CC$(2,1,3)$ code, it has a higher data throughput. In general we can exchange coder complexity for data throughput.

By using soft-decision Viterbi decoding [VD-SD], we obtained a set of theoretical and simulation results for the half rate convolutional codes CC$(2,1,5)$ and CC$(2,1,7)$, and for the two-third rate punctured convolutional codes PCC$(3,2,5)$ and PCC$(3,2,7)$ as shown in Figure 4.32. The soft-decision decoding was assumed to utilise infinite quantisation levels from the demodulator. Again, the theoretical and simulation results were in close agreement. By comparing the decoding methods of soft and hard decision decoding, the CC$(2,1,5)$ code required an E_b/N_0 of 5.7 dB to achieve a BER of 10^{-6} using soft-decision, whereas it needed 7.7 dB (as indicated in Figure 4.30) to acquire the same BER employing hard-decision decoding. The performance gain of soft-decision decoding was 2 dB. In the case of the two-thirds rate punctured PCC$(3,2,5)$ code, it required E_b/N_0 values of 6.2 dB (Figure 4.32) and 7.8 dB (Figure 4.31) to yield a BER of 10^{-6} for soft and hard decision, respectively. There was a 1.6 dB gain in E_b/N_0 by using soft-decision decoding. An interesting comparison in Figure 4.32 is that there is a 1 dB gain in E_b/N_0 if the constraint length of either CC$(2,1,5)$ or PCC$(3,2,5)$ code is increased from five to seven binary stages. Also, only 0.5 dB is gained in E_b/N_0 by reducing the coding rate of

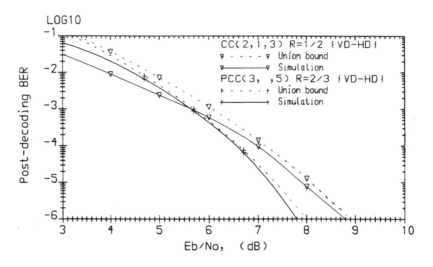

Figure 4.31: Post-decoding BER of the CC(2, 1, 3) and PCC(3, 2, 5) codes with [VD-HD] via Gaussian channel.

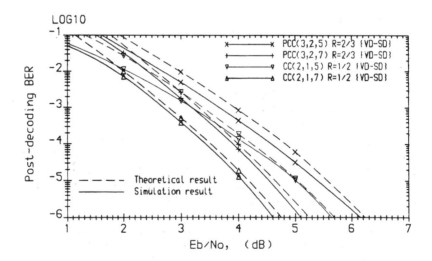

Figure 4.32: Post-decoding BER of various soft-decision Viterbi-decoded [VD-SD] convolutional codes via AWGN channel.

either the PCC(3, 2, 5) or the PCC(3, 2, 7) code from a two-thirds to a half rate CC(2, 1, 5) or CC(2, 1, 7) code, respectively.

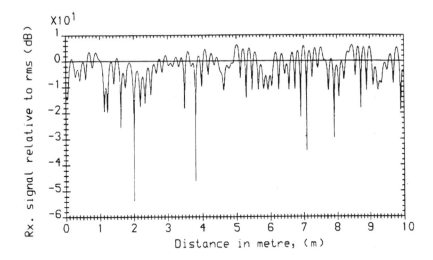

Figure 4.33: Received signal strength relative to its RMS value over Rayleigh-fading channel.

4.3.8.2 Convolutional Code Performance via Rayleigh Channels

When the transmissions were over a mobile radio channel to a MS in a vehicle travelling at 60 mph, the signal envelope was subjected to Rayleigh fading as shown in Figure 4.33. A deep fade of $-20\,$dB relative to its root mean square value was common and led to the occurrence of burst errors. Interleaving techniques were employed to transform the bursty channel into a near memoryless channel. The theoretical union and upper bounds of the post-decoding BER of the $CC(2,1,5)$ code, decoded by using the hard-decision Viterbi algorithm, for the memoryless channel were evaluated by substituting the channel BER p_b into Equations 4.41 and 4.42, respectively. These bounds were displayed in Figure 4.34. The simulation results for inter-block bit interleaving with $B = 24$ and $N = 7$, i.e., IBI/B(24,7), showed a good approximation to the theoretical calculations. Inter-block bit interleaving with a delay of 4032 bits converted the channel into a memoryless one, and an $E_b/N_0 =18.5\,$dB was required to achieve a BER of 10^{-6}. By reducing the interleaving delay to 2880 bits, the inter-block bit interleaving IBI/B(24,5) provided an approximately memoryless channel, but the performance was 1.5 dB inferior to that of IBI/B(24,7). The penalty of shorter delay was a reduction in the BER performance.

If the signal was not protected by coding and was not interleaved for transmission over the Rayleigh fading channel, then the simulations showed that an E_b/N_0 value of 52 dB was required to achieve a BER of 10^{-6}. After introducing the convolutional code $CC(2,1,5)$ without interleaving, the required E_b/N_0 value was reduced to 37.5 dB for a BER of 10^{-6}, as shown

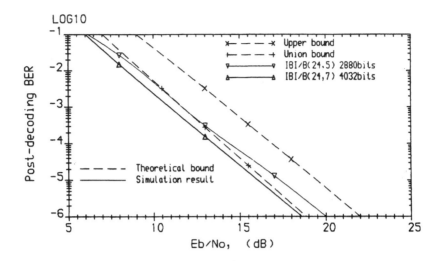

Figure 4.34: Theoretical bounds for the memoryless channel and simulation results using various interleavers for the post-decoding BER of the CC(2, 1, 5) [VD SD] code.

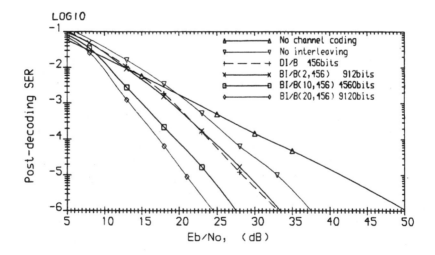

Figure 4.35: Effect of diagonal and block interleaving on the post-decoding BER of the CC(2, 1, 5) [VD SD] code over Rayleigh-fading channel.

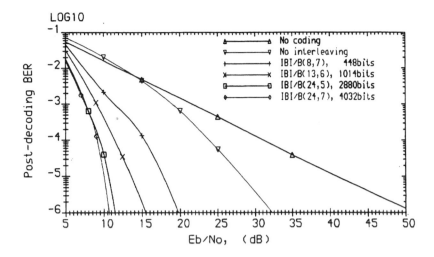

LOG10

Post-decoding BER

Eb/No, (dB)

No coding
No interleaving
IBI/B(8,7),   448bits
IBI/B(13,6),   1014bits
IBI/B(24,5),   2880bits
IBI/B(24,7),   4032bits

Figure 4.36: Effect of interblock interleaving on post-decoding BER for the CC(2, 1, 5) [VD SD] code over Rayleigh-fading channel.

in Figure 4.35, a coding gain of 14.5 dB where hard-decision decoding was used. Then in addition to the error protection, diagonal bit interleaving over a delay of 456 bits was applied. The burst errors were divided into smaller segments and were dispersed into adjacent blocks. Smaller segments of errors had a better chance of being corrected and that reduced the E_b/N_0 value to 33 dB. However, the interleaving depth of this method was only two and the error segments in the adjacent block were still bursty as illustrated by the PDF in Figure 4.10. This E_b/N_0 value was also obtained using block bit interleaving BI/B(2,456), but with the same depth. When the depth was increased to 10 and 20, the burst errors were more randomly distributed and the required E_b/N_0 values dropped to 27.5 and 24.5 dB, respectively. When comparing the E_b/N_0 at a BER of 10^{-6}, we found that for the CC(2, 1, 5) code the BI/B(20,456) scheme with 9120 bits delay (displayed in Figure 4.35) required 6 dB more SNR than the IBI/B(24,7) interleaver with 4032 bits delay (see Figure 4.34). The inter-block bit interleaving dispersed the burst errors more randomly with a smaller delay penalty compared to block bit interleaving.

In Figure 4.36, the results for the CC(2, 1, 5) code using soft-decision decoding and different inter-block bit interleaving delays is presented. With no interleaving the E_b/N_0 value of 32.5 dB gave a BER of 10^{-6} having a gain of 5 dB compared to the hard-decision version (see Figure 4.35). When IBI/B(8,7) with 448 bits delay was introduced, the E_b/N_0 value was reduced to 19.5 dB as illustrated in Figure 4.36. This delay was acceptable for speech transmissions and the performance of the code provided a guide-

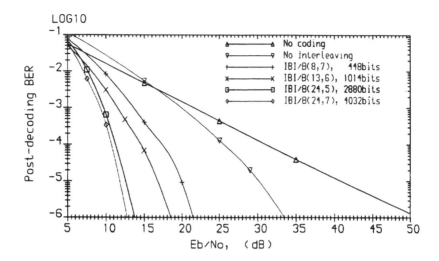

Figure 4.37: Effect of interblock interleaving on post-decoding BER for the PCC(3, 2, 5) [VD SD] code over Rayleigh-fading channel.

line for speech codec design. If the speech quality of the codec requires it to operate with a BER$> 10^{-3}$, a minimum value of E_b/N_0 of 11.5 dB must be guaranteed. When the delay was increased to 1024 and 2880 bits, the required E_b/N_0 values were reduced to 15.5 and 11.5 dB at a BER of 10^{-6}. With the longer delay of 4032 bits, a further reduction of 1 dB was achieved. As the channel became random, the E_b/N_0 was reduced to the minimum of 10.5 dB for the BER of 10^{-6}.

When the code was changed from half to two-thirds rate by keeping the same constraint length K of 5, a new set of results for the PCC(3, 2, 5) code was obtained, see Figure 4.37. The E_b/N_0 values were 33.5, 21.5, 18.5, 13.7, and 12.6 dB for a BER of 10^{-6} for the delays of 0, 448, 1014, 2880, and 4032 bits, respectively. Despite the higher coding rate, the values were only about 2 to 3 dB inferior to that of the CC(2, 1, 5) code. When speech was transmitted with an interleaving delay of 448 bits, a channel E_b/N_0 value of 13.5 dB was required for a BER of 10^{-3} that required 2 dB more compared with the half rate code with the same constraint length. It is interesting to note that at low E_b/N_0 values the BER with coding is even higher than that without coding. More bit errors were introduced at the decoder output because substantial burst errors could not be corrected. The Viterbi decoder selected the incorrect path and that precipitated more errors than if coding had not been used. The code had no error detection capability enabling the incorrect decoding to be identified. In addition, the non-systematic code was superior as the received bit sequence did not contain a copy of the information data. As a result, there was no remedy

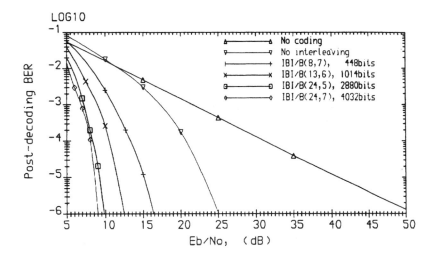

Figure 4.38: Effect of interblock interleaving on post-decoding BER for the CC(2,1,7) [VD SD] code over Rayleigh-fading channel.

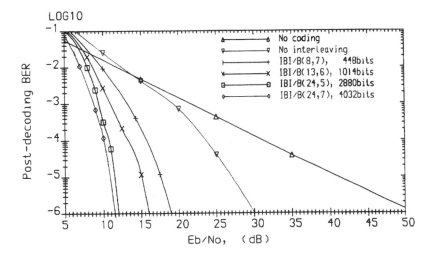

Figure 4.39: Effect of interblock interleaving on post-decoding BER for the PCC(3,2,7) [VD SD] code over Rayleigh-fading channel.

if incorrect decoding occurred.

When the constraint length of the codes was extended to $K = 7$, the half rate CC(2,1,7) code and the two-thirds rate PCC(3,2,7) code yielded the results displayed in Figure 4.38 and Figure 4.39, respectively. The gains in

E_b/N_0 for the CC(2, 1, 7) code compared to the CC(2, 1, 5) were 7, 3, 3, 2, and 1.5 dB for delays of 0, 448, 1014, 2880, and 4032 bits, respectively. The gradually decreasing gain in E_b/N_0 with increasing delay was because the behaviour of the channel became more random and approached a Gaussian channel. The gain of the CC(2, 1, 7) code over the CC(2, 1, 5) code for the case of the AWGN channel was as small as 1 dB as shown in Figure 4.32. Similarly, the gains of E_b/N_0 for the PCC(3, 1, 7) code in Figure 4.39 over the PCC(3, 1, 5) were 3.5, 2.5, 2.5, 1.7, and 1.3 dB for delays of 0, 448, 1014, 2880, and 4032 bits, respectively. The decreasing gain in E_b/N_0 with increasing delay was again observed. The gain in E_b/N_0 due to the increased constraint length for the two-thirds rate code was comparatively smaller than for the half rate code.

4.3.9   Conclusions on Convolutional Coding

For transmissions over the AWGN channel, the performance of convolutional codes is enhanced by 0.5 dB if the constraint length K is increased by 1. A degradation of at most 1 dB occurs when the coding rate is increased from half to two-thirdss. As a result, if the data throughput is increased by increasing the coding rate from half to two-thirdss, the constraint length of the code can also be increased in order to obtain the same performance. If soft-decision is applied, a further 1.5–2.0 dB gain in E_b/N_0 can be achieved. The coding gains of convolutional codes for transmissions over AWGN channel are summarised in Table 4.1.

In the Rayleigh fading channel, inter-block interleaving over 2880 bits can approximately render the fading channel into a memoryless one. For an acceptable delay of 448 bits introduced by the inter-block bit interleaving, digital speech transmission is possible with a BER of 10^{-3} or less, provided the E_b/N_0 value is above 11.5 dB. The soft-decision Viterbi decoding in the fading channel achieves a gain of 5 dB at BER of 10^{-6} compared to the hard-decision decoder. This gain is more than that for the AWGN channel (2 dB), and means that soft-decision decoding is more effective in the fading environment. A degradation of 2–3 dB occurs if the coding rate is increased from half to two-thirdss. If the constraint length of the code is increased from 5 to 7, the gain in E_b/N_0 for the half rate code ranges from 7 to 1.5 dB for a delay ranging from 0 to 4032 bits, respectively. For the long delay, the channel tends to be random and therefore the gain is only 1.5 dB. This is equivalent to the gain (1.5–2 dB) achieved for transmission over the AWGN channel. The improvement for the two-thirds rate code ranges from 3.5 to 1.3 dB over the delay range from 0 to 4032 bits, respectively. The coding gains of convolutional codes in the Rayleigh fading channel are tabulated in Table 4.2. For the high rate codes, the BER is higher at the low E_b/N_0 values than when channel coding is not used. This is due to the lack of error detection capability, and to the use of non-systematic codes.

	E_b/N_0 at BER of		Cod.-gain at BER of	
	10^{-3}	10^{-6}	10^{-3}	10^{-6}
No coding	6.8dB	10.5dB	0dB	0dB
$CC(2,1,5)$ $R = 1/2$ $[VD - HD]$	5.3dB	7.7dB	1.5dB	2.8dB
$PCC(3,1,5)$ $R = 2/3$ $[VD - HD]$	5.6dB	7.8dB	1.2dB	2.7dB
$CC(2,1,5)$ $R = 1/2$ $[VD - SD]$	3.2dB	5.6dB	3.6dB	4.9dB
$CC(2,1,7)$ $R = 1/2$ $[VD - SD]$	2.7dB	4.6dB	4.1dB	5.9dB
$PCC(3,1,5)$ $R = 2/3$ $[VD - SD]$	3.7dB	6.1dB	3.1dB	4.4dB
$PCC(3,1,7)$ $R = 2/3$ $[VD - SD]$	3.2dB	5.1dB	3.6dB	5.4dB

Table 4.1: Coding gain of convolutional codes over AWGN channels.

	E_b/N_0 at BER of		Cod.-gain at BER of	
	10^{-3}	10^{-6}	10^{-3}	10^{-6}
No coding	21.5dB	52.0dB	0dB	0dB
$CC(2,1,5)$ $R = 1/2$ $[VD - HD]$				
No interleaving	21.3dB	37.5dB	0.2dB	14.5dB
DI/B 456bits	19.5dB	33.5dB	2.0dB	18.5dB
BI/B(2,456) 912bits	19.0dB	33.5dB	2.5dB	18.5dB
BI/B(10,456) 4560bits	15.0dB	27.5dB	6.5dB	24.5dB
BI/B(20,456) 9120bits	13.5dB	24.5dB	8.0dB	27.5dB
IBI/B(24,5) 2880bits	11.7dB	20.0dB	9.8dB	32.0dB
IBI/B(24,7) 4032bits	10.9dB	18.6dB	10.6dB	33.4dB
$CC(2,1,5)$ $R = 1/2$ $[VD - SD]$				
No interleaving	19.0dB	32.2dB	2.5dB	19.8dB
IBI/B(8,7) 448bits	11.5dB	19.7dB	10.0dB	32.3dB
IBI/B(13,6) 1014bits	9.0dB	15.5dB	12.5dB	36.5dB
IBI/B(24,5) 2880bits	7.6dB	11.5dB	13.9dB	40.5dB
IBI/B(24,7) 4032bits	7.5dB	10.6dB	14.0dB	41.4dB
$PCC(3,1,5)$ $R = 2/3$ $[VD - SD]$				
No interleaving	20.0dB	33.5dB	1.5dB	18.5dB
IBI/B(8,7) 448bits	13.7dB	21.5dB	7.8dB	30.5dB
IBI/B(13,6) 1014bits	11.5dB	18.5dB	10.0dB	33.5dB
IBI/B(24,5) 2880bits	9.7dB	13.8dB	11.8dB	38.2dB
IBI/B(24,7) 4032bits	9.3dB	12.7dB	12.2dB	39.3dB
$CC(2,1,7)$ $R = 1/2$ $[VD - SD]$				
No interleaving	17.3dB	25.0dB	4.2dB	27.0dB
IBI/B(8,7) 448bits	11.0dB	16.5dB	10.5dB	35.5dB
IBI/B(13,6) 1014bits	8.8dB	12.5dB	12.7dB	39.5dB
IBI/B(24,5) 2880bits	7.2dB	9.8dB	14.3dB	42.2dB
IBI/B(24,7) 4032bits	6.8dB	9.0dB	14.7dB	43.0dB
$PCC(3,1,7)$ $R = 2/3$ $[VD - SD]$				
No interleaving	19.2dB	30.0dB	2.3dB	22.0dB
IBI/B(8,7) 448bits	13.2dB	19.0dB	8.3dB	33.0dB
IBI/B(13,6) 1014bits	11.0dB	16.0dB	10.5dB	36.0dB
IBI/B(24,5) 2880bits	9.5dB	12.0dB	12.0dB	40.0dB
IBI/B(24,7) 4032bits	8.7dB	11.5dB	12.8dB	40.5dB

Table 4.2: Coding gain of convolutional codes over Rayleigh fading channels.

4.4   Block Codes

The history of block codes began in 1950 when a class of single error correcting block codes was introduced by Hamming [1]. The correcting capability of Hamming codes was, however, very weak and of limited practical value. A major breakthrough came when Hocquenghem [35] in 1959 and Bose and Chaudhuri [36, 37] in 1960 discovered a large class of multiple error correcting codes which are named after them as the Bose-Chaudhuri-Hocquenghem (BCH) binary codes. Soon after this pioneering work, the cyclic structure of these codes was discovered by Peterson [38]. The limitation of the theory to binary codes was removed by Gorenstein and Zierler in 1961 [39] providing coverage of both binary and non-binary codes. An important subclass of BCH codes was discovered by Reed and Solomon [40], and these codes known as the Reed-Solomon (RS) codes achieve maximum separable distance between their codewords. The first decoding algorithm for binary BCH codes was suggested by Peterson in 1960 [38], followed by techniques [39, 41, 42] of how to practically implement the decoder. An efficient decoding algorithm proposed by Berlekamp [4, 43] and Massey [44, 45] became available for correcting a large number of errors. With the advance of digital integration circuit technology, the deployment of block codes became practical for a wide range of applications. Powerful Reed-Solomon (RS) block decoders [46, 47] have reportedly been built that operate at data rates above 120 Mbit/s. They have also become important as the outer layer code for use in concatenation with an inner layer convolutional code in both deep space communications [26] and in mobile satellite as well as radio applications [48].   Cyclic block codes are basically described by two parameters n and k, and a generator polynomial. A block of k information symbols at the input to the encoder is encoded into a block of n symbols. A characteristic of block codes is that each n symbol codeword is uniquely determined by a block of k input symbols. The ratio of k/n is the coding rate of the code and determines the amount of added redundancy.

4.4.1   The Structure of Block Codes

We begin to describe the algebraic structure that is fundamental in understanding the theory of block codes. Although both real and complex numbers are commonly employed in engineering applications, the algebraic theory for block coding requires the algebraic construction of fields. The operations for fields include addition, subtraction, multiplication and division, but their definitions are not the same as those of elementary arithmetic. The first step in understanding block codes is to grasp the concept of the fields with finite number of elements, and the arithmetic operations that can be performed. From this introduction, we will move to extension fields, polynomials and into coding algorithms.

4.4.1.1 Finite Fields

A finite field, also called a *Galois field*[3], is denoted by $GF(q)$. It describes a finite set of q elements with two defined operations, addition and multiplication. These operations performed on the inverse elements implicitly imply two further operations, that of subtraction and division. The rules of these operations are not significantly different from those employed in arithmetic operations with real and complex numbers. The rules of finite fields are illustrated as follows.

1. There are two operations, addition and multiplication, for operating on elements.

1) The field is closed. That is, the sum of the addition or the product of the multiplication results in a third element which is contained within the field.

2) The field always contains a unique additive identity element 0, and a unique multiplicative identity element 1, such that $u + 0 = u$ and $u \cdot 1 = u$ for any element u.

3) For every element u, there is a unique additive inverse element $-u$ such that $u + (-u) = 0$, and for $u \neq 0$, there is a unique multiplicative inverse element, denoted by u^{-1}, such that $u \cdot u^{-1} = 1$. The existence of the inverse elements implies the inverse operation, subtraction and division.

4) For operation on elements u, v and w, the following laws apply

$$
\begin{array}{llrcl}
associative: & & u + (v + w) & = & (u + v) + w \\
& & u \cdot (v \cdot w) & = & (u \cdot v) \cdot w \\
commutative: & & u + v & = & v + u \\
& & u \cdot v & = & v \cdot u \\
distributive: & & u \cdot (v + w) & = & u \cdot v + u \cdot w \ .
\end{array}
$$

In the ordinary arithmetic system, we observe examples of fields which conform to the above definitions of addition and multiplication. These fields are the set of all real numbers, the set of all complex numbers and the set of all rational numbers. Their elements obey the rules of ordinary addition and multiplication, and the fields are closed and contain an infinite number of elements. The additive identity element 0 and the multiplicative identity element 1 are among the elements in the fields, and every element has a unique additive and multiplicative inverse element. Furthermore, associative, commutative and distributive laws apply. In contrast, the set

---

[3]Galois fields are named in honour of the French mathematician Évariste Galois (1811-1832) who was killed in a duel at the age of 20. On the eve of his death, he wrote a letter to a friend in which he gave the results of his theory of algebraic equations, already presented to the Paris Academy.

+	0	1	2	3	4
0	0	1	2	3	4
1	1	2	3	4	0
2	2	3	4	0	1
3	3	4	0	1	2
4	4	0	1	2	3

•	0	1	2	3	4
0	0	0	0	0	0
1	0	1	2	3	4
2	0	2	4	1	3
3	0	3	1	4	2
4	0	4	3	2	1

Table 4.3: Arithmetic tables for $GF(5)$ operations.

of all integers is not a field because integers other than unity have no multiplicative inverses in the set. These fields all have an infinite number of elements. The number of elements in a field is called the *order* of the field and it may be finite or infinite. A field with a finite number of elements is called a *finite field* and is denoted by $GF(q)$, where q is the number of elements in the field. For example, $GF(5)$ is a finite field containing a set of five integer elements {0,1,2,3,4} operating under modulo-5 addition and multiplication. The addition and multiplication tables for the elements in $GF(5)$ are shown in Table 4.3. Notice that the $GF(5)$ contains the additive identity element 0 and the multiplicative identity element 1 in the field. Also, for each element in $GF(5)$, a unique additive inverse, and a unique multiplicative inverse, except for 0, always exist in the field. Thus, the addition to an inverse element implies the subtraction such that $2 - 3 = 2 + (-3) = 2 + 2 = 4$, and similarly the multiplication by an inverse element implies the division such that $2/3 = 2 \cdot 3^{-1} = 2 \cdot 2 = 4$.

We have stated that a finite field $GF(q)$ consists of q integer elements. Suppose q is not a prime number, but a multiple of u and v. If $GF(q)$ is a finite field, u and v are the elements of the field, where u^{-1} and v^{-1} are their inverse elements, respectively. Hence,

$$u = R_q[u] = R_q[v^{-1}vu] = R_q[v^{-1}q] = 0 \qquad (4.53)$$

where $R_q[\bullet]$ is the modulo-q operation of $[\bullet]$. The proof shows a contradiction because $u \neq 0$. This means that $GF(q)$ consisting of integer elements is not a field if q is not a prime number. So, is it possible to have a $GF(q)$ with q not a prime number, such as $GF(8)$ or $GF(25)$? The answer is yes. In general, the finite field exists for $GF(q^m)$, where q is a prime number greater than 1 and m is an integer. The simplest field of $GF(q^m)$ with $m = 1$ is called the *prime field $GF(q)$*. The prime field consists of the set of all integer elements having values from zero to less than q. The operations on the integer elements are modulo-q addition and multiplication. For example, $GF(3)$ has a set of integer elements {0, 1, 2} and similarly the set of integer elements of $GF(5)$ is {0, 1, 2, 3, 4}. The smallest prime field is $GF(2)$ which only consists of the additive and multiplicative identity elements 0 and 1, respectively. If m is greater than 1, $GF(q^m)$ is constructed as an extension of the prime field and is refered to as an *extension* *field.* Thus, $GF(8) = GF(2^3)$ is the extension field of $GF(2)$ and similarly $GF(25) = GF(5^2)$ is the extension field of $GF(5)$. The construction of the extension field is explained in Section 4.4.1.3 after the important algebraic concepts of vector spaces are described in Section 4.4.1.2.

4.4.1.2   Vector Spaces

The concept of vector space is closely related to the ideas of linear algebra and matrix theory in mathematics. The representation of an n-dimensional vector **v** is an enumeration of its coordinates $(v_1, v_2, \ldots, v_n)$. In two- or three-dimensional Euclidean space, the coordinates are simply the projections of **v** onto coordinate axes and the vector can be visualised geometrically as a directed line in a two- or three-dimensional plane. The properties of geometric vectors in ordinary coordinate systems provide an intuitive concept. The addition of two vectors is the addition of corresponding coordinates of the two vectors, and the multiplication of a vector by a real number is done by multiplying each coordinate by the number. These definitions can be extended mathematically to an n-dimensional vector space.

Having defined the concept of geometric vectors, we introduce the analogue of vector space over a *field F*. A set **V** is called a *vector space* and its elements are called *vectors*. The field elements of F are called *scalars*. A vector space **V** over a field F is structured by a set of vectors and a set of scalars under the operations of addition and multiplication in a mathematical system very much like a system of geometric vectors, real numbers and ordinary algebra. The defined operations are the addition of vectors called *vector addition*, and the multiplication of a vector by a scalar called *scalar multiplication*. The addition of any two vectors **u** and **v** in **V** i.e., **u** + **v**, results in a vector that is also in **V**. The multiplication of a scalar a in F by a vector **v** in **V**, i.e., a**v**, also gives a vector in **V**. The results of both vector addition and scalar multiplication are always vectors in **V** as the operations are subjected to the constraints imposed by the closure properties of fields.

For any vectors **u** and **v** in **V** and any scalars a and b in F, the following conditions must be satisfied:

1. A vector space **V** over a field F is a commutative group under vector addition.

1) The distributive laws apply, such that $a(\mathbf{u} + \mathbf{v}) = a\mathbf{u} + a\mathbf{v}$ and $(a + b)\mathbf{u} = a\mathbf{u} + b\mathbf{u}$.

2) The associative law applies, such that $(ab)\mathbf{u} = a(b\mathbf{u})$.

3) Let 1 be the multiplicative identity element in F. Then for any **u** in **V**, $1\mathbf{u} = \mathbf{u}$.

For vector addition in **V**, an additive identity element, called the *origin* of **V**, exists and is denoted by **0** such that $\mathbf{u} + \mathbf{0} = \mathbf{0} + \mathbf{u} = \mathbf{u}$ for all

u in **V**. Also, for scalar addition in F, an additive identity element exists in F called the zero scalar element and is denoted by 0. The two identity elements **0** and 0 are closely related. For all a in F and all **u** in **V**, we have $a\mathbf{0} = \mathbf{0}$ and $0\mathbf{u} = \mathbf{0}$.

So far we have decribed the concept and the definition of vector space over a field. Let us now confine our interest to the application to the error control codes. A set of n elements $(u_1, u_2, \ldots, u_n)$, where u_i is a field element in F, is called an *n-tuple* over F. The addition of any two n-tuples is defined by an element-by-element addition, such as

$$(u_1, u_2, \ldots, u_n) + (v_1, v_2, \ldots, v_n) = (u_1 + v_1, u_2 + v_2, \ldots, u_n + v_n) \quad (4.54)$$

where the addition of $u_i + v_i$ is done in F and their sum is another element in F. The multiplication of an element from F by an n-tuple over F is the element-by-element scalar multiplication,

$$a(u_1, u_2, \ldots, u_n) = (au_1, au_2, \ldots, au_n) \quad (4.55)$$

where each multiplication au_i is performed in F. Under the operations of elementwise addition and elementwise scalar multiplication, the distributive and associative laws apply, and the set of n-tuples constitutes a vector space over the field F. A vector space can be constructed in this way with any field F, but the main feature in error control coding is that in a vector space over finite fields, the scalars represent code symbols and the n-tuples represent codewords.

If $\mathbf{v}_1, \mathbf{v}_2, \ldots, \mathbf{v}_k$ are vectors in a vector space **V** over a field F, any sum of the form

$$\mathbf{u} = a_1\mathbf{v}_1 + a_2\mathbf{v}_2 + \cdots + a_k\mathbf{v}_k \quad (4.56)$$

where the scalars a_i are in the field F, is called a *linear combination* of $\mathbf{v}_1, \mathbf{v}_2, \ldots, \mathbf{v}_k$. A set of k vectors $\{\mathbf{v}_1, \mathbf{v}_2, \ldots, \mathbf{v}_k\}$ is said to be *linearly independent* if there is not a single set of scalars $\{a_1, a_2, \ldots, a_k\}$, except all a_i zero, such that

$$a_1\mathbf{v}_1 + a_2\mathbf{v}_2 + \cdots + a_k\mathbf{v}_k = \mathbf{0} \quad (4.57)$$

where **0** is the zero vector. If only one set of scalars, not all equal to zero, is found to cause the linear combination equal to the zero vector, the set of vectors $\{\mathbf{v}_1, \mathbf{v}_2, \ldots, \mathbf{v}_k\}$ is said to be *linearly dependent*. For example, the vectors (0,0,1), (0,1,0) and (1,0,0) are linearly independent over any field, whereas the vectors (1,1,1), (0,1,1) and (1,0,0) are linearly dependent over $GF(2)$ as their sum is equal to the zero vector.

In any vector space **V**, there is at least one set of linearly independent vectors that may generate any vector in **V** by means of linear combinations. The set of vectors $\{\mathbf{v}_i\}$ is said to *span* the vector space **V** and any such set of vectors is called the *basis* of the vector space. For example, the three binary vectors (0,0,1), (0,1,0) and (1,0,0) are linearly independent and constitute the basis of the vector space $\mathbf{V}_3$ which consists of all eight binary vectors (0,0,0), (0,0,1), ..., (1,1,1) formed by the linear combinations of the basis over $GF(2)$. If we use the two vectors (0,0,1) and (0,1,0) to form linear combinations over $GF(2)$, the resulting vectors (0,0,0), (0,0,1), (0,1,0) and (0,1,1) constitute a vector space $\mathbf{V}_2$ which is a *subspace* of $\mathbf{V}_3$. The vectors (0,0,1) and (0,1,0) are said to span $\mathbf{V}_2$. The concept of the vector space is closely related to the familiar x, y and z axes of 3-dimensional coordinate systems in Euclidean space. The basis vectors in the vector space are represented by the unit vectors. However, the unit vectors are not the only basis vectors for $\mathbf{V}_3$. The vectors (1,0,0), (0,1,0) and (0,1,1) are the alternative basis for $\mathbf{V}_3$. Similarly, the vectors (0,0,1) and (0,1,1) also form an alternative basis for subspace $\mathbf{V}_2$. Furthermore, the vector (1,1,1) is a basis for another subspace $\mathbf{V}_1$, composed of only the vectors (0,0,0) and (1,1,1). The *dimension* of the vector space is the number of spanning vectors that can be used to generate the space by linear combinations. In the example, the vector spaces $\mathbf{V}_3$, $\mathbf{V}_2$, and $\mathbf{V}_1$ have dimension three, two and one, respectively.

In summary, we have highlighted the important concept of vector spaces, including the linear dependency among vectors, linear combinations of vectors, basis vectors of forming vector spaces, and dimensionality of a vector space. These concepts lay the foundations for algebraic operations in a finite field for error control coding.

4.4.1.3 Extension Fields

We now extend the concept of vectors, or m-tuples, in vector spaces to polynomial representations in algebraic systems. In the previous section, the m-dimensional vector space over $GF(q)$ had q^m vectors, each constituted by m-tuples of elements in the field $GF(q)$. That is, for a two-dimensional vector space over $GF(2)$, the vectors include a set of four 2-tuples (0,0), (0,1), (1,0) and (1,1). The addition and subtraction operations on the vectors are performed element-by-element over $GF(q)$ and the result of the operation is another vector in the vector space. For example, $(0,1) + (1,0) = (1,1)$, where (1,1) is also a vector in the vector space. The multiplication and division operations on the vectors are not obvious. Let us associate each vector with a polynomial having coefficients corresponding to the elements in the vector. The set of all 2-tuples defined on $GF(2)$ is replaced by the set of all degree-1 polynomials defined on $GF(2)$. That is, the set of four 2-tuples over $GF(2)$ can be represented by 0, 1, z, $z + 1$, corresponding to (0,0), (0,1), (1,0) and (1,1), respectively.

The addition and subtraction on polynomials are performed on their coefficients over $GF(q)$. In our example, the addition of $(0z + 1)$ and $(z + 0)$ results in another polynomial, $(z + 1)$, in the set. This shows the closure property under addition. Similarly, the closure property under multiplica-

+	0	1	z	$z+1$
0	0	1	z	$z+1$
1	1	0	$z+1$	z
z	z	$z+1$	0	1
$z+1$	$z+1$	z	1	0

Table 4.4: Addition table for $GF(4)$.

tion applies if the product of any two polynomials is another polynomial in the set. We notice that all the polynomials in the set are of degree $(m-1)$ or less. The multiplication performed in a finite field can therefore be defined by taking the remainder of the product with respect to a fixed polynomial of degree m. By this definition, we always achieve a remainder of degree $(m-1)$ or less that must therefore be another polynomial in the set.

The fixed polynomial is denoted by $p(z)$ and must be a *prime polynomial* such that $p(z)$ is irreducible, that is, a degree-m polynomial that has no factors of degree less than m and greater than 0. This requirement is demonstrated by the following proof. Suppose that $p(z)$, whose degree is at least 2, is not prime. Then $p(z) = u(z)v(z)$ for some $u(z)$ and $v(z)$ in the set, each of degree at least 1, and their inverse polynomials are $u^{-1}(z)$ and $v^{-1}(z)$, respectively. Hence,

$$u(z) = R_{p(z)}[u(z)] = R_{p(z)}[v^{-1}(z)v(z)u(z)] = R_{p(z)}[v^{-1}(z)p(z)] = 0$$
$$(4.58)$$

where $R_{p(z)}[\bullet]$ represents the remainder of $[\bullet]$ upon division by $p(z)$. As $p(z)$ is a multiple of $u(z)$ and $v(z)$, $u(z)$ cannot be equal to 0. This contradicts Equation 4.58, whence our initial assumption that $p(z)$ is not a prime was wrong. The proof then demonstrates that $p(z)$ must be a prime number in order to perform multiplication in a finite field.

Having defined the addition and multiplication on polynomials, we now illustrate the relationship between polynomials and fields. A set of polynomials of degree $(m-1)$ with coefficients defined over $GF(q)$ constitutes a finite field $GF(q^m)$ with a total of q^m polynomials. As an example, $GF(4)$ is constituted by four polynomials of degree-1 defined over $GF(2)$, i.e., {0, 1, z, $z+1$}. The addition of polynomials is performed on their coefficients over $GF(2)$. By using the prime polynomial, $p(z) = z^2 + z + 1$, the multiplication of polynomials is the remainder of their product divided by $p(z)$. The addition and multiplication tables for $GF(4)$ are tabulated in Table 4.4 and Table 4.5. The $GF(4)$ consists of four elements including the additive identity element 0 and the multiplicative identity element 1. We observe from the addition table that each polynomial is its own additive inverse. Also, from the multiplication table, each non-zero polynomial has a unique multiplicative inverse, z being the inverse of $z + 1$ and vice versa, while 1 is its own inverse, as always. Thus, $GF(4)$ is a finite field constructed from

•	0	1	z	$z+1$
0	0	0	0	0
1	0	1	z	$z+1$
z	0	z	$z+1$	1
$z+1$	0	$z+1$	1	z

Table 4.5: Multiplication table for $GF(4)$.

Exponential notation	Polynomial notation	Binary notation
0	0	00
α^0	1	01
α^1	z	10
α^2	$z+1$	11

Table 4.6: Representations of $GF(4)$.

$GF(2)$.

In general, a finite field $GF(q^m)$ exists for any number q^m, where q is a prime and m is a positive integer. The relationship between $GF(q)$ and $GF(q^m)$ is that $GF(q)$ is a *subfield* of $GF(q^m)$ such that the elements of $GF(q)$ are a subset of the elements in $GF(q^m)$. Equivalently, $GF(q^m)$ is called the *extension field* of $GF(q)$. For example, $GF(2)$ is a subfield of $GF(4)$ such that the elements, $\{0,1\}$, of $GF(2)$ are a subset of the elements, $\{0,1,z,z+1\}$, of $GF(4)$. Also $GF(4)$ is an extension field of $GF(2)$.

4.4.1.4 Primitive Polynomials

Every Galois field has at least one *primitive element*, denoted by α, which can represent every field element, except zero, as a power of α. For example, in the $GF(5)$, we have $2^1 = 2$, $2^2 = 4$, $2^3 = 3$ and $2^4 = 1$, where the results of 2^3 and 2^4 are their modulo-5 values. Thus $\alpha = 2$ is a primitive element of $GF(5)$. Similarly, $\alpha = 3$ is also a primitive element of $GF(5)$ such that $3^1 = 3$, $3^2 = 4$, $3^3 = 2$ and $3^4 = 1$. Consider the example of $GF(4)$. We try $\alpha = z$, the consecutive powers of α give $\alpha^1 = z$, $\alpha^2 = z+1$ and $\alpha^3 = 1$, where the results of α^2 and α^3 are their modulo $p(z) = z^2 + z + 1$ values. By employing $\alpha = z+1$, we again generate all the field elements by raising the power of α, i.e., $\alpha^1 = z+1$, $\alpha^2 = z$, and $\alpha^3 = 1$. In both examples of $GF(4)$ and $GF(5)$, we have found two primitive elements of which either one can generate a list of the non-zero field elements as powers of α. Once all the field elements have been found, we can adopt different notations to represent the elements. As an example of $GF(4)$, we can associate a binary 2-tuple with the polynomial, as shown in Table 4.6. For the binary notation, the addition of two elements is implemented by a bitwise exclusive-OR operation. We show in Table 4.6 that the addition of 1 and z is equal to $(z+1)$. This occurs because we represent two elements 1 and z by 01 and 10, and the bitwise exclusive-OR of 01 and 10 is 11, corresponding to $(z + 1)$.

Degree	Primitive polynomials	Degree	Primitive polynomials
2	$z^2 + z + 1$	14	$z^{14} + z^{10} + z^6 + z + 1$
3	$z^3 + z + 1$	15	$z^{15} + z + 1$
4	$z^4 + z + 1$	16	$z^{16} + z^{12} + z^3 + z + 1$
5	$z^5 + z^2 + 1$	17	$z^{17} + z^3 + 1$
6	$z^6 + z + 1$	18	$z^{18} + z^7 + 1$
7	$z^7 + z^3 + 1$	19	$z^{19} + z^5 + z^2 + z + 1$
8	$z^8 + z^4 + z^3 + z^2 + 1$	20	$z^{20} + z^3 + 1$
9	$z^9 + z^4 + 1$	21	$z^{21} + z^2 + 1$
10	$z^{10} + z^3 + 1$	22	$z^{22} + z + 1$
11	$z^{11} + z^2 + 1$	23	$z^{23} + z^5 + 1$
12	$z^{12} + z^6 + z^4 + z + 1$	24	$z^{24} + z^7 + z^2 + z + 1$
13	$z^{13} + z^4 + z^3 + z + 1$	25	$z^{25} + z^3 + 1$

Table 4.7: Primitive polynomials over $GF(2)$ (Blahut [6].)

For the exponential notation, the field elements are represented by the successive powers of the primitive element. The advantage of this notation is that the multiplication of two elements is equivalent to the addition of their exponents. For example, the multiplication of z and $(z + 1)$ gives $(z^2 + z)$ which is then taken modulo-$p(z)$ to yield the product, namely 1. Equivalently, the multiplication in exponential notation of these two elements α^1 and α^2 also results in α^3, namely 1.

We know that the field elements of prime field $GF(q)$ are a set of integer elements $\{0, 1, 2, \ldots, q - 1\}$. However, for the extension field $GF(q^m)$, we would like to represent the polynomial elements as the successive powers of the primitive element, where multiplication of two polynomials can easily be done by the addition of the exponents of their corresponding exponential notation. This is usually convenient if the polynomial z corresponds to a primitive element of the field enabling the field elements to be found by computing the successive powers of z. A special prime polynomial, called a primitive polynomial, is selected to construct the field. A *primitive polynomial* $p(z)$ over $GF(q)$ is a prime polynomial over $GF(q)$ with z being a primitive element for constructing the field elements in the extension field. Table 4.7 lists the primitive polynomials of degree 2 to degree 25 over $GF(2)$ that enable us to construct a field from $GF(2^2)$ to $GF(2^{25})$.

Let us now summarise the construction of an extension field $GF(q^m)$, where q is a prime and m is an integer. We first generate all the field elements by using the primitive element and the primitive polynomial and then construct the addition and multiplication tables. As the $GF(q^m)$ is an extension field of $GF(q)$, the elements in the $GF(q^m)$ are represented by q^m polynomials of degree $(m-1)$ or less with coefficients in $GF(q)$. To generate the field elements, a degree-m primitive polynomial over $GF(q)$ is selected and the primitive element is $\alpha = z$. The power of α is raised successively, i.e., $\{\alpha^0, \alpha^1, \ldots, \alpha^{q^m-2}\}$ until all the elements except zero are generated. If the prime field is $GF(2)$, the coefficients of the polynomials are binary and therefore the degree-$(m - 1)$ polynomial elements can be represented by

Exponential notation	Polynomials notation	Binary notation	Hexadecimal notation
0	0	0000	0
$\alpha^0 \equiv \alpha^{15}$	1	0001	1
α^1	z	0010	2
α^2	z^2	0100	4
α^3	z^3	1000	8
α^4	$z + 1$	0011	3
α^5	$z^2 + z$	0110	6
α^6	$z^3 + z^2$	1100	C
α^7	$z^3 + z + 1$	1011	B
α^8	$z^2 + 1$	0101	5
α^9	$z^3 + z$	1010	A
α^{10}	$z^2 + z + 1$	0111	7
α^{11}	$z^3 + z^2 + z$	1110	E
α^{12}	$z^3 + z^2 + z + 1$	1111	F
α^{13}	$z^3 + z^2 + 1$	1101	D
α^{14}	$z^3 + 1$	1001	9

Table 4.8: Field elements of $GF(16)$ generated by $p(z) = z^4 + z + 1$.

$(m - 1)$ binary digits notation. The addition of two polynomial elements is done by adding coefficients over $GF(q)$ of corresponding powers of z. The multiplication of two elements is the addition of the powers of their corresponding exponential notation.

We now illustrate the example of constructing $GF(16)$, which can be written as $GF(2^4)$ where $q = 2$ and $m = 4$. The primitive polynomial is therefore defined on $GF(2)$ of degree-4. From Table 4.7, $p(z) = z^4 + z + 1$ is used to generate the field elements. The power of the primitive element $\alpha = z$ is raised to represent all the non-zero field elements $\{\alpha^0, \alpha^1, \ldots, \alpha^{14}\}$ in $GF(16)$ and their modulo-$p(z)$ value is computed to give the polynomial representations as shown in Table 4.8. For instance,

$$\alpha^6 = z^6 \pmod{p(z)} = z^3 + z^2 .$$

The addition of two elements requires us to add coefficients over $GF(2)$ of corresponding powers of z in each of the polynomials, and then to express the sum in its exponential representation. For example,

$$\begin{aligned} \alpha^6 + \alpha^7 &= (z^3 + z^2) + (z^3 + z + 1) \\ &= z^2 + z + 1 \\ &= \alpha^{10} . \end{aligned}$$

An alternative approach of directly adding two elements together in an exponential representation is done using the Zech logarithm, $Z(j)$, which is defined by

$$\alpha^{Z(j)} = 1 + \alpha^j . \tag{4.59}$$

j	$Z(j)$
$-\infty$	0
0	$-\infty$
1	4
2	8
3	14
4	1
5	10
6	13
7	9
8	2
9	7
10	5
11	12
12	11
13	6
14	3

Table 4.9: Zech's logarithms in $GF(16)$.

Two elements α^i and α^j can be added by

$$\begin{aligned}
\alpha^i + \alpha^j &= \alpha^i(1 + \alpha^{j-i}) \\
&= \alpha^{i+Z(j-i)} \; .
\end{aligned}$$

Using this technique, we tabulate the Zech logarithms for $GF(16)$ in Table 4.9. Let us illustrate the addition by using the following example:

$$\alpha^3 + \alpha^7 = \alpha^{3+Z(4)} = \alpha^4 \; .$$

By using Table 4.9, the exponential representations of the sums are tabulated in Table 4.10. The product of two elements is a new element with the power equal to the sum of the powers of the corresponding elements. For example,

$$\alpha^6 + \alpha^{12} = \alpha^{6+12} = \alpha^{15} \cdot \alpha^3 = \alpha^3 \; .$$

The multiplication table is shown in Table 4.11.

4.4.1.5   Minimal Polynomials

In the last section the concept of an extension field was introduced. We now investigate the relationship between the extension field and the prime field, and this will lead to the introduction of the minimal polynomials. It is these polynomials that play a cardinal role in the formation of the generator polynomials for BCH codes to be described in Section 4.4.3. A special case of minimal polynomials is the primitive polynomial which we used as the prime polynomial in constructing the extension field in Section 4.4.1.4.

In ordinary algebraic arithmetic, a polynomial of degree n with real

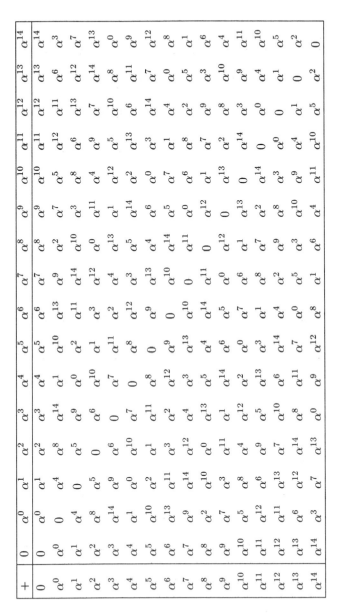

+	0	α^0	α^1	α^2	α^3	α^4	α^5	α^6	α^7	α^8	α^9	α^{10}	α^{11}	α^{12}	α^{13}	α^{14}
0	0	α^0	α^1	α^2	α^3	α^4	α^5	α^6	α^7	α^8	α^9	α^{10}	α^{11}	α^{12}	α^{13}	α^{14}
α^0	α^0	0	α^4	α^8	α^{14}	α^1	α^{10}	α^{13}	α^9	α^2	α^7	α^5	α^{12}	α^{11}	α^6	α^3
α^1	α^1	α^4	0	α^5	α^9	α^0	α^2	α^{11}	α^{14}	α^{10}	α^3	α^8	α^6	α^{13}	α^{12}	α^7
α^2	α^2	α^8	α^5	0	α^6	α^{10}	α^1	α^3	α^{12}	α^0	α^{11}	α^4	α^9	α^7	α^{14}	α^{13}
α^3	α^3	α^{14}	α^9	α^6	0	α^7	α^{11}	α^2	α^4	α^{13}	α^1	α^{12}	α^5	α^{10}	α^8	α^0
α^4	α^4	α^1	α^0	α^{10}	α^7	0	α^8	α^{12}	α^3	α^5	α^{14}	α^2	α^{13}	α^6	α^{11}	α^9
α^5	α^5	α^{10}	α^2	α^1	α^{11}	α^8	0	α^9	α^{13}	α^4	α^6	α^0	α^3	α^{14}	α^7	α^{12}
α^6	α^6	α^{13}	α^{11}	α^3	α^2	α^{12}	α^9	0	α^{10}	α^{14}	α^5	α^7	α^1	α^4	α^0	α^8
α^7	α^7	α^9	α^{14}	α^{12}	α^4	α^3	α^{13}	α^{10}	0	α^{11}	α^0	α^6	α^8	α^2	α^5	α^1
α^8	α^8	α^2	α^{10}	α^0	α^{13}	α^5	α^4	α^{14}	α^{11}	0	α^{12}	α^1	α^7	α^9	α^3	α^6
α^9	α^9	α^7	α^3	α^{11}	α^1	α^{14}	α^6	α^5	α^0	α^{12}	0	α^{13}	α^2	α^8	α^{10}	α^4
α^{10}	α^{10}	α^5	α^8	α^4	α^{12}	α^2	α^0	α^7	α^6	α^1	α^{13}	0	α^{14}	α^3	α^9	α^{11}
α^{11}	α^{11}	α^{12}	α^6	α^9	α^5	α^{13}	α^3	α^1	α^8	α^7	α^2	α^{14}	0	α^0	α^4	α^{10}
α^{12}	α^{12}	α^{11}	α^{13}	α^7	α^{10}	α^6	α^{14}	α^4	α^2	α^9	α^8	α^3	α^0	0	α^1	α^5
α^{13}	α^{13}	α^6	α^{12}	α^{14}	α^8	α^{11}	α^7	α^0	α^5	α^3	α^{10}	α^9	α^4	α^1	0	α^2
α^{14}	α^{14}	α^3	α^7	α^{13}	α^0	α^9	α^{12}	α^8	α^1	α^6	α^4	α^{11}	α^{10}	α^5	α^2	0

Table 4.10: Addition table for $GF(16)$.

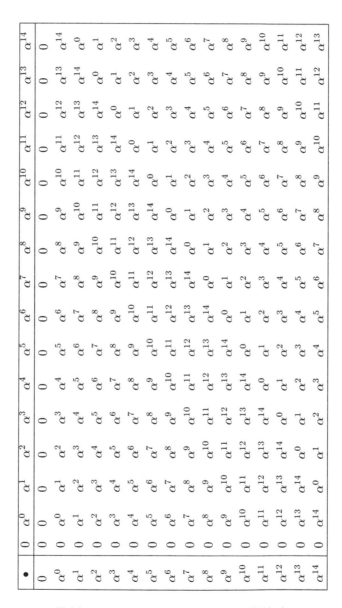

$\bullet$	0	α^0	α^1	α^2	α^3	α^4	α^5	α^6	α^7	α^8	α^9	α^{10}	α^{11}	α^{12}	α^{13}	α^{14}
0	0	0	0	0	0	0	0	0	0	0	0	0	0	0	0	0
α^0	0	α^0	α^1	α^2	α^3	α^4	α^5	α^6	α^7	α^8	α^9	α^{10}	α^{11}	α^{12}	α^{13}	α^{14}
α^1	0	α^1	α^2	α^3	α^4	α^5	α^6	α^7	α^8	α^9	α^{10}	α^{11}	α^{12}	α^{13}	α^{14}	α^0
α^2	0	α^2	α^3	α^4	α^5	α^6	α^7	α^8	α^9	α^{10}	α^{11}	α^{12}	α^{13}	α^{14}	α^0	α^1
α^3	0	α^3	α^4	α^5	α^6	α^7	α^8	α^9	α^{10}	α^{11}	α^{12}	α^{13}	α^{14}	α^0	α^1	α^2
α^4	0	α^4	α^5	α^6	α^7	α^8	α^9	α^{10}	α^{11}	α^{12}	α^{13}	α^{14}	α^0	α^1	α^2	α^3
α^5	0	α^5	α^6	α^7	α^8	α^9	α^{10}	α^{11}	α^{12}	α^{13}	α^{14}	α^0	α^1	α^2	α^3	α^4
α^6	0	α^6	α^7	α^8	α^9	α^{10}	α^{11}	α^{12}	α^{13}	α^{14}	α^0	α^1	α^2	α^3	α^4	α^5
α^7	0	α^7	α^8	α^9	α^{10}	α^{11}	α^{12}	α^{13}	α^{14}	α^0	α^1	α^2	α^3	α^4	α^5	α^6
α^8	0	α^8	α^9	α^{10}	α^{11}	α^{12}	α^{13}	α^{14}	α^0	α^1	α^2	α^3	α^4	α^5	α^6	α^7
α^9	0	α^9	α^{10}	α^{11}	α^{12}	α^{13}	α^{14}	α^0	α^1	α^2	α^3	α^4	α^5	α^6	α^7	α^8
α^{10}	0	α^{10}	α^{11}	α^{12}	α^{13}	α^{14}	α^0	α^1	α^2	α^3	α^4	α^5	α^6	α^7	α^8	α^9
α^{11}	0	α^{11}	α^{12}	α^{13}	α^{14}	α^0	α^1	α^2	α^3	α^4	α^5	α^6	α^7	α^8	α^9	α^{10}
α^{12}	0	α^{12}	α^{13}	α^{14}	α^0	α^1	α^2	α^3	α^4	α^5	α^6	α^7	α^8	α^9	α^{10}	α^{11}
α^{13}	0	α^{13}	α^{14}	α^0	α^1	α^2	α^3	α^4	α^5	α^6	α^7	α^8	α^9	α^{10}	α^{11}	α^{12}
α^{14}	0	α^{14}	α^0	α^1	α^2	α^3	α^4	α^5	α^6	α^7	α^8	α^9	α^{10}	α^{11}	α^{12}	α^{13}

Table 4.11: Multiplication table for $GF(16)$.

coefficients has exactly n roots, some of which may be repeated. If the roots are not from the field of real numbers, they are from the field of complex numbers that contains the field of real numbers as a subfield. For example, the polynomial $f(z) = z^2 + 4z + 13$ defined in the field of real numbers is irreducible. It does not have real roots, but instead it has two complex conjugate roots, $-2 \pm 3\imath$, where $\imath = \sqrt{-1}$. Similarly, in finite field arithmetic, if the polynomial defined in the subfield is irreducible, it has no roots in the subfield, only in the extension field. Every polynomial $f(z)$ of degree n has n roots, and if $f(z)$ is irreducible over the subfield then all n roots are in the extension field. For example, $f(z) = z^4 + z^3 + z^2 + z + 1$ is irreducible over $GF(2)$ and it has no roots from $GF(2)$. Instead, it has four roots, $\alpha^3, \alpha^6, \alpha^9$ and α^{12}, from the $GF(2^4)$, which is the extension field of $GF(2)$. By using the addition and the multiplication tables in Table 4.10 and Table 4.11, we can verify these roots by substituting into the polynomial. For α^3 we have,

$$
\begin{aligned}
f(\alpha^3) &= (\alpha^3)^4 + (\alpha^3)^3 + (\alpha^3)^2 + \alpha^3 + 1 \\
&= \alpha^{12} + \alpha^9 + \alpha^6 + \alpha^3 + 1 \\
&= 0
\end{aligned}
$$

and hence, α^3 is a root of $f(z)$. The other roots α^6, α^9 and α^{12} can also be verified by the same procedure. As $f(z)$ has a degree of four, with roots, $\alpha^3, \alpha^6, \alpha^9$ and α^{12}, then $(z+\alpha^3)(z+\alpha^6)(z+\alpha^9)(z+\alpha^{12})$ must be equal to $z^4 + z^3 + z^2 + z + 1$. Again, by using the addition and the multiplication tables in Table 4.10 and Table 4.11, we evaluate

$$
\begin{aligned}
&(z + \alpha^3)(z + \alpha^6)(z + \alpha^9)(z + \alpha^{12}) \\
&= (z^2 + \alpha^2 z + \alpha^9)(z^2 + \alpha^8 z + \alpha^6) \\
&= z^4 + (\alpha^2 + \alpha^8)z^3 + (\alpha^9 + \alpha^6 + \alpha^{10})z^2 + (\alpha^{17} + \alpha^8)z + \alpha^{15} \\
&= z^4 + z^3 + z^2 + z + 1 \ .
\end{aligned}
$$

The properties of these roots in extension fields are important in finite fields. Let $f(z)$ be an irreducible polynomial with coefficients from $GF(2)$, and β be a root of $f(z)$ such that $f(\beta) = 0$. As $f(z)$ is irreducible over $GF(2)$, it has no roots in $GF(2)$, and therefore β must be an element in some extension field $GF(2^m)$. The additions and multiplications required for the evaluation of the polynomial are performed in the extension field $GF(2^m)$, as $GF(2)$ is contained in any of its extension. Now let us describe the characteristics of these roots by the following key properties.

1. If $f(z)$ is an irreducible polynomial of degree n over $GF(2)$ and has a root β from $GF(2^m)$, then for any $l \geq 0$, β^{2^l} is also a root of $f(z)$, i.e., $\beta, \beta^2, \beta^4, \beta^8, \ldots, \beta^{2^{n-1}}$ are all roots of $f(z)$. This property can

be verified by the following proof. Let us consider

$$
\begin{aligned}
f^2(z) &= (f_0 + f_1 z + \cdots + f_n z^n)^2 \\
&= [f_0 + (f_1 z + f_2 z^2 + \cdots + f_n z^n)]^2 \\
&= f_0^2 + f_0 \cdot (f_1 z + f_2 z^2 + \cdots + f_n z^n) \\
&\quad + f_0 \cdot (f_1 z + f_2 z^2 + \cdots + f_n z^n) \\
&\quad + (f_1 z + f_2 z^2 + \cdots + f_n z^n)^2 \\
&= f_0^2 + (f_1 z + f_2 z^2 + \cdots + f_n z^n)^2 \ .
\end{aligned}
$$

Repeating the expansion of the above equation, we obtain,

$$
f^2(z) = f_0^2 + (f_1 z)^2 + (f_2 z^2)^2 + \cdots + (f_n z^n)^2 \ .
$$

As $f(z)$ is defined on $GF(2)$, the coefficient f_i is either 0 or 1. Therefore $f_i^2 = f_i$ and the equation becomes

$$
\begin{aligned}
f^2(z) &= f_0 + f_1 z^2 + f_2 (z^2)^2 + \cdots + f_n (z^2)^n \\
&= f(z^2) \ .
\end{aligned}
\tag{4.60}
$$

From Equation 4.60, we deduce that for any $l \geq 0$,

$$
f^{2^l}(z) = f(z^{2^l})
\tag{4.61}
$$

and for $z = \beta$,

$$
f^{2^l}(\beta) = f(\beta^{2^l}) \ .
\tag{4.62}
$$

As β is a root of $f(z)$, it implies that $f(\beta) = 0$. The powers of $f(\beta)$, i.e., $f^{2^l}(\beta)$, are also equal to zero. From Equation 4.62 we see that $f^{2^l}(\beta) = 0$, $f(\beta^{2^l}) = 0$, and therefore β^{2^l} is also a root of $f(z)$. This shows that if β (an element from $GF(2^m)$) is a root of the polynomial $f(z)$ over $GF(2)$, then all the β^{2^l} (elements from $GF(2^m)$) for $l \geq 0$ are also roots of $f(z)$. The element β^{2^l} is called a *conjugate* of β. For example, the polynomial $f(z) = z^4 + z^3 + 1$ is irreducible over $GF(2)$, and has four roots. One of the roots is α^7 which is an element in $GF(2^4)$. This can be verified by substituting α^7 into $f(z)$,

$$
\begin{aligned}
f(\alpha^7) &= (\alpha^7)^4 + (\alpha^7)^3 + 1 \\
&= \alpha^{13} + \alpha^6 + 1 \\
&= 0 \ .
\end{aligned}
$$

The conjugates of α^7 are $(\alpha^7)^2 = \alpha^{14}$, $(\alpha^7)^{2^2} = \alpha^{28} = \alpha^{13}$ and $(\alpha^7)^{2^3} = \alpha^{56} = \alpha^{11}$. It should be noted that for $l > (m - 1) = 3$, the conjugates repeat again such as $(\alpha^7)^{2^4} = \alpha^{112} = \alpha^7$, $(\alpha^7)^{2^5} = \alpha^{224} = \alpha^{14}$, and so on. We recall that the conjugates α^{14}, α^{13} and

α^{11} are also the roots of $f(z)$.

1) If β is a non-zero element in $GF(2^m)$, then β^{2^m-1} is always equal to 1. Adding 1 to both sides of the equation $\beta^{2^m-1} = 1$ gives

$$\beta^{2^m-1} + 1 = 0 . \tag{4.63}$$

β is an element, and in the above equation it is seen to be a root of the polynomial $(z^{2^m-1} + 1)$ over $GF(2)$. As the polynomial has degree of $2^m - 1$, it has $2^m - 1$ roots which are all the non-zero elements in $GF(2^m)$. As the zero element 0 of $GF(2^m)$ is the root of z, it then follows that the elements of $GF(2^m)$ form all the roots of $(z^{2^m} + z)$. As every element β in an extension field $GF(2^m)$ is a root of the polynomial $(z^{2^m} + z)$, there is a polynomial in $GF(2)$, called the *minimal polynomial* $\psi_\beta(z)$ of β. This polynomial is the smallest degree monic polynomial having β as a root, where a *monic* polynomial is defined as a polynomial with a leading coefficient of 1. In the case of $GF(2)$ the coefficient is either 0 or 1 and therefore all polynomials are monic. For example, a polynomial $(z^{2^4} + z)$ of degree-16 defined on $GF(2)$ has 16 roots which are all the elements in $GF(2^4)$.

Let us express the polynomial $(z^{2^4} + z)$ over $GF(2)$ as the product of the smallest degree monic polynomials,

$$
\begin{aligned}
&z^{2^4} + z \\
&\quad = z(z+1)(z^2+z+1)(z^4+z+1)(z^4+z^3+1)(z^4+z^3+z^2+z+1).
\end{aligned}
\tag{4.64}
$$

Each factor of the polynomial $(z^{2^4} + z)$ represents a minimal polynomial $\psi_\beta(z)$ over $GF(2)$ of some element β in $GF(2^4)$. The minimal polynomial $\psi_0(z)$ of zero element 0 from $GF(2^4)$ is the factor z and the minimal polynomial $\psi_{\alpha^0}(z)$ of unit element $\alpha^0 = 1$ is the factor $(z + 1)$. Also, the minimal polynomial of element α^3 is $(z^4 + z^3 + z^2 + z + 1)$. According to Property 1, the conjugates of α^3 are also the roots of the minimal polynomial. Thus, the elements α^3 and its conjugates, α^6, α^9 and α^{12}, have the same minimal polynomial. This can be verified as follows:

$$
\begin{aligned}
\psi_{\alpha^3}(\alpha^3) &= (\alpha^3)^4 + (\alpha^3)^3 + (\alpha^3)^2 + \alpha^3 + 1 = 0 \\
\psi_{\alpha^6}(\alpha^6) &= (\alpha^6)^4 + (\alpha^6)^3 + (\alpha^6)^2 + \alpha^6 + 1 = 0 \\
\psi_{\alpha^9}(\alpha^9) &= (\alpha^9)^4 + (\alpha^9)^3 + (\alpha^9)^2 + \alpha^9 + 1 = 0 \\
\psi_{\alpha^{12}}(\alpha^{12}) &= (\alpha^{12})^4 + (\alpha^{12})^3 + (\alpha^{12})^2 + \alpha^{12} + 1 = 0 .
\end{aligned}
$$

The minimal polynomials of all the elements in $GF(2^4)$ are tabulated in Table 4.12. Notice that the minimal polynomial of β is unique,

Conjugate roots	Minimal polynomial
0	z
α^0	$z + 1$
$\alpha^1, \alpha^2, \alpha^4, \alpha^8$	$z^4 \qquad\qquad + z + 1$
$\alpha^3, \alpha^6, \alpha^9, \alpha^{12}$	$z^4 + z^3 + z^2 + z + 1$
α^5, α^{10}	$z^2 + z + 1$
$\alpha^7, \alpha^{11}, \alpha^{13}, \alpha^{14}$	$z^4 + z^3 \qquad\qquad + 1$

Table 4.12: Minimal polynomials of the elements in $GF(2^4)$.

that is, for every β there is one and only one minimal polynomial. However, different elements of $GF(2^4)$ can have the same minimal polynomial. Moreover, for every element in $GF(2^m)$, the degree of the minimal polynomial over $GF(2)$ is at most m.

2) From Property 2, we understand that the minimal polynomial of the element β from $GF(2^m)$ is defined as the smallest degree polynomial over $GF(2)$ with the root of β. The minimal polynomial is therefore irreducible. Also, Property 1 states that if the element β from $GF(2^m)$ is a root of an irreducible polynomial, then all the other roots of the polynomial are the conjugates of β. Hence, the element β and its conjugates form all the roots of the minimal polynomial, and the total number of roots determines the degree of the minimal polynomial. Let e be the degree of the minimal polynomial of β from $GF(2^m)$, and e be defined as the smallest integer such that,

$$\beta^{2^e} = \beta \ . \tag{4.65}$$

As the element β and all its conjugates are all the roots of the minimal polynomial, the minimal polynomial of β is formed by

$$\psi_\beta(z) = \prod_{i=0}^{e-1} (z + \beta^{2^i}) \ . \tag{4.66}$$

For example, the conjugates of $\beta = \alpha^3$ in $GF(2^4)$ are

$$\beta^2 = \alpha^6, \quad \beta^{2^2} = \alpha^{12}, \quad \text{and} \quad \beta^{2^3} = \alpha^{24} = \alpha^9 \ .$$

The minimal polynomial of $\beta = \alpha^3$ is then formed as

$$\begin{aligned} \psi_{\alpha^3}(z) &= (z + \alpha^3)(z + \alpha^6)(z + \alpha^9)(z + \alpha^{12}) \\ &= (z^2 + \alpha^2 z + \alpha^9)(z^2 + \alpha^8 z + \alpha^6) \\ &= z^4 + z^3 + z^2 + z + 1 \ . \end{aligned}$$

The minimal polynomial of α^3 in $GF(2^4)$ can be verified with the aid of Table 4.12.

3) The minimal polynomial of a primitive element of $GF(2^m)$ has degree m and is a primitive polynomial. In the construction of the Galois field $GF(2^m)$, we use a primitive polynomial $p(z)$ of degree m and the primitive element which is a root of $p(z)$. The primitive element is α and the successive powers of α represent all the non-zero elements of $GF(2^m)$. They form a commutative group under multiplication and the group is closed as $\alpha^{2^m-1} = 1$. That is, if $l > (2^m - 1)$, the element $\alpha^l = \alpha^{(2^m-1)i+j} = \alpha^j$, $j \leq (2^m - 1)$ and therefore α^j is an element in the group. For example, in $GF(2^4)$, the minimal polynomial $z^4 + z + 1$ over $GF(2)$ of degree 4 of the primitive element α is used as the primitive polynomial to construct all the non-zero elements in $GF(2^4)$.

So far, we have studied the structure of the Galois field which introduces the finite field arithmetic in the cyclic codes. We now concentrate on the encoding and decoding algorithms of different error control codes.

4.4.2 Cyclic Codes

Cyclic codes were first introduced by Prange in 1957 [49]. They can be easily implemented by shift register circuits. For an (n, k) linear code C, k information symbols are encoded into an n-symbol codeword. This is a *cyclic code* if every cyclic shift of a vector in C is also a code vector in C. Thus if the elements of an n-tuple $\mathbf{v} = (v_{n-1}, \ldots, v_1, v_0)$ are cyclically shifted one place to the left, we obtain another n-tuple,

$$\mathbf{v}^{(1)} = (v_{n-2}, \ldots, v_0, v_{n-1}) \ .$$

This process is called a cyclic shift of $\mathbf{v}$. If the elements of $\mathbf{v}$ are cyclically shifted by i places to the left, the resultant n-tuple is

$$\mathbf{v}^{(i)} = (v_{n-i-1}, v_{n-i-2}, \ldots, v_1, v_0, v_{n-1}, \ldots, v_{n-i}) \ .$$

In order to explore the algebraic properties of the cyclic code, we express the code vector $\mathbf{v} = (v_{n-1}, \ldots, v_1, v_0)$ using the polynomial representation where the coefficients of the polynomial correspond to the elements of the vector, namely

$$\mathbf{v}(z) = v_{n-1}z^{n-1} + \cdots + v_2z^2 + v_1z + v_0 \ .$$

Hence, an n-tuple code vector is represented by a polynomial of degree $(n - 1)$ or less. If $v_{n-1} \neq 0$, the degree of $\mathbf{v}(z)$ is $(n - 1)$; if $v_{n-1} = 0$, the degree of $\mathbf{v}(z)$ is less than $(n - 1)$. The corresponding polynomial representation of the cyclically shifted code vector $\mathbf{v}^{(i)}(z)$ can be written as

$$\mathbf{v}^{(i)}(z) \; = \; \underbrace{v_{n-i-1}z^{n-1} + \cdots + v_1 z^{i+1} + v_0 z^i}_{(n-i)\text{ terms}}$$

$$+ \underbrace{v_{n-1}z^{i-1} + \cdots + v_{n-i+1}z + v_{n-i}}_{i\text{ terms}} \; .$$

The equivalent operation of the cyclic shift in terms of polynomial representation can be achieved by the following manipulation. We first observe that the multiplication of $\mathbf{v}(z)$ by z^i is

$$z^i \mathbf{v}(z) = \underbrace{v_{n-1}z^{n+i-1} + \cdots + v_{n-i}z^n}_{i\text{ terms}} + \underbrace{v_{n-i-1}z^{n-1} + \cdots + v_1 z^{i+1} + v_0 z^i}_{(n-i)\text{ terms}}$$

and that the order of the first i terms in the above equation exceeds the degree $(n-1)$, while those terms with degree less than i are absent. Due to the cyclic property, those i terms with degree higher than $(n-1)$ are shifted to the lower order part of the polynomial. This cyclic arithmetic is done by

$$z^i \mathbf{v}(z) \; = \; \underbrace{v_{n-i-1}z^{n-1} + \cdots + v_0 z^i}_{(n-i)\text{ terms}} + \underbrace{v_{n-1}z^{i-1} + \cdots + v_{n-i+1}z + v_{n-i}}_{i\text{ terms}}$$

$$+ \underbrace{v_{n-1}z^{i-1}(z^n+1) + \cdots + v_{n-i+1}z(z^n+1) + v_{n-i}(z^n+1)}_{i\text{ terms}}$$

$$= \; q(z)(z^n+1) + \mathbf{v}^{(i)}(z) \qquad\qquad (4.67)$$

where $q(z) = v_{n-1}z^{i-1} + \cdots + v_{n-i+1}z + v_{n-i}$. This means that if $\mathbf{v}(z)$ is a code polynomial, $\mathbf{v}^{(i)}(z)$ is also a code polynomial for any cyclic shift i. From Equation 4.67, we note that the cyclically shifted code polynomial $\mathbf{v}^{(i)}(z)$ is the remainder resulting from dividing the polynomial $z^i \mathbf{v}(z)$ by (z^n+1). That is,

$$\mathbf{v}^{(i)}(z) = R_{z^n+1}\left[z^i \mathbf{v}(z)\right] \qquad\qquad (4.68)$$

where $R_{f(z)}[\bullet]$ is the remainder of the modulo-$f(z)$ of $[\bullet]$.

So far we have defined cyclic codes, and now we will focus on their properties.

1. For an (n,k) linear code C, there are 2^k codeword polynomials $c(z)$. The codeword polynomials of degree $(n-1)$ are encoded by a generator polynomial $g(z)$ of degree $(n-k)$. As all the codeword polynomials of a cyclic code must be multiples of a generator polynomial $g(z)$, it then follows from Equation 4.68 that a codeword polynomial can be described by

$$c(z) = R_{z^n+1}\left[a(z)g(z)\right] \qquad\qquad (4.69)$$

where $a(z)$ is an arbitrary polynomial. A code polynomial $c(z)$ is modulo $(z^n + 1)$ and this implies that the block length is n.

1) Another property of the cyclic code is that the generator polynomial $g(z)$ of an (n, k) code is a factor of $(z^n + 1)$. Let r be the degree of the generator polynomial, where $r = n - k$. Multiplying $g(z)$ by z^k results in a polynomial $z^k g(z)$ of degree n. Dividing $z^k g(z)$ by $(z^n + 1)$, we obtain

$$z^k g(z) = (z^n + 1) + g^{(k)}(z) \qquad (4.70)$$

where $g^{(k)}(z)$ is the remainder. From Equation 4.68, we note that $\mathbf{v}^{(i)}(z)$ is a code polynomial given by cyclically shifting $\mathbf{v}(z)$ i times. Similarly, $g^{(k)}(z)$ is the code polynomial obtained by shifting $g(z)$ to the left cyclically k times. Hence, $g^{(k)}(z)$ is a multiple of $g(z)$, say $g^{(k)}(z) = a(z)g(z)$. Substituting $g^{(k)}(z)$ into Equation 4.70 yields

$$z^n + 1 = g(z) \left[z^k + a(z) \right] \ . \qquad (4.71)$$

Hence, $g(z)$ is a factor of $(z^n + 1)$. Consequently, for any cyclic code having the generator polynomial $g(z)$,

$$z^n + 1 = g(z)h(z) \qquad (4.72)$$

where the polynomial $h(z)$ is the *parity-check polynomial*. Then for any codeword polynomial $c(z)$

$$
\begin{aligned}
R_{z^n+1} \left[c(z)h(z) \right] &= R_{z^n+1} \left[a(z)g(z)h(z) \right] \\
&= R_{z^n+1} \left[a(z)(z^n + 1) \right] \\
&= 0 \ .
\end{aligned}
\qquad (4.73)
$$

Having presented the definition and the properties of cyclic codes, we now highlight their encoding. Suppose the information sequence is represented by a polynomial $i(z)$ of degree $(k - 1)$. The set of information polynomials $i(z)$ is mapped into the set of codeword polynomials $c(z)$ using the generator polynomial $g(z)$. A simple encoding method is

$$c(z) = i(z)g(z) \ . \qquad (4.74)$$

This method is called *non-systematic* encoding because the codeword polynomial does not contain a copy of $i(z)$. Alternatively *systematic* encoding is where the information polynomial $i(z)$ is inserted into the high-order coefficients of the codeword $c(z)$, and the parities are appended to the low-order coefficients. The codeword polynomial for systematic codes is

$$c(z) = i(z)z^{n-k} + b(z) \qquad (4.75)$$

where $b(z)$ is evaluated so that

$$R_{g(z)}\left[c(z)\right] = 0 \ .$$

It follows that,

$$R_{g(z)}\left[i(z)z^{n-k}\right] + R_{g(z)}\left[b(z)\right] = 0$$

and the degree of $b(z)$ is less than $(n-k)$, the degree of $g(z)$. Therefore,

$$b(z) = -R_{g(z)}\left[i(z)z^{n-k}\right] \ . \tag{4.76}$$

The systematic and non-systematic encoding procedures are unique one-to-one mappings from a set of information polynomials to a set of codeword polynomials, but the mappings are different for the two methods.

4.4.3   Bose-Chaudhuri-Hocquenghem Codes

The Bose-Chaudhuri-Hocquenghem (BCH) codes constitute a prominent class of cyclic block codes that have multiple-error detection and correction capabilities. In this section their theory and structure is studied. The class of binary and non-binary BCH codes is considered in Section 4.4.3.1 and in Section 4.4.3.2, respectively. For the non-binary BCH codes, an important subclass is that of the Reed-Solomon (RS) codes which achieve the maximum separable distance between codewords as will be detailed in Section 4.4.3.2.1.

A BCH code accepts k information symbols and produces an n-symbol codeword. If a codeword is designed to correct t random errors, the code is called a *t-error-correcting* code and is denoted as a $BCH(n, k, t)$ code. A BCH code is a cyclic code and therefore can be constructed by its generator polynomial $g(z)$. According to the second property of cyclic codes given in Section 4.4.2 and Equation 4.71, the generator polynomial is a factor of $(z^n + 1)$. That is,

$$z^n + 1 = a(z)g(z)$$

where $a(z)$ is an arbitrary polynomial. Also, the second property of minimal polynomials in Section 4.4.1.5 states that the polynomial $(z^{q^m-1} + 1)$ is the least common multiple (LCM) of the minimal polynomials of all the non-zero elements in $GF(q^m)$. That is,

$$z^{q^m-1} + 1 = \mathrm{LCM}[\psi_{\alpha^0}(z), \psi_{\alpha^1}(z), \dots, \psi_{\alpha^{q^m-2}}(z)].$$

On observing these two properties, we assign

$$n = q^m - 1 \tag{4.77}$$

such that

$$z^n + 1 = z^{q^m - 1} + 1 .$$

The generator polynomial, which is a factor of $(z^n + 1)$, can be constructed by a product of the minimal polynomials over $GF(q)$ of elements from $GF(q^m)$. As the minimal polynomial $\psi_{\alpha^i}(z)$ over $GF(q)$ is defined to have roots of element α^i and its conjugates from $GF(q^m)$, the roots of the generator polynomial are also the element of α^i and its conjugates. For a t-error-correcting code, the generator polynomial $g(z)$ is defined by the least common multiple of the minimal polynomials over $GF(q)$ having $2t$ consecutive powers of α, i.e., $\alpha^{j_0}, \alpha^{j_0+1}, \ldots, \alpha^{j_0+2t-1}$, as their roots, where j_0 is an integer, and hence,

$$g(z) = \mathrm{LCM}\left[\psi_{\alpha^{j_0}}(z), \psi_{\alpha^{j_0+1}}(z), \ldots, \psi_{\alpha^{j_0+2t-1}}(z)\right]. \qquad (4.78)$$

The degree of the generator polynomial $g(z)$, regardless of non-systematic encoding or systematic encoding described by Equations 4.74 and 4.76 respectively, determines the number of redundancy symbols in a codeword. The $2t$ roots of $g(z)$ allow us to correct t error symbols. We require the degree of $g(z)$ to be as small as possible by keeping $2t$ roots so as to correct t errors with minimum redundancy. When j_0 is chosen to be 1, the first root is the primitive element α and this usually gives the $g(z)$ of smallest degree. These codes are described as the *primitive BCH codes* and their blocklength is $(q^m - 1)$. The distance between codewords required for a t-error-correcting code is

$$d = 2t + 1 \qquad (4.79)$$

where d is called the *designed distance* of the code. But the minimum separable distance between codewords in some BCH codes may be greater, i.e., $d_{min} \geq 2t + 1$ and this is detailed in reference [7].

4.4.3.1 Binary BCH Codes

A BCH code is binary if the codeword symbols are binary, i.e., the symbol is defined on $GF(q)$, where $q = 2$. To represent a codeword as a polynomial, the symbols of a codeword are the coefficients of the polynomial. In the case of a binary code, the symbol is either 0 or 1, and the coefficients of the polynomial have binary values. The blocklength n of the code, as defined in Equation 4.77, is $(q^m - 1)$ symbols. The number of symbols in a codeword determines the extension field $GF(q^m)$ where the roots of the generator polynomial reside. As a consequence, the generator of the code is a binary polynomial, defined on $GF(2)$, having roots in the extension field $GF(q^m)$.

We now present examples of how to construct the generator polynomial of a binary BCH code. This generator polynomial is defined on $GF(2)$. As our first example, we let $m = 4$, giving a blocklength of 15 bits and a

+	0	1	2	3
0	0	1	2	3
1	1	0	3	2
2	2	3	0	1
3	3	2	1	0

•	0	1	2	3
0	0	0	0	0
1	0	1	2	3
2	0	2	3	1
3	0	3	1	2

Table 4.13: Arithmetic tables for $GF(4)$.

generator polynomial having roots from $GF(2^4) = GF(16)$. Table 4.8 gives $GF(16)$ as an extension field of $GF(2)$. The table is constructed using the primitive polynomial $p(z) = z^4 + z + 1$. The minimal polynomials over $GF(2)$ of all field elements in $GF(16)$ are listed in Table 4.12, where $\alpha = z$ is the primitive element. According to Equation 4.78, the generator polynomial for the double error correcting BCH code requires four consecutive roots in $GF(16)$ and can be constructed by

$$
\begin{aligned}
g(z) &= \text{LCM}[\psi_\alpha(z), \psi_{\alpha^2}(z), \psi_{\alpha^3}(z), \psi_{\alpha^4}(z)] \\
&= \text{LCM}[z^4 + z + 1, z^4 + z + 1, z^4 + z^3 + z^2 + z + 1, z^4 + z + 1] \\
&= (z^4 + z + 1)(z^4 + z^3 + z^2 + z + 1) \\
&= z^8 + z^7 + z^6 + z^4 + 1 .
\end{aligned}
\tag{4.80}
$$

The degree of $g(z)$ is eight, i.e., $n - k = 8$, and therefore $k = 7$. This means that 7 information bits are encoded into a codeword with a blocklength of 15 bits. The code can correct up to 2 error bits occurring at any position in a codeword, and the code is denoted as $BCH(15, 7, 2)$.

Another example of constructing a generator polynomial for a triple error correcting binary code is by multiplying six minimal polynomials having consecutive roots in $GF(16)$,

$$
\begin{aligned}
g(z) &= \text{LCM}[\psi_\alpha(z), \psi_{\alpha^2}(z), \psi_{\alpha^3}(z), \psi_{\alpha^4}(z), \psi_{\alpha^5}(z), \psi_{\alpha^6}(z)] \\
&= (z^4 + z + 1)(z^4 + z^3 + z^2 + z + 1)(z^2 + z + 1) \\
&= z^{10} + z^8 + z^5 + z^4 + z^2 + z + 1 .
\end{aligned}
\tag{4.81}
$$

The generator polynomial is of degree 10, and therefore 10 parity bits are included in a codeword in order to correct 3 error bits. The code with blocklength of 15 bits thus carries 5 information bits and is therefore described as $BCH(15, 5, 3)$.

4.4.3.2 non-binary BCH Codes

A non-binary BCH code has symbols of more than one bit defined on $GF(q)$, where $q > 2$. The coefficients of the polynomial are also non-binary and are elements of $GF(q)$. From Equation 4.77, the blocklength of the code is $(q^m - 1)$ symbols and the number of symbols in a codeword

determines the extension field $GF(q^m)$. The generator of the code is defined
as a polynomial over $GF(q)$ having roots in the extension field $GF(q^m)$.

Let us consider an example of a non-binary BCH code. A symbol consisting of two bits is defined on $GF(4)$. If the blocklength of the code is
selected to have $4^2 - 1$ symbols, i.e., $q = 4$ and $m = 2$, the generator
polynomial over $GF(4)$ has roots in the extension field $GF(q^m) = GF(16)$.
Table 4.13 is a decimal representation of the arithmetic tables of $GF(4)$
in Tables 4.4 and 4.5. We also show in Table 4.14 the minimal polynomials over $GF(4)$ of all the field elements in $GF(16)$, where α is primitive.
For the double error correcting BCH code, the generator polynomial over
$GF(4)$ requires four consecutive roots in $GF(16)$ and can be constructed
as

$$
\begin{aligned}
g(z) &= \mathrm{LCM}[\psi_\alpha(z), \psi_{\alpha^2}(z), \psi_{\alpha^3}(z), \psi_{\alpha^4}(z)] \\
&= \mathrm{LCM}[z^2 + z + 2, z^2 + z + 3, z^2 + 3z + 1, z^2 + z + 2] \\
&= (z^2 + z + 2)(z^2 + z + 3)(z^2 + 3z + 1) \\
&= z^6 + 3z^5 + z^4 + z^3 + 2z^2 + 2z + 1 .
\end{aligned}
$$

The degree of the generator polynomial is 6, the remainder of the information polynomial after division by $g(z)$ results in the parity polynomial.
The degree of the parity polynomial is 5 which therefore determines 6 parity
symbols in a codeword. Therefore the code is described as a non-binary
BCH(15,9,2) code over $GF(4)$ as it encodes 9 information symbols, each
of 2 bits, into a codeword of 15 symbols. The code is able to correct
2 quaternary error symbols in a codeword with a blocklength of 15 quaternary symbols (i.e., 30 bits). Furthermore, for a triple error correcting
code, the generator polynomial can be constructed by

$$
\begin{aligned}
g(z) &= \mathrm{LCM}[\psi_\alpha(z), \psi_{\alpha^2}(z), \psi_{\alpha^3}(z), \psi_{\alpha^4}(z), \psi_{\alpha^5}(z), \psi_{\alpha^6}(z)] \\
&= (z^2 + z + 2)(z^2 + z + 3)(z^2 + 3z + 1)(z + 2)(z^2 + 2z + 1) \\
&= z^9 + 3z^8 + 3^7 + 2z^6 + z^5 + 2z^4 + z + 2 .
\end{aligned}
$$

This code is described as a BCH(15,6,3) triple error correcting code over
$GF(4)$.

4.4.3.2.1 Reed-Solomon Codes

In Section 4.4.3.2, the non-binary
BCH codes define the symbols over $GF(q)$, where $q > 2$, and the roots of
the generator polynomial are over $GF(q^m)$. The Reed-Solomon codes are
a special case of non-binary BCH codes with $m = 1$. The coefficients of the
generator polynomial are from $GF(q)$ and the roots of the generator are also
elements from $GF(q)$. Hence, the minimal polynomials for constructing the
generator polynomial are defined on $GF(q)$ with roots from the same field.
Notice that the minimal polynomial over $GF(q)$ of an element β in the
same $GF(q)$ is

Conjugate roots	Minimal polynomial
0	z
α^0	$z + 1$
α^1, α^4	$z^2 + z + 2$
α^2, α^8	$z^2 + z + 3$
α^3, α^{12}	$z^2 + 3z + 1$
α^5	$z + 2$
α^6, α^9	$z^2 + 2z + 1$
α^7, α^{13}	$z^2 + 2z + 2$
α^{10}	$z + 3$
α^{11}, α^{14}	$z^2 + 3z + 3$

Table 4.14: Minimal polynomials of the elements in $GF(4^2)$.

$$\psi_\beta(z) = z - \beta .$$

For a t-error-correcting RS code, the generator polynomial is

$$g(z) = (z - \alpha^{j_0})(z - \alpha^{j_0+1}) \cdots (z - \alpha^{j_0+2t-1}) . \qquad (4.82)$$

As all the minimal polynomials of any element in $GF(q)$ are of degree 1, the choice of j_0 will not optimise the degree of the generator polynomial. The convention is to set $j_0 = 1$ to produce a generator polynomial of degree $2t$. Hence, for an RS code

$$n - k = 2t . \qquad (4.83)$$

The blocklength of the RS codes is determined by substituting $m = 1$ into Equation 4.77, whence

$$n = q - 1 .$$

An important characteristic of Reed-Solomon codes is their maximum minimum distance property. Observe that the designed distance d of the code is given by Equation 4.79 and the minimum separable distance d_{min} between codewords may be actually greater than the designed distance. The lower bound of d_{min} is therefore,

$$d_{min} \geq d = 2t + 1 = n - k + 1 \qquad (4.84)$$

because $2t = n - k$. In addition, systematic codewords exist with only one non-zero information symbol and $(n - k)$ parity symbols. This codeword has a maximum symbol distance of $(n - k + 1)$ from the all-zero codeword. Hence, the minimum distance of the code cannot be greater than $(n - k + 1)$ and the upper bound of d_{min} is therefore

$$d_{min} \leq n - k + 1 . \qquad (4.85)$$

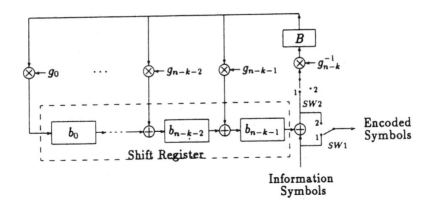

Figure 4.40: Systematic encoder for block codes.

Equation 4.85 is known as the Singleton bound. Combining both Equation 4.84 and 4.85 yields d_{min} for the RS codes as

$$d_{min} = n - k + 1 . \tag{4.86}$$

The Reed-Solomon code is a maximum-distance code, and the minimum distance is $(n - k + 1)$.

As an example, we will find the generator polynomial $g(z)$ for a double error correcting RS code, i.e., for $t = 2$. Suppose the symbols of the code are chosen to have four bits, defined on $GF(q)$, where $q = 16$. The roots of $g(z)$ are also defined on $GF(q)$, i.e, $GF(16)$. Consequently, for a double error correcting RS code, the generator polynomial $g(z)$ over $GF(16)$ has a set of four roots $\{\alpha, \alpha^2, \alpha^3, \alpha^4\}$ from $GF(16)$. From Equation 4.82, $g(z)$ can be constructed over $GF(16)$ as

$$\begin{aligned} g(z) &= (z - \alpha)(z - \alpha^2)(z - \alpha^3)(z - \alpha^4) \\ &= z^4 + \alpha^{13}z^3 + \alpha^6 z^2 + \alpha^3 z + \alpha^{10} . \end{aligned} \tag{4.87}$$

The degree of $g(z)$ is four, $n - k = 4$. The blocklength of the code is given by Equation 4.84 such that $n = 15$ and so $k = 11$. An information sequence of 11 hexadecimal symbols (44 bits) is encoded into a codeword with a blocklength of 15 hexadecimal symbols (60 bits). The Reed-Solomon code is denoted as RS(n, k, t), and, in this example, it is an RS$(15, 11, 2)$ code.

4.4.4 Encoding of Block Codes

The structure of cyclic codes has the advantage that their encoders and decoders can be implemented using shift-register circuits. The cyclic codes can be encoded either non-systematically or systematically. For the non-systematic codes, the encoder multiplies an arbitrary information polynomial $i(z)$ with the generator polynomial $g(z)$ to obtain the codeword $c(z)$, as shown in Equation 4.74. Whereas, for the systematic codes, the encoder evaluates and then appends the redundancy symbols to the arbitrary information polynomial (see Equation 4.75). The systematic code thus contains a copy of the information symbols. Suppose the codeword is transmitted over a noisy channel, and the number of error symbols in the received codeword exceed the correcting capability of the code. For the non-systematic codes, the decoder is required to correct the received codeword and then divide the corrected codeword by the generator polynomial $g(z)$ to regenerate the information polynomial $i(z)$, namely

$$i(z) = c(z)/g(z) \ . \tag{4.88}$$

However, if the corrected codeword is erroneous, the regenerated information polynomial can have any arbitrary value, which may result in disastrous symbol errors. By constrast to the non-systematic codes, the systematic RS decoder corrects the received codeword and then copies the information symbols from the corrected codeword to its output. As we are considering here the situation where the number of channel errors exceeds the correcting capability of the code, the information polynomial has error symbols due to both the channel noise and the correction error. However, the correctly received symbols in the information polynomial remain intact and are undisturbed by the decoder. The result is that systematic RS codes perform better than the non-systematic ones.

For systematic codes the generator polynomial $g(z) = g_{n-k}z^{n-k} + \cdots + g_1 z + g_0$ formulates a codeword by appending $(n-k)$ parity symbols $b_{n-k-1}, \ldots, b_1, b_0$ to k information symbols. The encoder employs a shift register (SR) having $(n-k)$ stages as depicted in Figure 4.40. At the beginning of the encoding process, the SR is cleared and both switches $SW1$ and $SW2$ are placed in position 1. The first k number of information symbols are passed directly to the encoder output forming the information part of the codeword, and they are also multiplied over $GF(q^m)$ by the coefficient g_{n-k}^{-1}, where $q = 2$ and $m = 1$ for binary codes. This product is buffered in register B. The value of register B is multiplied by g_{n-k-1}, followed by $GF(q^m)$ addition with b_{n-k-2} to form a new parity symbol of b_{n-k-1}. Again, the new b_{n-k-2} value is calculated by multiplying the content of register B with g_{n-k-2} and then adding b_{n-k-3} to this product. Similar multiplications and additions are performed to achieve new values from b_{n-k-3} to b_0. The second information symbol enters the encoder and the cycle of multiplications and additions is repeated to yield a new set of $(n-k)$ parity symbols. After the kth information symbol has entered the encoder and has produced the parity symbols, the switches are turned to position 2, preventing data from entering the SR. The $(n-k)$ parity

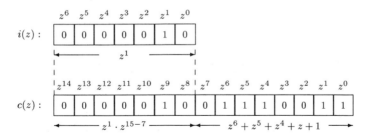

Figure 4.41: Binary representation of polynomials.

symbols are removed serially from the encoder.

4.4.4.1 Binary BCH Encoder

We now present an example of encoding systematically a binary BCH-$(15, 7, 2)$ double error correcting code. The numerical calculation of the encoding will first be demonstrated, followed by its implementation using a shift register circuit. The generator polynomial $g(z)$ of this code is given by Equation 4.80. Let us take an arbitrary information polynomial for our example as

$$i(z) = z^1 \ . \tag{4.89}$$

According to the systematic encoding described in Equation 4.75, the parity polynomial $b(z)$ is obtained using Equation 4.76 as,

$$b(z) = R_{(z^8 + z^7 + z^6 + z^4 + 1)}[z^1 \cdot z^{15-7}] \ .$$

By performing long division over $GF(2)$ we have

$$
\begin{array}{r}
z + 1 \\
\hline
z^8 + z^7 + z^6 + z^4 + 1 \ \overline{\smash{\big)}\, z^9} \\
z^9 + z^8 + z^7 + z^5 + z \\
\hline
z^8 + z^7 + z^5 + z \\
z^8 + z^7 + z^6 + z^4 + 1 \\
\hline
z^6 + z^5 + z^4 + z + 1 \qquad (490)
\end{array}
$$

where $b(z)$ is given by the remainder $(z^6 + z^5 + z^4 + z + 1)$. From Equation 4.75, the codeword $c(z)$ is

$$c(z) = z^9 + z^6 + z^5 + z^4 + z + 1 \ . \tag{4.91}$$

Observe that the binary representation of the codeword $c(z)$ in Figure 4.41 has a blocklength of 15 bits, where each bit represents a coefficient of

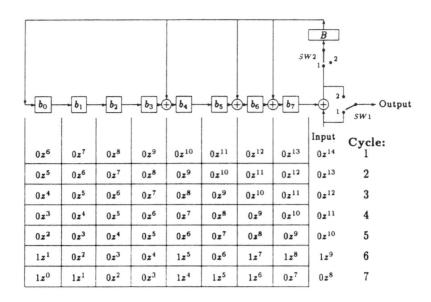

								Input	Cycle:
$0z^6$	$0z^7$	$0z^8$	$0z^9$	$0z^{10}$	$0z^{11}$	$0z^{12}$	$0z^{13}$	$0z^{14}$	1
$0z^5$	$0z^6$	$0z^7$	$0z^8$	$0z^9$	$0z^{10}$	$0z^{11}$	$0z^{12}$	$0z^{13}$	2
$0z^4$	$0z^5$	$0z^6$	$0z^7$	$0z^8$	$0z^9$	$0z^{10}$	$0z^{11}$	$0z^{12}$	3
$0z^3$	$0z^4$	$0z^5$	$0z^6$	$0z^7$	$0z^8$	$0z^9$	$0z^{10}$	$0z^{11}$	4
$0z^2$	$0z^3$	$0z^4$	$0z^5$	$0z^6$	$0z^7$	$0z^8$	$0z^9$	$0z^{10}$	5
$1z^1$	$0z^2$	$0z^3$	$0z^4$	$1z^5$	$0z^6$	$1z^7$	$1z^8$	$1z^9$	6
$1z^0$	$1z^1$	$0z^2$	$0z^3$	$1z^4$	$1z^5$	$1z^6$	$0z^7$	$0z^8$	7

Figure 4.42: Systematic encoder for the BCH$(15, 7, 2)$ code.

the codeword polynomial. By using systematic encoding, the information polynomial z^1 is multiplied by z^{15-7} to give the information part positioned at the high-order part $(z^{14}-z^8)$ of the codeword. The low-order part (z^7-z^0) of the codeword is constituted by the parity polynomial $z^6 + z^5 + z^4 + z + 1$.

The systematic encoder implemented by the shift register for the BCH-$(15, 7, 2)$ code is shown in Figure 4.42. It can be seen to be a derivative of Figure 4.40. The generator polynomial is $g(z) = z^8 + z^7 + z^6 + z^4 + 1$. The coefficient g_8 of the z^8 term is 1, the inverse of g_8, i.e., g_8^{-1} is also 1. In the case of binary codes, the coefficient of the generator polynomial is either 0 or 1 and the g_8^{-1} is always 1. Observe that all the multipliers illustrated in Figure 4.40 are absent in Figure 4.42. If the coefficient is 1, the multiplier is replaced by a direct hard-wire connection as shown in Figure 4.42, whereas if the coefficient is 0, no connection is made.

The shift register consists of 8 storage elements, $\{b_0, b_1, \ldots, b_7\}$, where each element buffers a parity bit. The feedback connections to these elements are determined by the coefficients of the generator polynomial. This circuit arrangement is actually performing a polynomial division similar to the division shown in Equation 4.90. The input to the shift register is the information polynomial $i(z)$ which is divided by the generator polynomial $g(z)$. The remainder of the division is buffered in the storage elements. In Figure 4.42, each row of the grid represents the contents stored in the elements for a particular cycle. Initially the shift register is cleared and all its elements are zeros. Both switches $SW1$ and $SW2$ are in position 1. At the first cycle, the zero coefficient of order z^6 in $i(z)$ is inserted into the shift register. According to the systematic code, this input bit corresponds to the order z^{14} in $c(z)$ and therefore passes directly to the output. The adder sums the input bit and b_7 over $GF(2)$ to produce the result of 0 which is buffered in register B. The value of register B is added to b_6 to form a new parity bit of b_7. Similarly, the new b_6 value is calculated by adding the content of register B to b_5. Furthermore, the value of b_4 is shifted to b_5 by forming its new value. Similar additions performed sequentially yield new values from b_4 to b_0. As a consequence, the storage elements contain the first partial remainder of the division of $i(z)$ by $g(z)$. The partial remainder polynomial has degree of z^{13} and its coefficients are all zero. For the rest of the cycles, similar additions and shifts of the parity bits are made and the partial remainder polynomial in the storage elements is always one degree less than in the previous cycle. At the end of the seventh cycle, the storage devices contain the remainder of the division of $i(z)z^{n-k}$ by $g(z)$. The switches $SW1$ and $SW2$ are turned to position 2 and all the parity bits are removed serially from the encoder.

It is interesting to note the similarity of the long division calculation demonstrated in Equation 4.90 and the division implemented by the shift register circuit. At the sixth cycle, the non-zero coefficient of order z^9 input bit is inserted into the shift register. The partial remainder polynomial is $z^8 + z^7 + z^5 + z$ which corresponds to the partial remainder of the first step in the long division process. Similarly, at the next cycle, the content of the storage elements is the remainder of the long division.

4.4.4.2 Reed-Solomon Encoder

We will explain the operation of a RS encoder by considering the specific example of a double error correcting code with four bits per symbol. The arithmetic operations are over $GF(16)$ and the generator polynomial for the double error correcting code is described by Equation 4.87. The result is an $RS(15, 11, 2)$ code over $GF(16)$, encoding 11 information symbols into a codeword of 15 symbols. Given that the information polynomial over $GF(16)$ is

$$
\begin{aligned}
i(z) &= 0z^{10} + 0z^9 + 0z^8 + \alpha^{12}z^7 + \alpha^{12}z^6 + \alpha^{12}z^5 + \alpha^{12}z^4 \\
&\quad + \alpha^{12}z^3 + \alpha^{12}z^2 + \alpha^{12}z^1 + \alpha^{12}
\end{aligned} \tag{4.92}
$$

consisting of 11 symbols, we will assume that the three high-order terms $(z^{10} - z^8)$ contain all-zero symbols and all the low-order terms $(z^7 - z^0)$ contain all-one (i.e., α^{12}) symbols. By using systematic encoding described in Equation 4.75, the parity polynomial $b(z)$ can be obtained with the aid

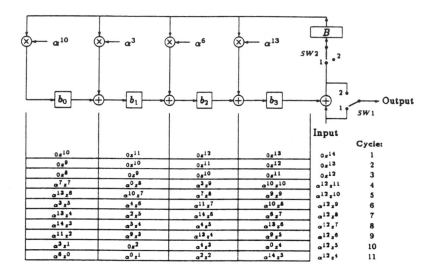

b_0	b_1	b_2	b_3	Input	Cycle:
$0z^{10}$	$0z^{11}$	$0z^{12}$	$0z^{13}$	$0z^{14}$	1
$0z^9$	$0z^{10}$	$0z^{11}$	$0z^{12}$	$0z^{13}$	2
$0z^8$	$0z^9$	$0z^{10}$	$0z^{11}$	$0z^{12}$	3
$\alpha^7 z^7$	$\alpha^0 z^8$	$\alpha^3 z^9$	$\alpha^{10} z^{10}$	$\alpha^{12} z^{11}$	4
$\alpha^{13} z^6$	$\alpha^{10} z^7$	$\alpha^7 z^8$	$\alpha^9 z^9$	$\alpha^{12} z^{10}$	5
$\alpha^3 z^5$	$\alpha^4 z^6$	$\alpha^{11} z^7$	$\alpha^{10} z^8$	$\alpha^{12} z^9$	6
$\alpha^{13} z^4$	$\alpha^2 z^5$	$\alpha^{14} z^6$	$\alpha^6 z^7$	$\alpha^{12} z^8$	7
$\alpha^{14} z^3$	$\alpha^3 z^4$	$\alpha^4 z^5$	$\alpha^{13} z^6$	$\alpha^{12} z^7$	8
$\alpha^{11} z^2$	$\alpha^9 z^3$	$\alpha^{13} z^4$	$\alpha^9 z^5$	$\alpha^{12} z^6$	9
$\alpha^3 z^1$	$0z^2$	$\alpha^4 z^3$	$\alpha^0 z^4$	$\alpha^{12} z^5$	10
$\alpha^6 z^0$	$\alpha^0 z^1$	$\alpha^2 z^2$	$\alpha^{14} z^3$	$\alpha^{12} z^4$	11

Figure 4.43: Systematic encoder for the RS(15, 11, 2) code.

of Equation 4.76 as,

$$b(z) = R_{(z^4 + \alpha^{13} z^3 + \alpha^6 z^2 + \alpha^3 z + \alpha^{10})} \left[(\alpha^{12} z^7 + \alpha^{12} z^6 + \alpha^{12} z^5 + \alpha^{12} z^4 \right.$$
$$\left. + \alpha^{12} z^3 + \alpha^{12} z^2 + \alpha^{12} z^1 + \alpha^{12}) \cdot z^{15-11} \right].$$

By performing long division over $GF(16)$, $b(z)$ is,

$$b(z) = \alpha^{14} z^3 + \alpha^2 z^2 + \alpha^0 z + \alpha^6 . \tag{4.93}$$

The polynomial $b(z)$ contains four parity symbols which are systematic encoded according to Equation 4.75. The codeword polynomial $c(z)$ becomes

$$
\begin{aligned}
c(z) = \ & 0z^{14} + 0z^{13} + 0z^{12} + \alpha^{12} z^{11} + \alpha^{12} z^{10} + \alpha^{12} z^9 + \alpha^{12} z^8 + \alpha^{12} z^7 \\
& + \alpha^{12} z^6 + \alpha^{12} z^5 + \alpha^{12} z^4 + \alpha^{14} z^3 + \alpha^2 z^2 + \alpha^0 z + \alpha^6
\end{aligned} \tag{4.94}
$$

and is of degree 14 with 15 terms. The coefficient of each term represents a four-bit symbol in the codeword. The blocklength of the codeword is 15 four-bit symbols, and therefore the codeword has 60 bits. The 11 information symbols reside in the high-order (z^{14}—z^4) terms of $c(z)$, while the low-order (z^3—z^0) terms contain four parity symbols.

Reed-Solomon codes belong to the class of cyclic codes and their cyclic property enables the RS encoder to be implemented by shift register circuits. The systematic encoder for the RS(15, 11, 2) code is shown in Figure 4.43. Its circuitry is similar to that for the binary BCH encoder shown in Figure 4.42. For Figure 4.43 we see that the shift register consists of four storage elements, $\{b_0, b_1, b_2, b_3\}$, each element buffers a parity symbol and each symbol has four bits. The feedback connections to these devices are determined by the coefficients of the generator polynomial. The generator polynomial of this code is $g(z) = z^4 + \alpha^{13}z^3 + \alpha^6 z^2 + \alpha^3 z + \alpha^{10}$, which is a monic polynomial such that the coefficient g_4 of the z^4 term is 1. The inverse of g_4 is also 1 and therefore the multiplier of g_4^{-1} (i.e., corresponding to the multiplier of g_{n-k}^{-1} in Figure 4.40) is a direct hard-wire connection. The multipliers of α^{13}, α^6, α^3 and α^{10} in the circuit correspond to the coefficients of z^3, z^2, z^1 and z^0, respectively. All the multiplications and additions are performed over $GF(16)$.

The input to the shift register is the information polynomial $i(z)$ which is divided by the generator polynomial $g(z)$. The partial remainder of the division is then buffered in the storage devices. As there are 11 information symbols in a codeword, the division is completed at the end of the eleventh cycle. The four parity symbols are the remainder of the division and are buffered in the shift register.

The cyclic code has a defined blocklength of n symbols of which k are information symbols, but systematic cyclic codes can be shortened by dropping s information symbols from each codeword. Now $(k - s)$ information symbols are combined with $(n - k)$ parity symbols to form a codeword with a blocklength of $(n - s)$ symbols. The code is converted from an (n, k) code to an $(n - s, k - s)$ code which is known as a *shortened cyclic code*. Those deleted symbols must be the ones in the high-order positions of the information sequence. As the deleted symbols in a shortened code are always set to zero and they are not transmitted, the receiver reinserts them and decodes just as if the code were not shortened. If the symbol of the code is q-ary, the original code consists of a set of q^k possible codewords and has minimum distance d_{min}. After the codeword has been shortened by s symbols, the set of possible codewords is reduced to q^{k-s} and therefore the minimum distance is d_{min} or larger. As the number of parity symbols of the codeword remains the same after shortening, the code is still able to correct t error symbols. For example, the RS(15, 11, 2) code consists of 11 information symbols and 4 parity symbols. If the number of information symbols is reduced by three, i.e., $s = 3$, such that the three high-order terms $(z^{14} - z^{12})$ are always set to zeros, then the blocklength of the shortened code is reduced by three and the code becomes RS(12, 8, 2). The generator polynomial $g(z)$ of the shortened code is given by Equation 4.87, and the four parity symbols in a codeword are able to correct two error symbols.

4.4.5 Decoding Algorithms for Block Codes

When a codeword is transmitted, errors can occur at any symbol position in a codeword. For binary codes, the magnitude of the error symbol is 1. For the non-binary code defined on $GF(q)$, where $q > 2$, the error magnitude can be any value from 1 to $(q - 1)$. If the positions of the errors and their magnitudes are found, the original codeword can be recovered by correcting the magnitude in each error position of the received codeword. In Section 4.4.5.1, we derive the error magnitudes and positions as the solution of a set of non-linear syndrome equations. The set of equations is then turned into a matrix with its coordinate elements arranged in a special structure known as a Vandermonde matrix. The direct inversion of the Vandermonde matrix was first suggested by Peterson [38] to obtain the solution for binary codes and then extended for non-binary codes by Gorenstein and Zierler [39]. The decoding algorithm is known as the Peterson-Gorenstein-Zierler method and is described in Section 4.4.5.2. Berlekamp [4] and Massey [44, 45] recognised that the best way to derive the decoding algorithm was as a solution to a problem in designing linear-feedback shift registers. Their method described in Section 4.4.5.3, is conceptually more complex than the direct matrix inversion, but is computationally much simpler. Finally, in Section 4.4.5.4, we describe the Forney algorithm for evaluating the error magnitudes for non-binary codes.

4.4.5.1   The Syndrome Equations

Consider a t error correcting BCH(n, k, t) code constructed by a generator polynomial over $GF(q)$ having $2t$ consecutive roots in $GF(q^m)$. Notice that this is an RS code if $m = 1$. To simplify the equations we let $j_0 = 1$ in Equation 4.78 whereby the roots of the generator polynomial are $\alpha^1, \alpha^2, \ldots, \alpha^{2t}$. Suppose that the codeword $c(z)$ is expressed in polynomial form over $GF(q)$, namely

$$c(z) = c_{n-1}z^{n-1} + c_{n-2}z^{n-2} + \ldots + c_1 z + c_0$$

where the coefficient c_i represents the ith symbol in the codeword. As errors can occur at any symbol position in a codeword, the error polynomial

$$e(z) = e_{n-1}z^{n-1} + e_{n-2}z^{n-2} + \ldots + e_1 z + e_0$$

is also of degree $(n-1)$, where the coefficients e_i with non-zero value indicate that the ith symbol of the codeword is in error and that its error magnitude is e_i. The coefficient with zero value means that there is no error in the ith position. For a t error correcting code, the maximum number of non-zero coefficient terms is t. If this condition is not met, the number of errors exceeds the correcting capability of the code and the received word cannot be corrected. The received word is therefore described as

$$r(z) = c(z) + e(z) \ . \tag{4.95}$$

Now suppose that there are v error symbols in a received codeword, where $0 \le v \le t$. These errors occur in unknown positions $[i_1, i_2, \ldots, i_v]$, with error magnitudes $[e_{i_1}, e_{i_2}, \ldots, e_{i_v}]$. By ignoring the zero coefficient terms in $e(z)$, we may express the error polynomial as

$$e(z) = e_{i_1} z^{i_1} + e_{i_2} z^{i_2} + \ldots + e_{i_v} z^{i_v} \qquad (4.96)$$

with its v number of unknown error magnitudes and v number of unknown positions. Moreover the value of v itself is also unknown as the number of error symbols v in a correctable codeword can be any value $\leq t$. If all these unknowns are found, the received word can be corrected.

We now define a syndrom S_l as the received polynomial $r(z)$ evaluated at $z = \alpha^l$, where α^l is a root of the generator polynomial. This implies that $c(\alpha^l) = 0$ as α^l is also a root of the codeword polynomial $c(z)$, and that the syndrome S_l depends only on the error polynomial $e(z)$, i.e.,

$$S_l = r(\alpha^l) = c(\alpha^l) + e(\alpha^l) = e(\alpha^l) \ .$$

Let us now evaluate the received polynomial $r(z)$ at α to obtain the syndrome S_1, where α is a root of the generator polynomial. Thus for $l = 1$, we have,

$$\begin{aligned} S_1 &= r(\alpha) = c(\alpha) + e(\alpha) = e(\alpha) \\ &= e_{i_1}\alpha^{i_1} + e_{i_2}\alpha^{i_2} + \cdots + e_{i_v}\alpha^{i_v} \ . \end{aligned} \qquad (4.97)$$

Notice that the position i_x of the error symbol can be any position in a codeword, namely $0 \leq i_x < (n-1)$. From Equation 4.77, $n = q^m - 1$ for BCH codes, and the element α^{i_x} representing the error position is in the extension field $GF(q^m)$. To simplify the notation, we assign P_x to be the error position α^{i_x}, and M_x to be its error magnitude e_{i_x}. Hence, Equation 4.97 becomes

$$S_1 = M_1 P_1 + M_2 P_2 + \cdots + M_v P_v \ .$$

Similarly, by evaluating the received polynomial $r(z)$ at the set of $2t$ roots $\{\alpha, \alpha^2, \ldots, \alpha^{2t}\}$ of the generator polynomial $g(z)$, we obtain a set of syndromes,

$$\begin{aligned} S_1 &= M_1 P_1 + M_2 P_2 + \cdots + M_v P_v &= \textstyle\sum_{i=1}^{v} M_i P_i \\ S_2 &= M_1 P_1^2 + M_2 P_2^2 + \cdots + M_v P_v^2 &= \textstyle\sum_{i=1}^{v} M_i P_i^2 \\ \vdots \qquad & \qquad\qquad \vdots & \qquad \vdots \\ S_{2t} &= M_1 P_1^{2t} + M_2 P_2^{2t} + \cdots + M_v P_v^{2t} &= \textstyle\sum_{i=1}^{v} M_i P_i^{2t} \end{aligned} \qquad (4.98)$$

This set of syndrome equations can be expressed in the following matrix form:

$$
\begin{bmatrix} S_1 \\ S_2 \\ \vdots \\ S_{2t} \end{bmatrix}
=
\begin{bmatrix}
P_1 & P_2 & \cdots & P_v \\
P_1^2 & P_2^2 & \cdots & P_v^2 \\
\vdots & & & \vdots \\
P_1^{2t} & P_2^{2t} & \cdots & P_v^{2t}
\end{bmatrix}
\begin{bmatrix} M_1 \\ M_2 \\ \vdots \\ M_v \end{bmatrix} .
\tag{4.99}
$$

The syndromes $S_1, S_2, \ldots, S_{2t}$ are computed from the received polynomial. They are constants in the set of simultaneous non-linear equations in which the error positions $P_1, P_2, \ldots, P_v$ and their magnitudes $M_1, M_2, \ldots, M_v$ are unknowns.

We have to find $2 \cdot v$ unknowns using $2 \cdot t$ syndromes by evaluating a set of $2t$ non-linear equations which is difficult to solve. Instead, we now describe a method in Section 4.4.5.2 used by Peterson [38] for direct solution of these non-linear equations.

4.4.5.2   Peterson-Gorenstein-Zierler Decoding

Here we use Peterson's method [38] for converting the syndrome equations into linear equations from which the error positions can be computed. Let us define an *error-locator polynomial* to be the polynomial with zeros at the inverse error positions P_i^{-1} for $i = 1, 2, \ldots, v$:

$$
L(z) = (1 - zP_1)(1 - zP_2) \cdots (1 - zP_v)
\tag{4.100}
$$

and upon multiplying out the product terms we obtain,

$$
L(z) = L_v z^v + \cdots + L_1 z + 1 .
\tag{4.101}
$$

If the coefficients of Equation 4.101 are found, we can find the zeros of $L(z)$ to obtain the error positions. Therefore we concentrate on finding the values of $L_1, L_2, \ldots, L_v$ from the given syndromes.

Let us now multiply both sides of Equation 4.101 by $M_i P_i^{j+v}$ and substitute $z = P_i^{-1}$ to give

$$
M_i P_i^{j+v} L(P_i^{-1}) = M_i P_i^{j+v}(L_v P_i^{-v} + \cdots + L_1 P_i^{-1} + 1) .
\tag{4.102}
$$

As P_i^{-1} is a root of the error-locator polynomial, $L(P_i^{-1}) = 0$, and hence,

$$
0 = M_i(L_v P_i^j + \cdots + L_1 P_i^{j+v-1} + P_i^{j+v}) .
\tag{4.103}
$$

The equation is valid for each i, and on summing over $i = 1, 2, \ldots, v$ for a given j, we have

$$
\sum_{i=1}^{v} M_i(L_v P_i^j + \cdots + L_1 P_i^{j+v-1} + P_i^{j+v}) = 0
$$

or

$$L_v \sum_{i=1}^{v} M_i P_i^j + \cdots + L_1 \sum_{i=1}^{v} M_i P_i^{j+v-1} + \sum_{i=1}^{v} M_i P_i^{j+v} = 0. \qquad (4.104)$$

The summation of each term in Equation 4.104 is recognised as a syndrome in Equation 4.98, and therefore

$$L_v S_j + \cdots + L_1 S_{j+v-1} + S_{j+v} = 0 . \qquad (4.105)$$

The syndromes in Equation 4.105 are all described in Equation 4.98 and they are written as $S_1, S_2, \ldots, S_{2t}$. As $v \leq t$, the subscript of S_{j+v} suggests that the value of j is in the interval $1 \leq j \leq v$, and Equation 4.105 is for a particular value of j, whence,

$$L_v S_j + \cdots + L_2 S_{j+v-2} + L_1 S_{j+v-1} = -S_{j+v} \qquad j = 1, \ldots, v. \quad (4.106)$$

In matrix form the coefficients $L_1, L_2, \ldots, L_v$ of the error-locator polynomial $L(z)$ and the syndromes $S_1, S_2, \ldots, S_{2t}$ are related by

$$\begin{bmatrix} S_1 & S_2 & S_3 & \ldots & S_{v-1} & S_v \\ S_2 & S_3 & S_4 & \ldots & S_v & S_{v+1} \\ S_3 & S_4 & S_5 & \ldots & S_{v+1} & S_{v+2} \\ \vdots & & & & \vdots & \\ S_v & S_{v+1} & S_{v+2} & \ldots & S_{2v-2} & S_{2v-1} \end{bmatrix} \begin{bmatrix} L_v \\ L_{v-1} \\ L_{v-2} \\ \vdots \\ L_1 \end{bmatrix} = \begin{bmatrix} -S_{v+1} \\ -S_{v+2} \\ -S_{v+3} \\ \vdots \\ -S_{2v} \end{bmatrix} ,$$

$$(4.107)$$

which can also be written as

$$\vec{\vec{S}} \vec{L} = \vec{S} . \qquad (4.108)$$

The coefficients of the error-locator polynomial $L(z)$ can be computed by inverting the matrix $\vec{\vec{S}}$, namely,

$$\vec{L} = \vec{\vec{S}}^{-1} \vec{S} \qquad (4.109)$$

provided that the matrix is non-singular. The structure of the matrix $\vec{\vec{S}}$ is recognised as a Vandermonde matrix [8] which is non-singular if its dimension corresponds to the number of unknowns. Hence, $\vec{\vec{S}}$ can be inverted if its dimension is $v \times v$, but it is singular and cannot be inverted if its dimension is greater than v [9, 39], where v is the actual number of errors that occurred. This theorem provides the basis of determining the actual number of errors v and $L_1, L_2, \ldots, L_v$.

Let us now summarise the method of obtaining the error positions and subsequently their error magnitudes as a series of steps.

1. Obtain a set of $2t$ syndromes $S_1, S_2, \ldots, S_{2t}$, by substituting the roots of the generator into the received polynomial $r(z)$.

1) We have to determine v in order to invert $\vec{S}$ for the solution of Equation 4.109. We first assign $v = t$, as t is the maximum number of correctable errors.

2) The $(v \times v)$ matrix $\vec{S}$ is formulated from the syndromes and its determinant, $\det(\vec{S})$ is computed.

3) If $\det(\vec{S}) = 0$, we cannot invert $\vec{S}$. Therefore v is reduced by 1 and the procedure reverts to Step 3. However, if $\det(\vec{S}) \neq 0$, $\vec{S}$ can be inverted and we proceed with the matrix inversion to obtain $\vec{S}^{-1}$.

4) The coefficients $L_1, L_2, \ldots, L_v$ of the error-locator polynomial $L(z)$ are evaluated from Equation 4.109, i.e., $\vec{L} = \vec{S}^{-1} \vec{S}$.

5) As we define $L(z)$ to have zeros at the inverse error positions P_1^{-1}, P_2^{-1}, ..., P_v^{-1}, these are the elements of $GF(q^m)$. There is usually only a moderate number of field elements and the simplest way to find the zeros of $L(z)$ is by substituting every field element into $L(z)$ in turn. The element α^i is the inverse of the error position if $L(\alpha^i) = 0$, and hence the error symbol is at position $(\alpha^i)^{-1}$. This process of finding the zeros by trial and error is known as a *Chien search*.

6) For binary BCH codes, the symbol of the code is a single bit and its value at the error positions is toggled for error correction. For nonbinary codes, the error magnitudes $M_1, M_2, \ldots, M_v$ are determined from Equation 4.99. Although Equation 4.98 represents a set of $2t$ equations, the actual number of errors v is now known and the v unknown error magnitudes can be solved by a set of v equations. The error magnitudes $M_1, M_2, \ldots, M_v$ are given by

$$\begin{bmatrix} M_1 \\ M_2 \\ \vdots \\ M_v \end{bmatrix} = \begin{bmatrix} P_1 & P_2 & \cdots & P_v \\ P_1^2 & P_2^2 & \cdots & P_v^2 \\ \vdots & & & \vdots \\ P_1^v & P_2^v & \cdots & P_v^v \end{bmatrix}^{-1} \begin{bmatrix} S_1 \\ S_2 \\ \vdots \\ S_v \end{bmatrix} . \tag{4.110}$$

We have located the positions and magnitudes of the errors. This enables the error polynomial $e(z)$ to be formulated, and when it is added to the received polynomial $r(z)$ the original codeword $c(z)$ is recovered.

Let us now illustrate the decoding algorithm by considering a triple error correcting binary BCH(15, 5, 3) code. Suppose the information polynomial

$$i(z) = z^1$$

is systematically encoded by the generator $g(z)$ given by Equation 4.81 to produce a codeword,

$$c(z) = z^{11} + z^9 + z^6 + z^5 + z^3 + z^2 + z .$$

If three or less errors occur, the received word $r(z)$ can be corrected. Suppose that double error symbols occur causing the received polynomial to be,

$$r(z) = z^{13} + z^{11} + z^9 + z^5 + z^3 + z^2 + z .$$

To rectify the errors, we proceed through the steps of the decoding algorithm. First we compute the syndromes over $GF(16)$ with the aid of Table 4.10 and Table 4.11. The six syndromes are

$$\begin{array}{lll}
S_1 = r(\alpha^1) = \alpha^0 & S_2 = r(\alpha^2) = \alpha^0 & S_3 = r(\alpha^3) = \alpha^1 \\
S_4 = r(\alpha^4) = \alpha^0 & S_5 = r(\alpha^5) = \alpha^{10} & S_6 = r(\alpha^6) = \alpha^2 .
\end{array}$$
$$(4.111)$$

Assigning $v = t = 3$, $\vec{S}$ is formulated with a dimension of (3×3),

$$\vec{S} = \begin{bmatrix} S_1 & S_2 & S_3 \\ S_2 & S_3 & S_4 \\ S_3 & S_4 & S_5 \end{bmatrix} = \begin{bmatrix} \alpha^0 & \alpha^0 & \alpha^1 \\ \alpha^0 & \alpha^1 & \alpha^0 \\ \alpha^1 & \alpha^0 & \alpha^{10} \end{bmatrix} .$$

The determinant of $\vec{S}$ is zero and therefore cannot be inverted. Next we assign $v = 2$, and formulate $\vec{S}$ again with a dimension of (2×2),

$$\vec{S} = \begin{bmatrix} S_1 & S_2 \\ S_2 & S_3 \end{bmatrix} = \begin{bmatrix} \alpha^0 & \alpha^0 \\ \alpha^0 & \alpha^1 \end{bmatrix} .$$

The $\det(\vec{S}) = \alpha^4$ and hence double errors are recognised. The inverse of $\vec{S}$ is calculated as

$$\vec{S}^{-1} = \begin{bmatrix} \alpha^{12} & \alpha^{11} \\ \alpha^{11} & \alpha^{11} \end{bmatrix} .$$

From Equation 4.109, we can evaluate the coefficients L_1, L_2 of $L(z)$,

$$\begin{bmatrix} L_2 \\ L_1 \end{bmatrix} = \begin{bmatrix} \alpha^{12} & \alpha^{11} \\ \alpha^{11} & \alpha^{11} \end{bmatrix} \begin{bmatrix} \alpha^1 \\ \alpha^0 \end{bmatrix} = \begin{bmatrix} \alpha^4 \\ \alpha^0 \end{bmatrix}$$

to give the error-locator polynomial,

$$L(z) = L_2 z^2 + L_1 z + 1 = \alpha^4 z^2 + \alpha^0 z + 1 .$$

By the Chien search, we find the zeros of the error-locator polynomial by substituting each field element of $GF(16)$ into $L(z)$,

$$
\begin{array}{llll}
L(\alpha^0) = \alpha^4 & L(\alpha^1) = \alpha^{12} & L(\alpha^2) = 0 & L(\alpha^3) = \alpha^{11} \\
L(\alpha^4) = \alpha^{13} & L(\alpha^5) = \alpha^{11} & L(\alpha^6) = \alpha^{12} & L(\alpha^7) = \alpha^1 \\
L(\alpha^8) = \alpha^1 & L(\alpha^9) = 0 & L(\alpha^{10}) = \alpha^6 & L(\alpha^{11}) = \alpha^0 \\
L(\alpha^{12}) = \alpha^4 & L(\alpha^{13}) = \alpha^{13} & L(\alpha^{14}) = \alpha^6 & .
\end{array}
$$

The elements α^2 and α^9 are the zeros of $L(z)$, and the inverses of these zeros are the error positions α^{13} and α^6, respectively. The error magnitudes of binary codes at these positions must be 1. Armed with the knowledge of the error positions and error magnitudes we formulate the error polynomial,

$$
e(z) = z^{13} + z^6 .
$$

The original codeword $c(z)$ is then recovered by adding the error polynomial $e(z)$ to the received word $r(z)$:

$$
\begin{aligned}
c(z) &= r(z) + e(z) \\
&= (z^{13} + z^{11} + z^9 + z^5 + z^3 + z^2 + z) + (z^{13} + z^6) \\
&= z^{11} + z^9 + z^6 + z^5 + z^3 + z^2 + z .
\end{aligned}
$$

As another example, consider a triple error correcting non-binary $RS(15,9,3)$ code over $GF(16)$. Suppose the information polynomial $i(z)$ consisting of all-zero symbols,

$$
i(z) = 0
$$

is systematically encoded by the generator polynomial

$$
\begin{aligned}
g(z) &= (z - \alpha)(z - \alpha^2)(z - \alpha^3)(z - \alpha^4)(z - \alpha^5)(z - \alpha^6) \\
&= z^6 + \alpha^{10} z^5 + \alpha^{14} z^4 + \alpha^4 z^3 + \alpha^6 z^2 + \alpha^9 z + \alpha^6 .
\end{aligned}
$$

The encoded polynomial $c(z)$ is also an all-zero codeword,

$$
c(z) = 0 .
$$

Suppose three error symbols occur such that the received polynomial $r(z)$ is

$$
r(z) = \alpha^2 z^{14} + \alpha^{13} z^8 + \alpha^6 z^2 .
$$

In the case of the all-zero encoded polynomial $c(z)$, all the non-zero symbols in the received polynomial $r(z)$ are the error symbols, and therefore $e(z) = r(z)$. To proceed with the decoding steps, we evaluate the six syndromes over $GF(16)$ by using Tables 4.10 and 4.11:

$$S_1 = r(\alpha^1) = \alpha^7 \quad S_2 = r(\alpha^2) = \alpha^{12} \quad S_3 = r(\alpha^3) = \alpha^{13}$$
$$S_4 = r(\alpha^4) = \alpha^8 \quad S_5 = r(\alpha^5) = \alpha^3 \quad S_6 = r(\alpha^6) = \alpha^2 \ .$$

Next we assign $v = t = 3$, formulate $\vec{S}$ with a dimension of (3×3):

$$\vec{S} = \begin{bmatrix} S_1 & S_2 & S_3 \\ S_2 & S_3 & S_4 \\ S_3 & S_4 & S_5 \end{bmatrix} = \begin{bmatrix} \alpha^7 & \alpha^{12} & \alpha^{13} \\ \alpha^{12} & \alpha^{13} & \alpha^8 \\ \alpha^{13} & \alpha^8 & \alpha^3 \end{bmatrix} \ .$$

As $\det(\vec{S}) = \alpha^8$, triple errors are recognised. The inverse of $\vec{S}$ is

$$\vec{S}^{-1} = \begin{bmatrix} 0 & \alpha^5 & \alpha^{10} \\ \alpha^5 & \alpha^6 & \alpha^{12} \\ \alpha^{10} & \alpha^{12} & \alpha^{13} \end{bmatrix} \ .$$

The coefficients L_1, L_2 and L_3 of $L(z)$ can be found by using Equation 4.109,

$$\begin{bmatrix} L_3 \\ L_2 \\ L_1 \end{bmatrix} = \begin{bmatrix} 0 & \alpha^5 & \alpha^{10} \\ \alpha^5 & \alpha^6 & \alpha^{12} \\ \alpha^{10} & \alpha^{12} & \alpha^{13} \end{bmatrix} \begin{bmatrix} \alpha^8 \\ \alpha^3 \\ \alpha^2 \end{bmatrix} = \begin{bmatrix} \alpha^9 \\ \alpha^{11} \\ \alpha^3 \end{bmatrix}$$

to give the error-locator polynomial,

$$L(z) = L_3 z^3 + L_2 z^2 + L_1 z + 1 = \alpha^9 z^3 + \alpha^{11} z^2 + \alpha^3 z + 1 \ .$$

By means of the Chien search we find the zeros of the error-locator polynomial on substituting each field element of $GF(16)$ into $L(z)$,

$$\begin{array}{llll}
L(\alpha^0) = \alpha^{13} & L(\alpha^1) = 0 & L(\alpha^2) = \alpha^3 & L(\alpha^3) = \alpha^0 \\
L(\alpha^4) = \alpha^8 & L(\alpha^5) = \alpha^1 & L(\alpha^6) = \alpha^0 & L(\alpha^7) = 0 \\
L(\alpha^8) = \alpha^3 & L(\alpha^9) = \alpha^7 & L(\alpha^{10}) = \alpha^2 & L(\alpha^{11}) = \alpha^{12} \\
L(\alpha^{12}) = \alpha^{10} & L(\alpha^{13}) = 0 & L(\alpha^{14}) = \alpha^4 &
\end{array}$$

The elements α^1, α^7 and α^{13} are the zeros of $L(z)$, and therefore the inverses of the zeros are the error positions α^{14}, α^8 and α^2, respectively. The error magnitudes of non-binary codes at these positions have to be determined. From Equation 4.110, the error magnitudes M_1, M_2 and M_3 are

$$\begin{bmatrix} M_1 \\ M_2 \\ M_3 \end{bmatrix} = \begin{bmatrix} P_1 & P_2 & P_3 \\ P_1^2 & P_2^2 & P_3^2 \\ P_1^3 & P_2^3 & P_3^3 \end{bmatrix}^{-1} \begin{bmatrix} S_1 \\ S_2 \\ S_3 \end{bmatrix}$$

$$= \begin{bmatrix} \alpha^2 & \alpha^8 & \alpha^{14} \\ (\alpha^2)^2 & (\alpha^8)^2 & (\alpha^{14})^2 \\ (\alpha^2)^3 & (\alpha^8)^3 & (\alpha^{14})^3 \end{bmatrix}^{-1} \begin{bmatrix} \alpha^7 \\ \alpha^{12} \\ \alpha^{13} \end{bmatrix}$$

$$= \begin{bmatrix} \alpha^7 & \alpha^6 & \alpha^0 \\ \alpha^2 & \alpha^{14} & \alpha^1 \\ \alpha^7 & \alpha^{12} & \alpha^{12} \end{bmatrix} \begin{bmatrix} \alpha^7 \\ \alpha^{12} \\ \alpha^{13} \end{bmatrix} = \begin{bmatrix} \alpha^6 \\ \alpha^{13} \\ \alpha^2 \end{bmatrix} .$$

The error positions and error magnitudes provide the error polynomial,

$$e(z) = M_3 P_3 + M_2 P_2 + M_1 P_1 = \alpha^2 z^{14} + \alpha^{13} z^8 + \alpha^6 z^2 .$$

As before, the original codeword $c(z)$ is then recovered by adding the error polynomial $e(z)$ to the received word $r(z)$:

$$\begin{aligned} c(z) &= r(z) + e(z) \\ &= (\alpha^2 z^{14} + \alpha^{13} z^8 + \alpha^6 z^2) + (\alpha^2 z^{14} + \alpha^{13} z^8 + \alpha^6 z^2) = 0 . \end{aligned}$$

Hence the received word is corrected and the all-zero codeword $c(z)$ is recovered.

4.4.5.3   Berlekamp-Massey Algorithm

In Section 4.4.5.2, we showed that the error positions can be computed by direct matrix inversion. For v error symbols in a codeword, the number of computations required to invert a v by v matrix is proportional to v^3. Consequently the matrix inversion approach is only practical for moderate values of v. In practice, long codes with large t are usually preferable as they have the capability of correcting a large number of errors. For these codes, Berlekamp and Massey introduced an efficient method of decoding without recourse to matrix inversion. Their method of inverting the matrix is inspired by the analogy of designing a linear-feedback shift register.

The first row of the matrix equations in Equation 4.107 expresses S_{v+1} in terms of S_1, S_2, ..., S_v, the second row expresses S_{v+2} in terms of S_2, S_3, ..., S_{v+1}, and so on. If the vector $\vec{L}$ in the equation is known, the recursive expression of the syndromes in terms of themselves suggests the equation of an autoregressive filter,

$$S_j = -\sum_{i=1}^{v} L_i S_{j-i}, \qquad j = v+1,\dots,2v . \tag{4.112}$$

This filter is implemented by a linear-feedback shift register as shown in Figure 4.44. The taps are represented by a polynomial $L(z)$ to give a sequence of syndromes. Now, the problem of finding L_1, L_2, ..., L_v in Equation 4.107 is converted to a filter design problem by evaluating its tapping coefficients L_i, $i = 1,\dots,v$ given in Equation 4.112 to produce the

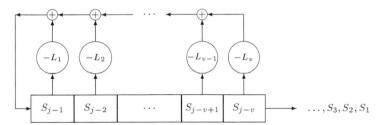

Load the shift register with $S_v, \ldots, S_2, S_1$

Figure 4.44: Autoregressive filter implemented by linear-feedback shift register.

sequence of known syndromes S_j.

There are many linear-feedback shift registers with different lengths and different connection polynomials that are capable of generating the known sequence of syndromes. The length of the shift register may be greater than the degree of $L(z)$ such that some rightmost shift register stages may not be connected. However, there are two criteria for designing the shift register. Firstly, for the syndromes of a correctable error pattern, the solution of $\vec{L}$ expressed in Equation 4.107 has a dimension of v and is unique because only the v by v matrix is invertible. Hence, the polynomial $L(z)$ is of degree v. From Figure 4.44, we note that the register is designed to contain a row of the v by v matrix $\vec{\vec{S}}$, and therefore the register length is v. As a consequence, the connection polynomial $L(z)$ has the same length v as the linear-feedback shift register and the rightmost tap of the register is non-zero. Secondly, if the dimension of the matrix $\vec{\vec{S}}$ is smaller than v, $\vec{\vec{S}}$ cannot be inverted and $\vec{L}$ has no solution. This means that the smallest dimension of $\vec{L}$ is v. From the above, we choose the shift register having the shortest length such that the polynomial $L(z)$ has the smallest degree.

The problem becomes one of designing the filter with the shortest length to predict a given sequence of syndromes. If the register length corresponds to the degree of the connection polynomial $L(z)$, the coefficients of $L(z)$ are the solution of the matrix equations expressed in Equation 4.107. Let us now illustrate the design procedure by deriving a minimum length linear-feedback shift register to produce the binary sequence 11101011000110 shown in Figure 4.45. Our approach is inductive, progressively modifying the register at each step. Let $LFSR_i$ denote the linear-feedback shift register design at the ith iteration. Also, l_i and $L^{(i)}(z)$ denote the register length and the connection polynomial, respectively.

Initially, at step $i = 1$, we use the simplest linear-feedback shift register $LFSR_1$ consisting of a single stage with a direct feedback from its output

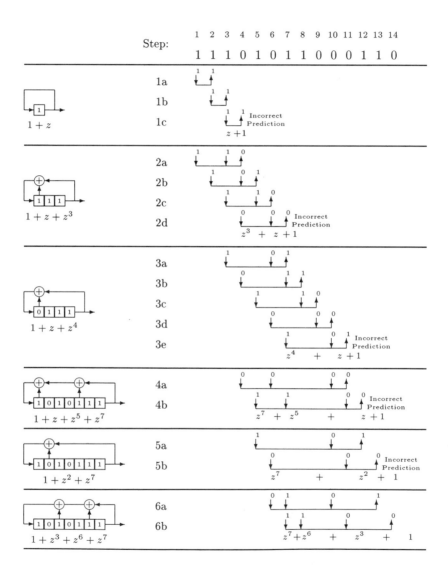

Figure 4.45: Constructing a minimum length shift register by the Berlekamp-Massey algorithm to generate the binary sequence.

to its input giving $l_1 = 1$ and $L^{(1)}(z) = z + 1$. The register design $LFSR_1$ is initialised by the first bit of the sequence and generates the first three bits but fails on the fourth. This failure prompts us to modify $LFSR_1$ either in its length or in its tapping positions. It is obvious that a length-2 shift register cannot combine two ones (first and second bits) to produce a one (third bit) and then combine two ones (second and third bits) on the next shift to produce a zero (fourth bit). This suggests that the length of the register must be increased to 3 with an even number of feedback connections.

At step $i = 2$ in Figure 4.45, we use a length-3 shift register $LFSR_2$ (i.e., $l_2 = 3$) with connection polynomial $L^{(2)}(z) = z^3 + z + 1$. It can generate the 4th, 5th and 6th bits correctly, but fails on the 7th bit. At this point, we modify $LFSR_2$ such that the new register $LFSR_3$ continues to generate from the 7th bit onward without reproducing the previous correct bits. Suppose l_3 is the length of the shift register at the next step $i = 3$, the register is modified in a way that the correction values added to the $(l_3 + 1)$th, $(l_3 + 2)$th, ..., (7-1)th bits are zero, but the value added to the 7th bit is one so as to change the generated bit from 0 to 1. In the example shown in Figure 4.45, $l_3 = 4$ and the correction values to the 5th and 6th bits are zero, but to the 7-th bit it is a one.

We observe that the linear combination (bitwise exclusive-OR for binary codes) of the generated bit from the shift register and the next bit to be predicted in the sequence is always zero if the prediction is correct, and is one if the prediction is wrong. For example, at steps 1a and 1b in Figure 4.45, the generated bit is 1 and the next bit in the sequence is also 1. The prediction of either steps is correct and therefore the linear combination of these two bits is zero. However, at step 1c, the generated bit is 1 and the fourth bit in the sequence is 0. The prediction is wrong and their linear combination is 1. The result of the non-zero linear combination is essential to generate the correction value about to be added to the output of the $LFSR_2$ design to compensate for its failure to generate the right prediction when a new $LFSR_3$ is formed.

As the bit generated from a shift register is the linear combination of its stages defined by the connection polynomial, the $LFSR_2$'s output can be modified by adding the appropriately shifted linear combination defined by the connection polynomial $L^{(1)}(z)$ of the $LFSR_1$ to form a new $LFSR_3$. The length of the register is increased to $l_3 = 4$ and the connection polynomial becomes

$$\begin{aligned} L^{(3)}(z) &= L^{(1)}(z)z^3 + L^{(2)}(z) \\ &= (z+1)z^3 + (z^3 + z + 1) = z^4 + z + 1 \ . \end{aligned}$$

Now, at step 3, the linear combinations due to the tapping defined by the connection polynomial of the $LFSR_1$ at steps 1a and 1b are zero and the next symbol is predicted with zero error. Shifted by three positions

according to the multiplication by z^3 this contributes a zero correction to
the prediction by $LFSR_3$ at the fifth and sixth bits. However, when
the $LFSR_3$ predicts the seventh bit, the correction value (observed from
the $LFSR_1$ at step 1c) is one which changes the previous prediction by
the $LFSR_2$ to the current prediction by the $LFSR_3$, i.e., from 0 to 1.
Hereafter, the $LFSR_3$ produces the 8th, 9th and 10th bits in the sequence
properly, but fails on the 11th. At this point, the $LFSR_3$ is again modified
for step 4.

To proceed with step 4, the $LFSR_3$ can be modified by adding the
appropriately shifted non-zero output of either the $LFSR_1$ or the $LFSR_2$
to form a new $LFSR_4$. The length of the $LFSR_4$ will be either 8 or
7, according to the positioning shifts of z^7 or z^4, respectively, and their
corresponding connection polynomials are

$$
\begin{aligned}
L^{(4)}(z) &= L^{(1)}(z)z^7 + L^{(3)}(z) \\
&= (z+1)z^7 + (z^4 + z + 1) = z^8 + z^7 + z^4 + z + 1
\end{aligned}
$$

or

$$
\begin{aligned}
L^{(4)}(z) &= L^{(2)}(z)z^4 + L^{(3)}(z) \\
&= (z^3 + z + 1)z^4 + (z^4 + z + 1) = z^7 + z^5 + z + 1 \ .
\end{aligned}
$$

As the criterion of designing the linear-feedback shift register is that it has
a minimum length, we choose the $LFSR_2$ to modify the $LFSR_3$. The
$LFSR_4$ generates the 11th bit in the sequence, but fails on the 12th.

At step 5 in Figure 4.45, the $LFSR_5$ is basically constructed from
$LFSR_4$ with modification by $LFSR_3$. The length of $LFSR_5$ is still 7
and the connection polynomial is

$$
\begin{aligned}
L^{(5)}(z) &= L^{(3)}(z)z + L^{(4)}(z) \\
&= (z^4 + z + 1)z + (z^7 + z^5 + z + 1) = z^7 + z^2 + 1 \ .
\end{aligned}
$$

The register now predicts correctly until it reaches bit 13.

At this point, the $LFSR_5$ can be modified by one of the previous shift
registers, $LFSR_1$, $LFSR_2$, $LFSR_3$ and $LFSR_4$. But we choose the one to
form the minimum length shift register for step 6. That is $LFSR_3$ which
forms a new $LFSR_6$ of length 7. The connection polynomial is

$$
\begin{aligned}
L^{(6)}(z) &= L^{(3)}(z)z^2 + L^{(5)}(z) \\
&= (z^4 + z + 1)z^2 + (z^7 + z^2 + 1) = z^7 + z^6 + z^3 + 1 \ .
\end{aligned}
$$

The new $LFSR_6$ is able to predict the 13th bit and also the last bit 14.
Finally, a length-7 $LFSR_6$ is the correct register. If it is initialised with the
first 7 bits of the sequence, it will then generate the subsequent bits. If the
bits in the sequence are the syndromes, the connection polynomial $LFSR_6$

is the error-locator polynomial, and the coefficients of the polynomial constitute the solution for $\vec{L}$ in Equation 4.107. We note that the length of the $LFSR_6$ corresponds to the degree of the connection polynomial $L^{(6)}(z)$ that satisfies the criteria of filter design.

Let us now summarise the design procedure of the last example and also generalise it for the design of non-binary codes. During the process of formulating the shift register, we always determine the register length l_i and the connection polynomial $L^{(i)}(z)$. A correction term $C^{(i)}(z)$ is updated for use at the next step whenever the linear-feedback shift register fails to predict the next symbol correctly. At each step, the shift register predicts the next symbol in the sequence of syndromes. If the prediction is correct, the connection polynomial remains intact and the correction term is multiplied by z. However, if the prediction is wrong, the connection polynomial is modified by adding the correction term. As all the previous connection polynomials are candidates for the correction term, we choose the one to give the minimum register length.

If the length of the register has increased after modification, the previous connection polynomial is multiplied by z and becomes the current correction term. For example, the prediction of the shift register at step 3e in Figure 4.45 is wrong. The connection polynomial $L^{(3)}(z)$ is therefore modified by adding the correction term $L^{(2)}(z)z^4$ to form $L^{(4)}(z)$ for the new register at the next step 4. The length of the register has extended from 4 to 7 stages, and the previous connection polynomial $L^{(3)}(z)$ is multiplied by z and is maintained as the correction term for the next step 4b. However, if the length of the register remains the same after the modification, the current correction term is multiplied by z and is kept for the next step. For example, the correction term at step 4b in Figure 4.45 is $L^{(3)}(z)z$. The next prediction of the shift register is wrong, and the connection polynomial $L^{(4)}(z)$ is modified by adding the correction term $L^{(3)}(z)z$ to form the $L^{(5)}(z)$ for step 5. The length of the new register remains the same with 7 stages, and the correction term is maintained and updated by multiplying by z to become $L^{(3)}(z)z^2$ for step 5b.

In the example, the bit sequence is binary. The prediction of the shift register is either 0 or 1. If the prediction is wrong, the discrepancy is always one. However, for the sequence of syndromes defined on $GF(q)$, where $q > 2$, the discrepancy can be any value greater than zero and less than q. In this case, the connection polynomial of the shift register produces a wrong prediction and has a discrepancy d with the next symbol in the sequence. If the connection polynomial is selected as the correction term it is normalised such that the discrepancy is equal to one. When the shift register is modified at the subsequent step, this normalised term is then multiplied by the value of the discrepancy for the current prediction.

From our deliberations, we see that the Berlekamp-Massey algorithm is an iterative process for predicting $2t$ syndroms. Each iteration predicts one syndrom and the process lasts for $2t$ iterations. Let $L^{(i)}(z)$ and $C^{(i)}(z)$

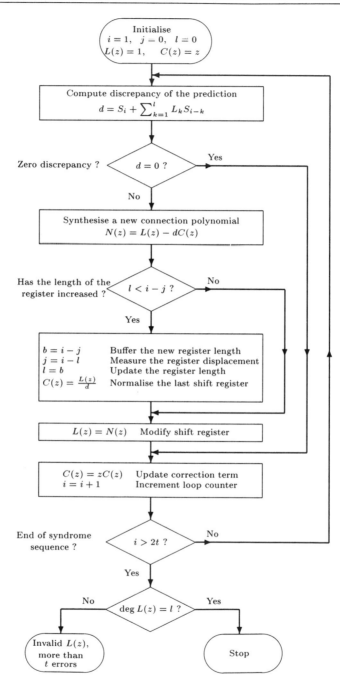

Figure 4.46: Computer flow diagram of the Berlekamp-Massey algorithm.

be the connection polynomial and the correction polynomial, respectively, at the ith iteration, $l^{(i)}$ be the length of the linear-feedback shift register, $j^{(i)}$ be the displacement of the correction polynomial from the beginning of the syndrome sequence (i.e., the location of the oldest symbol of the polynomial in the sequence), and the discrepancy in the prediction is $d^{(i)}$.

1. Initially, the values are set as

$$i = 1, \quad j^{(0)} = 0, \quad l^{(0)} = 0, \quad L^{(0)}(z) = 1 \ \text{and} \ C^{(0)}(z) = z .$$

1) The discrepancy of the prediction is evaluated as

$$d^{(i)} = S_i + \sum_{k=1}^{l^{(i)}} L_k^{(i)} S_{i-k} .$$

2) If the discrepancy is zero, the prediction at the current iteration i is correct and the correction polynomial is updated by

$$C^{(i+1)}(z) = z C^{(i)}(z) .$$

The register is left unchanged so that

$$L^{(i+1)}(z) = L^{(i)}(z), \quad l^{(i+1)} = l^{(i)} \ \text{and} \ j^{(i+1)} = j^{(i)} .$$

The register then continues to predict the next symbol and goes back to step 1 to calculate the discrepancy.

3) However, if the discrepancy is non-zero, the prediction is wrong. The connection polynomial of the shift register is modified according to the following expression

$$L^{(i+1)}(z) = L^{(i)}(z) - d^{(i)} C^{(i)}(z)$$

and the length of the new register becomes

$$l^{(i+1)} = \max\left(l^{(i)}, i - j^{(i)}\right) .$$

(a) If the length of the new register has not increased, i.e., $l^{(i+1)} = l^{(i)}$, the last correction polynomial is retained and is updated as

$$C^{(i+1)}(z) = z C^{(i)}(z)$$

and its displacement remains the same,

$$j^{(i+1)} = j^{(i)} .$$

i	$d^{(i)}$	$L^{(i)}(z)$	$C^{(i)}(z)$	$l^{(i)}$
1	α^0	$1 + \alpha^0 z$	z	1
2	0	$1 + \alpha^0 z$	z^2	1
3	α^4	$1 + \alpha^0 z + \alpha^4 z^2$	$\alpha^{11} z + \alpha^{11} z^2$	2
4	0	$1 + \alpha^0 z + \alpha^4 z^2$	$\alpha^{11} z^2 + \alpha^{11} z^3$	2
5	0	$1 + \alpha^0 z + \alpha^4 z^2$	$\alpha^{11} z^3 + \alpha^{11} z^4$	2
6	0	$1 + \alpha^0 z + \alpha^4 z^2$	$\alpha^{11} z^4 + \alpha^{11} z^5$	2

Table 4.15: Computing the $L(z)$ polynomial of the BCH(15, 5, 3) code using the Berlekamp-Massey algorithm.

(b) If the length of the new register has increased, i.e., $l^{(i+1)} > l^{(i)}$, the last connection polynomial is normalised and becomes the new correction polynomial,

$$C^{(i+1)}(z) = z \frac{L^{(i)}(z)}{d^{(i)}}$$

and its displacement is,

$$j^{(i+1)} = i - l^{(i)} .$$

The new register then predicts the next symbol and repeats step 1 to calculate the discrepancy of the prediction.

The algorithm is depicted in the form of the computer flow diagram illustrated in Figure 4.46. Notice that the calculation of the discrepancy $d^{(i)}$ requires not more than t multiplications for each iteration, and also that the linear-feedback shift register and the correction polynomial update require at most $2t$ multiplications per iteration. The procedure iterates $2t$ times and therefore the algorithm requires at most $(t + 2t)2t = 6t^2$ multiplications. By constrast, the direct matrix inversion method derived from the Peterson-Gorenstein-Zierler method requires on the order of t^3 multiplications. The low complexity of the Berlekamp-Massey algorithm makes it preferable to the direct matrix inversion approach suggested by Peterson, if $t \geq 6$.

Let us now consider the previous examples in Section 4.4.5.2 to illustrate the procedure of the Berlekamp-Massey algorithm. The first example is the triple error correcting BCH(15, 5, 3) code whose syndromes are given in Equation 4.111, while the procedures for determining the error-locator polynomial by the Berlekamp-Massey algorithm are shown in Table 4.15. The resulting $L(z)$ polynomial of the BCH(15, 5, 3) code obtained by the Berlekamp-Massey algorithm can be verified by that of the Peterson method in Section 4.4.5.2. The second example is the non-binary triple error correcting RS(15, 9, 3) code over $GF(16)$. The procedure is illustrated in Table 4.16.

i	$d^{(i)}$	$L^{(i)}(z)$	$C^{(i)}(z)$	$l^{(i)}$
1	α^7	$1 + \alpha^7 z$	$\alpha^8 z$	1
2	α^5	$1 + \alpha^5 z$	$\alpha^8 z^2$	1
3	α^{14}	$1 + \alpha^5 z + \alpha^7 z^2$	$\alpha^1 z + \alpha^6 z^2$	2
4	α^{11}	$1 + \alpha^{14} z + \alpha^{12} z^2$	$\alpha^1 z^2 + \alpha^6 z^3$	2
5	α^2	$1 + \alpha^{14} z + \alpha^{10} z^2 + \alpha^8 z^3$	$\alpha^{13} z + \alpha^{12} z^2 + \alpha^{10} z^3$	3
6	α^2	$1 + \alpha^3 z + \alpha^{11} z^2 + \alpha^9 z^3$	$\alpha^{13} z^2 + \alpha^{12} z^3 + \alpha^{10} z^4$	3

Table 4.16: Computing the $L(z)$ polynomial of the RS$(15,9,3)$ code using the Berlekamp-Massey algorithm.

4.4.5.4 Forney Algorithm

The Peterson-Gorenstein-Zierler decoder described in Section 4.4.5.2, provides a method of decoding BCH codes. The method requires two v by v matrix inversions which consumes substantial computational power. The first matrix inversion evaluates the error-locator polynomial, and this can be circumvented by employing the computationally efficient Berlekamp-Massey algorithm. The second matrix inversion involves the evaluation of the error magnitudes, which can be carried out more efficiently using the Forney algorithm. In this section, we commence by deriving the Forney algorithm and then illustrate the procedure with an example.

We recall from Equation 4.100 that the error-locator polynomial $L(z)$ is defined to have zeros at the inverse error positions P_i^{-1} for $i = 1, 2, \ldots, v$, namely,

$$L(z) = \prod_{i=1}^{v} (1 - zP_i) = (1 - zP_1)(1 - zP_2) \cdots (1 - zP_v) . \qquad (4.113)$$

We now define the *syndrome polynomial $S(z)$* as

$$S(z) \stackrel{\triangle}{=} \sum_{j=1}^{2t} S_j z^j = S_1 z + S_2 z^2 + \cdots + S_{2t} z^{2t} . \qquad (4.114)$$

The coefficient S_j of the polynomial is the syndrome, and from Equation 4.98 the syndrome polynomial can also be expressed as

$$S(z) = \sum_{j=1}^{2t} \sum_{i=1}^{v} M_i P_i^j z^j . \qquad (4.115)$$

We also define the *error-evaluator polynomial $E(z)$* as the product of the syndrome polynomial $S(z)$ and the error-locator polynomial $L(z)$,

$$E(z) \stackrel{\triangle}{=} S(z)L(z) \pmod{z^{2t}} \qquad (4.116)$$

and its expanded form is given by

$$E(z) = E_{2t-1} z^{2t-1} + \cdots + E_2 z^2 + E_1 z \ . \tag{4.117}$$

Substituting Equation 4.115 and 4.113 into Equation 4.116, yields

$$
\begin{aligned}
E(z) &= \sum_{j=1}^{2t} \sum_{i=1}^{v} M_i P_i^j z^j \prod_{l=1}^{v} (1 - z P_l) \quad (\mathrm{mod} \ z^{2t}) \\
&= \sum_{i=1}^{v} M_i P_i z \sum_{j=1}^{2t} (P_i z)^{j-1} \prod_{l=1}^{v} (1 - z P_l) \quad (\mathrm{mod} \ z^{2t}) \\
&= \sum_{i=1}^{v} M_i P_i z \left[(1 - z P_i) \sum_{j=1}^{2t} (P_i z)^{j-1} \right] \prod_{l \neq i} (1 - z P_l) \quad (\mathrm{mod} \ z^{2t})
\end{aligned}
$$

By extending and simplifying the square bracketed term we get:

$$
\begin{aligned}
E(z) &= \sum_{i=1}^{v} M_i P_i z \left[1 - (P_i z)^{2t} \right] \prod_{l \neq i} (1 - z P_l) \quad (\mathrm{mod} \ z^{2t}) \\
&= \sum_{i=1}^{v} M_i P_i z \prod_{l \neq i} (1 - z P_l) \ . \tag{4.118}
\end{aligned}
$$

This error-evaluator polynomial has the following values at the inverse error positions P_l^{-1}:

$$
\begin{aligned}
E(P_l^{-1}) &= \sum_{i=1}^{v} M_i P_i P_l^{-1} \prod_{j \neq i} (1 - P_l^{-1} P_j) \\
&= M_1 P_1 P_l^{-1} \prod_{j \neq 1} (1 - P_l^{-1} P_j) + \cdots + M_l \prod_{j \neq l} (1 - P_l^{-1} P_j) \\
&\quad + \cdots + M_v P_v P_l^{-1} \prod_{j \neq v} (1 - P_l^{-1} P_j) \\
&= M_l \prod_{j \neq l} (1 - P_l^{-1} P_j) \ . \tag{4.119}
\end{aligned}
$$

Using the product rule for differentiation, we evaluate the derivative of the error-locator polynomial $L(z)$ given by Equation 4.113 as

$$L'(z) = \sum_{i=1}^{v} -P_i \prod_{j \neq i} (1 - z P_j) \tag{4.120}$$

$$
\begin{array}{r|l}
S_6z^6 \ + & S_5z^5 \ + \quad S_4z^4 \ + \quad S_3z^3 \ + \quad S_2z^2 \ + \quad S_1z \\
& \hphantom{S_5z^5 \ + \quad} L_3z^3 \ + \quad L_2z^2 \ + \quad L_1z \ + \quad 1 \\
\hline
S_6z^6 \ + & S_5z^5 \ + \quad S_4z^4 \ + \quad S_3z^3 \ + \quad S_2z^2 \ + \quad S_1z \\
L_1S_6z^7 \ + \ L_1S_5z^6 \ + & L_1S_4z^5 \ + \ L_1S_3z^4 \ + \ L_1S_2z^3 \ + \ L_1S_1z^2 \\
L_2S_6z^8 \ + \ L_2S_5z^7 \ + \ L_2S_4z^6 \ + & L_2S_3z^5 \ + \ L_2S_2z^4 \ + \ L_2S_1z^3 \\
L_3S_6z^9 \ + \ L_3S_5z^8 \ + \ L_3S_4z^7 \ + \ L_3S_3z^6 \ + & L_3S_2z^5 \ + \ L_3S_1z^4 \\
\hline
& E_5z^5 \ + \quad E_4z^4 \ + \quad E_3z^3 \ + \quad E_2z^2 \ + \quad E_1z
\end{array}
$$

Figure 4.47: Long multiplication of $S(z)$ and $L(z)$.

and at the inverse error position $z = P_l^{-1}$,

$$
\begin{aligned}
L'(P_l^{-1}) &= -\sum_{i=1}^{v} P_i \prod_{j \neq i}(1 - P_i^{-1}P_j) \\
&= -P_1 \prod_{j \neq 1}(1 - P_1^{-1}P_j) - \cdots - P_l \prod_{j \neq l}(1 - P_l^{-1}P_j) \\
&\quad - \cdots - P_v \prod_{j \neq v}(1 - P_v^{-1}P_j) \\
&= -P_l \prod_{j \neq l}(1 - P_l^{-1}P_j) .
\end{aligned} \tag{4.121}
$$

From Equations 4.119 and 4.121,

$$
\frac{E(P_l^{-1})}{L'(P_l^{-1})} = \frac{M_l}{-P_l} . \tag{4.122}
$$

Rearranging Equation 4.122, we find the magnitude of the error symbol at position P_l to be,

$$
M_l = \frac{-E(P_l^{-1})}{P_l^{-1}L'(P_l^{-1})} = \frac{E(P_l^{-1})}{\prod_{j \neq l}(1 - P_l^{-1}P_j)} . \tag{4.123}
$$

As the error-evaluator polynomial $E(z)$ is defined to be the product of the error-locator polynomial $L(z)$ and the syndrome polynomial $S(z)$, the coefficients of $E(z)$ can be expressed as the product of L_i and S_i. Consider the example of a triple error correcting code having a degree-5 error-evaluator polynomial whose coefficients are given in Figure 4.47:

$$
E(z) = E_5z^5 + E_4z^4 + E_3z^3 + E_2z^2 + E_1z \tag{4.124}
$$

Similarly, for a t error correcting code in general, the error-evaluator polynomial $E(z)$ is formulated with coefficients

$$
E_j = S_j + \sum_{k=1}^{t} S_{j-k}L_k \qquad j = 1, \ldots, (2t - 1) \tag{4.125}
$$

where $S_j = 0$ for $j \leq 0$.

Let us now illustrate the Forney algorithm by evaluating the error magnitudes of the non-binary triple error correcting RS$(15,9,3)$ code over $GF(16)$, as previously discussed in Section 4.4.5.2. The syndrome polynomial $S(z)$ is written as

$$S(z) = \alpha^2 z^6 + \alpha^3 z^5 + \alpha^8 z^4 + \alpha^{13} z^3 + \alpha^{12} z^2 + \alpha^7 z$$

and the error-locator polynomial $L(z)$ is given by

$$L(z) = \alpha^9 z^3 + \alpha^{11} z^2 + \alpha^3 z + 1 \ .$$

By using the Equation 4.125, the coefficients of the error-evaluator polynomial $E(z)$ are evaluated as,

$$
\begin{aligned}
E_1 &= S_1 = \alpha^7 \\
E_2 &= S_2 + L_1 S_1 = \alpha^{12} + \alpha^3 \alpha^7 = \alpha^3 \\
E_3 &= S_3 + L_1 S_2 + L_2 S_1 = \alpha^{13} + \alpha^3 \alpha^{12} + \alpha^{11} \alpha^7 = \alpha^2 \\
E_4 &= S_4 + L_1 S_3 + L_2 S_2 + L_3 S_1 = \alpha^8 + \alpha^3 \alpha^{13} + \alpha^{11} \alpha^{12} + \alpha^9 \alpha^7 = 0 \\
E_5 &= S_5 + L_1 S_4 + L_2 S_3 + L_3 S_2 = \alpha^3 + \alpha^3 \alpha^8 + \alpha^{11} \alpha^{13} + \alpha^9 \alpha^{12} = 0
\end{aligned}
$$

giving

$$E(z) = \alpha^2 z^3 + \alpha^3 z^2 + \alpha^7 z \ .$$

The error positions are located by using either the Peterson-Gorenstein-Zierler method described in Section 4.4.5.2 or the Berlekamp-Massey algorithm presented in Section 4.4.5.3, and are

$$P_1 = \alpha^2, \qquad P_2 = \alpha^8 \ \text{and} \ P_3 = \alpha^{14}$$

and their corresponding inverse values are

$$P_1^{-1} = \alpha^{13}, \qquad P_2^{-1} = \alpha^7 \ \text{and} \ P_3^{-1} = \alpha^1 \ .$$

By using Equation 4.123, the error magnitudes are evaluated using the Forney algorithm as

$$
\begin{aligned}
M_1 &= \frac{E(P_1^{-1})}{(1 - P_2 P_1^{-1})(1 - P_3 P_1^{-1})} = \frac{\alpha^2 \alpha^{39} + \alpha^3 \alpha^{26} + \alpha^7 \alpha^{13}}{(1 - \alpha^8 \alpha^{13})(1 - \alpha^{14} \alpha^{13})} = \alpha^6 \\
M_2 &= \frac{E(P_2^{-1})}{(1 - P_1 P_2^{-1})(1 - P_3 P_2^{-1})} = \frac{\alpha^2 \alpha^{21} + \alpha^3 \alpha^{14} + \alpha^7 \alpha^7}{(1 - \alpha^2 \alpha^7)(1 - \alpha^{14} \alpha^7)} = \alpha^{13} \\
M_3 &= \frac{E(P_3^{-1})}{(1 - P_1 P_3^{-1})(1 - P_2 P_3^{-1})} = \frac{\alpha^2 \alpha^3 + \alpha^3 \alpha^2 + \alpha^7 \alpha^1}{(1 - \alpha^2 \alpha^1)(1 - \alpha^8 \alpha^1)} = \alpha^2 \ .
\end{aligned}
$$

i	$d^{(i)}$	$\varepsilon^{(i)}(z)$	$D^{(i)}(z)$	$l^{(i)}$
1	α^7	α^7	0	1
2	α^5	α^7	0	1
3	α^{14}	α^7	$\alpha^8 z$	2
4	α^{11}	$\alpha^7 + \alpha^4 z$	$\alpha^8 z^2$	2
5	α^2	$\alpha^7 + \alpha^4 z + \alpha^{10} z^2$	$\alpha^5 z + \alpha^2 z^2$	3
6	α^2	$\alpha^7 + \alpha^3 z + \alpha^2 z^2$	$\alpha^5 z^2 + \alpha^2 z^3$	3

Table 4.17: Computing the $\varepsilon(z)$ polynomial of the RS(15, 9, 3) code using Berlekamp-Massey algorithm.

The error-evaluator polynomial $E(z)$ can be found from the error-locator polynomial $L(z)$ by multiplying it by the syndrome polynomial $S(z)$. Notice however, that a sequence of recursively updated polynomials $E^{(i)}(z)$ can be defined to obey the same recursive relationship as $L^{(i)}(z)$. In this way, one could use the Berlekamp-Massey algorithm to obtain $E(z)$ in parallel to $L(z)$. Let us therefore define ε by:

$$E(z) \overset{\triangle}{=} z\varepsilon(z) . \qquad (4.126)$$

The Berlekamp-Massey algorithm applied to $L(z)$ is also deployed to obtain $\varepsilon(z)$. The initial value of $\varepsilon^{(0)}(z) = 0$, and the correction term $D^{(0)}(z) = -1$. Then, $\varepsilon(z)$ is updated to

$$\varepsilon^{(i+1)}(z) = \varepsilon^{(i)}(z) - d^{(i)} D^{(i)}(z) .$$

If $l^{(i+1)} = l^{(i)}$, the correction polynomial is updated to

$$D^{(i+1)}(z) = z D^{(i)}(z)$$

and if $l^{(i+1)} > l^{(i)}$, it becomes

$$D^{(i+1)}(z) = z \frac{\varepsilon^{(i)}(z)}{d^{(i)}} .$$

An example of finding the error-evalutor polynomial by the Berlekamp-Massey algorithm is illustrated with reference to Table 4.17. This is again an example of the RS(15, 9, 3) code over $GF(16)$. At the end of $2t = 6$ iterations, $\varepsilon(z)$ is found, and from Equation 4.126 the error-evaluator polynomial is obtained as

$$E(z) = \alpha^7 z + \alpha^3 z^2 + \alpha^2 z^3 .$$

4.4.6   Trellis Decoding for Block Codes

In our quest for decoding algorithms making use of the channel measurement information [50], we will now investigate the performance of trellis decoding [51, 52] of both binary and non-binary block codes. We commence by considering the trellis construction for binary BCH codes, and then applying the Viterbi algorithm to decode them. The trellis construction can easily be extended to non-binary codes. A t error correcting code is represented by $\text{BCH}(n, k, t)$, where k information bits are encoded into n-bit codewords. The code is defined over $GF(2)$. In the case of systematic codes, the generator polynomial $g(z) = g_{n-k}z^{n-k} + \cdots + g_1 z + g_0$ formulates a codeword by appending $(n - k)$ parity bits $b_{n-k-1}, \ldots, b_1, b_0$, to k information bits.

4.4.6.1   Trellis Construction

To illustrate the trellis construction we commence by considering the example of the $\text{BCH}(15, 11, 1)$ code over $GF(2)$. The systematic encoder, displayed in Figure 4.48, is derived from Figure 4.40, and the trellis diagram of this code is shown in Figure 4.49. The generator polynomial, $g(z) = z^4 + z + 1$, produces parity bits b_0, b_1, b_2, and b_3 which are buffered in the shift register (SR). The values of the set $\{b_0, b_1, b_2, b_3\}$ represent the states of the register. The four parity bits result in 2^4 different states in the SR. The sequential change of states during the process of encoding a codeword can be catalogued as a particular path through the trellis. There are 2^{11} unique paths in the trellis and each path represents a particular codeword. The trellis has 2^{15-11} rows and $(15 + 1)$ columns. The nodes on the same row represent the same state, whereas the nodes on the same column illustrate all the possible states a, b, ..., p with their corresponding values 0000, 0001, ..., 1111. The state-changes between adjacent columns in the trellis are marked by the transition vector. In Figure 4.49, the vectors are drawn either by a solid line or by a dashed line according to whether the encoder input is a 0 or 1, respectively.

Initially all the parity bits in the SR are set to zero. The number of encoder states increases as each new information bit is inserted into the encoder. The symbol signalling instants corresponding to the column positions in the trellis, shown in Figure 4.49, are indexed by the integer J. On inserting the first information bit into the encoder, $J = 0$, and two different nodes are possible at the next instant. The arrival of the second information bit when $J = 1$ causes the number of possible nodes at the next instant to increase to 2^2. The number of possible nodes continues to increase with J until the maximum number 2^4 is reached. This maximum number of states is reached when $J = 4$, and from then on the number of possible states is constant. The state transitions pattern from instant $J = i$ to $J = i + 1$, shown in Figure 4.49, repeats for $J = 4, 5, \ldots, 10$, until the last information bit to be encoded has entered the SR at $J = 11$. At this

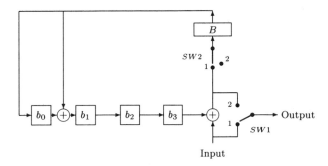

Figure 4.48: Systematic encoder for the BCH(15, 7, 2) code.

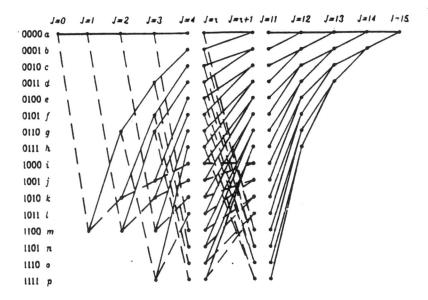

Figure 4.49: Trellis diagram for the binary BCH(15, 11, 1) code.

moment, the switches $SW1$ and $SW2$ are turned to position 2, and the SR is cleared one parity bit at a time as the bits are removed from the encoder to leave behind the all-zero state in the SR. The number of possible states is thus divided by two at every column in the trellis merging towards the all-zeros state, which is reached after clocking the encoder 12 times.

The trellis for the BCH(15, 11, 1) code is suitable for maximum likeli-

hood decoding by the Viterbi algorithm [20, 33] in order to improve the
BER performance. A further improvement in performance can be achieved
by using soft-decision decoding.

4.4.6.2   Trellis Decoding

The method of trellis decoding of block codes is similar to Viterbi decoding
of convolutional codes. The block decoder selects the path in the trellis
having the smallest distance from the received word and thereby identifies
the recovered codeword. The distance properties of the code determine its
error correcting capability.

The distance properties of the BCH$(15, 7, 2)$ binary block code are derived in a similar way to those of convolutional codes, which suggests that
this code can correct any combination of two bit errors. For convolutional
codes the number of incorrect paths increases exponentially and indefinitely
with the number of columns j in the trellis. However, block codes have a
fixed codeword length, and consequently the trellis is truncated after n
columns. The paths in the trellis initially diverge from, and finally converge to, the all-zero state at the codeword boundaries. All the paths have
the same length of n bits. There is a total number of 2^k possible paths in
the trellis and the weight distribution, $W_{\text{BCH}_{1572}}(d)$, of the BCH$(15, 7, 2)$
code was found by computer search through all of these paths as

$$W_{\text{BCH}_{1572}}(d) = 42d^5 + 84d^6 + 49d^7 + 56d^8 + 126d^9 + 84d^{10} + 7d^{15} , \quad (4.127)$$

giving a total of 42 error bits for all distance-5 paths, 84 for all distance-6
paths, etc. The minimum separable distance, d_{min}, between the codewords
is 5 bits. We will represent the total number of information bit errors for
all those paths having a distance d by the coefficient W_d. Following the
procedure of Section 4.3.6, the union bound on the post-decoding bit error
probability p_{bp} for binary block codes is

$$p_{bp} \leq \sum_{d=d_{min}}^{n} W_d P_{ICD}(d) \quad (4.128)$$

where $P_{ICD}(d)$ is the probability of incorrect decoding, i.e., the probability
that the decoder selects a path at distance d from the correct path. The
value of $P_{ICD}(d)$ is found from Equations 4.35 and 4.36 for hard-decision
decoding, or from Equation 4.45 for soft-decision decoding. Reed-Solomon
codes have a maximum separable distance of $(n - k + 1)$ symbols amongst
codewords. Optimum symbol-by-symbol decoding methods [53, 54] that
minimise the symbol error rate result in increased complexity compared
to Viterbi decoding, while their performance is essentially the same. Consequently, we decode using a bit oriented Viterbi algorithm and describe
the properties of the RS code in terms of bit-distance measures. For the

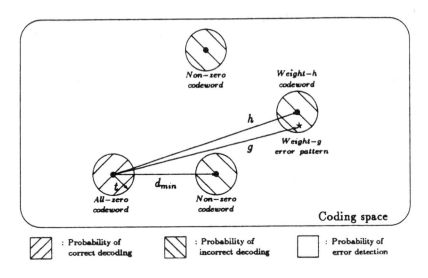

Figure 4.50: Representation of codewords in coding space.

RS$(4, 2, 1)$, $GF(16)$ code the weight distribution $W_{\text{RS}_{421}}(d)$ is found by computer search to be

$$W_{\text{RS}_{421}}(d) = 14d^4 + 58d^5 + 86d^6 + 134d^7 + 210d^8 + 218d^9 + 170d^{10}$$
$$+ 82d^{11} + 24d^{12} + 20d^{13} + 8d^{14}. \tag{4.129}$$

The union bound on the post-decoding bit error probability p_{bp} for this code is obtained by substituting $W_{\text{RS}_{421}}(d)$ into Equation 4.128.

4.4.7 Block Decoding Theory

In this section we consider the abilities of block codes to combat transmission errors. As the linear block code is considered here, we may analyse its behaviour by assuming that a particular codeword is transmitted, say the all-zero codeword, knowing that our findings are applicable for the transmission of any other codeword. The concept of geometric coding space is particularly useful for visualising the decoding situation. Figure 4.50 represents the coding space containing $(q^m)^n$ words of which $(q^m)^k$ are legitimate codewords. If the received word contains $\leq t$ errors, it lies within the all-zero codeword sphere and can be corrected. The probability of correct decoding is represented by the probability of codewords received within the all-zero codeword's decoding sphere. However, if the received word has $> t$ errors, it is uncorrectable and subsequently results in either incorrect decoding or error detection. In the former case [55], the received word falls within one of the non-zero codeword decoding spheres and is therefore erroneously decoded to be a legitimate non-zero codeword. The probability of incorrect decoding is constituted by the probability of codewords being received in the union of the non-zero codeword spheres. The probability of error detection is that of codewords being received outside the union of all the legitimate decoding spheres. Thus we may express the error detection probability of a codeword as

$$P_{ED} = 1 - P_{CD} - P_{ICD} \qquad (4.130)$$

where P_{CD} and P_{ICD} are the probabilities of correctly decoding and incorrectly decoding into another valid codeword, respectively. Considering that ratio of those uncorrectable error words which are detectable, we express this relative error detection probability as

$$P_{EDR} = \frac{P_{ED}}{P_{ED} + P_{ICD}} \; . \qquad (4.131)$$

Observe that for codes with high error correcting capability t the minimum coding space separation $(2t + 1)$ amongst legitimate codewords is high, consequently $P_{ICD} \ll 1$ and $P_{EDR} \sim 1$.

The block codes considered here are over $GF(q^m)$, where $q = 2$ and m is the number of bits in a symbol. The m-bit symbols are transmitted sequentially over a binary channel, which is modelled as an asymmetric memoryless channel as discussed in Section 4.2.5. The probability of receiving an error symbol for this channel is given by Equation 4.3 and the probability of receiving a symbol with i bit errors is expressed by Equation 4.2. We now derive analytical expressions for the probability of correct decoding, incorrect decoding, error detection, and subsequently the probability of post-decoding bit and symbol errors.

4.4.7.1   Probability of Correct Decoding

An (n, k, t) block code defined over $GF(q^m)$ with minimum distance $d_{min} = n - k + 1$ is able to correct t symbol errors. Hence, the probability of correct decoding P_{CD} is the probability of receiving an n-symbol word having t or fewer symbol errors and is given by

$$P_{CD} = \sum_{i=0}^{t} \binom{n}{i} [1 - (1 - p_b)^m]^i [(1 - p_b)^m]^{n-i} . \qquad (4.132)$$

The index i is the number of error symbols in a codeword and ranges from zero to t. There are $\binom{n}{i}$ possible error patterns, and $[1 - (1 - p_b)^m]^i$ is the probability of i symbols being received in error, while $[(1 - p_b)^m]^{n-i}$ is the probability of $(n - i)$ symbols being received correctly. Notice that for channels having a high SNR, p_b is very small and P_{CD} approaches unity.

4.4.7.2   Probability of Incorrect Decoding

If the received word contains more than t error symbols, it is either decoded incorrectly, or an error detection flag is raised after identifying uncorrectable errors in the received word. The probability of incorrect decoding [10] is

$$P_{ICD} = \sum_{h=d_{min}}^{n} P_{ICD}(h) \tag{4.133}$$

where $P_{ICD}(h)$ is the probability of incorrect decoding to a codeword having a symbol distance h from the all-zero codeword. In order to determine
the probability of incorrectly decoding to a weight-h codeword, we define
$N_{g,s}(h)$ as the number of weight-g error patterns that are at a distance s
from a particular weight-h codeword. When $s \leq t$ and $h \geq d_{min}$, each such
error pattern is decoded incorrectly into the weight-h codeword, as seen in
Figure 4.50. Suppose $P(g)$ is the probability of occurrence of a particular
weight-g error pattern and A_h is the number of weight-h codewords, then
the probability of incorrectly decoding as a weight-h codeword is the sum
of all the probabilities $P(g)$ of weight-g error patterns which lie inside the
decoding sphere of a particular weight-h codeword, that is,

$$P_{ICD}(h) = A_h \sum_{s=0}^{t} \sum_{g=h-s}^{h+s} N_{g,s}(h)P(g), \qquad 2t+1 \leq h \leq n. \tag{4.134}$$

Clearly, we have to determine $A_h, P(g)$ and $N_{g,s}(h)$ to be able to compute $P_{ICD}(h)$ in Equation 4.134. As we have seen in Equation 4.132, the
probability p_g of having exactly g number of erroneous symbols out of the
n symbols of a codeword is given by:

$$p_g = \binom{n}{g} [1 - (1 - p_b)^m]^g [(1 - p_b)^m]^{n-g}. \tag{4.135}$$

All the weight-g error polynomials are equiprobable in the case of memoryless channels, since the errors can occur in any arbitrary position with
the same probability. Also, as expected for any probability:

$$\sum_{g=0}^{n} p_g = 1.$$

The number of weight-g error patterns is found to be

$$N(g) = \binom{n}{g} (2^m - 1)^g. \tag{4.136}$$

Whence the probability of a specific weight-g error polynomial $P(g)$ is yielded as:

$$P(g) = \frac{p_g}{N(g)} = \frac{1}{(2^m - 1)^g}[1 - (1 - p_b)^m]^g[(1 - p_b)^m]^{n-g}. \qquad (4.137)$$

An alternative expression for $P(g)$ is derived as follows. Suppose an all-zero (n, k, t) codeword is transmitted. The probability of occurrence of a single error symbol, which we will refer to as producing a weight-1 error pattern, is

$$p_1 = \binom{n}{1} \cdot \sum_{i_1=1}^{m} \binom{m}{i_1} p_b^{i_1} (1 - p_b)^{mn-i_1} . \qquad (4.138)$$

The probability of a weight-2 error pattern is:

$$p_2 = \binom{n}{2} \sum_{i_1=1}^{m} \sum_{i_2=1}^{m} \binom{m}{i_1} \binom{m}{i_2} p_b^{i_1+i_2} (1 - p_b)^{mn-(i_1+i_2)} . \qquad (4.139)$$

Therefore the probability of having an arbitrary weight-g error pattern is given as:

$$p_g = \binom{n}{g} \sum_{i_1=1}^{m} \sum_{i_2=1}^{m} \cdots \sum_{i_g=1}^{m} \left[\binom{m}{i_1} \binom{m}{i_2} \cdots \binom{m}{i_g} \right.$$
$$\left. p_b^{i_1+i_2+\cdots+i_g} (1 - p_b)^{mn-(i_1+i_2+\cdots+i_g)} \right] . \qquad (4.140)$$

The probability of receiving a particular weight-g error pattern out of the possible $N(g) = \binom{h}{g} \cdot (2^m - 1)^g$ such patterns is found upon dividing p_g, by $N(g)$, yielding:

$$P(g) = \frac{1}{(2^m - 1)^g} \sum_{i_1=1}^{m} \sum_{i_2=1}^{m} \cdots \sum_{i_g=1}^{m}$$
$$\left[\binom{m}{i_1} \binom{m}{i_2} \cdots \binom{m}{i_g} p_b^{i_1+i_2+\cdots+i_g} \right.$$
$$\left. (1 - p_b)^{mn-(i_1+i_2+\cdots+i_g)} \right] . \qquad (4.141)$$

Having determined $P(g)$ to be used in Equation 4.134 we now evaluate the number of received words corrupted by a weight-g error pattern, having a distance s to the weight-h codeword, i.e. $N_{g,s}(h)$. First we define e_i and c_i to represent the symbol at the ith position of the error pattern and the weight-h codeword, respectively. By observing the values of the symbols between the error pattern and the codeword, the symbols e_i and c_i can have the following relationships,

$$
\begin{aligned}
c_i &= e_i = 0 \\
c_i &= e_i \neq 0 \\
c_i &= 0, \quad e_i \neq 0 \\
c_i &\neq 0, \quad e_i = 0 \\
c_i &\neq 0, \quad e_i \neq 0, \quad c_i \neq e_i .
\end{aligned}
$$

We define $[10, 56, 57]$ the following new variables.

1. w is the number of symbols when $c_i = e_i \neq 0$.

1) x is the number of symbols when $c_i \neq e_i$ and $e_i \neq 0$, $c_i \neq 0$.

2) y is the number of symbols when $c_i \neq 0$ and $e_i = 0$.

3) z is the number of symbols when $c_i = 0$ and $e_i \neq 0$.

Furthermore, we know that

1) g is the number of symbols when $e_i \neq 0$.

2) h is the number of symbols when $c_i \neq 0$.

From the above definitions, the variables obey the following equations:

$$
\begin{aligned}
g &= w + x + z & (4.142) \\
h &= w + x + y & (4.143) \\
s &= x + y + z. & (4.144)
\end{aligned}
$$

Consider now the total number of weight-g error words having a distance s from the weight-h codeword in terms of w, x, y and z. Observe from Figure 4.51 that there are $\begin{pmatrix} h \\ w \end{pmatrix}$ ways of $c_i = e_i \neq 0$; $\begin{pmatrix} h - w \\ x \end{pmatrix}$ ways of $c_i \neq e_i$, $e_i \neq 0$ and $c_i \neq 0$ where every e_i may take one of $(2^m - 2)$ possible values, excluding $e_i \neq 0$ and $e_i \neq c_i$; and finally $\begin{pmatrix} n - h \\ z \end{pmatrix}$ ways of being $c_i = 0$ and $e_i \neq 0$, where every e_i may take one of $(2^m - 1)$ $e_i \neq 0$ possible values. As a result,

$$
N_{g,s}(h \mid w, x, y, z) = \begin{pmatrix} h \\ w \end{pmatrix} \begin{pmatrix} h - w \\ x \end{pmatrix} (2^m - 2)^x \begin{pmatrix} n - h \\ z \end{pmatrix} (2^m - 1)^z .
$$
$$(4.145)$$

To simplify the conditional probability at the left hand side of Equation 4.145 the expressions of w and x in Equations 4.142, 4.143 and 4.144 are written in terms of h, g, s and z as

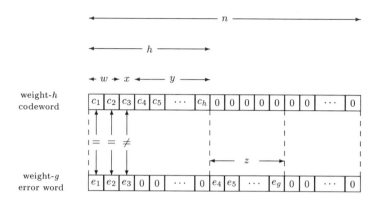

Figure 4.51: Representation of a weight-g error word and a weight-h codeword.

$$w = h - s + z \tag{4.146}$$

$$x = g - h + s - 2z \tag{4.147}$$

and on substituting Equations 4.146 and 4.147 into Equation 4.145 we have

$$N_{g,s}(h \mid z) = \binom{h}{h-s+z}\binom{s-z}{g-h+s-2z}\binom{n-h}{z}$$
$$(2^m - 2)^{g-h+s-2z}(2^m - 1)^z . \tag{4.148}$$

The value of z is lower and upper limited by h and g. Consider the weight, h, of the codeword to be greater than g, such as $h = n$, then z can be as small as zero. When $h < g$, the smallest value of z is $g - h$. Therefore we define z_{min} to be the minimum value of z,

$$z_{min} \geq 0 \quad \text{if } h \geq g$$
$$z_{min} \geq g - h \quad \text{if } g > h , \tag{4.149}$$

that is,

$$z_{min} = \max\{0, g - h\} . \tag{4.150}$$

The expression of z can be found from Equations 4.142, 4.143 and 4.144 as,

$$z = \frac{g - h + s - x}{2} . \tag{4.151}$$

Weight-h h	Number of weight-h codewords A_h
5	18
6	30
7	15
8	15
9	30
10	18
15	1

Table 4.18: Number of weight-h codewords of the BCH$(15, 7, 2)$ code.

By setting $x = 0$, we define z_{max} to be the maximum value of z

$$z_{max} = \left\lfloor \frac{g - h + s}{2} \right\rfloor, \tag{4.152}$$

where $\lfloor (\bullet) \rfloor$ is the lower truncated integer value of $(\bullet)$. By summing over z, the number of weight-g error patterns at a distance s from the weight-h codeword is expressed by

$$
\begin{aligned}
N_{g,s}(h) &= \sum_{z=z_{min}}^{z_{max}} N_{g,s}(h \mid z) \\
&= \sum_{z=z_{min}}^{z_{max}} \binom{h}{h - s + z} \binom{s - z}{g - h + s - 2z} \binom{n - h}{z} \\
&\quad (2^m - 2)^{g-h+s-2z} (2^m - 1)^z .
\end{aligned}
\tag{4.153}
$$

4.4.7.2.1 Number of Weight-h Codewords To evaluate $P_{ICD}(h)$ from Equation 4.134 we must compute A_h, the number of weight-h codewords. For example, the number of weight-h codewords of the BCH$(15, 7, 2)$ code is tabulated in Table 4.18.

For any maximum distance code, such as a Reed-Solomon code defined over $GF(q^m)$ with codeword length n and minimum distance d, the weight distribution A_h is given by [4]

$$A_h = \binom{n}{h} (q^m - 1) \sum_{j=0}^{h-d} (-1)^j \binom{h - 1}{j} (q^m)^{h-d-j} . \tag{4.154}$$

By substituting $P(g)$, $N_{g,s}(h)$ and A_h from Equations 4.141, 4.153 and 4.154 into Equation 4.134 and then substituting $P_{ICD}(h)$ into Equation 4.133, we have,

$$
\begin{aligned}
P_{ICD} &= \sum_{h=d}^{n} \left[\binom{n}{h} (q^m - 1) \sum_{j=0}^{h-d} (-1)^j \binom{h-1}{j} (q^m)^{h-d-j} \right] \\
&\quad \sum_{s=0}^{t} \sum_{g=h-s}^{h+s} \left\{ \left[\sum_{z=z_{min}}^{z_{max}} \binom{h}{h-s+z} \binom{s-z}{g-h+s-2z} \right. \right. \\
&\quad \left. \binom{n-h}{z} (2^m - 2)^{g-h+s-2z} (2^m - 1)^z \right] \frac{1}{(2^m - 1)^g} \\
&\quad [1 - (1 - p_b)^m]^g [(1 - p_b)^m]^{n-g}.
\end{aligned}
\tag{4.155}
$$

Observe that P_{ICD} tends to be zero when the SNR is high as p_b is approximately zero. Armed with P_{CD} and P_{ICD} the probabilty of error detection P_{ED} is computed using Equation 4.130, while the relative error detection probability, P_{EDR} is determined using Equation 4.131.

4.4.7.3   Post-decoding Bit and Symbol Error Probabilities

When the received codewords contain more than t symbol errors they are either known to be incorrect, or they are decoded into another codeword and the error is unknown. We will now determine the probability of the symbols and the bits in the regenerated codeword being in error for the cases when the errors are known and unknown.

When errors are detected, a systematic RS code conveys the information part of the codeword directly to the decoder output as decoded information. Consequently, the post-decoding error probability is equal to the pre-decoding error probability. Hence the contribution to the respective bit and symbol error probabilities p_{bp1} and p_{sp1} is

$$
p_{bp1} = p_b P_{ED}
\tag{4.156}
$$

$$
p_{sp1} = \sum_{i=1}^{m} \binom{m}{i} p_{bp1}^i (1 - p_{bp1})^{m-i}.
\tag{4.157}
$$

For error patterns that are undetectable the received word is decoded as a weight-h codeword, i.e., it contains h error symbols. The post-decoding symbol error probability p_{sp2} is given by

$$
p_{sp2} = \frac{1}{n} \sum_{h=d}^{n} h P_{ICD}(h)
\tag{4.158}
$$

where $P_{ICD}(h)$ is given by Equation 4.155. The post-decoding bit error probability due to incorrect decoding can be evaluated from p_{sp2} as,

$$p_{sp2} \; = \; 1 - (1 - p_{bp2})^m \qquad\qquad (4.159)$$

$$p_{bp2} \; = \; 1 - e^{\frac{1}{m}ln(1-p_{sp2})}. \qquad\qquad (4.160)$$

Then the total post-decoding symbol error probability p_{sp} and the total post-decoding bit error probability p_{bp} can be expressed in terms of the probability of pre-decoding bit errors p_b, the probability of error detection P_{ED} and the probability of incorrect decoding P_{ICD}, namely

$$
\begin{aligned}
p_{sp} \; &= \; p_{sp1} + p_{sp2} \\
&= \; \sum_{i=1}^{m} \left[\binom{m}{i} (p_b P_{ED})^i (1 - p_b P_{ED})^{m-i} \right] \\
&\quad + \frac{1}{n} \sum_{h=d}^{n} h P_{ICD}(h) \qquad\qquad (4.161)
\end{aligned}
$$

and

$$
\begin{aligned}
p_{bp} \; &= \; p_{bp1} + p_{bp2} \\
&= \; p_b P_{ED} + \left[1 - e^{\frac{1}{m}ln\left(1 - \frac{1}{n}\sum_{h=d}^{n} h P_{ICD}(h)\right)} \right]. \qquad (4.162)
\end{aligned}
$$

4.4.8   Block Coding Performance

In this section, we investigate the performance of BCH and RS codes operating at approximately half rate, but having different block lengths. The codes were transmitted via MSK modulation over AWGN or Rayleigh fading channels, and were decoded by either the hard-decision Berlekamp-Massey [BM-HD] algorithm, or by the soft-decision trellis decoding [TD-SD] method. For Rayleigh fading channels, interleaving was introduced at the transmitter in order to disperse burst errors at the receiver into the adjacent blocks resulting in an improvement in BER performance. The theoretical calculations of the probabilities of correct decoding, incorrect decoding, and relative error detection of each code were obtained from Equations 4.132, 4.133 and 4.131, respectively. These probabilities were compared with the results from simulations. Only systematic codes were used. When an incorrectable codeword was detected by syndromcheck, the information part of the codeword was passed to the decoder output. The probabilities of post-decoding symbol and bit errors were examined by simulation and compared with the theoretical results obtained from Equations 4.161 and 4.162.

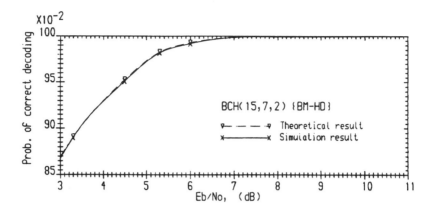

Figure 4.52: Probability of correct decoding for the $BCH(15, 7, 2)$ code using [BM-HD] decoding over AWGN channel

4.4.8.1 Block Coding Performance via Gaussian Channels

Figure 4.52 and Figure 4.53 illustrate the probabilities of correct decoding and incorrect decoding, respectively, as a function of E_b/N_0 for the short binary BCH(15, 7, 2) code decoded using the hard-decision Berlekamp-Massey algorithm [BM-HD]. Observe that for E_b/N_0 in excess of 7 dB $P_{CD} \sim 1$ and $P_{ICD} \sim 0$. The probabilities of detecting the incorrectable words calculated from Equation 4.131 were 0.57 for E_b/N_0 of 2 dB, and 0.60 for E_b/N_0 of 10 dB. Clearly, for practical E_b/N_0 values more than 57 % of the incorrectable words were detected, and 43 % were incorrectly decoded into other valid codewords. The comparatively low ability to detect code-overload is due to the densely packed, low correcting power ($t = 2$) code. The probability of post-decoding bit errors in Figure 4.54 was similar in its nature to P_{ICD} in Figure 4.53. The probability of incorrect decoding decreased to 10^{-6} when E_b/N_0 exceeded 9.7 dB, while the post-decoding bit error probability was similarly reduced to 10^{-6} at E_b/N_0 of 9.6 dB. This demonstrated that the post-decoding bit errors were mainly contributed by incorrect decoding. Although the coding rate of this code was close to half, a block length of 15 bits allowed only 8 parity bits. The small number of parity bits meant that the separable distance between codewords was small, and as a result the transmitted codeword was easily corrupted into an incorrectable and undetectable word. Thus the small distance code gave a high percentage of incorrect decoding decisions. Figures 4.52-4.54 also contain simulation results which were in close agreement to the theoretical ones.

We now compare the theoretical and simulation results for a number of non-binary Reed-Solomon codes having different codeword lengths, but all operating at a coding rate of 1/2. The Berlekamp-Massey hard-decision

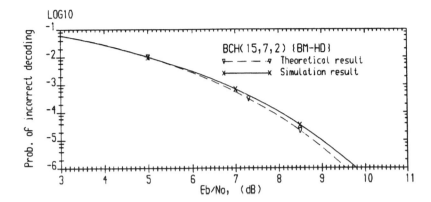

Figure 4.53: Probability of incorrect decoding for the BCH(15, 7, 2) code using [BM-HD] decoding over AWGN channel.

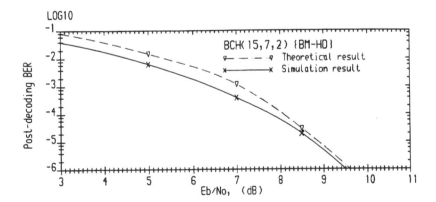

Figure 4.54: Post-decoding BER for the BCH(15, 7, 2) code using [BM-HD] decoding over AWGN channel.

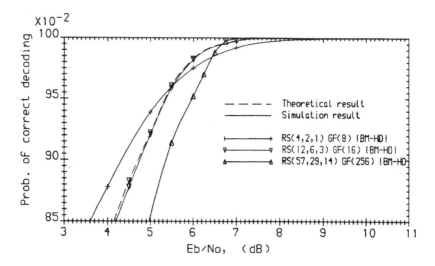

Figure 4.55: Probability of correct decoding for various RS codes over AWGN channel.

decoding method was used. Figure 4.55 shows the probability of correct decoding as a function of E_b/N_0 for the RS(4, 2, 1) code over $GF(8)$, the RS(12, 6, 3) code over $GF(16)$ and the RS(57, 29, 14) code over $GF(256)$. The probability of correct decoding P_{CD} using shorter RS codes was found to be higher than for the longer codes for E_b/N_0 values below approximately 6 dB. This was because the codes with a larger number of bits per symbol suffered higher symbol error probability, as exemplified earlier in the interleaving section by Figure 4.12. This situation was reversed for higher E_b/N_0 values because the longer codes had larger t values, and therefore were able to correct more errors in a codeword. The simulation and theoretical results coincided.

Figure 4.56 shows the probability of incorrect decoding P_{ICD} as a function of E_b/N_0. The corresponding P_{ICD} values at E_b/N_0 of 3 dB for RS(4, 2, 1) and RS(12, 6, 3) codes were 0.065 and 0.012, respectively. Although not displayed in Figure 4.56, the theoretical P_{ICD} value for the RS(57, 29, 14) code was 10^{-25}. This small P_{ICD} value was anticipated from the fact that the designed distances $(2t + 1)$ of RS(4, 2, 1), RS(12, 6, 3), and RS(57, 29, 14) codes, are 3, 7, and 29, respectively. The P_{ICD} value was significantly lower for longer codes because the separable distances between codewords were much larger. The probability of decoding an incorrectable word into another valid codeword was reduced as the larger distances restricted the occurrence of the incorrectable words. Again the simulation and theoretical values coincided.

Another property of RS codes is their capability to detect incorrectable

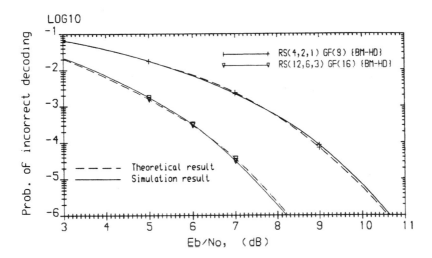

Figure 4.56: Probability of incorrect decoding of various RS codes over AWGN channel.

error patterns, the probability of which is characterised in terms of the relative error detecting probability P_{EDR} as defined in Equation 4.131. As we experienced in the case of the BCH(15, 7, 2) code, P_{EDR} was practically independent of E_b/N_0 for a particular code. For the short RS(4, 2, 1) code, the average value of P_{EDR} was 0.73; for the RS(12, 6, 3) code it was 0.98; whereas for the longest code the average P_{EDR} was $1 - 10^{-28}$. This suggested that the longer code protected by a larger number of parity symbols for the same coding rate, offered more reliable error detection. This is an important feature when transmitting computer data.

The post-decoding bit error probability p_{bp} was a function of both P_{ICD} and P_{ED} (see Equation 4.162). The simulation and theoretical results are displayed in Figure 4.57 for the three codes. The p_{bp} curves exhibited a cross-over region for E_b/N_0 between 5 dB and 6 dB. Below this region the bit error rate of the RS(57, 29, 14) code was the highest as it had a higher channel symbol error rate than the other two codes, as shown in Figure 4.12. For E_b/N_0 in excess of 6 dB the RS(57, 29, 14) code had the best p_{bp} performance as the noise-induced corruption of the transmitted codeword was restricted for most of the time to the confines of the decoding sphere of that codeword. As a result, the slope of the curve is much sharper for the longer codes than for the shorter ones. The same tendency can be observed in Figure 4.58, where the post-decoding symbol error probability p_{sp} is displayed for the same conditions as those in Figure 4.57.

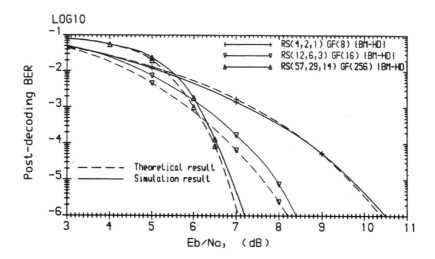

Figure 4.57: Post-decoding BER of various RS codes over AWGN channel.

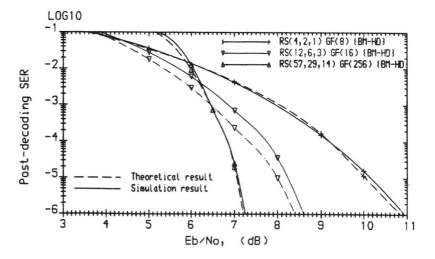

Figure 4.58: Post-decoding SER of various RS codes over AWGN channel.

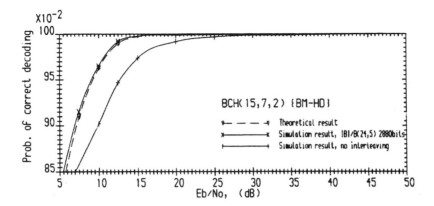

Figure 4.59: Probability of correct decoding of the BCH(15, 7, 2) code over Rayleigh-fading channel.

4.4.8.2 Block Coding Performance via Rayleigh Fading Channels

In this section, we investigate the performance of block codes transmitted via MSK modulation over Rayleigh fading channels. Figure 4.59 and Figure 4.60 show the corresponding probabilities of correct and incorrect decoding as a function of E_b/N_0 for the binary BCH(15, 7, 2) code. Without interleaving, burst errors occurred in the mobile channel, overloading the 2 bits per codeword correcting capability of the code, causing incorrect decoding. When inter-block bit interleaving IBI/B(24,5) with $B = 24$ and $N = 5$ was introduced, the channel became essentially memoryless. The burst errors were dispersed to the adjacent blocks, reducing the probability of incorrect decoding, and at the same time increasing the probability of correct decoding compared to the case without interleaving. The theoretical calculations were for the memoryless channel and were in a good agreement with the simulation results. The average probability of detecting the incorrectable words was calculated to be 0.58. This value was close to that obtained on the AWGN channel. This implies that the P_{EDR} depended on the distance separation between codewords, regardless of the channel error statistics. In Figure 4.61, the probability of post-decoding bit error p_{bp} is shown as a function of E_b/N_0. By using interleaving, the p_{bp} was reduced from 39 dB to 25 dB at a BER of 10^{-6}, this coding gain of 14 dB was achieved at a price of 2880 bits of delay.

We now consider the performance of the RS(12, 6, 3) and RS(57, 29, 14) codes. In Figure 4.62, we display the probability of correctly decoding a codeword as a function of E_b/N_0. When bit or symbol interleaving was deployed the P_{CD} was increased for both codes, and the longer the interleaving period the better was the performance. However, a bit interleaving

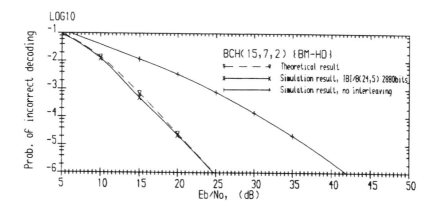

Figure 4.60: Probability of incorrect decoding of the BCH(15, 7, 2) code over Rayleigh-fading channel.

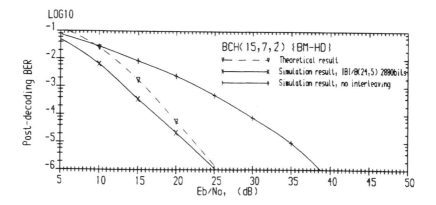

Figure 4.61: Post-decoding BER of the BCH(15, 7, 2) code over Rayleigh-fading channel.

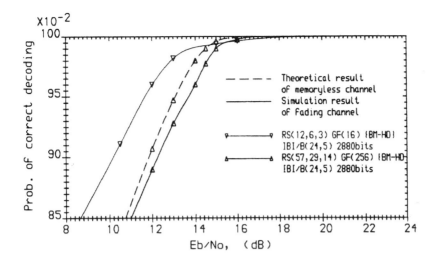

Figure 4.62: Probability of correct decoding of various RS codes with identical interleaving memory over Rayleigh-fading channel.

period of 2880 bits appeared to be sufficient to randomise the error statistics of the channel, and yielded a performance very similar to the theoretical one. For an E_b/N_0 of 11 dB, the probability of correct decoding P_{CD} was approximately 0.86 for the RS(57, 29, 14) code, and 0.92 for the RS(12, 6, 3) code. The longer code performed better for high E_b/N_0 values than the shorter code, while for low E_b/N_0 values the situation was reversed.

The probability of incorrect decoding P_{ICD} is displayed in Figure 4.63 for the RS(12, 6, 3) code with inter-block bit interleaving having a period of 2880 bits. The theoretical P_{ICD} curve is also displayed for that case when the interleaving period was infinitely long, representing the performance for the memoryless channel. For the RS(57, 29, 14) code, the P_{ICD} was less than 10^{-23} for an E_b/N_0 of 10.5 dB, and therefore its performance curve cannot be shown in Figure 4.63.

The BER performance of the RS(12, 6, 3) and RS(57, 29, 14) codes is compared in Figure 4.64 in terms of their post-decoding bit error probability with 2880 bits inter-block bit interleaving, as well as with two different symbol interleaving periods. Firstly, we focus our attention on the RS(12, 6, 3) code, where we observe that even if the bit interleaving period was 2880 bits long, the BER performance was worse than that of the theoretical curve for the memoryless channel. However, if symbol interleaving was used with the same delay, practically the same BER performance was achieved as for the memoryless channel. The same tendency was noted in case of the longer RS(57, 29, 14) code, where even a shorter symbol interleaving delay of 1368 bits resulted in a higher performance compared to

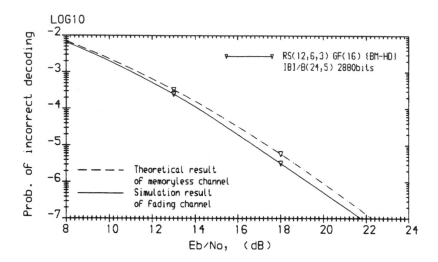

Figure 4.63: Probability of incorrect decoding of the RS(12, 6, 3) code over Rayleigh-fading channel.

when bit interleaving was used. This observation confirms that the symbol interleaving gives a better performance than bit interleaving with RS codes. Once again, the BER curves gave a sharper slope for the longer codes. Similar results were observed in terms of symbol error rates (SER) as shown in Figure 4.65.

In Figure 4.66 and Figure 4.67, the performance of the RS(12, 6, 3), and the RS(57, 29, 14) code with various interleaving delays is presented. The width of the block interleaver was determined by the blocklength. Consecutive error symbols on the channel were diverted into the adjacent blocks on deinterleaving. Observe that the performance of the shorter RS(12, 6, 3) code hardly improved when the interleaving delay increased beyond 912 bits, while for the longer RS(57, 29, 14) code 2280 bits was adequate. The longer delay for the longer code was incurred due to the width of the interleaver. However, as the delays were sufficiently long to randomise the channel to a memoryless one, the performances depended entirely on the correcting capabilities of the codes. This is evident by noting that the RS(57, 29, 14) code achieved a BER of 10^{-6} at E_b/N_0 of 13 dB, while for the RS(12, 6, 3) code 17.5 dB was necessary.

4.4.8.3   Soft/Hard Decisions via Gaussian Channels

In Figure 4.68, the post-decoding bit error probabilities of the RS(4, 2, 1) code over $GF(16)$ and the BCH(15, 7, 2) code are depicted both for hard-decision decoding and for soft-decision trellis decoding. The transmissions are over an AWGN channel. For both soft and hard decisions the

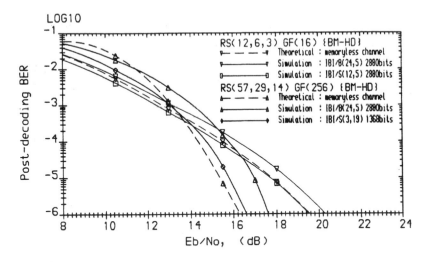

Figure 4.64: Post-decoding BER of various RS codes over Rayleigh-fading channel.

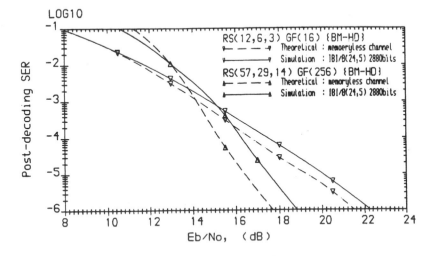

Figure 4.65: Post-decoding SER of various RS codes over Rayleigh-fading channel.

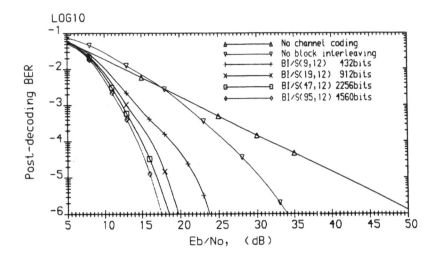

Figure 4.66: The effect of block interleaving on the post-decoding BER of the RS$(12, 6, 3)GF(16)$ code using [BM-HD] decoding over Rayleigh-fading channel.

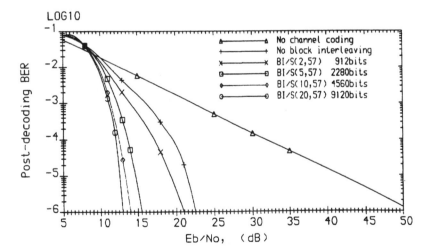

Figure 4.67: The effect of block interleaving on the post-decoding BER of the RS$(57, 29, 14)GF(256)$ code using [BM-HD] decoding over Rayleigh-fading channel.

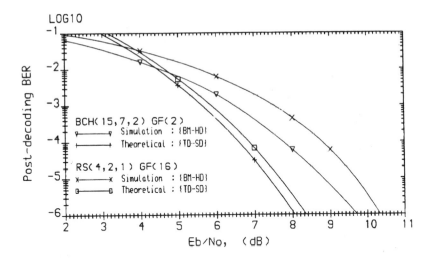

Figure 4.68: The effect of hard and soft decisions on the post-decoding BER of various block codes over AWGN channel.

BCH(15, 7, 1) code performed better than the RS(4, 2, 1) code as it corrected any combination of two bit errors. By constrast the RS code coped only with those bit errors which occurred in the same four bit symbols, as it can correct only one symbol error in a codeword. By using soft-decision, an improvement of about 2 dB was achieved over the hard-decision in both codes at a BER of 10^{-6}. The trellis decoding made use of the maximum likelihood method to select the path in the trellis that most resembled the received codeword. However, this probabilistic decoding did not have error detection capability, which is a serious disadvantage.

4.4.9   Conclusions on Block Coding

The performance of binary and non-binary block codes has been investigated theoretically, and simulation results have been presented. In the case of the Rayleigh fading channel, the interleaver selected for block codes is the block interleaver, with bit interleaving deployed for binary codes and symbol interleaving for non-binary codes. The block codes are decoded by either the hard-decision Berlekamp-Massey algorithm, or by the soft-decision trellis decoding method.

When the Berlekamp-Massey decoding algorithm is deployed, the data reliability increases with the amount of redundancy introduced into the codeword. For the same coding rate, longer codes contain a higher number of parity symbols than the shorter codes and the probability of detecting incorrectable words is very high. For example, the probability of detecting the incorrectable errors of the RS(12, 6, 3) code is 0.98, as compared with

	E_b/N_0 at BER of		Cod.-gain at BER of	
	10^{-3}	10^{-6}	10^{-3}	10^{-6}
No coding	6.8dB	10.5dB	0dB	0dB
RS(4, 2, 1) $GF(8)$ [BM-HD]	7.2dB	10.5dB	-0.4dB	0dB
RS(12, 6, 3) $GF(16)$ [BM-HD]	6.2dB	8.5dB	0.6dB	2.0dB
RS(57, 29, 14) $GF(256)$ [BM-HD]	6.1dB	7.2dB	0.7dB	3.3dB
RS(4, 2, 1) $GF(16)$ [BM-HD]	7.5dB	10.4dB	-0.7dB	0.1dB
BCH(15, 7, 2) $GF(2)$ [BM-HD]	6.5dB	9.7dB	0.3dB	0.8dB
RS(4, 2, 1) $GF(16)$ [TD-SD]	5.9dB	8.3dB	-0.9dB	2.2dB
BCH(15, 7, 2) $GF(2)$ [TD-SD]	5.6dB	8.0dB	1.2dB	2.5dB

Table 4.19: Coding gain of block codes over AWGN channel.

the RS(57, 29, 14) code whose is $1 - 10^{-28}$. This feature is valuable in automatic repeat request (ARQ) protocols, especially for computer data transfer. However, the penalty of using long codes is a large delay which may not be acceptable for speech transmissions. Also the complexity of the decoder increases with the amount of redundancy. The number of multiplications required by the Berlekamp-Massey algorithm (which is at most $6t^2$) increases exponentially with the number of correctable errors. As a result, the complexity increases with t.

When the soft-decision trellis decoding method is used to decode an (n, k) block code over $GF(2^m)$, the number of bits representing the state of the encoder is equal to $(n - k) \times m$ used by the parities in a codeword. The number of states is therefore $2^{(n-k) \times m}$. The complexity of the decoder increases with the number of states. Low complexity codes operate at high coding rates or they are short codes. In either case, the separation between the codewords is small, making them weak error correcting codes. Nevertheless, it is the complexity of the decoder that limits the application of the trellis decoding. If trellis decoding is applied, the soft-decision decoding achieves a gain of 2 dB at BER of 10^{-6} for transmissions over an AWGN channel.

When the block symbol interleaving has a depth of 9 with a delay of 432 bits suitable for speech transmissions, the RS(12, 6, 3) code over $GF(16)$ achieves an E_b/N_0 of 14.5 dB at a BER of 10^{-3}. Whereas the RS(57, 29, 14) code over $GF(256)$ operating without interleaving encounters a similar delay of 456 bits, and needs an E_b/N_0 of 16 dB to acquire a BER of 10^{-3}. This suggests that the short codes require less decoder complexity, and perform slightly better at the target BER of 10^{-3} than the longer codes, and therefore might be preferable when speech signals are transmitted.

Table 4.19 and Table 4.20 tabulate the performance of the block codes for specific BERs when the transmissions are over an AWGN or a Rayleigh fading channel, respectively.

	E_b/N_0 at BER of		Cod.-gain at BER of	
	10^{-3}	10^{-6}	10^{-3}	10^{-6}
No coding	23.0dB	52.0dB	0dB	0dB
BCH$(15,7,2)$ [BM-HD]				
No interleaving 15bits	22.5dB	39.0dB	0.5dB	13.0dB
IBI/B(24,5) 2880bits	13.0dB	25.0dB	10.0dB	27.0dB
RS$(12,6)$ $GF(16)$ [BM-HD]				
No interleaving 48bits	20.5dB	34.0dB	2.5dB	18.0dB
BI/S(9,12) 432bits	14.5dB	24.0dB	8.5dB	28.0dB
BI/S(19,12) 912bits	13.0dB	20.0dB	10.0dB	32.0dB
BI/S(47,12) 2256bits	12.5dB	19.0dB	10.5dB	33.0dB
BI/S(95,12) 4560bits	12.0dB	17.0dB	11.0dB	35.0dB
RS$(57,29)$ $GF(256)$ [BM-HD]				
No interleaving 456bits	16.0dB	22.5dB	7.0dB	29.5dB
BI/S(2,57) 912bits	14.0dB	21.0dB	9.0dB	31.0dB
BI/S(5,57) 2280bits	12.5dB	15.5dB	10.5dB	36.5dB
BI/S(10,57) 4560bits	11.5dB	14.0dB	11.5dB	38.0dB
BI/S(20,57) 9120bits	11.0dB	13.0dB	12.0dB	39.0dB

Table 4.20: Coding gain of block codes over Rayleigh fading channel.

4.5   Concatenated Codes

Concatenated coding was first introduced by Forney [58] to utilise multiple levels of coding. Figure 4.69 illustrates two levels of coding and two levels of interleaving to combat the channel's burst errors. The level of the coding and interleaving closer to the channel is called the inner layer, whereas the level outside the inner layer is known as the outer layer. The inner and outer FEC codes can be convolutional codes or block codes. At the receiving end, the demodulator may produce either hard or soft decisions. In either case, these decisions are fed to the inner interleaver. The inner deinterleaver disperses the channel burst errors into random patterns. If the channel being considered is Gaussian, the inner interleaving is not required. The inner FEC decoder is designed to remove the random errors. If the inner FEC decoder cannot correct the word or erroneously decodes the word, the decoding errors are bursty in nature and the outer interleaver is used to disperse the errors into adjacent codewords of the outer code. The outer FEC decoder then attempts to correct the remaining errors. Essentially there are two types of concatenated codes, depending on the way the two layers are amalgamated: nested codes and product codes.

4.5.1   Nested Codes

Suppose the outer code and the inner code are denoted as (N, K, T) and (n, k, t) respectively. The outer coder encodes K outer symbols into N outer symbols as shown in Figure 4.70, where each outer symbol consists of k inner symbols. The inner coder then encodes each outer symbol of k inner symbols into n inner symbols. If the inner symbol is defined on $GF(q)$, the outer symbol consisting of k inner symbols is defined on $GF(q^k)$. Hence,

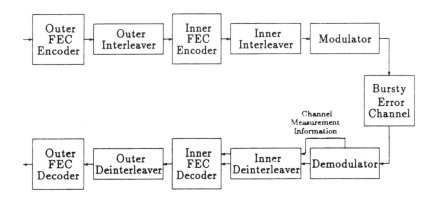

Figure 4.69: Concatenated coding system block diagram.

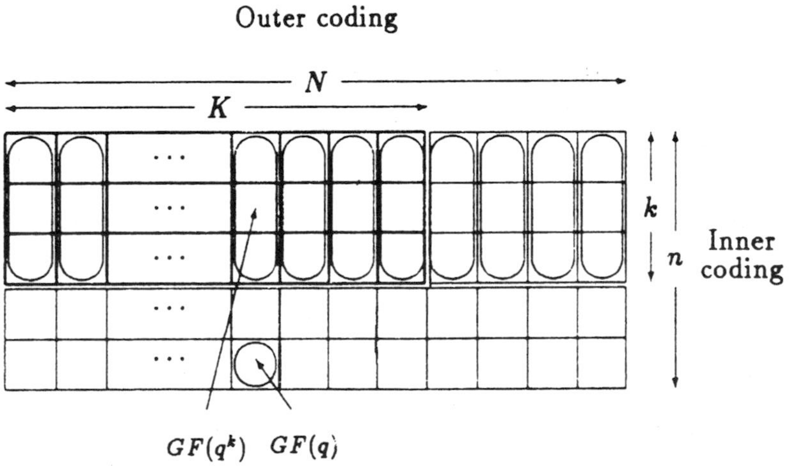

Outer coding

$GF(q^k)$ $GF(q)$

Figure 4.70: Structure of nested codes.

the outer layer is described as the (N, K, T) code over $GF(q^k)$, and the inner layer is the (n, k, t) code over $GF(q)$. The combined layers of coding accept a sequence of Kk information symbols and produce a sequence of Nn symbols. An example of a nested code is the single error correcting shortened RS(5, 3, 1) code over $GF(8) = GF(512)$ for the inner layer, and the double error correcting RS(511, 507, 2) code over $GF(8^3)$ for the outer layer. Both layers of coding jointly form a nested code of (2555, 1521) over $GF(8)$.

In Figure 4.70, the outer coder encodes a row at a time, and the inner coder encodes a column at a time. If the symbols are transmitted in rows

Outer coding

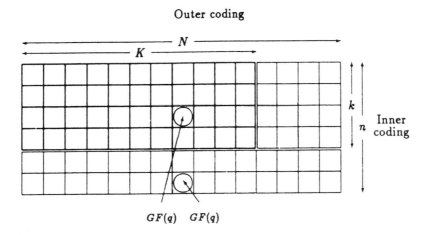

Figure 4.71: Structure of product codes.

to the channel, the burst errors corrupt consecutive symbols which are received and arranged in rows in the Figure. As a consequence, the bursts of error symbols are distributed over adjacent words of the inner codes. The dispersion of burst errors is equivalent to interleaving. The inner decoder corrects the burst errors and outputs N outer symbols to the outer decoder. The outer decoder corrects the decoding errors left by the inner layer.

A powerful nested code is formed by using convolutional codes as the inner layer and Reed-Solomon codes as the outer layer. This is because a convolutional code can achieve a high performance gain in correcting random errors. The errors at the convolutional decoder output due to erroneous decoding tend to be bursts and the concatenated outer Reed-Solomon code is used to correct them.

4.5.2 Product Codes

Product codes are obtained by encoding a matrix of information symbols (defined on $GF(q)$) in two dimensions, namely, in rows and in columns. If each row is encoded by an (N, K, T) code, and each column is encoded by an (n, k, t) code, the information matrix with dimensions $(k \times K)$ is encoded to the dimension of $(n \times N)$. Figure 4.71 illustrates an example of a product code by using RS(15, 11, 2) code over $GF(32)$ for the rows, and the shortened RS(6, 4, 1) code over $GF(32)$ for the columns.

In a mobile radio channel, the signal fading occurs intermittently. Most of the codewords between fadings have unnecessary protection resulting in low data throughtput. A method proposed [59] is to use an adaptive product coding. The information protected with relatively low overhead error detection codes is transmitted via the channel. If the decoder detects errors, automatic repeat request (ARQ) is activated and the encoder transmits more and more redundancy until the decoder is able to correct the errors. By using this method, the amount of data protection adapts to the channel condition improving the data throughput. A similar adaptive coding scheme was proposed in references [60] and [61] as well.

4.6   Comparison of Error Control Codes

Every type of error control code such as convolutional codes, Reed-Solomon codes and concatenated codes have unique properties. Convolutional codes have an advantage in correcting random errors, whereas Reed-Solomon codes are good at correcting both random as well as burst errors and have reliable error detection capability. Concatenated codes possess high error correcting capability due to their long blocklengths. It is difficult to select 'the' best code for all the different channel conditions and system requirements, such as combined coding and interleaving delays, coding rate, data throughput, data integrity, coder complexity, types of channels, etc. In this section, we investigate the performance of using different codes for speech and data signals for transmissions over mobile radio channels. In general, speech codecs are robust to moderate channel errors, but cannot tolerate long transmission delays. Hence, for speech signals short coding and interleaving delays, say about 500 bits are used, and moderate data protection is provided. For data transmission, data integrity is vital. The codes used for data signals must have a high error detection capability. When an incorrectable word is detected at the receiver, usually the retransmission of the codeword is necessary. As data transmission does not usually require realtime processing, coding and interleaving delays of about 2000 bits are long enough to randomise the Rayleigh fading channel's bursty error statistics even for low vehicular speeds. We have arranged for speech and data signals that the codecs operate with the same half rate in order to compare their bit error rate (BER) performance with the same amount of protection.

We experimented with both non-concatenated coding and concatenated coding of digital speech transmitted via MSK over Rayleigh fading channels. The BER results are shown in Figure 4.72 and Figure 4.73 respectively.

For non-concatenated codes, the RS$(57, 29, 14)$ code over $GF(256)$ deploying no interleaving had a coding delay of 456 bits. By stacking up nine RS$(12, 6, 3)$ codewords defined on $GF(16)$ in the interleaving matrix, the block symbol interleaver BI/S$(9,12)$ achieved a delay of 432 bits. Both RS codes were decoded using the hard-decision Berlekamp-Massey algorithm. We observe from Figure 4.72 that the BER performance curves cross-over at an E_b/N_0 of 21 dB. Below 21 dB, the shorter RS$(12, 6, 3)$ code had a better performance than the longer RS$(57, 29, 14)$ code. The longer code having a larger symbol field, i.e., 8 bits per symbol for the RS$(57, 29, 14)$

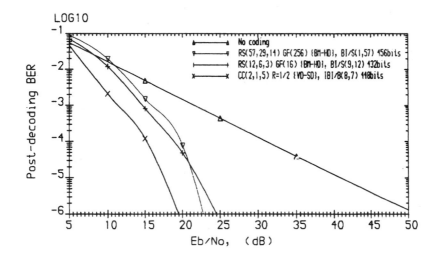

Figure 4.72: Post-decoding BER performance of half-rate non-concatenated block and convolutional codes for speech transmission over Rayleigh-fading channel.

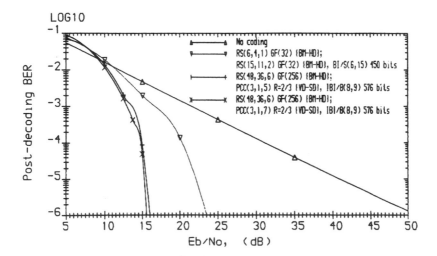

Figure 4.73: Post-decoding BER performance of half-rate concatenated codes for speech transmission over Rayleigh-fading channel.

code compared with 4 bits per symbol for the RS$(12, 6, 3)$ code, suffered a higher channel symbol error rate that was likely to cause more symbol errors in a decoded word. This tended to induce incorrect decoding and incorrectable errors, resulting in a worse performance. Above 21 dB, the longer RS$(57, 29, 14)$ code having more parity symbols in a codeword and better burst-dispersing properties had a higher correcting capability, and therefore had a better performance than the shorter RS$(12, 6, 3)$ code. Inter-block bit interleaving IBI/B(8,7) of the convolutional code CC$(2, 1, 5)$ introduced 448 bits of delay. By using soft-decision Viterbi decoding, the BER performance of the CC$(2, 1, 5)$ code was 3 dB to 5 dB better than for both the RS codes for BERs from 10^{-2} to 10^{-6}. The soft-decision decoding yielded a better performance by making use of the channel information, especially for transmissions over Rayleigh fading channels. For speech codecs able to tolerate a BER of 10^{-3}, the CC$(2, 1, 5)$ code, the RS$(12, 6, 3)$ code, and the RS$(57, 29, 14)$ code required E_b/N_0 values of 11.5 dB, 14.5 dB, and 15.5 dB, respectively. The shorter RS code was better than the longer code for the speech transmission. The CC code is the best among all three candidates.

Figure 4.73 compares the performance of the concatenated product code and the nested code. The product code employed RS$(15, 11, 2)$ code over $GF(32)$ for the outer coding and RS$(6, 4, 1)$ over $GF(32)$ for the inner coding. The code was block interleaved BI/S(6,15) with a delay of 450 bits. This code achieved a BER of 10^{-3} at 16.5 dB which was 1 dB and 2 dB more than that required by the RS$(57, 29, 14)$ code and the RS$(12, 6, 3)$ code, respectively. The inferior performance was mainly due to the weak correcting code in both dimensions, i.e., double error correcting code in the rows, and single error correcting codes in the columns. The half-rate nested code was constructed by concatenating the outer three-quarter rate RS$(48, 36, 6)$ code over $GF(256)$ and the inner two-thirds rate PCC$(3, 1, 5)$ code, and the inter-block bit interleaving IBI/B(8,9) was over 576 bits. The restricted delay did not allow us to use the outer interleaver. At a BER of 10^{-3}, the required E_b/N_0 was 13.5 dB which was 1 dB less than that of the RS$(12, 6, 3)$ code. If the constraint length of the inner CC code was increased from five to seven binary stages, the performance was improved by 0.5 dB. Notice that the improved nested code was still inferior to the half-rate CC$(2, 1, 5)$ code by some 1.5 dB. The poor performance was due to the weak inner two-thirds rate CC code at the low E_b/N_0 values. The nested code performed better at higher E_b/N_0 values. As an example, it achieved a BER of 10^{-6} at 15.5 dB, whereas the CC$(2, 1, 5)$ code required 19.5 dB. However, for speech codecs, any further reduction in BER below 10^{-3} is imperceptible. Hence, the CC$(2, 1, 5)$ is the most effective code amongst our benchmarkers for speech transmission.

Figure 4.74 and Figure 4.75 illustrate the performance of non-concatenated codes and concatenated codes, respectively, for the transmission of data. With a delay of over 2000 bits, the channel was randomised to a near memoryless one, and the codes performed better than those used

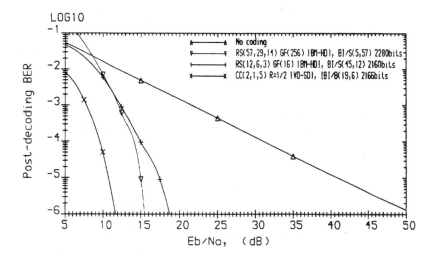

Figure 4.74: Post-decoding BER performance of half-rate non-concatenated block and convolutional codes for data transmission over Rayleigh-fading channel.

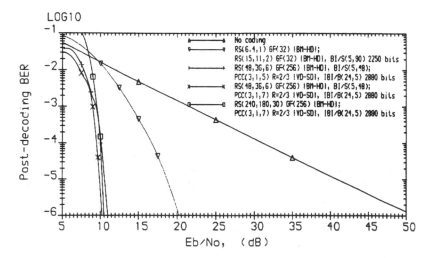

Figure 4.75: Post-decoding BER performance of half-rate concatenated codes for data transmission over Rayleigh-fading channel.

for the speech signals. For data transmission, we measure the performance of the codes at a BER of 10^{-6}. In Figure 4.74, the RS$(57, 29, 14)$ code and the RS$(12, 6, 3)$ code required 15.5 dB and 18.5 dB, respectively, yielding gains of 7 dB and 8 dB compared with their performance for speech data. However, the CC$(2, 1, 5)$ code decoded by soft-decision Viterbi algorithm required an E_b/N_0 of 11.5 dB at a BER of 10^{-6}. Although the CC$(2, 1, 5)$ achieved the best performance, it had no error detection capability, which is often a serious disadvantage. The RS$(57, 29, 14)$ code detected the incorrectable error words with a successful rate of $1 - 10^{-28}$, a value sufficiently reliable to invoke retransmission in ARQ systems. Hence, the RS$(57, 29, 14)$ code was more appropriate for data protection.

In Figure 4.75, the product code constructed by the RS$(15, 11, 2)$ and RS$(6, 4, 1)$ codes again demonstrated lack of correcting power. For the nested code constructed by the RS$(48, 36, 6)$ and PCC$(3, 1, 5)$ codes, the E_b/N_0 value was 10.5 dB at a BER of 10^{-6}. If the constraint length of the PCC$(3, 1, 5)$ code increased from 5 to 7, a gain of 0.5 dB was achieved. The RS$(48, 36, 6)$ code was used for the outer layer providing a successful error detection rate of $1 - 10^{-13}$. If the RS$(48, 36, 6)$ code was replaced by the RS$(240, 180, 30)$ code, the successful rate was virtually unity. However, this longer RS code had a higher probability of having an incorrectable error codeword, and therefore its performance deteriorated at low E_b/N_0 values.

The subject of this chapter was the range of forward error correction codecs employed in wireless communications systems. Our discussions initially concentrated on various interleaving techniques, which are powerful in terms of randomising the bursty channel error statistics at the cost of imposing additional interleaving delay. Hence the interleaver depth in interactive mobile radio speech and video systems must be low. These discussions were followed by a brief historical perspective on convolutional coding. The Viterbi algorithm was introduced through a series of worked examples, before its analytical performance was characterised. The performance of a range of coding schemes was summarised in coding gain tables over both Gaussian and Rayleigh fading channels. The family of block codes was then invoked, which operate over the mathematical construction of finite Galois fields. Reed-Solomon and Bose-Chaudhuri-Hocquenghem codes were introduced with the aid of a range of worked examples and then treated in depth also analytically. Similarly to the class of convolutional codes, they were characterised over both Gaussian and Rayleigh fading channels in coding gain terms. The chapter concluded with a discourse on concatenated codes. The next chapter focuses our attention on quaternary frequency shift keying.

Bibliography

[1] **R.W.Hamming**. "Error detecting and error correcting codes". *Bell Sys. Tech. J.,29*, pp.147–160, 1950.

[2] **P.Elias**. "Coding for noisy channels". *IRE Conv. Rec. pt.4*, pp.37–47, 1955.

[3] **J.L.Massey**. *Threshold decoding*. MIT Press, Cambridge, Mass., 1963.

[4] **E.R.Berlekamp**. *Algebraic Coding Theory*. McGraw-Hill, New York, 1968.

[5] **S.Lin** and **D.J.Costello**. *Error Control Coding Fundamentals and Applications*. Prentice-Hall, Inc., New Jersey, 1983.

[6] **R.E.Blahut**. *Theory and Practice of Error Control Codes*. Addison-Wesley, 1983.

[7] **F.J.MacWilliams** and **J.A.Sloane**. *The Theory of Error-Correcting Codes*. North-Holland, Amsterdam, 1977.

[8] **V.Pless**. *Introduction to the theory of error-correcting codes*. John Wiley and Sons, 1982.

[9] **W.W.Peterson** and **E.J.Weldon**. *Error correcting codes*. MIT Press, 1972.

[10] **A.M.Michelson** and **A.H.Levesque**. *Error-control techniques for digital communication*. John Wiley & Sons, 1985.

[11] **G.D.Forney**. "Burst-correcting codes for the classic burst channel". *IEEE Trans. Commun. Technol., vol.COM-19*, pp.772–781, October 1971.

[12] **GSM Recommodation 05.03**. "Channel coding". *Draft Version 3.1.0*, February 1988.

[13] **K.Y.Liu** and **J.J.Lee**. "Recent results on the use of concatenated Reed-Solomon/Viterbi channel coding and data compression for space communications". *IEEE Trans. Commun., vol.COM-32, no.5*, pp.518–523, May 1984.

[14] **J.L.Ramsey**. "Realisation of optimum interleavers". *IEEE Trans. Info. Theory, vol.IT-16, no.3*, pp.338–345, May 1970.

[15] **R.Steele**. "Towards a high-capacity digital cellular mobile radio system". *IEE Proc., Part F, 132, no.5*, pp.405–415, August 1985.

[16] **W.F.Bodtmann** and **H.W.Arnold**. "Fade-duration statistics of a Rayleigh-distributed wave". *IEEE Trans. Commun., vol.COM-30, no.3*, pp.549–553, March 1982.

[17] **J.M.Wozencraft**. "Sequential decoding for reliable communication". *IRE Natl. Conv. Rec., vol.5, pt.2*, pp.11–25, 1957.

[18] **J.M.Wozencraft** and **B.Reiffen**. *Sequential decoding*. MIT Press, Cambridge, Mass., 1961.

[19] **R.M.Fano**. "A heuristic discussion of probabilistic coding". *IEEE Trans. Info. Theory, vol.IT-9*, pp.64–74, April 1963.

[20] **A.J.Viterbi**. "Error bounds for convolutional codes and an asymphtotically optimum decoding algorithm". *IEEE Trans. Info. Theory, vol.IT-13*, pp.260–269, April 1967.

[21] **G.D.Forney**. "The Viterbi algorithm". *Proc. of the IEEE, vol.61, no.3*, pp.268–278, March 1973.

[22] **J.A.Heller** and **I.M.Jacobs**. "Viterbi decoding for satellite and space communication". *IEEE Trans. Commun. Technol., vol.COM-19, no.5*, pages pp.835–848, October 1971.

[23] **K.H.H.Wong** and **R.Steele**. "Transmission of digital speech in highway microcells". *Journal of Instn. of Electronic & Radio Engrs., vol.57, no.6 (supplement)*, pp.S246–S254, November-December 1987.

[24] **E.A.Bucher** and **J.A.Heller**. "Error probability bounds for systematic convolutional codes". *IEEE Trans. Inform. Theory, vol.IT-16*, pp.219–224, March 1970.

[25] **J.P.Odenwalder**. *Optimal decoding of convolutional codes*. PhD thesis, Ph.D. dissertation, Dept. of Systems Sciences, School of Engineering and Applied Sciences, University of California, Los Angeles, 1970.

[26] **Consultative Committee for Space Data Systems**. "Blue book". *Recommendations for Space Data System Standards: Telemetry Channel Coding*, May 1984.

[27] **T.Yamada, H.Harashima**, and **H.Miyakawa**. "A new maximum likelihood decoding of high rate convolutional codes using a trellis". *Trans. Inst. Electron. & Commun. Eng. Jpn. 66A*, pages pp.611–616, 1983.

[28] **L.H.C.Lee** and **P.G.Farrell**. "Error performance of maximum-likelihood trellis decoding of $(n, n-1)$ convolutional codes: a simulation study". *IEE Proc., vol.134, Pt.F, no.7*, pp.673–680, December 1987.

[29] **K.J.Larsen**. "Short convolutional codes with maximal free distance for rate 1/2, 1/3 and 1/4". *IEEE Trans. Info. Theory, vol.IT-19*, pp.371–372, May 1973.

[30] **J.B.Cain**, **G.C.Clark**, and **J.M.Geist**. "Punctured convolutional codes of rate $(n-1)/n$ and simplified maximum likelihood decoding". *IEEE Trans. Info. Theory, vol.IT-25, no.1*, pp.97–100, January 1979.

[31] **Y.Yasuda**, **K.Kashiki**, and **Y.Hirata**. "High-rate punctured convolutional codes for soft decision Viterbi decoding". *IEEE Trans. Commun., vol.COM-32, no.3*, pp.315–319, March 1984.

[32] **D.G.Daut**, **J.W.Modestino**, and **L.D.Wismer**. "New short constraint length convolutional code construction for selected rational rates". *IEEE Trans. Info. Theory, vol.IT-28*, pp.793–799, September 1982.

[33] **A.J.Viterbi**. "Convolutional codes and their performance in communication systems". *IEEE Trans. Commun. Technol., vol.COM-19, no.5*, pages pp.751–772, October 1971.

[34] **S.T.S.Chia**, **R.Steele**, **E.Green**, and **A.Baran**. "Propagation and bit error rate measurement for a microcellular system". *Journal of Instn of Electronic & Radio Engrs, vol.57, no.6 (supplement)*, pp.S255–S266, November-December 1987.

[35] **A. Hocquenghem**. "Codes correcteurs d'erreurs". *Chiffres (Paris), vol.2*, pp.147–156, September 1959.

[36] **R.C.Bose** and **D.K.Ray-Chaudhuri**. "On a class of error correcting binary group codes". *Information and Control, vol.3*, pp.68–79, March 1960.

[37] **R.C.Bose** and **D.K.Ray-Chaudhuri**. "Further results on error correcting binary group codes". *Information and Control, vol.3*, pp.279–290, September 1960.

[38] **W.W.Peterson**. "Encoding and error correction procedures for the Bose-Chaudhuri codes". *IRE Trans. Inform. Theory, vol.IT-6*, pp.459–470, September 1960.

[39] **D.Gorenstein** and **N.Zierler**. "A class of cyclic linear error-correcting codes in p^m synbols". *J. Soc. Ind. Appl. Math., 9*, pp.107–214, June 1961.

[40] **I.S.Reed** and **G.Solomon**. "Polynomial codes over certain finite fields". *J. Soc. Ind. Appl. Math., vol.8*, pp.300–304, June 1960.

[41] **R.T.Chien**. "Cyclic decoding procedure for the Bose-Chaudhuri-Hocquenghem codes". *IEEE Trans. Info. Theory, vol.10*, pp.357–363, 1964.

[42] **G.D.Forney**. "On decoding BCH codes". *IEEE Trans. Info. Theory, vol.11*, pp.549–557, 1965.

[43] **E.R.Berlekamp**. "On decoding binary Bose-Chaudhuri-Hocquenghem codes". *IEEE Trans. Info. Theory, vol.11*, pp.577–579, 1965.

[44] **J.L.Massey**. "Step-by-step decoding of the Bose-Chaudhuri-Hocquenghem codes". *IEEE Trans. Info. Theory, vol.11*, pp.580–585, 1965.

[45] **J.L.Massey**. "Shift-register synthesis and BCH decoding". *IEEE Trans. Info. Theory, IT-15*, pp.122–127, January 1969.

[46] **E.R.Berlekamp**, **R.E.Peile**, and **S.P.Pope**. "The application of error control to communications". *IEEE Commun. Magazine, vol.25, no.4*, pp.44–57, April 1987.

[47] **E.R.Berlekamp**. "The technology of error-correcting codes". *Proc. of the IEEE, vol.68, no.5*, pp.564–592, May 1980.

[48] **K.H.H.Wong**, **L.Hanzo**, and **R.Steele**. "Channel coding for satellite mobile channels". *International Journal on Satellite Comm., accepted for publication*, 1989.

[49] **E.Prange**. "Cyclic error-correcting codes in two symbols". *AFCRC-TN-57, 103, Air Force Cambridge Research Center, Cambridge, Mass.*, 1972.

[50] **D.Chase**. "A class of algorithms for decoding block codes with channel measurement information". *IEEE Trans. on Info. Theory, vol.IT-18, no.1*, pages pp.170–182, January 1972.

[51] **J.K.Wolf**. "Efficient maximum likelihood decoding of linear block codes using a trellis". *IEEE Trans. Info. Theory, vol.IT-24, no.1,*, pp.76–80, January 1978.

[52] **T.Matsumoto**. "Trellis decoding of linear block codes in digital mobile radio". *38 th IEEE Vehicular Technology Conf., Philadelphia, Pennsylvania*, pp.6–11, 15–17 June 1988.

[53] **C.R.P.Hartmann** and **L.D.Rudolph**. "An optimum symbol-by-symbol decoding rule for linear codes". *IEEE Trans. Info. Theory, vol.IT-22, no.5,*, pp.514–517, September 1976.

[54] **L.R.Bahl**, **J.Cocke**, **F.Jelinek**, and **J.Raviv**. "Optimum decoding of linear codes for minimising symbol error rate". *IEEE Trans. Info. Theory, vol.IT-20*, pp.284–287, March 1974.

[55] **T.Kasami** and **S.Lin**. "On the probability of undetected error for the maximum distance separable codes". *IEEE Trans. on Commun., vol.COM-32, no.9*, pp.998–1006, September 1984.

[56] **Z.McC.Huntoon** and **A.M.Michelson**. "On the computation of the probability of post-decoding error events for block codes". *IEEE Trans. on Info. Theory*, pp.399–403, May 1976.

[57] **A.M.Michelson**. "The calculation of post-decoding bit-error probabilities for binary block codes". *Nat. Telecomm. Conf. Rec.*, pp.24.3.1–24.3.4, 1976.

[58] **G.D.Forney**. *Concatenated codes*. MIT Press, Cambridge, Massachusetts, 1966.

[59] **S.D.Bate**, **B.Honary**, and **P.G.Farrell**. "Error control techniques applicable to HF channels". *IEE Proc., vol.136, Pt.I, no.1*, pp.57–63, February 1989.

[60] **U.H.-G.KreBel** and **P.A.M.Buné**. "Adapative forward error correction for fast data transmission over the mobile radio channel". *8 th European Conf. on Electrotechnics, Conf. Proc., on Area Commun., Sweden*, pp.170–173, June 1988.

[61] **P.A.M.Buné**. "A fast and secure data transmission scheme for the GSM system". *CEPT/GSM/WP2, Document 278, Stockholm, Sweden*, October 1987.

Chapter **5**

Quaternary Frequency Shift Keying

I.J.Wassell[1] and R.Steele[2]

For the mobile radio system considered in this chapter the system signalling rates are sufficiently low for the mobile channel to exhibit flat fading.
The modulated signal bandwidth is therefore less than the channel coherence bandwidth for a significant proportion of the time. Consequently
there is no need to employ relatively expensive and power hungry adaptive
equalisers to counteract the effects of time dispersive and frequency selective channels. The flat fading experienced by narrowband systems may be
combatted by employing frequency hopping or space diversity. An effective
way of decreasing the channel occupancy of a modulated signal, and to
decrease the probability of intersymbol interference, is to use more than
one bit per symbol. In this chapter we consider quaternary frequency shift
keying where two bits per symbol are transmitted.

5.1 An S900-D Like System

The deployment of multilevel modulation reduces the symbol rate compared to binary modulation and therefore the channel bandwidth can be
decreased. However, a consequence of transmitting more than one bit per
symbol is that the signal power must be commensurately increased for the
same channel noise if the symbol error rate is not to increase. A particular

---

[1]University of Southampton and Multiple Access Communications Ltd
[2]University of Southampton and Multiple Access Communications Ltd

advantage accrues if the TDMA multilevel signal has a symbol rate sufficiently low that the mobile radio channel exhibits flat fading rather than frequency selective fading. Thus by ensuring that the modulation bandwidth is less than the coherence bandwidth delay dispersion of the spectral components in the received signal is avoided and there is no need to employ equalisers to remove intersymbol interference (ISI). The flat fading can be combatted by means of diversity techniques.   Perhaps the most simple of all the multilevel modulation methods is quaternary frequency shift keying (QFSK), and a NB-TDMA system using QFSK has been studied by Ketterling, Pfitzmann and Tietgen [1–3]. In their QFSK/TDMA system they employ narrowband TDMA (NB-TDMA) and accommodate the channel induced ISI by restricting the number of TDMA channels per carrier to 10 and by employing 4-level FSK, i.e., QFSK. Their system is designed on the assumption that the time delays of the multipath signals do not exceed 10 μs and the fade depths are rarely more than 10 dB. The TDMA rate used is 128 kb/s or 64 k symbols/s, resulting in a symbol length of 15.6 μs, a duration not particularly long compared to the assumed excess delay spread of 10 μs. The system would have a better performance if the cellular clusters are reduced in size to ensure that the excess delay spread is significantly less than 10 μs, and if the symbol rate is decreased by arranging for fewer users per carrier. Alternatively higher level FSK can be used to reduce the symbol rate and thereby reduce the risk of ISI. However, the channel SNR would then have to be increased. Nevertheless the QFSK/NB-TDMA system [1–3] does have the virtue of simplicity. The systems described in Chapter 6, which are also designed to operate in large cells, have a better performance than QFSK/NB-TDMA but at the expense of greater complexity. We may speculate that eventually large cells will only exist in rural areas and sparsely populated countries. In the densely populated countries the large cell will be relatively rare, rendering this simple QFSK/NB-TDMA system adequate for many types of mobile communications.

Let us consider the QFSK/NB-TDMA system in more detail. The speech signals are residual excited linear prediction (RELP) encoded at 9.6 kb/s and converted to 11 kb/s by repeating essential and already protected parts of the information. The TDMA frame lasting 32 ms contains 10 voice band channels and a frame synchronisation word to yield a TDMA bit rate of 128 kb/s. Each consecutive two bits in the TDMA signal are formed into a symbol having four possible levels depending on the logical levels of the two-bits. This baseband waveform is applied to a voltage controlled oscillator to give the modulated signal. The dotted lines in Figure 5.1 show the waveforms of three sets of five consecutive arbitrary symbols. The sharp transitions at the symbol boundaries cause the eye pattern at a receiver with a non-coherent frequency discriminator to be wide open, but the spectral spillage of the RF modulated signal is unacceptably large. A compromise for containing the bandwidth of the modulated signal while

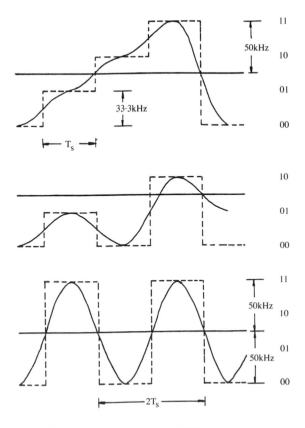

Figure 5.1: Baseband QFSK waveforms.

providing an acceptable eye pattern at the receiver is to filter the symbols prior to modulation. By this approach the smooth waveforms shown in Figure 5.1 are obtained.

To generate the smooth transitions the current and previous symbols address a read-only memory (ROM) containing the 16 possible intersymbol transitions associated with the 4 possible symbols. These transitions are displayed in Figure 5.2 for a cosine square function, where 90° of the function is generated in a symbol period T_s. The sub-figures (a), (b), (c) and (d) show all the possible transitions from symbols 00, 01, 10 and 11, to the other symbol values, respectively. The filter action of the cosine squared shaping reduces the spectral spillage into adjacent RF channels, while providing an acceptable eye pattern. Figure 5.3 shows all the transitions from 00 (solid lines) and 11 (dotted lines) to all the possible symbols and then back again over a two symbol period. The corresponding transitions for 10 and 01 are shown in Figure 5.4. The eye pattern shown in Figure 5.5 is

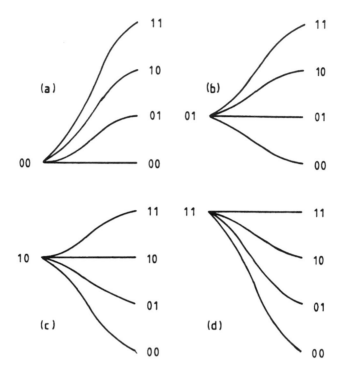

Figure 5.2: Baseband QFSK transition waveforms.

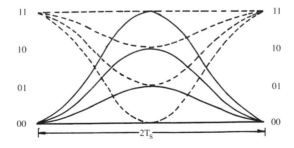

Figure 5.3: Baseband QFSK 00 to 11 transition waveforms.

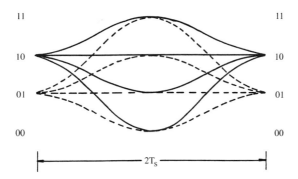

Figure 5.4: Baseband QFSK 10 to 01 transition waveforms.

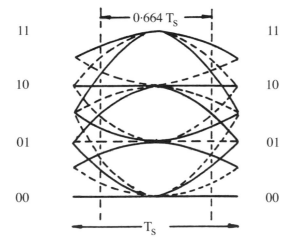

Figure 5.5: Baseband QFSK eye diagram.

formulated by overlaying Figures 5.3 and 5.4. Notice that because of the raised cosine filtering the greatest width of the eye is 0.664 T_s, instead of T_s which would be available in the absence of filtering.

The smooth symbol waveforms of Figure 5.1 frequency modulate an RF carrier, and the frequency deviation between adjacent steps in the symbol levels is set at 33.3 kHz. The highest modulation bandwidth occurs when the data are ... 0011001100 ... resulting in a peak deviation of ±50 kHz as shown in Figure 5.1, and a modulation signal that is nearly a pure tone of frequency $1/2T_s$ = as $1/T_s$ = 64 kBd. The resulting line spectrum of the QFSK signal is displayed in Figure 5.6. Because it is considered that the adjacent RF carrier power should be attenuated by 70 dB relative to the unmodulated carrier, the bandwidth is seen to be 320 kHz. By acknowledging that the two zeros-two ones periodic sequence is statistically rare, the bandwidth of the transmitted signal is in reality of the order of 250 kHz. The modulation index is given by [4]

$$h_f = 2f_d T_s$$

where f_d is the frequency deviation from the carrier. The frequencies nearest the carrier are located at $f_c \pm f_d$, and the farthest ones at $f_c \pm 3f_d$ corresponding to $h_f = 0.52$. Applying a pseudo random binary sequence (PRBS) to the modulator results in the power spectrum of Figure 5.7. This Figure shows a 40 dB down bandwidth of only 240 kHz. To reduce spectral occupancy further one could consider reducing the modulation index, for example, with $h_f = 0.25$ the 40 dB bandwidth is now only 160 kHz. Unfortunately the power efficiency of the modulation scheme (in terms of bit error rate (BER)) has now been reduced. The carrier spacing recommended in references [1–3] is 320 kHz which gives good adjacent channel interference protection, while a receiver bandwidth of 250 kHz gives a reasonable compromise between signal distortion and noise band-limitation.

At the receiver, the frequency variations, i.e., the derivative of the phase of the received signal, are converted into voltage variations to yield the eye pattern of Figure 5.5. The decision levels to regenerate the symbols are set at the centre of the eye openings. This eye pattern is drawn for ideal equipment and transmission channels.

The TDMA channel burst power is 10 Watts, giving approximately 1 Watt mean power for the MS. This power level should accommodate cell sizes of 10 to 20 km radius. The BS antenna is approximately 150 m, and the receiver sensitivity is of the order 1.0 μV_{emf} at 50 Ω.

Figure 5.8 shows the basic TDMA structure. The MS to BS transmissions are in the frequency band 890–915 MHz, while the BS to MS are from 935–960 MHz. Each duplex band is divided into 100 sub-bands of 250 kHz, and as each carrier carries 10 traffic channels, the total number of duplex channels is 1000. For a seven-cell cluster there are 140 radio channels per cell. A ten frame super-frame is used, where each of its frames contains 10 TDMA traffic channels (TCHs) on a single carrier. A frame synchronisation word with a frame number is inserted at the beginning of each TDMA frame. The frame number is of importance as it identifies the position of the frame in the super-frame. Provided no call connection is established the MS is able to determine the reception quality of the other nine BSs in its super-frame, and it can request to be switched to another BS if the reception is unacceptable.

Each TCH contains 408 bits of which 352 constitute the digital speech. At the beginning of each time slot or packet is a 12-bit synchronisation word, followed by a 4-bit frame number. Next comes a 10-bit signalling

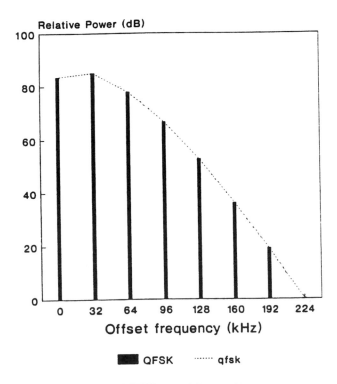

Figure 5.6: Line spectrum of QFSK signal for a ... 0011001100 ... data signal.

word made up of 8 bits of information and two bits for error detection. It is repeated as an error protection measure after the 352 bits of speech data have been processed. Because the MSs vary their distances from their BS as they travel the propagation times of the packets varies. The BS notes these variations, normalises them on a time basis and adjusts the synchronisation accordingly. It is able to achieve this by using the last 20 bits in a time-slot for transmission from MS to BS. Because each bit has a duration of 7.8 μs, the maximum propagation delay that can be accommodated is $20 \times 7.8 = 156$ μs. As the propagation velocity is 3.3 μ s/km, a distance of $156/3.3 = 47.3$ km can be allowed. For BS to MS transmissions the last 20 bits in the time slot contain system information for the MS.

In each 10 channel frame, one channel is used as a common control channel (CCCH) to convey commands relating to MS registration, roaming, call set-up, and so forth. The CCCH is shown in Figure 5.8. The CCCH organisation is dependent on the direction of transmission. For BS to MS transmission the packet is divided into two identical sub-packets, each consisting of 12 synchronisation bits, followed by 172 information bits, and

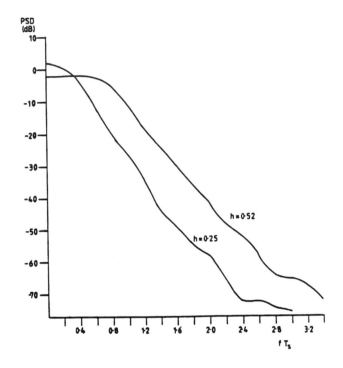

Figure 5.7: Power spectral density of the QFSK signal for a PRBS data signal.

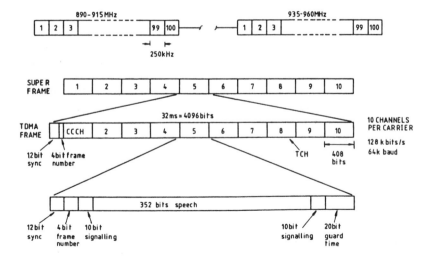

Figure 5.8: TDMA structure.

20 BS identification bits. The use of the two sub-packets builds in redundancy to mitigate the effects of channel errors. Alternate packet structures are used for transmission from MS to BS. During odd-numbered frames four sub-packets, each having 102 bits are used. The first 12 bits in the sub-packet relate to synchronisation, the next 56 bits are message bits and the last 34 bits equalise the propagation delay. These sub-packets are repeated as they are used in an ALOHA procedure [5]. The even numbered CCCH packets contain two sub-packets having 102 bits and one of 204 bits.

We have focused on a particular QFSK system that was a contender for the pan-European digital cellular mobile radio system. It was not a successful candidate because it had to compete with more complex systems that were more appropriate for large cell sizes. However, the QFSK/NB-TDMA system has the virtue of relative simplicity, and has applicability in other situations. We will, therefore, now embark on a closer inspection of QFSK, determining its theoretical performance in a variety of situations, leaving the reader to decide on its suitability for his or her application. We commence by considering the simplest channel, namely an additive white Gaussian noise (AWGN) channel. Later we will examine the performance of QFSK transmissions over Rayleigh fading channels.

5.2   QFSK Transmissions Over Gaussian Channels

The transmitted frequency in QFSK depends on the logical values of the two-bit symbols. A suitable symbol-to-carrier frequency mapping is specified in Table 5.1, where the signalling frequencies f_0, f_1, f_2 and f_3 are the

Quaternary symbols		Transmitted
Natural binary	Gray code	carrier frequency
0 0	0 0	f_0
0 1	0 1	f_1
1 0	1 1	f_2
1 1	1 0	f_3

Table 5.1: Symbol to carrier frequency mapping for QFSK.

values of the carrier frequency over the two-bit symbol interval. The frequencies are orthogonal and the spacing between adjacent tones is $2f_d$. These signalling frequencies, or elements, may be expressed as

$$s_i(t) = \sqrt{\frac{2E_s}{T_s}} \cos\left(2\pi f_i t + \phi_0\right) \qquad 0 \leq t \leq T_s \qquad (5.1)$$

where i is 0, 1, 2 or 3 depending on the logical values of the two bits being transmitted, E_s is the symbol energy, T_s is the symbol period and ϕ_0 is an arbitrary phase which can be set to zero with coherent demodulation. We will simplify our analysis by assuming square modulating waveforms; no bandlimiting by the channel; and the absence of intersymbol, adjacent channel and cochannel interference. Now because the channel contains additive white Gaussian noise (AWGN) $n(t)$, the signal at the input to the demodulator at the receiver during the symbol period from 0 to T_s is

$$r(t) = \alpha s_i(t) + n(t) \tag{5.2}$$

where α is an attenuation factor. For a Gaussian channel, α is a constant for a MS at a given distance from its BS, and for simplicity we will set it to unity, unless otherwise stated. The AWGN signal may be represented as

$$n(t) = n_I(t) \cos 2\pi f_i t - n_Q(t) \sin 2\pi f_i t \tag{5.3}$$

where $n(t), n_I(t)$ and $n_Q(t)$ are all zero mean Gaussian random processes having the same average power level. The quadrature amplitude signals $n_I(t)$ and $n_Q(t)$ occupy the frequency band from $-B/2$ to $B/2$, where B is the bandwidth of $n(t)$ over positive frequencies. The two-sided PSD of $n_I(t)$ and $n_Q(t)$ is $N_o/2$.

5.2.1   Demodulation in the Absence of Cochannel Interference

We will now present expressions for the probability of symbol error as a function of channel SNR for QFSK transmissions over Gaussian channels. Coherent demodulation is considered first, followed by non-coherent demodulation. In Sections 5.2.2 and 5.2.3 we deal with the effects of cochannel interference.

5.2.1.1   Coherent Demodulation

The QFSK coherent demodulator is shown in Figure 5.9, and an analysis, following the approach of Clark [6], will now be given. The received signal $r(t)$ is multiplied by a set of i coherent RF tones, $\sqrt{2} \cos 2\pi f_i t$, to give $m_i(t), i = 0, 1, 2$ and 3. The outputs from the four multipliers are

$$
\begin{aligned}
m_i(t) &= \left((2E_s/T_s)^{\frac{1}{2}} \cos 2\pi f_i t + n(t) \right) \sqrt{2} \cos 2\pi f_i t \\
&= (E_s/T_s)^{\frac{1}{2}} + (E_s/T_s)^{\frac{1}{2}} \cos 4\pi f_i t + \sqrt{2} n(t) \cos 2\pi f_i t. \tag{5.4}
\end{aligned}
$$

The final term in Equation 5.4 is the noise component, and from Equation 5.3 it can be expressed as

$$(2)^{-\frac{1}{2}} \left\{ n_I(t) + n_I(t) \cos 4\pi f_i t - n_Q(t) \sin 4\pi f_i t \right\}$$

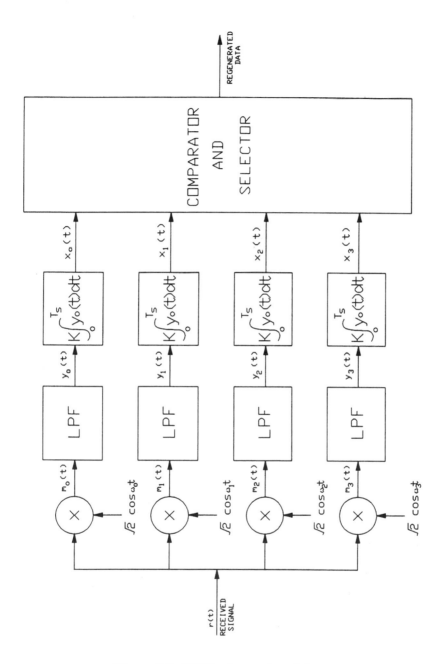

Figure 5.9: QFSK coherent demodulator.

and consequently when $m_i(t)$ is low pass filtered it becomes

$$y_i(t) = (E_s/T_s)^{\frac{1}{2}} + u(t) \tag{5.5}$$

where

$$u(t) = n_I(t)/\sqrt{2}. \tag{5.6}$$

The signal $u(t)$ is a Gaussian random process with zero mean and a variance of $N_o/2$ over the frequency band of the baseband signal. Each $y_i(t)$ signal, $i = 0, 1, 2, 3$ is integrated to give

$$
\begin{aligned}
x_i(t) &= K \int_0^{T_s} y_i(t)dt \\
&= K (E_s/T_s)^{\frac{1}{2}} + v_i
\end{aligned}
\tag{5.7}
$$

where K is the gain of the integrator, and

$$v_i = K \int_0^{T_s} u(t)dt \tag{5.8}$$

is the noise signal at the output of the integrator.

We digress at this stage to observe that the variance of a signal (here v_i) at the output of a linear filter (here an integrator) having a transfer function $G(f)$ and an input power spectral density (PSD) of $N_o/2$ is

$$N = \frac{N_o}{2} \int_{-\infty}^{\infty} |G(f)|^2 \, df \tag{5.9}$$

where

$$G(f) = \int_{-\infty}^{\infty} g(t)e^{-j2\pi ft}dt \tag{5.10}$$

and $g(t)$ is the impulse response of the filter. By Parsaval's theorem

$$\int_{-\infty}^{\infty} |G(f)|^2 \, df = \int_{-\infty}^{\infty} g^2(t)dt \tag{5.11}$$

and so

$$N = \frac{N_o}{2} \int_{-\infty}^{\infty} g^2(t)dt. \tag{5.12}$$

We now return to the main discourse, and observe that the integrators in the coherent demodulator of Figure 5.9 have an impulse response of

$$g(t) = \begin{cases} K; & 0 \le t \le T_s \\ 0; & \text{elsewhere} \end{cases} \tag{5.13}$$

Integrator outputs x_i	Quaternary symbols			
	0	1	2	3
x_0	$K\sqrt{E_sT_s} + v_0$	v_0	v_0	v_0
x_1	v_1	$K\sqrt{E_sT_s} + v_1$	v_1	v_1
x_2	v_2	v_2	$K\sqrt{E_sT_s} + v_2$	v_2
x_3	v_3	v_3	v_3	$K\sqrt{E_sT_s} + v_3$

Table 5.2: Relationship between the quaternary symbols and the integrator outputs.

and therefore from Equation 5.12

$$N = \frac{N_o}{2}K^2T_s \qquad (5.14)$$

is the variance of v_i.

The relationship between the quaternary symbols denoted by $i = 0, 1, 2, 3$ and the corresponding integrator outputs $x_i(t)$ are given in Table 5.2. Each of the quaternary symbols is orthogonal in that the transmitted carrier frequency f_i causes the ith path in the demodulator to give a signal $K\sqrt{T_sE_s}$ plus noise, while all the other paths only pass the noise component. Let us consider the case when the symbol '00' was transmitted using the frequency f_o. The signals $x_o(t), x_1(t), x_2(t)$ and $x_3(t)$, are $K\sqrt{T_sE_s} + v_o, v_1, v_2$ and v_3, respectively. In order to decide which symbol was transmitted the $x_i(t)$ signals are compared, the largest $x_i(t)$ is selected, and the bit word associated with it is regenerated.

For an erroneous two-bit symbol to be regenerated one or more of the $x_i(t), i = 1, 2, 3$ must exceed $x_o(t)$. The probability that one or more of these three conditions are satisfied is the union

$$
\begin{aligned}
P_e &= P\left[(x_1 > x_o) \cup (x_2 > x_o) \cup (x_3 > x_o)\right] \\
&= P(x_1 > x_o) + P(x_2 > x_o) + P(x_3 > x_o) \\
&\quad -P\left[(x_1 > x_o) \cap (x_2 > x_o)\right] - P\left[(x_1 > x_o) \cap (x_3 > x_o)\right] \\
&\quad -P\left[(x_2 > x_o) \cap (x_3 > x_o)\right] \\
&\quad +P\left[(x_1 > x_o) \cap (x_2 > x_o) \cap (x_3 > x_o)\right] \qquad (5.15)
\end{aligned}
$$

and so

$$P_e < P(x_1 > x_o) + P(x_2 > x_o) + P(x_3 > x_o) \qquad (5.16)$$

where $\cup$ and $\cap$ are the union and intersection of the events. The probability of a symbol error is, therefore, less than the sum of the individual probabilities of the three events. However, $x_i, i = 1, 2, 3$ are likely to exceed x_o with equal probability, and hence

$$P(x_i > x_o) = P\left(v_i > K\sqrt{T_sE_s} + v_o\right)$$

$$= P\left(v > K\sqrt{T_s E_s}\right) \tag{5.17}$$

where

$$v = v_o + v_i. \tag{5.18}$$

The v_i are Gaussian distributed, and hence v is also a Gaussian variable with a variance equal to the sum of the variances of v_o and v_i, namely $2N$, and a mean of zero. Consequently the PDF of v is

$$P(v) = \frac{1}{\sqrt{4\pi N}} exp\left(\frac{v^2}{4N}\right). \tag{5.19}$$

Hence

$$
\begin{aligned}
P(x_i > x_o) &= \int_{K\sqrt{T_s E_s}}^{\infty} \frac{1}{\sqrt{4\pi N}} exp - \left(\frac{v^2}{4N}\right) dv \\
&= \int_{\frac{K\sqrt{T_s E_s}}{\sqrt{2N}}}^{\infty} \frac{1}{\sqrt{2\pi}} exp - \left(\frac{v^2}{2}\right) dv \\
&= Q\left(K\sqrt{\frac{T_s E_s}{2N}}\right)
\end{aligned} \tag{5.20}
$$

where the Q-function is defined by

$$Q(\lambda) \triangleq \int_{\lambda}^{\infty} \frac{1}{\sqrt{2\pi}} exp\left(-\frac{x^2}{2}\right) dx. \tag{5.21}$$

From Equations 5.15, 5.16, and 5.20,

$$P_e < 3Q\left(K\sqrt{\frac{T_s E_s}{2N}}\right). \tag{5.22}$$

Substituting K from Equation 5.14 into Equation 5.22 and replacing the less than sign by an equals sign on the assumption of high signal-to-noise ratios (SNR), yields

$$P_e = Q\left(K\sqrt{\frac{T_s E_s}{2N}}\right) = Q\left(\sqrt{\frac{E_s}{N_o}}\right) \tag{5.23}$$

where E_s/N_o is the carrier-to-noise ratio C/N, assuming the receiver bandwidth is $1/T_s$. As we are transmitting two bits per symbol, the energy per bit is

$$E_b = E_s/2 \tag{5.24}$$

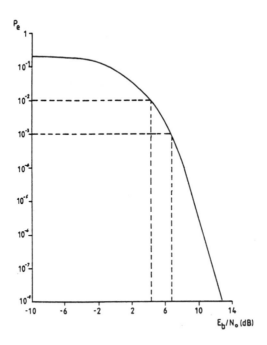

Figure 5.10: Coherent QFSK demodulation in the presence of AWGN.

giving a probability of symbol error in terms of bit energy as

$$P_e = Q\left(\sqrt{\frac{2E_b}{N_o}}\right). \qquad (5.25)$$

Observe that P_e is also the symbol error rate (SER). The variation of the probability of symbol error P_e as a function of E_b/N_o is shown in Figure 5.10. Now there are speech encoders that can operate with near toll quality performance with a bit error rate (BER) of 10^{-3}. As the BER is lower than the SER (often it is half), we observe from Figure 5.10 that toll quality speech is obtainable for E_b/N_o ratios exceeding 7 dB. If channel coding is employed, the SER before channel decoding can be of the order of 10^{-2}, necessitating a $E_b/N_o > 5$ dB. Notice that a SER of 10^{-8} can be achieved for an E_b/N_o of 12 dB.

5.2.1.2 Non-coherent Demodulation

We now turn our attention to a simpler type of demodulation known as non-coherent demodulation. Unlike coherent demodulation, there is no need to accurately acquire the transmitted carrier frequency in order to

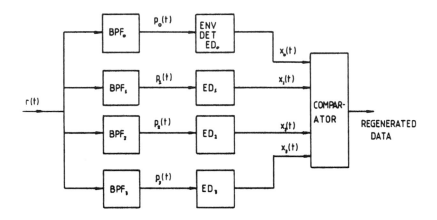

Figure 5.11: Non-coherent QFSK demodulator.

demodulate the received signal. Instead the received signal $r(t)$ is filtered by a bank of bandpass filters (BPFs) whose centre frequencies are those frequencies f_o, f_1, f_2 and f_3 that correspond to the frequencies used to convey the data symbols, see Table 5.1. As the received signal $r(t)$ is composed of a tone representing a data symbol plus Gaussian noise, the filtered signals are bandlimited noise signals, except for one filter whose output is the transmitted tone plus noise. Envelope detection of the filtered signals ensues to give signals x_0, x_1, x_2 and x_3. At a sampling instant these signals are sampled to give X_0, X_1, X_2 and X_3 respectively, and the largest X_i is noted. The value of i associated with X_i is used to regenerate the most probable transmitted symbol. The block diagram of the non-coherent QFSK demodulator is shown in Figure 5.11.

In our analysis we will assume that the bandpass filters have a bandwidth B about a centre frequency f_i, and that B is the reciprocal of T_s. We will also assume that the signal components are filtered without introducing waveform distortion and that ideal envelope detection occurs. The probability of a correct symbol detection given that the symbol s_o was transmitted is

$$P_{co} = P\{X_1 < X_0, X_2 < X_0, X_3 < X_0 | s_o \text{ sent }\}$$

which we may express as

$$= \int_{-\infty}^{\infty} \int_{-\infty}^{x_o} \int_{-\infty}^{x_o} \int_{-\infty}^{x_o} [f(x_1, x_2, x_3 | s_o) dx_1 dx_2 dx_3] f(x_o | s_o) dx_o \qquad (5.26)$$

where the conditional PDF of x_o given the transmission of s_o is $f(x_o | s_o)$, and the joint PDF of X_1, X_2 and X_3 may be simplified to

$$f(x_1, x_2, x_3 | s_o) = f(x_1) f(x_2) f(x_3). \qquad (5.27)$$

As

$$\int_{-\infty}^{x_o} f(x_i) dx_i \qquad (5.28)$$

is the same for $i = 1, 2, 3$, the probability of the QFSK non-coherent demodulator making a correct decision is

$$P_c = P_{co} = \int_{-\infty}^{\infty} \left[\int_{-\infty}^{x_o} f(x) dx \right]^3 f(x_o | s_o) dx_o \qquad (5.29)$$

and consequently the probability of making an erroneous decision is

$$P_e = 1 - P_c. \qquad (5.30)$$

To evaluate P_e, we must find expressions for the PDFs of $f(x)$ and $f(x_o|s_o)$. In order to do this we consider the effect of passing narrowband noise through an envelope detector. Now the quadrature representation of a narrowband noise signal $n(t)$ given in Equation 5.3 may be transformed into an equivalent envelope and phase form

$$n(t) = R(t) \left[\cos(2\pi f_c t + \theta(t) \right] \qquad (5.31)$$

where the envelope $R(t)$ and the phase $\theta(t)$ are given by

$$R(t) = \sqrt{(n_I(t))^2 + (n_Q(t))^2} \qquad (5.32)$$

and

$$\theta(t) = \tan^{-1} \left(\frac{n_Q(t)}{n_I(t)} \right). \qquad (5.33)$$

To find the PDFs of $R(t)$ and $\theta(t)$, which we write for convenience as R and θ, we apply the PDF transform theorem, namely

$$f(R, \theta) = f(n_I, n_Q) |J_1| \qquad (5.34)$$

where J_1 is the Jacobian matrix. As n_I and n_Q are independent Gaussian random variables, the joint PDF of n_I and n_Q is

$$f(n_I, n_Q) = \frac{1}{2\pi N} \exp\left(-(n_I^2 + n_Q^2)/2N \right) ; -\infty < n_I, n_Q < \infty. \qquad (5.35)$$

Solving for n_I and n_Q in terms of R and θ gives a unique solution,

$$n_I = R \cos \theta \qquad (5.36)$$

and

$$n_Q = R \sin \theta. \tag{5.37}$$

The Jacobian is

$$|J_1| = \begin{vmatrix} \frac{\partial n_I}{\partial R} & \frac{\partial n_I}{\partial \theta} \\ \frac{\partial n_Q}{\partial R} & \frac{\partial n_Q}{\partial \theta} \end{vmatrix} \tag{5.38}$$

and from Equations 5.36 and 5.37,

$$|J_1| = \begin{vmatrix} \cos \theta & -R \sin \theta \\ \sin \theta & R \cos \theta \end{vmatrix} = R. \tag{5.39}$$

Substituting $|J_1|$, n_I and n_Q from Equations 5.39, 5.36 and 5.37 into Equations 5.34 and 5.35 yields

$$f(R, \theta) = \frac{R}{2\pi N} \exp(-R^2/2N) \tag{5.40}$$

where

$$0 \leq R \leq \infty \tag{5.41}$$

and

$$-\pi \leq \theta < \pi. \tag{5.42}$$

Integrating $f(R, \theta)$ over the range of θ and R gives the Rayleigh PDF

$$f(R) = \frac{R}{N} \exp(-R^2/2N) \tag{5.43}$$

for the envelope R, and the uniform PDF

$$f(\theta) = \frac{1}{2\pi} \tag{5.44}$$

for the phase θ.

Let us now return to Figure 5.11. If the message symbol transmitted is '00', say, the signal $x_0(t)$ is the envelope of the sum of a single tone f_0 and Gaussian noise. Signals x_1, x_2, and x_3 are the envelopes of band limited white Gaussian noise. However, we have just shown in Equation 5.43 that x_1, x_2 and x_3 have Rayleigh PDFs. Exchanging R in Equation 5.43 for x_i ; $i = 1, 2, 3$, we obtain the PDF of the envelope of the noise signals as

$$f(x_i) = \frac{x_i}{N} \exp \left[-\frac{x_i^2}{2N} \right]. \tag{5.45}$$

We are now in a position to evaluate the cube term in Equation 5.29.

Our next objective is to determine $f(x_o|s_o)$. To do this we consider the receiver path that is carrying the data symbol '00' in the form of a carrier frequency f_o. Because of the receiver noise, the input to the envelope detector is a sinusoid plus the additive narrow band noise. From Equations 5.1

and 5.3 the signal applied to the envelope detector is

$$
\begin{aligned}
p_o(t) &= \sqrt{2S}\cos 2\pi f_o t + n_I(t)\cos 2\pi f_o t - n_Q(t)\sin 2\pi f_o t \\
&= \left(\sqrt{2S} + n_I(t)\right)\cos 2\pi f_o t - n_Q(t)\sin 2\pi f_o t
\end{aligned}
\tag{5.46}
$$

where for convenience we have set E_s/T_s to S. The envelope of $p_o(t)$ is

$$
R(t) = \sqrt{\left(\sqrt{2S} + n_I(t)\right)^2 + n_Q^2(t)}
\tag{5.47}
$$

and to determine its PDF we apply the following notation,

$$
\chi_1 = \sqrt{2S} + n_I(t)
\tag{5.48}
$$

and

$$
\chi_2 = n_Q(t)
\tag{5.49}
$$

where χ_1 and χ_2 are independent Gaussian random variables with

$$
E[\chi_1] = \sqrt{2S}, \quad E[\chi_2] = 0
\tag{5.50}
$$

and

$$
\sigma_{\chi_1}^2 = \sigma_{\chi_2}^2 = N.
\tag{5.51}
$$

Consequently the joint PDF of χ_1 and χ_2 can be written as

$$
f(\chi_1, \chi_2) = \frac{1}{2\pi N}\exp\left(-\frac{(\chi_1 - \sqrt{2S})^2 + \chi_2^2}{2N}\right)
\tag{5.52}
$$

and the phase of $p_o(t)$ is

$$
\theta(t) = \tan^{-1}(\chi_2/\chi_1).
\tag{5.53}
$$

Applying the PDF transformation of Equation 5.34 and the approach of determining polar PDFs used for receiver channels 1, 2, 3, we have

$$
f(R, \theta) = \frac{R}{2\pi N}\exp\left(-\frac{(r^2 + 2S - 2\sqrt{2S}R\cos\theta)}{2N}\right).
\tag{5.54}
$$

The PDF of the envelope R is obtained by integrating over all values of θ,

$$
\begin{aligned}
f(R) &= \int_{-\pi}^{\pi} f(R, \theta)d\theta \\
&= \frac{R}{N}\exp\left(-\frac{2S + R^2}{2N}\right)\left[\frac{1}{2\pi}\int_{-\pi}^{\pi}\exp-\left(\frac{\sqrt{2S}R\cos\theta}{N}\right)d\theta\right].
\end{aligned}
\tag{5.55}
$$

The second term in Equation 5.55 is the modified Bessel function of the first kind and zero order defined by

$$I_o(a) \triangleq \frac{1}{2\pi} \int_{-\pi}^{\pi} \exp(a \cos u) du$$

enabling the PDF of R to be written as

$$f(R) = \frac{R}{N} I_o \left(\frac{\sqrt{2S}R}{N} \right) \exp \left(-\frac{R^2 + 2S}{2N} \right) \tag{5.56}$$

where $R \geq 0$. This is the Rician PDF.

As an aside, it is interesting to make the connection with the Rician PDF of fading signals in radio propagation channels described in Chapter 2. In the propagation case the Rician PDF of the fading envelope is due to a strong component (could be a line-of-sight) and a collection of scattered components whose combination have a Rayleigh envelope. In the QFSK demodulator, the envelope detector operates on the signal element tone $s_i(t)$ and an AWGN signal that has a Rayleigh envelope. The envelopes in both situations are the same, although the physical situations are radically different.

We are now able to formulate the conditional PDF $f(x_o|s_o)$ for the signal at the output of the envelope detector in the receiver path carrying the data symbol by merely replacing R in Equation 5.56 by the envelope signal $x_o(t)$, namely

$$f(x_o|s_o) = \frac{x_o}{N} I_o \left(\frac{\sqrt{2S}x_o}{N} \right) \exp \left(-\frac{x_o^2 + 2S}{2N} \right) ; 0 \leq x_o \leq \infty. \tag{5.57}$$

Having obtained expressions for the PDFs $f(x_i)$ in Equation 5.45 and $f(x_o|s_o)$ in Equation 5.57, we now continue with the evaluation of P_e. To determine the cube term in Equation 5.29 we first integrate $f(x_i)$ over the range 0 to x_o, namely

$$\int_0^{x_o} \frac{x}{N} \exp \left(-\frac{x^2}{2N} \right) dx = 1 - \exp \left(-\frac{x_o^2}{2N} \right) \tag{5.58}$$

to yield

$$\left[1 - \exp \left(-\frac{x_o^2}{2N} \right) \right]^3 = \sum_{j=0}^{3} \binom{3}{j} (-1)^j \exp \left(-j \frac{x_o^2}{2N} \right) \tag{5.59}$$

after applying the Binomial theorem. Substituting this result, and $f(x_o|s_o)$ from Equation 5.57 into Equation 5.29 yields the probability of making a correct symbol regeneration in the non-coherent demodulator as

$$
P_c = \int_0^\infty \sum_{j=0}^{3} \binom{3}{j} (-1)^j \exp\left(-j\frac{x_o^2}{2N}\right) \frac{x_o}{N}
$$

$$
\cdot I_o\left(\frac{\sqrt{2S}x_o}{N}\right) \exp\left(-\frac{x_o^2 + 2S}{2N}\right) dx_o. \tag{5.60}
$$

Reversing the order of the summation and integration gives

$$
P_c = \frac{1}{N} \sum_{j=0}^{3} \binom{3}{j} (-1)^j \exp(-2S/2N)
$$

$$
\cdot \int_0^\infty x_o \exp\left(-\frac{(1+j)x_o^2}{2N}\right) I_o\left(\frac{\sqrt{2S}x_o}{N}\right) dx_o. \tag{5.61}
$$

As

$$
\int_0^\infty x_o \exp\left(-\frac{(1+j)x_o^2}{2N}\right) I_o\left(\frac{\sqrt{2S}x_o}{N}\right) dx_o
$$

$$
= \frac{N}{1+j} \exp\left(\frac{2S}{2N(1+j)}\right) \tag{5.62}
$$

we may express Equation 5.61 as

$$
P_c = \exp(-2S/2N) \sum_{j=0}^{3} \binom{3}{j} \frac{(-1)^j}{1+j} \exp\left(\frac{S}{N(1+j)}\right). \tag{5.63}
$$

Hence the probability of a symbol error becomes

$$
P_e = 1 - P_c = \sum_{j=1}^{3} \binom{3}{j} \frac{(-1)^{j+1}}{1+j} \exp\left(-\frac{j}{j+1} \cdot \frac{S}{N}\right) \tag{5.64}
$$

as the summation term is unity for $j = 0$.

Replacing S in Equation 5.64 by E_s/T_s, and as

$$
N = N_o/T_s \tag{5.65}
$$

then

$$
P_e = \sum_{j=1}^{3} \binom{3}{j} \frac{(-1)^{j+1}}{1+j} \exp\left(-\frac{j}{j+1} \cdot \frac{E_s}{N_o}\right). \tag{5.66}
$$

Because there are two bits in each symbol the energy per bit is given by Equation 5.24 enabling us to write the probability of symbol error in terms

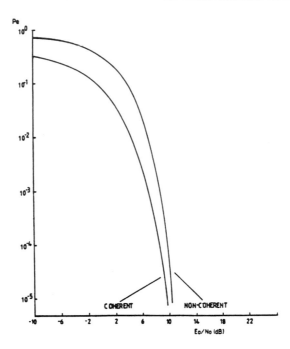

Figure 5.12: Non-coherent QFSK demodulation in the presence of AWGN.

of energy per bit as

$$P_e = \frac{3}{2}\exp\left(-\frac{E_b}{N_o}\right) - \exp\left(-\frac{4}{3}\frac{E_b}{N_o}\right) + \frac{1}{4}\exp\left(-\frac{3}{2}\frac{E_b}{N_o}\right). \tag{5.67}$$

The variation of P_e as a function of E_b/N_o is shown in Figure 5.12, with the curve for coherent QFSK included as a bench mark.

5.2.2   Single Cochannel Interferer with Non-coherent Demodulation

An individual transmitted element of a QFSK signal is given by Equation 5.1, and therefore the signal from a single cochannel interferer has the form

$$c_i(t) = \sqrt{\frac{2E_I}{T_s}}\cos(2\pi f_i t + \phi) \tag{5.68}$$

where E_I is the symbol energy of the interferer, and ϕ is its phase whose fluctuations are uniformly distributed between zero and 2π. Suppose that each symbol is equally likely to be transmitted, and that the interference falls into only one demodulation channel at a time. As a consequence there are two mutually exclusive possibilities regarding the signal and its interference, namely, they can either fall into the same, or into different demodulation channels. An analysis utilising this approach is adopted in reference [7]. When the interference and desired signal are both transmitting the symbol '00', say, the input to the non-coherent demodulator is

$$r_o(t) = s_o(t) + c_o(t) + n(t) \tag{5.69}$$

as the index i is zero. To ease the nomenclature we let

$$E_s/T_s = S \tag{5.70}$$

and

$$E_I/T_s = I \tag{5.71}$$

so the sum of the desired and interfering signal is

$$a(t) = \sqrt{2S} \cos 2\pi f_i t + \sqrt{2I} \cos(2\pi f_i t + \phi). \tag{5.72}$$

By re-writing Equation 5.72 as a phasor

$$a(t) = A(t) \cos(2\pi f_i t + \alpha(t)) \tag{5.73}$$

where

$$\alpha(t) = \tan^{-1} \left(\frac{-\sqrt{2T} \sin \phi}{\sqrt{2S} + \sqrt{2I} \cos \phi} \right) \tag{5.74}$$

is the phase angle and

$$A(t) = \sqrt{2S + 2I + 2\sqrt{2S}\sqrt{2I} \cos \phi} \tag{5.75}$$

is the envelope, the received signal is again composed of the sum of a sinusoidal input and additive noise. As a consequence the PDF of the envelope of the signal $x_o(t)$ at the output of the envelope detector in Figure 5.11 is Rician and is given by Equation 5.57 after exchanging $\sqrt{2S}$ for $A(t)$, namely

$$f(x_o|s_o, \phi) = \frac{x_o}{N} I_o \left(\frac{A(t) x_o}{N} \right) \exp \left(-\frac{x_o^2 + A(t)^2}{2N} \right). \tag{5.76}$$

The cochannel interference signal $c_o(t)$ is unable to pass through the BPFs having centre frequencies f_1, f_2, f_3. Only the noise signal $n(t)$ is filtered by these BPFs, to give signals whose envelopes have Rayleigh PDFs specified by Equation 5.45. As a consequence the conditional symbol error probability can be expressed as

$$P_{eo}|X_o, \phi = 1 - P(X_1 < X_o, X_2 < X_o, X_3 < X_o : X_o, \phi). \tag{5.77}$$

Following the procedure used in Section 5.2.1, we write Equation 5.77 in the form

$$P_{eo}|X_o, \phi = 1 - \int_0^\infty \left\{ \int_0^{x_o} \int_0^{x_o} \int_0^{x_o} \prod_{i=1}^3 f_{x_i}(x_i) dx_i \right\} f(x_o|s_o, \phi) dx_o \quad (5.78)$$

which can be expressed as

$$P_{eo}|X_o, \phi = 1 - \int_0^\infty \left[\int_0^{x_o} f_{x_j}(x_j) dx_j \right]^3 f(x_o|s_o, \phi) dx_o \quad (5.79)$$

to yield

$$P_{eo}|X_o, \phi = \sum_{k=1}^3 \left(\begin{array}{c} 3 \\ k \end{array} \right) \frac{(-1)^{k+1}}{1+k} \exp\left(-\frac{k}{k+1} \cdot \frac{A^2(t)}{2N} \right). \quad (5.80)$$

The average symbol error probability is determined by substituting for $A(t)$ and multiplying the above equation by the uniform PDF of ϕ and integrating over all possible values of ϕ to give

$$P_{eo} = \sum_{k=1}^3 (-1)^{k+1} \left(\begin{array}{c} 3 \\ k \end{array} \right) \frac{1}{1+k}$$

$$\cdot I_o \left(\frac{2(SI)^{\frac{1}{2}}}{N} \left(\frac{k}{k+1} \right) \right) \exp\left(-\frac{(S+I)k}{N(1+k)} \right). \quad (5.81)$$

The probability of an error when the wanted signal and cochannel interference are passed by different bandpass filters in the demodulator will now be determined. The approach we use is similar to that employed in Section 5.2.1, and consequently many of the same equations are applicable. Assuming that the desired signal element has a carrier frequency f_o, while the interfering element has a carrier frequency f_1, then the envelopes of the signals at the outputs of envelope detectors ED_o and ED_1 in Figure 5.11 will have Rician PDFs. The envelope detectors ED_2 and ED_3 respond to Gaussian noise only and their outputs X_2 and X_3 have Rayleigh PDFs. The conditional probability of a symbol error is

$$P_{e1}|X_o = 1 - P\{X_1 < X_o, X_2 < X_o, X_3 < X_o|X_o\} \quad (5.82)$$

and so

$$P_{e1}|X_o = 1 - \int_0^\infty \left[\sum_{k=0}^1 \left(\begin{array}{c} 2 \\ k \end{array} \right) (-1)^k \exp\left(-\frac{kx_o^2}{2N} \right) \right] \quad (5.83)$$

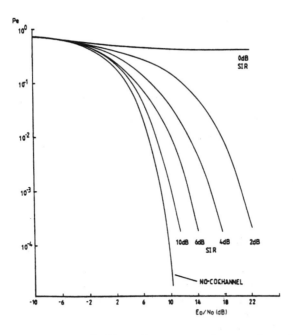

Figure 5.13: Non-coherent QFSK in the presence of one co-channel interferer.

$$\left[\int_0^{x_o} \frac{x_1}{N} I_o\left(\frac{\sqrt{2I}x_1}{N}\right) \exp\left(-\frac{x_1^2 + 2I}{2N}\right) dx_1\right] f(x_o|s_o) dx_o.$$

Substituting for $f(x_o|s_o)$ from Equation 5.57 and integrating gives

$$P_{e1} = 1 + \sum_{k=0}^{2} (-1)^{k+1} \binom{2}{k} \frac{1}{1+k} \exp(-k(2+k)b).$$

$$\cdot \left(1 - Q[(2a)^{\frac{1}{2}}, (2b)^{\frac{1}{2}}] + \frac{1}{2+k} \exp[-(a+b)] I_o\{2(ab)^{\frac{1}{2}}\}\right) \tag{5.84}$$

where

$$a = \frac{1}{N}\left(\frac{1+k}{2+k}\right) \quad \text{and} \quad b = \frac{S}{N}\frac{1}{(1+k)(2+k)} \tag{5.85}$$

and the Marcum Q function is defined as

$$Q(a,b) = \int_b^{\infty} I_o(at) \exp\left(-\frac{t^2 + a^2}{2}\right) dt. \tag{5.86}$$

On the assumption that the interfering element is equally likely to have any f_i ; $i = 0, 1, 2, 3$, the average probability of symbol error is

$$P_e = \frac{1}{4}P_{eo} + \frac{3}{4}P_{e1}. \qquad (5.87)$$

The variation of P_e as a function of E_b/N_o for signal-to-interference ratios of 0, 2, 4, 6 and 10 dB are plotted in Figure 5.13. The curve for no cochannel interference is also displayed as a bench mark. With the exception of $E_b/N_o = 0$ dB, i.e., the interfering signal power is equal to the wanted signal power, all the curves exhibit a rapid fall in P_e with E_b/N_o. When the SIR exceeds 10 dB, E_b/N_o is always less than 2 dB from the curve for no cochannel interference for $P_e > 10^{-4}$.

5.2.3   Multiple Cochannel Interferers

Let us assume that there are a large number of interferers and that their effect is equivalent to an increase in the additive white Gaussian noise (AWGN) power at the receiver input. As a consequence P_e no longer tends towards zero as the SNR is increased. Instead it approaches an asymptotic error probability determined by the level of the AWGN that corresponds to the multiple cochannel interference level. Results obtained by making this Gaussian noise assumption for the cochannel interference are pessimistic for small numbers of equal power interferers. However, if the number of interferers is greater than six, the assumption becomes more realistic.

5.2.3.1   Coherent Demodulation

The probability of symbol error P_e for a coherently detected QFSK system in the absence of cochannel interference is given by Equation 5.25. In order to express P_e in terms of carrier-to-noise power ratio C/N we note that

$$C = \frac{E_b}{T} \qquad (5.88)$$

and

$$N = N_o B \qquad (5.89)$$

where B is the receiver bandwidth. For the situation where

$$B = 1/T_s \qquad (5.90)$$

we have the equivalence

$$\frac{C}{N} = \frac{E_b}{N_o}. \qquad (5.91)$$

Consequently Equation 5.25 becomes,

$$P_e = Q\left(\sqrt{\frac{2C}{N}}\right). \tag{5.92}$$

We may model the effects of cochannel interference by adding an equivalent power C_I to the noise power N to give the probability of symbol error as

$$P_e = Q\left(\sqrt{\frac{2C}{N + C_I}}\right). \tag{5.93}$$

Defining the signal-to-interference power ratio as

$$\text{SIR} \triangleq \frac{C}{C_I} \tag{5.94}$$

and substituting into Equation 5.93 gives

$$
\begin{aligned}
P_e &= Q\left(\sqrt{\frac{2}{\frac{N}{C} + \frac{1}{\text{SIR}}}}\right) \\
&= Q\left(\sqrt{\frac{2E_b/N_o}{1 + \frac{1}{\text{SIR}}E_b/N_o}}\right).
\end{aligned}
\tag{5.95}
$$

Figure 5.14 shows the graphical representation of Equation 5.102. Higher values of P_e occur at a given E_b/N_o compared to those in Figure 5.13, and the asymptotic nature of P_e at high E_b/N_o is apparent. The SIR must be above 7 dB to ensure that P_e can exceed 10^{-3} at high channel SNR values.

5.2.3.2   Non-Coherent Demodulation

The probability of symbol error P_e for a non-coherently detected QFSK system in the absence of cochannel interference is given by Equation 5.67. For convenience we define

$$\Upsilon \triangleq E_b/N_o \tag{5.96}$$

and so we can express the probability of error as

$$P_e = \frac{3}{2}\exp(-\Upsilon) - \exp\left(-\frac{4}{3}\Upsilon\right) + \frac{1}{4}\exp\left(-\frac{3}{2}\Upsilon\right). \tag{5.97}$$

As before we add an equivalent noise power C_I to account for the effect of cochannel interference. Noting the equivalence relation of Equation 5.91 we modify Υ to yield

$$\Upsilon = \frac{C}{N + C_I}. \tag{5.98}$$

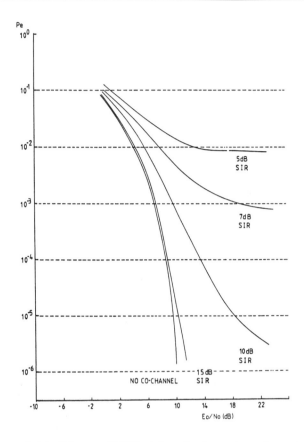

Figure 5.14: Coherent QFSK with multiple co-channel interferers.

Using the definition for SIR of Equation 5.94 allows us to write

$$\Upsilon = \frac{E_b/N_o}{1 + \left(\frac{1}{SIR}\right)(E_b/N_o)}. \tag{5.99}$$

The probability of symbol error as a function of channel SNR is displayed in Figure 5.15. The performance is seen to be worse than for coherent demodulation, amounting to an order of magnitude in P_e when the SIR is 7 dB.

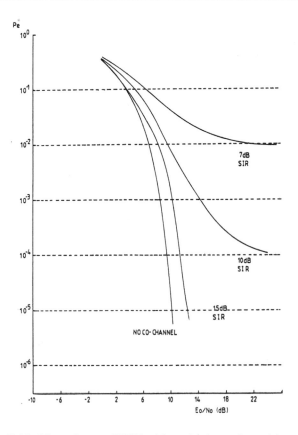

Figure 5.15: Non-coherent QFSK with multiple co-channel interferers.

5.3 QFSK Transmission Over Rayleigh Fading Channels

In this section we derive the symbol error rate performance when QFSK signals are transmitted over a non-frequency selective, slowly fading Rayleigh channel. This channel results in multiplicative distortion of the transmitted signal, and slow fading implies that the multiplicative process may be regarded as constant over at least one symbol period. The fading appears as a multiplicative factor α on the transmitted signal $s_i(t)$, [8, 9] and the received signal becomes

$$r(t) = \alpha s(t) + n(t). \tag{5.100}$$

We have already calculated the error probability for various modulation schemes with a time invariant channel, i.e., a channel where α is a constant, which for convenience we set to unity. When α is no longer a constant we must include it in our equation for the probability of symbol error. In Section 5.2 the channel SNR, i.e., the carrier-to-noise ratio C/N, is E_s/N_o, but in fading conditions it becomes

$$\gamma = \alpha^2 \frac{E_s}{N_o} = \alpha^2 \frac{2E_b}{N_o}. \tag{5.101}$$

We compute the symbol error probability when α is a random variable by averaging $P_e(\gamma)$ over the PDF of γ, namely $f(\gamma)$: namely

$$P_e = \int_0^\infty P_e(\gamma) f(\gamma) d\gamma. \tag{5.102}$$

To obtain the PDF of γ we utilise the following transformation

$$f(\gamma) = f(\alpha_1) \left| \frac{d\alpha_1}{d\gamma} \right| + f(\alpha_2) \left| \frac{d\alpha_2}{d\gamma} \right| \tag{5.103}$$

where $f(\alpha)$ is a Rayleigh PDF,

$$f(\alpha) = \frac{\alpha}{\alpha_o^2} \exp(-\alpha^2/2\alpha_o^2) \tag{5.104}$$

and the average value of α^2 is,

$$E[\alpha^2] = 2\alpha_o^2. \tag{5.105}$$

From Equation 5.101

$$\alpha_1 = \sqrt{\gamma N_o/E_s} \tag{5.106}$$
$$\alpha_2 = -\sqrt{\gamma N_o/E_s} \tag{5.107}$$

and furthermore,

$$\frac{d\alpha_1}{d\gamma} = \frac{1}{2}\sqrt{\frac{1}{\gamma}\frac{N_o}{E_s}} \tag{5.108}$$

$$\frac{d\alpha_2}{d\gamma} = -\frac{1}{2}\sqrt{\frac{1}{\gamma}\frac{N_o}{E_s}}. \tag{5.109}$$

However, the Rayleigh distribution does not exist for negative values and therefore from Equations 5.103 - 5.109:

$$f(\gamma) = f(\alpha_1) \left| \frac{d\alpha_1}{d\gamma} \right|. \tag{5.110}$$

Substituting $f(\alpha_1)$ and $d\alpha_1/d\gamma$ from Equations 5.104 and 5.108 into Equation 5.110 yields

$$f(\gamma) = \frac{N_o}{2\alpha_o^2 E_s} \exp\left\{-\left(\frac{N_o}{E_s}\frac{\gamma}{2\alpha_o^2}\right)\right\}.$$ (5.111)

Defining the average channel signal-to-noise ratio as

$$\Lambda \triangleq 2\alpha_o^2 \frac{E_s}{N_o}$$ (5.112)

gives the PDF of γ as

$$f(\gamma) = \frac{1}{\Lambda} \exp-\left(\frac{\gamma}{\Lambda}\right).$$ (5.113)

This function is known as the chi-square probability distribution.

5.3.1 Coherent Demodulation

The probability of symbol error for QFSK with coherent demodulation in a fading environment becomes with the aid of Equations 5.23 and 5.101

$$
\begin{aligned}
P_e(\gamma) &= Q\left(\sqrt{\alpha^2 \frac{E_s}{N_o}}\right) \\
&= Q\left(\sqrt{\gamma}\right)
\end{aligned}
$$ (5.114)

and so from Equations 5.113 and 5.114,

$$P_e(\Lambda) = \int_0^\infty Q\left(\sqrt{\gamma}\right)\frac{1}{\Lambda}\exp(-\gamma/\Lambda)d\gamma.$$ (5.115)

Substituting for the Q function gives

$$P_e(\Lambda) = \frac{1}{\Lambda}\int_0^\infty \left\{\frac{1}{\sqrt{2\pi}}\int_{\sqrt{\gamma}}^\infty \exp\left[-\frac{1}{2}\nu^2\right]d\nu \exp(-\gamma/\Lambda)\right\}d\gamma.$$ (5.116)

The inner integral can be evaluated using a series expansion, while $P_e(\Lambda)$ is obtained by numerical integration using Simpson's rule.

5.3.2 Non-Coherent Demodulation

The probability of symbol error for non-coherent demodulation as a function of the received SNR is given by Equation 5.67 for transmissions over Gaussian channels. We may rewrite this equation for the Rayleigh fading channel as

$$P_e(\gamma) = \frac{3}{2}\exp\left(-\frac{\gamma}{2}\right) - \exp\left(-\frac{2}{3}\gamma\right) + \frac{1}{4}\exp\left(-\frac{3}{4}\gamma\right). \qquad (5.117)$$

Substituting $P_e(\gamma)$ and $f(\gamma)$ into Equation 5.102 gives

$$\begin{aligned}
P_e(\Lambda) &= \frac{1}{\Lambda}\int_0^\infty \exp\left(-\frac{\gamma}{\Lambda}\right)\left[\frac{3}{2}\exp(-\frac{\gamma}{2})- \right.\\
&\quad \left. \exp\left(-\frac{2}{3}\gamma\right) + \frac{1}{4}\exp\left(-\frac{3}{4}\gamma\right)\right]d\gamma \qquad (5.118)
\end{aligned}$$

and upon integrating

$$P_e(\Lambda) = \frac{3}{2+\Lambda} - \frac{3}{2\Lambda+3} + \frac{1}{3\Lambda+4}. \qquad (5.119)$$

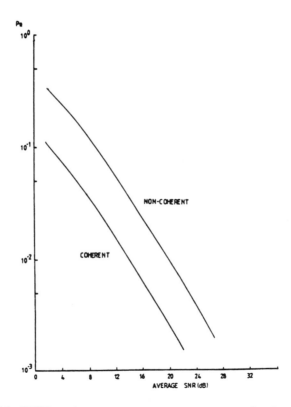

Figure 5.16: QFSK performance for transmissions over a flat Rayleigh fading channel.

Curves of $P_e(\Lambda)$ against average signal-to-noise ratio are given for coherent QFSK and non-coherent QFSK in Figure 5.16. When compared to the curves shown in Figure 5.12 for the Gaussian channel we see the devastating effect of fading on the performance of the QFSK modulation. This occurs because even at high *average* values of channel SNR, a deep fade means that the received signal is swamped by the channel noise giving a bit error rate of 0.5 for a short time. Nevertheless, P_e values of 10^{-3} can be achieved for channel SNR values above 30 dB. By deploying channel coding the reduction in channel SNR for this P_e will decrease by 10 dB or more.

This chapter has examined QFSK for mobile radio systems where the signalling rates are sufficiently low to ensure that the received signal experiences flat fading rather than frequency selective fading for the majority of the time. The S900-D system conceived for the conventional large cell system, has its effective transmission rate (and hence bandwidth) reduced by the use of multilevel modulation. In this way the coherence bandwidth of the channel is not exceeded for a significant proportion of the time. The S900-D system is shown to be workable in static channels, flat fading channels and in the presence of co-channel interference when coherently demodulated. Non-coherent demodulation is not power efficient and its use with four-level FM modulation requires high signal-to-interference ratios at the receiver to achieve an acceptable performance. We note that the use of an S900-D like system may be more appropriate in a cordless telecommunications environment, where its low complexity would be an advantage.

Bibliography

[1] **K-H.Tietgen**: "Numerical Modulation Methods Applied in the FD/TDMA-System S900-D," *Proc. 2nd Nordic Seminar on Digital Land Mobile Radio Communications*, Stockholm, Paper No.33B, Oct 1986.

[2] **D.E.Pfitzmann and H-P.Ketterling**: "A New CP-4FSK Sampling Demodulator for S 900-D," *Proc. 2nd Nordic Seminar on Digital Land Mobile Radio Communications*, Stockholm, Paper No.33A, Oct 1986.

[3] **H-P.Ketterling**: "The Digital Mobile Radio Telephon System S 900-D. *Proc. 2nd Nordic Seminar on Digital Land Mobile Radio Communications*, Stockholm, Paper No.32, Oct 1986.

[4] **R.R.Anderson and J.Salz**: "Spectra of Digital FM," *BSTJ*, pp.1165-1189, Jul/Aug 1965.

[5] **G.L.Choudhury and S.S.Rappaport**: "Diversity ALOHA–A Random Access Scheme for Satellite Communications," *IEE Trans. Commun.*, Vol 31, No.3, pp.450–457, Mar 1983.

[6] **A.P.Clark**: *"Principles of Digital Data Transmission,"* Pentech Press, 1976.

[7] **M.J.Massaro**: "Error Performance of M-ary Non-coherent FSK in the Presence of CW Tone Interference," *IEEE Trans. Commun.*, pp.1363-1369, Nov 1975.

[8] **S.Stein** and **J.J.Jones**: *"Modern Communication Principles,"* McGraw-Hill, 1967.

[9] **J.G.Proakis**: *"Digital communications"*, McGraw-Hill, New York, 1982.

Chapter 6

Partial-response Modulation

I.J.Wassell[1] and R.Steele[2]

6.1 Generalised Phase Modulation

Binary phase shift keying (BPSK) and quaternary phase shift keying (QPSK) were established by the mid-1960s. During the next decade new concepts in modulation resulted in the modulating data being shaped to yield smooth phase transitions in the carrier waveform at symbol boundaries. Further, the phase was constrained to change in a continuous fashion. An important example of this type of modulation is minimum shift keying (MSK) [1] in which the phase changes linearly over a symbol period. MSK is also called fast FSK [2] (FFSK) as well as continuous phase FSK (CPFSK) [3] with a modulation index of a 0.5. Its sidelobe spectral energy relative to non-continuous phase modulation methods, such as QPSK, is significantly reduced. MSK modulation belongs to the class of continuous phase modulation (CPM) [4] and as its phase response is shaped over a symbol period it is referred to as a full response CPM.

As pressure mounted to increase the spectral efficiency of narrow band radio communications it became necessary to devise methods of reducing the spectral spillage of the CPM signal into adjacent channels. A basic technique evolved whereby the phase response to a symbol was spread over

---

[1]University of Southampton and Multiple Access Communications Ltd
[2]University of Southampton and Multiple Access Communications Ltd

a number of symbol periods. By deliberately introducing intersymbol inter-ference (ISI) the spectrum of the modulated signal became more compact. Although the CPM signal could be demodulated in the presence of ISI, an enhanced performance was achieved by removing ISI at the receiver by means of equalisation. The equalisation methods employed were essen-tially derived from those used in high bit rate transmissions over telephone networks where ISI occurred.

CPM schemes that deliberately introduce ISI are generally known as partial response modulations as only part of the symbol shaping is over a symbol period. By increasing the spectral energy in the channel compared to the out-of-channel energy, the power efficiency is increased. Thus for a given BER the transmitted power can be reduced. Further, all CPM sys-tems have constant carrier envelopes enabling non-linear amplifiers to be used thereby providing a high dc-power-to-RF-power conversion ratio. Ex-amples of partial response CPM are tamed frequency modulation (TFM), generalised TFM (GTFM), Gaussian MSK (GMSK) and multi-h CPFSK.

6.1.1   Digital Phase Modulation

Digital phase modulation (DPM) [5,6] is a form of CPM that is well suited to VLSI implementation. The main distinction between DPM and most other forms of CPM is that it is essentially a phase modulation technique in which the shaped symbol pulses are applied directly to a phase modulator. The demodulation of DPM is easier to implement than with other forms of CPM which integrate the data signal prior to phase modulation, see Section 6.1.2.

In DPM the data signal $\alpha(t)$ is applied to the modulator which houses a digital phase shaping filter having an impulse response $q(t)$. For simplicity we will consider the data to be binary, although CPM can accommodate M-ary data symbols. The convolution of $\alpha(t)$ with $q(t)$ yields the phase signal $\phi(t, \alpha)$, a phase that is dependent on both the data α and time t. This $\phi(t, \alpha)$ signal addresses two ROMs to yield $\cos \phi(t, \alpha)$ and $\sin \phi(t, \alpha)$ as shown in Figure 6.1. The front-end of the DPM modulator is therefore completely digital. To produce the modulated signal for radio transmission, digital-to-analogue conversion (DAC) of both $\cos \phi(t, \alpha)$ and $\sin \phi(t, \alpha)$ en-sues, and after passing through anti-aliasing filters, the resulting analogue signals modulate the quadrature carriers $\cos 2\pi f_o t$ and $\sin 2\pi f_o t$, where f_o is the carrier frequency. The radiated signal is formulated as

$$
\begin{aligned}
\tilde{s}(t, \alpha) &= A \cos \phi(t, \alpha) \cos 2\pi f_o t - A \sin \phi(t, \alpha) \sin 2\pi f_o t \\
&= A \cos \left(2\pi f_o t + \phi(t, \alpha) \right)
\end{aligned}
\tag{6.1}
$$

where the amplitude A of $\tilde{s}(t, \alpha)$ is constant, independent of the data. This constant envelope feature is an advantage when efficient class-C amplifiers are used. The power in $\tilde{s}(t, \alpha)$ is

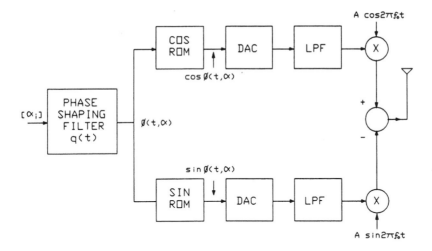

Figure 6.1: Block diagram of the bandpass DPM modulator.

$$\frac{A^2}{2} = \frac{E_b}{T} \tag{6.2}$$

where E_b is the energy per bit and T is the duration of a bit. Hence the bandpass RF signal becomes

$$\tilde{s}(t, \alpha) = \sqrt{\frac{2E_b}{T}} \cos(2\pi f_o t + \phi(t, \alpha)) \tag{6.3}$$

which we may write as

$$
\begin{aligned}
\tilde{s}(t, \alpha) &= Re\left[\sqrt{\frac{2E_b}{T}} \exp\left\{j(2\pi f_o t + \phi(t, \alpha))\right\}\right] \\
&= Re\left[s(t, \alpha) \exp(j2\pi f_o t)\right]
\end{aligned} \tag{6.4}
$$

where $s(t, \alpha)$ is the complex baseband signal

$$s(t, \alpha) = \sqrt{\frac{2E_b}{T}} \exp j\phi(t, \alpha). \tag{6.5}$$

The inphase (I) and quadrature (Q) components of $s(t, \alpha)$ are

$$s_I(t, \alpha) = \sqrt{\frac{2E_b}{T}} \cos \phi(t, \alpha) \tag{6.6}$$

and

$$s_Q(t, \alpha) = \sqrt{\frac{2E_b}{T}} \sin \phi(t, \alpha) \qquad (6.7)$$

respectively.

The filter impulse response $q(t)$ spans a number of bit periods and as a consequence the output $\phi(t, \alpha)$ consists of partially overlapping pulses. We allow the filter $q(t)$ to intentionally introduce intersymbol interference (ISI) in order to contain the spectral spillage of the transmitted signal $\tilde{s}(t, \alpha)$ into adjacent channels. As a consequence of the modulator filter an equaliser is employed at the receiver to remove these ISI effects, even if the transmission channel is ideal. For a Gaussian channel the design of the equaliser is a relatively straightforward procedure as we know exactly how the ISI was introduced. We will see that the task is more daunting in the presence of the multipath effects experienced in mobile radio channels.

The information carrying phase in Equation 6.3 is

$$\phi(t, \alpha) = \sum_{i=-\infty}^{\infty} \alpha_i q(t - iT) \qquad (6.8)$$

where $\alpha_i = \ldots \alpha_{-2}, \ \alpha_{-1}, \ \alpha_0, \ \alpha_1, \ \ldots$ is an infinitely long sequence of uncorrelated data bits. The phase shaping filter of Figure 6.1 has the discrete time FIR filter arrangement shown in Figure 6.2. The data are applied to the filter at a rate $1/T$, and the delay D in each stage of the filter is less than T. Thus D is the sample period of the filter, and

$$\eta = T/D \qquad (6.9)$$

is the oversampling ratio. Accordingly $1/D$ is the sampling rate of the filter. We may, therefore, express $\phi(t, \alpha)$ as a sequence of samples at instants $t = nD$, where n is the sampling instant number, namely

$$\phi(n, \alpha) = \sum_{i=-\infty}^{\infty} \alpha_i q_{n-i\eta} \qquad (6.10)$$

and $\{q_n\}$ is the weighting sequence of the filter. The duration of the impulse response of the phase shaping filter is L symbol periods, and hence the number of filter coefficients is

$$K = \eta L. \qquad (6.11)$$

The restriction on $(n - i\eta)$ is clearly 0 to $K - 1$.

During each bit period the phase shaping filter is excited by the sampled value of the bit and $\eta - 1$ consecutive zeros. The modulation index is defined [5] in terms of the maximum possible phase change during one bit

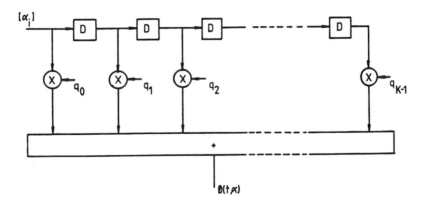

Figure 6.2: Phase shaping FIR filter

interval T, namely,

$$h_p \triangleq \max\{\phi(t+T,\alpha) - \phi(t,\alpha)\}/\pi \quad . \tag{6.12}$$

If an impulse sequence $\ldots 0,0,1,0,0,0,0, \ldots$ is applied to the filter its output is the filter weighting sequence $\{q_n\}$ whose maximum value is $\max\{q_n\}$. When the input sequence is $\ldots 0,0,-1,0,0,0,0, \ldots$ the minimum value of the filter output is $-\max\{q_n\}$. However, when the DPM modulator is transmitting data there are a number of symbols passing through the FIR filter at any one time, and as α can be either $+1$ or -1, the value of h_p becomes

$$h_p = 2\max\{q_n\}/\pi \; ; \; n = 0, \, 1, \ldots, \, K-1. \tag{6.13}$$

The selection of the filter coefficients is critical as it determines both the power spectrum of the transmitted signal and the BER performance of the modem. It has been found [5] that raised cosine and similarly shaped pulses are suitable. Of particular interest is the raised cosine (RC) impulse response that spans L symbol periods, abbreviated to 'L-RC', and defined by

$$q(t) = \left\{ \begin{array}{ll} \frac{\beta}{LT}\left(1 - \cos\frac{2\pi}{LT}t\right) & ; \quad 0 < t < LT \\ 0 & ; \quad \text{elsewhere} \end{array} \right. \tag{6.14}$$

where β is a system parameter. The phase shaping filter is generally implemented in a digital form, and the weighting sequence $\{q_n\}$ of this FIR filter has coefficients [7] that are the samples of $q(t)$ weighted by the sampling period D, namely

$$q_n = Dq(nD). \tag{6.15}$$

By replacing t by nD in Equation 6.14, and with the aid of Equation 6.9

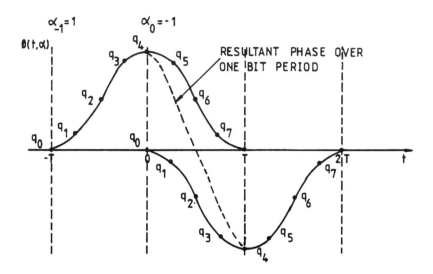

Figure 6.3: DPM phase response.

we may express Equation 6.15 as

$$q_n = \frac{\beta}{L\eta}\left[1 - \cos\frac{2\pi}{L\eta}n\right] \tag{6.16}$$

where the parameter β is adjusted to achieve the desired modulation index. The maximum value of q_n is $2\beta/L\eta$ and hence

$$\max\{q_n\} = \frac{2\beta}{L\eta}. \tag{6.17}$$

From Equations 6.13 and 6.17

$$\beta = \frac{L\eta\pi h_p}{4} \tag{6.18}$$

and on substituting β into Equation 6.9 gives the weighting sequence

$$q_n = \frac{\pi h_p}{4}\left(1 - \cos\frac{2\pi}{L\eta}n\right). \tag{6.19}$$

Example

Consider an FIR impulse response extending over two symbol periods, with four samples per data bit, i.e., $L = 2$, $\eta = 4$. Figure 6.3 shows the separate contributions of α_{-1} and α_o to $\phi(t, \alpha)$, where α_{-1} and α_o are $+1$ and -1, respectively. We see that α_{-1} and α_o generate phase waveforms

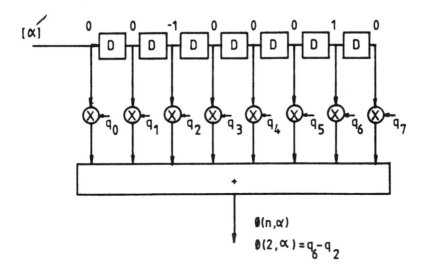

Figure 6.4: DPM FIR filter with the data sequence that yields the response in Figure 6.3.

having amplitudes q_o to q_7 originating at the instants when α_{-1} and α_o are applied. The resulting phase waveform over the interval 0 to 4 D, i.e., over a one bit period is shown by the dotted line, where it is assumed that no other bits have activated the filter. From Equation 6.5,

$$\phi(n, \alpha) = \alpha_{-1} q_{n+4} + \alpha_o q_n \qquad (6.20)$$

and for a particular instant, say when $n = 2$,

$$\phi(2, \alpha) = q_6 - q_2 \qquad (6.21)$$

and Figure 6.4 shows the condition of the FIR filter that yields this $\phi(2, \alpha)$.

6.1.2 Digital Frequency Modulation

We designate forms of CPM where the integration of the phase occurs before phase modulation as digital frequency modulation (DFM) [8,9]. A DFM signal can be produced by applying the data sequence $\{\alpha_i\}$ to a filter having an impulse response $g(t)$ that spreads each data bit over a number of bit intervals. The resulting signal is multiplied by $2\pi h_F$, where h_F is the DFM modulation index, and applied to a voltage control oscillator (VCO), see Figure 6.5. The filtered data sequence changes the frequency of the VCO directly thereby producing DFM.

An alternative approach that is easier to implement is to integrate the data sequence $\{\alpha_i\}$ prior to filtering, and apply the resultant signal to a

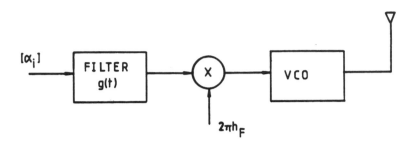

Figure 6.5: Direct FM modulator.

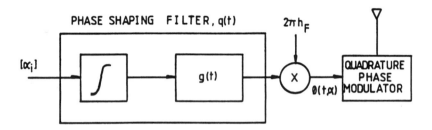

Figure 6.6: Production of FM via phase modulation.

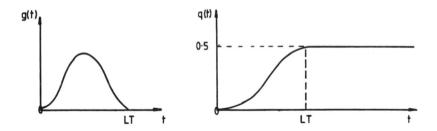

Figure 6.7: Frequency and phase shaping filter responses.

phase modulator. The arrangement is shown in Figure 6.6, where the phase signal is

$$\phi(t, \alpha) = 2\pi h_F \int_{-\infty}^{t} \sum_{i=-\infty}^{\infty} \alpha_i g(\tau - iT) d\tau. \tag{6.22}$$

The impulse response $q(t)$ is given by

$$q(t) = \int_{-\infty}^{t} g(\tau) d\tau \tag{6.23}$$

where for a causal system

$$g(t) = 0 \; ; \; LT \leq t \leq 0 \tag{6.24}$$

and the impulse response $g(t)$ is normalised such that

$$q(t) = 0.5 \; ; \; t \geq LT \; . \tag{6.25}$$

Stylised responses of $g(t)$ and $q(t)$ are shown in Figure 6.7.

The phase of the DFM signal in the nth symbol interval can be represented with the aid of Equations 6.22 and 6.23 as

$$\phi(t, \alpha) = 2\pi h_F \sum_{i=-\infty}^{n} \alpha_i q(t - iT) \; ; \quad nT \leq t \leq (n+1)T \tag{6.26}$$

and upon rearranging and employing Equation 6.25,

$$\begin{aligned} \phi(t, \alpha) &= 2\pi h_F \sum_{i=n-L+1}^{n} \alpha_i q(t - iT) + \pi h_F \sum_{i=-\infty}^{n-L} \alpha_i \\ &= \theta(t, \alpha) + \theta_n \end{aligned} \tag{6.27}$$

where $\theta(t, \alpha)$ is the correlative state vector that depends on the L most recent symbols currently in the filter.

In Figure 6.8 we display the $q(t)$ response for successive bits applied to the filter when $L = 2$. The second term θ_n in Equation 6.27 is the accumulated phase of all the previous symbols that have passed through the filter and it is referred to as the phase state. It is the elimination of this second term that gives DPM an implementation advantage over DFM, or over any similar modulation technique such as GMSK [10] or TFM [11].

Minimum Shift Keying Minimum shift keying (MSK) can be produced in a similar way to off-set quadrature phase shift keying (OQPSK) except that sinusoidal rather than rectangular shaping of the RF quadrature signal envelopes is performed. A good description of this approach is given by Pasupathy [1]. An alternative method, and the one that fits appropriately

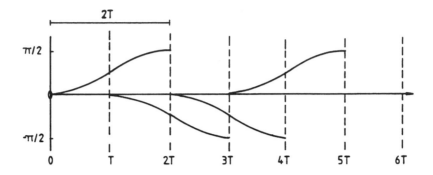

Figure 6.8: Output of filter $q(t)$ in response to a sequence of data, $L = 2$.

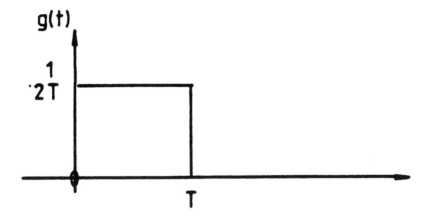

Figure 6.9: Impulse response $g(t)$ of filter for MSK modulation.

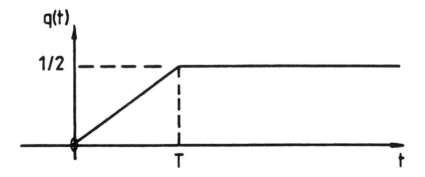

Figure 6.10: Impulse response $q(t)$ of filter for MSK modulation.

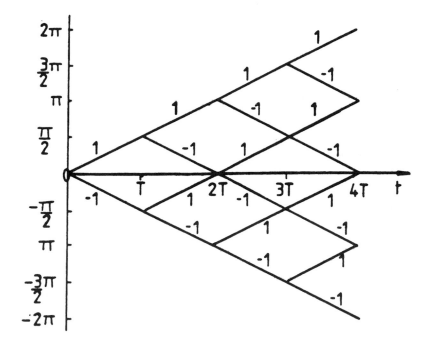

Figure 6.11: MSK phase tree.

in our discourse, is to simply select $g(t)$ in Figure 6.6 to be

$$g(t) = \begin{cases} 1/(2T) & ; \quad 0 \le t \le T \\ 0 & ; \quad \text{elsewhere} \end{cases}. \qquad (6.28)$$

The impulse response $g(t)$ is shown in Figure 6.9, and the corresponding $q(t)$ in Figure 6.10. From Equation 6.26 the linear variation of $q(t)$ from 0 to T yields a linear phase transition of $\pi/2$ because h_F for MSK is defined to be a half, i.e., $2\pi h_F \alpha_i$ can take the values $\pm\pi$ and q_{t-iT} changes linearly from 0 to 0.5 during a symbol period.

The phase tree for MSK is displayed in Figure 6.11, where logical ones and logical zeros are represented by 1 and -1 respectively. When the input data bit is a logical one the phase increases by $\pi/2$, while it decreases by $\pi/2$ when the data is a logical zero. Notice that after two bit periods the phase tree is fully developed as any phases in excess of $\pm\pi$ wrap around to $\mp\pi/2$. Only four possible phases exist enabling us to represent the tree structure by the trellis arrangement shown in Figure 6.12. If in the tree diagram a logical 1 increases the phase from π to $3\pi/2$, the change in the trellis is from $-\pi$ (as this is identical to $+\pi$) to $-\pi/2$, and so forth.

When viewed as a form of frequency modulation the filter $g(t)$ in Fig-

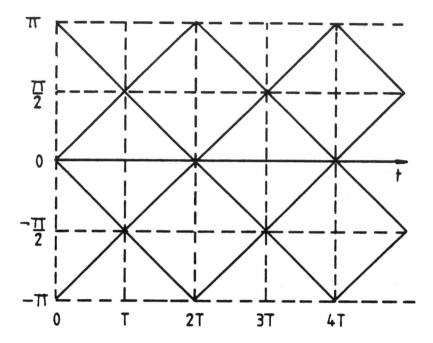

Figure 6.12: MSK phase trellis.

ure 6.5 with the impulse response shown in Figure 6.9 implies a step change in frequency upon receipt of a data bit of opposite polarity. Thus a logical '1' ($\alpha_i = 1$) at the input of the filter in Figure 6.5 will cause the modulated signal to have a frequency of $\omega_c + \omega_d$, where ω_c is the nominal centre frequency and ω_d is the frequency deviation. From Figure 6.5 it can be seen that the frequency deviation of the VCO is given by the expression

$$\omega_d = 2\pi h_f g(t). \tag{6.29}$$

For MSK

$$g(t) = \frac{1}{2T} \tag{6.30}$$

over a symbol interval, and $h_f = 0.5$, giving

$$\omega_d = 2\pi \cdot \frac{1}{2} \cdot \frac{1}{2T} = \frac{\pi}{2} f_b \tag{6.31}$$

where $f_b = 1/T$ is the bit rate. Expressing Equation 6.31 in hertz gives

$$f_d = \frac{f_b}{4}. \tag{6.32}$$

So upon receipt of a logical '1' the VCO frequency is increased by $f_b/4$ Hz from the nominal centre frequency. Similarly upon the receipt of a logical '0' ($\alpha_i = -1$) the VCO frequency is decreased by $f_b/4$ Hz from the nominal centre frequency. Because we have a simple frequency shift keyed system (albeit without phase discontinuities at bit intervals) it is possible to employ non-coherent frequency discriminator demodulation, as well as coherent phase demodulation. The loss in performance incurred by using a frequency discriminator is balanced by somewhat easier implementation.

Gaussian Minimum Shift Keying: A particular form of DFM where the impulse response $g(t)$ is Gaussian shaped is known as Gaussian minimum shift keying (GMSK) [10] having an impulse response $g(t)$ of

$$g(t) = \frac{1}{2T} \left[Q \left(2\pi B_b \frac{t - T/2}{\sqrt{\ell n2}} \right) - Q \left(2\pi B_b \frac{t + T/2}{\sqrt{\ell n2}} \right) \right] \qquad (6.33)$$

for

$$0 \leq B_b T \leq \infty$$

where $Q(t)$ is the Q-function

$$Q(t) = \int_t^\infty \frac{1}{\sqrt{2\pi}} \exp(-\tau^2/2) d\tau \qquad (6.34)$$

B_b is the bandwidth of a low pass filter having a Gaussian shaped spectrum, T is the bit period, and

$$B_N = B_b T \qquad (6.35)$$

is the normalised bandwidth. It may be shown [12] that $g(t)$ is the result of convolving a non return to zero (NRZ) data stream of unity amplitude with a Gaussian low pass filter whose impulse response is

$$h_t(t) = \sqrt{\frac{2\pi}{\ell n2}} B_b \exp \left(-\frac{2\pi^2 B_b^2}{\ell n2} t^2 \right). \qquad (6.36)$$

To give the desired discrete time values of the Gaussian impulse response Equation 6.33 is multiplied by D and the time variable t is replaced by nD, where n is an integer. Upon using Equations 6.9 and 6.35, Equation 6.33 becomes

$$g_n = \frac{1}{2\eta} \left[Q \left(\frac{2\pi}{\sqrt{\ell n2}} B_N \left(\frac{n}{\eta} - \frac{1}{2} \right) \right) - Q \left(\frac{2\pi}{\sqrt{\ell n2}} B_N \left(\frac{n}{\eta} + \frac{1}{2} \right) \right) \right]. \qquad (6.37)$$

The response $g(t)$ is shown in Figure 6.13 for different B_N values.

To enable g_n to be represented in terms of a power series, the Q-functions are first replaced by error functions, i.e.,

$$Q(\theta) = \frac{1}{2} - \frac{1}{2} \text{ erf } \left(\theta/\sqrt{2} \right) \qquad (6.38)$$

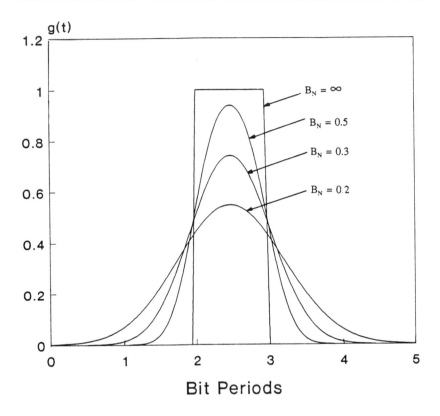

Figure 6.13: Impulse response $g(t)$ for GMSK for different values of B_N.

enabling us to rewrite Equation 6.37 as

$$g_n = \frac{1}{4\eta}\left[\,\mathrm{erf}\,\left(\pi\sqrt{\frac{2}{\ell n2}}B_N\left(\frac{n}{\eta}+\frac{1}{2}\right)\right) - \,\mathrm{erf}\,\left(\pi\sqrt{\frac{2}{\ell n2}}B_N\left(\frac{n}{\eta}-\frac{1}{2}\right)\right)\right].$$
(6.39)

The error functions are now expressed by the following series expansion

$$\mathrm{erf}(z) = \frac{2}{\sqrt{\pi}}\sum_{k=0}^{\infty}\frac{(-1)^k z^{2k+1}}{k!(2k+1)}$$
(6.40)

yielding after simplification

$$g_n = c\left[\sum_{k=0}^{\infty}\frac{(-1)^k b^a}{ak!}\left(\left(n+\frac{\eta}{2}\right)^a - \left(n-\frac{\eta}{2}\right)^a\right)\right]$$
(6.41)

where

$$a = 2k + 1, \; b = \pi\sqrt{\frac{2}{\ln 2}} \frac{B_N}{\eta} \qquad (6.42)$$

and

$$c = \frac{1}{2\sqrt{\pi\eta}}. \qquad (6.43)$$

The use of 100 terms in the expansion of Equation 6.41 and quadruple precision arithmetic are satisfactory for generating g_n.

In implementing a GMSK modulator the time impulse response $g(t)$ of the Gaussian filter must clearly be limited. Specifically, it is symmetrically truncated to L symbol intervals. When the modulation index for GMSK is 0.5, the phase state θ_n in Equation 6.27 can, like MSK, only assume four values 0, $\pi/2$, π and $3\pi/2$. By using $h_F = 0.5$, a GMSK signal can be demodulated by a parallel coherent minimum shift keying (MSK) demodulator, and other sub-optimal receivers.

When viewed as a form of frequency modulation the effect of the Gaussian filter with impulse response $g(t)$ is to prevent the instantaneous changes of frequency inherent in MSK. Consequently, a modulated signal power spectrum with much lower levels of side-lobe energy than that of MSK results. The instantaneous frequency changes due to spreading each bit over more than one bit interval is similar to that of the phase signal in the DPM modulator. Thus if $g(t)$ has a duration of 2 bits and a Gaussian shape, then Figure 6.3 describes the instantaneous frequency of the GMSK modulator output signal. As the modulator filter bandwidth decreases, the duration of its impulse response increases. Normalised bandwidths, B_N, of 0.5 and 0.3 correspond to impulse response durations of approximately 2 and 3 bits, respectively. Examples of instantaneous frequency variations are shown in Figure 6.14. It can be seen that as B_N is decreased, the number of possible frequency trajectories increases. This means that a frequency discriminator with simple threshold decisions has significantly degraded performance in demodulating GMSK when B_N is decreased to 0.3 [13]. For $B_N < 0.3$, discriminator detection can be employed provided additional post-discriminator processing is used. The advantage of smaller normalised filter bandwidths is the reduction in spectral occupancy of the modulated signal.

Tamed Frequency Modulation: In 1978 de Jager and C B Dekker [11] proposed a method of frequency modulation that resulted in the modulated signal having a compact power spectrum without sidelobes. They called the technique tamed frequency modulation (TFM). In TFM, the spectral compactness of the transmitted signal is achieved by careful control of its phase transitions. The premodulation filter $g(t)$ consists of a cascaded 3-tap transversal filter, and a low pass filter $g_o(t)$, see Figure 6.15.

The phase shifts of the modulated carrier over a one bit period are

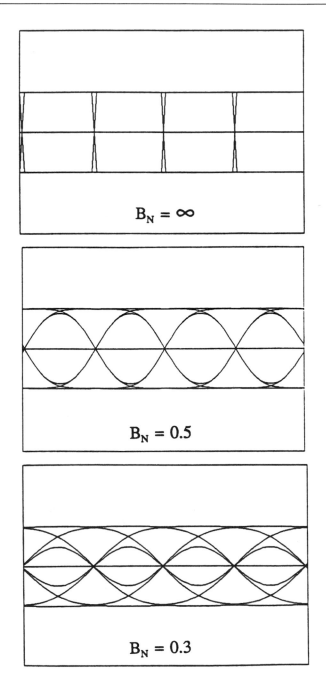

Figure 6.14: Instantaneous frequency variations of GMSK.

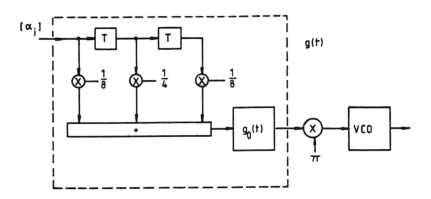

Figure 6.15: TFM modulator.

restricted to either 0, $\pm\pi/4$ or $\pm\pi/2$, and are determined by the three latest consecutive input binary data bits. The frequency shaping function $g(t)$ is

$$g(t) = \frac{1}{8}[g_o(t - T) + 2g_o(t) + g_o(t + T)] \tag{6.44}$$

where

$$g_o(t) \approx \frac{1}{T}\left[\frac{\sin\left(\frac{\pi t}{T}\right)}{\frac{\pi t}{T}} - \frac{\pi^2}{24}\left\{\frac{2\sin\left(\frac{\pi t}{T}\right) - \frac{2\pi t}{T}\cos\left(\frac{\pi}{T}t\right) - \left(\frac{\pi t}{T}\right)^2\sin\left(\frac{\pi T}{2}\right)}{\left(\frac{\pi t}{T}\right)^3}\right\}\right]$$

$$\approx \sin\left(\frac{\pi t}{T}\right)\left[\frac{1}{\pi t} - \frac{2 - \frac{2\pi t}{T}\cos\left(\frac{\pi t}{2}\right) - \frac{\pi^2 t^2}{T^2}}{\frac{24\pi t^3}{T^2}}\right].$$

$$\tag{6.45}$$

An extension [14] of TFM, called generalised tamed frequency modulation (GTFM), provides flexibility in selecting $g(t)$. The tap coefficients of the transversal filter and the roll-off of the low pass filter response can be chosen to trade increase spectrum spillage into neighbouring bands for lower bit error rate (BER), and vice versa. Both the GTFM and TFM modulators can generate their signals by means of look-up tables.

6.1.3   Power Spectra

The spectral spillage of a modulated signal into adjacent channels is of prime importance in digital cellular mobile radio. Figures 6.16, 6.17 and 6.18 show the spectra of DPM, GMSK and MSK, respectively, for pseudo random modulating data at a rate of 250 kbit/s and a carrier frequency of 910 MHz [15]. The DAC in the modulator has 256 levels. The DPM spectrum of Figure 6.16 is for raised cosine pulse shaping over 3-bit periods

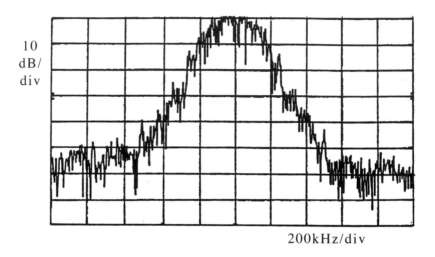

10
dB/
div

200kHz/div

Figure 6.16: Measured DPM spectrum.

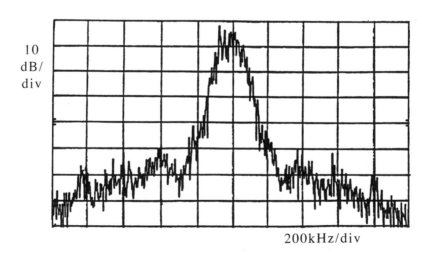

10
dB/
div

200kHz/div

Figure 6.17: Measured GMSK spectrum.

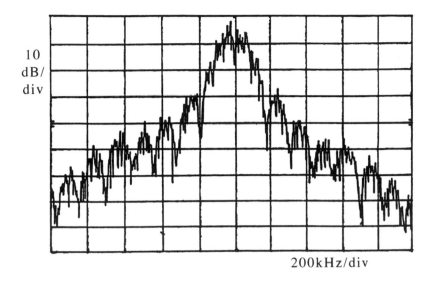

10
dB/
div

200kHz/div

Figure 6.18: Measured MSK spectrum.

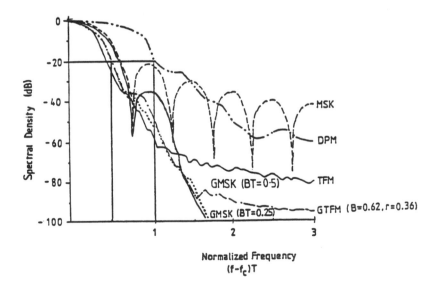

Figure 6.19: Stylised power spectral density curves for different types of modulation signals.

(3RC), a modulation index h_p of 1.08, and 8 samples of the phase per bit period. The GMSK spectrum of Figure 6.17 is for $B_bT = 0.25$, a modulation index h_F of 0.5 and an oversampling ratio of 8. The spectrum of Figure 6.18 is for MSK with a modulation index h_F of 0.5. Stylised spectra for these modulations, in addition to those of TFM and generalised TFM (GTFM), are displayed in Figure 6.19. The rectangular shaped impulse response $g(t)$ used in MSK manifests itself as the sinc-like function with its deep narrow troughs and broad peaks. The narrow main-lobe and the low side-lobe levels of GMSK and GTFM are obtained at the expense of considerable ISI, and this increases the complexity of the equaliser located at the receiver. DPM introduces less ISI than the other DFM methods, and for a given excess delay in the mobile radio channel it is able to operate with fewer states in its Viterbi equaliser compared to GMSK. However, DPM does have considerable out-of-band energy. Nevertheless when DPM operates in the presence of co-channel interference this out-of-band energy is not the limiting factor and is acceptable.

6.1.3.1 Modulated Signal Power Spectral Density Estimation

The PSDs shown in the previous section were obtained from prototype hardware modulators. Instead of building new hardware each time we wish to investigate the PSD of the modulated signal, spectral estimation [16] can be performed instead. To achieve this, pseudo-random data are used and the baseband in-phase and quadrature signals analyzed on the computer to yield the PSD of the modulated signal. One way of computing the PSD is to decompose the complex modulator output sequence into k subsequences, each of M samples. These subsequences are spaced B samples apart, where $B = M/2$, as shown in Figure 6.20. Each subsequence is multiplied by a Hanning window and its FFT computed. Next the periodogram (normalised magnitude squared) of each FFT is calculated, and the k periodograms are averaged to give the PSD estimate. To perform the spectral estimate, an FFT block size of $M = 256$ is appropriate, and $k = 100$ periodograms should be averaged. Theoretical treatments for the evaluation of the PSD of FM signals are given in references [17–19].

6.1.4 TDMA Format for DPM and DFM Transmissions

Having described the basic principles of DPM and DFM, we now consider how these modulation methods are arranged for TDMA transmissions. The speech signal is digitally encoded, and the resulting data stream is channel coded followed by interleaving. Because the channel data are to be transmitted via TDMA in a time slot, the channel data are conveyed to a packetiser. Other vital information, such as a propagation sounding sequence and system control information, is also inserted into the packetiser

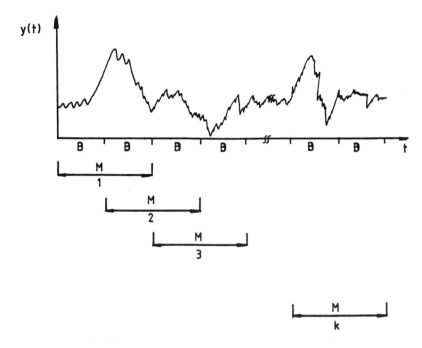

Figure 6.20: Spectral estimation process.

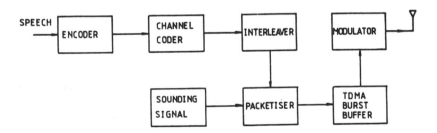

Figure 6.21: Basic TDMA transmitter for mobile radio.

to yield packets that are forwarded at a constant continuous rate into the TDMA burst buffer. The arrangement is shown in Figure 6.21. For a mobile station the packet is removed from the buffer at the TDMA rate during an appropriate time slot. No data are removed from the buffer until a further TDMA frame period has elapsed. By contrast a base station may continuously provide packets for its numerous mobile stations at the TDMA rate in every frame slot.

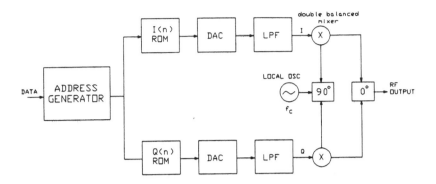

Figure 6.22: CPM modulator structure.

6.1.5   Hardware Aspects

There are two basic ways to perform digital frequency modulation: either direct frequency modulation as shown in Figure 6.5, or via phase modulation shown in Figure 6.1. The disadvantage of direct frequency modulation is that it is difficult to keep the centre frequency within the allowable bounds, while maintaining linearity and constant deviation sensitivity. Consequently a modulator employing this approach must adopt extra measures, such as closed loop control of the phase, to overcome these problems [11].

We will now consider DFM modulators employing the phase modulation approach shown in Figure 6.1. A practical realisation is shown in Figure 6.22. The phase shaping filter having impulse response $q(t)$, and the cos and sin read-only memories (ROMS) are combined into two ROMS, $I(n)$ and $Q(n)$. Binary data applied to the modulator are used to generate appropriate addresses for the $I(n)$ and $Q(n)$ ROMS whose outputs are 8 bit sample values of $\cos \phi(t, \alpha)$ and $\sin \phi(t, \alpha)$ respectively. After D/A conversion and anti-alias low-pass filtering, the resulting I and Q signals are each applied to a double balanced mixer. The local oscillator output and a 90° phase shifted version provide the orthogonal signals which drive the two balanced mixers (shown as multipliers). The outputs from the mixers form the inputs to a 0° combiner, i.e., an adder, from which the modulated signal emerges. The major difficulty associated with this type of modulator is achieving amplitude and phase matching of the signals in the I and Q paths. Deviation from 90° phase shift in the quadrature splitter, deviation from 0° addition in the combiner, or different phase shifts through the balanced mixers cause the modulated signal to have unwanted amplitude and phase variations. These distortions in the modulated signal are compounded if the modulator is followed by stages of non-linear amplification. The consequence is that at the receiver the eye pattern of the demodulated signal will be severely impaired making clock recovery and bit regeneration difficult.

Obtaining propriety RF components with the required close tolerances becomes increasingly difficult as the carrier frequency increases, rendering operation over a wide frequency band (greater than 200 MHz) in the vicinity of 2 GHz difficult to achieve at the time of writing. For wide-band operation it may be necessary to perform the modulation at a fixed intermediate frequency and up-convert the modulated signal to the required operating frequency. This approach, however, involves considerable extra complexity, including additional filtering, mixers and local oscillators, and may be inappropriate for a hand portable.

Another form of modulator structure is based on a simple PSK modulator followed by a phase locked loop. However, there are limits on the usable modulation index and the duration of the filter impulse response $g(t)$. A description of this type of modulator is provided in reference [4].

6.2 CPM Receivers

The mobile radio communications considered in this section are concerned with CPM transmissions over channels that may exhibit frequency selective fading, i.e., the modulated signal bandwidth will be greater than the channel coherence bandwidth for a significant proportion of the time. However, we commence our deliberations by considering what form the optimal receiver takes when the transmissions are over Gaussian channels. This will lead us to the notion of maximum likelihood sequence detection and then to the use of the Viterbi algorithm. The use of the Viterbi algorithm to equalise the effects of inter-symbol interference due to frequency selective fading will then be addressed.

6.2.1 Optimal Receiver

Let us consider a transmitter having an alphabet of M unique messages $m_i, i = 1, 2, \ldots, M$ which are transmitted as a signal $s_i(t); i = 1, 2, \ldots, M$ with exact correspondence, over an additive white Gaussian noise (AWGN) channel to a maximum likelihood (ML) receiver [4, 20]. Figure 6.23 shows the arrangement. The receiver selects from the corrupted received signal $r(t)$ what it deems to be the most likely transmitted signal $s_i(t)$, given the channel noise is $n(t)$. For convenience it will be assumed all signals in this subsection are bandpass; however, later sections will draw distinctions between bandpass and low-pass signals. The receiver achieves this by maximising the *a posteriori* probability $P[s_i(t)|r(t)]$. This procedure gives rise to the name, maximum a posteriori (MAP) receiver. If no prior knowledge of the value of $s_i(t)$ is known to the receiver, the MAP process is equivalent to the ML detection. Such a process provides the minimum probability of message error when all the transmitted messages are equally probable. As

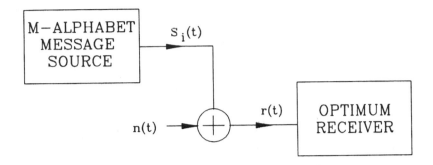

Figure 6.23: Maximum likelihood reception.

$P[s_i(t)|r(t)]$ is unknown, we express it using the mixed form of Bayes rule as

$$P[s_i(t)|r(t)] = \frac{f_R[r(t)|s_i(t)]P[s_i(t)]}{f[r(t)]} \quad (6.46)$$

where $f[r(t)]$ is the probability density function (PDF) of $r(t)$ and $P[s_i(t)]$ is the probability that $s_i(t)$ was transmitted. As $f[r(t)]$ is independent of the signal transmitted, the optimum receiver attempts to maximise the numerator in Equation 6.46. The probabilities $P[s_i(t)]$ are often unknown to the receiver, and hence the term $f_R[r(t)|s_i(t)]$ is maximised.

The reason $r(t)$ differs from $s_i(t)$ is due to the channel noise

$$n(t) = r(t) - s_i(t) \quad (6.47)$$

which is statistically independent of $s_i(t)$. We wish to express these signals as a convergent series of so called basis functions [21]. The coefficients (or weights) applied to these basis functions are components of vectors existing in a vector space known as the signal space. To show that these coefficients do indeed lie in a vector space, and that these coefficients allow the calculation of the maximum likelihood probability requires detailed mathematical treatment which we do not feel is justified in this text.

In setting up the vector space, we have intimated that a set of basis functions must be specified, for example

$$\Psi_i(t) \; ; \; i = 1, 2, ..., n. \quad (6.48)$$

The inner product of two vectors is defined as

$$< h(t), g(t) >= \int h(t)g^*(t)dt \quad (6.49)$$

and an orthonormal basis set is one for which

$$< \Psi_i(t)\Psi_j(t) >= \left\{ \begin{array}{ll} 1 & ; \quad i = j \\ 0 & ; \quad i \neq j \end{array} \right. . \tag{6.50}$$

The range of integration for the inner product is over the range defined for that vector space. The components in the vector space for a function $h(t)$ are given by

$$h_j =< h(t), \Psi_j(t) > . \tag{6.51}$$

To express $h(t)$ in terms of these components we write

$$h(t) = \sum_{j=1}^{n} h_j \Psi_j(t). \tag{6.52}$$

We now define a basis set that forms a suitable basis for all the transmitted messages $s_i(t)$. Thus

$$s_i(t) = \sum_{j=1}^{N} s_{ij} \Psi_j(t) \tag{6.53}$$

where N is the dimension of the signal space which has a maximum value equal to the number of transmitted messages. The coefficients are given by

$$s_{ij} =< s_i(t), \Psi_j(t) > . \tag{6.54}$$

We may also express the noise signal $n(t)$ using the same basis set, i.e.,

$$n(t) = \sum_{j=1}^{N} n_j \Psi_j(t). \tag{6.55}$$

Similarly the coefficients are

$$n_j =< n(t), \Psi_j(t) > . \tag{6.56}$$

Finally, the received signal $r(t)$ may be expressed as

$$r(t) = \sum_{j=1}^{N} r_j \Psi_j(t). \tag{6.57}$$

We note that in terms of signal space components

$$r_j = s_{ij} + n_j. \tag{6.58}$$

The coefficients $\{r_j\}$ are Gaussian random variables because they are produced by linear operations on other Gaussian random variables. The mean of r_j for the ith transmitted message is

$$
\begin{aligned}
E[R_j|s_i(t)] &= E[s_{ij} + N_j] \\
&= s_{ij} + E[N_j] \\
&= s_{ij} \\
&= \eta \; .
\end{aligned}
\tag{6.59}
$$

Note that the use of capital letters denote random variables. We now wish to evaluate the conditional variance of r_j,

$$
\begin{aligned}
E[(R_j - \eta)^2|s_i(t)] &= E[(s_{ij} + N_j - s_{ij})^2] \\
&= E[N_j^2] \; .
\end{aligned}
\tag{6.60}
$$

Substituting for N_j yields

$$
\begin{aligned}
E[N_j^2] &= E[< n(t), \Psi_j(t) >< n(\tau)\Psi_j(\tau) >] \\
&= E\left[\int n(t)\Psi_j(t)dt \int n(\tau)\Psi_j(\tau)d\tau\right] \\
&= E\left[\int \int n(t)n(\tau)\Psi_j(t)\Psi_j(\tau)dtd\tau\right] \\
&= \int \int E[n(t)n(\tau)]\Psi_j(t)\Psi_j(\tau)dtd\tau \\
&= \frac{N_o}{2}\int \int \delta(t-\tau)\Psi_j(t)\Psi_j(\tau)dtd\tau \\
&= \frac{N_o}{2}\int \Psi_j^2(t)dt \\
&= \frac{N_o}{2}
\end{aligned}
\tag{6.61}
$$

where $N_o/2$ is the power spectral density of the Gaussian noise $n(t)$. The multivariate density function of $r_1, r_2, \ldots, r_N$ conditional on $s_i(t)$ is therefore

$$
\begin{aligned}
f[r_1, r_2, \ldots, r_N|s_i(t)] &= \prod_{j=1}^{N} \frac{\exp[-(r_j - s_{ij})^2/N_o]}{\sqrt{\pi N_o}} \\
&= \frac{\exp\left[-\sum_{j=1}^{N}(r_j - s_{ij})^2/N_o\right]}{(\pi N_o)^{N/2}} \; .
\end{aligned}
\tag{6.62}
$$

From Equation 6.46 we noted that maximising the term $f_R[r(t)|s_i(t)]$ or equivalently the term $f[r_1, r_2, \ldots, r_N|s_i(t)]$ over all possible messages $s_i(t)$ yields maximum likelihood reception. The conditional density of Equation 6.62 depends only upon the term

$$\sum_{j=1}^{N}(r_j - s_{ij})^2. \tag{6.63}$$

It is possible to use Parseval's identity to show that

$$\sum_{j=1}^{N}(r_j - s_{ij})^2 = \int (r(t) - s_i(t))^2 dt , \tag{6.64}$$

which is the square Euclidean distance between the signals $r(t)$ and $s_i(t)$. Noting the monotonicity of the exponential function, the density function f_R is maximised by choosing $s_i(t)$ that is closest to $r(t)$ in terms of Euclidean distance. The maximum likelihood receiver may then be implemented by calculating

$$\int (r(t) - s_i(t))^2 dt = \int r^2(t) dt + \int s_i^2(t) dt - 2 \int r(t) s_i(t) dt \tag{6.65}$$

for each i. We note that the first term is constant with respect to i, hence the receiver needs to perform only the correlation $\int r(t) s_i(t) dt$ and subtract it from the second term. The second term is the energy of $s_i(t)$ and if all the transmitted messages contain equal energy, then only the correlation need be formed. The integrals required to evaluate the expression may be implemented using linear filters. This technique is known as matched filtering. For DFM and DPM systems, the phase of the transmitted signal in any particular symbol interval usually depends on previous symbols as well as on the latest symbol. Consequently an optimum maximum likelihood (ML) receiver must observe many symbol intervals before reaching a decision on the value of a specific symbol. In theory one must find the most likely sequence at the receiver corresponding to the whole of the transmitted sequence. This procedure is known as maximum likelihood sequence detection (MLSD) [3, 4] and is complex in terms of implementation and analysis. However, the Viterbi Algorithm (VA) [20, 22] provides a recursive optimal solution to the problem of estimating the sequence of states in a phase modulated signal. As the transition between phase states corresponds to a unique data sequence, the VA also provides MLSD. It will be shown later that it is the recursive application of the VA that gives the reduction in complexity required to produce an optimal receiver.

6.2.2 Probability of Symbol Error

The probability of symbol error is directly determined by the distance properties of the signal space. We will commence by examining the relationship between signal space and the probability of symbol error. We will then dis-

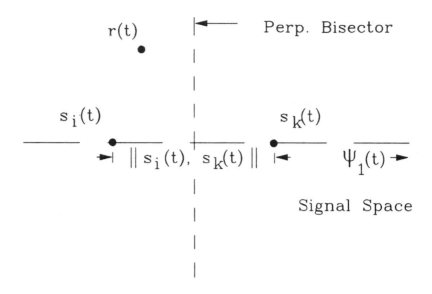

Figure 6.24: Signal space probability of error.

cuss factors affecting modulation distance properties. Suppose the signal $s_i(t)$ is transmitted and a maximum likelihood receiver recovers the signal $s_k(t)$ with a probability $P_e(k; i)$. This event will only occur if the received signal $r(t)$ is closer in *distance* to $s_k(t)$ than to $s_i(t)$. Consider the signal space representation of Figure 6.24 where one of the orthogonal signal space axes is seen to join $s_i(t)$ with $s_k(t)$. As stated previously, white, zero mean Gaussian noise will appear on all of the orthogonal axes. The signal space distance D between the two signals is given by

$$
\begin{aligned}
D &= ||s_i(t) - s_k(t)|| \\
&= \sqrt{\int (s_i(t) - s_k(t))^2 \, dt}.
\end{aligned}
\tag{6.66}
$$

The probability that $r(t)$ lies nearer to $s_k(t)$ than $s_i(t)$ is the probability that the first vector space component $r_1 - s_{i1}$ exceeds $D/2$. Thus the probability of wrongly identifying $s_i(t)$ is

$$
P_e(k; i) = \int_{\frac{D}{2}}^{\infty} \frac{1}{\sqrt{\pi N_o}} \exp\left(\frac{-u^2}{N_o}\right) du
\tag{6.67}
$$

as $r_1 - s_{i1}$ is Gaussian noise with zero mean and variance $N_o/2$.

To find the probability of detecting any other 'incorrect' signal we utilise the 'union bound'. This states that the probability of one or more events occurring is overbounded by the summation of the individual probabilities. Consequently the probability of error, given signal $s_i(t)$ was sent, is

$$P_e(i) \leq \sum_{k \neq i} P_e(k, i)$$

$$\leq \sum_{k \neq i} \int_{\frac{D}{2}}^{\infty} \frac{1}{\sqrt{\pi N_o}} \exp\left(-\frac{u^2}{N_o}\right) du \qquad (6.68)$$

and the total probability of error is

$$P_e = \sum_i P_e(i) P\{s_i(t) \text{ sent }\}. \qquad (6.69)$$

If all messages are equally likely and if $P_e(i)$, is identical for every i then P_e and $P_e(i)$ are identical. Equation 6.68 can also be expressed using Q-functions to yield

$$P_e(i) \leq \sum_{k \neq i} Q\left(\frac{\|s_i(t) - s_k(t)\|}{\sqrt{2N_o}}\right). \qquad (6.70)$$

The concept of signal space will now be applied to phase modulated signals. Suppose that two signals $s_i(t)$ and $s_k(t)$ are different over a duration of N intervals, where each interval has a duration of T and energy E. The square Euclidean distance between the signals is from Equation 6.66

$$D^2 = \int_0^{NT} s_i^2(t) dt + \int_0^{NT} s_k^2(t) dt - 2 \int_0^{NT} s_i(t) s_k(t) dt. \qquad (6.71)$$

The first term is,

$$\int_0^{NT} s_i^2(t) dt = \int_0^{NT} \left[\sqrt{\frac{2E}{T}} \cos(w_c t + \phi_i(t))\right]^2 dt$$

$$= \frac{E}{T} \int_0^{NT} [1 + \cos 2(w_c t + \phi_i(t))] \, dt$$

$$= NE + \text{ term depending on } \left(\frac{1}{w_c}\right). \qquad (6.72)$$

The second term in Equation 6.72 can be neglected for systems of interest here. The energy contributed by the second squared term in Equation 6.71 will also be NE. The cross signal term in Equation 6.71 is,

$$-2 \int_0^{NT} s_i(t) s_k(t) dt = \frac{4E}{T} \int_0^{NT} \cos(w_c t + \phi_i(t)) \cos(w_c t + \phi_k(t)) dt$$

$$= \frac{2E}{T} \int_0^{NT} \cos(2w_c t + \phi_i(t) + \phi_k(t))$$

$$+ \frac{2E}{T} \int_0^{NT} \cos(\phi_i(t) - \phi_k(t)) dt \qquad (6.73)$$

after expressing this equation in terms of complex exponentials, rearranging and changing back to cosine terms. The first term depends on $1/w_c$ and will be neglected and so the cross-correlation term can be expressed as

$$\frac{2E}{T} \int_0^{NT} \cos \Delta\phi(t) dt \qquad (6.74)$$

where $\Delta\phi(t) = \phi_i(t) - \phi_k(t)$. Consequently the square Euclidean distance of Equation 6.71 can be expressed as

$$D^2 = 2NE - \frac{2E}{T} \int_0^{NT} \cos \Delta\phi(t) dt \qquad (6.75)$$

or equivalently

$$D^2 = \frac{2E}{T} \int_0^{NT} [1 - \cos \Delta\phi(t)] dt. \qquad (6.76)$$

For binary systems, $E = E_b$, where E_b is the energy per bit. With an M-ary system then,

$$E_b = \frac{E}{\log_2 M}. \qquad (6.77)$$

Thus the square Euclidean distance is now

$$D^2 = \frac{2E_b \log_2 M}{T} \int_0^{NT} [1 - \cos \Delta\phi(t)] dt. \qquad (6.78)$$

We now define the normalised Euclidean distance function as

$$d^2(s_i(t), s_k(t)) \triangleq \frac{\log_2 M}{T} \int_0^{NT} [1 - \cos \Delta\phi(t)] dt. \qquad (6.79)$$

Thus the square Euclidean distance may be written as

$$D^2 = 2E_b d^2(s_i(t), s_k(t)). \qquad (6.80)$$

Equation 6.70 enables us to express the probability of error $P_e(i)$ in terms of the signal space distance D as follows:

$$P_e(i) \leq \sum_{k \neq i} Q\left(\frac{D}{\sqrt{2N_o}}\right). \qquad (6.81)$$

Substituting for D from Equation 6.80 yields

$$P_e(i) \le \sum_{k \ne i} Q\left(\sqrt{\frac{E_b}{N_o}}d(s_i(t), s_k(t))\right). \qquad (6.82)$$

When the ratio of signal energy to noise energy is reasonably high, say in excess of 10 dB, then one inter-signal distance completely dominates the expression for $P_e(i)$. Indeed we can go further and say that the worst case combination of $s_i(t)$ and $s_k(t)$ will eventually dominate the total error probability P_e as the SNR increases. The worst case distance for any observation interval N is called the minimum distance d_{min}. Consequently the probability of error may be expressed as

$$P_e \cong Q\left(\sqrt{\frac{E_b}{N_o}d_{min}^2}\right). \qquad (6.83)$$

In practice an upper bound to d_{min}^2 called d_B^2, is easier to evaluate. This term can be evaluated for both full and partial response systems with arbitrary pulse shapes and modulation indices.

We now consider a practical implementation of the optimal receiver using the recursive solution provided by the Viterbi algorithm.

6.2.3 Principle of Viterbi Equalisation

In the presence of AWGN, the bandpass signal at the receiver is given by

$$\tilde{r}(t) = \tilde{s}(t, \alpha) + \tilde{n}(t) \qquad (6.84)$$

where $\tilde{s}(t, \alpha)$ is the bandpass (signified by a raised tilda $\sim$) transmitted signal at time t and for data α, and $\tilde{n}(t)$ is the bandpass AWGN signal. As demonstrated in Section 6.2.1, the maximum likelihood (ML) receiver minimises Equation 6.65. The possible received signals $\tilde{s}(t, \bar{\alpha})$ now depend on the infinitely long estimated sequence $\{\bar{\alpha}\}$, and accordingly the ML receiver minimises the function

$$\int (\tilde{r}(t) - \tilde{s}(t, \bar{\alpha}))^2 dt \qquad (6.85)$$

with respect to the estimated data sequence $\{\bar{\alpha}\}$, i.e., it minimises the Euclidean distance.

From Equation 6.65 minimising Equation 6.85 is equivalent to maximising the correlation

$$C(\bar{\alpha}) = \int_{-\infty}^{\infty} \tilde{r}(t)\tilde{s}(t, \bar{\alpha})dt. \qquad (6.86)$$

It would be possible to construct a receiver based on Equation 6.86 in which

State	α_{n-1}, α_{n-2}
Γ_o	-1, -1
Γ_1	-1, 1
Γ_2	1, -1
Γ_3	1, 1

Table 6.1: State table.

all the possible transmitted sequences are correlated with the received signal. The sequence $\bar{\alpha}$ chosen would be that which maximised $C(\bar{\alpha})$. However, even with short bursts this structure becomes unmanageable as the number of comparisons increases exponentially with sequence length. To overcome this problem we define [4, 20, 22]

$$C_n(\bar{\alpha}) \triangleq \int_{-\infty}^{(n+1)T} \tilde{r}(t)\tilde{s}(t,\bar{\alpha})dt \qquad (6.87)$$

and so

$$C_n(\bar{\alpha}) = C_{n-1}(\bar{\alpha}) + Z_n(\bar{\alpha}) \qquad (6.88)$$

where $C_n(\bar{\alpha})$ and $Z_n(\bar{\alpha})$ are referred to as a metric and an incremental metric, respectively, for a particular $\bar{\alpha}$. The incremental metric is given by

$$Z_n(\bar{\alpha}) = \int_{nT}^{(n+1)T} \tilde{r}(t)\tilde{s}(t,\bar{\alpha})dt. \qquad (6.89)$$

Thus $C_n(\bar{\alpha})$ can be evaluated recursively using Equation 6.88, where $Z_n(\bar{\alpha})$ is an incremental metric generated by correlating the received signal with an estimated signal over the nth symbol interval.

Example

In order to clarify how the optimal receiver works in terms of Equation 6.88, we will consider a modulator whose output waveform over a bit interval is dependent on the current and previous two bits, namely α_n, α_{n-1}, and α_{n-2}, yielding $2^3 = 8$ possible waveform segments. We may represent this situation by the state transition diagram shown in Figure 6.25.

Table 6.1 displays the four states that are associated with the values of α_{n-1}, α_{n-2}. In both the Figure and the Table, logical 0 and logical 1 data bits are represented by -1 and $+1$, and by solid and dotted lines, respectively.

Assuming the system is initially in state Γ_o, and $\alpha_n = -1$, the state is unchanged, and the modulator generates a waveform segment $\tilde{S}_o$. However, if $\alpha_n = 1$, the state changes to Γ_2 as the new values of α_{n-1} and α_{n-2} are 1 and -1. The modulator output is waveform segment $\tilde{S}_2$. Should the next data bit be another logical 1 the system moves to Γ_3 and the output waveform is $\tilde{S}_6$. Any further logical ones applied to the modulator generate

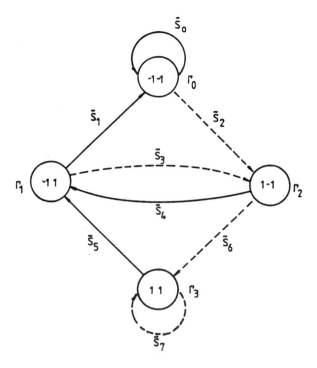

Figure 6.25: State transition diagram.

waveform segment $\tilde{S}_7$ as the transitions are from Γ_3 back to Γ_3. A logical 0 will cause a transition to Γ_1 and $\tilde{S}_5$ to occur, and we note that it is impossible to move directly from Γ_3 to Γ_2.

An alternative representation of the state transition diagram is the trellis diagram of Figure 6.26. The two columns of circles represent the four states Γ_0, Γ_1, Γ_2, Γ_3, at instants $n-1$ and n, while $\tilde{S}_i$; $i = 0, 1, \ldots, 7$, are the waveform segments generated by the modulator during a bit period. Observe that each state is connected to two other states as we are dealing with binary modulation, and one connection is due to the presence of a logical 1 and the other with a logical 0 data bit.

Knowing how the transmitter generates the waveform segments $\tilde{S}_i$ based on the present and previous two input data bits, we now consider how the optimal receiver regenerates the bit sequence in the presence of channel noise. As we know, the receiver will correlate the received signal waveform $\tilde{r}$ over one bit interval with all the possible known transmitted waveforms $\tilde{S}_i$; $i = 0, 1, \ldots, 7$. Now this cross-correlation process yields the eight incremental metrics $Z_n(\bar{a}) = Z_{ni}$; $i = 0, 1, \ldots, 7$, as shown in Figure 6.27. Although the largest Z_{ni} implies that $\tilde{S}_i$ was the most likely transmitted waveform, we refrain from making a decision on a single transmitted

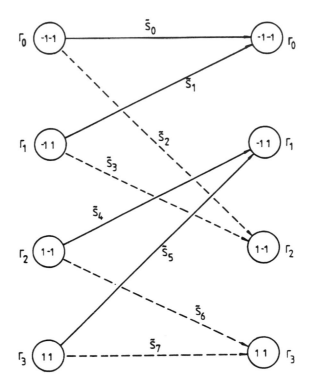

Figure 6.26: Trellis diagram, where _____ -1 and - - - 1.

bit. Instead we compute each of the eight values of $C_n(\bar{\alpha})$ given by Equation 6.88 from a knowledge of the four values of $C_{n-1}(\bar{\alpha})$ and the eight values of $Z_{n,i}$. For example, the two metrics for state Γ_o at instant n are represented by $C_n(\alpha_n, \Gamma_j)$ where $\alpha_n = \pm 1$, and Γ_j are the states at $n-1$, namely

$$C_n(-1, \Gamma_o) = C_{n-1}(\Gamma_o) + Z_{no}$$

and

$$C_n(-1, \Gamma_1) = C_{n-1}(\Gamma_1) + Z_{n1}.$$

Only the larger value of these two metrics is retained and designated $C_n(\Gamma_o)$, along with the logical value of α_n, namely -1. This procedure is repeated for each of the four states to yield four metrics that will be employed in the next bit period.

This recursive process continues for each successive bit interval, namely the path in the trellis associated with the larger C_n at each state is retained, and the logical value of the bit is stored along with the other bits associated with the path leading to that state. At the end of a data sequence we inspect

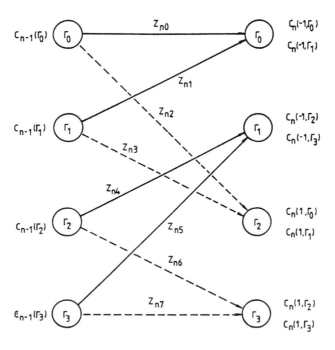

Figure 6.27: Trellis diagram showing metrics, where ———— -1 and - - - 1.

the final metrics for each of the four states. The largest metric identifies the optimum path through the trellis, and the sequence of bits associated with this path is deemed to be the most probable transmitted one.

Figure 6.28 shows the development of the trellis for the 4-state modem described above over a ten bit period. The reader should refrain from calculating the numbers shown on the Figure as much data has been omitted to avoid obfuscation. Starting with an all-zero sequence, the sub-figure for the first bit period, i.e., when $k = 1$ shows the effect of a logical 0 and a logical 1. By $k = 3$ all the states are in use. The numbers assigned to each state are in terms of the path having the *lowest* cumulative metric, as the computer program that generated these metrics operated on the basis of minimum Euclidean distance rather than the equivalent maximum cross-correlation, see Equation 6.65. Thus at each node, or state, the lowest Euclidean distance metric and the bit sequence from $k = 1$ to the current k are stored. For a 10 bit sequence we examine the accumulated Euclidean distance metrics for each of the four states, and select the lowest, namely zero in Figure 6.28. Thus we trace the path back from state Γ_3 to the beginning of the sequence. When the path is a dotted line a logical 1 occurs, and when it is a solid line a logical 0 is formed. The regenerated sequence is seen to be 0000111011, as -1 signifies a logical 0.

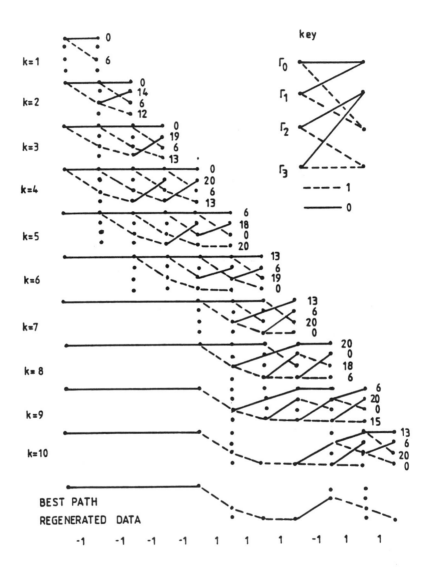

Figure 6.28: Trellis development of a four state VE.

Discussion: We have seen that maximising the cross-correlation (of Equation 6.87) is equivalent to minimising the Euclidean distance between the received and transmitted signals. The procedures described in this section for recovering the data are known as the Viterbi algorithm (VA) [22]. Later we will employ it to equalise the effects of a fading channel, when we will refer to it as a Viterbi equaliser (VE).

In order to decrease the probability of generating errors in the regenerated bit stream, i.e., to avoid selecting the wrong path through the trellis, $v - 1$ dummy zero bits are inserted prior to the data in order to initialise the trellis of the Viterbi equaliser to the all-zero state. This is done in Figure 6.28. As a consequence all the possible paths to be traced through the trellis diagram by the Viterbi algorithm can be assumed to originate from the zero state node. After $v - 1$ data bits have been received the full recursion of the algorithm is reached. After processing all the information data in a manner previously described, the Viterbi processor receives v postcursor dummy zero bits, which ensures that the final state can be assumed to be the zero state. Subsequently the output data sequence for the whole burst, corresponding to the path with the minimum accumulated metric and which passes through the zero state terminal mode, is read out.

We have confined our discussions to binary data as that is our concern in this text. Multilevel modulation can be handled using the VA but at a considerable increase in complexity. Even with binary data the implementation of the Viterbi algorithm is not trivial. However, the power of the algorithm resides in that only 2^{v-1} paths need to be stored, rather than the vast number of paths associated with a tree structure, and consequently the hardware complexity required for the VA does not increase with the length of the data sequence, except for a linear increase in storage requirements. The number of computations is proportional to the data sequence. These remarks can be appreciated with reference to Figure 6.28. However, the situation is radically different if each bit's influence is associated with more waveform segments, i.e., if the ISI is deliberately increased to improve the spectral compactness of the modulated signal. For example, if the output waveform segment is dependent on α_n, α_{n-1}, α_{n-2}, α_{n-3}, α_{n-4} ; $v = 5$, the number of states becomes $2^{v-1} = 16$, and the number of incremental metrics increases to 32. The hardware complexity of the Viterbi algorithm to remove the ISI therefore increases exponentially with v.

Having described the principle of optimum reception of bursts of TDMA data, we will now commence our description of the regeneration of CPM data signals transmitted over mobile radio channels. A prerequisite to the complex baseband signal processing required in the recovery of the data is the demodulation of the received RF signal, and this topic is now addressed.

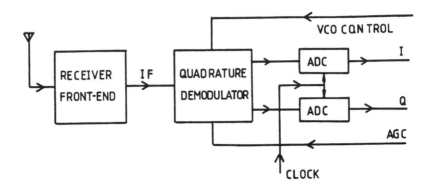

Figure 6.29: RF to baseband conversion.

6.2.4   RF to Baseband Conversion

The NB-TDMA radio frequency (RF) signal conveys the data in bursts occupying one TDMA slot. The receiver at either a MS or BS tunes to the appropriate carrier and using a conventional receiver front-end down-converts the RF signal to the intermediate frequency (IF). Quadrature demodulation follows, see Figure 6.29, to yield baseband analogue in-phase and quadrature signals.

Figure 6.30 shows the quadrature demodulator in more detail. The level of the IF signal is adjusted by the automatic gain control (AGC) signal via a variable gain amplifier. If the phase off-set between the voltage controlled oscillator (VCO) signal and the IF signal is zero the demodulation is coherent. Generally it is not necessary to force the phase off-set to zero because it can be accommodated as a channel imperfection when estimating the channel impulse response. This estimation procedure is described in Section 6.2.5. The low pass filters in Figure 6.30 remove the second harmonic of the IF at the outputs of the multipliers to leave the quadrature baseband signals. Thus the action of the quadrature demodulator is to accept the high frequency bandpass IF signal whose magnitude spectrum is of the form displayed in Figure 6.31 and to yield a baseband quadrature spectrum of the type shown in Figure 6.29.

Returning to Figure 6.29, we see that the baseband analogue quadrature signals are analogue-to-digitally converted (ADC). This process can be viewed as sampling the input signal at a rate

$$f_s = \eta_R/T \qquad (6.90)$$

where η_R is the receiver oversampling ratio, and T is the bit duration. Typically, $\eta_R \geq 2$, thereby ensuring that the degree of spectral aliasing is tolerable. Each sample is encoded into n bits in the ADC. The value of n is

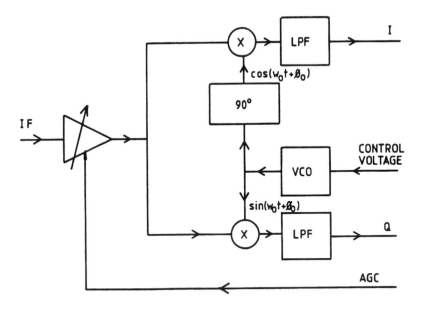

Figure 6.30: Quadrature demodulator.

selected to provide sufficient waveform integrity without making excessive demands on the subsequent digital signal processing. Experiments [15] have shown that η_R values of 2 and 4 are satisfactory for GMSK and DPM demodulation, respectively.

6.2.5 Baseband Processing

After generating the digital I and Q signals, and assuming slot synchronisation has been achieved, the data in the slot, i.e., the packet, are rate converted down to one that allows digital signal processing to be performed within a TDMA frame duration. The data in the packet are separated into the received baseband sounding signal $\hat{c}(t)$, and the received baseband traffic data $r(t)$. The arrangement is shown in Figure 6.32, where the double connecting lines represent complex baseband signals, i.e., inphase and quadrature, and the single lines represent real signals.

The principle of data regeneration is that a channel estimate $h^w(t)$ is formed from $\hat{c}(t)$, and all the possible baseband signals $\bar{s}(t)$ over a one bit period are convolved with $h^w(t)$ to yield the signal estimates $\bar{x}(t)$. The waveforms of $\bar{s}(t)$ over a bit period are all those generated at the transmitter, plus additional ones to allow for multipath propagation, as described later in this section. The traffic data $r(t)$ are convolved with the ambiguity function to allow for imperfect channel estimation and the resulting

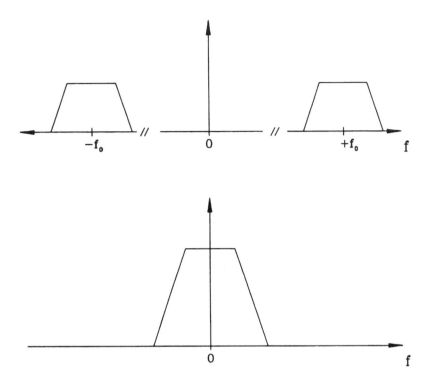

Figure 6.31: Bandpass and low pass spectra.

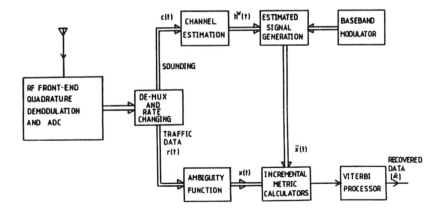

Figure 6.32: Block diagram of the receiver having a Viterbi equaliser.

signal $x(t)$, along with $\bar{x}(t)$, enable the incremental metrics required by the Viterbi algorithm (VA) to be calculated. Viterbi processing is performed, according to the description given in Section 6.2.3, and the traffic data $\{\hat{\alpha}\}$ are recovered.

As we are now concerned with baseband processing we will redraw Figure 6.32 entirely at baseband, removing the RF part. Further we will do the same for the transmitter, and convert the real mobile radio channel to its complex baseband equivalent as described in Chapter 2. The resulting diagram is displayed in Figure 6.33, although system controls are not shown. We will now commence a detailed description of how the regenerated sequence $\{\hat{\alpha}\}$ is produced.

Channel Estimation: In order to obtain an estimate of the channel impulse response the data corresponding to the sounding sequence are processed. Let us momentarily digress to consider the case where the baseband sounding sequence is conveyed over an ideal channel and applied to a matched filter in the receiver. The resulting filter output signal $\rho(t)$ is an approximation to an impulse function. Now suppose we return to reality and replace the ideal channel with a mobile radio channel. In a baseband representation, and assuming linearity, we may remove the baseband radio channel to after the matched filter, and as a consequence the impulse-like signal at the output of the matched filter is convolved with the channel impulse response $h(t)$ of the mobile radio channel to give an estimate of the baseband channel impulse response $h'(t)$. In other words the convolution of the sounding sequence, the impulse response of the channel and the matched filter impulse response provide us with a good estimate of $h(t)$.

Let us consider the situation in more detail. The matched filtering of the received complex baseband sounding signal $\hat{c}(t)$, a corrupted version of the transmitted sounding signal $c(t)$, provides an estimate of the channel baseband impulse response, namely

$$h'(t) = \hat{c}(t) * p(t). \tag{6.91}$$

The equivalent low-pass impulse response of the matched filter, $p(t)$ is a time reversed version of the complex conjugate of the sounding signal, delayed to make it causal, i.e.,

$$p(t) = c^\circ(T_c - t). \tag{6.92}$$

The duration of $c(t)$ is T_c, and the raised zero denotes the complex conjugate operation. To clarify the operation of the channel estimator, we initially consider that the channel is noiseless. The baseband sounding signal at the output of the baseband mobile channel is

$$\hat{c}(t) = c(t) * h(t) \tag{6.93}$$

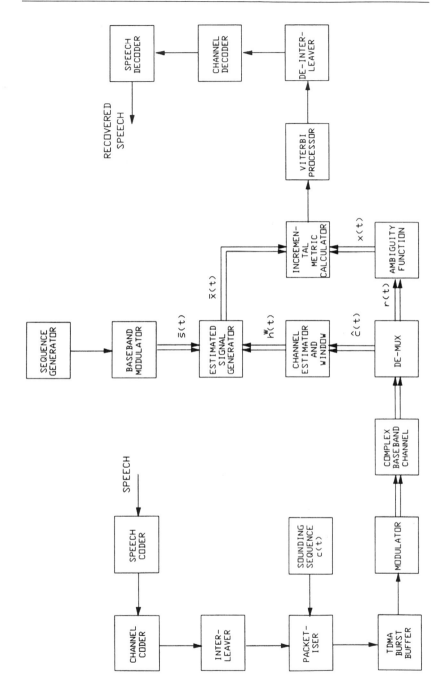

Figure 6.33: Baseband system.

and after matched filtering the resulting signal is

$$h'(t) = c^\circ(T_c - t) * \hat{c}(t) * h(t) \tag{6.94}$$

or

$$h'(t) = \rho(t) * h(t) \tag{6.95}$$

where $\rho(t)$ is known as the ambiguity function [6]. Thus the estimated channel response $h'(t)$ is the actual response $h(t)$ convolved with the ambiguity function

$$\rho(t) = c^\circ(T_c - t) * c(t). \tag{6.96}$$

It should be noted that due to the even symmetry of $c(t)$ about its midpoint

$$c^\circ(T_c - t) = c^\circ(t) \tag{6.97}$$

so that we can put

$$
\begin{aligned}
\rho(t) &= c^\circ(t) * c(t) \\
&= [c_I(t) - jc_Q(t)] * [c_I(t) + jc_Q(t)] \\
&= c_I(t) * c_I(t) + c_Q(t) * c_Q(t).
\end{aligned}
\tag{6.98}
$$

Consequently the ambiguity function is wholly real, which enables simplified processing to be performed in the receiver.

Channel estimation for both DPM and DFM can be performed using either a swept frequency signal (a chirp), or by transmitting a pseudo random binary sequence (PRBS) via phase (or frequency) modulation.

Chirp Sounding

The instantaneous frequency of the transmitted chirp signal is given by

$$f(t) = f_c + \frac{\lambda}{\pi}t \tag{6.99}$$

where f_c is the nominal centre frequency, λ is the sweep parameter and T_c is the chirp duration. The bandpass transmitted chirp signal is

$$\tilde{c}(t) = \cos 2\pi \left(f_c + \frac{\lambda t}{\pi} \right) t \tag{6.100}$$

which can also be expressed in the form

$$
\begin{aligned}
\tilde{c}(t) &= \Re \left[\exp j2\pi(f_c + \frac{\lambda t}{\pi})t \right] \\
&= \Re \left[\exp(j2\lambda t^2) \exp(j2\pi f_c t) \right] \\
&= \Re[c(t) \exp(j2\pi f_c t)]
\end{aligned}
\tag{6.101}
$$

where

$$
\begin{aligned}
c(t) &= \exp(j2\lambda t^2) \\
&= \cos 2\lambda t^2 + j \sin 2\lambda t^2
\end{aligned}
\tag{6.102}
$$

is the complex baseband representation of the bandpass chirp signal $\tilde{c}(t)$. If we define the maximum allowable frequency sweep in terms of the bit rate T, then from Equation 6.99

$$
\frac{\lambda T_c}{\pi} = a \cdot \frac{1}{T}
\tag{6.103}
$$

where a is the new sweep parameter. Letting

$$
T_c = nT,
\tag{6.104}
$$

i.e., the chirp duration is an integer number of bit periods, we can write

$$
\lambda = \frac{a\pi}{nT^2}.
\tag{6.105}
$$

Substituting for λ in Equation 6.102 yields

$$
c(t) = \cos\left(\frac{2a\pi}{nT^2}\right) t^2 + j \sin\left(\frac{2a\pi}{nT^2}\right) t^2.
\tag{6.106}
$$

To express $c(t)$ in discrete time we have

$$
T = \eta D
\tag{6.107}
$$

where η is the oversampling ratio, and we express time as

$$
t = kD \quad ; \quad k = -\infty \text{ to } \infty,
\tag{6.108}
$$

to yield

$$
c(k) = \cos\left(\frac{a2\pi}{n\eta^2}\right) k^2 + j \sin\left(\frac{a2\pi}{n\eta^2}\right) k^2.
\tag{6.109}
$$

Due to the symmetry of $c(t)$ about its midpoint the ambiguity function is entirely real as shown in Figure 6.34, where $a = 0.5$. In order to ensure that the matched filtering (channel estimation) at the receiver is performed correctly, a period of zero carrier follows the frequency chirp. This period should be long enough, e.g. a duration of 6 bits, to accommodate the largest expected channel delay spread. The channel estimate is produced by performing matched filtering with a filter whose impulse response is given by $c^o(T_c - t)$.

Sequence Sounding: To reduce hardware complexity the channel estimation process can be facilitated by transmitting a signal which has undergone phase (or frequency) modulation by a pseudo random binary

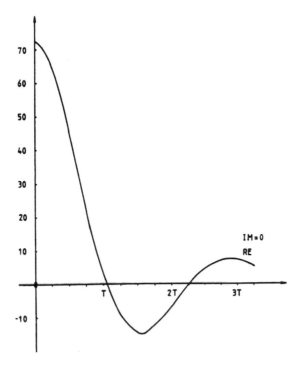

Figure 6.34: Chirp ambiguity function, a=0.5.

code. This code is selected to exhibit good autocorrelation properties, i.e.,
a peaked response with low sidelobes and preferably no imaginary components, by means of a computer search. As an example, for both DPM
(3RC) and DFM (GMSK, $B_N = 0.3$), 16-bit codewords can be used, and
to ensure accurate channel estimation at the receiver, the first 6 bits of
a codeword can be appended to the end of the codeword and the final
6 bits to the start of the codeword. Thus the final 28-bit channel sounding
preamble can tolerate 6 bits of delay spread before the channel estimate
is corrupted by the following section of the TDMA burst. Suitable codewords for DPM (3RC) and GMSK ($B_N = 0.3$) are 1001101011001000 and
1110000111010011, respectively. Their ambiguity functions are shown in
Figures 6.35 and 6.36, respectively. The DPM ambiguity function can be
seen to possess a much narrower main lobe than that of the GMSK ambiguity function (1 bit duration as opposed to 2 bits). As a consequence
the resolution of the channel estimation will be higher in the DPM case.
This is expected because the bandwidth occupied by the DPM signal is
considerably greater than that of the GMSK signal. The GMSK ambiguity
function also possesses a small imaginary component. However, the imag-

Normalised amplitude

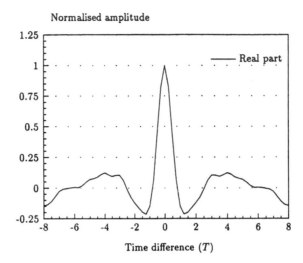

Time difference (T)

Figure 6.35: DPM sequence ambiguity function.

Normalised amplitude

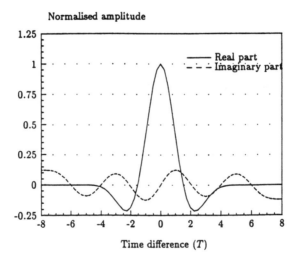

Time difference (T)

Figure 6.36: GMSK sequence ambiguity function.

inary component can be neglected without significantly affecting the link performance.

The ambiguity function $\rho(t)$, i.e., the autocorrelation function of the sounding signal $c(t)$, can be made more impulse-like. This can be achieved in the case of a chirp signal by increasing its frequency sweep, an approach that is usually unacceptable because of the need to occupy a greater channel

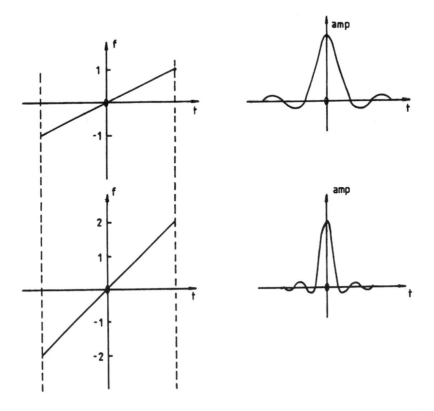

Figure 6.37: Chirp sounding.

bandwidth. Figure 6.37 displays the frequency-time characteristics and the autocorrelation function of a chirp signal for two different frequency sweeps. By increasing the range of the swept frequency the autocorrelation signal has a narrower mainlobe and therefore the ability to resolve multipaths is enhanced. Typically in narrowband TDMA (NB-TDMA) applications, the width of the mainlobe of $\rho(t)$ is approximately of two bits duration. In the case of sequence sounding, increasing the length of the sequence will also narrow the main lobe of the autocorrelation function

In general terms, chirp soundings have autocorrelation functions with narrower main lobes than those produced by sequence sounding. However, the small performance difference between chirp and sequence sounding does not justify the extra complexity involved in chirp sounding.

Channel Windowing: A Viterbi channel equaliser of a given complexity can only cope with a certain delay spread of signal paths, and therefore the full $h'(t)$ cannot be used in the decoding process. Consequently $h'(t)$ is windowed to give a shortened $h'(t)$, say $h^w(t)$, and it is $h^w(t)$ that is generated in the channel estimator shown in Figure 6.33. In addition to truncation, the window function may involve amplitude weighting of $h'(t)$ near the edges of the window.

The need to shorten the duration of $h'(t)$ can be illustrated as follows. Suppose that in the modulator, the duration of one data pulse is spread over $L = 3$ bit periods. The corresponding number of states required in the Viterbi equaliser to remove ISI (when no channel multipath is present) is 2^{L-1} and the number of incremental metrics to be calculated is 2^L, i.e., there are eight possible received waveforms in one bit period. When a multipath is present each bit is effectively spread over additional bit periods, and consequently the number of possible received waveforms per bit period increases beyond eight. To remove the dispersive effects of the multipath channel as well as the ISI deliberately introduced in the modulator, the number of states in the Viterbi processor must be increased. Let the number of states in the Viterbi processor be 2^{v-1}, where $v > L$. Correspondingly the number of incremental metrics is now 2^v, as there are two incremental metrics associated with each state. When the sum of the duration L of the modulator filter impulse response and the duration of the estimated channel response L_p is greater than v, full equalisation is not possible. The receiver is now obliged to operate on a segment of the channel impulse response which must be selected to maximise the BER performance. This arrangement is the so-called reduced-state Viterbi equaliser, in which none of the estimated waveforms will match exactly the received waveforms; nevertheless good BER performance can still be achieved.

As we have said, a reduced-state equaliser selects an appropriate segment of the estimated channel impulse response. The duration of the response is limited to L_T bits, where

$$L_T = v - L \qquad (6.110)$$

because we can only accommodate L_T bits of $h'(t)$ before we exceed the number of states in the equaliser. We will call this selection process rectangular windowing. A reasonable basis on which to select the appropriate segment of the estimated channel impulse response $h'(t)$ is to slide the rectangular window of length L_T bits over the whole of the estimated response, calculate the energy contained within the window at each point and then to identify the window position where the energy is maximum. Hopefully, the window resides on that part of $h'(t)$ which enables the VE to regenerate the data with the lowest BER. Thus for a channel estimate of L_p bits total duration and with an oversampling ratio η_R the energy is calculated at the ith sample position as

$$E_i = \sum_{k=0}^{\eta L_T} |h'_{i+k}|^2 \quad ; \quad i = 0 \text{ to } \eta(L_p - L_T). \qquad (6.111)$$

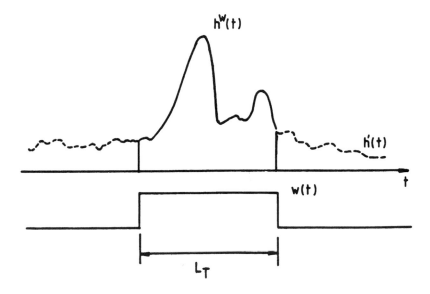

Figure 6.38: Channel windowing of $h'(t)$ to give $h^w(t)$.

Figure 6.38 shows an arbitrary $h'(t)$, the movement of the sliding window and the selected segment $h^w(t)$.

Estimated Signal Generation: Before the information data in the packet is processed, an attempt is made to determine the possible signals that would emanate from a channel having an impulse response $h^w(t)$. Specifically, all possible v-bit sequences are generated by the local modulator within the Viterbi equaliser (VE). The length, v, of the sequence is determined by the number of states (2^{v-1}) that can be accommodated in terms of the acceptable complexity in the Viterbi channel equaliser. Thus a baseband local modulator is supplied with all the possible combinations of the v-bit sequence to yield phase estimates, $\bar{\phi}(t)$. The quadrature estimates $(2E/T)^{\frac{1}{2}}\cos\bar{\phi}(t)$ and $(2E/T)^{\frac{1}{2}}\sin\bar{\phi}(t)$ are computed and it is these terms that are convolved with the estimated wideband baseband channel impulse response $h^w(t)$ to give the estimates of the possible received waveforms over a one bit interval. However, in reality all the $\cos\bar{\phi}(t)$ and $\sin\bar{\phi}(t)$ estimates are stored in a ROM, i.e., the filtering and the trigonometric functions are reduced to a set of stored numbers, addressed by the possible v-bit data sequences. Notice in Figure 6.33 that

$$\bar{s}(t) = (2E/T)^{\frac{1}{2}}[\cos\bar{\phi}(t) + j\sin\bar{\phi}(t)]. \tag{6.112}$$

The Role of the Ambiguity Function: From Equation 6.95 the estimate of the channel impulse response $h'(t)$ is the actual channel impulse response $h(t)$ convolved with $\rho(t)$. Further, $h'(t)$ is windowed to give $h^w(t)$. Consequently the estimated signals in the receiver will be based on $h^w(t)$, not on the unknown $h(t)$. During the reception of information, as distinct from the sounding sequence $c(t)$, the received signal is

$$r(t) = s(t) * h(t), \tag{6.113}$$

where $s(t)$ is the information signal. The receiver estimates that the received signals are

$$
\begin{aligned}
\bar{x}(t) &= \bar{s}(t) * h^w(t) \\
&= \bar{s}(t) * (w(t)h'(t)) \\
&= \bar{s}(t) * w(t)(h(t) * \rho(t)) \\
&= w(t)(h(t) * \rho(t)) * \bar{s}(t) \\
&= w(t)(\rho(t) * h(t) * \bar{s}(t)) \tag{6.114}
\end{aligned}
$$

where $\bar{s}(t)$ is the local modulator output, and $w(t)$ is the windowing function applied to the estimated channel $h'(t)$, see Figure 6.38. In order to obtain the closest match between the received and estimated signals, $r(t)$ in Equation 6.113 is convolved with $w(t)\rho(t)$. Consequently we can write the baseband received signal as

$$
\begin{aligned}
x(t) &= r(t) * (w(t)\rho(t)) \\
&= h(t) * s(t) * w(t)\rho(t) \\
&= w(t)\rho(t) * h(t) * s(t). \tag{6.115}
\end{aligned}
$$

The use of the weighted ambiguity function allows better gain matching between the received signal $x(t)$ and the estimated signals $\bar{x}(t)$ compared to that achieved using $\rho(t)$ directly. We will call the weighted ambiguity function $w(t)\rho(t) = \rho_w(t)$ and because $\rho(t)$ is entirely real we may write

$$
\begin{aligned}
x(t) &= [r_I(t) + jr_Q(t)] * \rho_w(t) \\
&= r_I(t) * \rho_w(t) + jr_Q(t) * \rho_w(t) \\
&= x_I(t) + jx_Q(t) \tag{6.116}
\end{aligned}
$$

where $x_I(t)$ and $x_Q(t)$ are the inphase and quadrature components of the baseband signal. We may write the quadrature received baseband components explicitly in the presence of additive channel noise as

$$x_I(t) = \sqrt{\frac{2E}{T}}[h_I(t) * s_I(t) - h_Q(t) * s_Q(t)] * \rho_w(t) + n_I(t) * \rho_w(t) \tag{6.117}$$

and

$$x_Q(t) = \sqrt{\frac{2E}{T}}[h_I(t) * s_Q(t) + h_Q(t) * s_I(t)] * \rho_w(t) + n_Q(t) * \rho_w(t) \quad (6.118)$$

where $h_I(t)$ and $h_Q(t)$ and $n_I(t)$ and $n_Q(t)$ are the complex components of $h(t)$ and the additive baseband channel noise $n(t)$, respectively.

Notice that although we have ignored the effect of channel noise in producing the channel estimate, we have included noise in the above equations. This may be justified by considering the high energy of the sounding sequence compared with the energy per bit. The difference means that the effective SNR of the channel estimate is much higher (approximately 12 dB greater for a 14-bit preamble) than that experienced during data reception. Consequently when the SNR becomes too low to ensure reasonable channel estimation, the data are unusable anyway.

Incremental Metric Calculator: Initially, as an aid to understanding, we will dispense with quadrature representation, and consider real partial response baseband signals enabling us to express the received signal as

$$x(t) = \alpha(t) * q(t) * h(t) * \rho_w(t) \quad (6.119)$$

where $\alpha(t)$ is a sequence of impulses representing the data vector at the transmitter, $q(t)$ is the impulse response of the modulator filter, and $h(t)$ is the channel impulse response. We will also ignore the effects of channel noise. Similarly, the estimated signal sequences from the local modulator in the receiver after convolution with the estimated windowed channel response are given by

$$\bar{x}(t) = \bar{\alpha}(t) * q(t) * h^w(t) \quad (6.120)$$

where $h^w(t)$ in this case is a truncated channel estimate obtained by applying a rectangular window to $h'(t)$.

To generate these signal sequences over each bit period we feed all the possible v-bit patterns $\bar{\alpha}_{-v+1} \ldots, \bar{\alpha}_o$, into a local baseband modulator filter and convolve its output with the truncated channel estimate $h^w(t)$, see Figure 6.39. Let us represent the combined effect of the modulator filter and the estimated windowed channel by a network having an impulse response

$$d(t) = q(t) * h^w(t) \quad (6.121)$$

such that $\bar{x}(t)$ is the convolution of the data sequence $\bar{\alpha}(t)$ with $d(t)$ as shown in Figure 6.39.

The estimated, locally generated baseband waveforms $\bar{x}(t)$ are seen to be dependent on the number of data bits v, the modulator filter impulse response $q(t)$, and the windowed estimated channel impulse response $h^w(t)$. However, in generating the digital representation of the received signal $x(t)$ it is necessary to sample at a rate η_R times the bit rate in order to prevent

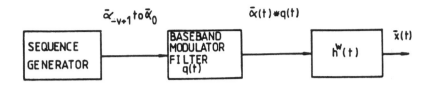

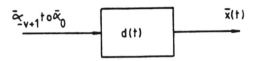

Figure 6.39: Locally generated signal estimates.

excessive aliasing.

Consider the case where $\eta_R = 2$ and $v = 5$. The estimates $\bar{x}(t)$ are formulated using five bits, namely $\bar{\alpha}_{-4}, \bar{\alpha}_{-3}, \bar{\alpha}_{-2}, \bar{\alpha}_{-1}$ and $\bar{\alpha}_0$. As $\eta_R = 2$, there are two samples per bit, achieved by introducing a zero between the $\bar{\alpha}$ values. Figure 6.40 shows an arbitrary response $d(t)$ together with its sampled values $d_0, d_1, \ldots, d_9$. Also displayed is the transversal arrangement of the network having the weighting sequence $\{d_n\}$ showing the data during the first sampling interval. The output of this transversal filter is seen to be

$$\bar{x}_{u,0} = \bar{\alpha}_0 d_0 + \bar{\alpha}_{-1} d_2 + \bar{\alpha}_{-2} d_4 + \bar{\alpha}_{-3} d_6 + \bar{\alpha}_{-4} d_8. \qquad (6.122)$$

In the next sampling interval the data are shifted one delay stage to the right, and a zero is arranged to follow $\bar{\alpha}_0$ to yield

$$\bar{x}_{u,1} = \bar{\alpha}_0 d_1 + \bar{\alpha}_{-1} d_3 + \bar{\alpha}_{-2} d_5 + \bar{\alpha}_{-3} d_7 + \bar{\alpha}_{-4} d_9. \qquad (6.123)$$

As $v = 5$, there are 32 possible received values of $\bar{x}_{u,0}$ and $\bar{x}_{u,1}$, i.e., $u = 0, 1, ..31$ and these are used in the computation of the incremental metrics employed in the Viterbi Algorithm (VA).

At the first sampling instant of the kth bit the estimated signals $\bar{x}_{uo}$ are subtracted from the received signal x_{ko} and the results squared to give the Euclidean distances $(x_{ko} - \bar{x}_{uo})^2$. At the second sampling instant the distances $(x_{k1} - \bar{x}_{u1})^2$ are formed. Notice that the second subscript of x and $\bar{x}$ signifies the first or second sampling instant during a bit period. The incremental metrics for a one bit interval are therefore

$$m_{uk} = (x_{ko} - \bar{x}_{uo})^2 + (x_{k1} - \bar{x}_{u1})^2; \quad u = 0 \text{ to } 31. \qquad (6.124)$$

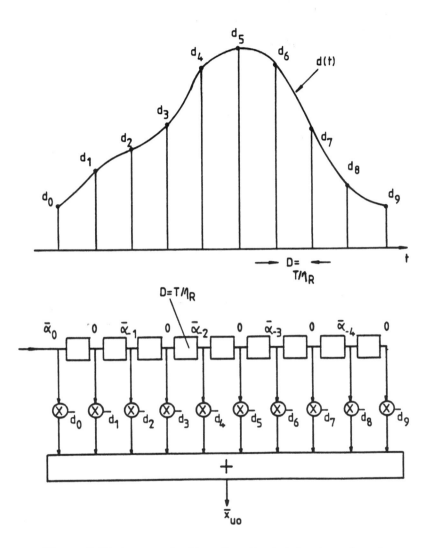

Figure 6.40: An example of how the estimated signals are formed.

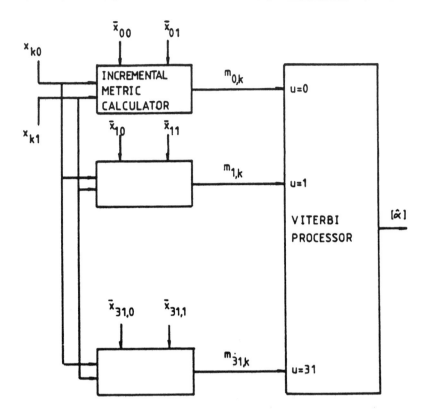

Figure 6.41: Incremental metric formation.

For general values of η_R and v this equation becomes

$$m_{uk} = \sum_{i=0}^{\eta_R - 1} (x_{ki} - \bar{x}_{ui})^2; \quad u = 0, 1, \ldots, 2^v - 1. \qquad (6.125)$$

Having formed the 32 incremental metrics they are applied to the Viterbi processor in Figure 6.41, and the new metrics for each state are established according to the description given in Section 6.2.3. At the next bit interval, $k + 1$, the inputs are $x_{k+1,0}$ and $x_{k+1,1}$. However, the $\bar{x}_{u,o}$ and $\bar{x}_{u,1}; u = 0, 1, \ldots, 31$, do not change as they are constant for the TDMA burst. They will only change when an updated windowed estimate of the channel impulse response is formulated, and that does not occur until the next packet is received. Observe that the samples applied to the incremental metric calculators are essentially sampled values of the analogue waveform segments $x(t)$ and $\bar{x}(t)$, and that the Viterbi processor favours the minimum values of the incremental metrics, as the Euclidean distance rather than the correlation criterion is used.

Having described how the Viterbi equalisation operates with real signals, we now address our actual problem where the I/Q network furnishes us with inphase and quadrature signals. If x_{Iki} and x_{Qki} represent the inphase and quadrature components of the received signal at the ith sample of the kth bit interval, and $\bar{x}_{Iui}$ and $\bar{x}_{Qui}$ are the corresponding values for the locally generated signal estimates, $u = 0, 1, \ldots, 2^v - 1$, then the square of the Euclidean distance is

$$(x_{Iki} - \bar{x}_{Iui})^2 + (x_{Qki} - \bar{x}_{Qui})^2. \tag{6.126}$$

As i ranges from 0 to $\eta_R - 1$, the incremental metric becomes the sum of the square of the Euclidean distances at each sampling instant during a bit period; namely

$$m_{uk} = \sum_{i=0}^{\eta_R-1} \{(x_{Iki} - \bar{x}_{Iui})^2 + (x_{Qki} - \bar{x}_{Qui})^2\}; u = 0, 1, \ldots 2^v - 1. \tag{6.127}$$

6.2.6 Viterbi Equalisation of Digital Phase Modulation

Having described the basic principles of baseband processing we now consider the specific case of Viterbi equalisation of DPM signals [5, 6]. From Chapter 2 it is shown that for DPM the local modulator complex baseband output signal at the receiver is

$$\bar{s}(t) = \exp j \left[\sum_p \bar{\alpha}_p q(t - pT) \right] \tag{6.128}$$

where to avoid cluttering the text we have set the amplitude $(2E/T)^{\frac{1}{2}}$ to unity and ϕ_0 to zero. The discrete time version of $\bar{s}(t)$ derived from Equation 6.128 is

$$\bar{s}(n, \bar{\alpha}) = \exp j \left[\sum_p \bar{\alpha}_p q_{n-p\eta_R} \right] \tag{6.129}$$

and as $\bar{x}(n, \bar{\alpha})$ is the convolution of $\bar{s}(n, \bar{\alpha})$ with $h^w(n)$, see Equation 6.114, we will write $\bar{x}(n)$ as

$$\bar{x}_{ui} = \sum_{n=i-\eta_R L_T}^{i} h_{i-n}^w \cdot \exp j \left[\sum_{p=\lceil \frac{n+1}{\eta_R} \rceil - L}^{\lfloor \frac{n}{\eta_R} \rfloor} \bar{\alpha}_p q_{n-p\eta_R} \right]; u = 0, 1, \ldots, 2^v - 1 \tag{6.130}$$

where $\lceil \cdot \rceil$ and $\lfloor \cdot \rfloor$ denote the nearest integers above and below $\cdot$, respectively, i is the index of the sample during a per bit period and u is the incremental metric index. The limits of the first summation ensure that h^w is limited to the range 0 to $\eta_R L_T$, where L_T is given by Equation 6.110. The coefficients in the modulator filter span the range from q_o to $q_{\eta_R L - 1}$, as shown in Chapter 3, and hence the limits on the second summation.

We are interested in all the possible samples $\bar{x}_{u,i}$ of the estimated received waveforms in a one bit interval. Firstly we will reconsider the earlier example given in Section 6.2.5, but with $\eta_R = 1$. In this example the estimated channel response is truncated to a duration equivalent to $L_T = 2$ bits. With $v = 5$, $L = 3$, the valid modulator filter coefficients are q_o, q_1, q_2 and the truncated channel response is h^w_o, h^w_1, h^w_2. The estimated received signals for a 5-bit sequence $\bar{\alpha}_{-4}$ to $\bar{\alpha}_0$ are, upon substituting into Equation 6.130,

$$\bar{x}_{ui} = \sum_{n=i-L_T}^{i} h^w_{i-n} \exp j \left[\sum_{p=n-2}^{n} \bar{\alpha}_p q_{n-p} \right]$$

$$= \sum_{n=i-2}^{i} h^w_{i-n} \exp j [\bar{\alpha}_{n-2} q_2 + \bar{\alpha}_{n-1} q_1 + \bar{\alpha}_n q_0]. \qquad (6.131)$$

Now i is zero as $\eta_R = 1$, and thus we can write

$$\begin{aligned} \bar{x}_{uo} = \; & h^w_2 \exp j [\bar{\alpha}_{-4} q_2 + \bar{\alpha}_{-3} q_1 + \bar{\alpha}_{-2} q_0] \\ & + h^w_1 \exp j [\bar{\alpha}_{-3} q_2 + \bar{\alpha}_{-2} q_1 + \bar{\alpha}_{-1} q_0] \\ & + h^w_0 \exp j [\bar{\alpha}_{-2} q_2 + \bar{\alpha}_{-1} q_1 + \bar{\alpha}_0 q_0]. \end{aligned} \qquad (6.132)$$

Conceptually the process of producing samples of the estimated received waveforms can be viewed as the local modulator and channel model being cycled through all of the 32 possible 5-bit sequences every bit period.

To ensure a good performance the oversampling ratio η_R should be at least 2, which means that the receiver must produce $\bar{x}_{u,i}, i = 0, 1$; and $u = 0, 1, \ldots 31$ each bit period. For this case we can write the ith sample as

$$\bar{x}_{ui} = \sum_{n=i-4}^{i} h^w_{i-n} \exp j \left[\sum_{p=\lceil \frac{n+1}{2} \rceil - 3}^{\lfloor n/2 \rfloor} \bar{\alpha}_p q_{n-2p} \right] \qquad (6.133)$$

where the valid modulator filter coefficients are q_o to $q_{\eta_R L - 1}$, i.e., q_o to q_5; the truncated channel response is h^w_o to h^w_4, and i can be 0 or 1. The two samples are therefore given by

$$\begin{aligned} \bar{x}_{uo} = \; & h^w_4 \exp j [\bar{\alpha}_{-4} q_4 + \bar{\alpha}_{-3} q_2 + \bar{\alpha}_{-2} q_0] \\ & + h^w_3 \exp j [\bar{\alpha}_{-4} q_5 + \bar{\alpha}_{-3} q_3 + \bar{\alpha}_{-2} q_1] \end{aligned}$$

$$+h_2^w \exp j[\bar{\alpha}_{-3}q_4 + \bar{\alpha}_{-2}q_2 + \bar{\alpha}_{-1}q_0]$$
$$+h_1^w \exp j[\bar{\alpha}_{-3}q_5 + \bar{\alpha}_{-2}q_3 + \bar{\alpha}_{-1}q_1]$$
$$+h_0^w \exp j[\bar{\alpha}_{-2}q_4 + \bar{\alpha}_{-1}q_2 + \bar{\alpha}_0 q_0] \qquad (6.134)$$

and

$$
\begin{aligned}
\bar{x}_{u1} &= h_4^w \exp j[\bar{\alpha}_{-4}q_5 + \bar{\alpha}_{-3}q_3 + \bar{\alpha}_{-2}q_1] \\
&+ h_3^w \exp j[\bar{\alpha}_{-3}q_4 + \bar{\alpha}_{-2}q_2 + \bar{\alpha}_{-1}q_0] \\
&+ h_2^w \exp j[\bar{\alpha}_{-3}q_5 + \bar{\alpha}_{-2}q_3 + \bar{\alpha}_{-1}q_1] \\
&+ h_1^w \exp j[\bar{\alpha}_{-2}q_4 + \bar{\alpha}_{-1}q_2 + \bar{\alpha}_0 q_0] \\
&+ h_0^w \exp j[\bar{\alpha}_{-2}q_5 + \bar{\alpha}_{-1}q_3 + \bar{\alpha}_0 q_1].
\end{aligned}
\qquad (6.135)
$$

The set of signal estimates are complex quantities so we can write

$$\bar{x}_{ui} = x_{Iui} + jx_{Qui} \quad \text{for } u = 0 \text{ to } 31; \text{ and } i = 0, 1. \qquad (6.136)$$

The incremental metrics for the Viterbi demodulator are given by the Euclidian distances between the received signal and the estimated signals over a one bit period and are from Equation 6.127:

$$
\begin{aligned}
m_{uk} &= [(x_{Iko} - \bar{x}_{Iuo})^2 + (x_{Qko} - \bar{x}_{Quo})^2] \\
&+ [(x_{Ik1} - \bar{x}_{Iu1})^2 + (x_{Qk1} - \bar{x}_{Qu1})^2]
\end{aligned}
\qquad (6.137)
$$

where x_{Iki} and x_{Qki} are the inphase and quadrature components of the complex baseband received signal during the kth bit. The set of signal estimates $\bar{x}_{ui}$ is only updated once per burst, although it is used once per bit period in order to evaluate the set of incremental metrics.

Non-rectangular Weighting Functions: In order for the receiver to handle long multipath delays, the truncation length (L_T) of the estimated channel impulse response must be of the order of the maximum expected excess delay. From Equation 6.130, increasing L_T requires more bits in $\{\bar{\alpha}_k\}$ to produce each signal estimate $\bar{x}_{ui}$, and each additional bit in $\{\bar{\alpha}_k\}$ doubles the number of states in the Viterbi equaliser and hence the hardware complexity. An alternative is to employ a non-rectangular weighted truncation of the channel estimate.

To illustrate the concept, consider an example where $\eta_R = 1$, $L = 3$, $v = 5$, and where the estimated channel impulse response is truncated to a length of $(v + 1)$ bits. As stated earlier, 2^{v-1} is the number of states in the demodulator. Modifying Equation 6.130 by setting $i = 0$ as $\eta_R = 1$, and n ranges from 0 to $v + 1$ as there are $v + 2$ samples in the estimated channel impulse response, the signal estimates are

$$\bar{x}_{uo} = \sum_{n=-(v+1)}^{0} h_{-n}^w \exp j \left[\sum_{p=n+1-L}^{n} \bar{\alpha}_p q_{n-p} \right] \qquad (6.138)$$

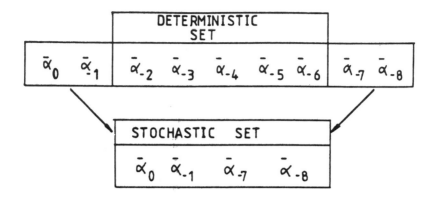

Figure 6.42: Deterministic and stochastic sets.

and on expanding,

$$
\begin{aligned}
\bar{x}_{uo} = \; & h_6^w \exp j[\bar{\alpha}_{-8}q_2 + \bar{\alpha}_{-7}q_1 + \bar{\alpha}_{-6}q_0] \\
& + h_5^w \exp j[\bar{\alpha}_{-7}q_2 + \bar{\alpha}_{-6}q_1 + \bar{\alpha}_{-5}q_0] \\
& + h_4^w \exp j[\bar{\alpha}_{-6}q_2 + \bar{\alpha}_{-5}q_1 + \bar{\alpha}_{-4}q_0] \\
& + h_3^w \exp j[\bar{\alpha}_{-5}q_2 + \bar{\alpha}_{-4}q_1 + \bar{\alpha}_{-3}q_0] \\
& + h_2^w \exp j[\bar{\alpha}_{-4}q_2 + \bar{\alpha}_{-3}q_1 + \bar{\alpha}_{-2}q_0] \\
& + h_1^w \exp j[\bar{\alpha}_{-3}q_2 + \bar{\alpha}_{-2}q_1 + \bar{\alpha}_{-1}q_0] \\
& + h_0^w \exp j[\bar{\alpha}_{-2}q_2 + \bar{\alpha}_{-1}q_1 + \bar{\alpha}_0 q_0].
\end{aligned}
\tag{6.139}
$$

It can be seen that the sampled values of the signal estimates depend on $\bar{\alpha}_{-8}$ to $\bar{\alpha}_0$. A rectangular window would require the Viterbi processor to have 2^8 states. However, as we are limited to 16 states we can only choose a subset of five consecutive $\bar{\alpha}$'s to generate the sampled values of the signal estimates. It seems reasonable from consideration of symmetry, and also to use the longest possible impulse response, that $\bar{\alpha}_{-2}$ to $\bar{\alpha}_{-6}$ be selected as the 'deterministic' state subset and $\bar{\alpha}_0$, $\bar{\alpha}_{-1}$, $\bar{\alpha}_{-7}$, $\bar{\alpha}_{-8}$ be designated as the 'stochastic' subset, as shown in Figure 6.42. The elements in this 'stochastic' subset may have the values of ± 1 with equal probability. One way to generate the sampled value of the signal estimate $\bar{x}_{uo}$ for a given deterministic set is to produce sampled values of the signal estimates for all possible combinations of the stochastic subset and take their average value. Thus to compute $\bar{x}_{uo}$ in Equation 6.139 for a particular deterministic set of $\bar{\alpha}_{-2}$, $\bar{\alpha}_{-3} \ldots$, $\bar{\alpha}_{-6}$, we consider all 16 combinations of $\bar{\alpha}_0$, $\bar{\alpha}_{-1}$, $\bar{\alpha}_{-7}$, $\bar{\alpha}_{-8}$, to give 16 values of $\bar{x}_{uo}$. The value of $\bar{x}_{uo}$ adopted for the metric is the average of these 16 values of $\bar{x}_{uo}$, namely $\bar{X}_{uo}$.

This procedure is time-consuming and it is possible to produce the same effect by weighting the estimated channel response. For example, consider evaluating the average value of the final term in Equation 6.139, namely,

$$h_0^w \exp j[\bar{\alpha}_{-2}q_2 + \bar{\alpha}_{-1}q_1 + \bar{\alpha}_0 q_0].\qquad(6.140)$$

This term has two stochastic elements $\bar{\alpha}_0$ and $\bar{\alpha}_{-1}$, and its average value on expanding the four possible combinations is,

$$
\begin{aligned}
&(h_0^w/4) \exp j(\bar{\alpha}_2 q_2)[\exp j(q_0 + q_1) + \exp j(q_1 - q_0) + \\
&\exp j(-q_1 + q_0) + \exp j(-q_1 - q_0)] \\
=\;& (h_0^w/4) \exp j(\bar{\alpha}_2 q_2)(\exp j q_0 + \exp j(-q_0)) \\
&(\exp j q_1 + \exp j(-q_1)) \\
=\;& (h_0^w/4) \exp j(\bar{\alpha}_2 q_2) 4 \cos q_0 \cos q_1 \\
=\;& h_0^w w_0^0 \exp j \bar{\alpha}_{-2} q_2 \qquad\qquad\qquad (6.141)
\end{aligned}
$$

where the superscript zero implies $i = 0$. Thus it can be appreciated that multiplying h_o^w by a weighting factor

$$w_o^o = \cos q_1 \cos q_o \qquad(6.142)$$

is equivalent to averaging when $\bar{\alpha}_0$ and $\bar{\alpha}_{-1}$ are allowed their possible ± 1 values.

The above procedure is repeated for the other terms in Equation 6.139 having elements in the stochastic set. Notice that the middle four terms in Equation 6.139 will produce weights of unity as they contain none of the stochastic set $\bar{\alpha}_0$, $\bar{\alpha}_{-1}$, $\bar{\alpha}_{-7}$, $\bar{\alpha}_{-8}$. The sampled values of the signal estimates in Equation 6.139 may be expressed as

$$\bar{X}_{uo} = \sum_{n=-6}^{0} h_{-n}^w w_{-n}^0 \exp j \left[\sum_{p=n-2}^{n} \bar{\alpha}_p q_{n-p} \right] \qquad(6.143)$$

where p is from -2 to -6 as the stochastic elements have already been taken into consideration in terms of the weights.

The sampled values of the signal estimates produced in this manner will now no longer exactly match any of the possible received signals and as a result, static BER performance will suffer. The advantage of this scheme is that it enables tolerable performance to be achieved over a much larger range of excess delay for a given receiver complexity.

When the oversampling ratio η_R is increased from unity a different weighting function is required for each sampling instant i. However, this approach is impractical, and so an average of the weighting factors over a one bit interval,

$$W_n = \frac{1}{\eta_R} \sum_{i=0}^{\eta-1} w_n^i \qquad(6.144)$$

is performed and W_n replaces w_n^o in Equation 6.143.

For explanatory purposes we consider a slightly different situation where the system parameters are $L = 3$, $v = 3$ and $\eta_R = 4$. The Viterbi equaliser has 4 states, and the estimated channel response has $\eta_R(v + 1) + 1 = 17$ samples. The deterministic subset has elements $\bar{\alpha}_{-2}$, $\bar{\alpha}_{-3}$ and $\bar{\alpha}_{-4}$ and the stochastic sub-set elements are $\bar{\alpha}_0$, $\bar{\alpha}_{-1}$, $\bar{\alpha}_{-5}$ and $\bar{\alpha}_{-6}$. From Equations 6.130 and 6.138 the signal estimates are given by

$$\bar{x}_{ui} = \sum_{n=i-(v+1)\eta_R}^{i} h_{i-n}^w \exp j \left[\sum_{p=\lceil \frac{n+1}{\eta_R} \rceil - L}^{\lfloor n/\eta_R \rfloor} \bar{\alpha}_p q_{n-p\eta_R} \right] \quad (6.145)$$

upon changing the summation limit $n = i - \eta_R L_T$ to $i - (v + 1)\eta_R$. The index i goes from 0 to $\eta_R - 1$. When $i = 0$,

$$\bar{x}_{ui} = h_0^w \exp j[\bar{\alpha}_{-2}q_8 + \bar{\alpha}_{-1}q_4 + \bar{\alpha}_0 q_0] + \dots$$

$$\cdot$$
$$\cdot$$
$$\cdot$$

$$h_{16}^w \exp j[\bar{\alpha}_{-6}q_8 + \bar{\alpha}_{-5}q_4 + \bar{\alpha}_{-4}q_0] \quad (6.146)$$

and similar expressions are obtained for sampling instants $i = 1$, 2 and 3.

The weighting factor w_o^o associated with the first term in Equation 6.143 is due to the stochastic subset elements $\bar{\alpha}_{-1}$, $\bar{\alpha}_0$, and can be shown to be

$$w_o^o = \cos q_o \cos q_4. \quad (6.147)$$

Similarly the weighting factor for the final term in Equation 6.143 is

$$w_{16}^o = \cos q_4 \cos q_8. \quad (6.148)$$

When there are no stochastic elements within a term in Equation 6.146 the weight function is unity. Sets of weights $\{w_n^i\}$ are also produced at the other sampling instants of $i = 1$, 2 and 3. By employing Equation 6.146 the weighting factors for each of the estimated channel samples are generated. These weights are given in Table 6.2, where for convenience $Q(\cdot)$ means $\cos q(\cdot)$. Clearly to evaluate the average weight, W_n, the sum of the terms in each row of Table 6.2 must be divided by η_R as in Equation 6.144. Thus when $\eta_R = 4$, then

$$W_n = \frac{1}{4} \sum_{i=0}^{3} w_n^i. \quad (6.149)$$

The weight function for this example of 3RC DPM having $h_p = 1$ is plotted in Figure 6.43. Computer simulations indicate that when $v = 3$ the

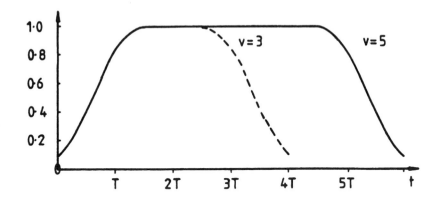

Figure 6.43: Weight functions.

n	w_n^0		w_n^1		w_n^2		w_n^3
0	$Q_0 \ Q_4$	+	$Q_1 \ Q_5$	+	$Q_2 \ Q_6$	+	$Q_3 \ Q_7$
1	Q_3	+	$Q_0 \ Q_4$	+	$Q_1 \ Q_5$	+	$Q_2 \ Q_6$
2	Q_2	+	Q_3	+	$Q_0 \ Q_4$	+	$Q_1 \ Q_5$
3	Q_1	+	Q_2	+	Q_3	+	$Q_0 \ Q_4$
4	Q_0	+	Q_1	+	Q_2	+	Q_3
5	1	+	Q_0	+	Q_1	+	Q_2
6	1	+	1	+	Q_0	+	Q_1
7	1	+	1	+	1	+	Q_0
8	1	+	1	+	1	+	1
9	Q_{11}	+	1	+	1	+	1
10	Q_{10}	+	Q_{11}	+	1	+	1
11	Q_9	+	Q_{10}	+	Q_{11}	+	1
12	Q_8	+	Q_9	+	Q_{10}	+	Q_{11}
13	$Q_{11} \ Q_7$	+	Q_8	+	Q_9	+	Q_{10}
14	$Q_{10} \ Q_6$	+	$Q_{11} \ Q_7$	+	Q_8	+	Q_9
15	$Q_9 \ Q_5$	+	$Q_{10} \ Q_6$	+	$Q_{11} \ Q_7$	+	Q_8
16	$Q_8 \ Q_4$	+	$Q_9 \ Q_5$	+	$Q_{10} \ Q_6$	+	$Q_{11} \ Q_7$

Table 6.2: Values of w_n for $n = 0$ to 16.

receiver can handle up to two bit periods excess path delay when weighting is used. With rectangular weighting, all the available states are required to remove ISI introduced by the modulator, and negligible excess delay can be accommodated. As a 16-state Viterbi equaliser is often used in practice, we also present in Figure 6.43 the weighting window for $v = 5$.

The optimum position of the weighted window is found by an approach similar to that described in Section 6.2.5 for the rectangular window. The energy values used to locate the position of the window are computed according to

$$E_i = \sum_{k=0}^{\eta_R(v+1)} W_k^2 |h_{i+k}^w|^2 \text{ for } i = 0, \text{ to } \eta(L_p - (v+1)) \qquad (6.150)$$

where L_p is the duration of the estimated channel impulse response in bit periods prior to windowing. The positioning of the centre of the window is where the value of i that results in the largest value of E_i resides.

This method of producing a so-called reduced state VE is not the only possible approach. References [23] - [26] show other schemes which attempt to approximate the received signal set using fewer states.

6.2.7   Viterbi Equalisation of GMSK Signals

The use of a frequency modulation scheme (in this case GMSK) [9, 15] requires the Viterbi processor to have extra states compared to DPM. These extra states are a consequence of the p possible values of the phase state term θ_n. Considering the situation at the modulator, we note that DPM requires 2^{L-1} states while $p2^L$ states are required when frequency modulation is used. The integer p is related to the modulation index by

$$h_f = 2k/p \qquad (6.151)$$

where k and p are integers and h is rational [8]. The phase states θ_n have values

$$\theta_n \in \{0, \ 2\pi/p, \ 2.2\pi/p \ldots (p-1)2\pi/p\} \qquad (6.152)$$

and each trellis state is now defined by a phase state θ_n and one of 2^{L-1} possible α sequences $\{\alpha_{n-1}, \alpha_{n-2} \ldots \alpha_{n-v+1}\}$. For GMSK we have $h_f = 0.5$, $k = 1$, $p = 4$ and so from Equation 6.152 there are four phase states with values $0, \pi/2, \pi$ and $3\pi/2$.

Consider the case of GMSK, $h_f = 0.5$ and $L = v = 3$ (i.e., an ideal channel) whose trellis diagram is shown in Figure 6.44. The four phase states, $0, \pi/2, \pi$ and $3\pi/2$ are shown, as are the bits defining the correlative state vector at time instants $n - 1$ and n. Figure 6.45 shows the relationship between the correlative state vector and the phase state at time instants n and $n+1$. With the arrival of the latest bit, α_{n+1}, it may be observed that α_{n-2} is no longer part of the correlative state vector and contributes a value

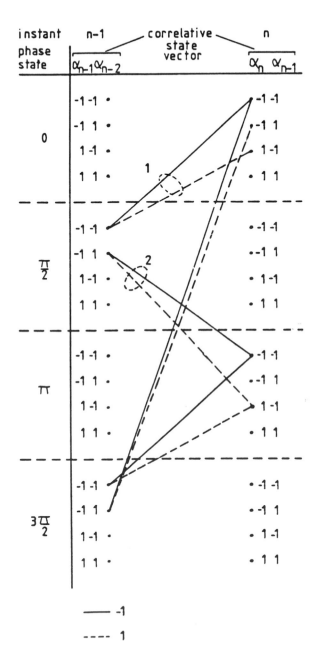

Figure 6.44: GMSK trellis diagram, where ——— -1 and - - - 1.

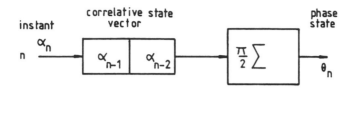

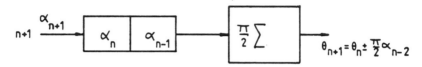

Figure 6.45: Relationship between the correlative state vector and phase state.

of either $\frac{\pi}{2}$ or $\frac{-\pi}{2}$ to the previous value of the phase state θ_n. Some of the transitions within the trellis are shown for explanatory purposes. Consider phase state $\pi/2$ and correlative state $\alpha_{n-1} = -1, \alpha_{n-2} = -1$. If the latest bit $\alpha_n = 1$, then the new correlative state becomes $\alpha_n = 1, \alpha_{n-1} = -1$; alternatively, if $\alpha_n = -1$, the correlative state is $\alpha_n = -1, \alpha_{n-1} = -1$. The new phase state is determined by bit α_{n-2}. It is the oldest correlative state bit and so with the arrival of the latest bit α_n, it will now contribute a constant value of phase ($\pm\pi/2$) to the phase state vector, depending on whether its value is ± 1. Returning to our example transitions, both of them will end up in phase state 0, because $\alpha_{n-2} = -1$, i.e., a subtraction of $\pi/2$ from the current phase state. These transitions are labelled 1 in Figure 6.44. Further example transitions, this time with $\alpha_{n-2} = 1$ are labelled 2 in Figure 6.44. In this case the transitions end in phase state π.

It can be appreciated from Figure 6.44 that the trellis structure for GMSK is four times more complicated than that required for DPM with an equivalent value of L.

The Viterbi Equaliser structure for GMSK is very similar to that proposed for DPM. To generate the estimated received signals at the receiver, we must convolve all the possible local modulator outputs with the estimated channel impulse response $h^w(t)$. The signal estimates will be used to generate $p2^v$ incremental metrics per received bit by calculating the Euclidean distances in a similar manner to that employed for DPM. Once the Viterbi algorithm has been applied recursively at each bit interval over the whole of the received TDMA burst, the path through the trellis with the smallest total metric is chosen as the most likely path, and the data sequence associated with it is regenerated.

One way to simplify the Viterbi equalisation of GMSK is to remove the effects of the additional phase states by tracking the phase states at each

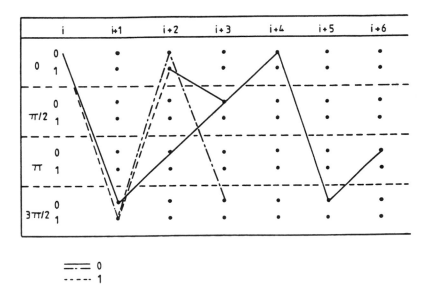

Figure 6.46: Demonstration of double errors in GMSK.

trellis node. The equaliser stores the present phase state at each trellis node to allow the signal estimates corresponding to the appropriate phase state to be chosen during the next bit interval. After updating each trellis node by storing the new cumulative metric and the updated information sequence, the latest phase state is also recorded for use in the following recursion. With only the additional complexity needed to store the phase state at each node in the trellis, the number of equaliser states for GMSK can be reduced to that required for a similar DPM scheme.

A characteristic of the Viterbi equalisation (VE) of GMSK is that single bit errors do not occur. Either double or multiples of double errors occur. The production of double errors can be demonstrated by considering the simple trellis of Figure 6.46 which has one correlative state and four phase states. The path corresponding to the transmission of an all logical zero (-1) sequence is shown, along with a diverging erroneous path occurring at instant i. It can be seen that two logical ones will be generated before the erroneous path rejoins the correct path at instant $i + 3$. Also shown in Figure 6.46 is another erroneous path, again diverging from the correct path at instant i. We will now investigate the effect of allowing only one error to occur. Consequently a path giving a logical 1 output is chosen for the transition from instant i to $i+1$, and a path giving logical 0 outputs for the remainder of the burst. It can be seen that the correct and erroneous paths will never converge and so it is not possible for single burst errors to occur in GMSK.

Input	Delayed input	TX	Error TX	No errors		Two errors	
				Delay output	Output	Delay output	Output
1	0	1	1	0	1	0	1
1	1	0	0	1	1	1	1
1	1	0	0	1	1	1	1
0	1	1	1	1	0	1	0
1	0	1	0*	0	1	0	0*
1	1	0	1*	1	1	0	1
1	1	0	0	1	1	1	1
1	1	0	0	1	1	1	1
0	1	1	1	1	0	1	0
0	0	0	0	0	0	0	0
0	0	0	0	0	0	0	0

Table 6.3: Differential coding and decoding.

An improvement in BER can be achieved if the double errors are converted to single errors. This can be achieved by differential encoding the data stream prior to the GMSK modulator, and differential decoding after the VE. The effect is demonstrated in Table 6.3, which shows various serial bit streams associated with a differential encoder/decoder pair of Figure 6.47. The first column shows the input bit stream, and the delayed bit stream is shown in the second column. The delay element is initialised to a logical '0'. The differential encoder output is given in the third column. When this bit stream is not corrupted, then the differentially decoded bit stream shown in column six is identical to the input bit stream. If we now introduce two errors into the transmitted bit stream TX as shown in column 4, and proceed to decode it as before, then there is only one bit in error when we compare the decoder output shown in column eight, with the encoder input. Note that bits in error are denoted * in Table 6.3. Consequently double bit errors are converted into single bit errors by adopting differential data encoding.

6.2.8   Simulation of DPM Transmissions

We now present simulation results for the transmission of data over radio channels via DPM. In our simulations the bandpass system elements were replaced by their complex baseband equivalents. Our experiments were conducted using 13448 data bits, and as a consequence the experiments are valid for BERs down to the order of 10^{-3}. The TDMA burst format is shown in Figure 6.48 for chirp sounding, and in Figure 6.49 for sequence sounding. The DPM had a 3-RC phase pulse shape and $h_p = 1.0$. Four times oversampling was used in both the modulator and the VE. A reduced 16-state VE was employed during all the simulations. No channel coding

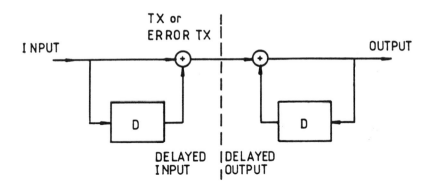

Figure 6.47: Differential coding and decoding.

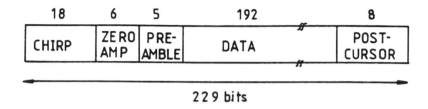

Figure 6.48: Format of the data packet using chirp sounding.

was used.

6.2.8.1 DPM Transmissions over an AWGN Channel

If we assume there is a single propagation path between a static transmitter and static receiver, then the received signal does not experience fading or Doppler phenomena. The only sources of impairment are the transmitter and receiver filters (which for the purpose of the simulation we assumed introduced no band limiting of the signal) and the system noise, which we modelled as additive white Gaussian noise (AWGN). To simulate the AWGN channel we made use of the complex baseband representation of narrowband noise, adding sample values of zero-mean independent Gaussian processes to the in-phase and quadrature complex baseband signals. By manipulating the variance of the Gaussian sources we determined the BER as a function of carrier to noise (C/N) ratio, or alternatively E_b/N_o, where E_b is the energy per bit and N_o the one sided noise power spectral density (PSD). The relationship between C/N and E_b/N_o is simply

$$\frac{C}{N} \;=\; \frac{E_b}{T} \cdot \frac{1}{BN_o} \qquad\qquad (6.153)$$

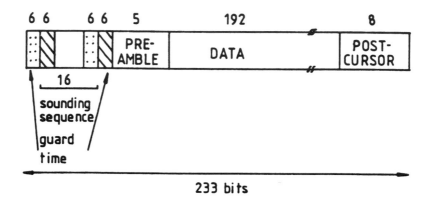

Figure 6.49: Format of the data packet using sequence sounding.

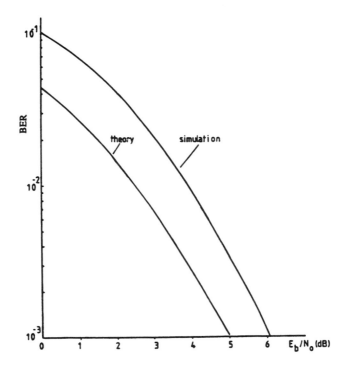

Figure 6.50: BER Performance of DPM in an AWGN channel.

$$= \frac{E_b}{N_o} \cdot \frac{1}{BT}. \tag{6.154}$$

where B is the receiver bandwidth and T is the bit duration. It was convenient to introduce a parameter a that specified the relationship between B and T, namely

$$B = a \left(\frac{1}{T} \right) \tag{6.155}$$

giving

$$\frac{E_b}{N_o} = a \frac{C}{N}. \tag{6.156}$$

In order to calibrate the system, the carrier power was measured. Its value remained constant for DPM because of its constant envelope property. Next the Gaussian noise generators were set to give a PDF with variance N. Finally, knowledge of the transmitted DPM signal bandwidth gave the parameter $a = 3$. Figure 6.50 shows the simulated system's BER as a function of E_b/N_o.

The theoretical determination of BER as a function of E_b/N_o is a difficult problem. One approach [4] is to utilise a bound based on the minimum free distance of the transmitted signal set. For 3RC DPM, $h_p = 1$, the value of $d_{min} = 3$ [5] and at high values of E_b/N_o the probability of bit error may be approximated by

$$P_e \approx Q \left(\sqrt{d_{min}^2 \frac{E_b}{N_o}} \right). \tag{6.157}$$

This theoretical result is also plotted in Figure 6.50 for comparison. The simulation curve is roughly 1 dB poorer in performance than that given by theory.

6.2.8.2 DPM Transmissions over Non-Frequency Selective Rayleigh and Rician Channels

The frequency non-selective Rayleigh fading channel is modelled by multiplying the transmitted signal envelope by an attenuation factor α, chosen from a Rayleigh PDF, and shifting the phase angle of the received signal by an angle θ that can have values in the range $(0-2\pi)$ with equal probability. For ease of simulation, the fading parameters were held constant for the duration of one TDMA burst, thereby providing slightly optimistic results.

The BER as a function of E_b/N_o is shown in Figure 6.51 for different values of the Rician parameter K in dB, along with the results for the AWGN situation which we use as a bench marker. The BER performance was very much worse for the Rayleigh fading channel ($K = -\infty$dB) compared to that for the AWGN channel ($K = +\infty$dB). This was because even when E_b/N_o had a high value, the received signal was on occasions in a deep

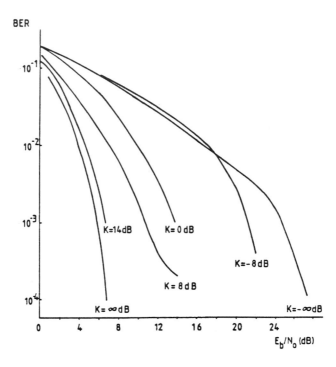

Figure 6.51: BER performance of DPM in frequency non-selective Rayleigh and Rician channels.

fade, enabling the noise to induce errors. The Rayleigh channel is the most severe type of non-frequency selective fading channel.   A study of propagation in microcells [27] revealed that in the majority of cases a dominant path exists between the transmitter and the receiver. The PDF of the fading envelope is Rician in this situation. To model a Rician channel a direct path, concurrent in time with the Rayleigh fading path, was added to the channel model. The Rician parameter K is the ratio of power in the dominant path to the power in the Rayleigh fading path. Figure 6.51 presents a family of Rician BER curves for which the AWGN channel and Rayleigh channel appear as special cases.   Inclusion of a direct path, even for $K = 0$ dB, caused a substantial improvement in performance, some 10 dB reduction in channel SNR compared to a Rayleigh fading channel at a BER of 10^{-3}. The improvement continued with increasing K, and at $K=14$ dB the performance approached that of the AWGN channel. Reference [28] suggests that a Rician channel with $K > 13$ dB behaves predominantly like an AWGN channel, an observation substantiated by this simulation. A Rayleigh fading channel was produced when the direct path was overwhelmed by the Rayleigh fading component [29]. It cannot be over-stressed

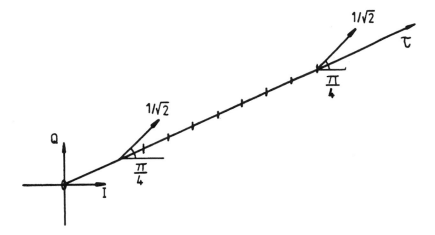

Figure 6.52: Typical complex impulse response of a frequency selective two-ray
static channel.

that by opting for microcellular structures, rather than large conventional
cells, the PDFs of the received signal envelopes become Rician and may
have high values of K, yielding significant gains in system performance.

6.2.8.3 DPM Transmissions over Frequency Selective Two-Ray Static Channels

This channel gives an insight into VE performance over a non-time-varying
frequency selective channel. The channel was modelled as a complex baseband finite impulse response (FIR) filter, having complex coefficients at
delay intervals given by the reciprocal of the system sampling frequency.
Only two of the tap coefficients were non-zero, giving rise to the two-ray
channel. A typical complex impulse response where the delay between the
two equal amplitude rays was eight sampling intervals (i.e., 2 bits with four
times oversampling) is shown in Figure 6.52. This complex response was
resolved into its inphase and quadrature components and complex baseband convolution performed. Figure 6.53 shows the BER as a function of
delay between the paths at an E_b/N_o of 5 dB. The degradation in BER as
a result of the frequency selective channel is evident.

6.2.8.4 DPM Transmissions over Frequency Selective Two-Ray Fading Channels

The mobile radio channel is dynamic in nature and as a consequence its
impulse response varies with time. When there is no direct path between
the transmitter and receiver, the amplitude distributions on each channel

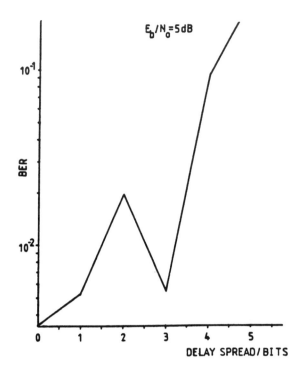

Figure 6.53: BER performance of DPM in a frequency selective two-ray static channel.

path (or ray) have Rayleigh statistics. Where a direct path existed, as may be the case in a microcell, or in air-to-ground radio communications, then it was assumed that only one of the paths experienced Rayleigh fading; produced as a result of many scatterers. For ease of simulation the channel impulse response was assumed to be constant during a TDMA burst.

Consider the situation where both paths possessed equal average power and experienced independent Rayleigh fading. Figure 6.54 shows the BER as a function of path delay separation when $E_b/N_o = 16.5$ dB. The value of E_b/N_o was chosen to ensure that with zero path separation (i.e., the two paths coalesced into one path), the BER was of the order of 10^{-2}. For path separations in the range from one to four bit periods, the BER improved by an order of magnitude compared with that achieved with zero path separation. This indicates that the VE uses the delayed path to provide a form of diversity. When the delay spread was less than one bit period the VE was unable to resolve the paths and so no improvement resulted. From Figure 6.54 it is clear that the reduced 16-state VE can cope with delay spreads of up to four bit periods duration. Delay spreads in excess of four bit periods caused the equalizer performance to degrade seriously

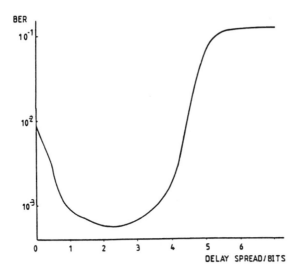

Figure 6.54: BER performance of DPM in a frequency selective two-ray Rayleigh fading channel.

and the saturated BER was much worse than that achieved with zero path separation. With four bit periods of path separation, the Viterbi equaliser could not accommodate both paths within its impulse response window. In the situation where one path dominated and the impulse response window was selected to accommodate it, then the interference caused by the second path lying outside the window was not removed. The result was a higher error rate than if both paths had been accommodated within the impulse response window.

The second part of this experiment involved only one of the two paths experiencing Rayleigh amplitude fluctuations. Figure 6.55 shows the BER as a function of path separation for values of $K = 0$ dB, 8 dB and 14 dB respectively. Again the values of E_b/N_o in each case were chosen to ensure a BER of 10^{-2} for zero path separation. When $K = 0$ dB, the results were similar to those produced when both paths experienced Rayleigh variations (see Figure 6.54), except for a small improvement in the saturated BER. The results for $K = 8$ dB were more interesting because the saturated BER was now only marginally inferior to that measured at zero path separation. It appears that the diffuse path was sufficiently large to provide diversity gain in the one to four bit duration path separation region, but not large enough to cause significant problems at greater path separations. With $K = 14$ dB the channel was virtually AWGN and in this situation the VE could not improve the BER. Notice that as the channels go from Rayleigh, to Rician to virtually Gaussian, the required E_b/N_o to achieve a BER of 10^{-2} at zero separation decreased from 16.5 dB to 4.6 dB. In practice, the

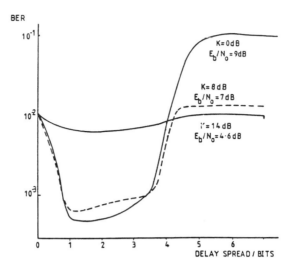

Figure 6.55: BER performance of DPM in a frequency selective two-ray Rician fading channel.

channel impulse response will not remain constant over the duration of a burst and consequently the channel estimates used in the VE will not be as accurate towards the end of a burst. However, the simulation results will not be changed dramatically, except in the rare cases where the channel is changing rapidly (i.e., significant Doppler phenomena).

6.2.9   Simulations of GMSK Transmissions

Initially we simulated the VE for GMSK (B_N=0.3, h_f=0.5) with a trellis where all four phase states were explicitly represented. This meant the VE possessed $4 \times 16 = 64$ states. The BER performance of the VE was evaluated over both static AWGN channels and Rayleigh fading channels. Next, we incorporated phase state memory into the simulation of the VE, reducing the number of states in the VE by a quarter of the previous value. Simulations showed that the BER performances of the equalisers were identical. This result was significant because a factor of four reduction in complexity gave no performance penalty.

6.2.9.1   GMSK Transmissions over an AWGN Channel

The simulations were similar to those with DPM, except that the band-width parameter a was reduced to 1.5 in order to account for the reduced spectral occupancy of GMSK. The BER as a function of E_b/N_o is shown in Figure 6.56. At a BER of 10^{-3} there was a degradation of about 1.5 dB in AWGN performance using GMSK ($B_N = 0.3$) as compared to DPM

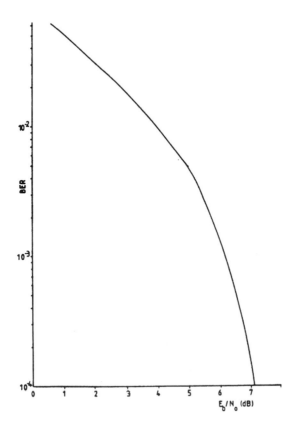

Figure 6.56: BER performance of GMSK in an AWGN channel.

(3RC). This was expected because GMSK ($B_N = 0.3$, $h_f = 0.5$) had a free distance d_f of two compared with three for DPM (3RC, $h_p = 1.0$) [4].

6.2.9.2   GMSK Transmissions over Frequency Selective Rayleigh Fading Channels

The BER performance of the GMSK Viterbi equaliser for frequency selective Rayleigh fading channels is shown in Figure 6.57 for $E_b/N_o = 15$ dB. The frequency selective channel had two independent equal power paths each with Rayleigh distributed amplitudes and phases uniformly distributed between 0 and 2π. The curve in Figure 6.57 is similar for DPM, see Figure 6.54. Again in the delay range from half a bit to four bits, there was a substantial improvement in BER performance due to multipath diversity. For delays in excess of four bits, the windowed estimated channel impulse response formed within the VE could no longer accommodate both

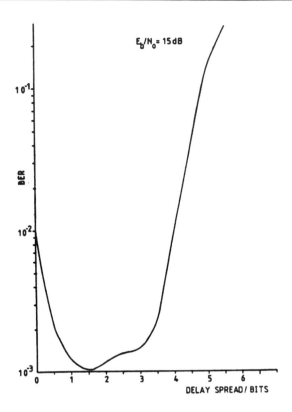

Figure 6.57: BER performance of GMSK in a frequency selective two-ray Rayleigh fading channel.

paths and consequently BER performance was seriously degraded.

6.2.9.3   Comment

We have described how moderately wideband systems known as NB-TDMA employing Viterbi equalisation and either DPM or DFM modulation are suitable for use with dispersive mobile radio channels. There is scope for employing sophisticated reduced state equalisers which can cope with greater excess path delays, without needing additional states. Further refinements include adaption of the equaliser within the TDMA burst, and the use of non-linear decision feedback equalisers (DFE). The DFE potentially has a performance similar to that achieved by Viterbi equalisation, but with a considerable reduction in the processing load. The reduction in complexity and power consumption gained in using a DFE would be valuable for hand-held portables.

$$*\qquad\qquad*$$

The focus of this chapter was on partial response modulation, where intentional inter-symbol interference is introduced in order to enhance the spectral efficiency of the modem. A prominent representative of this class of modems is Gaussian-filtered Minimum Shift Keying, which can be shown to possess the most compact spectrum possible. However, due to the inter-symbol interference the modulated signal can only be demodulated with the aid of a channel equaliser. In many implementations of the GSM system a Viterbi equaliser is used, which was detailed in this chapter. Our next chapter is focuses on frequency hopping, which is a powerful means of mitigating the effects of fading in wireless systems.

Bibliography

[1] **S.Pasupathy**: "Minimum shift keying: A spectrally efficient modulation," *IEEE Trans. Commun. Mag.*, pp.14-22, June 1979.

[2] **R.de Buda**: "Coherent demodulation of frequency shift keying with low deviation ratio," *IEEE Trans. Commun.*, pp.196-209, Jun 1972.

[3] **W.P.Osborne** and **M.B.Luntz**: "Coherent and non-coherent detection of CPFSK," *IEEE Trans. Commun.*, Vol. 22, pp.1023-1036, Aug 1974.

[4] **J.B.Anderson, T.Aulin** and **C-E. Sundberg**: "Digital phase modulation," *Plenum Press*, New York, 1985.

[5] **T.Maseng**: "Digitally phase modulated (DPM) signals," *IEEE Trans. Commun.*, Vol. 33, pp.911-918, Sept 1985.

[6] **T.Maseng** and **O.Trandem**: "Adaptive digital phase modulation," *Nordic Seminar on Digital Land Mobile Communications,"* Stockholm, Paper 19, Oct 1986.

[7] **A.Papoulis**: "Circuits and Systems — A modern approach," *Holt, Rinehart, Winston Inc.*, 1980.

[8] **T.Aulin** and **C-E.Sundberg**: "Continuous phase modulation — Part 1: Full response signalling," *IEEE Trans. Commun.*, Vol. 29, pp.196-209, Mar 1981.

[9] **T.Aulin** and **C-E.Sundberg**: "Continuous phase modulation — Part 2: Partial response signalling," *IEEE Trans. Commun.*, Vol. 29, pp.210-225, Mar 1981.

[10] **K.Murota** and **K.Hirade**: "GMSK modulation for digital mobile radio telephony," *IEEE Trans. Commun.*, Vol. 29, pp.1044-1050, Jul 1981.

[11] **F.de Jager** and **C.B.Dekker**: "Tamed frequency modulation — A novel method to achieve spectrum economy in digital transmission," *IEEE Trans. Commun.*, Vol. 26, pp.534-542, May 1978.

[12] **M.K.Simon** and **C.C.Wang**: "Differential detection of Gaussian MSK in a mobile radio environment," *IEEE Trans. Vehicular Technology*, Vol. 33, No. 4, pp.307-320, Nov 1984.

[13] **M.Hirono, T.Miki** and **K.Murota**: "Multilevel decision method for bandlimited digital FM with limiter-discriminator detection," *IEEE J. Sel. Areas Commun.*, Vol. 2, No. 4, pp.498-506, July 1984.

[14] **K-S.Chung**: "Generalised tamed frequency modulation and its application for mobile radio communications," *IEEE J. Sel. Areas Commun.*, Vol. 2, No. 4, pp.487-497, July 1984.

[15] **S.W.Wales, P.H.Waters** and **M.L.Streeton**: "Experimental program phase II final report," Report No: 72/87/R/515/U, *Plessey Research Roke Manor*, Dec 1987.

[16] **P.D.Welch**: "The use of fast Fourier transform for the estimation of power spectra: A method based on time averaging over short modified periodograms," *IEEE Trans. Audio and Elect.*, Vol. 15, No. 2, pp.70-73, June 1967.

[17] **R.R.Anderson** and **J.Salz**: "Spectra of digital FM," *BSTJ*, pp.1165-1189, July/Aug 1965.

[18] **H.E. Rowe** and **V.K.Prabhu**: "Power spectrum of a digital frequency modulation signal," *BSTJ*, pp.1095-1125, July/Aug 1975.

[19] **T.Maseng**: "The power spectrum of digital FM as produced by digital circuits," *Signal Processing*, Vol. 9, No. 4, pp.253-261, Dec 1985.

[20] **J.G.Proakis**: "Digital communications," *MacGraw-Hill*, New York, 1982.

[21] **J.M.Wozencraft** and **I.M.Jacobs**: "Principles of Communication Engineering," *Wiley*, 1965.

[22] **G.D.Forney, Jr**: "The Viterbi algorithm," *Proc. IEEE*, Vol. 61, No. 3, pp.268-278, March 1973.

[23] **A.P.Clark, J.D.Harvey** and **J.P.Driscoll**: "Near maximum likelihood detection processes for distorted radio signals," *The Radio and Electronic Engineer*, Vol. 48, No. 6, pp.301-307, June 1978.

[24] **A.Svensson, C-E.Sundberg** and **T.Aulin**: "A class of reduced complexity Viterbi detectors for partial response CPM," *IEEE Trans. on Commun.*, Vol. 32, No. 10, pp.1079-1087, Oct 1984.

[25] **A.D.Fagan** and **F.D.O'Keane**: "Performance comparison of detection methods derived from maximum-likelihood sequence estimation," *IEE Proc.*, Vol. 133, Pt. F, No. 6, pp.535-542, Oct 1986.

[26] **J.C.S.Cheung** and **R.Steele**: "Modified Viterbi equaliser for mobile radio channel having large multipath delays," *Electronics Letters*, Vol. 25, No. 19, pp.1309-1311, Sept 1989.

[27] **S.T.S.Chia, R.Steele, E.Green** and **A.Baran**: " Propagation and BER measurements for a microcellular system," *J. IERE*, Vol. 57, No. 6 (supplement), pp.S255-S266, Nov/Dec 1987.

[28] **F.Davarian**: "Fade margin calculations for channels impaired by Ricean fading," *IEEE Trans. Vehic. Tech.*, pp.41-44, Feb 1985.

[29] **M.Schwartz**: "Communication systems and techniques," *McGraw-Hill*, 1966.

Chapter 7

Frequency Hopping

D.G. Appleby[1], and Y.F. Ko[2]

7.1 Introduction

In this chapter we will consider frequency hopping in the context of narrowband mixed time and frequency division multiple access (TD/FDMA) for digital cellular mobile radio systems. In the type of system under consideration, such as for example the GSM Pan-European system, traffic channels are designated as combinations of time slot position and carrier frequency. However, the cellular multiple access protocol is based primarily on frequency division multiplexing (FDM), since the orthogonal sets of radio channels, needed to avoid cochannel interference between neighbouring cells, are obtained by assigning distinct sets of frequencies to all the cells in a reuse cluster, as in a pure FDMA system. Employing time division multiplexing (TDM) as well as FDM results in increased transmission symbol rates and fewer more widely spaced carrier frequencies are required in a given allocated band, which leads to less stringent transceiver design specifications on selectivity and frequency drift. In addition, higher spectral efficiency is possible with TDM, because of the greater precision achievable with digital switching technology. Narrowband operation is broadly defined by the provision that the transmitted symbol duration in most urban situations should be much greater than the delay spread caused by multipath propagation. This is widely regarded as a desirable condition, since it avoids the necessity for complex equalisation techniques to remove the severe intersymbol interference which would otherwise occur. Firstly we

---

[1]University of Southampton
[2]University of Southampton

must differentiate between 'slow' and 'fast' frequency hopping techniques. In the former case, which is of primary interest here, the hop rate is much less than the information symbol rate and thus many symbols are sent on the same carrier frequency during each hop, maintaining narrowband transmission conditions within each hop, provided of course that the symbol modulation bandwidth does not exceed the coherence bandwidth. On the other hand fast frequency hopping, in which the hop rate is equal to, or often greater than, the symbol rate, has been advocated for mobile radio [1] because of its spread spectrum properties. It can thus be classified as a wideband code division multiple access (CDMA) technique and so suffers many of the disadvantages of such techniques, in particular, implementation would be more difficult because of the requirement for very fast frequency synthesisers.   Slow frequency hopping multiple access (SFHMA), using code division multiplexing as the main multiple access mechanism has been the subject of considerable research effort in recent years [2–4] leading to a viable system design, the SFH900 [5], which was a very strong contender for the GSM Pan-European digital mobile radio network. Not only does SFHMA provide inherent frequency diversity, but it has the property of randomising cochannel interference, referred to as 'interferer diversity' [5], which allows error correction coding to be applied effectively to correct errors caused by interference as well as signal fading. Slow frequency hopping without CDMA has also been proposed for use in various narrowband TDMA systems [6,7], including the GSM Pan-European system, often as an optional 'add-on' feature. In all these applications it is only the frequency diversity advantage of frequency hopping which is being exploited to avoid the problem of stationary or slowly moving mobile stations being subjected to prolonged deep fades.   In the following section we discuss the principles and the characteristics of SFHMA and then in Section 7.3 an exemplary SFHMA system based on the SFH900 proposal [5,8] is described in detail. Sections 7.4 and 7.5 are devoted to analyses of the error rate performance of the SFHMA system in AWGN and in cochannel interference respectively. Estimates of spectral efficiency of the system are presented in Section 7.6 followed by a summary of the main conclusions drawn.

7.2   Principles of Slow Frequency Hopping Multiple Access

In this section we shall describe three multiple access techniques based on SFH. Possible reuse cellular structures which may be employed for SFHMA systems are discussed in Section 7.2.2, followed by a study of the factors that affect propagation in Section 7.2.3.

7.2.1   SFHMA Protocols

Three code division multiple access techniques employing SFH with pseudo random sequences have been defined in [2], they are namely, **orthogonal**, **random**, and **mixed**;

Orthogonal: orthogonal hopping sequences of length N can be assigned to active users within a cell to which N hop frequencies have been allocated to ensure that during any hop only one user can transmit on a particular frequency. In neighbouring cells orthogonality of transmissions is attained by allocating distinct sets of frequencies, as in FDMA. A given set of hop frequencies and sequences can be reused in cells which are at or beyond the reuse distance. This protocol assumes complete synchronisation of sequences so that any users with the same sequence of hop frequencies will experience continuous cochannel interference on every hop and thus the situation is the same as for FDMA, except for the inherent frequency diversity.

Random: each active user is assigned a unique hop sequence which is uncorrelated with, but not necessarily orthogonal to, all other sequences. The constraint of low cross correlation of hopping sequences is less stringent than that of orthogonality and thus allows a much larger set of sequences to be selected. Cochannel interference will occur in this case, but it will be caused by a different subset of the other active users on each hop and in consequence the interference intensity will be subjected to random hop-to-hop variations. This behaviour, which is known as 'interferer diversity', is an important feature of frequency hopping, since it spreads the interference effects evenly, on average, over all the available frequencies. It also enables the effects to be counteracted by powerful error correction coding. Reuse cellular structures are not necessary.

Mixed: as in the orthogonal protocol, a set of N orthogonal sequences of the N available hop frequencies is assigned to each cell, but in this case the sets of sequences assigned to reuse cells operating with the same frequencies are distinct and are selected to ensure that any two sequences from different cells are uncorrelated. Thus cochannel interference is caused only by transmissions in reuse cells, and in addition, its effects are mitigated by interferer diversity, as in the random protocol. It is assumed that a pair of uncorrelated sequences of length N hop to the same frequency in only one of the N hops.

Frequency diversity is the only advantage of the orthogonal protocol, whereas the random protocol has the additional important advantage of interferer diversity and was the preferred SFHMA protocol in early investigations [2]. However, later studies [5, 8] of SFHMA system design have opted for the mixed protocol, because when used in conjunction with fractional reuse structures, as described in the following subsection, it offers more freedom for system design optimisation and hence may be expected to yield higher spectral efficiency. Therefore in the remainder of this chapter

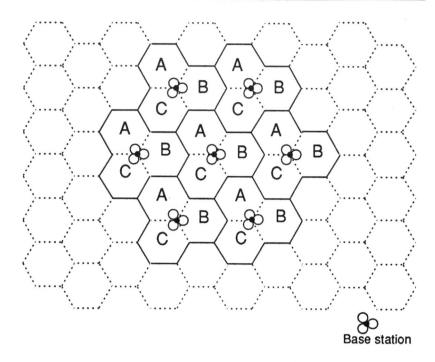

Base station

Figure 7.1: Three-colour cellular frequency reuse structure.

we shall consider only the mixed protocol.

7.2.2 Reuse Cellular Structures

The basic form of the reuse structure proposed for SFHMA systems [5,8] is the three-colour cluster, as shown in Figure 7.1, where the colours, indicated by the letters A, B and C, represent distinct sets of N frequencies assigned to the three cells. Also shown in the diagram is the collocated corner base station configuration, with directional antennas pointing into the respective cells, which is favoured primarily because of the obvious economies in the number of base station sites required and in the connections to the fixed cable network. Some authors refer to this configuration as a single sectored cell, but we prefer to describe it as a cluster of three separate cells.

Choosing a reference cell with colour A, then all other cells marked A in Figure 7.1 are full reuse cells using the same frequencies. It is assumed that it is possible to assign uncorrelated sets of hopping sequences to all the reuse cells near enough to produce significant levels of cochannel interference. An active user in the reference cell will experience interference from an active user in a reuse cell on only one of each sequence of N hops, when both transceivers are tuned to the same frequency. Each additional active user

Frequency group	A1	A2	A3	A4	A5	A6	A7
			S				hade
f1	*	*	*				
f2	*			*	*		
f3	*					*	*
f4		*		*		*	
f5		*			*		*
f6			*	*			*
f7			*		*	*	

Table 7.1: Shade allocation scheme for a 21/3 cellular frequency reuse structure.

in the same or another reuse cell will cause similar frequency collisions, but on a different frequency in the reference user's sequence. The dependence on the number of active users in reuse cells can be expressed in terms of the proportion of the hop sequence occupied by frequency collisions per active user. This quantity is known as the frequency collision rate and for the basic three-colour cluster under consideration it has the value of $1/N$.

Fractional reuse structures, which offer greater flexibility for design trade-offs, are obtained by dividing each colour into M overlapping sub-sets of L groups of frequencies, which henceforth we will refer to as 'shades' of that colour (Verhulst uses the term 'pseudo-colour' in [5]). The resulting reuse cluster contains $3M$ cells arranged as M sub-clusters of size 3, each using a different shade of the 3 colours and centred on a common base station site. This is termed a $3M/L$ fractional structure, where L/M is the fraction of the N frequencies making up a colour, which is contained in each shade. As an example take $M=7$ and $L=3$, giving a 21/3 structure as shown in Figure 7.2. A shade allocation scheme for colour A is presented in Table 7.1 which indicates that the fractional overlap of hop frequencies is 1/3. Other schemes can be obtained from the one shown by interchanging pairs of rows in the allocation matrix.

Figure 7.3 shows the positions of both full and partial reuse cells for colour A, i.e., the potential sources of cochannel interference, out to the first ring of full reuse cells. For full reuse cells, as in the case of the three-colour cluster, the frequency collision rate is simply the inverse of the sequence length, i.e. $M/LN = 7/3N$. However, for partial reuse cells we must allow for the fact that, because of the incomplete frequency overlap, certain pairs of sequences will produce no collisions, with a probability equal to the fractional overlap, k. Thus a general expression for the frequency collision rate, y, in this case is given by:

$$y = \frac{7k}{3N} \qquad (7.1)$$

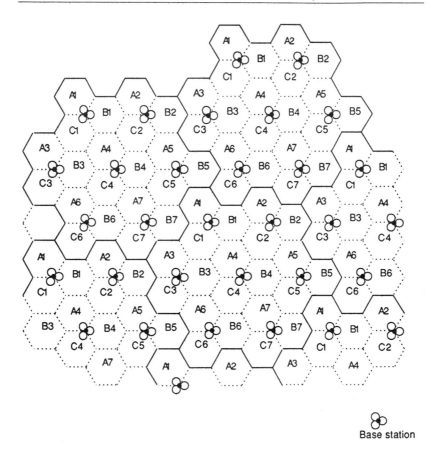

Base station

Figure 7.2: 21/3 cellular frequency reuse structure.

where k can have the following values:

$k = 0$ - for cells of different colour
$k = 1/3$ - for cells of the same colour but different shade, i.e. partial reuse
$k = 1$ - for cells of the same colour and shade, i.e. full reuse

For comparison, consider a second example in which we make $M = 4$ and $L = 3$, giving a 12/3 reuse structure as shown in Figure 7.4. One possible shade allocation matrix is presented in Table 7.2. Figure 7.5 shows the reuse cells for this case, for which $k = 2/3$ in cells with partial frequency reuse. The frequency collision rate now becomes

$$y = \frac{4k}{3N} \tag{7.2}$$

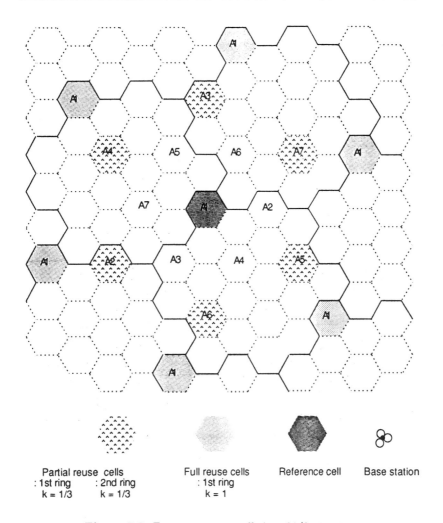

Figure 7.3: Frequency reuse cells in a 21/3 structure.

	S		hade	
Frequency group	A1	A2	A3	A4
f1	*	*	*	
f2	*	*		*
f3	*		*	*
f4		*	*	*

Table 7.2: Shade allocation scheme for a 12/3 cellular frequency reuse structure.

Type of	y	
reuse cell	21/3	12/3
Partial	7/9N	8/9N
Full	7/3N	4/3N

Table 7.3: Frequency collision rates of partial and full frequency reuse cells.

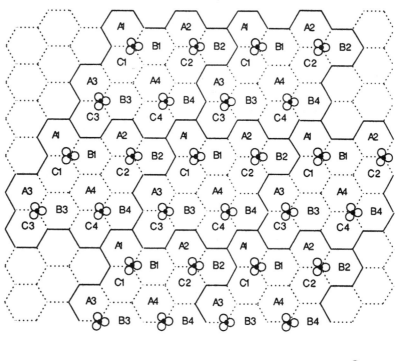

Figure 7.4: 12/3 cellular frequency reuse structure.

Values of y for both structures are given in Table 7.3.

7.2.3   Propagation Factors

In this section we will discuss briefly a set of factors which can be used to characterise the radio channel for studies of the SFHMA system [5, 8]. More detailed treatments of propagation characteristics are presented in Chapters 1 and 2. For our purpose here it is convenient to assume a

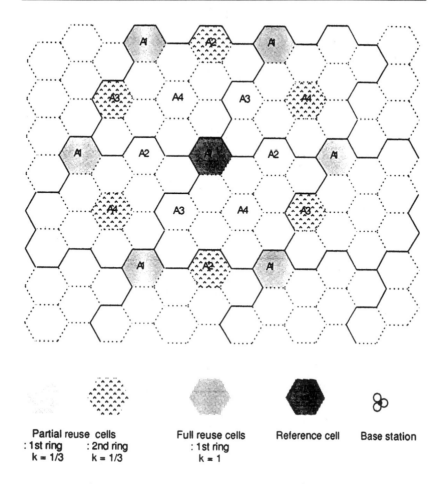

Figure 7.5: Frequency reuse cells in a 12/3 structure.

narrowband model for multipath distortion, even though in the system to be considered the normal narrowband criterion that the symbol duration should be much greater than the delay spread is not always satisfied.

Using the narrowband model, the power levels of both the wanted signal and interfering signals can be estimated as the product of three principal factors:

- mean received level A_i

- shadowing attenuation B_i

- multipath fading F_i.

A_i is deterministic and may include factors such as antenna radiation patterns, transmitter power control and receiver adjacent channel rejection, in addition to the mean path loss, which, based on the usual simplified model, can be assumed to be proportional to $d^{-\alpha}$, where d is the distance from the relevant transmitter and α is the propagation exponent, which depends on the environment. For all the calculations reported here α has been taken as 3.5.

B_i depends on large scale variations in the locations of the transmitter and receiver relative to the respective local topographies, and in practice it is unpredictable. Thus this factor is usually taken to be a random variable with lognormal distribution having a mean value of 0 dB and a standard deviation in the range 6 to 8 dB for typical urban environments. For normal vehicle speeds B_i may be assumed to be constant in frequency over a few MHz and in time over a few hundred ms.

F_i represents the rapid and severe fluctuations in signal level experienced on a radio link with a moving vehicle, often referred to as fast fading. These fluctuations are caused by the changes in relative phases between the multipath components due to small scale variations in vehicle location (of the order of half the wavelength). It can usually be assumed, especially in medium to large cells, that the resulting signal envelope has a Rayleigh amplitude distribution and a negative exponential pdf of power. Thus a negative exponential pdf with a mean of unity is appropriate for this factor. If the spacing of consecutive hop frequencies is greater than the coherence bandwidth (of the order of a few hundred kHz) and if the transmission duration during each hop is less than the coherence time (of the order of a few ms) then the F_i can be assumed to be constant during each hop and to be statistically independent from hop to hop.

Interfering signals are characterised by a fourth factor E_i, which is a discrete random variable having two values 0 and 1 with a Bernoulli distribution, representing the effective on/off status of the interference source. This factor accounts for the random nature of frequency collisions inherent in the mixed SFHMA protocol.

An important aspect in evaluating the effects of these propagation factors is the rate at which they are likely to change. A factor can be broadly categorised as 'fast' if it changes independently from hop to hop, as noted in the case of the fading factor, F_i. On the other hand a factor which remains highly correlated over many hop periods can be categorised as 'slow'. For the wanted signal the factors A_o and B_o are slow, but for interfering signals all the A_i and B_i are fast in the mobile-base direction, i.e. uplink, because of the changes in the locations of sources of interference from hop to hop, although they are slow in the other direction. F_i and E_i are always fast.

The value zero for the index i refers to conditions relevant to the wanted signal within the reference cell and non-zero values refer to interfering signals from other cells. Hence the received carrier-to-interference ratio (CIR) is given by:

$$\lambda = \frac{A_o B_o F_o}{\displaystyle\sum_{i=1}^{M} A_i B_i F_i E_i} \tag{7.3}$$

7.3 Description of an SFHMA System

The system described in this section is closely related to the SFH900 system proposed by the French collaboration LCT and Sagatel [5,8], as a candidate for the GSM Pan-European digital mobile radio system.

7.3.1 Multiple Access Protocol

A mixed SFHMA cellular protocol is assumed, in which, as explained in Sub section 7.2.1, strictly orthogonal hopping sequences are used for CDMA within each cell, but in neighbouring reuse cells different uncorrelated sets of sequences are deployed. Thus there will be no interference between users within a cell since in any time slot only one user should be transmitting on any given carrier frequency. Contention with immediate neighbour cells is prevented by using orthogonal sets of hopping frequencies. Interference from further cells within the main reuse cluster, in which partial overlap of frequency sets is allowed, is reduced because of interferer diversity, i.e., on each hop there is a new subset of possible interferers with statistically independent propagation characteristics.

7.3.2 Time Division Multiplexing

The hop duration of 4 ms is divided into 3 time slots, each 1.23 ms long, separated by guard times of 100 μs. These slots are used in sequence for transmission, reception and frequency switching. The users in a cell are divided into 3 approximately equal subsets which use the 3 possible phases of this sequence. Apart from other advantages in simplifying the design of the transceivers, this feature allows the number of FH channels in a cell to be trebled for a given set of FH sequences and also results in further randomisation of the interference generated by transmissions from the mobile stations.

We consider that the slot duration is short enough to justify the assumption that the transmission channel impulse response is static during each hop, even for fast vehicles.

7.3.3 Modulation and Equalisation

Binary GMSK modulation with a normalised premodulation filter bandwidth $B_t T = 0.3$ is assumed, together with quasi-coherent demodulation

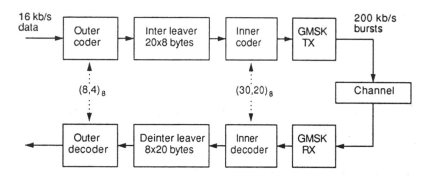

Figure 7.6: Concatenated RS channel coding.

using a Viterbi Algorithm equalisation technique to counteract intersymbol interference caused by the pre-modulation Gaussian LPF and multipath distortion. Reference should be made to Chapter 6, for further details of GMSK modulation and Viterbi equalisation.

Transmission of an 8-bit training pattern during each slot enables the demodulator to derive the complex baseband impulse response of the transmission channel and also to determine the carrier phase. Hence there is no need for a phase-locked loop, but the carrier frequency error must be limited to less than about 30 Hz to ensure that there is negligible degradation due to the consequent phase drift during the slot. Adequate frequency control is achieved by measurements on the unmodulated carrier transmissions on certain hops assigned to the Master Channel (see Section 7.3.5 below) or alternatively by sending two training patterns during each data hop (at 1/4 and 3/4 of the slot duration) so that the phase difference can be estimated.

7.3.4   Speech and Channel Coding

A 16 kbit/s speech encoder is used with three levels of protection against digital errors. An important feature of the system under consideration is the highly redundant forward error correction (FEC) channel coding to enable correction of long error bursts extending over complete hops caused by fading or frequency collisions. The overall average code rate is 1/3 on speech traffic channels, which is achieved by concatenating two shortened Reed-Solomon (RS) codes. Figure 7.6 shows the cascaded arrangement of the two coders and two decoders. The inner code is a (30,20) RS code with 8-bit symbols which codes a data block of length 160 bits to be transmitted in one hop. This code is applied to all traffic including control signals. The outer RS code has three levels of redundancy to give different levels of protection to various parts of the encoded speech frame. These codes are (8,3), (8,4) and (8,5), also with 8-bit symbol length, but the net rate is maintained at 1/2. Symbols from a given codeword are transmitted on separate hops so that errors due to fading and interference are decorrelated. This is achieved by interleaving symbols from blocks of 20 codewords (bytes) at the input to the inner coder, as shown in Figure 7.6.

It is possible to use different cosets of the inner code in cells assigned the same hop frequencies so that captures by strong interfering cochannel signals can be identified and hence rejected. In the analyses of error probability described in Sections 7.4 and 7.5, a single (8,4) outer code has been assumed.

7.3.5 Transmitted Signal Structure

A hop frame comprises 60 hops, which are divided into 3 subsets as follows:

H1 - Common Control Channels including   4 hops
     the Master Channel
H2 - Associated Control Channels   8 hops
H3 - Traffic Channels   48 hops

The hop frame duration is $60 \times 4 = 240$ ms. Allowing for the proportion of the frame allocated to speech traffic, the time division multiplexing (including guard times), the channel coding redundancy and the dual training patterns the transmitted bit-rate required becomes very close to 200 kbit/s. For this bit rate the optimum hop frequency spacing is about 150 kHz, which provides 17 dB adjacent channel rejection.

7.3.6 Frequency Reuse

Base stations are assumed to be located at the junctions of three cells into which they would radiate via 120° sectorial antennas. Following [5], we represent the base station antenna radiation pattern with a back-to-front ratio of -20 dB by the expression

$$g(\theta) = \max[\cos(1.1\theta), 0.01] \qquad (7.4)$$

A reuse cluster size of 3 is appropriate for the primary 6 MHz band in which the Master Channel is located. The Master Channel is designed to have a simple and regular time-frequency structure to enable rapid acquisition of network control information by a mobile entering the system. In the remainder of the allocated system bandwidth the available frequencies will be divided into groups to be allocated as 'shades' in fractional reuse structures, as described in Section 7.2.2. In practice some terrain dependent variations of reuse structures would be required to cope with the consequent propagation problems.

7.4   BER Performance in the Absence of Co-channel Interference

In this section the performance of the SFHMA system described above is
investigated in AWGN channels. A pure AWGN channel is a highly idealised channel condition in mobile radio environments, in which the system
is expected to perform well. In order to achieve spectral efficiency cellular
mobile radio systems are designed to be limited by cochannel interference
rather than noise. However, it is often useful to be able to compare systems on the basis of performance in the ideal AWGN channel, which is
further characterised here according to whether the received signal level is
assumed to be either static, or varying due to Rayleigh fading.   GMSK belongs to the class of modulation schemes called partial response Continuous
Phase Modulation (CPM). BER performance is of course dependent on the
particular type of demodulator. A comprehensive discussion on various demodulator structures for CPM signals can be found in [9]. The two classes
of demodulator considered in this section are the sub-optimal MSK-type
orthogonal coherent detector as suggested by Murota and Hirade in their
original paper on GMSK [10], and the optimum Maximum-Likelihood Sequence Estimation (MLSE) detector, which is closely approximated by the
quasi-coherent demodulator and equaliser based on the Viterbi Algorithm
assumed for the SFHMA system. The MSK-type receiver is of interest here
partly because of its simplicity in practical implementation with only a minor degradation in power efficiency compared to the MLSE detector, but
primarily because its performance in more realistic mobile radio conditions
is easier to analyse than the MLSE performance. Furthermore the SFHMA
receiver is expected to perform similarly in these channel conditions, while
the MLSE detector upper bounds the performance of GMSK.

As many published modulation scheme performances are evaluated at
a typical coded speech data rate of 16 kbit/s, we start our analysis by
considering the two types of GMSK receiver structure operating at this
data rate with no FEC coding in Sections 7.4.1 and 7.4.2. In Section 7.4.3
some of the key assumptions on the mobile radio channels and system
operations are discussed. The basis for comparison between the SFHMA
and uncoded systems is also addressed in this section. The effects on BER
performance resulting from the FEC coding and TDM specified for the
SFHMA system are investigated in Section 7.4.4 for the case of the static
AWGN channel. Finally we evaluate and compare the BER performances
in the more realistic Rayleigh fading AWGN channel in Section 7.4.5.

7.4.1   BER Performance of the MLSE Detector

General discussions on the principle of MLSE detectors can be found in
Chapter 6 and also in [11], while MLSE receiver structures suitable for
GMSK are reviewed in [12] and [13]. Although the structure of the optimal

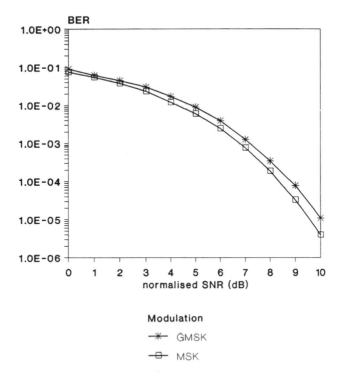

Figure 7.7: BER versus normalised SNR for GMSK ($B_t T = 0.3$) and for MSK using an ideal MLSE detector, in a static AWGN channel with continuous uncoded transmission at 16 kbit/s.

MLSE is known, it is difficult to evaluate its BER performance analytically. However, it is generally accepted [9, 10] that at fairly high SNR the BER performance is asymptotically dominated by the minimum Euclidean distance between the symbols in the signal space. For the type of modulation being considered here, the probability of bit errors for the ideal coherent MLSE detector is bounded in the high SNR condition as [10]:

$$P_e = Q \left(d_{\min} \sqrt{\frac{E_b}{N_o}} \right) \tag{7.5}$$

where $d_{\min}$ is the normalised minimum Euclidean distance, E_b is the mean energy per data bit, N_o is the single-sided power spectral density of the AWGN and $Q()$ is the normalised Gaussian integral, defined as:

$$Q(x) = \frac{1}{\sqrt{2\pi}} \int_x^\infty \exp(-u^2/2) du \tag{7.6}$$

In general the minimum squared Euclidean distance in the signal space is defined as:

$$D_{min}^2 = \min_{\underline{\alpha}, \underline{\beta}} \left\{ \int_0^{LT} \left[v(t, \underline{\alpha}) - v(t, \underline{\beta}) \right]^2 dt \right\} \qquad (7.7)$$

where T is the symbol duration, L is the number of symbols observed and $\underline{\alpha}$ and $\underline{\beta}$ are digital sequences of length L, that differ in at least the first digit, and which ensure that the modulator finishes in a common state having started from a common state.

The normalised minimum squared Euclidean distance is given by:

$$d_{min}^2 = \frac{D_{min}^2}{2E_b} \qquad (7.8)$$

For constant envelope CPM signals, which can be expressed in the form:

$$v(t, \underline{\alpha}) = \sqrt{2S} \cos[\omega_0 t + \theta(t, \underline{\alpha})] \qquad (7.9)$$

where S is the mean signal power, the normalised minimum Euclidean distance squared can be shown [9] to be given by:

$$d_{min}^2 = \min_{\underline{\alpha}, \underline{\beta}} \left\{ \frac{1}{T} \int_0^{LT} \left[1 - \cos[\theta(t, \underline{\alpha}) - \theta(t, \underline{\beta})] \right] dt \right\} \qquad (7.10)$$

From the calculated data presented by Murota and Hirade [10] for GMSK we find that for the normalised premodulation bandwidth $B_t T = 0.3$

$$d_{min} = \sqrt{1.78} = 1.334 \qquad (7.11)$$

which may be compared with the value of $\sqrt{2}$ for MSK or BPSK.

Figure 7.7 shows bit error probability versus the normalised signal-to-noise ratio E_b/N_o for GMSK with $B_t T = 0.3$ using an ideal MLSE detector, obtained by evaluating Equation 7.5. This curve represents the performance of continuous uncoded transmission of data at 16 kbit/s. Also included is a graph of error probability for MSK, which when compared with the GMSK graph, shows that there is only a small penalty in performance in using GMSK, which is more than outweighed by the superior spectral efficiency of GMSK.

7.4.2 BER Performance of the MSK-Type Detector

In the original paper on GMSK [10] Murota and Hirade described an orthogonal coherent MSK-type detector and presented experimental results showing that a BER performance close to that of an ideal MLSE detector can be obtained. Although the MSK-type detector is theoretically suboptimum for partial response CPM signals and is suitable only for schemes

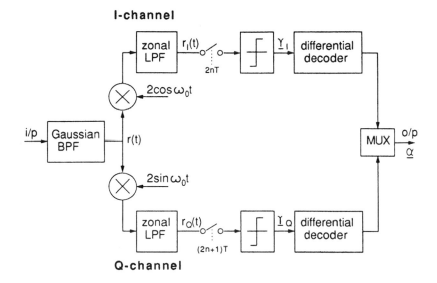

Figure 7.8: Conceptual structure of a parallel MSK-type receiver for GMSK
signals.

with a modulation index of 0.5, it is of special interest because of the relative ease of implementation compared with an MLSE detector. MSK-type
receivers can be implemented in serial form as well as the more familiar
parallel form, but since it has been shown [14] that, assuming ideal phase
and symbol timing recovery, both forms have equal performances, we shall
analyse in this section the performance in a static AWGN channel of a
parallel MSK-type detector only.

Figure 7.8 shows the conceptual structure of a parallel MSK-type receiver, in which the main demodulation noise filtering is provided by a
Gaussian band-pass filter (BPF) at the receiver input, as in the experimental circuit described in [10]. Perfect synchronisation of both carrier
phase and symbol timing is assumed in the following analysis. The impulse
response of the Gaussian BPF can be expressed in the form:

$$h_r(t) = 2Re\{h_{rb}(t)\exp(j\omega_o t)\} \tag{7.12}$$

where ω_o is the centre frequency and $h_{rb}(t)$ is the complex envelope of
the impulse response of the equivalent low-pass filter, which assuming zero
delay is given by the real Gaussian function:

$$h_{rb}(t) = \frac{\mu_r}{\sqrt{\pi}}\exp(-\mu_r^2 t^2) \tag{7.13}$$

where the parameter μ_r is related to the 3 dB bandwidth of the BPF, B_r, as:

$$\mu_r = \frac{\pi B_r}{\sqrt{2\ell n 2}} \tag{7.14}$$

Thus the filter output waveform can be obtained as the sum of the GMSK input signal convolved with $h_r(t)$ and the filtered AWGN, which can be expressed in the form:

$$r(t) = \mathrm{Re}\{[s(t) * h_{rb}(t)]\exp(j\omega_o t)\} + n_I(t)\cos\omega_o t - n_Q(t)\sin\omega_o t \tag{7.15}$$

where $n_I(t)$ and $n_Q(t)$ are the zero mean low-pass Gaussian processes of the inphase and quadrature components of the band-pass noise output and $s(t)$ is the complex envelope of the input signal which is given by:

$$s(t) = \sqrt{2S}\exp[j\theta(t,\underline{\alpha})] \tag{7.16}$$

After multiplying by the local inphase and quadrature carrier waveforms, $2\cos\omega_o t$ and $-2\sin\omega_o t$, and then zonal filtering to remove the sum frequency components, the baseband signals in the I and Q channels can be expressed as:

$$r_I(t) = \sqrt{2S}\cos\theta(t,\underline{\alpha}) * h_{rb}(t) + n_I(t) \tag{7.17}$$

and

$$r_Q(t) = \sqrt{2S}\sin\theta(t,\underline{\alpha}) * h_{rb}(t) + n_Q(t) \tag{7.18}$$

The output binary data sequence $\underline{\alpha}$ is determined by in effect differentially decoding the offset binary quadrature symbol sequences γ_I and γ_Q, which are obtained symbol by symbol after simple threshold detection of samples of $r_I(t)$ and $r_Q(t)$ taken at even and odd multiples of T, respectively. Since the processes of filtering and symbol detection are identical for the two quadrature channels, apart from a time shift of T, it is sufficient to consider only the I channel in evaluating the probability of quadrature symbol errors.

For a given data sequence $\underline{\alpha}$ the amplitude of the sample of $r_I(t)$ at the optimum sampling instant $(t = 0)$ is a Gaussian distributed random variable, y, which, putting $S = E_b/T$, has a mean value given by:

$$M_y(\underline{\alpha}) = \sqrt{\frac{2E_b}{T}}\int_{-\infty}^{\infty}\cos\theta(\tau,\underline{\alpha})h_{rb}(-\tau)d\tau \tag{7.19}$$

and a variance given by:

$$\begin{aligned} \sigma_y^2 &= \ <n_I^2(t)> \\ &= \ N_o B_{rn} \end{aligned} \tag{7.20}$$

where B_{rn} is the noise bandwidth of the BPF. Noting that the BPF has unity gain at the centre frequency ω_o and using Parseval's theorems, B_{rn}

can be determined as:

$$B_{rn} = \int_{-\infty}^{\infty} h_{rb}^2(t)dt$$

$$= \frac{B_r}{2}\sqrt{\frac{\pi}{\ell n2}} \tag{7.21}$$

As a function of the data sequence, the probability of a quadrature symbol error is given by:

$$P_{qe}(\underline{\alpha}) = Q\left\{\frac{M_y(\underline{\alpha})}{\sigma_y}\right\} \tag{7.22}$$

It is of course the ISI in the baseband signal waveforms, produced by both the premodulation filter and the predetection filter, which causes the above error probability to vary with the data sequence. The overall probability can be obtained by averaging over the set of possible relevant sequences, having a length equal to the effective duration of the ISI, which is dependent on the normalised bandwidths of the filters. However, at high values of SNR, as for the MLSE detector, the overall error probability will be dominated by the contribution due to the minimum value of M_y. By analogy with the treatment of the MLSE detector we introduce a squared distance parameter due to Svensson and Sundberg [14] defined as:

$$\delta^2 = \left[\frac{M_y(\underline{\alpha})}{\sigma_y}\right]^2 \cdot \frac{N_o}{E_b} \tag{7.23}$$

Hence the bound on the quadrature symbol error probability can be expressed in the form:

$$P_{qe} = Q\{\delta_{\min}\sqrt{E_b/N_o}\} \tag{7.24}$$

where using the convolution integral $\delta_{\min}$ is obtained from:

$$\delta_{\min}^2 = \min_{\underline{\alpha}}\left\{\frac{2}{TB_{rn}}\left[\int_{-\infty}^{\infty}\cos\theta(\tau,\underline{\alpha})h_{rb}(-\tau)d\tau\right]^2\right\} \tag{7.25}$$

Because of the differential decoding of the detected quadrature symbols each isolated symbol error will result in two output bit errors, but a group of consecutive quadrature symbol errors, occurring alternately on both the I and Q channels, will also only give rise to two bit errors at the beginning and end of the group. Consequently it can be shown that the probability of bit errors is given by:

$$P_b = 2P_{qe}(1 - P_{qe})$$

$$\approx 2P_{qe} \quad \text{if } P_{qe} << 1 \tag{7.26}$$

Thus for high SNR we can approximate the bit error probability as:

$$P_b = 2Q\{\delta_{\min}\sqrt{E_b/N_o}\} \tag{7.27}$$

It has been found by Murota [15], and confirmed during computations made by the authors, that the maximum ISI is caused by the sequence $\underline{\alpha} = \cdots 0,0,0,1,1,1\cdots$ which can therefore be used to evaluate $\delta_{\min}$. This sequence produces the following phase waveform [15]:

$$\theta(t) = \frac{\pi}{2T}\int_0^t erf(\mu_t\tau)d\tau + \frac{\sqrt{\pi}}{2T\mu_t} \tag{7.28}$$

where μ_t is a parameter of the premodulation Gaussian low pass filter related to the 3 dB bandwidth, B_t, as:

$$\mu_t = \pi B_t\sqrt{\frac{2}{\ell n2}} \tag{7.29}$$

and $erf()$ is the error function defined as:

$$erf(x) = \frac{2}{\sqrt{\pi}}\int_0^x \exp(-u^2)du \tag{7.30}$$

Equation 7.27 was evaluated for normalised bandwidths of $B_tT = 0.3$ and $B_rT = 0.63$, the latter value being the empirical optimum value found by Murota and Hirade [10] and which has been also confirmed during these computations by the authors. The convolution integral in Equation 7.25 was determined numerically, with the limits reduced to $\pm T$, since outside this range $h_{rb}(t)$ becomes negligible.

Figure 7.9 presents graphs of bit error rate (BER) probability versus E_b/N_o for the MSK-type detector and for the MLSE detector (from Section 7.4.1) for comparison. Again in the context of the analysis of the SFHMA system this performance may be taken as being representative of the continuous transmission of uncoded data at 16 kbit/s. It will be seen that the degradation in performance compared to the optimum MLSE detector is only about 1 dB in SNR. This computed performance for the MSK-type detector is also commensurate with the experimental results reported in [10].

7.4.3   Channel Models and System Assumptions

In the previous subsections we have analysed bit error rates in a static AWGN channel to provide an indication of the performance of the GMSK modulation scheme achievable under idealised conditions. In the following subsections we shall determine the BER performances of the SFHMA system, described in Section 7.3, and of an equivalent uncoded FDMA system in both static and fading AWGN channels. However, prior to starting the

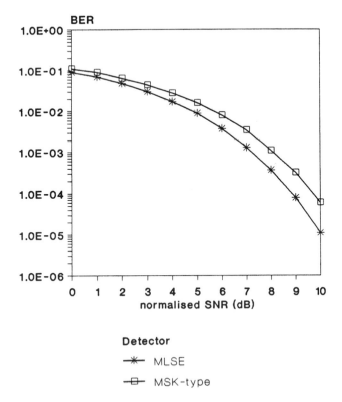

Figure 7.9: BER versus normalised SNR for GMSK ($B_t T = 0.3$) using an ideal MSK-type detector ($B_r T = 0.63$) and also using an ideal MLSE detector, in a static AWGN channel with continuous uncoded transmission at 16 kbit/s.

system analysis we wish to present the dynamic channel model and the principal assumptions affecting system performance.

As indicated previously in Section 7.2.3 we assume a narrowband flat Rayleigh fading channel model to account for multipath propagation distortion, although we realise that the specified SFHMA system is not strictly narrowband because the transmitted bit duration (5μ s) is comparable with the maximum expected delay spread. In [8], the use of the Rayleigh channel model is justified on the grounds that propagation tests at 900 MHz in urban areas showed that most of the multipath distortion energy occurs at short delay values and hence flat fading is the dominant channel impairment. Thus the complex envelope of the received GMSK signal can be expressed as:

$$s(t) = \rho(t) \exp\{j\phi(t)\} \exp\{j\theta(t, \underline{\alpha})\} \qquad (7.31)$$

where both $\rho(t)$, which is the Rayleigh distributed real envelope, and $\phi(t)$, which is a uniformly distributed phase perturbation, vary slowly enough to justify the following assumptions:

(1) the receiver can estimate the phase perturbations without error and hence can fully maintain the signal phase coherence;

(2) the magnitude of the signal envelope remains constant over the receiver observation interval of L_r bits.

The baseband signal in the I channel now becomes:

$$r_I(t) = \rho(t) \cos \theta(t, \underline{\alpha}) * h_{rb}(t) + n_I(t) \tag{7.32}$$

where the Rayleigh distributed envelope $\rho(t)$ has a long term mean squared value of $2S$. It can be shown that the normalised short term signal-to-noise ratio $\gamma_b = E_b/N_o$ averaged over L_r bits has an exponential pdf given by:

$$
\begin{aligned}
p(\gamma_b) &= \tfrac{1}{\Gamma_b} \exp\left(-\tfrac{\gamma_b}{\Gamma_b}\right) && \text{for } \gamma_b \geq 0 \\
&= 0 && \text{for } \gamma_b < 0
\end{aligned}
\tag{7.33}
$$

where Γ_b is the long term mean value.

In the SFHMA system coded speech data signals are transmitted in bursts at a rate of 200 kbit/s. In our analysis, we shall compare the SFHMA system performance with the performance of a conventional FDMA system with no error control coding and with continuous transmission at a rate of 16 kbit/s, which is the information bit rate of the SFHMA system. The principal performance measure we shall use is the probability of output data bit errors. Furthermore, for fair comparisons of the coded and uncoded systems, we assume that the same amount of energy is used to transmit the same number of information bits in the same fixed duration. As illustrated in Figure 7.10, the increase in bit rate due to FEC coding implies that the transmitted bit period in the coded system, T_c, is less than the data bit period, T, of the uncoded continuous transmission case and the transmitted power S is assumed constant. This means that the energy per bit is decreased by the use of coding because $E_c = ST_c$, and therefore the channel bit error probability is increased. Note that the further increase of transmitted bit rate due to the burst mode of operation does not lead to a reduction of energy per channel bit because there is a compensating increase of signal power during the burst to S_b.

In all the following analysis subsections we have assumed MSK-type detectors in both systems, because this greatly reduces the computational complexity, especially for the analysis of cochannel interference. Furthermore, we make the assumption that hard decision symbol detection is used. In the SFHMA system no allowance has been made for signal validity information being supplied by the demodulator to the inner channel decoder.

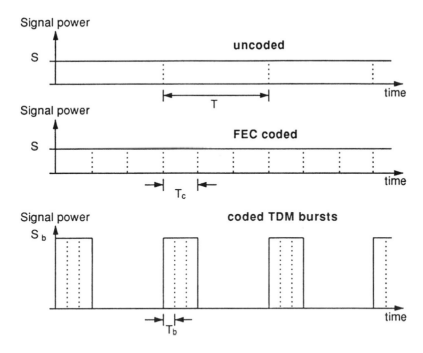

Figure 7.10: Basis of comparison between SFHMA and uncoded FDMA systems.

7.4.4 BER Analysis of the SFHMA System in a Static AWGN Channel

We shall use the MSK-type detector performance as the basis of our analysis in this section. In order to establish the overall bit error probability performance of the SFHMA system in AWGN channels, we must first determine the channel BER by applying Equation 7.27 modified to take into account the redundancy due to the overall 1/3 rate channel coding and the inclusion of an 8-bit training pattern in each transmitted burst. Then the probability of uncorrected errors in the output from the cascaded decoders can be evaluated. Note that to simplify the notation the inner and outer decoders will be referred to henceforth as decoders 1 and 2 respectively.

In each 60-hop frame the speech coder of an active user will produce a total of 16 kbit/s $\times$ 240 ms = 3840 data bits, which are carried as equal length blocks by the 48 data hops in the frame. Thus, allowing for the coding and the training patterns, the total number of bits transmitted over the channel per frame to convey this speech data will be $3 \times 3840 + 48 \times 8 = 11904$. If we apply the constraint of equal total energies per frame, i.e., equal average signal powers, then the energies per channel bit and per data bit are related as:

$$E_b = \frac{11904}{3840} E_c = 3.100 \, E_c \qquad (7.34)$$

From Equation 7.27 the channel BER at the input to decoder 1 is given

by:

$$\begin{aligned} P_{ib1} &= 2Q\{\delta_{\min}\sqrt{E_b/(3.1 \, N_o)}\} \\ &= 2Q\{0.692 \sqrt{E_b/N_o}\} \end{aligned} \qquad (7.35)$$

where $\delta_{\min}$ has been evaluated for $B_t T = 0.3$ and $B_r T = 0.63$ as described in Section 7.4.2.

The next step of this analysis is to derive the BER performance improvement resulting from the channel coding. Consider initially the analysis of a single RS decoder capable of correcting up to t symbol errors in an (n,k) RS coded signal, often denoted as a (n,k,t) RS decoder. We will assume that:

(1) the RS code used is systematic, i.e. the input data symbols appear unchanged in the codeword

(2) the decoder is capable of detecting all decoding failures, i.e. where more than t symbol errors occur in a codeword

(3) in such events the decoder will output the received data symbols containing the original channel errors without attempting error correction.

In practice of course some severely corrupted codewords may lie close to a completely different codeword and thus be apparently successfully, but in fact incorrectly, decoded. A rigorous analysis of the post-decoding symbol and bit error probabilities for RS codes is presented in Section 4.4.7. However, for the codes used in the SFHMA system it has been found that the approximate method described below, based on the above assumptions, gives adequate accuracy in the output bit error probability calculations with much less computational complexity.

Neglecting incorrect decoding events, decoder output symbol errors can be considered to be due solely to input data symbol errors in those codewords containing more than t symbol errors in total. Thus the mean number of data symbol errors per received codeword is obtained by averaging the total number of symbol errors in such cases of decoding failure and multiplying by the proportion of data symbols in each codeword, i.e.:

$$N_{ds} = \frac{k}{n} \sum_{j=t+1}^{n} j \binom{n}{j} P_{is}^j (1 - P_{is})^{n-j} \qquad (7.36)$$

where P_{is} is the probability of input symbol errors. Expressing the probability of output symbol errors as the mean number per data symbol gives:

$$P_{os} = \frac{N_{ds}}{k}$$

$$= \frac{1}{n} \sum_{j=t+1}^{n} j \binom{n}{j} P_{is}^{j} (1 - P_{is})^{n-j} \qquad (7.37)$$

To determine the output bit error probability, P_{ob}, we postulate that the output symbol errors are caused by a sequence of independent bit errors with probability P_{ob}, giving the relationship:

$$P_{os} = 1 - (1 - P_{ob})^{m} \qquad (7.38)$$

where m is the number of bits per symbol. Thus by a simple rearrangement of the above expression we obtain:

$$P_{ob} = 1 - (1 - P_{os})^{1/m} \qquad (7.39)$$

A check was made on the justification of applying the approximate method to the performance analysis of the (30,20,5) and (8,4,2) decoders in the SFHMA system by using the exact expressions given in Section 4.4.7 to compute the relative probability of incorrect decoding P_{IDR}, given by the ratio

$$P_{IDR} = \frac{P_{ICD}}{1 - P_{CD}} \qquad (7.40)$$

where P_{ICD} is the probability of incorrect decoding, P_{CD} is the probability of correct decoding and $1 - P_{CD}$ is therefore the probability of decoding failure. Graphs of P_{IDR} vs P_{ib} for the two SFHMA decoders are presented in Figure 7.11. The low values of P_{IDR} obtained fully justify the use of the approximate method in all the evaluations of the bit error rate performance of the SFHMA system described in the remainder of this chapter. Note that the maximum value of t is assumed for each decoder, which for an RS code is given by

$$t = \lfloor (n - k)/2 \rfloor \qquad (7.41)$$

where $\lfloor \; \rfloor$ denotes the next lower integer. Choosing a value of t less than the maximum will increase the reliability of detecting decoding failures and decrease the probability of incorrect decoding, but this does not seem to be required in these cases. Consider now the cascaded decoders in the SFHMA receiver. At the input to the first (inner) decoder one or more channel bit errors in one 8-bit symbol will constitute an input symbol error with probability given by:

$$P_{is1} = 1 - \mathcal{P}\{\text{no bit error}\}$$

$$= 1 - (1 - P_{ib1})^{8} \qquad (7.42)$$

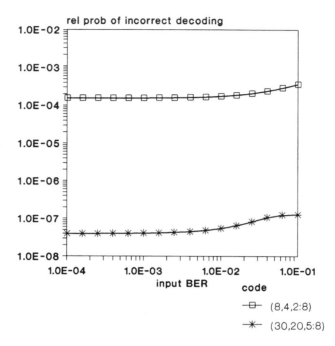

Figure 7.11: Relative probability of incorrect decoding versus input BER for the RS codes used in the SFHMA system.

Before the second decoding stage each output block of 20 data symbols from the first decoder are deinterleaved into 20 different 8-symbol codewords, each of which will contain symbols from 7 other hops. Hence it is reasonable to assume that symbol errors at the input to the second (outer) decoder are independent, even when the first decoder is dealing with bursts of errors due to fading.

The input symbol error rate (SER) of the second decoder is equal to the output SER of the (30,20,5) decoder, which from Equation 7.37 can be expressed as:

$$
\begin{aligned}
P_{is2} &= P_{os1} \\
&= \frac{1}{30} \sum_{j=6}^{30} j \binom{30}{j} P_{is1}^j (1 - P_{is1})^{30-j}
\end{aligned}
\tag{7.43}
$$

Applying Equation 7.37 again we obtain the output SER of the (8,4,2) (second) decoder as:

$$P_{os2} = \frac{1}{8} \sum_{j=3}^{8} j \binom{8}{j} P_{is2}^{j} (1 - P_{is2})^{8-j} \qquad (7.44)$$

Finally the output BER is obtained from Equation 7.39 as:

$$P_{ob2} = 1 - (1 - P_{os2})^{1/8} \qquad (7.45)$$

The SFHMA system BER performance is obtained by evaluating Equations 7.35 and 7.42 to 7.45. The results are presented in Figure 7.12 as a graph of BER versus E_b/N_o, where they are compared with the performance of an uncoded system (from the analysis in Section 7.4.2). Somewhat surprisingly, this comparison shows that the powerful FEC coding gives no improvement in performance of the SFHMA system, in fact just the opposite in the critical BER range 10^{-3} to 10^{-2}, where the speech decoder input BER threshold is most likely to be located.

7.4.5   BER Analysis in a Rayleigh Fading Channel

In this section, we shall first develop the BER performance for continuous transmission of uncoded speech data with an MSK-type detector in a Rayleigh fading channel with AWGN. Next the SFHMA system performance is derived for a similar channel condition and compared with the continuous uncoded case.

For the analysis of the continuous transmission uncoded reference system, the effective receiver observation interval, $L_r T$, may be considered to be the duration of the total ISI caused by both the premodulation filters. Thus the value of L_r is inversely related to the normalised bandwidths of the filters. For $B_t T = 0.3$ and $B_r T = 0.63$ then $L_r = 6$ would be an appropriate value.

The dynamic bit error probability in the Rayleigh fading channel, as a function of the long term mean normalised signal-to-noise ratio, Γ_b, is found by averaging the error probability with respect to the short term normalised signal-to-noise ratio γ_b as:

$$P_b(\Gamma_b) = \int_0^{\infty} P_b(\gamma_b) p(\gamma_b) d\gamma_b \qquad (7.46)$$

where $P_b(\gamma_b)$ is determined by evaluating Equations 7.25 to 7.28 and $p(\gamma_b)$ is the pdf given by Equation 7.33.

For the analysis of the SFHMA system performance the observation interval is taken as one hop slot duration (1.23 ms), during which time one complete codeword of the inner RS code is received. The corresponding value of L_r is approximately 250. The block of 20 symbols at the output of the inner decoder is deinterleaved into 20 consecutive codewords at the in-

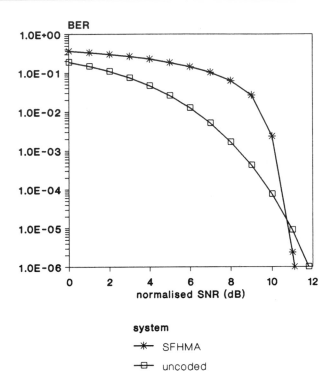

Figure 7.12: BER versus normalised SNR for both SFHMA and uncoded
FDMA systems in a static AWGN channel [MSK-type detector: $B_tT = 0.3$ and $B_rT = 0.63$].

put to the outer decoder. Thus the codewords at this point can be assumed
to contain independent symbol errors with a probability which is obtained
by using Equations 7.35, 7.42 and 7.43 to determine the input SER of the
outer (second) decoder P_{is2} as a function of γ_b and then averaging with
respect to γ_b as:

$$P_{is2}(\Gamma_b) = \int_0^\infty P_{is2}(\gamma_b)p(\gamma_b)d\gamma_b. \tag{7.47}$$

As before, the integral is evaluated numerically. The dynamic output bit
error probability can then be determined as a function of the long term
mean signal-to-noise ratio by means of Equations 7.44 and 7.45.

Figure 7.13 presents the results of the analysis of both the SFHMA
and the uncoded systems, as graphs of BER versus mean signal-to-noise
ratio Γ_b. For this case of the fading channel the benefits of the FEC coding incorporated in the SFHMA system are clearly demonstrated by the
superior performance of the system in the critical BER range of 10^{-3} to
10^{-2}. At the mid-range BER value of 3×10^{-3} the coding gain is approxi-

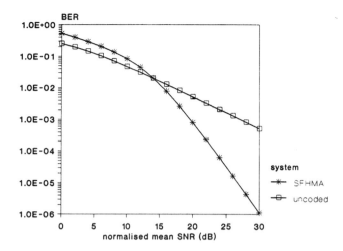

Figure 7.13: BER versus normalised mean SNR for both SFHMA and uncoded
FDMA systems in a Rayleigh fading AWGN channel [MSK-type
detector: $B_tT = 0.3$ and $B_rT = 0.63$].

mately 4.5 dB. It should be noted that the FEC coding scheme, comprising
concatenated RS codes with block interleaving, is specifically designed to
combat bursts of errors in a fading channel rather than the random errors
arising in a static AWGN channel. Thus some difference in the efficiency
of the FEC coding in the two channel conditions should be expected.

7.5 BER Performance in the Presence of Co-channel Interference

Cochannel interference arises from frequency reuse which is a feature of any
efficient mobile radio systems [16–18]. It is probably the most important
parameter in the design of cellular mobile radio systems, because it ultimately determines the system capacity. A unique advantage of SFHMA
systems is that they provide interferer diversity, i.e., the set of interfering
sources changes from hop to hop, however this feature makes the analysis of performance much more difficult. In this section we will examine
cochannel interference in the classical sense without taking account of interferer diversity. Two approaches to the analysis are presented, the first
is a simplified analysis based on Murota's treatment [15], which assumes
noise-free reception, whereas in the second approach by Ko [19] a more
elaborate analysis is entailed which also accounts for receiver noise. Both
treatments are based on the MSK-type receiver because of the greatly reduced computational complexity. Its BER performance is believed to be

only slightly inferior to the MLSE detector.

We shall first consider the cochannel interference in non-fading channels for both the uncoded continuous transmission system and for the SFHMA system. Then the analysis is extended to cover frequency non-selective, Rayleigh fading channels.

7.5.1 BER Analysis in a Noiseless Static Channel

In this analysis, following [15], the presence of cochannel interference from a single source in a static noiseless channel is considered for the uncoded continuous system. In Verhulst et al.'s original analysis of SFH digital cellular radio systems [2], and in subsequent analysis of the SFH900 system [3,5,8], this noiseless channel condition is also assumed. No account has been taken of the additional ISI caused by pre-demodulation filtering of the received GMSK signals. As in the previous analysis of the performance of the MSK-type detector in a static AWGN channel (see Section 7.4.2), we assume that the probability of decision errors in the receiver's quadrature channels is dominated by the worst case ISI condition which occurs for the data sequence $\cdots 0, 0, 0, 1, 1, 1 \cdots$. Therefore, we restrict our attention to the special case for which the signal phase waveform $\theta(t)$ is given by Equation 7.28.

The interference signal waveform can be expressed in a similar form to that of the desired signal except that the signal power, S, is replaced by the interference power, I. The interferer's phase is modelled by a random process, $\psi(t)$, uniformly distributed between $-\pi$ and π, resulting from the assumption that the signal and interference are not phase coherent and are modulated by independent data sequences. The total received signal $r_s(t)$ in this case is:

$$r_s(t) = \sqrt{2S}\cos[\omega_o t + \theta(t)] + \sqrt{2I}\cos[\omega_o t + \psi(t)] \qquad (7.48)$$

for which the equivalent low pass complex envelope is given by:

$$z(t) = \sqrt{2S}\exp[j\theta(t)] + \sqrt{2I}\exp[j\psi(t)] \qquad (7.49)$$

It is convenient to regard $z(t)$ as the complex sum of the desired signal and the interference signal phasors. Figure 7.14 shows a phasor diagram of the resultant complex envelope at the decision instant $t = 0$, the correct decision being obtained in the I channel when the resultant complex envelope remains in the right half plane. Furthermore, to allow for the inherent ISI in the GMSK signal the reference phase is defined as the modulation phase change at $t = 0$ of a classical MSK signal, i.e., $B_t T = \infty$. Therefore, substituting $t = 0$ into Equation 7.28 and substituting for μ_t from

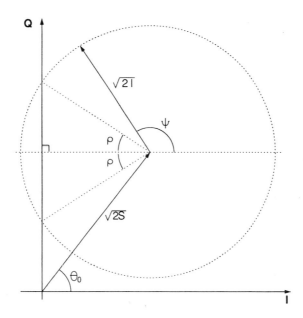

Figure 7.14: Phasor diagram for GMSK signal with cochannel interference signal at decision instant ($t = 0$).

Equation 7.29, the signal reference phase is obtained as:

$$\theta_o = \frac{\sqrt{\ell n 2}}{2 B_t T \sqrt{2\pi}} \tag{7.50}$$

Due to the symmetry of the two quadrature channels and to the uniform distribution of $\psi(t)$ it can be argued, with reference to Figure 7.14, that the probability of quadrature symbol decision errors is given by the ratio of the angle subtended by the intersection of the Q axis with the circular locus of the resultant envelope, 2ρ, to the overall range of the phase angle, i.e. 2π. Thus the quadrature symbol error probability can readily be determined in terms of the signal reference phase, θ_o, and the carrier signal-to-interference ratio (CIR), $\lambda = S/I$, as:

$$
\begin{aligned}
P_{qe}(\lambda) &= \tfrac{\rho}{\pi} = \tfrac{1}{\pi} \cdot \cos^{-1}\sqrt{\lambda \cos\theta_o} & \text{for } 0 \le \sqrt{\lambda} \le 1/\cos\theta_o \\
&= 0 & \text{for } \sqrt{\lambda} > 1/\cos\theta_o
\end{aligned}
\tag{7.51}
$$

From Equation 7.26 the output bit error probability is given by:

$$P_b(\lambda) = 2P_{qe}(\lambda)[1 - P_{qe}(\lambda)] \tag{7.52}$$

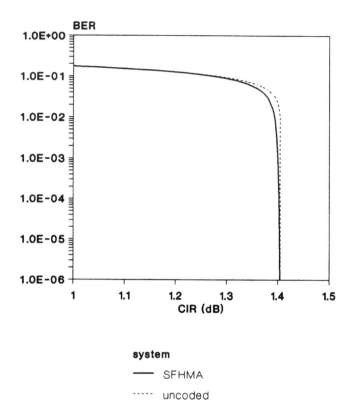

Figure 7.15: BER versus CIR for both the SFHMA and uncoded FDMA
systems with cochannel interference in a static noiseless channel
[MSK-type detector: $B_t T = 0.3$].

In Figure 7.15 we display the performance curve of BER against CIR, which
is obtained by evaluating Equations 7.50 to 7.52 for $B_t T = 0.3$.

For the SFHMA system we used Equations 7.50 to 7.52 to calculate
the channel bit error probability at the input to the inner decoder, P_{ib1},
then the effect of channel coding is evaluated by means of Equations 7.42
to 7.44, as described above in Section 7.4.4. The results obtained are also
displayed in the graph shown in Figure 7.15, for comparison with results
for the uncoded system. It can be seen that the two curves are very close
with a sharp knee at a CIR value of 1.4 dB, and then they fall extremely
steeply. This behaviour is entirely consistent with the assumed model,
which involves constant amplitudes of both the signal and the interference.

7.5.2 BER Analysis in a Static AWGN Channel

For the situation considered here, where the GMSK signal is accompanied by both cochannel interference and AWGN, we have employed an approximate analysis developed by Ko [19], which was shown to provide a very large saving in computation time, with only moderate degradation of accuracy, compared to more exact methods of analysis. Intersymbol interference due to the receiver filter is taken into account, but the treatment is simplified by considering only the data sequence $\underline{\alpha}_{min}$ giving the minimum Euclidean distance and by approximating multiple interfering signals as arising from a single source.

With some minor changes to be compatible with the previous analysis of the MSK-type detector in section 7.4.2, Ko's approximate expression (see Equation 2.67 in [19]) for the probability of quadrature symbol error becomes:

$$
\begin{aligned}
P_{qe} &= \frac{1}{2\pi} \int_0^{2\pi} Q \left\{ \frac{\sqrt{2S}}{\sigma_y} \int_{-\infty}^{\infty} h_{rb}(-\tau) \cos\theta(\tau, \underline{\alpha}_{\min}) d\tau \right. \\
&\quad \left. + \frac{\sqrt{2I}}{\sigma_y} \int_{-\infty}^{\infty} h_{rb}(-\tau) \cos\psi d\tau \right\} d\psi
\end{aligned}
\tag{7.53}
$$

where S is the input signal power, I is the total cochannel interference power, $h_{rb}(t)$ is the impulse response of the low-pass equivalent of the Gaussian band-pass filter in the MSK detector given by Equations 7.13 and 7.14, $\theta(t, \underline{\alpha}_{\min})$ is the signal phase waveform with the worst case intersymbol interference (ISI) due to the transmitter Gaussian filter (for the data sequence $\cdots 0, 0, 0, 1, 1, 1 \cdots$) given by Equations 7.20 and 7.21. and 7.29, ψ is the interference phase assumed to be random and uniformly distributed over 0 to 2π and σ_y^2 is the variance of the noise in the baseband quadrature channel given by Equations 7.20 and 7.21.

The signal power can be expressed in the form:

$$
S = E_b/T
\tag{7.54}
$$

where E_b is the mean per energy/bit. In terms of S and the CIR, λ, the interference power is given by:

$$
I = S/\lambda = E_b/\lambda T
\tag{7.55}
$$

Assuming that

(1) the interference phase ψ is constant over the effective duration of $h_{rb}(t)$,

(2) the zero frequency gain of the equivalent low-pass filter is unity, which

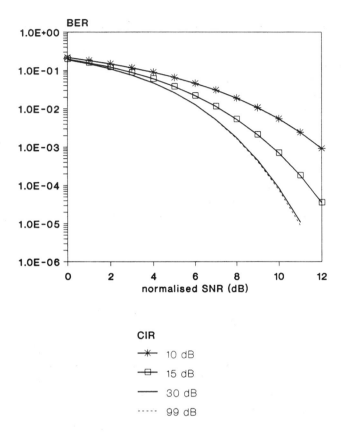

Figure 7.16: BER versus normalised SNR for an uncoded FDMA system with cochannel interference in a static AWGN channel [MSK-type detector: $B_t T = 0.3\ B_r T = 0.63$].

implies that:

$$\int_{-\infty}^{\infty} h_{rb}(-\tau)d\tau = \int_{-\infty}^{\infty} h_{rb}(\tau)d\tau = 1 \qquad (7.56)$$

(3) the noise bandwidth of the receiver band-pass filter is equal to its -3 dB bandwidth B_r,

and also noting that $h_{rb}(-\tau) = h_{rb}(\tau)$, Equation 7.53 can be simplified and expressed as a function of the normalised signal-to-noise ratio (SNR)

E_b/N_o and λ, in the form:

$$P_{qe} = \frac{1}{2\pi} \int_0^{2\pi} Q \left\{ \left(\Phi_s + \frac{\cos \psi}{\sqrt{\lambda}} \right) \sqrt{\frac{2}{B_r T}} \sqrt{\frac{E_b}{N_o}} \right\} d\psi \qquad (7.57)$$

where Φ_s is the signal phase integral, which is given by:

$$\Phi_s = \int_{-\infty}^{\infty} h_{rb}(\tau) \cos \theta(\tau, \underline{\alpha}_{\min}) d\tau \qquad (7.58)$$

which from Equations 7.13 and 7.28 can be expressed as:

$$\Phi_s = \frac{\mu_r}{\sqrt{\pi}} \int_{-\infty}^{\infty} \exp(-\mu_r^2 t^2) \cos \left[\frac{\pi}{2T} \int_0^t erf(\mu_t \tau) d\tau + \frac{\sqrt{\pi}}{2\mu_t T} \right] dt \qquad (7.59)$$

where the receiver and transmitter filter parameters, μ_r and μ_t, are related to the normalised filter bandwidths, $B_r T$ and $B_t T$ by Equations 7.14 and 7.29. In all the calculations described here a fixed value of ϕ_s was employed, which was obtained by putting $B_t T = 0.3$ and $B_r T = 0.63$.

Calculations of the bit error rate using Equations 7.57 and 7.52 have been made for the uncoded system with $B_t = 0.3$ and $B_r T = 0.63$. Graphs of BER vs SNR and of BER vs CIR are presented in Figures 7.16 and 7.17, respectively. As would be expected, the curve for the highest value of CIR (99 dB) in Figure 7.16 is very similar to the curve for the uncoded FDMA system in Figure 7.12 for the static AWGN channel. As the CIR is decreased below 30 dB significant deterioration in BER performance becomes apparent. Plotting the results against CIR, as in Figure 7.17, is of more interest when considering system performance under interference limited conditions, which will be our prime concern in the remainder of this chapter.

In the SFHMA system case the value of the SNR used in Equation 7.57 has been adjusted to allow for the overall FEC coding rate as in Section 7.4.4. Then having used Equation 7.52 to obtain the input BER at the inner decoder, Equations 7.42 to 7.45 were employed to obtain the output BER. A graph of BER vs CIR is shown in Figure 7.18. As in Figure 7.17, it will be seen that for very high values of SNR the BER falls extremely steeply for CIR greater than 3 dB.

The reason that this steep cut-off occurs at a higher value of CIR than the value of 1.4 dB obtained for the noiseless channel analysis, which can be seen in Figure 7.15, is that in this case there is additional ISI due to the receiver filter, which was not accounted for in the previous analysis. It can be shown that in the limit as the SNR $\rightarrow \infty$ the cut-off will occur at $\lambda = \phi_s^{-2}$ giving CIR values very close to 3 dB for $B_r T = 0.63$ and 1.4 dB for $B_r T >> 1$.

Further comparison of Figures 7.17 and 7.18 shows once more that the FEC coding employed in the SFHMA system is ineffective against the ran-

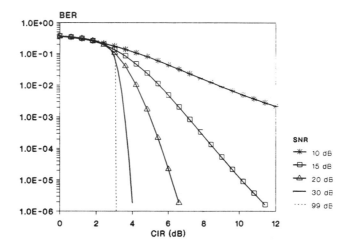

Figure 7.17: BER versus CIR for an uncoded FDMA system with cochannel interference in a static AWGN channel [MSK-type detector: $B_tT = 0.3$ and $B_rT = 0.63$].

dom channel errors produced in a static channel situation, the coding gain varying from low to negative values as the SNR changes from high to low.

7.5.3 BER Analysis in a Rayleigh Fading AWGN Channel

Again we base the analysis of system performance in a noiseless Rayleigh fading channel on [15]. Therefore we treat the multiple interferers as an equivalent single source producing a received power equal to the total power of the component signals and we can apply the expressions obtained in Section 7.5.1 to determine the channel bit error probability. It is assumed that the desired and interfering signals both fade independently, that the signal-to-interference ratio, λ, is constant over each hop burst duration and that there is no other form of diversity. Consequently, the pdf of the hop CIR can be shown to be given by [20]:

$$p(\lambda) = \frac{\Lambda}{(\lambda + \Lambda)^2} \tag{7.60}$$

where Λ is the mean CIR.

For the uncoded system the mean BER is obtained by averaging the channel bit error probability over λ, using the above pdf as:

$$P_b(\Lambda) = \int_0^\infty P_b(\lambda) p(\lambda) d\lambda \tag{7.61}$$

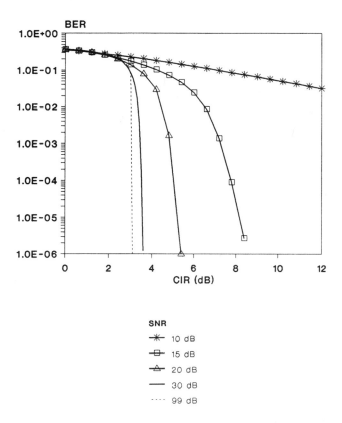

Figure 7.18: BER versus CIR for the SFHMA system with cochannel interference in a static AWGN channel [MSK-type detector: $B_tT = 0.3$ and $B_rT = 0.63$].

In the case of the SFHMA system the above expression is used together with Equations 7.42 and 7.43 to determine the probability of input symbol errors at the outer decoder, averaged over λ, as a function of the mean signal to interference ratio as:

$$P_{is2}(\Lambda) = \int_0^\infty P_{is2}(\lambda)p(\lambda)d\lambda \qquad (7.62)$$

Thereafter, the output bit error probability is evaluated by using Equations 7.44 and 7.45. The results for both systems are presented as graphs of BER versus mean CIR in Figure 7.19. They show that the FEC coding of the SFHMA system is even more effective against cochannel interference in a fading channel condition than it was against noise, as can be seen by comparison with Figure 7.13. In this case the coding gain at a BER of

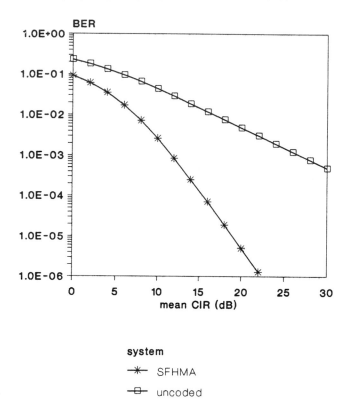

Figure 7.19: BER versus mean CIR for the SFHMA and uncoded FDMA systems with cochannel interference in a noiseless Rayleigh fading channel [MSK-type detector: $B_t T = 0.3$].

3×10^{-3} is approximately 11 dB.

7.5.4   BER Analysis of a Noiseless Rayleigh Fading Channel

This is a severe channel condition in which the wanted and the interfering signals are all subjected to independent fading and perturbed by AWGN. It has been shown in [19] that in a Rayleigh fading channel, the interference can be regarded as arising from an equivalent single source. Moreover, as in the previous analysis of fading, it is assumed that the instantaneous SNR and CIR are constant over the duration of a hop, but from hop to hop the wanted signal and the interference are subjected to mutually independent Rayleigh fading. Therefore before averaging it is necessary to modify Equation 7.57 to express the quadrature error probability as a function of the signal-to-noise voltage ratio $\rho_s = \sqrt{E_b/N_o}$ and of the interference-to-noise voltage ratio $\rho_i = \sqrt{IT/N_o} = \sqrt{E_b/N_o\lambda}$ in the form;

$$P_{qe}(\rho_s, \rho_i) = \frac{1}{2\pi} \int_0^{2\pi} Q\left\{(\rho_s\Phi_s + \rho_i\cos\psi)\sqrt{\frac{2}{B_rT}}\right\} d\psi \qquad (7.63)$$

from which the bit error probability as a function of ρ_s and ρ_i is obtained using Equation 7.52.

Then the mean BER as a function of the mean normalised SNR, Γ_b, and the mean CIR, Λ, is obtained by averaging over both ρ_s and ρ_i via the double integral:

$$P_b(\Gamma_b, \Lambda) = \int_0^\infty \int_0^\infty P_{be}(\rho_s, \rho_i)p(\rho_s)p(\rho_i)d\rho_s d\rho_i \qquad (7.64)$$

The two Rayleigh pdfs can be expressed as

$$p(\rho_s) = \frac{2\rho_s}{\Gamma_b}\exp\left(-\frac{\rho_s^2}{\Gamma_b}\right) \qquad (7.65)$$

and

$$p(\rho_i) = \frac{2\rho_i}{\Gamma_i}\exp\left(-\frac{\rho_i^2}{\Gamma_i}\right) \qquad (7.66)$$

where Γ_i is the mean normalised interference-to-noise power ratio, which is related to the mean CIR and the mean SNR by:

$$\Gamma_i = \frac{\Gamma_b}{\Lambda} \qquad (7.67)$$

Simpson's rule was used to evaluate the double integral in Equation 7.64. It was found that great care had to be taken with setting the upper limits in order to obtain sensible results without excessive computation time.

Computation of the SFHMA system performance proceeded in a similar way, using Equations 7.63 and 7.52 to give the channel error rate and then Equations 7.42 and 7.43 to determine the input symbol error rate of the outer decoder as a function of ρ_s and ρ_i. It was this probability which was averaged as indicated in Equation 7.64 and then used to compute the output BER by means of Equations 7.44 and 7.45. This procedure is essentially the same as used in previous calculations described in Section 7.4.5 and is justified by the same assumption, i.e. all the symbols in each codeword at the input to the outer decoder have been transmitted on different hops and so contain independent errors. Graphs of BER vs mean CIR for various values of mean SNR for the uncoded and SFHMA systems are shown in Figures 7.20 and 7.21. The curves for the highest SNR (99 dB) in both cases are very similar to those in Figure 7.19 for a fading channel without noise.

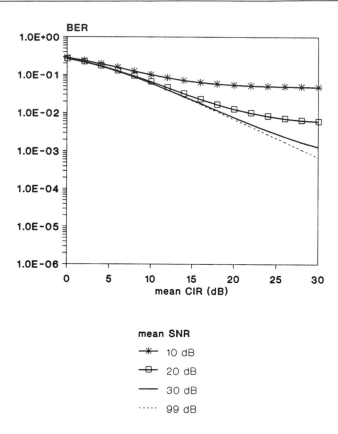

Figure 7.20: BER versus mean CIR for the uncoded FDMA system with cochannel interference in a Rayleigh fading AWGN channel [MSK-type detector: $B_t T = 0.3$ and $B_r T = 0.63$].

7.6   Estimation of Spectral Efficiency

Efficient utilisation of the limited radio spectrum available for mobile radio is one of the most important considerations in the design of digital cellular systems. In this section we will describe two methods for estimating the spectral efficiency of a SFHMA system and present results obtained by the application of both methods for the base-to-mobile (down) link. For comparison, spectral efficiency estimates are also determined by a more conventional method for a TDM/FDMA system, which is assumed to use the same digital modulation and demodulation schemes, but with no FEC coding.

There are a number of definitions of spectral efficiency in general use, but the one which seems most appropriate for our present purpose is the

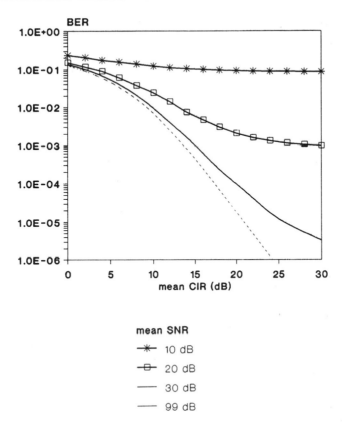

Figure 7.21: BER versus mean CIR for the SFHMA system with cochannel interference in a Rayleigh fading AWGN channel [MSK-type detector: $B_t T = 0.3$ and $B_r T = 0.63$].

maximum traffic, in Erlangs, carried per cell per unit of spectral width, which is conveniently expressed in units of erl/cell/MHz.

Interferer diversity, which is claimed to be one of the main advantages of SFHMA, requires a different approach in determining spectral efficiency since cochannel interference from a reuse cell is spread statistically over all hop frequencies as random frequency collisions with a probability dependent on the system traffic loading. However, in conventional FDMA (or TDMA) system analysis it is necessary to assume a worst case condition in which there is continuous interference from an active user on the same channel in each reuse cell, irrespective of the traffic loading.

7.6.1 Spectral Efficiency of the SFHMA System: Method A

This method is the one described by Dornstetter and Verhulst in their paper [8] on the analysis of the SFH900 system, which differs only in small details from that presented in an earlier paper by Verhulst [5]. It is assumed that the system performance is interference limited. In the interest of simplicity only the simpler down-link (base-to-mobile) situation will be considered here, but both directions are dealt with in [8].

The combined characteristics of the demodulation and inner (intra hop) FEC decoding are represented by a simple threshold comparison, i.e. a transmitted hop is assumed valid if the hop CIR $\lambda > \gamma$ and invalid if $\lambda < \gamma$, where γ is the value of the CIR such that the probability of correct reception of a hop is 0.5. A value of 7 dB is quoted for γ in [8], which is based on the measured BER vs SNR characteristics of an SFH system with no FEC coding using a quasi coherent GMSK receiver with a Viterbi Algorithm equaliser. The measurements were made in a simulated dispersive channel having static multipath distortion with AWGN, but in the absence of cochannel interference. Although a satisfactory explanation is not provided in [8] for the choice of this value of SNR, which yielded a BER close to 5×10^{-2}, as a suitable CIR threshold value, it can be argued that this choice is justified on the grounds that the amplitude distribution of the sum of several interfering signals with independent phases and subject to independent dynamic multipath distortion will be approximately Gaussian by application of the Central Limit Theorem.

Stipulation of a value for the probability of correct reception of a hop, q, is used to account for the effectiveness of the outer FEC coding and the speech coding to give an adequate decoded speech quality. A minimum value of 0.7 for q is suggested in [8] to ensure acceptable quality and in the case of the RS (8,4) outer code it is claimed that the corresponding output BER would be approximately 6×10^{-3}. Our analysis of the outer decoder indicates that 0.8 might be a better choice for the minimum q to ensure a similar value of output BER. Spectral efficiency is found by determining q as a function of the system traffic load, allowing for fast fading, interference diversity and shadowing, and then finding the load which ensures $q > 0.7$ with a high probability over the whole cell.

Using Equation 7.3 from Section 7.2.3, we can write:

$$q = \mathcal{P}\{\lambda \geq \gamma\}$$
$$= \mathcal{P}\left\{\frac{A_o B_o F_o}{\sum_{i=1}^{M} A_i B_i F_i E_i} \geq \gamma\right\} \tag{7.68}$$

where the probability is with respect to the 'fast' variables, i.e. the fading factors F_i and the interference indicators E_i, while the 'slow' variables, i.e. mean power levels A_i and shadowing factors B_i, are treated as constants.

Assuming that the F_i values are independent and exponentially distributed (Rayleigh fading) and that the binary E_i values are independent with a Bernoulli distribution characterised by the frequency collision probability $p_i = \mathcal{P}\{E_i = 1\}$ and the probability of no collision $1 - P_i = \mathcal{P}\{E_i = 0\}$, it is shown in Appendix A that q is given by:

$$q = \prod_{i=1}^{M} \left(1 - \frac{p_i \gamma}{\Lambda_i + \gamma} \right) \tag{7.69}$$

where $\Lambda_i = A_o B_o / A_i B_i$ is the short term mean CIR with respect to the ith interferer in the worst case for $p_i = 1$. The simple on/off model used above to represent hop frequency collisions implies a further assumption that the cellular system is fully synchronised. The collision probability p_i is given by the product of the frequency collision rate per active user in the ith cell, y_i, and the mean number of active users per TDM slot in that cell, where 'active' means that the mobile is engaged in a call and is currently receiving transmissions from its own BS. Assuming that all cells are carrying the same traffic load of a_{cell} Erlangs and that silence detection with no transmission during silent periods is implemented to reduce cochannel interference, the collision probability can be expressed as:

$$p_i = \frac{y_i a_{cell} r_a}{n_t} \tag{7.70}$$

where n_t is the number of slots per TDM frame ($= 3$ for the SFHMA system under consideration) and r_a is the speech activity ratio (for the down link r_a is taken to be 0.5 in [5]). As described in Section 7.2.2, the collision rate y_i depends on the reuse cell structure, having a value of $1/N$ for the basic 3-cell full reuse cluster, where N is the number of hop frequencies assigned to each of the three basic colours. In general for a full reuse cluster of size C the collision rate y_i becomes $C/3N$. In the case of a $3M/L$ fractional reuse structure the appropriate value of y_i is $k_i M/LN$, where k_i is the fractional overlap between the sets of frequencies, i.e. shades, assigned to the ith and to the reference cells. In terms of the total bandwidth W allocated to the downlink and the frequency spacing f_s, N is simply:

$$N = \frac{W}{3f_s} \tag{7.71}$$

Thus substituting in Equation 7.70, firstly for the case of the general full reuse structure, we can write:

$$p_i = \frac{C f_s a_{cell} r_a}{n_t W} \tag{7.72}$$

and secondly for fractional reuse structures:

$$p_i = \frac{3Mk_i f_s a_{Cell} r_a}{L n_t W} \tag{7.73}$$

Denoting the spectral efficiency by $\eta = a_{Cell}/W$ and rewriting Equations 7.72 and 7.73 gives η in terms of p_i:

$$\eta = \frac{n_t p_i}{C f_s r_a} \tag{7.74}$$

for full reuse and

$$\eta = \frac{L n_t p_i}{3M k_i f_s r_a} \tag{7.75}$$

for fractional reuse.

Dornsetter and Verhulst applied Equation 7.69 as an essential part of a simulation of a 19-cell network, the results of which are presented in [8]. The network comprised the central reference cell with the nearest ring of 6 reuse cells of the same colour plus two closer rings of different colours producing adjacent channel interference. Mobile locations within the reference cell were selected at random and for each location, having calculated the set of products of the deterministic A_i factors and the random B_i factors (with log-normal pdf), q was computed for various values of traffic load expressed as the mean number of users engaged in calls per MHz per cell, X. For each value of X the values of q for the various locations were analysed to determine the '90% worst case value', q_{90}, defined as that value of q which is exceeded with a probability of 90%. The spectral efficiency can then be found from a graph of q_{90} vs X at the point where $q_{90} = 0.7$. X is related to the collision probability and spectral efficiency as follows. The traffic per cell is given by:

$$a_{Cell} = WX \tag{7.76}$$

Hence the spectral efficiency becomes simply:

$$\eta = a_{Cell}/W = X \tag{7.77}$$

and from Equations 7.70 and 7.76 we have:

$$p_i = \frac{y_i r_a W X}{n_t} \tag{7.78}$$

Also described by Dornsetter and Verhulst in [8] is a simplified model based on their experiences with the simulation. They noted that there was a close association between sample locations with values of q close to q_{90} and those with values of mean CIR close to the '90% worst case value', Λ_{90}, i.e. $\mathcal{P}\{\Lambda > \Lambda_{90}\} = 90\%$. In addition, because of the directional BS antenna involved, only 3 cells out of the ring of 6 can produce significant interference levels in the reference cell and it was noticed that in most instances only 2

Size	3	9	12	21	27
Λ_{90} (dB)	5.2	13.0	14.5	18.2	20.6

Table 7.4: Values of Λ_{90} versus reuse cluster size [5]

of these were simultaneously giving significant levels. Thus for a network having K rings, the proposed model gives a simple analytical expression for q_{90} in terms of the sets of values of p_i and Λ_{90i}:

$$q_{90} = \prod_{i=1}^{K} \left(1 - \frac{p_i \gamma}{\gamma + 2\Lambda_{90i}} \right)^2 \qquad (7.79)$$

Curves of q_{90} vs X plotted in [8] show very close agreement with the simulation results. The crucial values of Λ_{90} used in these calculations are presented in Table 7.4. They were taken from the paper by Stjernvall [21] (NB referenced in [5] but not in [8]), which presents computed distributions of CIR in the down link direction due to the closest ring of 6 reuse cells, for various sizes of reuse clusters having corner BS sites and assuming correlated lognormal shadowing, with a sigma of 6 dB and a correlation coefficient of 0.7.

Our own calculations of q_{90} for the basic 3-cell structure, using the above values of γ and Λ_{90} and taking a frequency spacing of $f_s = 1/7$ MHz (also as in [5]), are presented in Figures 7.22 to 7.24 as a graphs of q_{90} vs X for $r_a = 0.5$, q_{90} vs p_i (for any r_a) and q_{90} vs the spectral efficiency η for $r_a = 1$, respectively. In each case curves are shown for the following conditions:

(1) Nearest reuse ring (size 3) only

(2) Nearest reuse ring plus the two nearer rings of cells of the other two colours which produce adjacent-channel interference, for which the values of Λ_{90} quoted in [5] are $R_{AC} + 2$ dB and $R_{AC} + 5$ dB, where R_{AC} is the adjacent-channel interference rejection, for which the value appropriate to $f_s \simeq 150$ kHz was taken as 17 dB.

(3) The second reuse ring (size 9) only.

Condition (2) was the one used for the computations reported in [5]; however, comparing the curves for conditions (1) and (2) it will be seen that adjacent-channel interference has only a small effect on spectral efficiency and can safely be neglected, as was done in all our other calculations.

In Figure 7.24 a fourth curve is plotted for the more accurate approximation obtained by using the first 5 reuse rings (up to size 27). The difference between this curve and that for condition (1) is of greater significance and thus we have used multiple reuse rings out to the largest size

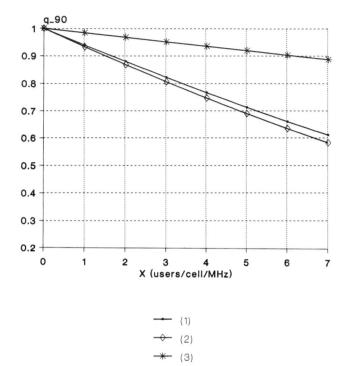

Figure 7.22: q_{90} versus X for $\gamma = 7$ dB, full reuse cluster size 3 and $r_a = 0.5$:

    (1) first reuse ring (size 3) only;

    (2) first reuse ring plus 2 nearest rings of other colours giving adjacent-channel interference;

    (3) second reuse ring (size 9) only.

possible in all subsequent computations to determine spectral efficiency.

For each reuse structure there is an upper limit on the value of η which can be attained, the theoretical value of which is determined as follows. The probability of frequency collisions, p_i, can be considered to be the product of three factors: the fractional overlap between the sets of hop frequencies in the reuse and reference cells, the speech activity ratio, and the mean channel utilisation, i.e. the probability of a channel being occupied by a call, which may also be regarded as the traffic carried per channel. Thus p_i can be expressed as:

$$p_i = k_i U r_a \qquad (7.80)$$

where U is the channel utilisation. The collision probability will be maxi-

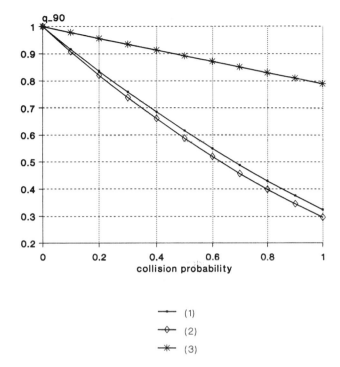

Figure 7.23: q_{90} versus collision probability p_i for $\gamma = 7$ dB and full reuse cluster size 3:

(1) first reuse ring (size 3) only;

(2) first reuse ring plus 2 nearest rings of other colours giving adjacent-channel interference;

(3) second reuse ring (size 9) only.

mum when all channels are occupied, i.e. $U = 1.0$, and so we can write:

$$p_{max} = r_a \qquad (7.81)$$

for full reuse structures, and

$$p_{max} = k_i r_a \qquad (7.82)$$

for fractional reuse structures. Substituting in Equations 7.74 and 7.75 we have for full reuse structures:

$$\eta_{max} = \frac{n_t}{C f_s} \qquad (7.83)$$

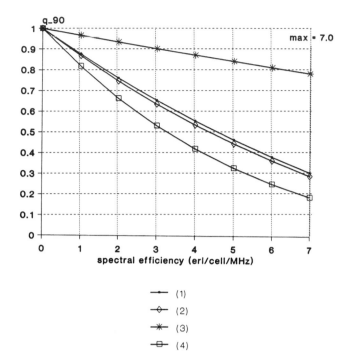

Figure 7.24: q_{90} versus η for $\gamma = 7$ dB, full reuse cluster size 3 and $r_a = 1.0$:

      (1) first reuse ring (size 3) only;

      (2) first reuse ring plus 2 nearest rings of other colours giving adjacent-channel interference;

      (3) second reuse ring (size 9) only;

      (4) first 5 reuse rings (sizes 3,9,12,21 and 27).

and for fractional reuse structures:

$$\eta_{textrmmax} = \frac{Ln_t}{3Mf_s} \tag{7.84}$$

A graph of q_{90} vs η for a full reuse structure of size 9 with $r_a = 1$ is shown in Figure 7.25. With the limited data available on Λ_{90} it was only possible to account for two reuse rings of sizes 9 and 27. However because of the increased distance between rings in this case we believe that two rings are sufficient to give an adequate accuracy.

Calculations of q_{90} as a function of η for various fractional reuse structures, all with $r_a = 1$, have been made using Equations 7.75 and 7.79 together with the data in Table 7.5. The graphs of Λ_{90} vs η shown in

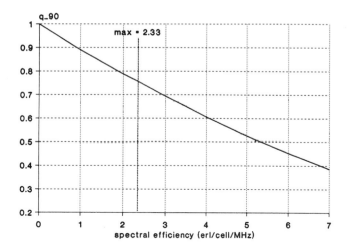

Figure 7.25: q_{90} versus η for $\gamma = 7$ dB, full reuse cluster size 9 and $r_a = 1.0$.

Structure	M	L	k	y fractional	full
9/2	3	2	1/2	9/4N	3/2N
12/2	4	2	1/2	1/N	2/N
12/3	4	3	2/3	8/9N	4/3N
21/3	7	3	1/3	7/9N	7/3N
21/4	7	4	1/2	7/8N	7/4N

Table 7.5: Fractional reuse structure characteristics.

Figures 7.26 and 7.27 are examples of the results of the above calculations.

Note that although in principle the maximum value of the channel utilisation, U, is 1.0, it is not possible to obtain this ideal value in practical channel assignment sub-systems. Achievable maximum values can be determined using the Erlang B formula and will depend on the number of channels per cell, n_{ch}, and the permissible blocking probability, P_B. Taking the system bandwidth (for downlink) to be $W = 24$ MHz, hop frequency spacing to be $f_s = 1/7$ MHz and putting $P_B = 2\%$, we have determined values of $U_{textrmmax}$ for each reuse structure and hence obtained more practical values of $\eta_{textrmmax}$. The results of the computations are listed in Table 7.6, where it will be seen that $\eta_{textrmmax}$ is reduced by a factor in the range 0.8 to 0.9.

Threshold values of spectral efficiency, η_{th}, for $r_a = 1$ have been determined, for which q_{90} becomes equal to threshold values of 0.7 and 0.8. These values are listed below in Tables 7.7 and 7.8, together with $\eta_{textrmmax}$ and

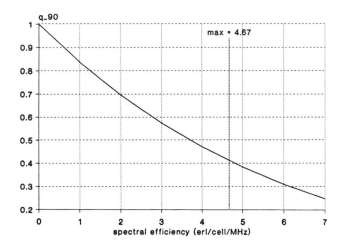

Figure 7.26: q_{90} versus η for $\gamma = 7$ dB, fractional reuse cluster size 9/2 and $r_a = 1.0$.

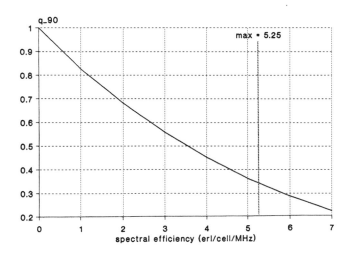

Figure 7.27: q_{90} versus η for $\gamma = 7$ dB, fractional reuse cluster size 12/3 and $r_a = 1.0$.

Structure	$\eta_{textrmmax}$ (U=1)	n_{ch}	$U_{textrmmax}$	$\eta_{textrmmax}$
3	7.00	168	0.901	6.31
9	2.33	56	0.803	1.87
9/2	4.67	112	0.872	4.07
12/2	3.50	84	0.846	2.96
12/3	5.25	126	0.881	4.63
21/3	3.00	72	0.831	2.49
21/4	4.00	96	0.859	3.44

Table 7.6: Maximum values of channel utilisation and spectral efficiency (erl/cell/MHz) : $P_B = 2\%$.

Cluster size	$\eta_{textrmmax}$	η_{th}	$\eta_{textrmsyst}$	$U\%$
3	6.31	1.76	1.76	25
9	1.87	2.93	1.87	80
9/2	4.07	1.98	1.98	42
12/2	2.96	1.58	1.58	45
12/3	4.63	1.87	1.87	36
21/3	2.49	2.05	2.05	68
21/4	3.44	1.91	1.91	48

Table 7.7: Spectral efficiency (erl/cell/MHz) for various reuse structures: $r_a = 1$; $P_B = 2\%$; $q_{90} = 0.7$.

the values of attainable system spectral efficiency, $\eta_{textrmsyst}$, given simply by taking the smaller of η_{th} and $\eta_{textrmmax}$, i.e.

$$\eta_{textrmsyst} = \min\{\eta_{th}, \eta_{textrmmax}\} \tag{7.85}$$

The last column of both Tables 7.7 and 7.8 is the channel utilisation computed from η_{Syst}. To obtain values of η_{Syst} for the case of $r_a = 0.5$ and $q_{90} = 0.7$ it is only necessary to multiply the values of η_{th} given in Table 7.7 by $1/r_a$, i.e. by 2, and then apply Equation 7.85. The results of doing this are presented in Table 7.9. The results for η_{Syst} summarised in Tables 7.7 and 7.8 suggest that for both values of q_{90} the optimum structure is the 9/2 cluster, with the 12/3 structure in second place, but having lower requirements on channel utilisation, which may provide greater network operational flexibility.

As might be expected, the effect of employing silence detection is shown

Cluster size	$\eta_{textrmmax}$	η_{th}	$\eta_{textrmsyst}$	$U\%$
3	6.31	1.12	1.12	16
9	1.87	1.88	1.87	80
9/2	4.07	1.25	1.25	27
12/2	2.96	1.00	1.00	29
12/3	4.63	1.19	1.19	23
21/3	2.49	1.30	1.30	43
21/4	3.44	1.21	1.21	30

Table 7.8: Spectral efficiency (erl/cell/MHz) for various reuse structures: $r_a = 1$; $P_B = 2\%$; $q_{90} = 0.8$.

Cluster size	$\eta_{textrmmax}$	η_{th}	η_{Syst}	$U\%$
3	6.31	3.52	3.52	50
9	1.87	5.86	1.87	80
9/2	4.07	3.96	3.96	85
12/2	2.96	3.16	2.96	85
12/3	4.63	3.74	3.74	71
21/3	2.49	4.10	2.49	83
21/4	3.44	3.82	3.44	86

Table 7.9: Spectral efficiency (erl/cell/MHz) for various reuse structures : $r_a = 0.5$; $P_B = 2\%$; $q_{90} = 0.7$.

by the results for $r_a = 0.5$, given in Table 7.9, to be a large increase in spectral efficiency and channel utilisation in most cases. Comparison of the values with these results for structures of size 3, 9, 12/2, 21/3 and 21/4 reported in [5] and [8] shows reasonable agreement, taking into account that Dornsetter and Verhulst do not allow for channel utilisation being limited to values less than 100%. In [8] it is suggested that trunking efficiency factors should not normally be applied to SFHMA systems, but this is only true if situations involving high channel utilisation can be avoided, which may involve some sacrifice of spectral efficiency.

7.6.2 Spectral Efficiency of the SFHMA System: Method B

An alternative approach has been developed by the authors, which has the merit of allowing the inclusion of the full analysis of the concatenated FEC coding in a channel degraded by both noise and cochannel interference, as described in Section 7.5. The key to this method is the derivation of an expression for the pdf of the hop CIR, λ, as a function of the mean CIR and the probability of frequency collisions, starting from the expression for the probability of correct reception of a hop given in [8] and also derived in Appendix A. Armed with this pdf the variation of output BER with mean CIR can be determined, as described previously in Section 7.5.4 for a fading channel, but with collision probability as a parameter. By these means it is possible to find the value of the collision probability, p_i, and hence the spectral efficiency, which causes the BER to equal or exceed a threshold value (specified to ensure adequate speech decoder output) at a mean CIR equal to the 90% worst case value for the reuse structure under consideration. From Equation 7.69 the cumulative distribution of the hop

CIR, λ, is given by:

$$
\begin{aligned}
F(\gamma) &= \mathcal{P}\left\{\lambda \le \gamma\right\} = 1 - \mathcal{P}\left\{\lambda > \gamma\right\} = 1 - q \\
&= 1 - \prod_{i=1}^{M}\left(1 - \frac{p_i\gamma}{\Lambda_i + \gamma}\right)
\end{aligned}
\tag{7.86}
$$

where Λ_i is the short term mean CIR due to the ith interferer. The pdf is then obtained by differentiating $F(\gamma)$ wrt γ and putting $\lambda = \gamma$ to give:

$$
f(\lambda) = \sum_{i=1}^{M} \frac{p_i\Lambda_i}{(\Lambda_i + \lambda)^2} \prod_{\substack{j=1 \\ j\neq i}}^{M}\left(1 - \frac{p_j\lambda}{\Lambda_j + \lambda}\right).
\tag{7.87}
$$

Because of the directionality of the BS antennas assumed for the corner BS configuration, not all of the 6 cells in a given reuse ring will produce a significant level of interference at a mobile location in the reference cell. Generally the number of significant interferers alternates between 3 and either 2 or 4 in successive rings, as shown diagrammatically in Figure 7.28. Rings giving 4 interferers, for example of size 21, have 12 cells in total and may be regarded as two interlaced rings of 6 reuse cells. To evaluate Equation 7.87 for a reuse cluster of size 3 we used only the first 3 rings with 8 significant interferers. To simplify the computations required it is assumed that Λ_i remains constant over the surface of the reference cell at the value estimated at the BS location. This value is given by

$$
\Lambda_i = \frac{\Lambda_M \sum_{j=1}^{8} I_{rj}}{I_{ri}}
\tag{7.88}
$$

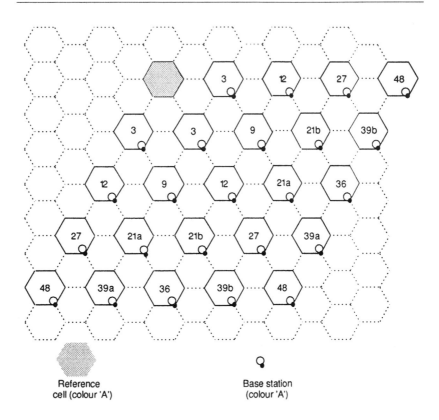

Figure 7.28: Positions of reuse cells producing significant cochannel interference for cluster size 3.

where Λ_M is the overall mean CIR and I_{ri} is the relative interference power received from the ith source, which is obtained as the product of the ith BS antenna power gain, G_{Ai}, given by Equation 7.4 and the path gain, G_{Pi}, relative to the path from the central source in the nearest ring, i.e.

$$I_{ri} = G_{Ai}G_{Pi} \qquad (7.89)$$

G_{Pi} is given by

$$G_{Pi} = \left[\frac{(D/R)_i}{3}\right]^{-3.5} \qquad (7.90)$$

where D/R is the reuse distance to cell radius ratio which is related to the ring size C by the well known expression

$$D/R = \sqrt{3C} \qquad (7.91)$$

Ring size	D/R	G_P	Source no.	G_A	I_r	Λ/Λ_M
3	3.00	1.000	1	0.407	0.407	5.45
			2	1.000	1.000	2.22
			3	0.407	0.407	5.45
9	5.20	0.146	4	0.839	0.122	18.2
			5	0.839	0.122	18.2
12	6.00	0.0884	6	0.407	0.0360	61.6
			7	1.000	0.0884	25.1
			8	0.407	0.0360	61.6
					====	
					2.218	
21	7.94	0.0332	9	0.738	0.0245	
			10	0.917	0.0304	
			11	0.917	0.0304	
			12	0.738	0.0245	
27	9.00	0.0214	13	0.407	0.0087	
			14	1.000	0.0214	
			15	0.407	0.0087	

Table 7.10: Data for calculation of Λ_i for reuse cluster size 3.

Numerical values for the above quantities are presented in Table 7.10. As in the previous analyses of performance the SFHMA system in a fading channel, it is necessary to determine the SER at the input to the second (outer) RS decoder as a function of the hop SNR normalised wrt the coded bit rate, γ_{cb}, and the hop CIR, λ, and then to evaluate a double integral to obtain the mean SER in terms of the mean CIR, Λ, and the mean SNR, Γ_{cb}, i.e.;

$$P_{is2}(\Lambda, \Gamma_{cb}) = \int_0^\infty \int_0^\infty P_{is2}(\lambda, \gamma_{cb}) f(\lambda) p(\gamma_{cb}) d\lambda d\gamma_{cb} \qquad (7.92)$$

where $f(\lambda)$ is the pdf of λ given by Equation 7.86 as above and $p(\gamma_{cb})$ is the pdf of γ_{cb}, which is assumed to have the negative exponential form associated with the Rayleigh fading model, as in Equation 7.33. Equations 7.57 (with SNR $= \gamma_{cb}$), 7.52, 7.42 and 7.43 are used to determine the first factor in the integrand. Finally the output BER from the second decoder is found by application of Equations 7.44 and 7.45. The mean SNR is normalised wrt the data bit rate by scaling Γ_{cb} as described in Section

7.4.4.

For all the calculations described here we put $r_a = 1$ and the SNR was kept constant at the relatively high, but arbitrarily chosen value of 30 dB, so that the system performance is primarily limited by cochannel interference rather than by noise.

A graph of output BER vs mean CIR, for cluster size 3 and $r_a = 1$, is presented in Figure 7.29 with η as parameter. From this data, for each value of η the threshold value of mean CIR, denoted by Λ_{th}, was found at which the BER crosses a threshold value of 3×10^{-3}, which is a representative input BER threshold for typical 16 kbits/s toll quality speech decoders. Figure 7.30 presents a graph of Λ_{th} vs η, from which the value of η can be determined, denoted by η_{th}, at which $\Lambda_{th} = \Lambda_{90}$ for cluster size 3 (5.2 dB).

Similar calculations were made for reuse cluster size 9, for which case 5 significant interferers in the two nearest reuse rings of size 9 and 27 were allowed for in evaluating the hop CIR pdf via Equation 7.87. The results are also plotted in Figure 7.30. In this case it will be seen that the curve does not extend to the point where $\Lambda_{th} = \Lambda_{90}$ for cluster size 9 (13.0 dB) before η reaches its maximum value of 2.33 (for $U = 1$).

The computations required for the fractional reuse structures are complicated by the need to allow for two different values of collision probability depending on whether the interferer being considered is in either a fractional or a full reuse cell. For a $3M/L$ structure the full reuse rings will be those appropriate for a full reuse structure of size $3M$, while the fractional reuse rings will be those appropriate to the basic 3-cell structure. In order to limit the computational complexity involved in the calculations reported here we used only 3 rings of size 3, 9 and 12 with 8 significant interferers, which allowed for the determination of the performance of the 9/2, 12/2 and 12/3 structures. In other respects the calculations were carried out in the same way as those for the full reuse structures described above. A graph of Λ_{th} vs η for the three structures is presented in Figure 7.31 and it will be seen that all three curves cross the relevant $\Lambda_{th} = \Lambda_{90}$ (= 5.2 dB) line for the basic 3-cell structure.

Table 7.11 summarises the values of η_{th} derived from Figures 7.30 and 7.31, together with the values of system spectral efficiency, η_{Syst}, found by applying Equation 7.85 and the corresponding values of channel utilisation. The values of η_{max} are taken from Table 7.6. Apart from the cluster size 9 case all the values of η_{Syst} and U lie between the corresponding values obtained using method A for $q_{90} = 0.7$ and $q_{90} = 0.8$, but closer to the former values.

7.6.3   Spectral Efficiency of the TD/FDMA System

In this section we shall determine the spectral efficiency of an equivalent TD/FDMA system employing the same basic features, such as modulation,

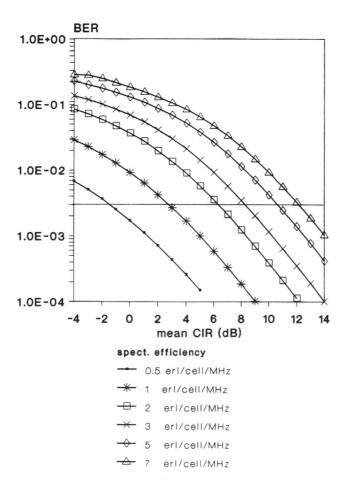

Figure 7.29: BER versus mean CIR for the SFHMA system with cochannel interference in a Rayleigh fading channel allowing for interferer diversity in full reuse cluster size 3 : normalised SNR = 30 dB and r_a = 1.0 [MSK-type detector: $B_t T$ = 0.3 and $B_r T$ = 0.63].

Cluster size	η_{max}	η_{th}	$\eta_{textrmsyst}$	U %
3	6.31	1.60	1.60	23
9	1.87	>2.33	1.87	80
9/2	4.07	1.92	1.92	41
12/2	2.96	1.47	1.47	42
12/3	4.63	1.74	1.74	33

Table 7.11: Spectral efficiency (erl/cell/MHz) for various reuse structures: $r_a = 1$; $P_B = 2\%$.

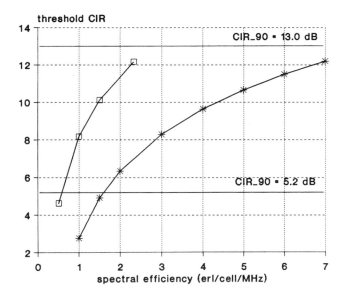

Figure 7.30: Λ_{th} versus η for full reuse clusters sizes 3 and 9 : $\mathrm{BER}_{th} = 3 \times 10^{-3}$, SNR = 30 dB and $r_a = 1.0$.

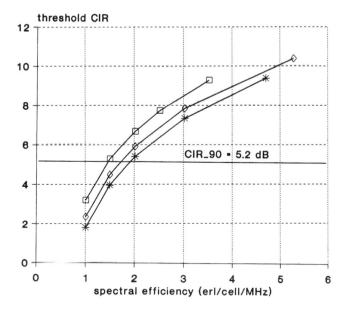

Figure 7.31: Λ_{th} versus η for fractional reuse clusters sizes 9/2, 12/2 and 12/3 : $\mathrm{BER}_{th} = 3 \times 10^{-3}$, SNR = 30 dB and $r_a = 1.0$.

type of receiver, 16 kbit/s speech coding and corner base station cell configuration, as in the SFHMA system, but with constant carrier frequencies and no FEC coding. The transmitted bit rate is set at 172 kbit/s, which allows for 8 TDM slots/frame, together with a guard time of almost 8% of the slot duration and a 25% overhead for network control signalling, as in the SFHMA system. For this bit rate the minimum frequency spacing is taken as 125 kHz.

To determine the threshold value of the mean CIR, Λ_{th}, which gives a BER threshold of 3×10^{-3} it was necessary to repeat the calculations described in Section 7.5.4 using Equations 7.63 - 7.67, but with the normalised SNR adjusted to allow for the total overhead of 34% in the transmitted bit rate. The results for a normalised SNR of 30 dB are shown in Figure 7.32 as a graph of BER vs mean CIR from which the value of λ_{th} is found to be 25 dB. The minimum cluster size is determined from the requirement

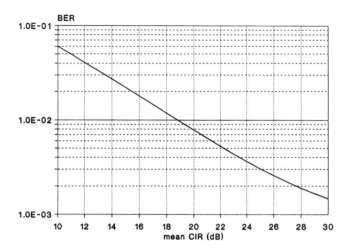

Figure 7.32: BER versus mean CIR for the uncoded TD/FDMA system with cochannel interference in a Rayleigh fading AWGN channel at normalised mean SNR = 30 dB [MSK-type detector : $B_tT = 0.3$ and $B_rT = 0.63$].

Size	36	39	48
Λ_{90} (dB)	22.9	23.5	25.1

Table 7.12: Extrapolated values of Λ_{90} versus reuse cluster size.

that the corresponding $\Lambda_{90} \geq \Lambda_{th}$. However the available data for Λ_{90} (see Table 7.4) only extends to a maximum value of 20.6 dB and so we have extrapolated the data to cover three additional cluster sizes, based on the assumption that Λ_{90} is proportional to $(D/R)^{-3.5}$ for larger clusters. The values obtained are presented in Table 7.12 and it will be seen that the minimum cluster size is 48. To determine the spectral efficiency we proceed to calculate the number of channels available per cell, the traffic carried per channel from the Erlang B formula for a blocking probability of 2% and finally the traffic carried per cell divided by the system bandwidth as indicated in Table 7.13. For comparison we also show the calculations for the SFHMA system, assuming the same worst case operation with no account taken of interferer diversity and for full reuse structures only. From the results of the analysis described in Section 7.5.4 for the SFHMA system in a fading AWGN channel, shown in Figure 7.21, at 30 dB SNR the CIR giving BER = 3×10^{-3} is 12.5 dB, which implies a minimum cluster size of 9.

These results demonstrate the superiority of the SFHMA system even

Quantity		TD/FDMA	SFHMA
System bandw.(MHz)	W	24	24
Cluster size	C	48	9
No. TDM slots	n_t	8	3
Frequency spacing(MHz)	f_s	0.125	0.143
No. freq./system	$n_{fs} = W/f_s$	192	168
No. channels/cell	$n_{ch} = n_{fs}n_t/C$	32	56
Traffic/channel (erl)	a_{ch}	0.727	0.803
Traffic/cell (erl)	a_{Cell}	23.3	45.0
Spectral efficiency (erl/cell/MHz)	$\eta_{\text{Syst}} = a_{textrmcell}/W$	0.971	1.88

Table 7.13: Calculation of spectral efficiency for SNR = 30 dB : $P_B = 2\%$.

with interferer diversity and silence detection discounted. Comparison with results for the SFHMA system obtained in the previous sections, which do allow for these effects, show good agreement for the cluster size 9. This is not as surprising as it might seem at first sight, since in this case the spectral efficiency is limited by the maximum channel utilisation, which implies that interferer diversity is not significant.

7.7 Conclusions

In this chapter we have described the principles and essential characteristics of SFHMA, concentrating on the mixed protocol for the basic code division multiplexing. A specification of a complete exemplary SFHMA system is provided as a basis for the performance analyses.

Details and results of BER performance analyses, mostly developed by the authors, are presented for the following conditions:

(1) a comparison of the ideal MLSE receiver for GMSK signals with a more practical MSK-type detector, which showed that the performance was only degraded by about 1 dB,

(2) system performance in static and Rayleigh fading channels,

(3) system performance with cochannel interference, both with and without noise and also with and without fading.

In the studies under the headings (2) and (3) comparative results are presented for an equivalent basic FDMA system which has no FEC coding.

No allowance was made initially in the cochannel interference analyses of the SFHMA system, described in Section 7.5, for the effects of interferer diversity, the main emphasis being placed on the performance of the

GMSK receiver with concatenated FEC coding. However interferer diversity is accounted for in the two methods presented in the following section for estimating spectral efficiency. Method A is essentially the simplified model reported by Verhulst [5], with some refinement to allow for non-ideal channel utilisation, while method B is a development of the authors' more rigorous analysis of SFHMA performance in a channel degraded by both cochannel interference and noise, but utilising an expression for the pdf of the hop CIR derived from Verhulst's analysis [5]. The results produced by the two methods agree surprisingly well, considering the different gross approximations involved.

Finally, in order to demonstrate the advantage of SFHMA we have estimated the spectral efficiency of a comparable TD/FDMA system. Comparing the results presented in Tables 7.7, 7.9 and 7.13, it can be seen that for the case of no silence detection ($r_a = 1$) the SFHMA system can provide a spectral efficiency just over twice that of the TD/FDMA system, while if silence detection is implemented the potential improvement is by a factor of almost four. Clearly there is a considerable advantage in spectral efficiency to be gained by employing SFHMA, under the assumptions used in the analyses, but too much reliance should not be placed on the absolute values of the improvement factors indicated above. Many crude approximations have been made in the underlying analyses and, moreover, the use of the mean input BER to the unspecified speech decoder as the basis of comparison of the systems can be criticised, because it does not allow for the temporal characteristics of the stream of errors.

SFHMA also offers the advantage of greater operational flexibility because of the variety of reuse structures available and because of the potential ability to cope with non-uniform traffic loading, which is implied by not needing to operate at high levels of the channel utilisation to achieve good spectral efficiency.

7.8 Appendix A:

Derivation of the Probability of Correct Reception of a Hop

A hop will be correctly received if the hop CIR, $\lambda = S/I$, is not less than the threshold value, γ, with a probability given by:

$$q = \mathcal{P}\{S/I \geq \gamma\} \qquad (7.93)$$

where $S = A_o B_o F_o$ is the received hop signal power and I is the total received hop interference power, obtained by summing the powers received from the sources in the reuse cells, i.e.:

$$I = \sum_{i=1}^{M} A_i B_i E_i F_i. \tag{7.94}$$

Substituting for S and rearranging, Equation 7.93 can be written as:

$$
\begin{aligned}
q &= \mathcal{P}\{F_o \geq \gamma I / A_o B_o\} \\
&= \int_0^{\infty} p(I) \left[\int_{\gamma I / A_o B_o}^{\infty} p(F_o) dF_o \right] dI
\end{aligned}
\tag{7.95}
$$

where $p(I)$ is the pdf of the total interference power and $p(F_o)$ is the pdf of the signal fading factor, which has an exponential pdf with a mean value of unity, i.e.:

$$p(F_o) = \exp(-F_o) \tag{7.96}$$

Thus we can determine the inner integral in the RHS of Equation 7.95 giving:

$$q = \int_0^{\infty} p(I) \exp(-\gamma I / A_o B_o) dI \tag{7.97}$$

This expression is of the form of a Laplace transform of $p(I)$ for $s = \gamma / A_o B_o$ so that we can write:

$$q = \mathcal{L}\{p(I)\} \mid_{s = \gamma / A_o B_o} \tag{7.98}$$

Since I is the sum of a set of independent random variables I_i, we can apply the known relationship that the transform, i.e. the characteristic function, of the pdf of the sum is the product of the transforms of the pdfs of the components:

$$\mathcal{L}\{p(I)\} = \prod_i \mathcal{L}\{p(I_i)\} \tag{7.99}$$

The discontinuous nature of the power received from the i_{th} interferer may be modelled as random on/off switching of a continuous exponentially distributed process, which has a mean value of $A_i B_i$. Hence the pdf $p(I_i)$ can be determined as the sum of two conditional pdfs for the 'on' ($E_i = 1$) and 'off' ($E_i = 0$) states, weighted by the state probabilities p_i and $(1 - p_i)$ respectively, where p_i is the frequency collision probability for the i_{th} reuse cell. Conditional on $E_i = 1$ the pdf of I_i is the same as that of the continuous process, i.e.:

$$p(I_i | E_i = 1) = \frac{1}{A_i B_i} \exp\left(\frac{-I_i}{A_i B_i}\right) \tag{7.100}$$

The alternative condition of $E_i = 0$ implies that I_i can only take the single discrete value of zero and so the pdf in this case becomes a unit impulse at

$I_i = 0$, i.e.:

$$p(I_i|E_i = 0) = \delta(I_i) \qquad (7.101)$$

Combining these conditional pdfs gives the overall pdf of I_i as:

$$p(I_i) = (1 - p_i)\delta(I_i) + \frac{p_i}{A_iB_i}\exp\left(\frac{-I_i}{A_iB_i}\right) \qquad (7.102)$$

Taking Laplace transforms gives:

$$\mathcal{L}\{p(I_i)\} = 1 - p_i + \frac{p_i}{1 + sA_iB_i} \qquad (7.103)$$

Substituting into Equation 7.99 and putting $s = \gamma/A_oB_o$ we obtain:

$$q = \prod_{i=1}^{M}\left(1 - p_i + \frac{p_i}{1 + \gamma A_iB_i/A_oB_o}\right) \qquad (7.104)$$

A somewhat different derivation of this expression is briefly described in [8]. We will define the ratio A_oB_o/A_iB_i as the short term mean CIR with respect to the i_{th} interference source, denoted by Λ_i. Note that this quantity relates the mean received signal power to the maximum mean interference power which could be received from the i_{th} source if it operated continuously. Finally, substituting in the above equation and rearranging gives:

$$\begin{aligned}
q &= \prod_{i=1}^{M}\left(1 - p_i + \frac{p_i\Lambda_i}{\Lambda_i + \gamma}\right) \\
&= \prod_{i=1}^{M}\left(1 - \frac{p_i\gamma}{\Lambda_i + \gamma}\right) \qquad (7.105)
\end{aligned}$$

In this chapter we have considered slow frequency hopping characterised the achievable performance using a variety of scenarios. This chapter concludes our discussions on various mobile radio system components. In the forthcoming chapter our aim is to compose a complete system on the basis of the previously introduced components. We invoke the example of the global system of mobile communications, known as GSM, in order to provide a discourse on the design of an amalgamated wireless system.

Bibliography

[1] **G.R.Cooper and R.W.Nettleton**, "A spread spectrum technique for high capacity mobile communications", *IEEE Trans. Veh. Technol.*, Vol. VT-27, No. 4, pp. 264-75, November 1978.

[2] **D.Verhulst, M.Mouly and J.Szpirglas**, "Slow frequency hopping multiple access for digital cellular radiotelephone", *IEEE Trans. Veh. Technol.*, Vol. VT-33, No. 3, pp. 179-190, August 1984.

[3] **D.Verhulst and M.Mouly**, "Spectrum efficiency evaluation techniques for future digital mobile systems", *Proc. of 13th Nordic Seminar Digital Land Mobile Radio*, Espoo, Finland, 5-7 February 1985.

[4] **E. A.Geraniotis and M. B.Pursley**, "Error probabilities for slow frequency-hopped spread-spectrum multiple-access communications over fading channels", *IEEE Trans. Commun.*, pp. 996-1009, May 1982.

[5] **D.Verhulst**, "Spectrum efficiency analysis of the digital system SFH900", *Proc. 2nd Nordic Seminar Digital Land Mobile Radio*, Stockholm, Sweden, 14-16 October 1986.

[6] **ANT/Bosch**, "Description of the experimental system S-900D for digital radiotelephone", GSM doc No 85/85, 1985.

[7] **J.Udenfeldt**, "TMS90-an experimental TDMA digital mobile telephone system", *Ericsson Radio System AB*, Stockholm, Sweden.

[8] **J-L.Dornstetter and D.Verhulst**, "Cellular efficiency with slow frequency hopping: Analysis of the digital SFH900 mobile system", *IEEE J. Select. Areas Commun.*, Vol. SAC-5, No. 5, pp. 835-848, June 1987.

[9] **J.B.Anderson, T.Aulin and C-E.Sundberg**, "Digital Phase Modulation", *Plenum Press*, New York 1986.

[10] **K.Murota and K.Hirade**, "GMSK modulation for digital mobile radio telephony", *IEEE Trans. Commun.*, Vol. COM-29, pp. 1044-1050, July 1981.

659

[11] **J. G.Proakis**, "Digital communications", *McGraw-Hill Book Company*, 1983.

[12] **T.Aulin, N.Rydbeck and C-E.Sundberg**, "Continuous phase modulation - Part II. Partial response signaling", *IEEE Trans. Commun.*, Vol. COM-29, No.3, pp. 210-225, March 1981.

[13] **S. G.Wilson and M.G.Mulligan**, "An improved algorithm for evaluating trellis phase codes", *IEEE Trans. Inform. Theory*, Vol. IT-30, No. 6, pp. 846-851, November 1984.

[14] **A.Svensson and C-E.Sundberg**, "Serial MSK-type detection of partial response continuous phase modulation", *IEEE Trans. Commun.*, Vol. COM-33, No. 1, pp. 44-52, January 1985.

[15] **K.Murota**, "Spectrum efficiency of GMSK land mobile radio", *IEEE Trans. on Veh. Technol.*, Vol. VT-34, No. 2, pp. 69-75, May 1985.

[16] **J.Oetting**, "Cellular mobile radio-an emerging technology", *IEEE Commun. Mag.*, pp. 10-15, November 1983.

[17] **V.H.MacDonald**, "Advanced mobile phone service: The cellular concept", *Bell Syst. Tech. Journal*, Vol. 58, No. 1, pp. 15-41, January 1979.

[18] **R.Steele and V.K.Prabhu**, "High-user-density digital cellular mobile radio systems", *IEE Proc. on Communications*, Radar and Signal Processing, Pt. F, Vol. 132, No. 5, pp. 396-404, August 1985.

[19] **Y.F.Ko**, "Digital cellular mobile radio links and networks", *PhD Thesis, Dept of Electronics and Computer Science, University of Southampton*, Section 2.4.2., December 1989.

[20] **W.Wr.Jakes**, "Microwave mobile communications", Wiley, New York, pp. 367.

[21] **J.E.Stjernvall**, "Calculation of capacity and co-channel interference in a cellular system", *Proc. of Nordic Seminar on Digital Land Mobile Radio*, Espoo, Finland, pp. 209-217, 5-7, February 1985.

Chapter **8**

Global System of Mobile Communications—known as GSM

L. Hanzo[1] and J. Stefanov[2]

8.1 Introduction

The first cellular radio system in Europe was installed in Scandinavia in 1981 and it served initially only a few thousand subscribers. At the time of writing there are six different cellular systems operating in 16 European countries and serving more than 1.2 million subscribers. There is however, a general incompatibility of systems and user equipment. A mobile station (MS) designed for one system cannot be used in another which makes it impossible for mobiles to roam across international borders while making calls. The low scale of equipment production also results in higher equipment cost and call charges. In 1982 CEPT (Conference Europeene des Postes et Telecommunication), the main governing body of the European PTTs, created the Groupe Speciale Mobile (GSM) Committee and tasked it with specifying a cellular pan-European public mobile communication system to operate in the 900 MHz band. This pan-European system became so widespread all over the globe that later it was renamed as the Global System of Mobile (GSM) communications.

The first important decision made by the GSM Committee was the se-

---

[1]University of Southampton and Multiple Access Communications Ltd
[2]University of Southampton and Multiple Access Communications Ltd

	Access type	Transm. bit rate Kbit/s	Carrier spacing kHz	Mod. type	Channels per carrier
CD-900	CDMA/TDMA	7980	4500	4-PSK	63
MATS-D/W	CDMA/TDMA	2496	1250	QAM	32
ADPM	TDMA	512	600	ADPM	12
DMS-90	TDMA	340	300	GMSK	10
MOBIRA	TDMA	252	250	GMSK	9
SFH-900	TDMA	200	150	GMSK	3
S900-D	TDMA	128	250	4-FSK	10
MAX II	TDMA	104.7	50	8-PSK	4
MATS-D/N	FDMA	19.5	25	GTFM	1

Table 8.1: Systems tested by GSM in Paris, 1986.

lection of a digital system. This was followed by the launch of experimental programmes of different types of digital cellular radio systems in a number of European countries. By the middle of 1986 nine proposals were received for the future pan-European system, and GSM organised a trial in Paris to identify the one having the best performance. The technical details of the candidate systems are described in references [1]- [4] and [9]. A short summary [6] of their salient features is listed in Table 8.1.   The first generation cellular radio systems use analogue frequency modulation. Single-channel-per-carrier (SCPC) Frequency Division Multiple Access (FDMA) is utilised, where the channel bandwidth is either 25 or 30 kHz. The proposed digital systems for GSM employed a variety of access methods, transmission rates and modulation schemes. Six systems adopted Time Division Multiple Access (TDMA), another two employed Code Division Multiple Access (CDMA) combined with TDMA, and one used FDMA on its mobile station to base station (MS to BS) uplink. The transmission bit rates for the various systems spanned the range from 20 kb/s to 8 Mbit/s. Altogether seven different modulation schemes were used in the nine systems.

In addition to the field trials in Paris, laboratory testing of the equipment was carried out using a propagation simulator. The test arrangements allowed for measurements with or without interference under static or dynamic conditions. The channel simulator had two independent Rayleigh fading paths, enabling five different propagation profiles to be used corresponding to various rural, suburban and urban propagation environments. Based on the field trials and laboratory tests the candidate systems were assessed in order of importance against the following criteria [8]: spectrum efficiency, subjective voice quality, mobile cost, hand-portable feasibility, base station cost, ability to support new services and co-existance with current systems.

The test results demonstrated that the spectrum efficiency of all the candidate systems was equal or better than that of the first generation analogue cellular systems. However, not all the systems provided an acceptable transmission quality over the range of propagation conditions. Furthermore, there were significant implementational and operational risk factors involved with some of the systems. The two 'hybrid' CDMA/TDMA systems proposed were wideband systems requiring broadband radio subsystems and very complex baseband signal processing. These systems had the highest degree of risk in terms of being implementable within the GSM time scale for the system to be operational in 1992 and that their costs might be commercially too high. The predicted performance of broadband receivers was considered to be difficult to achieve in practice [6] and broadband systems can be susceptible to spurious interference. The implementation of the complex VLSIs needed for baseband processing would be difficult and costly, although the resulting microcircuits were expected to be of low cost because of the large size of the market.

On the whole, the TDMA technique was preferred over FDMA and TDMA-CDMA. Partly, because for good transmission quality FDMA systems need antenna diversity not only at the base station but also at the mobile station, which in most cases is unacceptable. However, the high bit rate associated with TDMA implies dispersive wideband channel models and requires the use of channel coding, interleaving, as well as channel equalisation. In TDMA systems the lack of a duplexer in the mobile station and the multiplexing of several channels on one RF carrier in the base station leads to simpler, more compact and more cost-efficient design of both the MS and BS. TDMA systems are also more flexible than FDMA systems in accommodating new services in the future. The hand-over in TDMA systems can be performed more efficiently than in FDMA systems which is particularly important for high traffic density microcells.

Based upon the results of the tests in Paris, GSM decided at the beginning of 1987 to adopt a narrowband TDMA system with the basic system features specified as follows: 8 TDMA channels per carrier; regular pulse excited linear predictive (RPE-LPC) speech codec operating at 13 kb/s; half-rate ($R=1/2$), constraint length five ($K=5$) convolutional codecs CC(2,1,5); carrier spacing of 200 kHz; constant envelope Gaussian minimum shift keying (GMSK) modulation. By the end of 1988 the working parties of GSM and their supporting expert groups had substantially completed the specifications of the pan-European system. The specifications were released by GSM as 13 sets of recommendations [9] covering various aspects of the system, which are summarised in Table 8.2.

The following description of the GSM system, its basic elements and their functions, is based almost entirely on the GSM recommendations [9]. Since the system is still evolving at the time of writing, fine details of the material presented in this chapter are subject to changes.

The GSM system operates in two paired bands: 890–915 MHz for uplink transmission, where the mobile transmits and the base station receives, and 935–960 MHz for downlink transmission, where the base station transmits

R.00 - *Preamble* to the GSM Recommendations.

R.01 - *General structure* of the Recommendations, description of a GSM network, associated recommendations, vocabulary, etc.

R.02 - *Service aspects:* bearer-, tele- and supplementary services, use of services, types and features of mobile stations (MS), licensing and subscription, as well as transferred and international accounting, etc.

R.03 - *Network aspects,* including network functions and architecture, call routing to the MS, technical performance, availability and reliability objectives, handover and location registration procedures as well as discontinuous reception and cryptological algorithms, etc.

R.04 - *Mobile/base station (BS) interface and protocols,* including specifications for layer 1 and 3 aspects of the open systems interconnection (OSI) seven-layer structure.

R.05 - *Physical layer on the radio path,* incorporating issues of multiplexing and multiple access, channel coding and modulation, transmission and reception, power control, frequency allocation and synchronisation aspects, etc.

R.06 - *Speech coding specifications,* such as functional, computational and verification procedures for the speech codec and its associated voice activity detector (VAD) and other optional features.

R.07 - *Terminal adaptors for MSs,* including circuit and packet mode as well as voice-band data services.

R.08 - *Base station (BS) and mobile switching centre (MSC) interface,* and transcoder functions.

R.09 - *Network interworking* with the public switched telephone network (PSTN), integrated services digital network (ISDN) and packet data networks.

R.10 - *Service interworking, short message service.*

R.11 - *Equipment specification and type approval specification* as regards to MSs, BSs, MSCs, home (HLR) and visited location register (VLR) as well as system simulator.

R.12 - *Operation and maintenance,* including subscriber, routing tariff and traffic administration as well as BS, MSC, HLR and VLR maintenance issues.

Table 8.2: GSM Recommendations [R.01.01]

and the mobile receives. A guard band of 200 kHz is provided at the lower end of each duplex band and the remaining spectrum is divided into 124 paired duplex channels with 200 kHz channel spacing in each band. The spacing between the duplex bands is 45 MHz.

The access scheme is TDMA with 8 timeslots per radio carrier detailed later by referring to Figure 8.2. The duration of each timeslot is approximately 0.58 ms, which leads to a TDMA frame duration of approximately $8 \cdot 0.58 = 4.6$ ms. The information is transmitted in bursts at a rate of approximately 271 kbit/s using Gaussian Minimum Shift Keying (GMSK) - which was the topic of Chapter 6 - with a bandwidth-bit interval product of BT=0.3. For the channel spacing of 200 kHz the use of this type of modulation allows the carrier separation to be 18 dB for the first adjacent channel and 50 dB for the next one. A rudimentary description of GMSK is provided in Section 8.8, while a detailed treatment is given in Chapter 6.

At the data rate of 271 kbit/s the multipath propagation leads to deep fades and to uncontrolled intersymbol interference in addition to that introduced in a controlled manner by the GMSK modulator. Transmission errors are combated by using channel coding and channel equalisation. The channel coding method employs two concatenated codes. A block code provides error detection for the most significant 50 speech bits, followed by a half-rate convolutional code, while the 78 least significant speech bits are not protected at all. The number of coded bits per block and the convolutional code rate depend on the type of information transmitted (speech, data, signalling). Aspects of the channel coding and interleaving are provided in Section 8.7, while the equalisation of GMSK signals is summarised in Section 8.9 and described in detail in Chapter 6. Suffice to state here that the GSM equaliser is expected to handle excess path delays of up to 16 μs. Slow frequency hopping with 217 hops/s is used to provide diversity effect and to increase the efficiency of coding and interleaving for slow moving mobile stations, as explained in Section 8.3. It helps also to decrease the effects of the cochannel interference.

A general objective of the GSM system is to provide a wide range of services and facilities, both voice and data, that are compatible with those offered by the existing fixed Public Services Telephone Networks (PSTN), Public Data Networks (PDN) and Integrated Services Digital Networks (ISDN). Another objective is to give compatibility of access to the GSM network for any mobile subscriber in any country which operates the system, and these countries must provide facilities for automatic roaming, locating and updating the mobile subscriber's status.

8.2 Overview of the GSM System

Mobile radio communications in a GSM Public Land Mobile Network (GSM-PLMN) is facilitated by a series of network functions and proce-

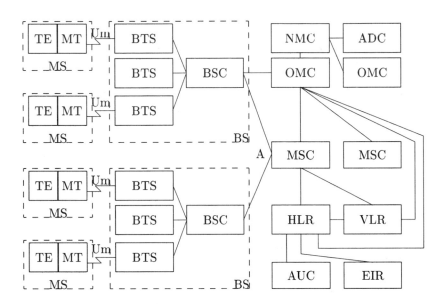

Figure 8.1: Simplified structure of GSM PLMN. ©ETT, Hanzo and Steele,
1994.

dures. Figure 8.1 shows the simplified structure of a typical GSM PLMN
with the functional entities of the system and their logical interconnections.
The Mobile Station (MS) is the equipment used by a subscriber to access
the services offered by the system. Functionally the MS includes a Mobile Termination (MT), and Terminal Equipment (TE) which may consist
of more than one piece of equipment such as a telephone set and Data
Terminal Equipment (DTE). If necessary, one or more Terminal Adapters
(TA) may also be included. The MT performs the functions needed to
support the physical channel between the MS and the base station, such as
radio transmissions, radio channel management, channel coding/decoding,
speech encoding/decoding, and so forth, where the MS-BS radio interface
is designated by U_m. There are various types of mobile stations such as
vehicle mounted stations, portable stations and hand-held stations.

The Base Station System (BSS), defined also as a Base Station (BS), is
divided functionally into a Base Transceiver Station (BTS) and Base Station Controller (BSC) and they are interconnected by the A-bis interface.
The BS is associated with the radio channel management including channel
allocation, link quality supervision, transmission of associated signalling information and broadcast messages, as well as controlling transmitted power
levels and frequency hopping. Its further functions entail error correction
encoding and decoding, digital speech transcoding or data rate adaptation,
intracell handover initiation to a 'better' RF channel, as well as data and

signalling encryption. The BTS is the transmission equipment used to give radio coverage for a traffic cell. All control functions in the base station are performed by the BSC. The radio equipment in a BS may serve more than one cell, in which case the BS will consist of several BTSs under the control of one BSC.

The Mobile Switching Centre (MSC) is linked to the BS via the A interface and performs all the switching functions needed for the operation of the mobile stations in the group of cells it services. The functions of an MSC include call routing and call control; procedures needed for interworking with other networks (e.g., PSTN, ISDN); procedures related to the mobile station's mobility management such as paging to receive a call, location updating while roaming and authentication to prevent unauthorised access; as well as procedures required to implement handovers. Handover (HO) is the process of re-assigning the mobile station's communications to a different base station, when the mobile moves outside the range of the serving base station. The GSM system supports also another type of handover, the intra-cell handover, which is a call transfer from one channel to another one within the same cell, when a channel cannot be used any longer due to interference disturbance or maintenance problems. Due to the high traffic demand expected, RF channels have to be frequently reused, which results in small microcells and increased probability of high cochannel interferences. To counteract the interference limitation, efficient handover algorithms based on intelligent and hence complex received signal quality evaluations are absolutely essential.

The Home Location Register (HLR) is a data base unit for the management of mobile subscribers. Part of the mobile location information is stored in the HLR, which allows the incoming calls to be routed to the MSC in command of the area where the MS roams. The MS has to periodically inform the PLMN about its geographic location by updating the contents of the HLR. To assist this process the PLMNs are divided into disjoint geographic areas characterised by unique identifiers broadcast regularly to all MSs via the so-called Broadcast Control Channels (BCCHs) conveyed over reserved RF carriers. Should the MS observe a change of identifier, it issues a location update request. The HLR contains the International Mobile Subscriber Identity (IMSI) number which is used for the authentication of the subscriber by his Authentication Centre (AUC). This enables the system to confirm that the subscriber is allowed to access it. Every subscriber belongs to a home network and the specific services which the subscriber is allowed to use are entered into his HLR. The Equipment Identity Register (EIR) allows for stolen, fraudulent or faulty mobile stations to be identified by the network operators.

The Visitor Location Register (VLR) is the functional unit that attends to a MS operating outside the area of its HLR. The visiting MS is automatically registered at the nearest MSC and the VLR is informed of the MSs arrival. A roaming number is then assigned to the MS and this enables calls to be routed to it. The Operations and Maintenance Centre (OMC), Network Management Centre (NMC) and ADministration Centre (ADC) are the functional entities through which the system is monitored, controlled, maintained and managed.

When a mobile user initiates a call the MS searches for a BS providing a sufficiently high received signal level on the BCCH carrier, it will synchronise to it, then the BS allocates a bidirectional signalling channel and also sets up a link with the MSC serving the area. The MSC uses the IMSI received from the mobile station to interrogate the subscriber's HLR. The subscriber's data obtained from the HLR are then sent to the local VLR. After the user is accepted by the network, the MS defines the type of service it requires and provides the destination number of its call. The BS serving the cell allocates a traffic channel and the MSC routes the call to its destination. If the MS moves to another cell, it is re-assigned to another BS and a handover occurs. If both BSs in the handover process are controlled by the same BSC, the handover takes place under the control of the BSC. If the base stations are controlled by different BSCs, then the handover is performed by the MSC.

The procedure of setting up a call connection from a fixed network to a MS is similar to the procedure described above. The main difference is that the MS must be paged by the BSC. A paging signal is transmitted on a paging channel (PCH) monitored continuously by all MSs and covers the location area in which the MS has registered. When the MS receives the paging signal, it starts an access procedure identical to that employed when the MS initiates a call.

8.3   Mapping Logical Channels onto Physical Channels

8.3.1   Logical Channels

The elaborate design of the MS-BS radio interface (U_m interface in GSM terminology) is motivated by the needs of providing appropriate signalling and traffic channels in the system. The traffic channels might take the form of circuit and packet switched bearer services at various synchronous and asynchronous speeds as well as teleservices, such as speech, short message, teletext or facsimile communications. Appropriate signalling and traffic communications are provided by the first three layers of the seven-layer Open Systems Interconnection (OSI) model. Accordingly, the physical layer (L1) interfaces with the data link layer (L2) via a number of logical channels constituted by speech and data traffic channels (TCHs) as well as signalling or control channels (CCHs). Our prime objective is to transmit the traffic channel's speech or data information; however, their transmission via the network requires a variety of control channels. A feasible solution to efficient networking is provided by the set of logical control channels defined by GSM. The need for each specific control channel provided arises from the system architecture discussed previously, although further alternative concepts can readily be contrived.

As seen in Table 8.3, there are two general forms of speech and data traffic channels: the full rate traffic channels (TCH/F), which carry information at a gross rate of 22.8 kbit/s, and the half rate traffic channels (TCH/H), which communicate at a gross rate of 11.4 kbit/s. A physical channel carries either a full rate traffic channel, or two half rate traffic channels. In the former the traffic channel occupies one timeslot, while in the latter the two half-rate traffic channels are mapped onto the same timeslot, but in alternate frames.

Encoded speech and user data can be conveyed on a variety of full or half rate traffic channels. These may be notated as TCH/$\alpha\beta$, where α is either F or H signifying full rate or half rate, respectively, and β is either S when speech is being carried, or 9.6, 4.8 or 2.4 representing the data rate in kbit/s. The bit rate in the full rate traffic channel for speech (TCH/FS) is 13 kbit/s, becoming 22.8 kbit/s after embedded channel coding. The channel coded data rate in the half rate speech channel TCH/HS is 11.4 kbit/s. This channel is envisaged for future evolution of the system and allows the traffic capacity to be doubled, when a low bit rate toll quality voice codec becomes available. The traffic channels may carry a wide variety of user information, but not signalling information. However, when user data are transmitted, they may carry protocols such as, e.g., in the case of packet switching services according to CCITT Recommendation X25.

The control channels carry signalling or synchronisation data. As summarised in Table 8.3, four categories of control channels are used, known as the broadcast control channel (BCCH), the common control channel (CCCH), the stand-alone dedicated control channel (SDCCH) and the associated control channel (ACCH). The broadcast control channels are used only in downlink communications from the base station to the mobile stations in its vicinity. There are three types of BCCHs. The frequency correction channel (FCCH) is provided to facilitate frequency synchronisation of the mobile station to the master radio frequency source in the base station. The transmitted information on the FCCH is equivalent to an unmodulated carrier with a fixed frequency offset from the nominal carrier frequency. The function of the synchronisation channel (SCH) is to enable frame synchronisation of the mobile station and the identification of the serving base station. Accordingly, the data transmitted on the SCH contain the TDMA frame number (FN) and the base station identity code (BSIC). When not acting as an FCCH or SCH, the BCCH carries general information, such as the number of common control channels, whether these common control channels are combined with stand-alone dedicated control channels and associated control channels on the same physical channels.

There are three types of CCCHs. The paging channel (PCH) contains

Logical channels					
Duplex BS ↔ MS Traffic channels: TCH		Control channels: CCH			
FEC-coded Speech	FEC-coded Data	Broadcast CCH BCCH BS → MS	Common CCH CCCH	Stand-alone Dedicated CCH SDCCH BS ↔ MS	Associated CCH ACCH BS ↔ MS
TCH/F 22.8 kbit/s	TCH/F9.6 TCH/F4.8 TCH/F2.4 22.8 kbit/s	Freq.Corr.Ch: FCCH	Paging Ch: PCH BS → MS	SDCCH/4	Fast ACCH: FACCH/F FACCH/H
		Synchron. Ch: SCH	Random Access Ch: RACH MS → BS	SDCCH/8	Slow ACCH: SACCH/TF SACCH/TH SACCH/C4 SACCH/C8
TCH/H 11.4 kbit/s	TCH/H4.8 TCH/H2.4 11.4 kbit/s				
		General Inf.	Access Grant Ch: AGCH BS → MS		

Table 8.3: GSM logical channels. ©ETT, Hanzo and Steele, 1994.

paging signals from the BS to the MSs in case of a network originated call. There is a random access channel (RACH), which is used only in the uplink communications whereby the MSs request the allocation of a bidirectional stand-alone dedicated control channel for BS-MS signalling. An access grant channel (AGCH) is provided for downlink communications for the purpose of allocating a stand-alone dedicated control channel or a traffic channel to MSs, which was previously requested via the RACH.

There are two types of SDCCHs. The stand-alone dedicated control channel having 4 sub-channels and notated by SDCCH/4, and SDCCH/8 which has 8 sub-channels. These SDCCH channels are used for setting up the services required by the user. This involves interrogation of the mobile station as to the services required, the availability response of the base station and the allocation of a free traffic channel.

The ACCHs, like the SDCCHs, are also bidirectional channels. In the downlink they carry, for example, control commands from the base station to the MS to set its transmitted power level, while in the uplink they convey the status of the mobile station, such as the received signal levels from various adjacent BSs, etc. An ACCH is always allocated in conjunction with either a traffic channel, or with a stand-alone dedicated control channel, as will be explained later. There are two types of ACCH. The fast associated control channel (FACCH) facilitates urgent actions, such as handover commands and channel reassignment in intra-cell handovers. This type of channel can be associated either with a full rate traffic channel (FACCH/F), or with a half rate traffic channel (FACCH/H), and it is provided by stealing bits from its traffic channels, hence degrading their performance. The slow associated control channel (SACCH) is sub-divided into 4 types, depending on what type of channel it is associated with. Thus SACCH/TF is associated with a full rate traffic channel; SACCH/TH is associated with a half rate traffic channel; SACCH/C4 is associated with SDCCH/4; while SACCH/C8 is associated with SDCCH/8. In downlink transmissions the SACCH carries commands from the base station to the mobile station for setting its output power level. The mobile station responds by informing the BS of the set level of its output power, the measured received RF signal strength and the quality of the signals from adjoining cells.

After this rudimentary characterisation of the logical channels used in the GSM system we are ready to describe the TDMA physical channels carrying the information of the logical channels and the way logical channels are mapped onto physical ones.

8.3.2 Physical Channels

A physical channel in a TDMA system is defined as a timeslot with a timeslot number TN in a sequence of TDMA frames. However, the GSM system deploys TDMA combined with frequency hopping and hence the physical channel is partitioned in both time and frequency. Consequently the physical channel is defined as a sequence of radio frequency channels and timeslots. Each carrier frequency supports 8 physical channels mapped onto 8 timeslots within a TDMA frame. A given physical channel uses always the same timeslot number TN in every TDMA frame. Therefore, a timeslot sequence is defined by a timeslot number TN and a TDMA frame number FN sequence.

8.3.2.1 Mapping the TCH/FS and its SACCH as well as FACCH onto Physical Channels

In our deductive approach we use the example of the full rate speech traffic channel (TCH/FS) to explain how this logical channel is mapped onto the physical channel constituted by a so-called Normal Burst (NB) of the TDMA frame structure. This mapping is explained in macroscopic terms by referring to Figures 8.2 and 8.3. Bit-level fine details of the individual mapping steps will be provided in later sections, as details of the speech coding, error correction coding, etc. become available. Then this example will be extended to other physical bursts such as the Frequency Correction (FB), Synchronisation (SB), Access (AB) and Dummy Burst (DB) carrying logical control channels, as well as to their TDMA frame structures, as seen in Figures 8.2 and 8.7.

The Regular Pulse Excited (RPE) speech encoder delivers 260 bits/20 ms at a bit rate of 13 kbit/s, which are divided into three significance classes: Class 1a (50 bits), Class 1b (132 bits) and Class 2 (78 bits). The Class 1a bits are encoded by a systematic (53,50) cyclic error detection code by adding three parity bits. Then the bits are reordered and four zero tailing bits are added, in order to periodically clear the memory of the subsequent half rate, constraint length five convolutional codec CC(2,1,5), as protrayed in Figure 8.3. Clearing the convolutional codec's memory after each transmission burst curtails the propagation of transmission errors across burst boundaries, which otherwise would be encountered due to diverging from the error-free trellis path in the Viterbi decoder. These diverging paths were visualised in the Viterbi decoding examples of Chapter 4.

Now the unprotected 78 Class 2 bits are concatenated to yield a block of 456 bits/20 ms, which implies an encoded bit rate of 22.8 kbit/s. This frame is partitioned into eight 57-bit subblocks that are block-diagonally interleaved before undergoing interburst interleaving, to be detailed later in Section 8.7. The flag bits hl and hu are included to classify whether the burst being transmitted is really a TCH/FS burst or if it has been 'stolen' by an urgent FACCH message. Now the bits are encrypted and positioned in a Normal Burst (NB), as depicted at the bottom of Figure 8.2, where three tailing bits (TB) - which are also often referred to as ramping symbols - are added at both ends of the transmission burst. The 3+3 ramping symbols are included in order to assist the transceiver to power up and

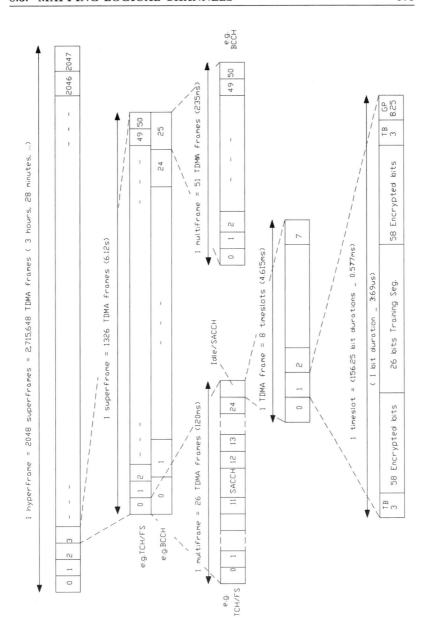

Figure 8.2: The GSM TDMA frame structure ©ETT, Hanzo and Steele, 1994

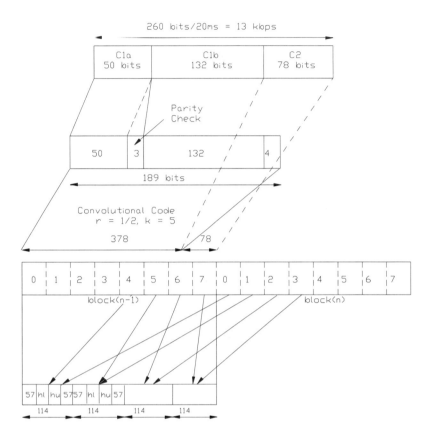

Figure 8.3: Mapping the TCH/FS logical channel onto a physical channel.
©ETT, Hanzo and Steele, 1994.

down smoothly, as will be highlighted in the context of the power ramping
mask of Figure 8.29, which is necessary for mitigating the spurious adjacent
channel emissions.

This mapping process is also summarised in the form of a hardware
oriented block diagram in Figure 8.4, where the Voice Activity Detector
(VAD) is included to enable or disable transmissions depending on whether
speech is deemed to be present at the input of the RPE speech encoder.
This allows a substantial reduction of the power consumption as well as that
of the interferences imposed on other users. The effects of subjectively
annoying silent periods are mitigated at the receiver by adding comfort
noise during these intervals.

The 8.25 bit-interval duration guard space (GP) at the bottom of Fig-

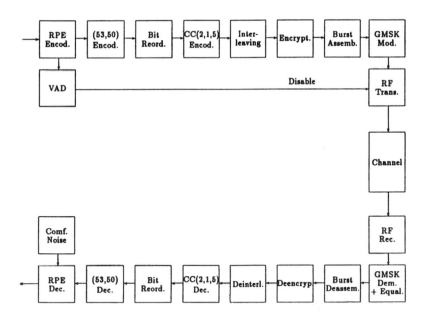

Figure 8.4: Block diagram of the TCH/FS channel.

ure 8.2 is provided to prevent burst overlapping due to delay fluctuations. Finally, a 26-bit equaliser training segment is included in the centre of the normal traffic burst. This segment is constructed by a 16-bit Viterbi channel equaliser training pattern surrounded by five quasi-periodically repeated bits on both sides. These bits provide a sufficiently long quasi-periodic extension of the training sequence before useful data are entered into the modulator, to keep the side-lobes of its autocorrelation function, as well as those of its spectra, sufficiently low. The 16-bit pattern was determined by evaluating the autocorrelation of the modulated signals due to all 2^{16} candidate training sequences and selecting the one with the highest auto-correlation peak, while maintaining a low main lobe to side-lobes ratio. For GMSK with BT=0.3 and modulation index of 0.5 several good sequences can be found. Since the MS has to be informed about which BS it communicates with, for neighbouring BSs different training patterns are used. Therefore, in the GSM system the eight best training patterns are used to be associated with eight different BS colour codes.

This 156.25 bit duration TCH/FS normal burst (NB) constitutes the basic timeslot of the TDMA frame structure, which is input to the GMSK modulator at a bit rate of approximately 271 kbit/s. Since the bit interval is 3.69 μs, the timeslot duration is $156.25 \cdot 3.69 \approx 0.577$ ms. Eight such normal bursts of eight appropriately staggered users are multiplexed onto one RF carrier giving a TDMA frame of $8 \cdot 0.577 \approx 4.615$ ms

duration, as we see in Figure 8.2. The physical channel as characterised above provides a physical timeslot with an effective information
throughput of 114 bits/4.615 ms=24.7 kbit/s, which is sufficiently high
to transmit the 22.8 kbit/s TCH/FS. It even has a 'reserved' capacity of
24.7$-$22.8=1.9 kbit/s, which can be exploited to transmit slow control information associated with this specific traffic channel, i.e., to construct a
so-called Slow Associated Control Channel (SACCH). The TCH/FS has a
repetition delay of 20 ms and an interleaving delay of $8 \cdot 4.615 = 37$ ms,
yielding a total speech delay of 20+37=57 ms.

To understand how we accommodate and access the SACCH we have
to proceed with the construction of the TDMA frame hierarchy, as highlighted in Figure 8.2. The TCH/FS TDMA frames of the eight users are
multiplexed into multiframes of 24 TDMA frames, but the 13th frame
will carry a SACCH message, rather than the 13th TCH/FS frame, while
the 26th frame will be an idle or dummy frame, as seen at the left hand
side of Figure 8.2 representing the traffic channel hierarchy. The general
control channel frame structure shown at the right of Figure 8.2 is discussed later. This way 24 TCH/FS frames are sent in a 26-frame multiframe during $26 \cdot 4.615 = 120$ ms. This reduces the traffic throughput to
$\frac{24}{26} \cdot 24.7 = 22.8$ kbps required by TCH/FS, allocates $\frac{1}{26} \cdot 24.7 = 950$ bps
to the SACCH and 'wastes' 950 bps in the idle frame. Observe that the
SACCH frame has eight timeslots to transmit the eight 950 bps SACCHs
of the eight users on the same carrier. The 950 bps idle capacity will be
used in the case of half rate channels, where 16 users will be multiplexed
onto alternate frames of the TDMA structure to increase system capacity,
when a half rate speech codec becomes available. Then sixteen 11.4 kbps
encoded TCH/HSs will be transmitted in a 120 ms multiframe, where also
sixteen SACCHs are available.

The construction of SACCH bursts is slightly different from TCH/FS
bursts in that only 184 control bits are transmitted during 20 ms, in contrast to 260 speech bits, as portrayed in Figure 8.5. The additional channel capacity is exploited to accommodate a (224,184) external block code,
extended by four zero tailing bits, in order to clear the memory of the subsequent half rate, constraint length five convolutional codec, which curtails
error propagation across transmission burst boundaries. The total number
of bits is now $(224 + 4) \cdot 2 = 456$ transmitted via four consecutive bursts,
each carrying 114 bits. Each of these bursts is accommodated in a new
120 ms multiframe, yielding a repetition delay of $4 \cdot 120 = 480$ ms. The 456
error protected bits are transmitted at an average rate of 950 bps, which
provides an unprotected information rate of $\frac{184}{456} \cdot 950 = 382$ bps for the
SACCH, when associated with a traffic channel.

As opposed to independent SACCH frames, Fast Associated Control
Channel (FACCH) messages are transmitted via the physical channels provided by bits 'stolen' from their own host traffic channels. The construction of the FACCH bursts from 184 control bits is identical to that of

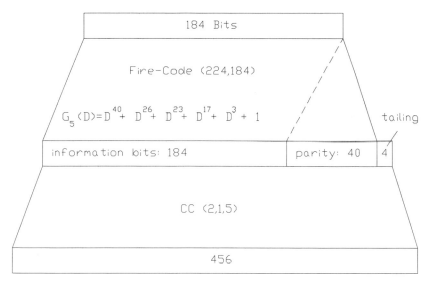

Figure 8.5: Mapping of SACCH, FACCH, BCCH, SDCCH, PCH and AGCH logical channels.

the SACCH, as also shown in Figure 8.5, but its 456-bit frame is mapped onto eight consecutive 114-bit TDMA traffic bursts, exactly as specified for TCH/FS. This is carried out by stealing the even bits of the first four and the odd bits of the last four bursts, which is signalled by setting $hu = 1$, $hl = 0$ and $hu = 0$, $hl = 1$ in the first and last bursts, respectively. The unprotected FACCH information rate is 184 bits/20 ms=9.2 kbps, which is transmitted after concatenated error protection at a rate of 22.8 kbps. The repetition delay is 20 ms and the interleaving delay is $8 \cdot 4.615 = 37$ ms, resulting in a total delay of 57 ms.

In a subsequent stage of Figure 8.2 51 TCH/FS multiframes are amalgamated into one superframe lasting $51 \cdot 120$ ms $= 6.12$ s, which contains $26 \cdot 51 = 1326$ TDMA frames. There would be no need for any further levels of TDMA hierarchy, if it was not for the encryption, which uses the TDMA frame number (FN) as a parameter in its algorithm. However, with 1326 FNs only the encryption rule is not sufficiently secure. Therefore 2048 superframes are concatenated to form a hyperframe of $1326 \cdot 2048 = 2\,715\,648$ TDMA frames lasting $2048 \cdot 6.12$ s ≈ 3 h 28 min, using a satisfactorily high number of FNs in the encryption algorithm. This step now concludes our example of mapping the TCH/FS and its SACCH logical channel onto an appropriate physical channel constituted by a specific timeslot dedicated to a specific user of a specific RF channel carrying the messages of eight TDMA users. To reduce the complexity of MSs they do not have to receive and transmit simultaneously, their receive and transmit timeslots carrying the same TN are shifted by three in the TDMA frame with respect to each other.

8.3.2.2   Mapping Broadcast and Common Control Channels onto Physical Channels

In our TCH/FS and SACCH example in the previous subsection the RF channel was shared amongst 8 or 16 TDMA users and a specific timeslot was dedicated to one TCH/FS or two TCH/HS users. In contrast, the BCCH and CCCH logical channels of all MSs roaming in a specific cell share the physical channel provided by timeslot zero of the so-called BCCH carriers available in the cell. Furthermore, all BCCHs and CCCHs are simplex channels operating in the uplink or downlink directions, as opposed to traffic channels, stand-alone dedicated control channels and fast or slow associated control channels, which are full duplex channels. Another difference is that in the case of BCCHs and CCCHs 51 TDMA frames are mapped onto a $51 \cdot 4.615 = 235$ ms duration multiframe, rather than on a 26-frame, 120 ms duration multiframe. To compensate for the extended multiframe length of 235 ms, 26 multiframes constitute a 1326-frame superframe of 6.12 s duration, as demonstrated by Figure 8.2. Also the allocation of the uplink and downlink frames is different, since these control channels exist only in one direction, as seen in Figure 8.6.

Specifically, the random access channel (RACH) is only used by the MSs in the uplink direction if they request, for example, a bidirectional stand-alone dedicated control channel (SDCCH) to be mapped onto an RF channel to register with the network and set up a call. The uplink RACH carries messages of eight bits per 235 ms multiframe, which is equivalent to an unprotected control information rate of 34 bps. These messages are concatenated FEC coded to a rate of 36 bits/235 ms=153 bps. They are not transmitted by the Normal Bursts (NB) derived for TCH/FS, SACCH or FACCH logical channels, but by the so-called Access Bursts (AB), depicted in Figure 8.7 in comparison to a NB and other types of bursts to be described when introduced. The tailing and synchronisation bits are given in Recommendation 05.02. The FEC coded, encrypted 36-bit messages, containing amongst other parameters also the encoded 6-bit BS identifier code (BSIC) constituted by the 3-bit PLMN colour code and 3-bit BS colour code for unique BS identification, are positioned after the 41-bit synchronisation sequence, which is extended to ensure reliable access burst recognition. These messages have no interleaving delay, while they are transmitted with a repetition delay of one control multiframe length, i.e. 235 ms.

The GSM system is specified to allow operation of mobile stations in cells when they are up to 35 km from their base station. The time a radio signal takes to travel the 70 km from the base station to the mobile station and back again is 233.3 μs. As signals from all the mobiles in the cell must

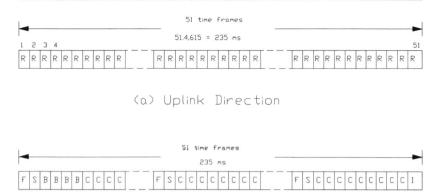

(a) Uplink Direction

(b) Downlink Direction

R: Random Access Channel

F: Frequency Correction Channel

S: Synchronisation Channel

B: Broadcast Control Channel

C: Access Grant/Paging Channel

I: Idle Frame

Figure 8.6: The control multiframe

reach the base station without overlapping each other, a guard period of 68.25 bits (252 μs) is provided in the access burst. This long guard period in the access burst is needed when the mobile station attempts its first access to the base station, or after handover has occurred.

To provide the same long guard period in the other bursts would be spectrally inefficient. The GSM system overcomes this problem by using adaptive frame alignment. When the base station detects a 41-bit random access synchronisation sequence with a long guard period, it measures the received signal delay relative to the expected signal from a mobile station of zero range. This delay, called the timing advance, is signalled using a 6-bit number to the mobile station, which advances its timebase over the range of 0 to 63 bits, i.e., in units of 3.69 μs. By this process the TDMA bursts arrive at the BS in their correct timeslots and do not overlap with adjacent ones. This process allows the guard period in all other bursts to be reduced to 8.25 $\cdot$ 3.69 μs $\approx$ 30.46 μs (8.25 bits) only. In normal

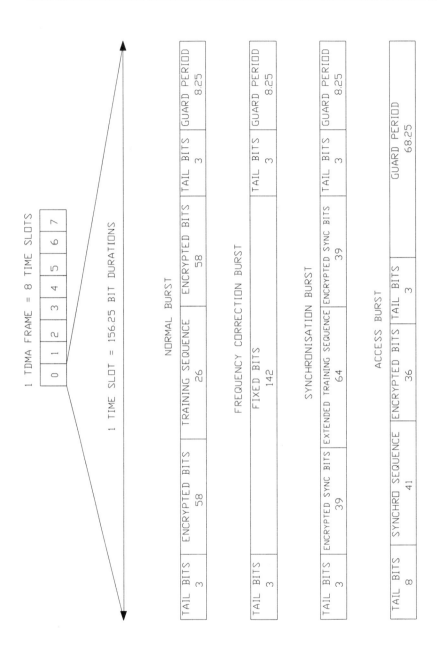

Figure 8.7: GSM burst structures. ©ETT, Hanzo and Steele, 1994.

operation the BS continuously monitors the signal delay from the MS and will instruct the MS to update its time advance parameter. In very large traffic cells there is an option to actively utilise every second timeslot only to cope with higher propagation delays, which is spectrally inefficient, but in these large, low-traffic rural cells admissible.

The downlink multiframe transmitted by the BS is shared amongst a number of BCCH and CCCH logical channels, as depicted in Figure 8.6. In particular, the last frame is an idle frame (I), while the remaining 50 frames are divided in five blocks of ten frames, where each block starts with a frequency correction channel (FCCH) followed by a synchronisation channel (SCH). In the first block of ten frames the FCH and SCH frames are followed by four broadcast control channel (BCCH) frames and by either four access grant control channels (AGCH) or four paging channels (PCH). In the remaining four blocks of ten frames the last eight frames are devoted to either PCHs or AGCHs, which are mutually exclusive for a specific MS being either paged or granted a control channel. Clearly, the downlink control multiframe hosts a total of four BCCHs, five FCCHs and SCHs, as well as $4 \cdot 8 + 4 = 36$ AGCHs or PCHs, constituting nine so-called paging blocks of four frames.

Each MS communicating via the downlink control multiframe is associated with one paging block of four AGCHs or PCHs out of the nine existing such blocks, although each block of four can be shared by several MSs. Therefore the bit rate of the physical channels provided for the concatenated error protected BCCH and AGCH or PCH logical channels is $4 \cdot 114 = 456$ bits per 235 ms, which is 1.94 kbps. The 456-bit error protected messages are derived from the 184-bit unprotected control messages exactly in the same way, as explained by referring to Figure 8.5 for the SACCHs and FACCHs in the previous subsection and have a repetition delay of 235 ms. The unprotected transmission rate available for BCCH, AGCH and PCH logical channels is $\frac{184}{456} \cdot 1.94$ kbps $= 782$ bps.

The FCCH uses the frequency correction burst (FCB) shown in Figure 8.7. The three tailing or ramping bits are used to facilitate smooth power ramping, as was demonstrated in the context of Figure 8.29, while the 8.25 bit length guard space prevents eventual burst overlapping due to propagation path-length differences, which were not equalised by the adaptive time frame alignment employed. The 142 fixed bits are chosen to yield a modulated signal which is equivalent to an unmodulated carrier with a fixed frequency offset above the nominal carrier frequency.

The SCH logical channel's control information is conveyed by the synchronisation burst (SB), seen also in Figure 8.7. The 25 synchronisation bits are concatenated FEC coded as described in Section 8.7 and encrypted to yield 78 protected bits, which are allocated both sides of the 64-bit extended training sequence specified in Recommendation 05.02. The role of the remaining bits is identical to those in other bursts.

There is a fifth burst type not shown in Figure 8.7, called the Dummy

Burst (DB), which has an identical structure to that of a NB, except for the fact that it carries no useful data. The 116 fixed encrypted bits have an equal probability of logical ones and zeros and are transmitted when no useful data are available, but the link has to be maintained to monitor the powers of the BCCH carriers of adjacent cells by the MSs.

It is important to note that a physical channel is specified by defining both the RF carrier and a TDMA timeslot number (TN). This is equivalent to saying that an RF channel carries eight physical channels. We emphasise that different timeslots of a specific RF carrier can carry both traffic and control logical channels, and therefore timeslots of the same RF channel can be assigned to different multiframe structures. It is plausible that multiframe types can only be altered at superframe boundaries, i.e., every 6.12 s, but with this proviso they are allowed to change arbitrarily in hyperframes. If the physical channels have to accommodate heavy control logical channel traffic, the allocation of timeslot zero only on the BCCH RF carrier would result in excessive access delays. In this case timeslots two, four and six can also be assigned to BCCH/CCCH logical channels. The 51-frame control multiframe structure is then slightly modified to transmit dummy bursts (DB) in the SCH and FCCH to maintain carrier transmissions, the power of which has to be monitored by the MSs. If there is only very little control information to be sent, timeslot zero of the BCCH carrier can be shared by BCCH/CCCH as well as by four SDCCH logical channels or by eight SDCCHs.

8.3.2.3   Broadcast Control Channel Messages

The BCCH transmits a variety of parameters to the MSs, of which those used to determine the legitimate combinations of control logical channels per physical channel are to be highlighted here. The number of basic physical channels supporting CCCHs is transmitted in the form of the 2-bit parameter BS_CC_CHANS. All CCCHs have to use timeslots on the BCCH carrier often referred to as C0 in GSM jargon. The first CCCH uses timeslot 0, the second CCCH will utilise timeslot 2 of C0 and for high control traffic demands timeslots 4 and 6 of C0 are allocated to further CCCHs. This means that $1 \leq$ BS_CC_CHANS ≤ 4.

A 1-bit flag indicating whether the CCCHs are combined with four SDCCHs and four SACCHs onto the same physical channel called BS_CCCH_-SDCCH_COMB is also broadcast on C0. Namely, if they are combined, the number of 'available' AGCH/PCH paging blocks must be reduced from nine. The AGCH and PCH share the physical channel on a 'block of ten frames' basis, which enables the MS to determine after deinterleaving and FEC decoding whether the block contains an AGCH or PCH message. As mentioned earlier, these two messages are mutually exclusive, since the MS is either being granted a channel or paged. The physical channel is assigned to AGCH or PCH messages on a demand basis, where PCH messages have higher priority than AGCH messages, since the former already represent partially established calls occupying parts of the system. However, to prevent long call set up delays, a number of the available blocks in each 51-frame multiframe can be reserved for AGCHs. The number of such reserved blocks (BS_AG_BLKS_RES) is also encoded using three bits and broadcast via the BCCH, while the number of 'available' PCH blocks has to be reduced by BS_AG_BLKS_RES.

Another parameter broadcast in the BCCH is the number of 51-frame multiframes between transmissions of paging messages to MSs of the same paging group, which is abbreviated as BS_PA_MFRMS. Since $2 \leq$ BS_PA_MFRMS ≤ 9, it is encoded by three bits. The total number of paging blocks 'available' on a CCCH logical channel is denoted by N and is computed as the product of 'available' blocks (ABL) in a multiframe and the number of multiframes between paging MSs of the same paging group, i.e.: N=ABL $\cdot$ BS_PA_MFRMS. Once N is known, the MSs are only required to monitor every Nth paging block of their CCCH. All MSs listening to a particular paging block out of the available total of N belong to a specific paging group (PAGING_GROUP). Hence, there are N paging groups and the MS computes which one it belongs to from the International Mobile Subscriber Identity (IMSI), N and from the assigned number of CCCH timeslots ($1 \leq$BS_CC_CHANS≤ 4) as given below:

$$\text{PAGING\_GROUP}(0 \ldots N-1) = ((\text{IMSI} \mod 1000) \mod (\text{BS\_CC\_CHANS} \cdot N)) \mod N. \tag{8.1}$$

The knowledge of the paging group is only required to pinpoint which paging block of which 51-frame multiframe has to be monitored by any MS. This determines for the MS, when it is supposed to be in active and dormant mode, thereby contributing towards the power-efficient MS operation. The required multiframe is encountered when:

$$\text{PAGING\_GROUP} \;\text{div}\; (N \;\text{div}\; \text{BS\_PA\_MFRMS}) = ((FN \;\text{div}\; 51) \mod (\text{BS\_PA\_MFRMS})) \tag{8.2}$$

where div represents integer division, mod is short for modulo and the paging block index (PBI) to be monitored is given by:

$$\text{PBI} = (\text{PAGING\_GROUP} \mod (N \;\text{div}\; \text{BS\_PA\_MFRMS}). \tag{8.3}$$

8.3.3 Carrier and Burst Synchronisation

The GSM Recommendations do not specify the BS-MS synchronisation algorithms to be used; these are left to the equipment manufacturers. However, a unique set of timebase counters is defined to ensure perfect BS-MS synchronism. The BS sends frequency correction (FCB) and synchronisation bursts (SB) on specific timeslots of the BCCH carrier to the MS to ensure that the MS's frequency standard is perfectly aligned with that of the BS, as well as to inform the MS about the required initial state of its

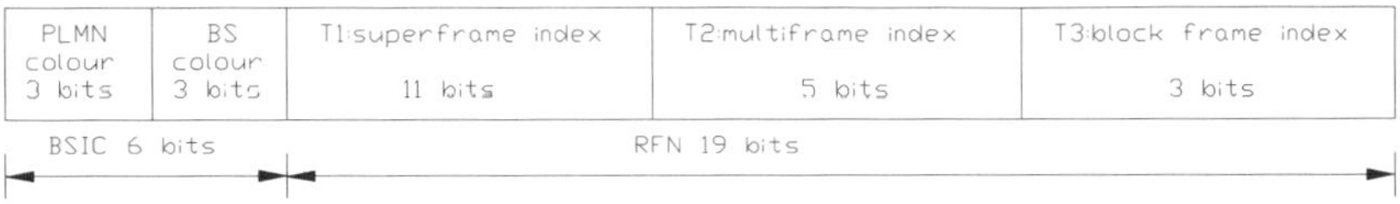

PLMN colour 3 bits	BS colour 3 bits	T1:superframe index 11 bits	T2:multiframe index 5 bits	T3:block frame index 3 bits
BSIC 6 bits		RFN 19 bits		

Figure 8.8: Synchronisation channel(SCH) message format.

internal counters. The MS sends its uniquely numbered traffic and control bursts staggered by three timeslots with respect to those of the BS to prevent simultaneous MS transmission and reception, and also takes into account the required timing advance (TA) to cater for different BS-MS-BS round-trip delays.

The timebase counters used to uniquely describe the internal timing states of BSs and MSs are the Quarter bit Number (QN=0...624) counting the quarter bit intervals in bursts, Bit Number (BN=0...156), Timeslot Number (TN=0...7) and TDMA Frame Number (FN=0...26·51·2048), given in the order of increasing interval duration. The MS sets up its timebase counters after receiving a SB by determining QN from the 64-bit extended training sequence in the centre of the SB, setting TN=0 and decoding the 78 encrypted, protected bits carrying the 25 SCH control bits. The SCH carries frame synchronisation information as well as BS identification information to the MS, as seen in Figure 8.8, and it is provided solely to support the operation of the radio subsystem. The first six bits of the 25-bit segment consist of three PLMN colour code bits and three BS colour code bits supplying a unique BS Identifier Code (BSIC) to inform the MS which BS it is communicating with. The second 19-bit segment is the so-called Reduced TDMA Frame Number (RFN) derived from the full TDMA Frame Number (FN), constrained to the range of $[0...(26·51·2048) - 1] = [0...2,715,647]$ in terms of three subsegments T1, T2 and T3. These subsegments are computed as:

$$T1(11 \text{ bits}) = (FN \text{ div } (26·51)) \tag{8.4}$$

$$T2(5 \text{ bits}) = (FN \text{ mod } 26) \tag{8.5}$$

$$T3'(3 \text{ bits}) = ((T3\text{-}1) \text{ div } 10), \text{ where } T3 = (FN \text{ mod } 5). \tag{8.6}$$

Here T1 determines the superframe index in a hyperframe, T2 the multiframe index in a superframe, T3 the frame index in a multiframe, while T3' the block index of a frame in a specific control multiframe. Their role is best understood by referring to Figure 8.2. Once the MS has received the Synchronisation Burst (SB), it readily computes the FN required in various control algorithms, such as encryption, handover, etc., as shown below:

$$FN = 51·((T3\text{-}T2) \text{ mod } 26) + T3 + 51·26·T1, \text{ where } T3 = 10·T3' + 1. \tag{8.7}$$

It is desirable to have perfect synchronism of all the channels under the

control of a BS, and hence all its RF carrier frequencies and timel
counter frequencies are derived from the same reference frequency. l
possible but not mandatory to synchronise different BSs together. Wl...
the BS detects a RACH message with its 41-bit synchronisation sequence,
its unique BSIC and 68.25-bit guard period, it will notice that the MS is
asking for random access to it. Using the 41-bit synchronisation sequence
in this decoded AB the BS can now evaluate the propagation delay, which
will be the timing advance to be signalled, rounded to the nearest integer
bit period, to the MS. As the timing advance is encoded by 6 bits, it is hard-
limited to 64·3.69=236 μs and it is kept constant for higher propagation
delays. This timing advance is continuously updated and the adjustment
error is less than half of a bit period.

8.3.4 Frequency Hopping

Frequency hopping combined with interleaving is known to be very efficient
in combating channel fading, and it results in near-Gaussian performance
even over hostile Rayleigh-fading channels. The principle of Frequency
Hopping (FH) is that each TDMA burst is transmitted via a different RF
Channel (RFCH). If the present TDMA burst happened to be in a deep
fade, then the next burst most probably will not be, as long as hopping is
carried out to a frequency sufficiently different from the present one, having a differently fading envelope. However, this is not easily ensured due to
the limited bandwidth available for GSM, since, for example, uplink transmissions are carried out in the 890-915 MHz band, where the maximum
relative hopping frequency is ~25 MHz/900 MHz≈2.8%. Nevertheless, FH
reduces the amount of time spent by the MS in a fade to 4.615 ms, the
duration of a TDMA burst, which brings substantial gains in the case of
slowly moving MSs, such as pedestrians. The GSM frequency hopping algorithm is shown in Figure 8.9. The algorithm's input parameters include
the TDMA Frame Number (FN) specified in terms of the indices FN(T1),
FN(T2) and FN(T3), as received in the synchronisation burst (SB) via
the Synchronisation Channel (SCH). A further parameter is the set of RF
channels called mobile allocation (MA) assigned for use in the MS hopping
sequence, which is limited to $1 \leq N \leq 64$ channels out of the legitimate
124 GSM channels. The Mobile Allocation Index Offset (MAIO) determines the minimum value of the Mobile Allocation Index (MAI), which is
the output variable of the FH algorithm determining the next RF channel
to which frequency hopping is required. Lastly, the Hopping Sequence generator Number $0 \leq HSN \leq 63$ is a further control parameter, which results
in cyclic hopping if HSN=0, as seen in Figure 8.9, and in pseudo-random
hopping patterns if $1 \leq HSN \leq 63$. This is because for HSN=0 the mobile
allocation index is computed as:

$$MAI = ((FN+MAIO) \mod N), \tag{8.8}$$

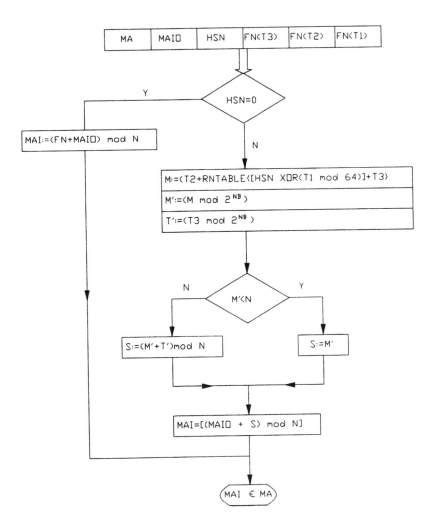

Figure 8.9: The GSM frequency hopping algorithm.

where (mod N) is taken to ensure that MAI remains an element of the set MA.

For $1 \leq \text{HSN} \leq 63$ somewhat more complex operations have to be computed, using a number of intermediate internal variables, as demonstrated by Figure 8.9. The only undefined variable in the figure is NB, representing the number of bits required for the binary encoding of N, the number of RF channels in the set MA. The function RNTABLE simply assigns one out of 114 pseudo-random numbers specified by GSM according to its argument, the XOR operator means bit-wise exclusive OR, while the remaining operations are self-explanatory. The result of the process is the mobile allocation index (MAI) specifying the next RF channel to be used by the MS. Note that frequency hopping is not allowed on timeslot zero of the BCCH carrier, which is ensured by using a single RFCH, i.e., setting $N=1$ and MAIO=0. In this case the FH sequence generation is uneffected by the value of HSN.

8.4 Full-rate 13 kbps Speech Coding

Speech compression foundations were the subject of Chapter 3, hence in this section a basic familiarity with speech coding and in particular the underlying background on Regular Pulse Excited (RPE) coding is assumed. Nonetheless, the treatment is conceptual, rather than mathematical, and hence it is readable for communications practitioners who are seeking a basic appreciation of the associated speech compression aspects. The selection of the most appropriate speech codec for the GSM system from the set of candidate codecs was based on extensive comparative tests among various operating conditions. The rigorous comparisons published in [7] are interesting and offer deep insights for system designers as regards the pertinent trade-offs in terms of speech quality, robustness against channel errors, complexity, system delay, etc.

8.4.1 Candidate Codecs

Originally the participating countries have proposed six different codecs with an overall channel coding and speech coding rate of 16 kbps for comparison. At a preliminary test the codecs were compared to the presently used companded FM system, and then two of the codecs were withdrawn. The remaining codecs were two different sub-band codecs and two pulse-excited codecs, which are detailed in Chapter 3.

SBC-APCM: Subband codec with block adaptive PCM. This codec used quadrature mirror filters (QMF) to split the input signal into 16 subbands of 250 Hz bandwidth, out of which the two highest bands were not transmitted. Adaptive bit allocation was used in the subbands on the basis of the power ratios of the various subbands, which constituted the side-information to be transmitted. The gross transmission rate of the subband

signals was 10 kbit/s, the side-information was 3 kbit/s, which was protected by 3 kbit/s forward error correction coding (FEC) redundancy.

SBC-ADPCM: Subband codec with adaptive delta PCM. In this scheme the speech input signal was split into 8 subbands, out of which only 6
were transmitted. The subband signals were encoded by differential coding
with backward estimation and adaptation, as opposed to the SBC-APCM
candidate, where forward estimation and adaptation were used. The bit
allocation of the subbands was fixed; hence no side-information was transmitted, which made the scheme more noise resilient, and hence no FEC
protection was required, and the bit-rate was 15 kbit/s only.

MPE-LTP: Multi-pulse excited LPC codec with long term predictor.
The particular speech codec implementation used in the comparisons required a 13.2 kbit/s transmission bit rate, and 2.8 kbit/s embedded FEC
coding was deployed to protect the most important bits of the speech codec.

RPE-LTP: Regular pulse excited LPC codec. A thorough theoretical
analysis of this method was given in Chapter 3 and the forthcoming subsection is devoted to implementational details of this scheme, since it was
selected for standardisation on the basis of the overall comparison tests.

These four codecs were compared in terms of speech quality, robustness, processing delays and computational complexity. From the experience
with the companded FM reference system two benchmarker bit error rates
(BER) were supposed, at which performance comparisons were carried out.
The pessimistic case was a BER of $10^{-2} = 1\%$, which would require a carrier to noise ratio (CNR) of approx. 18 dB, exceeded probably in 90% of
the reception area in the FM system. The optimistic channel exhibits a
BER of $10^{-3} = 0.1\%$, requiring approx. CNR=26 dB, guaranteed for at
least 50% of the coverage area for the reference FM system. The average mean opinion scores (MOS) on a five point scale over the various test
conditions were found to be [7]: MOS(RPE-LPC)= 3.54, MOS(MPE-
LTP)= 3.27, MOS(SBA-APCM)= 3.14, MOS(SBC-ADPCM)= 2.92,
MOS(FM)= 1.95. These results have emphasised the superiority of the
pulse-excited codecs and the importance of the long-term predictor (LTP).
The RPE codec exhibiting the most favourable properties was further improved by deploying a LTP, and the RPE-LTP codec guarantees an MOS
≈ 4.0 over a wide range of operating conditions.

8.4.2 The RPE-LTP Speech Encoder

The schematic diagram of the RPE-LTP encoder is shown in Figure 8.10,
where the following functional parts can be recognised [9], - [11]: 1. Pre-
processing, 2. STP analysis filtering, 3. LTP analysis filtering, 4. RPE
computation.

1. Pre-processing: Pre-emphasis can be deployed to increase the
numerical precision in computations by emphasising the high-frequency,
low-power part of the speech spectrum. This can be carried out by the

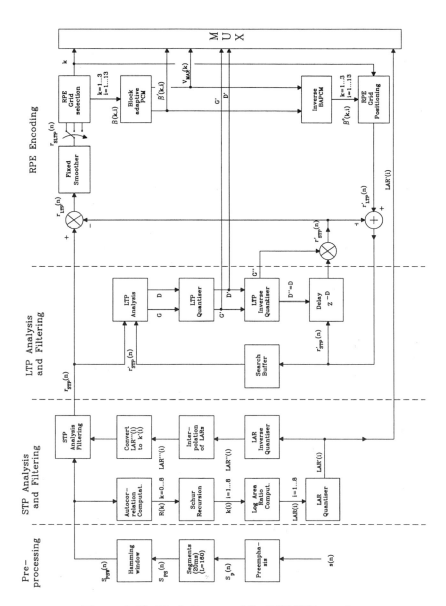

Figure 8.10: Block diagram of the RPE-LTP encoder.

help of a one-pole filter with the transfer function of:

$$H(z) = 1 - c_1 z^{-1},\qquad(8.9)$$

where $c_1 \approx 0.9$ is a practical value. The pre-emphasised speech $s_p(n)$ is segmented into blocks of 160 samples in a buffer, where they are windowed by a Hamming window to counteract the spectral domain Gibbs oscillation, caused by truncating the speech signal outside the analysis frame. The Hamming window has a tapering effect towards the edges of a block, while it has no influence in its middle ranges:

$$s_{psw}(n) = s_{ps}(n) \cdot c_2 \cdot \left(0.54 - 0.46 \cos 2\pi \frac{n}{L}\right)\qquad(8.10)$$

where $s_{ps}(n)$ represents the pre-emphasised, segmented speech, $s_{psw}(n)$ is its windowed version and the constant $c_2 = 1.5863$ is determined from the condition that the windowed speech must have the same power as the non-windowed.

2. STP analysis filtering: For each segment of $L=160$ samples nine autocorrelation coefficients $R(k)$ are computed from $s_{psw}(n)$ by:

$$R(k) = \sum_{n=0}^{L-1-k} s_{psw}(k) s_{psw}(n+k) \qquad k = 0 \ldots 8.\qquad(8.11)$$

From the speech autocorrelation coefficients $R(k)$ eight reflection coefficients k_i are computed according to the Schur recursion [12], which is an equivalent method to the Durbin algorithm used for solving the LPC key equations to derive the reflection coefficients k_i, as well as the STP filter coefficients a_i. However, the Schur recursion delivers the reflection coefficients k_i only. The reflection coefficients k_i are converted to logarithmic area ratios $(LAR(i))$, because the logarithmically companded LARs have better quantisation properties than the coefficients k_i:

$$LAR(i) = \log_{10}\left(\frac{1 + k(i)}{1 - k(i)}\right),\qquad(8.12)$$

where a piecewise linear approximation with five segments is used to simplify the real-time implementation:

$$LAR'(i) = \begin{cases} k(i), & \text{if } |k(i)| < 0.675 \\ \text{sign}\,[k(i)][2|k(i)| - 0.675], & \text{if } 0.675 < |k(i)| < 0.95 \\ \text{sign}\,[k(i)][8|k(i)| - 6.375], & \text{if } 0.975 < |k(i)| < 1.0 \end{cases}$$
$$(8.13)$$

The various $LAR(i)$ $i = 1 \ldots 8$ filter parameters have different dynamic ranges and differently shaped probability density functions (PDFs) as seen in Figure 3.25. This justifies the allocation of 6, 5, 4 and 3 bits to the first, second, third and fourth pairs of LARs, respectively. The quantised $LAR(i)$ coefficients $LAR'(i)$ are locally decoded into the set $LAR''(i)$, as well as transmitted to the speech decoder. So as to mitigate the abrupt changes in the nature of the speech signal envelope around the STP analysis frame edges, the LAR parameters are linearly interpolated, and towards the edges of an analysis frame the interpolated $LAR''(i)$ parameters are used. Now the locally decoded reflection coefficients $k'(i)$ are computed by converting $LAR''(i)$ back into $k'(i)$, which are used to compute the STP residual $r_{STP}(n)$ in a so-called PARCOR (partial correlation) structure. The PARCOR scheme directly uses the reflection coefficients $k(i)$ to compute the STP residual $r_{STP}(n)$, and it constitutes the natural analogy to the acoustic tube model of human speech production.

3. LTP analysis filtering: As we have seen in Chapter 3, the LTP prediction error is minimised by that LTP delay D which maximises the cross-correlation between the current STP residual $r_{STP}(n)$ and its previously received and buffered history at delay D, i.e., $r_{STP}(n-D)$. To be more specific, the $L=160$ samples long STP residual $r_{STP}(n)$ is divided into four $N=40$ samples long subsegments, and for each of them one LTP is determined by computing the cross-correlation between the presently processed subsegment and a continuously sliding $N=40$ samples long segment of the previously received 128 samples long STP residual segment $r_{STP}(n)$. The maximum of the correlation is found at a delay D, where the currently processed subsegment is the most similar to its previous history. This is most probably true at the pitch periodicity or at a multiple of the pitch periodicity. Hence the most redundancy can be extracted from the STP residual, if this highly correlated segment is subtracted from it, multiplied by a gain factor G, which is the normalised cross-correlation found at delay D. Once the LTP filter parameters G and D have been found, they are quantised to give G' and D', where G is quantised only by two bits, while to quantise D' seven bits are sufficient.

The quantised LTP parameters (G', D') are locally decoded into the pair (G'', D'') so as to produce the locally decoded STP residual $r'_{STP}(n)$ for use in the forthcoming subsegments to provide the previous history of the STP residual for the search buffer, as shown in Figure 8.10. Observe that since D is integer, we have $D = D' = D''$. With the LTP parameters just computed the LTP residual $r_{LTP}(n)$ is calculated as the difference of the STP residual $r_{STP}(n)$ and its estimate $r''_{STP}(n)$, which has been computed by the help of the locally decoded LTP parameters (G'', D) as shown below:

$$r_{LTP}(n) = r_{STP}(n) - r''_{STP}(n) \tag{8.14}$$

$$r''_{STP}(n) = G'' r'_{STP}(n - D). \tag{8.15}$$

Here $r'_{STP}(n - D)$ represents an already known segment of the past history of $r'_{STP}(n)$, stored in the search buffer. Finally, the content of the search buffer is updated by using the locally decoded LTP residual $r'_{LTP}(n)$ and the estimated STP residual $r''_{STP}(n)$ to form $r'_{STP}(n)$, as shown below:

$$r'_{STP}(n) = r'_{LTP}(n) + r''_{STP}(n) .\qquad(8.16)$$

4. RPE computation: The LTP residual $r_{LTP}(n)$ is weighted with the fixed smoother, which is essentially a gracefully decaying band limiting low-pass filter with a cut-off frequency of 4 kHz/3=1.33 kHz according to a decimation by three about to be deployed, as detailed in Chapter 3. The impulse response of this filter is also given in Chapter 3. The smoothed LTP residual $r_{SLTP}(n)$ is decomposed into three excitation candidates, by actually discarding the 40th sample of each subsegment, since the three candidate sequences can host 39 samples only. Then the energies E1, E2, E3 of the three decimated sequences are computed, and the candidate with the highest energy is chosen to be the best representation of the LTP residual. The excitation pulses are afterwards normalised to the highest amplitude $v_{max}(k)$ in the sequence of the 13 samples, and they are quantised by a three bit uniform quantiser, whereas the logarithm of the block maximum $v_{max}(k)$ is quantised with six bits. According to three possible initial grid positions k, two bits are needed to encode the initial offset of the grid for each subsegment. The pulse amplitudes $\beta(k,i)$, the grid positions k and the block maxima $v_{max}(k)$ are locally decoded to give the LTP residual $r'_{LTP}(n)$, where the 'missing pulses' in the sequence are filled with zeros.

8.4.3 The RPE-LTP Speech Decoder

The block diagram of the RPE-LTP decoder is shown in Figure 8.11, which exhibits an inverse structure, consituted by the functional parts of: 1. RPE decoding, 2. LTP synthesis filtering, 3. STP synthesis filtering, 4. Post-processing.

1. RPE decoding: In the decoder the grid position k, the subsegment excitation maxima $v_{max}(k)$ and the excitation pulse amplitudes $\beta'(k,i)$ are inverse quantised, and the actual pulse amplitudes are computed by multiplying the decoded amplitudes with their corresponding block maxima. The LTP residual model $r'_{LTP}(n)$ is recovered by properly positioning the pulse amplitudes $\beta(k,i)$ according to the initial offset k.

2. LTP synthesis filtering: Firstly the LTP filter parameters (G', D') are inverse quantised to derive the LTP synthesis filter. Then the recovered LTP excitation model $r'_{LTP}(n)$ is used to excite this LTP synthesis filter (G', D') to recover a new subsegment of length N=40 of the estimated STP residual $r'_{STP}(n)$. To do so, the past history of the recovered STP residual $r'_{STP}(n)$ is used, properly delayed by D' samples and multiplied by G' to deliver the estimated STP residual $r''_{STP}(n)$, according to:

$$r''_{STP}(n) = G'.r'_{STP}(n - D'),\qquad(8.17)$$

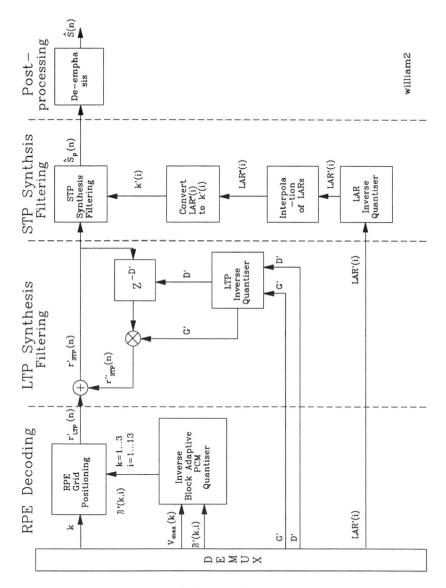

Figure 8.11: Block diagram of the RPE-LTP decoder.

Parameter to be encoded	No. of bits
8 STP LAR coefficients	36
4 LTP Gains G	4 x 2 = 8
4 LTP Delays D	4 x 7 = 28
4 RPE Grid-positions	4 x 2 = 8
4 RPE Block maxima	4 x 6 = 24
4x13=52 Pulse amplitudes	52 x 3 = 156
Total number of bits per 20 ms	260
Transmission bit-rate	13 kbit/s

Table 8.4: Summary of the RPE-LTP bit-allocation scheme.

and then $r''_{STP}(n)$ is used to compute the most recent subsegment of the recovered STP residual, as given below:

$$r'_{STP}(n) = r''_{STP}(n) + r'_{LTP}(n). \tag{8.18}$$

3. STP synthesis filtering: To compute the synthesized speech $\hat{s}(n)$ the PARCOR synthesis is used, where similarly to the STP analysis filtering the reflection coefficients $k(i)$ $i = 1...8$ are required. The $LAR'(i)$ parameters are decoded by using the LAR inverse quantiser to give $LAR''(i)$, which are again linearly interpolated towards the analysis frame edges between parameters of the adjacent frames to prevent abrupt changes in the character of the speech spectral envelope. Finally, the interpolated parameter set is transformed back into reflection coefficients, where filter stability is guaranteed, if recovered reflection coefficients, which fell outside the unit circle are reflected back into it, by taking their reciprocal values. The inverse formula to convert $LAR(i)$ back into $k(i)$ is given by:

$$k(i) = \frac{10^{LAR(i)} - 1}{10^{LAR(i)} + 1}. \tag{8.19}$$

4. Post-processing: The post-processing is constituted by the de-emphasis, using the inverse of the filter $H(z)$ in Equation 8.9.

The summarised RPE-LTP bit allocation scheme is tabulated in Table 8.4 for a period of 20 ms, which is equivalent to the encoding of L=160 samples, while the detailed bit-by-bit allocation is given in [9].

The 260 bits derived have to be reordered according to their subjective importance before error correction coding, as proposed by GSM, and classified into categories of Class 1a, Class 1b and Class 2 in descending order of prominence to facilitate a three-level error protection scheme, as will be highlighted in Section 8.7. Note that this sensitivity order is based on subjective tests. Objective bit-sensitivity analysis based on a combination of segmental signal-to-noise ratios and cepstrum distances results in a similar significance order [13].

Having described the 13 kbps full-rate GSM speech codec, let us now focus our attention on the features of the 5.6 kbps half-rate scheme in the next section.

8.5 The Half-rate 5.6 kbps GSM Speech Codec [14, 21]

8.5.1 Half-rate GSM Codec Outline and Bit Allocation

The in-depth theory of speech coding was the subject of Chapter 3. In this section we adopt a practical - rather than theoretical - approach to the portrayal of the half-rate GSM speech codec, in order to keep the level of treatment sufficiently conceptual for readers having a communications, rather than speech compression vein. Below we briefly highlight the techniques proposed by Gerson et al. [14], which led to the definition of the half-rate GSM standard codec employing a 5.6 kbps so-called Vector Sum Excited Linear Predictive (VSELP) codec [15, 16]. The codec's schematic is shown in Figure 8.12, where two different block-diagrams characterise its operation in four different operational modes. In Mode 0 the codec obeys the schematic portrayed at the top of Figure 8.12, while in the remaining three modes, Modes 1, 2 and 3, it is configured as seen at the bottom of Figure 8.12. The analysis synthesis filter's coefficients are determined every 20 ms and this interval is divided into four 5 ms excitation optimisation subsegments, corresponding to 160 and 40 samples, respectively, when using a sampling frequency of 8 kHz. In our forthcoming discussion we focus our attention on the above-mentioned different operating modes and the corresponding schematics.

The codec's bit allocation scheme is summarised in Table 8.5 for the synthesis modes of $0 - 3$. The speech spectral envelope is encoded by allocating 28 bits/20 ms synthesis frame for the vector quantisation of the reflection coefficients. A so-called soft interpolation bit is used to inform the decoder, whether the current frame's prediction residual energy was lower with or without interpolating the direct form LPC coefficients.

As mentioned before, there are four different synthesis modes, corresponding to different excitation modes, implying the presence of different grades of voicing in the speech signal. As seen in the table, two bits/frame are used for excitation mode selection. The decisions as to what amount of voicing is present and hence which excitation mode has to be used are based on the Long Term Predictor (LTP) gain, which is typically high for highly correlated voiced segments and low for noise-like, uncorrelated unvoiced segments.

In the unvoiced Mode 0 the schematic at the top of Figure 8.12 is used, where the speech is synthesised by superimposing the G_1 and G_2-

Parameter	Bits/frame
LPC coefficients	28
LPC Interpolation flag	1
Excitation Mode	2
Mode 0 :	
Codebook 1 index	$4 \times 7 = 28$
Codebook 2 index	$4 \times 7 = 28$
Modes 1, 2, 3	
LTPD (subframe 1)	8
Δ LTPD (subframes 2, 3, 4)	$3 \times 4 = 12$
Codebook 3 index	$4 \times 9 = 36$
Frame energy E_F	5
Excitation gain-related quantity $[E_s E_1]$	$4 \times 5 = 20$
Total no. of bits	112/20 ms
Bitrate	5.6 kbps

Table 8.5: Bit-allocation scheme of the 5.6 kbps VSELP half-rate GSM codec

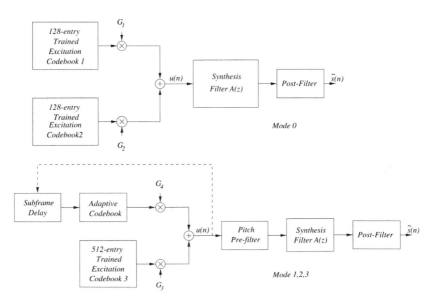

Figure 8.12: Schematic of the 5.6 kbps VSELP half-rate GSM codec, portraying the unvoiced Mode 0 and the voiced Modes 1, 2 and 3.

LTPD range	Resolution
$21 \ldots (22 + \frac{2}{3})$	$\frac{1}{3}$
$23 \ldots (34 + \frac{5}{6})$	$\frac{1}{6}$
$35 \ldots (49 + \frac{2}{3})$	$\frac{1}{3}$
$50 \ldots (89 + \frac{1}{2})$	$\frac{1}{2}$
$90 \ldots 142$	1

Table 8.6: Legitimate non-integer LTPD values and LTP resolution in the 5.6 kbps VSELP half-rate GSM codec.

scaled outputs of two 128-entry trained codebooks in order to generate the excitation signal, which is then filtered through the synthesis filter $A(z)$ and the spectral post-filter. Accordingly, both Excitation Codebooks 1 and 2 have a 7-bit address in each of the 4 subsegments, as shown in Table 8.5.

In Modes 1-3, where the input speech exhibits some grade of voicing, the schematic at the bottom of Figure 8.12 is used. The excitation is now generated by superimposing the G_3-scaled 512-entry trained codebook's output onto that of the G_4-scaled so-called adaptive codebook. The fixed codebook in these modes requires a 9-bit address, yielding a total of 4×9 = 36 coding bits for the 20 ms frame, as seen in Table 8.5. The adaptive codebook delay or long-term predictor delay (LTPD) is encoded in the first subsegment using 8 bits, allowing for 256 integer and non-integer delay positions. In consecutive subframes the LTPD is encoded differentially with respect to the previous subframe's delay, which we indicated as ΔLTPD in Table 8.5. The 4 encoding bits allow for a maximum difference of $[-8, +7]$ positions with respect to the previous LTPD value. The legitimate LTPD values are listed in Table 8.6.

Observe in the table that for low LTPD values a finer resolution is used and the highest resolution is assigned for the range $23 \ldots (34 + \frac{5}{6})$, corresponding to a pitch lag of between 2.875 and 4.35 ms or a pitch frequency of 230-348 Hz.

Returning to Table 8.5, the overall frame energy is encoded with 5 bits, which allows spanning a dynamic range of 64 dB, when using a stepsize of 2 dB and 32 steps. The excitation gains $G_1 - G_4$ are not directly encoded. Instead, the energy of each subframe E_s is expressed normalised by the frame energy E_F, which is then jointly vector quantised with another parameter about to be introduced. Specifically, it was found advantageous to express the relative contribution E_1 of the first excitation component constituted by Codebook 1 at the top of Figure 8.12 in Mode 0, and by the adaptive codebook at the bottom of Figure 8.12 in Modes 1-3 to the overall excitation. Clearly, this relative contribution must be limited to the range of $0 \ldots 1$. Then the parameter pair $[E_s, E_1]$ is vector-quantised using 5 bits/5 ms subsegment, which allowed for 32 possible combinations. Accordingly, Table 8.5 assigns a total of 20 bits/20 ms frame for the encoding of this gain-related information.

8.5.2   Spectral Quantisation in the Half-rate GSM Codec

According to Table 8.5, the codec employs 28-bit vector quantisation (VQ) of the so-called reflection coefficients, where the best set is deemed to be the one which minimises the prediction residual energy. A reduced-complexity version of the so-called Fixed Point Lattice Technique (FLAT) [17, 18] was proposed for the standard, which will be briefly highlighted below.

It would be impractical to use a 2^{28}-entry codebook for both search-complexity and storage-capacity reasons, whence a suboptimum 3-way split-vector implementation was proposed by Gerson [17], where the reflection coefficients k_1-k_3, k_4-k_6 and k_7-k_{10} are stored in separate codebooks. The number of quantisation or codebook address bits is $Q_1 = 11, Q_2 = 9$ and $Q_3 = 8$ bits, respectively. A particularly attractive property of the reflection coefficient based lattice-type predictors is that in the case of the above so-called split-vector quantisers the choice of the current acoustic tube model segment's reflection coefficient quantiser can partially compensate for the quantisation effects of the preceding tube section quantiser.

To elaborate on these issues Gerson, et al. [14] introduced the ingenious concept of pre-quantisation, where in each of the three split codebooks a so-called pre-quantiser using $P_1 = 6, P_2 = 5$ and $P_3 = 4$ bits is invoked. Each vector of the pre-quantiser is associated with a set of vectors in the actual quantiser. For example, each of the $P_1 = 6$ bit quantiser entries is associated with $n_1 = 2^{Q_1}/2^{P_1} = 2^{11}/2^6 = 2^5 = 32$ vectors in the first actual VQ codebook, etc. In order to reduce the overall complexity, the prediction residual error is computed for each of the pre-quantiser vectors at a given acoustic tube model segment and the four vectors resulting in the four lowest error energy values are earmarked. These four vectors are then used as pointers to identify four sets of vectors, which are associated with the earmarked pre-quantiser vectors. The four sets of actual quantised vectors are then exhaustively searched in order to find the set which minimises the prediction residual energy.

This technique results in a substantial complexity reduction. Specifically, instead of searching the $2^{Q_1} = 2^{11} = 2048$-entry codebook storing the reflection coefficients ($k_1 - k_3$), initially the $2^{P_1} = 2^6 = 64$-entry pre-quantiser codebook is searched to find the best four 'pointers', around each of which then the prediction residual is evaluated 32 times, requiring its computation 128 times. For simplicity, assuming an identical evaluation complexity for both steps, the complexity of the full-search was reduced by a factor of $2048/(64+ 128) \approx 10.67$. The corresponding factors for the ($k_4 - k_6$) and ($k_7 - k_{10}$) codebooks are $2^9/(32 + 64) \approx 5.3$ and $2^8/(16 + 64) = 3.2$, respectively.

The reflection coefficients themselves have been reported to have a

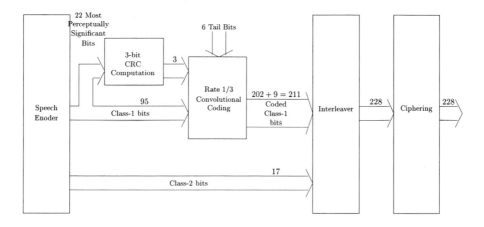

Figure 8.13: The 5.6/11.4 kbps GSM half-rate error protection schematic.

high spectral sensitivity in the vicinity of the unit circle, when $k_i \approx 1$. This may result in a large speech spectrum variation due to the quantisation of the reflection coefficients. Hence a very fine Max-Lloyd quantiser would be required for their quantisation in this domain, instead of uniform quantisation. Therefore two widely used non-linear transformations have been proposed for circumventing this problem, namely the log-area ratios (LAR) and the inverse sine transformation $S_i = \sin^{-1}(k_i)$, which are more amenable to uniform quantisation. The GSM half-rate codec uses the latter, employing an efficient 8-bit representation for the codebook entries, which were generated by uniformly sampling their so-called inverse-sine representations. Let us now breifly consider the error protection strategy used.

8.5.3   Half-rate GSM Error Protection

The error control strategy used is based on the schematic of Figure 8.13, which is quite similar in terms of its philosophy to that of the full-rate or the enhanced full-rate GSM schemes. The 112 bits/20 ms are divided in 95 more sensitive Class-1 bits and 17 more robust Class-2 bits. The most sensitive 22 Class-1 bits are assigned a 3-bit cyclic redundancy checking (CRC) pattern, which is then invoked by the decoder for initiating bad frame masking. Bad frames may be encountered due to channel errors, or due to fast associated control channel messages replacing a speech frame, for example in order to signal an urgent hand-over request. In this case the speech frame is wiped out by this fast associated control channel message and at the decoder it has to be replaced by a post-processed speech segment.

As displayed in Figure 8.13, the 17 robust Class-2 bits are unprotected, while the 95 Class-1 bits are 1/3-rate, constraint-length 7 convolutionally encoded. Here we note that the definition of constraint length in this case includes the current input bit of the encoder plus the six shift-register stages. Hence six tailing bits are necessary for flushing the encoder's shift-registers after each transmission burst, in order to prevent error propagation across transmission frame boundaries. We note, however, that a so-called punctured code was employed, where the effective coding rate becomes 1/2, due to puncturing, i.e. obliterating some of the encoded bits. The 95 Class-1 bits and the 6 tailing bits yield 101 bits, which generate 202 punctured convolutionally coded bits, while the 3 CRC bits are 1/3-rate coded, yielding a total of 211 bits. After concatenating the 17 unprotected bits the total rate becomes 228 bits/20 ms=11.4 kbps, which is exactly half of that of the full-rate and enhance full-rate systems.

Having highlighted the basic features of the 5.6 kbps half-rate GSM codec, we now describe some of the features of the 12.2 kbps so-called enhanced full-rate GSM speech codec in the next section.

8.6   The 12.2 kbps Enhanced Full-rate GSM Speech Codec [19–21]

8.6.1   Enhanced Full-rate GSM Codec Outline

This section gives a brief account of the operation of the 12.2 kbps enhanced full-rate GSM speech codec, which will replace the 13 kbps RPE speech codec. This scheme was standardised by the European Telecommunications Standardisation Institute (ETSI) in 1996. Here we follow the approach of Salami et al. [19, 20] and the interested reader is referred to [20] for a more indepth discussion. The codec employs the successful ACELP excitation model invented in 1987 by Adoul et al. at Sherbrooke University [22], which was detailed in Chapter 3. The enhanced full-rate GSM scheme uses a bit rate of 10.6 kbps for channel coding, resulting in a channel coded rate of 22.8 kbps, similar to the 13 kbps RPE GSM codec, which was the topic of Section 8.4 and was characterised by the schematics of Figures 8.10 and 8.11.

The enhanced full-rate GSM (EFR-GSM) encoder schematic is portrayed in Figure 8.14, while that of the decoder is displayed in Figure 8.17, both of which will be detailed below. Similarly to the RPE GSM encoder of Figure 8.10, the input speech is initially pre-emphasised using a high-pass filter, in order to augment the low-energy, high-frequency components, before the speech signal is processed. Observe in Figure 8.14 that as usual, the spectral quantisation is carried out on a frame-by-frame basis, while the excitation optimisation is carried out on a subsegment-by-subsegment

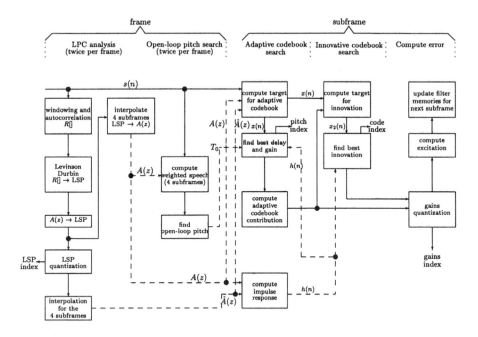

Figure 8.14: Enhanced full-rate 12.2 kbps GSM encoder schematic.

basis, although we note at this early stage that the spectral quantisation is quite original.

The codec's bit-allocation scheme is summarised in Table 8.7, while the rationale behind using the specified number of bits will be detailed during our forthcoming discourse. The 38 LSF quantisation bits per 20 ms constitute a 1.9 kbps bit rate contribution, which is typical for medium-rate codecs, although the quantisation scheme to be highlighetd below is unconventional. The fixed ACELP codebook (CB) gains are quantised using 5 bits/subframe, while the fixed ACELP codes are represented by 35 bits per subframe, which again will be justified below with reference to Table 8.7. The adaptive codebook index, corresponding to the pitch-lag, is represented by 9 bits, catering for 512 possible positions in the 1st and 3rd subframes using a very fine over-sampling by a factor of six in the low-delay region. In the 2nd and 4th subframes the pitch-lag is differentially encoded with respect to the odd subframes, again employing an oversampling by six in the low-delay domain.

We note that historically the DoD codec was the first scheme to invoke the above-mentioned differential coding of the pitch-lag and oversampling in the low-lag pitch domain. These measures became fairly widely employed in state-of-the-art codecs, despite the inherent error sensitivity of differen-

Parameter	1. & 3. subfr.	2. & 4. subfr.	No. of bits	Total (kbps)
Two LSF Sets			38	1.9
Fixed CB Gain	5	5	4·5=20	1
ACELP Code	35	35	4·35=140	7
Adaptive CB Index	9	6	2·9+2·6=30	1.5
Adaptive CB Gain	4	4	16	0.8
Total			244/20ms	12.2

Table 8.7: 12.2 kbps enhanced full-rate GSM codec bit allocation

tial coding. The high resolution pitch-lag coding of low values is important,
since it is beneficial to ensure a more-or-less constant relative pitch resolution, rather than a constant absolute resolution, as in the case of uniformly
applied 125 μs sample-spaced pitch encoding. Lastly, the pitch-gains are
encoded using four bits per subframe. Below we will consider most of the
above-mentioned operations of Figure 8.14 in more depth.

8.6.2   Enhanced Full-rate GSM Encoder

8.6.2.1   Spectral Quantisation and Windowing in the Enhanced Full-rate GSM Codec

Let us initially consider the spectral quantisation employed in the EFR-
GSM codec, where 10th-order LPC analysis is invoked twice for each 20
ms speech frame, upon using two different 30 ms-duration asymmetric windows, which will be justified below. In contrast to other state-of-the-art
speech codecs, such as the 8 kbps ITU G.729 ACELP codec's window function, where a 5 ms or 40-sample look-ahead was used, the EFR-GSM codec
employs no 'future speech samples' i.e. no look-ahead in the filter coefficient computation and both asymmetric window functions act on the
same set of 240 speech samples, corresponding to the 30 ms analysis interval. Whereas in the 10 ms-framelength G.729 codec an additional 5
ms look-ahead delay was deemed acceptable in exchange for a smoother
speech spectral envelope evolution, in the 20 ms-framelength EFR-GSM
scheme this was deemed unacceptable. This implies that there is a 10 ms
or 80-sample 'look-back' interval in the window functions.

Before specifying the shape of the window functions, let us state the
rationale behind using two LSF sets, which are used for the 2nd and 4th
subframes, respectively. Accordingly, the peak of the first window $w_1(n)$ of
Figure 8.15 is concentrated near the centre of the 2nd subframe, while that
of the second window $w_2(n)$ is near the centre of the 4th subframe. Hence
the latter has to exhibit a rapidly decaying slope, given that no look-ahead
is employed. For the 1st and 3rd subframes the LSFs are interpolated on
the basis of the surrounding subframes. Specifically, the first window $w_1(n)$

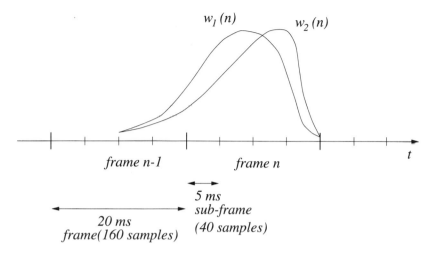

Figure 8.15: Stylised enhanced full-rate GSM window functions.

is constituted by two Hamming-window segments of different sizes, which is given below:

$$
w_1(n) = \begin{cases} 0.54 - 0.46 \cdot \cos \frac{\pi n}{L_1 - 1}, & n = 0, \ldots L_1 - 1 \\ 0.54 - 0.46 \cdot \cos \frac{\pi(n - L_1)}{L_2 - 1}, & n = L_1, \ldots L_1 + L_2 - 1 \end{cases} , \quad (8.20)
$$

where the parameters $L_1 = 160$ and $L_2 = 80$ were standardised. Although this window is asymmetric, it is gently decaying towards both ends of the current 20 ms frame, as seen in Figure 8.15. By contrast, since the centre of gravity of the second window is close to the beginning of the frame, it has to be tapered more abruptly, which is facilitated by using a short raised-cosine segment, as seen below:

$$
w_2(n) = \begin{cases} 0.54 - 0.46 \cdot \cos \frac{\pi n}{L_1 - 1}, & n = 0, \ldots L_1 - 1 \\ \cos \frac{2\pi(n - L_1)}{4L_2 - 1}, & n = L_1, \ldots L_1 + L_2 - 1 \end{cases} , \quad (8.21)
$$

where the parameters $L_1 = 232$ and $L_2 = 8$ were employed.

As seen in Figure 8.14, the autocorrelation coefficients are computed from the windowed speech and the Levinson-Durbin algorithm is employed in order to derive both the reflection and the linear predictive coefficients, which describe the speech spectral envelope with the help of the $A(z)$ polynomial. Further details of Figure 8.14 concerning for example the pitch lag search and excitation optimisation will be unravelled during our later discussions. The LPC coefficients are then converted to LSFs and quantised using the so-called split matrix quantiser (SMQ) of Figure 8.16, which is considered next.

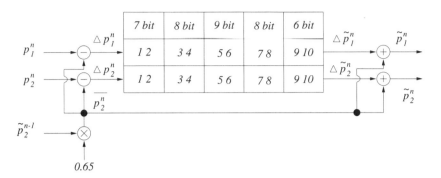

Figure 8.16: The 38-bit split matrix LSF quantisation of the sets generated using windows $w_1(n)$ and $w_2(n)$ of Figure 8.15 in the 12.2 kbps enhanced full-rate GSM codec.

First the long-term mean of both LSF vectors is removed, yielding the zero-mean LSF vectors p_1^n and p_2^n for frame n, corresponding to the two windows in Figure 8.15. Then both LSF sets of frame n are predicted from the previous quantised LSF set $\tilde{p}_2^{n-1}$, taking into account their long-term correlation of 0.65, as portrayed in Figure 8.16. Both LSF difference vectors are then input to the split matrix quantiser. Specifically, the LSFs of both vectors are paired, as suggested by Figure 8.16, creating a 2×2 submatrix from the first two LSFs of both LSF vectors and quantising them by searching through a 7-bit, 128-entry codebook. Similarly, the third and fourth LSFs of both LSF vectors are paired and quantised using the 8-bit, 256-entry codebook of Figure 8.16, etc. Observe that the most important LSFs corresponding to the medium frequency range are quantised using a larger codebook than those towards the lower and higher frequencies. Finally, after finding the best-matching codebook entries for all 2×2 submatrices the previous subtracted predicted values are added to them, in order to produce both quantised LSF vectors, namely $\tilde{p}_1^n$ and $\tilde{p}_2^n$, respectively.

8.6.2.2   Adaptive Codebook Search

A combined open- and closed-loop pitch analysis is used, which was summarised by Salami et al. [19] as follows:

- As seen in Figure 8.14, based on the weighted speech an open-loop pitch search is carried out twice per 20 ms frame or once every two subframes, favouring low pitch values in order to avoid pitch doubling. In this search an integer sample-based search is used and the open-loop lag T_o is identified.

- Then a closed-loop search for integer pitch values is conducted on a subframe basis. This is restricted to the range $[T_o \pm 3]$ in the 1st

Track	Pulses	Positions
1	p_0, p_1	0, 5, 10, 15, 20, 25, 30, 35
2	p_2, p_3	1, 6, 11, 16, 21, 26, 31, 36
3	p_4, p_5	2, 7, 12, 17, 22, 27, 32, 37
4	p_6, p_7	3, 8, 13, 18, 23, 28, 33, 38
5	p_8, p_9	4, 9, 14, 19, 24, 29, 34, 39

Table 8.8: 12.2 kbps enhanced full-rate GSM codec's ACELP pulse allocation. ©IEEE, Salami et al. [19].

and 3rd subframes, in order to maintain a low search complexity. As to the 2nd and 4th subframes, the closed-loop search is concentrated around the pitch values of the previous subframe, in the range of $[-5 \ldots + 4]$.

- Finally, fractional pitch delays are also tested around the best closed-loop lag value in the 2nd and 4th subframes, although only for the pitch delays below 95 in the 1st and 3rd subframes, corresponding to pitch frequencies in excess of about 84 Hz.

- Having determined the optimum pitch lag, the adaptive codebook entry is uniquely identified, while its gain is restricted to the range of $[0 \ldots 1.2]$ and quantised using four bits, as seen in Table 8.7.

Let us now consider the optimisation of the fixed codebook in the next subsection.

8.6.2.3   Fixed Codebook Search

Again, the principles of ACELP coding [22] were detailed in Chapter 3, hence here only a rudimentary overview is given. As shown in Table 8.7, 35 bits per subsegment are allocated to the ACELP code. The 5 ms, 40-sample excitation vector hosts 10 non-zero excitation pulses, each of which can take the values ±1. Salami et al. [19] subdivided the 40-sample subframe into five so-called tracks, each comprising two excitation pulses. The two pulses in each track are allowed to be co-located, potentially resulting in pulse amplitudes of ±2. The standardised pulse positions are summarised in Table 8.8. Since there are eight legitimate positions for each excitation pulse, three bits are necessary for signalling each pulse position. Given that there are ten excitation pulses, a total of 30 bits are required for their transmission. Furthermore, the sign of the first pulse of each of the five tracks is encoded using one bit, yielding a total of 35 bits per subsegment. The sign of the second pulse is inherently determined by the order of the pulse positions, an issue elaborated on in references [19] and [20]. The 3-bit pulse positions were also Gray-coded, implying that adjacent pulse positions are different only in one bit position. Hence a bit-error results in the closest

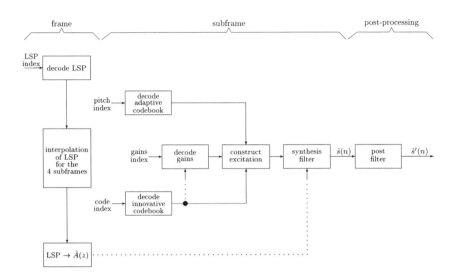

Figure 8.17: Enhanced full-rate GSM decoder schematic.

possible excitation pulse position to the one that was transmitted. This ACELP codebook is then invoked in order to generate the 20 ms synthetic speech frame, which is compared to the original speech segment in order to identify the best excitation vector.

At the decoder portrayed in Figure 8.17 the received codec parameters are recovered and the synthetic speech is reconstructed. Specifically, the decoded LSF parameters are interpolated for the individual subframes. Both the fixed and adaptive codebook vectors are regenerated and with the aid of the corresponding gain factors the excitation signal is synthesised. The excitation is then filtered through the synthesis filter and the post filter, in order to generate the synthetic speech.

Following the above brief description of the enhanced full-rate GSM codec we now consider the associated channel coding and interleaving aspects of the original full-rate GSM system in the next section.

8.7   Channel Coding and Interleaving

The TDMA frame-hierarchy used in the GSM system has been highlighted with the help of Figure 8.2 in Section 8.3, where we have seen how the timeslots are multiplexed into TDMA frames, multiframes, superframes and hyperframes. The major unknown is now how speech and data bits are mapped onto timeslots of the TDMA frames. First we focus our attention on the embedded forward error correction (FEC) and interleaving scheme used for speech transmissions.

8.7.1 FEC for the 13kbps Speech Channel

As explained in Section 8.4, the RPE-LTP codec delivers 260 bits/20 ms at a bit rate of 13 kbit/s according to the bit-allocation summarised in Table 8.4. The detailed mapping of the sensitivity-ordered Class 1a (C1a), Class 1b (C1b) and Class 2 (C2) speech bits onto a normal burst (NB) is given in Figures 8.18 and 8.19.

As seen in Figure 8.18(b) [23], the first 50 C1 bits $(d_0 \ldots d_{49})$ are protected by a weak error detecting block code. The code implemented is a (53,50) shortened, systematic, cyclic code with the generator polynomial $G_4(D) = D^3 + D + 1$. The corresponding encoder is displayed in Figure 8.20. Observe that the taps of the linear shift register encoder are allocated at the positions specified by the generator polynomial $G_4(D)$. Since a systematic encoding rule has been adopted, the switch SW is closed for the duration of the first fifty clock pulses, and the information bits enter the encoder as well as being passed on to the 'Reordering & Tail' block in Figure 8.4. After fifty clock pulses the switch SW is opened and the parity bits P_0, P_1 and P_2 exit the encoder, while the rest of the bits, i.e., the C1b and C2 bits are unaltered.

At this stage a first interleaving step depicted in Figure 8.18(c) is carried out. Namely, the C1a and C1b bits with even indices, i.e., $d_0, d_2 \ldots d_{180}$, are collected in the first part of the data word, followed by the three parity bits P_0, P_1 and P_2. Then the C1a and C1b bits with odd indices, i.e., $d_1, d_3 \ldots d_{179}$, are stored in the buffer, followed by the 78 uncoded C2 bits.

Observe also that four zero tailing bits have been added at the end of the C1 section, which are necessary to reset the constraint-length $K = 5$, rate=1/2 convolutional encoder CC(2,1,5) to be employed to protect the C1 bits. Now the 189 C1 bits $(u_0, u_1 \ldots u_{188})$ are encoded by the powerful half-rate convolutional code in the block designated by CC(2,1,5) in Figure 8.4. The CC(2,1,5) codec uses the generator polynomials $G_0 = 1 + D^3 + D^4$ and $G_1 = 1 + D + D^3 + D^4$ and the encoder structure displayed in Figure 8.21 in the full-rate speech channels. The total frame length amounts to $2 \cdot 189 + 78 = 456$ bits, as demonstrated by Figure 8.18(d).

Thereafter the 456 bits long encoded frame is partitioned into eight 57 bits long sub-blocks $(B_0 \ldots B_7)$, as shown in Figure 8.19a. The interleaving seen in Figure 8.4 is carried out by assigning the coded bits $c(n, k)$ into the interleaved sequence $i(B, j)$ as stated below:

$$i(B, j) = c(n, k), \tag{8.22}$$

where $k = 0, 1 \ldots 455$ is the bit index in the nth coded frame, displayed in

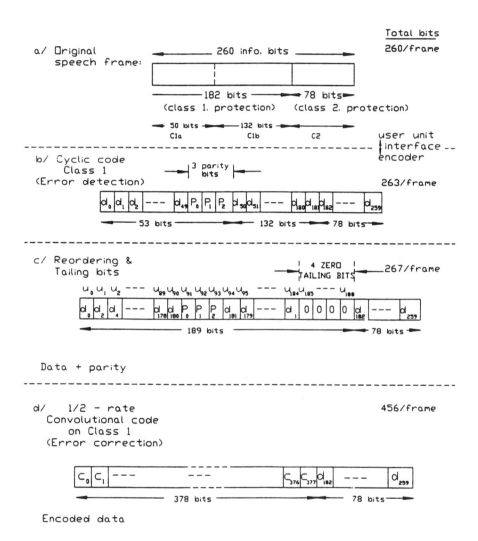

Figure 8.18: Forward error correction coding in TCH/FS.

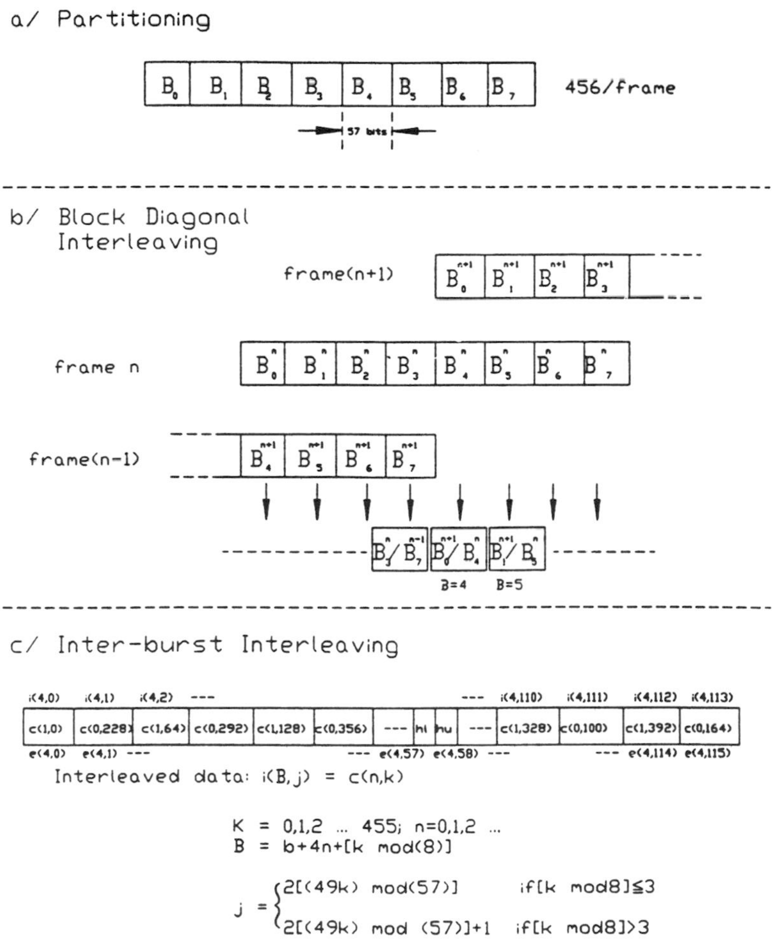

a/ Partitioning

b/ Block Diagonal Interleaving

c/ Inter-burst Interleaving

Interleaved data: $i(B,j) = c(n,k)$

$K = 0,1,2 \ldots 455; \quad n=0,1,2 \ldots$

$B = b+4n+[k \bmod(8)]$

$$j = \begin{cases} 2[(49k) \bmod(57)] & \text{if}[k \bmod 8] \leq 3 \\ 2[(49k) \bmod(57)]+1 & \text{if}[k \bmod 8] > 3 \end{cases}$$

Figure 8.19: Partitioning and interleaving in TCH/FS.

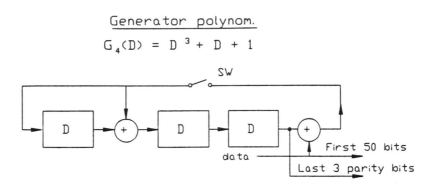

Figure 8.20: The C1a (53,50) systematic, cyclic block encoder.

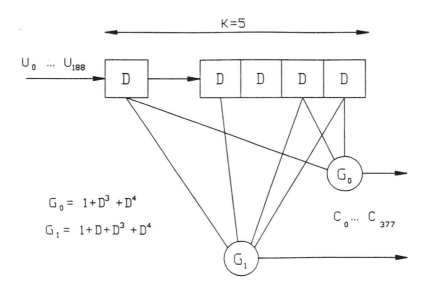

Figure 8.21: The CC(2,1,5) TCH/FS convolutional encoder.

Figure 8.18(d), $n = 0, 1, 2 \ldots$ is the frame index, $B = b + 4n + [k \bmod 8]$ is the subblock index with an initial value of b and

$$j = \begin{cases} 2[(49k) \bmod (57)] & \text{if } [k \bmod 8] \leq 3 \\ 2[(49k) \bmod (57)] + 1 & \text{if } [k \bmod 8] > 3 \end{cases} \qquad (8.23)$$

is the bit index in the interleaved 57-bit subblocks. Each of the $N = 57$ bits long subblocks disperses its bits over eight consecutive such subblocks. Each of the 57 bits being spread will be followed or preceded by the 'position-equivalent' bit of the subblock half-a-frame, i.e., four 57-bit subblocks apart. This is represented by the block diagonal interleaving shown in Figure 8.19(b) and facilitated by the shift $4n$ in $B = b + 4n + [k \bmod 8]$. Two originally adjacent $c(n, k)$ bits will be reordered so that their separation in the reordered frame becomes $2[(49k) \bmod (57)]$ or $2[(49k) \bmod (57)] + 1$ depending on the value of $[k \bmod 8]$. Since 49 and 57 are relative primes, this separation distance becomes pseudo-random, which is advantageous in terms of combating periodic error bursts. Observe also that for fixed b and n values B has a periodicity of eight in terms of k, which means that every eighth bit of $c(n, k)$ is directed to the same subblock B with a pseudo-random offset. Due to the term $4n$ in the definition equation $B = b + 4n + [k \bmod 8]$ four 114 bits long inter-block interleaved blocks are constructed from a 456 bits long speech frame. However, the 114-bit interleaved blocks are derived from two consecutive 456-bit speech frames by using additional block-diagonal interleaving, as displayed in Figure 8.19(b) and explained as follows.

Let us assume that frames $n = 0$ and $n = 1$ are being interleaved and $b = 0$. Then, for example, $c(0, 4), c(0, 12), c(0, 20), \ldots c(0, 452)$ will be mapped onto $i(4, 51), i(4, 37), i(4, 23), \ldots i(4, 65)$, respectively. Observe that even bits of block $n = 0$ are mapped onto odd positions of the fourth 114 bits long interleaved burst. It is also easily seen that $i(4, 1) = c(0, 228), i(4, 3) = c(0, 228+64), i(4, 5) = c(0, 228+128)$, etc. On the other hand, the even bits of the fourth interleaved burst $(B = 4)$ are supplied by the even bits of the speech block $n = 1$ as follows: $i(4, 0) = c(1, 0), i(4, 2) = c(1, 64), i(4, 4) = c(1, 128)$, etc. In other words, bits originally 228 positions apart follow each other in the interleaved bursts, as evidenced by Figure 8.19(c), where the final step of including the stealing flags $e(B, 57) = hl(B)$ and $e(B, 58) = hu(B)$ is portrayed as well. These bits are used in fast associated control channel (FACCH) signalling. Specifically, $hl = 0, hu = 0$ indicates that the current frame carries traffic channel (TCH) information, $hu = 1$ means that every even-numbered bit is stolen by the FACCH for signalling information, while $hl = 1$ represents that the odd bits are signalling information.

The inner working of the reordering and interleaving process is best understood by generating three consecutive coded frames and programming the interleaving formula in a spreadsheet. The 116-bit bursts are then encrypted, split into two 58 bits long sequences and amalgamated in a normal burst, as shown in Figure 8.2 to create a 577 μs, i.e., 156.25-bit duration long TDMA timeslot, which is transmitted at a burst rate of approximately 271 kHz.

8.7.2   FEC for Data Channels

As mentioned earlier, the full- and half-rate data traffic channels standardised in the GSM system are: TCH/F9.6, TCH/F4.8, TCH/F2.4, as well as TCH/H4.8, TCH/H2.4. As a representative example, here we only consider the format of the TCH/F9.6.

The user unit's interface defined in R.04.21 delivers blocks of 60 bits every 5 ms in accordance with the modified CCITT V.110 standard derived from the so-called 80 bits/5 ms frame or 16 kbit/s V.110 standard. The 80 bits of the original V.110 Recommendation entail a combination of 48 data bits (D1 ... D48), eight fixed zeros and nine fixed ones, i.e., a total of 17 synchronisation bits, three bits (E1, E2, E3) to transmit the code of the user data-rate, four bits (E4 ... E7) for network independent clocking and multiframe synchronisation and 11 channel control bits. The 60 bits/5 ms modified V.110 standard frame is yielded by discarding the 17 fixed synchronisation bits (ones and zeros), as well as the data-rate code specified by the bits (E1, E2, E3), while still keeping a channel capacity of (12 - 9.6) kbps=2.4 kbps for control channel information, such as RS-232 or V.24 standard signalling.

The burst structure of the TCH/F9.6 9.6 kbit/s full-rate traffic channel is explained by referring to Figure 8.22. Four consecutive 60-bit data blocks are arranged to constitute a 240-bit information block, which is followed by four zero tailing bits for resetting the subsequent punctured convolutional codec after the transmission of a frame. The 240-bit data blocks are encoded by the half-rate, constraint length $K = 5$ punctured convolutional code PCC(2,1,5) using the same generator polynomials $G_0 = 1 + D^3 + D^4$ and $G_1 = 1 + D + D^3 + D^4$, as in the speech channel. This PCC(2,1,5) convolutional code produces 488 encoded bits from the 244 data bits, but the following 32 coded bits are consistently punctured, i.e., not transmitted: $b(11 + 15j)$; $j = 0, 1, \ldots 31$. (On details of puncturing see Section 4.6.) The 456 encoded bits $b(0) \ldots b(455)$ are then mapped onto four consecutive 114-bit TDMA bursts $(K, K + 1, K + 2 and K + 3)$ using the mapping rule:

$$c(K, k) = b(k) \tag{8.24}$$
$$c(K + 1, k) = b(k + 114) \tag{8.25}$$
$$c(K + 2, k) = b(k + 228) \tag{8.26}$$
$$c(K + 3, k) = b(k + 342) \quad k = 0, 1 \ldots 113. \tag{8.27}$$

The stealing flags hl and hu used in the full-rate speech channel TCH/FS are included also here in the centres of the bursts, as seen in Figure 8.22(b) and have the same interpretation as in TCH/FS. The encoded bits are now

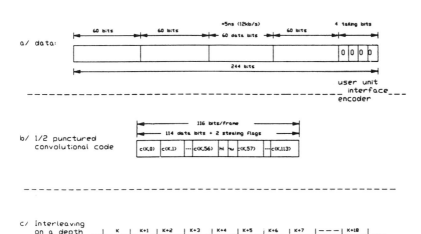

Figure 8.22: TCH/F9.6 FEC and burst structure.

reordered according to the following inter-burst interleaving rule:

$$i(K, j) = c(n, k) \tag{8.28}$$

$$K = K_0 + n + (k \bmod 19) \tag{8.29}$$

$$j = (k \bmod 19) + 19(k \bmod 6) \quad k = 0 \dots 113, \tag{8.30}$$

where k is the bit-index in the 114-bit encoded bursts, n is the encoded burst-index, K is the interleaved burst-index and j is the bit-index in the interleaved burst. Inter-burst interleaving is described in general in Section 4.9, and is known to possess good randomising properties in dispersing bursty channel errors, particularly if the interleaving memory is sufficiently long.

In the specific scheme highlighted above $N \cdot B = 19 \cdot 6 = 114$ bits of an encoded burst are dispersed over $N = 19$ consecutive interleaved bursts, including the current one, while donating $B = 6$ bits to each one of them. A representative example of mapping the encoded bits of the Kth encoded burst $c(K, k), k = 0 \dots 113$ onto bits of the subsequent 19 interleaved bursts is given in Figure 8.22(c). Viewing the mapping from a different angle, bits $0, 19, 38, 57$, etc. of the Kth encoded burst are mapped onto itself, then bits $1, 20, 39, 58$ etc. onto burst $(K + 1)$, etc. This arrangement disperses 114 bits of a burst over $19 \cdot 114 = 2166$ bits, which appears sufficiently long when combined with frequency hopping to randomise the bursty error statistics even for the slowly fading received signal envelopes of pedestrians as well.

8.7.2.1   Low-Rate Data Transmission

The 4.8 kbit/s full-rate data channel (TCH/F4.8) assumes 60 data bits from the user unit every 10 ms, as opposed to 5 ms in the previously considered TCH/F9.6. Furthermore, an $R=1/3$ rate convolutional encoder with the generator polynomials $G_1 = 1 + D + D^3 + D^4$, $G_2 = 1 + D + D^2 + D^4$, $G_3 = 1 + D + D^2 + D^3 + D^4$ is deployed combined with an inter-burst interleaver similar to that described for the TCH/F9.6 full-rate channel. The half-rate 4.8 kbit/s TCH/H4.8 channel also assumes 60 bits/10 ms from the user interface, but uses the generator polynomials and interleaving scheme of TCH/F9.6. Details of these channels along with those of other low-rate channels not discussed here are readily found in Recommendation 05.03.

8.7.3   FEC in Control Channels

As seen in Figure 8.5 of Section 8.3, the 184-bit control channel messages are delivered to the FEC encoder, where systematic shortened binary cyclic FIRE coding, using the generator polynomial $G_5(D) = (D^{23} + 1)(D^{17} + D^3 + 1) = D^{40} + D^{26} + D^{23} + D^{17} + D^3 + 1$ is carried out. Hence the 184 information bits are followed by 40 parity bits and 4 zero-valued so-called tailing bits, in order to periodically clear the memory of the subsequent convolutional codec, yielding a 228-bit sequence. Then the $R=1/2$, $K=5$ convolutional code CC(2,1,5) is employed using the generator polynomials $G_0 = 1 + D^3 + D^4$ and $G_1 = 1 + D + D^3 + D^4$ identical to those of TCH/FS, delivering 456 encoded bits. These blocks are then reordered and interleaved exactly as specified for TCH/FS, and the stealing flags are set to $hl = 1$, $hu = 1$ for SACCHs to indicate that no frame stealing is taking place. The block and convolutional encoding procedures, as well as the interleaving schemes used in BCCH, PCH, AGCH and SDCCH are perfectly identical to those described for the SACCH and are more powerful than those of speech channels due to the additional outer (224, 184) block coding deployed.

The 456 encoded bits of a FACCH are mapped onto 8 consecutive 114-bit bursts, as explained for TCH/FS, stealing even and odd bits in the first and last four bursts, respectively. This clearly implies that the present 20 ms, 456-bit speech frame is wiped out by an urgent FACCH message, such as a hand-over command. Hence, in contrast to SACCHs, for FACCHs $hl = 0$, $hu = 1$ is set in the first four 114-bit bursts, where even-indexed bits are stolen from the traffic channel, and $hl = 1$, $hu = 0$ is set in the last four 114-bit bursts of a traffic channel to signal that odd bits are stolen. This way an FEC-protected 456-bit link is created for the transmission of 184 FACCH bits. Again, since the control information requires higher

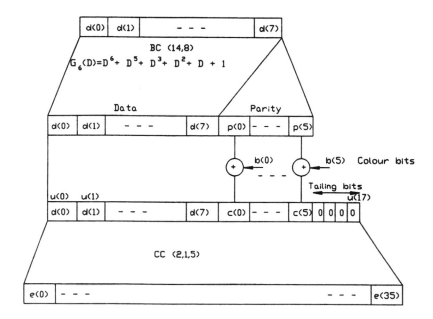

Figure 8.23: FEC in random access control channels.

integrity than speech channels, this link is enhanced by the deployment of the (224,184) FIRE code, which provides an additional layer of error correction when compared to speech and data channels.

Random Access Channels (RACHs) have different message and coding formats, as portrayed in Figure 8.23. The eight RACH information bits $d(0) \ldots d(7)$ are input to a simple systematic cyclic shift-register encoder characterised by the feedback generator polynomial $G_6(D) = D^6 + D^5 + D^3 + D^2 + D + 1$, yielding six parity bits $p(0) \ldots p(5)$. The six bits of the Base Station Identifier Code (BSIC) $b(0) \ldots b(5)$, to which random access is intended are bitwise modulo 2 added to $p(0) \ldots p(5)$ to deliver six so-called colour bits $c(0) \ldots c(5)$:

$$c(k) = b(k) \oplus p(k), \qquad k = 0 \ldots 5 \qquad (8.31)$$

where $b(0)$ is the most significant bit (MSB) of the PLMN colour code and $b(5)$ is the least significant bit (LSB) of the BS colour code. The sequence $d(0) \ldots d(7), c(0) \ldots c(5)$ is then followed by four zero tailing bits to periodically reset the subsequent convolutional codec, and this 18-bit codeword is then convolutionally encoded using the $R=1/2$, $K=5$ CC(2,1,5) code by means of the generator polynomials $G_0 = 1 + D^3 + D^4$ and $G_1 = 1 + D + D^3 + D^4$, as explained for TCH/FS and shown in Figure 8.23.

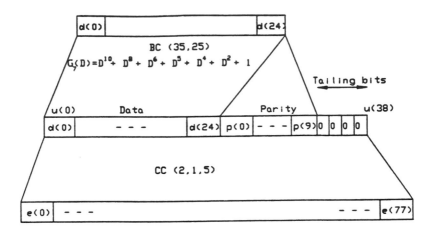

Figure 8.24: FEC in synchronisation channels.

The burst carrying the Synchronisation Channel (SCH) information on the downlink BCCH is constituted by the information bits $d(0) \ldots d(24)$, parity bits $p(0) \ldots p(9)$ and four resetting zero tailing bits for the subsequent CC(2,1,5) code, as demonstrated by Figure 8.24. The systematic cyclic shift-register encoder defined by the generator polynomial $G_7 = D^{10} + D^8 + D^6 + D^5 + D^4 + D^2 + 1$ outputs the parity bits $p(0) \ldots p(9)$ and the resulting 38 bits are encoded using the CC(2,1,5) code known from the TCH/FS. For a comprehensive catalogue of various FEC generator polynomials used in speech, data and control channels, see Table 8.9.

8.7.4 FEC Performance

In order to conclude the GSM error correction coding section, we conducted a number of experiments via narrowband flat-fading Rayleigh channels to assess the BER performance of the full-rate speech and data channels in contrast to a wide variety of bench mark schemes [23,24]. The simulations were carried out using minimum shift keying (MSK) and a Rayleigh-fading envelope with a propagation frequency of 900 MHz and vehicular speed of 30 mph. Our findings are summarised in Figures 8.25 and 8.26 for the speech and data channels, respectively. The overall coding and interleaving delay was close to those standardised for the speech and data channels, i.e., 456 bits and 2-3000 bits, respectively, to provide a fair basis for comparisons. The overall coding rate was $R=0.5$, but some of the bench markers had two layers of coding and interleaving to efficiently randomise the bursty error statistics and provide error detection as well, as explained in Section 4.23.

For the speech channel we have used the following bench mark systems:

$G_0 = D^4 + D^3 + 1$	TCH/FS, TCH/F9.6, TCH/H4.8, SACCH, FACCH, SDCCH, BCCH, PCH, AGCH, RACH, SCH
$G_1 = D^4 + D^3 + D + 1$	TCH/FS, TCH/F9.6, TCH/H4.8, SACCH, FACCH, SDCCH, BCCH, PCH, AGCH, RACH, SCH, TCH/F4.8, TCH/F2.4, TCH/H2.4
$G_2 = D^4 + D^2 + 1$	TCH/F4.8, TCH/F2.4, TCH/H2.4
$G_3 = D^4 + D^3 + D^2 + D + 1$	TCH/F4.8, TCH/F2.4, TCH/H2.4
$G_4 = D^3 + D + 1$	TCH/FS
$G_5 = D^{40} + D^{26} + D^{23} + D^{17} + D^3 + 1$	SACCH, FACCH, BCCH, PCH, AGCH, SDCCH
$G_6 = D^6 + D^5 + D^3 + D^2 + D + 1$	RACH (uplink)
$G_7 = D^{10} + D^8 + D^6 + D^5 + D^4 + D^2 + 1$	SCH (downlink BCCH)

Table 8.9: Summary of Generator Polynomials Used for Data, Speech and Control Channels.

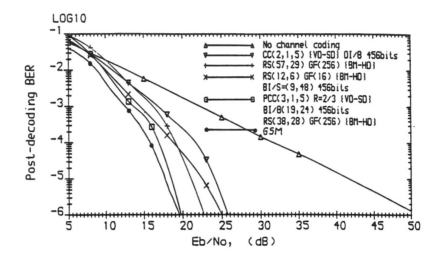

Figure 8.25: BER vs. E_b/N_0 performance of benchmark schemes for TCH/FS.

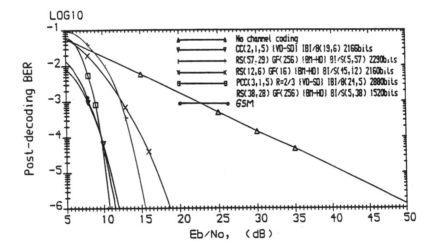

Figure 8.26: BER vs. E_b/N_0 performance of benchmark schemes for TCH/F9.6.

1. The single layer convolutional code CC(2,1,5) with parameters recommended in the GSM system, but the interleaving scheme used is a simple diagonal bit interleaver (DI/B) over the specified 456-bit interval. In the decoder Viterbi decoding with soft decisions (VD-SD) has been used.

1) Another single layer arrangement deployed is the Reed-Solomon RS-(57,29) codec over Galois Field GF(256) decoded by the Berlekamp-Massey hard decision (BM-HD) decoding method.

2) Also a short Reed-Solomon codec, the RS(12,6) codec over GF(16) associated with a symbol block interleaver (BI/S) has been investigated.

3) A powerful concatenated coding arrangement is constituted by the combination of the 2/3-rate punctured convolutional inner codec PCC(3,1,5) with constraint length $K=5$ and the outer layer RS(38,28) codec over GF(256). The convolutional code has been decoded by using VD-SD and the block code by BM-HD.

4) Finally the GSM 05.03 codec has been simulated and compared to our bench markers.

For speech channels the following observations can be made. It is clear that the GSM scheme is the most powerful one in terms of its BER versus E_b/N_0 performance, where E_b represents the bit-energy and N_0 the one-sided noise spectral density. Observe the dramatic performance increase due to the sophisticated interleaving scheme deployed, when compared to the curve representing our first bench marker using the same codec with less powerful interleaving. The RS(12,6) code performs better than the RS(57,29) code above $E_b/N_0 = 20$ dB, while below this cross-over point the situation is reversed. The best contender among our bench-mark systems is the concatenated PCC$(3, 1, 5)$/RS$(38, 28)$ scheme giving very similar performance to that of the GSM system around $E_b/N_0 = 20$ dB. It is worth emphasising that our concatenated scheme has the additional advantage of reliable error detections, which is very important when deploying speech post-enhancement algorithms [32]. In Figure 8.26 the performance of data communication channels has been compared. The coding schemes implemented are identical to those used for speech channels having merely longer interleaving periods. Since for the duration of 2500 bits the channel can be considered memoryless, a significantly increased performance is achieved for all the schemes studied. However, at this long interleaving period the performance differences between the two single layer convolutional codes are negligible. Furthermore, above $E_b/N_0 = 10$ dB the concatenated scheme has the highest performance with the favourable error detection capability, which can be exploited in automatic repeat request (ARQ) systems.

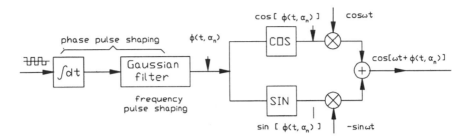

Figure 8.27: GMSK modulator schematic diagram.

8.8   Transmission and Reception

The family of constant envelope, continuous phase modulation schemes is widely used over fading mobile radio channels due to their robustness against signal fading and interference, while maintaining good spectral efficiency. High interference resistance is achieved if a high modulation index and moderate filtering are used, which keeps the phase changes engendered by interference relative to those due to modulation low. Unfortunately, a high modulation index requires a higher bandwidth, which means that a compromise has to be found. In other words, the slower and smoother are the phase changes, the better is the spectral efficiency. Spreading phase changes with zero initial and final slopes to three or four modulation intervals yields a partial response system. A representative of this family, called Gaussian Minimum Shift Keying (GMSK) is widely deployed via fading channels. It is derived from the full response Minimum Shift Keying (MSK) scheme, where phase changes between adjacent bit periods are piecewise linear, which results in a discontinuous phase derivative, i.e., instantaneous frequency. This clearly widens the spectrum, but by smoothing the phase using a Gaussian filter this problem is circumvented, as seen in Figure 8.27, where the GMSK signal is generated by modulating and adding two quadrature carriers.

The key parameter of GMSK in controlling both bandwidth and interference resistance is the 3 dB-bandwidth × bit interval product $(B \cdot T)$ referred to as normalised bandwidth. It was found by GSM that as the $B \cdot T$ product is increased from 0.2 to 0.5, the interference resistance is improved by approximately 2 dB at the cost of increased bandwidth occupancy, and best compromise was achieved for $B \cdot T = 0.3$. The surprising fact is that the spectral efficiency gain due to higher interference tolerance and hence smaller microcells was deemed to be more significant than the spectral loss caused by wider GMSK spectral lobes.

The GSM system will initially operate in the 890–915 MHz band for uplink transmission and 935–960 MHz band for downlink transmission. There are 124 paired duplex radio channels with a carrier spacing of 200 kHz and duplex spacing of 45 MHz between the uplink and downlink directions.

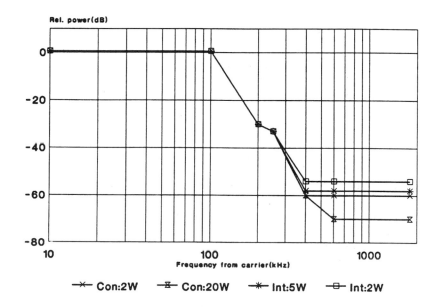

Figure 8.28: GMSK RF spectrum output mask.

A guard band of 200 kHz is left between the bottom edge of each band and the first RF carrier. Thus, the carrier frequencies in the two bands for the nth duplex radio channel will be (in MHz):

$$F_{nI} = 890.2 + 0.2(n-1) \text{ MHz} \tag{8.32}$$

$$F_{nII} = F_{nI} + 45 \text{ MHz}. \tag{8.33}$$

A TDMA system with 8 timeslots per RF carrier and 200 kHz channel spacing has the same spectral occupancy as an SCPC (Single Channel Per Carrier) system with 25 kHz channel spacing. When taking into account the 22.8 kbit/s channel coded data rate, the bandwidth of 25 kHz implies an approximate spectral efficiency of 1 bit/Hz. The actual RF output spectrum of the transmitted signals in a TDMA system, however, is determined by the modulation process and the switching transients occurring when bursts of RF signals are transmitted.

The recommendations for the output RF spectrum mask due to the GMSK modulation are given in graphic form in Figure 8.28. At the nominal bandwidth of 200 kHz the spectrum must have decayed by 30 dB with respect to the carrier frequency component. The specified relative power levels for frequency offsets from the carrier equal to or greater than 400 kHz depend on the output power level at which the transmitter is operating and the type of station. For transmitter output power levels below

43 dBm (20W) the specifications allow slightly higher levels in the spectrum at points 600 kHz to 1.8 MHz away from the carrier. The example of curve $Con : 2$ W representing a transmitted power of 2 W at the antenna connector in Figure 8.28 demonstrates that for 10 dB lower transmitted power a 10 dB higher modulation spectral mask is acceptable, when compared with a 20 W transmitter. Higher levels at points 400 kHz to 1.8 MHz away from the carrier are allowed also for equipment with integral antennas (e.g., portable sets) operating at power levels below 37 dBm (5 W), as represented by the curves $Int : 5$ W and $Int : 2$ W, respectively.

Switching transients and power ramping The switching transients caused by the transmission of bursts of RF energy widen the output RF spectrum. The RF spectrum due to the switching transients is required to be 23, 26, 32 and 36 dBm down relative to the level specified by the modulation mask at frequencies of 400, 600, 1200 and 1800 kHz measured from the carrier frequency, respectively. The switching transients can be reduced by ramping the output power up and down when transmitting a burst, instead of just keying the transmitter on and off. The information transmitted in the burst must not be affected by the process of power ramping, which is performed at the beginning and end of the timeslot using the mask illustrated in Figure 8.29. The timeslot in the figure corresponds to a duration of 156.25 bits, that of the burst length. In a normal burst, frequency correction burst or a synchronisation burst a guard period of 8.25 bit periods is inserted between adjacent ones. The remaining 148 bit periods form the 'active part' of the bursts. In an access burst the guard period after the burst is 68.25 bits long leaving an active part of 88 bit periods.

The 'useful part' of a burst in all cases is one bit period shorter than the active part and it begins half way through the first bit period as shown in Figure 8.30 for a burst with a 148 bit periods long active part. During that part of the burst when information is transmitted, the amplitude of the modulated RF signal must stay approximately constant. The power control of the transmitted signal exemplified by the ramping of the transmitted power occurs during the guard periods. Observe in Figure 8.29 that the approximately 70 dB power ramp-up occurs during 28 μs corresponding to 7.6 bit intervals, while ramp-down takes place in 18 μs, i.e., 4.9 bit intervals. When bursts are transmitted at the same frequency in consecutive timeslots, i.e., no frequency hopping is used, power ramping between the slots is not required and the signal transmitted in the guard times between the active slots may be any modulated signal. In this case the recommended time masks apply for the beginning and the end of the series of consecutive bursts.

In the GSM system the base and mobile stations are classified according to the transmitter output power. There are eight power categories of base station transmitted power, namely 2.5, 5, 10, 20, 40, 80, 160 as well as 320 W, and four for mobile stations given by 2, 5, 8, as well as 20 W. Adap-

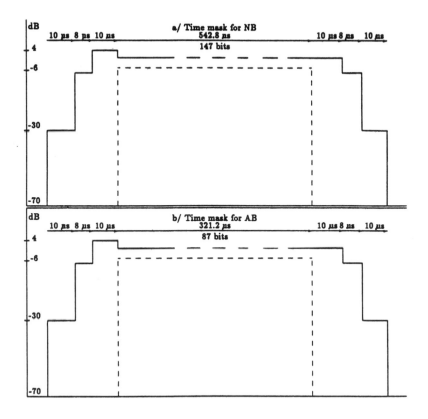

Figure 8.29: Power ramping time-masks for (a) Normal bursts and (b) Access bursts.

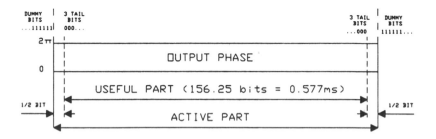

Figure 8.30: Active and useful parts of a normal burst.

tive RF power control is mandatory for the mobile stations, but optional for the base stations. This feature reduces cochannel interference whilst maintaining the quality of the radio channel. It also decreases the power consumption, which is important for hand-held mobile stations. Provisions are made for 16 different power control levels with 2 dB spacing between adjacent steps. The lowest power level for all mobile stations, regardless of their power class, is 13 dBm (20 mW) and the highest power level is equal to the maximum peak power corresponding to the class of the particular mobile station, as listed above. In the base stations the same 16 steps of 2 dB-spaced power levels are provided to achieve adaptive RF power control. Level '0' corresponds to the maximum peak power specified for the transmitter. For example, for a transmitter power class '3' the power level '0' is 80 W.

The output power of a base station transmitter can be reduced from its maximum level in at least six steps of 2 dB (with each step accurate to ± 0.5 dB) to adjust the radio coverage by the network operator. This RF output power adjustment is provided in every base station, and it is not connected in any way to the adaptive RF power control procedure.

The allowable spurious emissions from the base and mobile stations in the GSM system are higher than for the TACS analogue cellular mobile radio system. The spurious signal power from a base station transmission must be below 0.5 pW (-93 dBm), 250 nW (-36 dBm) and 1 μW (-30 dBm) in the frequency bands 890–915 MHz, 100 kHz–1 GHz and 1–12.75 GHz, respectively. For a mobile station transmitter the permissible levels of spurious emissions are 250 nW (-36 dBm) and 1 μW (-30 dBm) in the frequency bands 100 kHz–1 GHz and 1–12.75 GHz, respectively, when a channel is allocated to the mobile station. These permitted emission levels decrease to 2 nW (-57 dBm) and 20 nW (-47 dBm) in the corresponding bands when the MS is in its idle mode. In a base station two or more transmitters are often combined onto a single antenna. An undesirable consequence of the combining is the generation of intermodulation signals by the non-linearities in the transmitters. This effect occurs also when each transmitter feeds a separate antenna, but the transmitters are in close vicinity of each other. The peak power of any intermodulation product must not exceed 0.5 pW (-93 dBm) in the frequency band 890–915 MHz, nor 250 nW (-36 dBm) in the band 935–960 MHz. These limits apply not only to base station transmitters, but also to mobile PBX transmitters when operating in close vicinity of each other.

The receiver's performance is assessed in terms of a reference sensitivity level, which is the RF signal level at the receiver input assuming a specific propagation channel, for which the required error rate performance is achieved. For the hand portable receivers the reference sensitivity level is -102 dBm, and for all other types of mobile and base station receivers this level is -104 dBm. The receiver's blocking and intermodulation characteristics are specified assuming static propagation conditions for both wanted and unwanted signals, and a wanted RF signal level at the receiver input equals the reference sensitivity level. It is required that a mobile station receiver performance does not degrade by more than 3 dB when an interfering continuous wave signal is applied to the receiver at a level of -23 dBm or less, and that the frequency spacing between the wanted and unwanted signals is not less than 800 kHz and not more than 45 MHz. The requirements for the base station receivers are 10 dB higher, e.g., the level at which the 3 dB degradation in performance occurs is set at -13 dBm.

The intermodulation characteristics are measured by applying a signal with a frequency f_0 to the receiver input at the reference sensitivity level, plus two interfering signals of frequencies f_1 and f_2 such that $2f_1 - f_2 = f_0$ and $|f_2 - f_1| = 800$ kHz. The signal at frequency f_1 is a sinusoid, while the other interfering signal is a GMSK modulated carrier f_2 that has been modulated by any 148 bits subsequence of a 511 bits pseudo-random sequence. The receiver performance under these conditions must not degrade by more than 3 dB when the peak levels of both interfering signals are -43 dBm or less.

The error rate performance of the RF subsystem is specified in the GSM system for various propagation conditions, referred to as NAMEx, where NAME is the name of the propagation model and x is the vehicle speed in km/h. The models for rural area, hilly terrain, typical urban area and the profile for equalisation testing are referred to as RAx, HTx, TUx and EQx channels, respectively. An additive white Gaussian noise channel, i.e., a static channel, is also considered. Depending on the type of traffic or control channel, the performance is described in terms of frame erasure rate (FER), bit error rate (BER), or residual bit error rate (RBER). The RBER is defined as the ratio of the number of errors detected due to unprotected Class 2 bits over the frames defined as 'good', to the number of transmitted bits in the 'good' frames, where a frame is deemed to be good, for example, if the (53,50) cyclic error detecting block code protecting the Class 1a speech bits does not indicate code overload.

In GSM parlance there are three different types of error rates. One is concerned with the conditions of operation at reasonable signal levels and in the absence of interference. Another applies when the receiver is operating with signal levels close to its noise floor, and the third category applies for operation in the presence of interference. The nominal error rates (NER) apply to propagation conditions, when there is no interference and the received RF signal level is equal to -85 dBm. Under these conditions the chip error rate (channel BER), which is equivalent to the bit error rate of the non-protected C2 bits of a full rate traffic channel for speech (TCH/FS), is specified as $\leq 10^{-4}$ for the static channel, $\leq 4 \times 10^{-3}$ for the rural channel RA250, $\leq 4 \times 10^{-4}$ for the typical urban channel TU3 and $\leq 1\%$ for the channel used for the equaliser testing.

The reference sensitivity performance is the error rate performance when a received RF signal level is equal to the reference sensitivity level

Type of Channel		Propagation Conditions			
		Static	TU 50	RA 250	HT 100
SDCCH	(FER)	0.1 %	4 %	4 %	6 %
RACH	(FER)	0.1 %	10 %	10 %	10 %
SCH	(FER)	1 %	15 %	15 %	15 %
TCH/F9.6, H4.8	(BER)	10^{-5}	0.3 %	0.1 %	0.8 %
TCH/F4.8	(BER)		10^{-4}	10^{-4}	10^{-4}
TCH/F2.4	(BER)		10^{-5}	10^{-5}	10^{-5}
TCH/H2.4	(BER)		10^{-4}	10^{-4}	10^{-4}
TCH/FS	(FER)	10^{-3}	3 %	2 %	7 %
C1b	(RBER)	0.4 %	0.2 %	0.2 %	0.5 %
C2	(RBER)	2 %	8 %	7 %	8 %

Table 8.10: Reference sensitivity performance.

Type of Channel		Propagation Conditions			
		TU 3 (No FH)	TU 3 (FH)	TU 50	RA 250
SDCCH	(FER)	8 %	4 %	4 %	4 %
RACH	(FER)	12 %	12 %	12 %	10 %
SCH	(FER)	15 %	15 %	15 %	15 %
TCH/F9.6/H4.8	(BER)	1.5 %	0.3 %	0.3 %	0.2 %
TCH/F4.8	(BER)		10^{-4}	10^{-4}	10^{-4}
TCH/F2.4	(BER)		10^{-5}	10^{-5}	10^{-5}
TCH/H2.4	(BER)		10^{-4}	10^{-4}	10^{-4}
TCH/FS	(FER)	7 %	2.5 %	3.5 %	3 %
C1b	(RBER)	0.4 %	0.2 %	0.2 %	0.2 %
C2	(RBER)	8 %	8 %	8 %	8 %

Table 8.11: Reference interference performance.

of − 102 dBm for hand-portables and −104 dBm for all other mobile and base stations and when no interference is present. The reference sensitivity performance specifications are shown in Table 8.10, while the reference interference figures are summarised in Table 8.11 for the different types of channels and propagation conditions.

The reference interference performance is the error rate limit when the wanted input RF signal level is −85 dBm and a random GSM modulated interfering signal is present. The signal-to-interference ratio is ≤ 9 dB, while the signal-to-adjacent channel ratios at 200 and 400 kHz from the carrier are ≤ -9 dB and −41 dB, respectively. This interference ratio is called the 'reference interference ratio' and is identical for all types of base and mobile stations. In the tests, the wanted and interfering signals are subject to the same propagation profiles, and when frequency hopping is used, they have the same hopping sequence. Under these conditions the error rates for the various types of channels and propagation conditions satisfy the limits shown in Table 8.11.

8.9 Wideband Channels and Viterbi Equalisation

8.9.1 Channel Models

The understanding of the GSM wideband channel models and Viterbi equalisation assumes a sound appreciation of the mobile radio propagation phenomena, as detailed in Chapter 2. If the transmitted signal's bandwidth is narrow compared to the channel's coherence bandwidth (B_c), all transmitted frequency components encounter nearly identical propagation delays, i.e., the so-called narrow band condition is met and the signal is subjected to non-frequency-selective or flat envelope fading. When the signal bandwidth is increased, for example to accommodate several TDMA timeslots as in the GSM system, the channel becomes more dispersive which results in intersymbol interference. The channel's coherence bandwidth (B_c) is defined as the frequency where the correlation of two received signal components' attenuation becomes less than 0.5 and (B_c) is inversely proportional to the delay-spread (d), i.e., $B_c = 1/2\pi\ d$. Clearly, the wideband propagation channel is the superposition of a number of dispersive fading paths, suffering from various attenuations and delays, aggraevated by the phenomenon of Doppler shift caused by the MS's movement. The maximum Doppler shift (f_{Dmax}) is given by $f_{Dmax} = v/\lambda_c = v \cdot f_c/c$, where v is the vehicular speed, λ_c is the wavelength of the carrier frequency f_c and c is the velocity of light. The momentary Doppler shift f_D depends on the angle of incidence α, which is uniformly distributed, i.e., $f_D = f_{Dmax} \cdot \cos \alpha$, which hence has a random cosine distribution with a Doppler spectrum limited to $-f_{Dmax} < f < f_{Dmax}$. Due to time-frequency duality, this 'frequency dispersive' phenomenon results in 'time-selective' behaviour and the wider the Doppler spread, i.e., the higher the vehicular speed, the faster is the time-domain impulse response fluctuation.

In order to provide exactly specified, identical test conditions for different implementations of the GSM system, in particular for various Viterbi equalisers, a set of 12-tap and 6-tap typical channel impulse responses were defined, some of which are depicted in Figure 8.31. The Rural Area (RA) response is the least hostile amongst all standardised responses, decaying fast within a one bit interval and in terms of bit error rate performance it behaves as a single-path non-dispersive channel, where no Viterbi Equaliser (VE) is required. The Hilly Terrain (HT) model has a short-delay section due to local reflections and a long-delay part around 15 μs due to distant reflections, therefore in practical terms it can be considered a two- or three-path model, providing useful diversity gain, when using a VE. The Typical Urban (TU) impulse response spreads over a delay interval of 5 μs, which is almost two 3.69 μs bit intervals duration and therefore results in serious Inter Symbol Interference (ISI). Whence in simple terms it can be treated as a two-path model. The last standardised impulse response is artificially

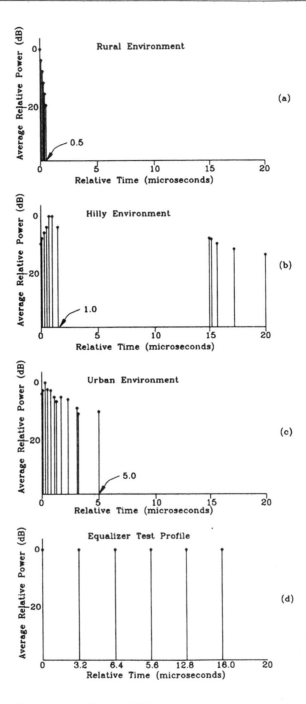

Figure 8.31: Typical GSM channel impulse responses.

contrived to test the VE's performance and is constituted by six equidistant unit-amplitude impulses representing six equal-powered independent Rayleigh-fading paths with a delay-spread over 16 μs. With these impulse responses in mind the required channel is simulated by summing the appropriately delayed and weighted received signal components. In all but one cases the individual components are assumed to have Rayleigh amplitude distribution. In the RA model the main tap at zero delay is supposed to have Rician distribution with the presence of a dominant path.

In summary, we highlighted four dispersive channel models, three of which represent realistic propagation environments, while the most hostile equaliser test response is worse than any practical channel. If reliable communications is expected, the uncontrolled ISI introduced by the mobile channel, as well as the controlled ISI introduced in the partial response modulator have to be removed, which requires a channel equaliser. The Bit Error Rate (BER) is minimised, if a 'Maximum Likelihood Sequence Estimator' (MLSE) is employed to decide upon the most likely transmitted sequence, rather than deciding on a 'maximum likelihood decoded symbol' basis. The Viterbi Algorithm (VA) detailed in Chapter 4 and augmented in Chapter 6 in the context of channel equalisation of GMSK signals is a well-suited efficient method for MLSE and is employed in most proposed implementations of the GSM system. A brief portrayal of Viterbi equalisation is given in the next section.

8.9.2   Viterbi Equaliser

Both during and after the definition of the GSM standards a number of VE implementations have been proposed in the literature, which have different complexities and performances [25],- [28], [30]. A simple general VE block diagram is shown in Figure 8.32. Once a call is set up, communications are maintained using Normal Bursts (NB) incorporating the 26-bit midamble in the centre of the burst, of which 16 bits constitute the frame synchronisation word and 5 bits are quasi-periodically repeated at both ends to keep the autocorrelation function and frequency domain oscillations low. As mentioned before, there are eight different, specially selected synchronisation words associated with eight different adjacent BS colour codes. These special synchronisation sequences have been found by computer search, evaluating the autocorrelation functions of all possible 2^{16} sequences. Favourable are those sequences which have the highest autocorrelation function main- to side-lobe ratio with near-zero values around the sampling instants $\pm T, \pm 2T, \pm 3T, \pm 4T$, etc., when quasi-periodically extended at both ends. It is highly desirable that both MSs and BSs use the same VE to keep development and production costs low, which additionally requires the recognition of synchronisation (SB) and access bursts (AB) as well, where 64- and 41-bit long synchronisation words are used, respectively. However, for the sake of simplicity, we only consider NBs with

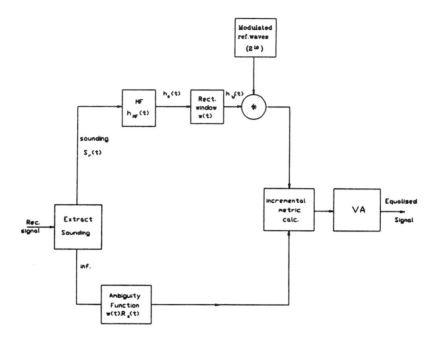

Figure 8.32: MLSE block diagram.

26-bit midambles.

The modulated NB with the channel sounding sequence $s(t)$ in its centre is convolved with the channel's impulse response $h_c(t)$ and corrupted by noise. Neglecting the noise for simplicity, the received sounding sequence becomes:

$$s_r(t) = s(t) * h_c(t), \qquad (8.34)$$

which is then matched-filtered using the causal impulse response $h_{MF}(t)$ to derive an estimate of the channel's impulse response:

$$h_e(t) = s_r(t) * h_{MF}(t) = s(t) * h_c(t) * h_{MF}(t) = R_s(t) * h_c(t), \qquad (8.35)$$

where $R_s(t)$ is the sounding sequence's autocorrelation function. Clearly, if $R_s(t)$ is a highly peaked Dirac impulse-like function, then its convolution with $h_c(t)$ becomes $h_e(t) \approx h_c(t)$. With $s(t)$ in the middle of the NB the estimated $h_c(t)$ is quasi-stationary for the 0.577 ms burst duration, and can be used to equalise the 114 bits of useful information on both sides of it, although the time-variant channel precipitates higher error rates towards the burst edges. Since the complexity of the VE grows exponentially with the number of signalling intervals in the legitimate modulated reference sequences generated from all possible transmitted sequences for metric comparisons, the estimated channel response $h_c(t)$ has to be windowed to a computationally affordable length using the rectangular function $w(t)$, while having sufficiently long memory to compensate for the typical GSM impulse responses of the previous section.

Specifically, in addition to the duration L_{CISI} of the controlled ISI, also the channel's delay-spread L_c has to be considered in calculating the required observation interval $L_o = L_{CISI} + L_c$ of the 2^{L_o-1}-state VE. In practical terms, using a bit interval of 3.69 μs and maximum channel impulse response durations of around 15–20 μs, a VE with a memory of 4-6 bit intervals is a good compromise, where $h_{\hat{c}}(t)$ is retained over that 4-6 bit interval of its total time domain support length, where it is exhibiting the highest energy. L_o consecutive transmitted bits give rise to 2^{L_o} possible transmitted sequences, which are first input to a local modulator to generate the modulated waveforms, and then convolved with the windowed estimated channel response $h_w(t)$ to derive the legitimate reference waveforms for metric calculation, as portrayed in Figure 8.32.

Recall that the condition $h_e(t) = h_c(t)$ is met only, i.e. the estimated impulse response is identical to the true channel impulse response only, if $R_s(t)$ is the Dirac delta function, which is not fulfilled when finite-length sounding sequences are used. The true channel response $h_c(t)$ could only be computed by deconvolution from Equation 8.35 upon neglecting the rectangular window $w(t)$. Alternatively, the received signal can be convolved for the sake of metric calculation with the known windowed autocorrelation function $w(t) \cdot R_s(t)$ often referred to as the ambiguity function, as seen in the lower branch of Figure 8.32 after extracting the sounding sequence from the received normal burst. Clearly, this way the received signal is 'predistorted' using the ambiguity function, identically to the estimated impulse response in Equation 8.35. This filtered signal is then compared to all possible reference signals and the incremental metrics m_i, $i = 0 \ldots (2^{L_o-1})$ are computed, which are utilised by the Viterbi algorithm (VA) to determine the maximum likelihood transmitted sequence, as explained in Chapter 6 and [29] and [30]. Let us now briefly consider the bit error rate performance of the GSM system.

8.9.3 GSM System Performance

The various implementations referenced have different complexities and performances. As a representative example we quote the BER versus SNR performance published in [28], as seen in Figure 8.33 for the TU50 channel (typical urban, vehicular speed of 50 km/h), the RA100, HT100 and RA250 GSM channel models. Best performance is achieved via the TU50 channel, which is due to the advantageous 'diversity' effect' introduced by the impulse response tap at 5 μs in Figure 8.31, since the probability of both paths having a deep fade simultaneously is fairly low. The worst performance is experienced via the HT100 channel, where the VE has apparent difficulties in combating excess delays above 15 μs. Interestingly, the

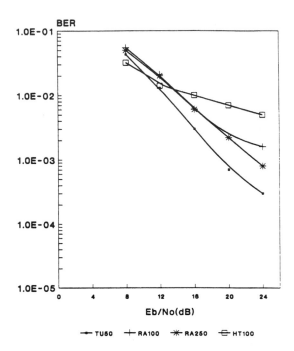

Figure 8.33: Viterbi equaliser BER versus E_b/N_0 performance.

RA100 and RA250 performances are worse than the TU50 integrity, since the RA models represent virtually single-path conditions with no 'diversity effect'. The BER is in all cases below 1%, if the E_b/N_0 ratio is in excess of about 12 dB. This residual BER can then be further reduced by the GSM concatenated error correction scheme, described in Section 8.7.   The performance of a complete GSM speech channel simulator has been reported in [31] for the various GSM channel models using vehicular speeds ranging from 0 km/h (AWGN) through pedestrians walking at 3 km/h to 250 km/h high-speed trains. For slowly walking pedestrians results are reported both with and without frequency hopping (FH). The concatenated coded C1 speech BER versus E_b/N_0 results are reproduced in Figure 8.34, where we observe virtually error free operation for the AWGN channel for E_b/N_0 in excess of 4 dB and for most of the fading channels above 12 dB. When using the TU3 channel the MSs are idling in deep fades and so the interleaving memory is not sufficiently long to randomise error bursts before channel decoding, which yields a high residual BER. This is seen being effectively combated by FH. The higher residual BER of the RA250 channel is due to the higher Doppler shift and lack of 'diversity effect'. The unprotected C2 bits have a high residual BER in Figure 8.35, which is uneffected by FH. In

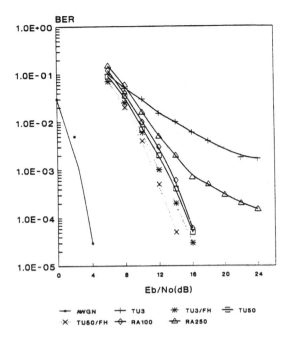

Figure 8.34: Speech C1 BER versus E_b/N_0 performance, Hodges et al. [31].

fact, this residual C2 BER is higher than that of the VE implementations proposed in [28] or [27]. Similar tendencies are recognised as regards to Frame Error Rates (FER) depicted in Figure 8.36. The interference resistance of the system expressed in carrier to interference ratio [C/I (dB)] is characterised in [31], which is again similar to the noise resistance, as seen in Figures 8.37, 8.38 and 8.39. In summary, all reported VE implementations reduce the channel BER to values sufficiently low for the concatenated channel coding/interleaving scheme to remove most of the errors for E_b/N_o and C/I ratios in excess of 12–14 dB, a value providing higher robustness and spectral efficiency than current analogue systems.

8.10 Radio Link Control

8.10.1 Link Control Concept

The radio sub-system link control in the GSM system involves procedures necessary for maintaining the link quality and managing traffic distribution, as well as for adaptive RF power control, handover, and to generate radio link failure responses. The call selection and re-selection procedures apply for a MS which is not engaged in communication with a BS, i.e., it is

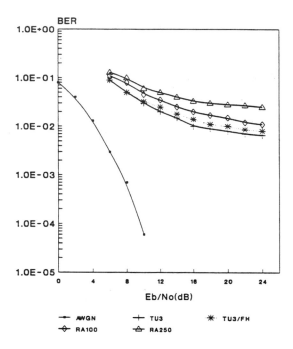

Figure 8.35: Speech C2 BER versus E_b/N_0 performance, Hodges et al. [31].

in its idle mode. These procedures allow for the MS to select a cell which provides the highest probability of reliable communications with the serving BS. The adaptive RF power control decreases interference with other cochannel users and, through dense frequency reuse, improves spectral efficiency, whilst maintaining an adequate communications quality. It also facilitates a reduction in power consumption, which is particularly important in hand-held MSs. The handover process maintains a call in progress as the MS moves between cells, or when there is an unacceptable degradation of quality caused by interference, in which case an intra-cell handover to another carrier in the same cell is performed. A radio link failure occurs when a call with an unacceptable voice or data quality cannot be improved either by RF power control or by handover. The reasons for the link failure may be loss of radio coverage or very high interference levels.

The radio sub-system link control procedures rely on measurements of the received RF signal strength (RXLEV), the received signal quality (RX-QUAL), and the absolute distance between base and mobile stations (DIS-TANCE). The received RF signal strength measurements are performed on the broadcast control channel (BCCH) carrier which is continuously transmitted by the BS on all timeslots and without variations of the RF level. A MS measures the received signal level from the serving cell and from the

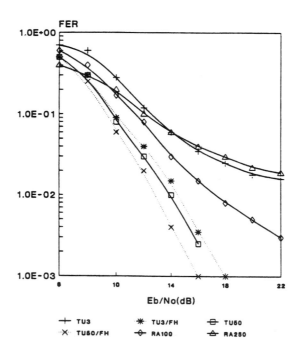

Figure 8.36: Speech FER versus E_b/N_0 performance, Hodges et al. [31].

BSs in all adjacent cells by tuning and listening to their BCCH carriers. The rms level of the received signal is measured over a dynamic range of -103 to -41 dBm for intervals of one SACCH multiframe (480 ms), with a relative accuracy of ±1 dB within any 20 dB section of this range, and an absolute accuracy of ±4 dB over the range from -103 to -70 dBm under normal conditions. The absolute accuracy over the full dynamic range and under both normal and extreme conditions is ±6 dB. The received signal level is averaged over at least 32 SACCH frames (≈15 s) and mapped to give RXLEV values between 0 and 63, where RXLEV = 0 if the received signal level (RSL) is less than -103 dBm, RXLEV = 1 if -103 dBm $\leq$ RSL < -102 dBm, ..., RXLEV = 63 if RSL> -41 dBm. The RXLEV parameters are then coded into 6-bit words for transmission to the serving BS via the SACCH.

The received signal quality (RXQUAL) is assessed by estimating the chip error rate, i.e., the BER before channel decoding, using the Viterbi channel equaliser's metrics and/or those of the Viterbi convolutional decoder. Eight values of RXQUAL span the BER range before channel decoding according to Table 8.12.

The absolute distance between base and mobile stations is measured

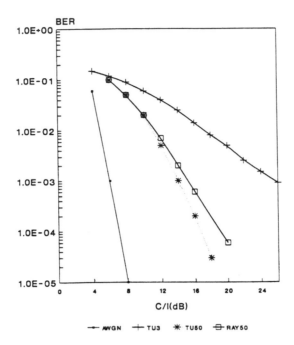

Figure 8.37: Speech C1 BER versus C/I performance, Hodges et al. [31].

RXQUAL 0	BER < 0.2%
RXQUAL 1	BER = 0.2% to 0.4%
RXQUAL 2	BER = 0.4% to 0.8%
RXQUAL 3	BER = 0.8% to 1.6%
RXQUAL 4	BER = 1.6% to 3.2%
RXQUAL 5	BER = 3.2% to 6.4%
RXQUAL 6	BER = 6.4% to 12.8%
RXQUAL 7	BER > 12.8%

Table 8.12: Received signal quality vs. channel bit error rate.

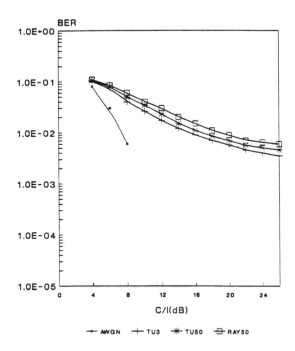

Figure 8.38: Speech C2 BER versus C/I performance, Hodges et al. [31].

using the 'timing advance' parameter. The timing advance is coded as a 6 bit number corresponding to a propagation delay from 0 to 63·3.69 μs = 232.6 μs. This allows measurements of an absolute distance from zero to almost 70 km with an accuracy of about 1 km.

The radio link control employs not only the parameters RXLEV, RX-QUAL and DISTANCE obtained by the measurements highlighted, but also other parameters transmitted by the BS. A MS needs to identify which surrounding BS it is measuring and the BCCH carrier frequency may not be sufficient for this purpose, since in small cluster sizes the same BCCH frequency may be used in more than one surrounding cell. To avoid ambiguity a 6-bit Base Station Identity Code (BSIC) is transmitted on each BCCH carrier in the SCH. Two other parameters represented by one-bit Boolean flags transmitted in the BCCH data provide additional information about the BS. Namely, PLMN_PERMITTED indicates whether the measured BCCH carrier belongs to a PLMN which the MS is permitted to access. The second flag, CELL_BAR_ACCESS, indicates whether the cell is barred for access by the MS, although it belongs to a permitted PLMN. The parameters BSIC, PLMN_PERMITTED and CELL_BAR_ACCESS, together with the RXLEV, are used in the cell selection and re-selection procedures, as seen in Figure 8.40. A MS in idle mode, i.e., after it has just

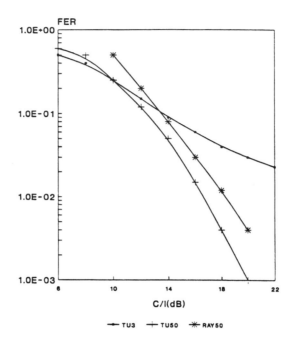

Figure 8.39: Speech FER versus C/I performance, Hodges et al. [31].

been switched on, or after it has lost contact with the network, searches all 124 RF channels and takes readings of RXLEV on each of them. The station then tunes to the carrier with the highest RXLEV and searches for frequency correction bursts (FCB) in order to determine whether or not the carrier is a BCCH carrier. If it is not, then the MS tunes to the next highest carrier, and so on, until it finds a BCCH carrier. The MS then finds a synchronisation burst (SB), synchronises to the BCCH carrier and decodes the parameters BSIC, PLMN_PERMITTED and CELL_BAR_ACCESS from the BCCH data and makes a decision to camp on the cell or to continue the search. The MS may have a BCCH carrier storage option, i.e., store the BCCH carrier frequencies used in the network accessed, in which case the search time would be reduced. The process described is summarised in the flowchart of Figure 8.40.

The RF power control procedures employ RXLEV measurement results. In every SACCH multiframe the BS compares the RXLEV readings reported by the MS, or obtained by the base station, with a set of thresholds. The exact strategy for RF power control is determined by the network operator with the aim of providing an adequate quality of service for speech and data transmissions and keeping interferences low. The criteria for determining the radio link failure are based on the measurements of RXLEV

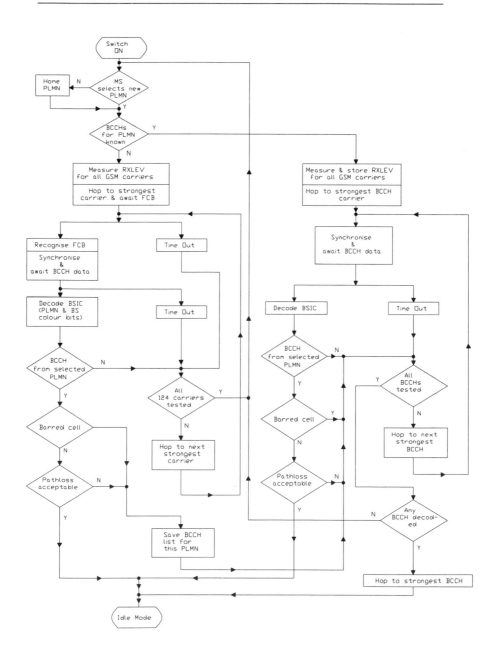

Figure 8.40: Initial cell selection by the MS. ©ETT, Hanzo and Steele, 1994

and RXQUAL performed by both the mobile and base stations. The procedures for handling radio link failures result in the re-establishment or the release of the call in progress. The network operator determines the exact criteria employed.

The handover process involves the most complex set of procedures in the radio link control. Handover decisions are based on results of measurements performed both by the base and mobile stations. The base station measures RXLEV, RXQUAL, DISTANCE, and also the interference level in unallocated timeslots, while the MS measures and reports to the BS the values of RXLEV and RXQUAL for the serving cell, and RXLEV for the adjacent cells. When the MS moves away from the BS, the RXLEV and RXQUAL parameters for the serving station become lower, while RXLEV for one of the adjacent cells increases.

8.10.2 A Link Control Algorithm

With the general link control concepts in mind GSM have devised an optional algorithm fulfilling all the system requirements. This algorithm provides an evolutionary basis from which PLMN operators can start developing their own procedures to meet special local criteria, which will be presented here in a number of steps.

8.10.2.1 BS Preprocessing and Averaging

The GSM HO algorithm explained with reference to Figures 8.41, 8.42 and 8.43 is based on the evaluation, storage and processing of a number of parameters. The MSs continuously measure the DownLink (DL) received level (RXLEV), downlink received quality (RXQUAL) from the serving cell and the downlink received levels from the nth adjacent cells (RXLEV_NCELL(n)), and report the measured values back to the BS via the SACCH. The new measurement 'samples' are conveniently generated for every new SACCH multiframe of 480 ms duration. If BS power budget (BS-PBGT) control is also implemented, a similar set of values is measured by the BS: UpLink (UL) RXLEV, uplink RXQUAL and RXLEV in unallocated timeslots, representing the interference level. Furthermore, the MS–BS distance is calculated from the timing advance (TA) parameter. It remains for the network operator to resolve how RXQUAL is determined. The options available are to monitor the metric statistics of the Viterbi channel equaliser, that of the Viterbi-type convolutional decoder and/or the code overload rate detected by the external block codes, used in data and speech traffic channels.

In possession of the above mentioned 480 ms based measurement 'samples', the BS has to evaluate their weighted or unweighted averages. Alternatively, median values can be utilised for further decisions, where the extreme outliers are ignored. To have sufficient confidence in the estimates derived from finite-sized measured sample sets, averaging is carried out for at least 32 samples, measured over $32 \cdot 0.48$ s ≈ 15 s durations. The actual timing of the processing is dictated by the OMC.

8.10.2.2 RF Power Control and HO Initiation

Now the averaged parameters are compared in the BS with their associated upper and lower RF power control thresholds, and if any of the UL&DL=(XX) RXLEV_XX & RXQUAL_XX parameters fall outside the required range, the BS attempts to rectify the shortfall by means of RF power control, as seen in Figure 8.41. More explicitly, if any of the four threshold comparisons fail, the BS and MS will attempt to appropriately increase or decrease the transmitted powers to meet the required conditions before initiating HO.

In the 'HO initiation' set of threshold comparisons the BS initiates HO if any of the comparisons fail to meet the corresponding condition, as portrayed in Figure 8.42. Observe that the system remembers the cause of the HO request and in most cases when the criteria are not satisfied HO is imperative.

8.10.2.3 Decision Algorithm

Upon HO requests due to any of the causes considered in the 'RF power control and HO initiation' phases, the BS sends a message with the 'preferred list of target cells' to the MSC. Alternatively, in the case of traffic-motivated HOs the MSC may send a 'HO-candidate enquiry message' to the BS, which responds with the same 'preferred list of target cells' message. This list is compiled using the average received signal levels RXLEV_DL and RXLEV_NCELL(n), as well as a few further system parameters. Also the so-called power budget parameter (PBGT(n)) can be evaluated for each connection taking into account each of the legitimate adjacent cells $(n=1 \ldots 16)$, using the following equation:

$$
\begin{aligned}
\text{PBGT}(n) \quad = \quad & [\min(\text{MS\_TXPWR\_MAX}, P) - \text{RXLEV\_DL}] \\
- \quad & [\min(\text{MS\_TXPWR\_MAX}(n), P) - \text{RXLEV\_NCELL}(n)].
\end{aligned}
$$

Here MS_TXPWR_MAX is the maximum allowed MS transmitted power on a traffic channel in the serving cell to control cell size, MS_TXPWR_MAX(n) is the same parameter in the nth adjacent channel, while P is the maximum transmitted power capability of the MS. This equation physically evaluates the power budget for each legitimate adjacent cell in contrast to the present serving cell, since the first square bracketed term represents the present pathloss, while the second term the pathloss of the nth candidate serving cell. However, the PBGT(n) parameter is evaluated only for those cells, which satisfy

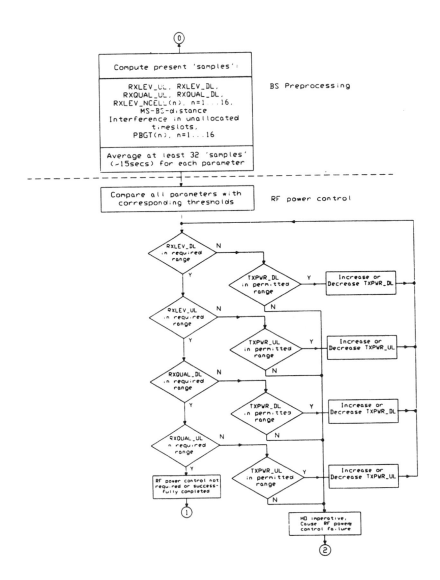

Figure 8.41: Handover preprocessing and RF power control.

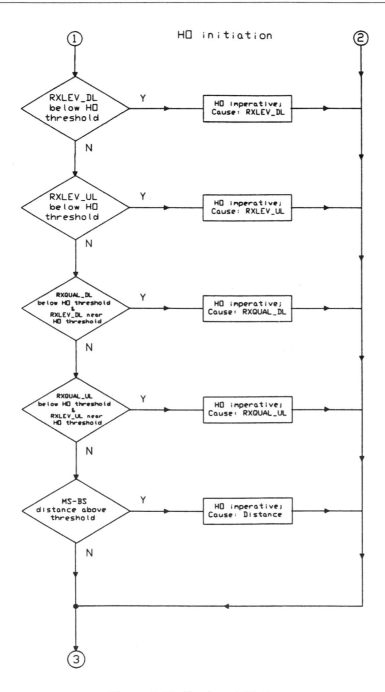

Figure 8.42: Handover initiation.

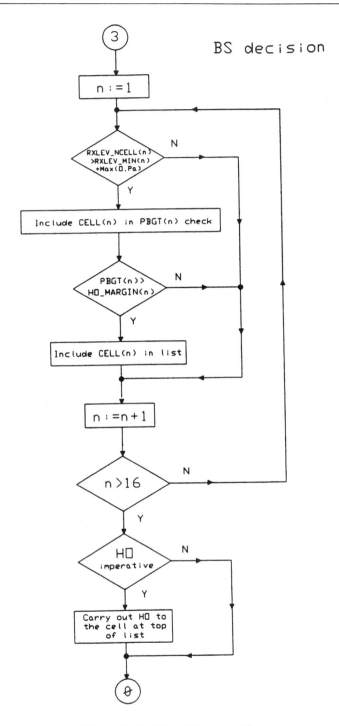

Figure 8.43: BS decision algorithm.

$$\text{RXLEV\_NCELL}(n) > \text{RXLEV\_MIN}(n) + \max(0, P_a), \qquad (8.36)$$

where

$$P_a = [\text{MS\_TXPWR\_MAX}(n) - P], \qquad (8.37)$$

with P being again the MS's maximum power capability, which is different for vehicle-mounted and for hand-held MSs, and $\text{MS\_TXPWR\_MAX}(n)$ being the maximum permitted MS power due to coverage area limitation in the nth adjacent cell. In other words, $\text{PBGT}(n)$ is evaluated for those candidate target cells where the received power $\text{RXLEV\_NCELL}(n)$ exceeds the corresponding minimum $\text{RXLEV\_MIN}(n)$ by the margin $\max(0, P_a)$.

The parameter HO_MARGIN seen in Figure 8.43 in the $\text{PBGT}(n)$ comparison is introduced to facilitate a hysteresis in the HO process by requiring the pathloss of adjacent cell n to be considerably more favourable than that of the present serving cell, before HO is requested to it. If HO_MARGIN=0, there is no hysteresis, while for HO_MARGIN$\neq$0 the HO from cell A to B occurs at a different point from the handover B to A. Clearly, a power budget or pathloss-motivated HO is possible to ensure communications with that BS which yields the lowest pathloss, even if all the other quality and received power threshold conditions are duly met in the serving cell.

Most of the HO request-causes are self-explanatory, but an interesting case is when the received signal level RXLEV_UL is high, typically $-80\ldots-40$ dBm, yet the received signal quality is low. This indicates the probability of high cochannel interference on the uplink, which can be eliminated by intracell HO. If the BS does not support intracell HO, then it sends a HO request to the MSC with the serving cell at the top of its preferred cell list. In some cases the OMC can initiate an intracell HO, for example due to resource management criteria. It is particularly important to avoid anomalous HO decisions, such as subsequent power increase and decrease commands. Therefore after a power control action the set of samples used in the decision have to be discarded.

8.10.2.4 HO Decisions in the MSC

During periods of peak traffic load the number of HO requests is often higher than that of the free traffic channels. In these cases the MSC sorts the HO requests in the following order of priority: RXQUAL, RXLEV, DISTANCE, PBGT. Then requests due to RXQUAL degradation enjoy the highest priority, followed by RXLEV, etc. A further classification principle is the priority order, which can be associated with each adjacent cell, where eight priority levels can be allocated. This allows, for example, umbrella cells to be given low priorities to handle calls only if no other suitable cell can carry the initiated traffic.

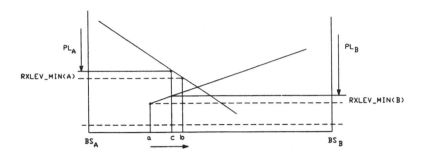

Figure 8.44: Handover without hysteresis.

8.10.2.5 Handover Scenarios

In this subsection a number of examples are used to support our discussions on HO algorithms [33]. Figure 8.44 represents a situation where the MS is travelling from cell A to cell B, the reference sensitivity level is -102 dBm and the minimum received power level to allow HOs to these cells are RXLEV_MIN(A) and RXLEV_MIN(B), respectively. When the received power levels exceed these thresholds at positions a and b, HO to these cells is possible. Observe that at point b the received signal level from BS_A falls below RXLEV_MIN(A) and therefore HO to BS_B is here imperative, although its received signal level is higher than that from BS_B. When the pathloss from B (PL_B) becomes lower than that from A (PL_A), HO to B is recommended. In this example cell B is probably allocated to an open area with low pathloss exponent and hence, in spite of the lower transmitted power seen, it has a larger coverage than cell A.

The size of cell B is readily reduced by increasing the minimum expected received level RXLEV_MIN(B). In parallel to this the possible HO region is shrinking and when point a moves past point b as RXLEV_MIN(B) is increased, HO is no longer possible at point c due to lack of received signal power, although the pathloss criterion $PL_B < PL_{textrmA}$ is met. Conversely, if the size of cell A is reduced by increasing RXLEV_MIN(A), point b keeps moving to the left. When point b moves past point c, HO to cell B becomes imperative due to lack of coverage, before the pathloss criterion $PL_B < PL_A$ becomes true.

In Figure 8.45 the HO hysteresis introduced by the HO_MARGIN is demonstrated. The simple principle is that the received signal level from BS_A must fall significantly, by the HO_MARGIN, below that from BS_B, before HO to cell B is carried out. This ensures that after HO the received level will be by HO_MARGIN(A) dB higher than from the current serving cell A. Observe that the hysteresis area is adjusted by appropriately selecting both HO_MARGINs, where HO from cell A to cell B occurs at point b, while HO from B to A happens at point a.

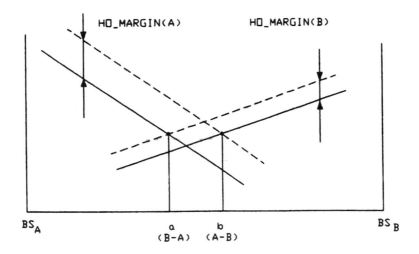

HO_MARGIN(A)   HO_MARGIN(B)

BS$_A$ a   b   BS$_B$
                      (B-A)   (A-B)

Figure 8.45: Handover with hysteresis.

8.11 Discontinuous Transmission

8.11.1 DTX Concept

The idea of discontinuous transmission (DTX) has long been known in bandwidth and power limited satellite systems, where spectral efficiency is improved using Digital Speech Interpolation (DSI). In mobile radio communications, however, GSM is the first system to use voice activity detection (VAD) and DTX to further reduce the MS's power consumption and increase spectral efficiency through reducing interference during silent periods. Assuming an average speech activity of 50% and a high number of interferers combined with frequency hopping to randomise the interference load, significant spectral efficiency gains can be scored. Due to the reduction in power consumption full DTX operation is mandatory for MSs, but in BSs only receiver DTX functions are compulsory. Earlier adaptive VAD designs were proposed for PCM speech codecs, stationary handsets and indoors background noise [34, 35]. The fundamental problem is how to differentiate between speech and noise, while keeping false noise triggering and speech spurt clipping as low as possible. In vehicle-mounted MSs the severity of the speech/noise recognition problem is aggravated by the excessive non-stationary vehicle background noise. This problem is resolved by deploying a combination of threshold comparisons and spectral domain techniques [36, 37]. Another important associated problem is the introduction of noiseless inactive segments, which is mitigated by introducing comfort noise in these segments at the receiver, which is also addressed in [36] and [37].

8.11.2   Voice Activity Detection

The basic function of the VAD is to differentiate between noisy speech and noise only under very high-noise conditions. Any VAD has to meet a compromise between the minimisation of false triggering due to high noise levels and the transmission of low level speech. Fast speech recognition is crucial to minimise initial talk-spurt clipping, and a short hangover delay reduces unwanted activity while preventing final talk-spurt clipping. The specific VAD implementation favoured must be in harmony with the speech codec selected, but differences between noise and speech properties can be exploited both in the time [34] and frequency domain [36]. In the GSM VAD a combination of spectral domain and energy differences is utilised in the decision process.

The VAD's schematic diagram is shown in Figure 8.46. In a first step the SNR is improved by adaptive noise filtering, the coefficients of which are determined during noise-only periods. Then the energy of the filtered signal is compared against an adaptive threshold computed in the 'Threshold adaptation' block for a speech/noise decision in the 'VAD decision' block. The adaptive noise filter coefficient- and noise threshold-update must take place during exclusively noise periods, which is ensured by additionally checking the signal stationarity and the lack of pitch frequencies with the help of the 'Periodicity detection' and 'Spectral comparison' blocks. A further fixed threshold is deployed to ensure that low level noise is not detected as speech. Finally, a hangover (HGO) mechanism is used to prevent mid-spurt and end-spurt clipping of speech bursts, as seen in in Figure 8.46.

More specifically, the VAD's fundamental function is to adaptively filter the input signal using the set of filter coefficients a_i, $i = 0 \ldots 8$ during noise-only periods. The filtered signal's energy P_{VAD} is compared against an adaptively adjusted threshold Th_{VAD} to derive a speech/noise indicator signal VAD, which after being subjected to hangover imposition, yields the transmit flag TXFL utilised by the transmitter's DTX handler to enable/disable transmissions. The rest of the block diagram is concerned with the adaptive adjustment of the filter coefficients a_i and that of the threshold Th_{VAD}.

The 'adaptive block filtering' operation of the 160 input signal samples $s(n)$ using an 8th order filter a_i yields 168 samples, as follows:

$$s_f(n) = \sum_{i=0}^{8} a_i s(n-i), \; n = 0 \ldots 167, \; 0 \le (n-i) \le 159. \qquad (8.38)$$

The energy of the current 20 ms (160 samples) filtered input signal extended

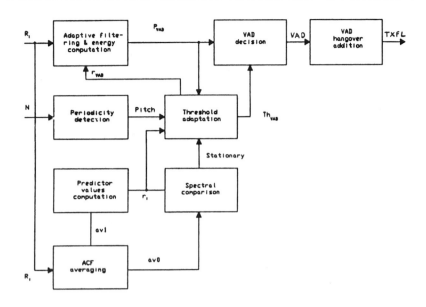

Figure 8.46: Functional block diagram of the VAD.

by the filter's 8-sample memory is given by:

$$P_{VAD} = \sum_{n=0}^{167} (\sum_{i=0}^{8} a_i s(n-i))^2, \ \ 0 \le (n-i) \le 159. \qquad (8.39)$$

After expanding the operations in the expression of P_{VAD} above and substituting the expressions

$$R_i = \sum_{n=0}^{159} s(n)s(n-i), \ \ i = 0 \ldots 8, \ \ 0 \le (n-i) \le 159 \qquad (8.40)$$

and

$$r_i = \sum_{k=0}^{8-i} a_k a_{k+i}, \ \ i = 0 \ldots 8 \qquad (8.41)$$

for the input signal's and the filter coefficient's autocorrelations, respectively, we have:

$$P_{VAD} = r_0 \cdot R_0 + 2 \sum_{i=1}^{8} r_i \cdot R_i. \qquad (8.42)$$

The result of the 'VAD decision' is the Boolean flag VAD, which is one if $P_{VAD} > Th_{VAD}$, zero otherwise. The HGO is implemented to prevent the VAD from prematurely curtailing the end of low-energy speech-spurts or removing short mid-speech silent gaps. The principle is that speech continues to be transmitted for four more 20 ms frames, even if VAD=0 indicates the presence of noise, in case at least three previous 20 ms speech segments were deemed to be present. Should VAD=1 be set during the HGO period, the hangover counter (HOCT) is reset to four.

With the principles of VAD known, we now embark upon a description of the adaptive adjustment of the VAD-threshold, that of the filter coefficients a_i and their correlations r_i. To get a stationary estimate of the input signal's statistics, each input signal autocorrelation coefficient R_i $i = 0 \dots 8$ is averaged over four frames, i.e., 80 ms to derive the averages $av0_i(n) = \sum_{j=0}^{3} R_i(n - j)$, $i = 0 \dots 8$, $av1_i(n) = av0_i(n - 4)$, $i = 0 \dots 8$, where n is the 20 ms frame index. The averaged autocorrelation coefficients $av1_i(n)$ are input to the Schur recursion [12] and the reflection coefficients $k_i(n)$ are computed exactly as in the RPE-LTP speech codec. In determining the noise envelope's reflection coefficients, the averages $av1_i(n)$, $i = 0 \dots 8$ are used, since $av0_i(n)$ might still contain the end of a speech burst. In a subsequent step the reflection coefficients k_i are converted to simple finite impulse response (FIR) LPC filter coefficients a_i, $i = 0 \dots 8$ and their autocorrelation is computed as follows:

$$r_i = \sum_{k=0}^{8-i} a_k \cdot a_{k+i}, \ i = 1 \dots 8. \tag{8.43}$$

The LPC filter coefficient autocorrelations r_i, as well as the averaged input signal autocorrelation coefficients $av0_i$ are then compared using the simple distance measure d_m defined as:

$$d_m = r_0 \cdot av0_0 + 2 \sum_{i=1}^{8} r_i \cdot av0_i / av0_0 \tag{8.44}$$

to derive a statistical similarity flag called 'Stationary' in the 'Spectral Comparison' block of Figure 8.46. The spectral distance of the consecutive 20 ms input segments is evaluated by computing $d = (d_m - d_{m-1})$, and Stationary = 1 is set if $d < 0.05$, i.e., the spectrum is deemed stationary, while Stationary = 0, if the spectral difference $d \geq 0.05$, i.e., the spectrum is non-stationary.

The 'Threshold Adaptation' process is based on the input parameters Stationary, r_{av1} derived so far, as well as on the Boolean flag 'Pitch', indicating the presence of voiced input and on the energy of the current adaptive filtered input signal P_{VAD}. This process has two output variables, the VAD decision threshold Th_{VAD} and the updated adaptive filter coefficient set a_i, $i = 0 \dots 8$ determining the updated set r_i.

The threshold adaptation updates Th_{VAD} every 20 ms using the flow-chart of Figure 8.47 in two basic scenarios. Whenever the signal energy is very low, i.e., $R_0 < P_{th}$, where the power threshold P_{th} is set by GSM to

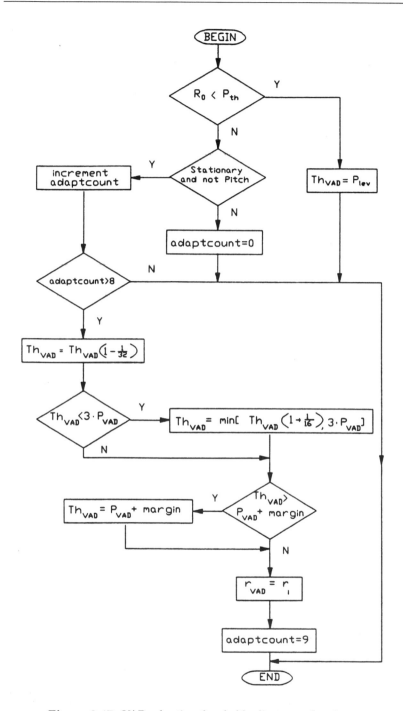

Figure 8.47: VAD adaptive threshold adjustment flowchart.

$300\,000$, $Th_{VAD} = P_{LEV} = 80\,000$ is selected, since any further tests would be unreliable due to the course quantisation at such low signal level. If, however, $R_0 \geq P_{th}$ is met, more elaborate tests are performed. If (Stat,Pitch) = (1,0) for the current input frame, i.e., the input signal appears stationary and unvoiced with no pitch periodicity, the adaptation counter (Adapt-count) is incremented and checked, whether it reached the value $adp = 8$ to allow threshold update. If not, no further action is taken before the next 20 ms frame arrives. Otherwise Th_{VAD} is decreased by the multiplicative factor $(1 - 1/32)$. Then the new Th_{VAD} value is compared with $3 \cdot P_{VAD}$ and in case it is larger than $3 \cdot P_{VAD}$, the threshold really had to be decreased, else it is set to $3 \cdot P_{VAD}$, unless this exceeds a multiplicative increase of $(1 + 1/16)$. The threshold is not permitted to be higher than $(P_{VAD}+$ margin), where margin$= 8 \cdot 10^6$, and after ensuring this the adaptation counter is set to $adp + 1 = 9$ to allow for continuous threshold adjustments. Finally, by forcing $r_{VAD,i} = r_i$, $i = 0 \ldots 8$, the set of filter coefficient correlations is updated for consecutive power computations.

The last parameter of the threshold adaptation is the Pitch-flag, updated also every 20 ms, which is true if a periodic input signal is detected. As seen in the flowchart of Figure 8.47, the threshold Th_{VAD} is only updated if the input is stationary but not periodic, which is characteristic of noise.

In summary, the GSM VAD strikes a good compromise between lowest possible on-air time, i.e., activity and unobjectionable talk-spurt clipping. The typical channel activities vary from 55% in quiet locations through 60% in office noise to 65–70% in strong airport or railway station noise.

8.11.3   DTX Transmitter Functions

The DTX transmitter's operation is explained by referring to Figure 8.48 and relies on the VAD differentiating between speech and noise. If speech is deemed to be present, i.e., the transmitter state machine is in its speech transmit (SPTX) state, the VAD flag is set to one, while for noise VAD=0. If the VAD stops detecting speech, it does not immediately disable the speech transmission by setting the transmit flag (TXFL) to zero, but first enters the so-called Hangover (HGO) state. The HGO state is designed to prevent negligibly short silence periods from disabling transmissions or to remove final talk spurt clipping, where only a fraction of the frame delivers speech and hence VAD=0 was detected. The HGO delay is of four speech frame durations, i.e., 4·20=80 ms long. Hence the hangover counter (HOCT) is initially set to HOCT=4 and in every subsequent noise frame, where VAD=0, it is decremented by one. When HOCT=0, the HGO delay has elapsed and the 'end of speech' (EOS) flag has to be set to one. Since the last four frames encoded by the speech encoder during the hangover interval were deemed to be noise, their spectral envelope parameters ($LARs$) as well as RPE subsegment maxima, averaged over four blocks, are used to form a so-called silence identifier (SID) frame to be passed to the speech decoder for comfort noise insertion to 'fill' the subjectively annoying 'deaf' periods introduced by disabled transmissions. Now the first averaged SID frame following the elapse of the hangover is scheduled for transmission, but if it happens to be stolen by the FACCH for example, then the EOS flag has to remind the DTX transmitter to send the subsequent frame instead. However, the frame immediately after the elapse of the HGO is the only one with EOS=1. Clearly, with the HGO elapsed the system enters the Comfort Noise Update (CNU) state and the SID frames are sent during further silent periods in each SACCH multiframe, i.e., at intervals of 480 ms, whenever it is asked for by the radio subsystem through setting the Noise Update Flag (NUFT) to logical one. In the simplest scenario NUFT=1 is aligned with the timeslots of the SACCH structure. From the CNU state, if VAD=1 is encountered, the DTX transmitter state-machine returns to its speech transmit (SPTX) state, otherwise it enters the Comfort Noise Computation (CNC) state, sets the transmit flag (TXFL) to zero and disables transmissions. Detecting VAD=1 forces the system to SPTX state, while on VAD=0 and NUFT=1 further SID frames have to be transmitted in CNU mode.

8.11.4 DTX Receiver Functions

The DTX receiver's operation is in close cooperation with the entire receiver, since it uses soft and hard decision information from the Viterbi channel equaliser, Viterbi channel decoder and cyclic error detecting block decoder to generate the so-called Bad Frame Indicator (BFI) flag. When the BFI flag signals a corrupted speech or SID frame, the Speech/noise Extrapolation (SE) functions are invoked to improve the perceived link quality. If, however, several adjacent frames are damaged, the received signal is gradually muted to zero. The interplay of system elements is completed by the comfort noise generator activated upon reception of SID frames for natural sounding Comfort Noise Insertion (CNI) in inactive speech intervals. The DTX receiver's operation is essentially conducted by the input flags BFI and SID, as evidenced by Figure 8.49.

Firstly, in 'Speech Received (SPRX)' state the SID frame detector decides whether the received frame is a speech or an SID frame and, after evaluating the received signal quality, forms the pair (BFI, SID). The SID detector is extremely reliable, since in SID frames all of the 95 C1b FEC coded RPE excitation bits are set to zero at the transmitter and the received frame is only deemed to be an SID sequence rendering SID=1 if at most 15 out of the 95 corresponding bits are non-zero. It will only be used for comfort noise insertion, however, if at most one bit of it is corrupted, i.e. BFI=0, while in the case of more than one but less than 16 corrupted bits (BFI,SID)=(1,1) is set, which requires noise extrapolation using gradually muted previous SID frames. The normal operation is described by

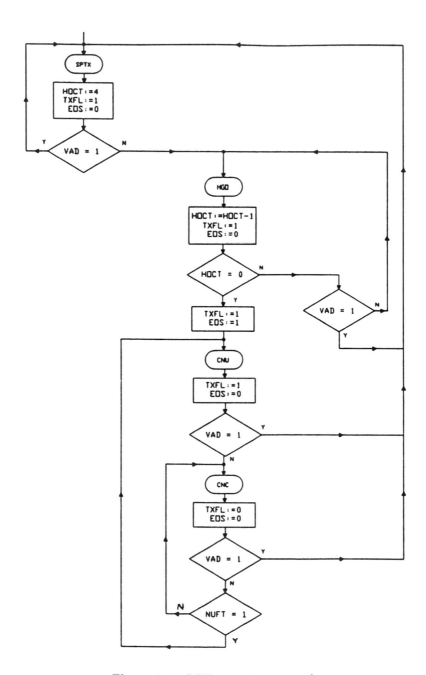

Figure 8.48: DTX transmitter operation.

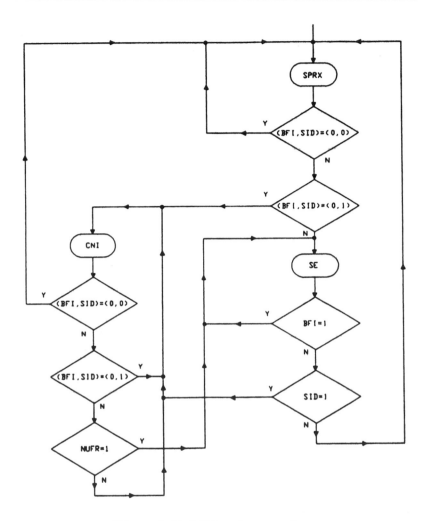

Figure 8.49: DTX receiver operation.

(BFI=0, SID=0), when an uncorrupted speech frame is received (SPRX). In this situation the received frame is simply decoded by the speech decoder. If a frame with a BFI=1 flag has arrived, irrespective of whether speech or noise is deemed to be present, the receiver switches into speech/noise extrapolation mode to improve the subjective link assessment, in case a single speech frame is corrupted or stolen by the FACCH. Several consecutive BFI=1 flags render the receiver to mute its output gradually to zero. When the link-quality improves, BFI=0 is encountered and upon SID=0 the receiver returns to its normal speech reception state SPRX.

If, however, the pair (BFI=0, SID=1) is detected, the DTX receiver inserts comfort noise to the speech decoder in its CNI state, based on the noise spectral parameters (*LARs*) received in the last SID frame. These noise spectral *LAR* parameters are updated by receiving fresh SID frames via the SACCH every 480 ms in each new multiframe. When detecting BFI=0, SID=0, normal speech decoding is invoked, while if this frame happens to be corrupted, i.e., BFI=1 and the receiver's noise update flag NUFR=1 indicates matching alignment with a SACCH time slot, then again the receiver enters the extrapolation mode (SE). In case several consecutive SID frames are corrupted, i.e. more than one of the 95 zero C1b bits is turned to one, the comfort noise is gradually muted to zero to inform the listener that the link is corrupted. The process described is highlighted also in Figure 8.49.

8.11.5   Comfort Noise Insertion and Speech/Noise Extrapolation

Experiments carried out by GSM have shown that silent gaps inserted by the DTX system are extremely annoying and degrade speech intelligibility. Best subjective and objective results are achieved if comfort noise of appropriately matched level and spectral envelope is inserted and updated via sending an SID frame at each 480 ms interval through the SACCH, when no speech is transmitted. The 456-bit SID frame is a 'speech-like' frame transmitted every 24th 20 ms speech frame to characterise the current background noise spectral envelope using the Logarithmic Area Ratio (*LAR*) parameters. The level of the noise is represented by the subsegment maxima computed, but the regularly spaced excitation pulses are set to zero at the transmitter to aid the SID frame recognition at the receiver. The Long Term Predictor (LTP) is disabled by setting its gain to zero, while erratic noise level changes are mitigated by limiting subsequent increases to 50% of the previous maximum value. Furthermore, the *LARs* and block maxima are averaged over the last four speech frames, before inclusion in an SID frame. At the receiver the decoded *LARs* and block maxima are used with locally injected uniformly distributed pseudo-random RPE samples and grid positions to represent the background noise at the transmitter.

Whenever the BFI flag signals a corrupted speech or noise frame, the previous 20 ms frame is input to the speech decoder. This repetition is hardly perceptible, if only one frame is lost in every ten, but becomes inadequate when encountered more frequently. In subsequent corrupted frames therefore their level is gradually muted to zero by decreasing the maximum 64-valued (6-bit logarithmically quantised) subsegment maxima each time by four. Hence it is set to zero in at most 16 subsequent 20 ms speechframes, i.e. in 320 ms.

8.12 Ciphering

The GSM communications security aspects are described in Recommendations 02.09, 02.17, 03.20 and 03.21, while an overview is given in [38]. The GSM security issues centre around the Subscriber Identity Module (SIM) received at subscription, which is preferably a removable plug-in module with a Personal Identification Number (PIN). These features facilitate the production of identical handsets with PIN protection against unauthorised use, while allowing GSM access through any GSM handset. The SIM contains, amongst a number of parameters, the International Mobile Subscriber Identity (IMSI), the Individual Subscriber Authentication Key (K_i) and the Authentication Algorithm (A3). On attempting to access the PLMN the MS identifies itself to the network, receives a random number (R), which together with K_i is used to calculate the Signed response (S) by invoking the confidential algorithm (A3): $S=[K_i(A3)R]$. The result S is sent back to the network and compared with the locally computed version to authorise access. In addition to the random number (R) the network sends a key number (K_n) to the MS, which is related to the ciphering key K_c and serves to avoid using different K_c keys at the receiver and transmitter. This key number K_n is then stored by the MS and is included in its first message to the network. Besides S, the MS computes the ciphering key (K_c) using another confidential algorithm (A8) stored in the SIM, and the input parameters K_i and R: $K_c=[K_i(A8)R]$. The ciphering key K_c is also computed in the network and hence no confidential information is sent unprotected via the radio path.

Once authentication is confirmed and both the network and the MS know K_c, the network issues a ciphering mode command and from now on all messages are ciphered at the transmitter and deciphered at the receiver, using the confidential algorithm (A5). Confidentiality is further enhanced by protecting the user's identity, when identification takes place assigning a Temporary Mobile Subscriber Identity (TMSI) valid for a specific location area. This TMSI uniquely describes the IMSI in a specific location area, but outside the area it must be associated with the Location Area Identity (LAI). The network, more precisely the Visitor Location Register (VLR) keeps track of the TMSI-IMSI association and allocates a new TMSI in each new location area update procedure, i.e., in each new VLR.

The following representative example is provided to describe one out of a variety of specific scenarios, where the authentication and ciphering algorithms described are utilised. We assume that the MS associated with a specific TMSI is registered in the VLR. All required MS characteristics are stored in the VLR and identification is based on the LAI and TMSI parameters. As mentioned, authentication is carried out upon each location updating and the set [IMSI, TMSI, K_c, K_n, R, S] is available in the VLR. The process is described with reference to Figure 8.50.

The MS stores its own set of [IMSI, TMSI, LAI, K_i, K_c, K_n] parameters

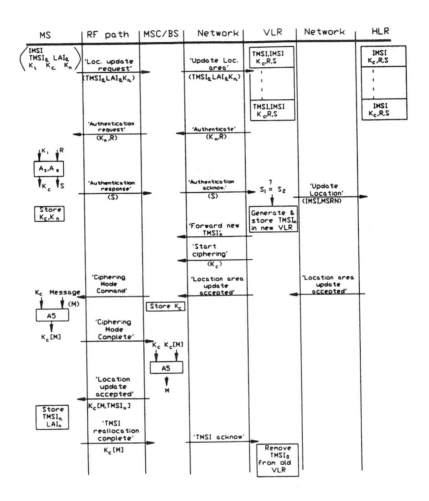

Figure 8.50: Location area update using the confidential algorithms A3, A5 and A8.

and requests location update via the radio path by sending its [TMSI_0, LAI_0, K_n] parameters to the MSC/BS. The MSC/BS forwards these via the network to the VLR, which issues an authentication request through the network to the MSC/BS and via the radio link to the MS by sending K_n and R. The MS computes K_c and S from K_i and R utilising the algorithms A3 and A8, which are stored in its SIM. The signed response S is transmitted back to the MSC/BS and from there via the network to the VLR, where authentication is performed by comparing the received and locally generated S parameters. The VLR updates the MS's location in its own HLR at the appropriate IMSI entry and assigns a MS Roaming Number (MSRN). The MSRN stored in the MS's HLR is then used by incoming calls to find the MS and route the calls to the appropriate VLR, where the momentary TMSI and LAI parameters locate and identify the called subscriber.

Simultaneously, the VLR also generates the new TMSI_n and forwards it to the MSC/BS and from now on TMSI_n, LAI_n and K_n are used to identify the MS. The HLR acknowledges the location update to the VLR and to the MSC/BS that, in turn, issues a 'ciphering mode command' to the MS. The MS responds with a 'ciphering mode complete' message and therefore ciphers all its messages by the algorthm A5 using K_c, while the MSC/BS deciphers and vice versa. The MSC/BS informs the MS using a ciphered message that 'location update is accepted' by the system and also sends out the new TMSI_n, which is acknowledged by the MS via sending the 'TMSI reallocation complete' message. Finally, this is accepted by the MSC/BS in that it sends a 'channel release' command to the MS and a 'TMSI acknowledge' to the old VLR to discard TMSI_0.

8.13   Telecommunication Services

The telecommunication services supported by a GSM PLMN are divided into two broad categories: bearer services and teleservices. The bearer services provide for the transmission of signals between access points (called user-network interfaces in ISDN), while the teleservices provide communications between users according to protocols established by the network operators. The teleservices thus include also the Terminal Equipment (TE) functions, see Figure 8.51.

The bearer services may include more than one transit network. The terminating network may include a GSM PLMN, either the originating one or another one. The terminal equipment (TE) may consist of one or more pieces of equipment—telephone set, Data Terminal Equipment (DTE), teletext terminal, etc. Both bearer services and teleservices are offered together with a set of supplementary services. A supplementary service modifies and/or supplements a basic telecommunication service and consequently it cannot be offered to a customer as a stand alone service.

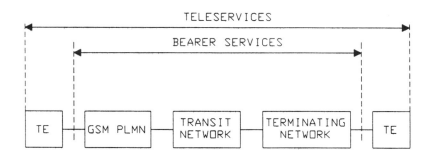

Figure 8.51: Bearer services and teleservices.

Bearer Services: The bearer services provide the user with the possibility of gaining access to various forms of communications. For example, information transfer between a user in a GSM PLMN and a user in a terminating network, including the same GSM PLMN, another GSM PLMN and other types of PLMNs. A bearer service involves only low layer attributes (layers 1–3 of the OSI model), such as information transfer, access, interworking, and general attributes. The information transfer capability is concerned with the transfer of different types of information, e.g., digitised speech, through a GSM PLMN and another network, or through a GSM PLMN only. The method of transfer, be it circuit switching or packet transportation is called the transfer mode, while the information transfer rate is the bit rate in circuit mode or the throughput rate in packet mode. The second group of attributes describes the access at the mobile station. The signalling access characterises the protocol on the signalling channel at the access point (V-series protocol, X-series protocol, etc.), while the information access describes the interface according to the protocol used to transfer user information at the access point (V-series interface, X-series interface, etc) and the bit rate in circuit mode, throughput rate in packet mode. The interworking attributes are concerned with the type of terminating network, such as GSM PLMN, PSTN, ISDN, etc., as well as with the terminal to terminating network interface, e.g., V-series interface, X-series interface, etc. Finally, the general attributes include the supplementary services, quality of service, service interworking, commercial and operational attributes. The following bearer services are supported by a GSM PLMN:

1. Asynchronous 300–9600 bit/s circuit switched data service interworking with the public switched telephone network (PSTN).

1) Circuit switched synchronous data transmission at 300–9600 bit/s, interworking with the PSTN, circuit switched public data networks (CSPDN) and ISDN.

2) Asynchronous 300–9600 bit/s packet assembler/disassembler (PAD)

access interworking with the packet switched public data network
(PSPDN).

3) Packet switched synchronous 2400–9600 bit/s data service interworking with the PSPDN.

The bearer services can be transparent or non-transparent. In a transparent service the error protection is provided only by Forward Error Correction (FEC). The non-transparent services have the additional protection of Automatic Repeat Request (ARQ) in the radio link protocol, which
results in higher data integrity. However, the extra error protection is
achieved at the expense of greater transmission delay and reduced throughput.

Teleservices: The teleservices provide the user with the possibility of
gaining access to various forms of applications, covering for example:

- Applications involving two terminals, which provide compatible or
  identical teleservice attributes at an access point in a GSM PLMN
  and an access point in a terminating network.

- Applications involving a terminal at one access point in a GSM PLMN
  and a system providing high layer functions (e.g., speech storage system, message handling system, etc.) located either within the GSM
  PLMN or in a terminating network.

A teleservice is characterised by a set of low level attributes and a set of
high level attributes. The low level attributes are the same as those used to
characterise the bearer services. High level attributes refer to functions and
protocols of layers 4 to 7 of the OSI model. They are concerned with the
transfer, storage and processing of user messages, provided by a subscriber
terminal, a retrieval centre or a network service centre. The high level
attributes include a variety of legitimate user information, such as speech,
short message, data, videotext, teletext, facsimile, as well as layers 4 to
7 protocol functions, which refer to the layer protocol characteristics of
the different teleservices. The teleservices supported by a GSM PLMN are
divided into six categories:

1. Transmission of speech information and voice band signalling tones
   of the PSTN/ISDN.

1) Short message service, which enables a user of a telecommunication
   network (e.g., PSTN) to send a short alphanumeric message (up to
   180 characters) to a mobile subscriber of the GSM network.

2) Message Handling System (MHS) access, providing the transmission
   of a short message from a message handling system in a fixed network
   (e.g., paging system) to a mobile station.

3) Videotext access.

4) Teletext transmission.

5) Facsimile transmission.

Supplementary Services: The supplementary services are divided into eight categories:

1. Number identification, entailing five services:

   - calling number identification presentation,
   - connected number identification presentation,
   - calling number identification restriction,
   - connected number identification restriction, and
   - malicious calls identification.

   The 'identification presentation' services provide for the ability to indicate the number of the calling/connected party, while the 'identification restriction' services offer to the calling/connected party the ability to restrict presentation of the party's number.

1) Call offering. A group of eight different services: six 'call forwarding' services, a 'call transfer' service and a 'mobile hunting access' service. The call forwarding services permit a called mobile subscriber to have the network send all incoming calls, or just those associated with a specific basic service, addressed to the called mobile subscriber's directory number to another directory number. The call forwarding can be: 1) unconditional, 2) on condition the mobile subscriber is busy, 3) when there is no reply, 4) if the subscriber cannot be reached due to radio congestion, 5) when there is no paging response and 6) whenever the mobile subscriber is not registered. The 'call transfer' service enables the served mobile subscriber to transfer an established incoming or outgoing call to a third party. The 'mobile hunting access' service can be used only by a mobile PABX and enables all incoming calls to be distributed over a group of accesses, belonging to the mobile PABX.

2) Call completion. This group is divided into three services:

   - call waiting,
   - call hold, and
   - completion of calls to busy subscribers.

   The 'call waiting' service allows the mobile subscriber to be notified of an incoming call whilst the termination is in a busy state. The subscriber can subsequently either answer, reject or ignore the incoming call. The 'call hold' service allows a served mobile subscriber to interrupt communication on an existing call and then subsequently to re-establish communication. The 'completion of calls to busy subscribers' service allows a calling mobile subscriber which encounters a busy called subscriber to be notified when the called subscriber becomes unengaged and have the call re-initiated.

3) Multi party. Two services are offered in this group:

   • three party, and
   • conference calling.

   The first one enables a mobile subscriber to establish a three party conversation, while the second service provides the mobile subscriber with the ability to have a multi-connection call, i.e., simultaneous communication between more than two parties.

4) Closed user group service: allows a group of subscribers, connected to the PLMN and/or the ISDN, to intercommunicate only amongst themselves. If required, one or more subscribers may be provided with incoming/outgoing access to subscribers outside this group.

5) Charging, including three services:

   • advice of charge, which allows the subscriber to receive charging information related to the used telecommunication services,
   • freephone service, allowing the served mobile subscriber to be reached with a freephone number and charged for these calls, and
   • reverse charging, which allows a called mobile subscriber to be charged for the usage-based calls.

6) Additional information transfer. A user-to-user signalling service allowing a mobile subscriber to send/receive a limited amount of information to/from another PLMN or ISDN subscriber over the signalling channel.

7) Call restriction. This group of services makes it possible for a mobile subscriber to prevent outgoing or incoming calls. There are seven different types of service:

   • barring of all outgoing calls,
   • barring of all outgoing international calls directed to non-CEPT countries,
   • barring of all outgoing international calls except those directed to the home PLMN country,
   • barring of all outgoing calls when roaming outside the home PLMN country,

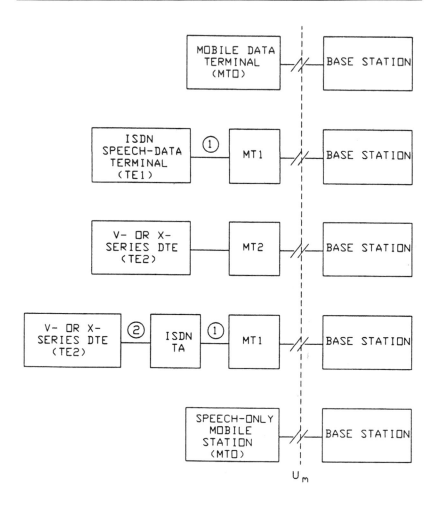

Figure 8.52: GSM PLMN access points.

- barring of all incoming calls, and
- barring of all incoming calls when roaming outside the home PLMN country.

Access to the GSM Network: The different access points in a GSM network are shown in Figure 8.52. At access points 1 and 2 bearer services may be accessed, while at access point 3 teleservices are accessed. All terminal equipment accessing a GSM PLMN interface at one of these access points must meet the specifications of the protocols at that interface.

The Mobile Station (MS) is shown to consist of Mobile Termination (MT) and Terminal Equipment (TE). The mobile termination supports functions associated with the management of the radio interface U_m and flow control of user data between interface U_m and access points 1 or 2. These functions include:

— radio transmission termination

— radio channel management

— error protection for information sent across the radio path

— speech encoding/decoding

— flow control and mapping of user data and signalling

— rate adaptation of user data between the radio channel rate and user rates

— multiple terminal support

— mobility management

There are three types of mobile termination: MT0, a fully integrated mobile station, including data terminal and its adaptation functions, MT1 which includes ISDN terminal adaptation functions, and MT2 which includes CCITT V- or X-series terminal adaptation functions.

The terminal equipment may consist of one or more pieces of equipment such as telephone set, Data Terminal Equipment (DTE), teletext terminal, etc. The TE1 type equipment represents an ISDN interface and the TE2 equipment presents a non-ISDN interface, e.g., CCITT V- or X-series interface. A non-ISDN terminal (TE2) may be connected to an MT1 type termination using Terminal Adapter (TA).

8.14 Summary

The GSM system's salient features were summarised in this chapter. Time Division Multiple Access (TDMA) with eight users per carrier is used at a multi-user rate of 271 kbit/s, demanding a channel equaliser to combat dispersion in large cell environments. The error protected chip-rate of the full-rate traffic channels is 22.8 kbit/s, while in half-rate channels it is 11.4 kbit/s. Apart from the full- and half-rate speech traffic channels there are five different-rate data traffic channels and 14 various control and signalling channels to support the system's operation. A moderately complex, 13 kbit/s Regular Pulse Excited speech codec with a long term predictor (LTP) is used, combined with an embedded three-class error correction codec and multi-layer interleaving to provide sensitivity-matched unequal error protection for the speech bits. An overall speech delay of 57.5 ms is maintained. Slow frequency hopping at 217 hops yields substantial performance gains for slowly moving pedestrians.

System feature	Specification
Uplink bandwidth	890-915=25 MHz
Downlink bandwidth	935-960=25 MHz
Total GSM bandwidth	50 MHz
Carrier spacing	200 kHz
No. of RF carriers	125
Multiple access	TDMA
No. of users/carrier	8
Total no. of channels	1000
TDMA burst rate	271 kbit/s
Modulation	GMSK with BT=0.3
Bandwidth efficiency	1.35 bit/s/Hz
Channel equaliser	yes
Speech coding rate	13 kbit/s
FEC coded speech rate	22.8 kbit/s
FEC coding	Embedded block/convolutional
Frequency hopping	217 hops/s
DTX and VAD	yes
Maximum cell radius	35 km

Table 8.13: Summary of GSM features.

Constant envelope partial response GMSK with a channel spacing of 200 kHz is deployed to support 125 duplex channels in the 890-915 MHz up link and 935-960 MHz down link bands, respectively. At a transmission rate of 271 kbit/s a spectral efficiency of 1.35 bit/s/Hz is achieved. The controlled GMSK-induced and uncontrolled channel-induced inter-symbol interferences are removed by the channel equaliser. The set of standardised wide-band GSM channels was introduced in order to provide bench-markers for performance comparisons. Efficient power budgeting and minimum co-channel interferences are ensured by the combination of adaptive power- and handover-control based on weighted averaging of up to eight uplink and downlink system parameters. Discontinuous transmissions assisted by reliable spectral-domain voice activity detection and comfort-noise insertion further reduce interferences and power consumption. Due to ciphering, no unprotected information is sent via the radio link. As a result, spectrally efficient, high-quality mobile communications with a variety of services and international roaming is possible in cells of up to 35 km radius for signal to noise- and interference-ratios in excess of 10-12 dB. The key system features are summarised in Table 8.13.

∗   ∗

Our aim in this chapter was to review the components of the GSM system and to consider their interconnections. In this system design study we considered mainly the system's physical layer functions, giving some cognisance to network-layer functions as well. We relied on all of the previous chapters, detailing the features of mobile channels, the speech codec, the channel codec, GMSK modulation, Viterbi equalisation, frequency hopping, etc. Over the past decade GSM has become the most wide spread international system of mobile communications across the globe. However, the services offered by GSM are limited to relatively low-rate data, speech, email and fax. The next chapter considers a range of powerful multimedia systems based on bandwidth-efficient multi-level modulation, which can accommodate more users in a given bandwidth than their second generation counterparts and hence may offer an evolutionary path for these existing systems.

Bibliography

[1] *Proceedings of the Nordic Seminar on Digital Land Mobile Radio Communication (DMR), Espoo, Finland,* February 1985.

[2] *Proceedings of the Second Nordic Seminar on Digital Land Mobile Radio Communication (DMRII), Stockholm, Sweden,* October 1986.

[3] *Proceedings of the International Conference on Digital Land Mobile Radio Communication (ICDMC) Venice, Italy,* June/July 1987.

[4] *Proceedings of Digital Cellular Radio Conference, Hagen, FRG,* October 12-14, 1988.

[5] **A. Moloberti.** "Definition of the radio subsystem for the GSM pan-European digital mobile communication system". *Proc. of ICDMC, Venice, Italy,* pp. 37-46, June/July 1987.

[6] **A.W.D. Watson.** "Comparison of the contending multiple access methods for the pan-European mobile radio systems". *IEE Colloquium, Digest No:1986/95,* pp. 2/1-2/6, 7 October 1986.

[7] **E. Natvig.** "Evaluation of six medium bit-rate coders for the pan-European digital mobile radio system". *IEEE Journal on Selected Areas in Communications, vol.6, no.2,* pp. 324-334, February 1988.

[8] **D.M. Balston.** "Pan-European cellular radio: or 1991 and all that". *Electronics and Communication Engineering Journal,* pp. 7-13, January/February 1989.

[9] *Group Speciale Mobile (GSM) Recommendation,* April 1988.

[10] **P. Vary and R.J. Sluyter.** "MATS-D speech codec: regular-pulse excitation LPC". *Proc. of the Second Nordic Seminar on Digital Land Mobile Radio Communication (DMRII), Stockholm, Sweden,* pp. 257-261, October, 1986.

[11] **P. Vary and R. Hoffmann.** "Sprachcodec für das europäische Funkfernsprechnetz". *Frequenz 42 (1988) 2/3,* pp. 85-93, 1988.

[12] **J. Schur.** "Über Potenzreihen, die im Innern des Einheitskreises beschränkt sind". *Journal für die reine und angewandte Mathematik, Bd 147*, pp. 205-232, 1917.

[13] **W. Webb, L. Hanzo, R. Salami and R. Steele.** "Does 16-QAM provide an alternative to a half-rate GSM speech codec?". *Proc. of IEEE-VT Conf., St.-Louis, Missouri, U.S.A.,* May 1991.

[14] **I.A. Gerson, M.A. Jasiuk, J-M. Muller, J.M. Nowack, E.H. Winter**, "Speech and channel coding for the half-rate GSM channel," *Proceedings ITG-Fachbericht,* vol. 130, pp. 225–233, November 1994.

[15] **I.A. Gerson and M.A. Jasiuk**, "Vector sum excited linear prediction (VSELP) speech coding at 8 kbps," pp. 461–464.

[16] **I.A. Gerson and M.A. Jasiuk**, "Techniques for improving the performance of CELP-type speech codecs," *IEEE JSAC,* vol. 10, pp. 858–865, June 1992.

[17] **I.A. Gerson**, "Method and means of determining coefficients for linear predictive coding." US Patent No 544,919, October 1985.

[18] **A. Cumain**, "On a covariance-lattice algorithm for linear prediction," *Proceedings of IEEE ICASSP '82*, pp. 651–654, May 1982.

[19] **R. Salami, C. Laflamme, B. Besette, J-P.Adoul, K. Jarvinen, J. Vainio, P. Kapanen, T. Hankanen and P. Haavisto**, "Description of the GSM enhanced full rate speech codec," *Proc. of ICC'97*, 1997.

[20] "PCS1900 Enhanced Full Rate Codec US1." SP-3612.

[21] **L. Hanzo, F.C.A. Brooks and J.P. Woodard**, Modern Voice Compression and Communications: Principles and applications for fixed and wireless channels, IEEE Press, in preparation [3]

[22] **J.P. Adoul, P. Mabilleau, M. Delprat and S. Morissette**, "Fast CELP Coding Based on Algebraic Codes," *Proc. ICASSP*, pp. 1957–1960, April 1987.

[23] **K.H.H. Wong.** "Transmission of channel coded speech and data over mobile radio channels". *PhD Thesis, Dept. of Electronics and Computer Science, University of Southampton,* 1989.

[24] **L. Hanzo, K.H.H. Wong and R. Steele.** "Efficient channel coding and interleaving schemes for mobile radio communications". *Proc. of IEE Colloq., Savoy Place, London, U.K.,* 22 February 1988.

[25] **L.B. Lopes.** "GSM radio link simulation". *IEE Colloquium, University research in Mobile Radio,* pp. 5/1-5/4, 1990.

---

[3] For detailed contents please refer to http://www-mobile.ecs.soton.ac.uk

[26] **J.C.S. Cheung and R. Steele.** "Modified Viterbi equaliser for mobile radio channels having large multi-path delay". *Electronics Letters, vol.25, no.19,* pp. 1309-1311, 14 Sept., 1989

[27] **N.S. Hoult, C.A. Dace and A.P. Cheer.** "Implementation of an equaliser for the GSM system". *Proc. of the 5th Int. Conf. on Radio Receivers Associated Systems, Cambridge, U.K.,* 24-26 July, 1990.

[28] **R.D'Avella, L. Moreno and M. Sant'Agostino.** "An adaptive MLSE receiver for TDMA digital mobile radio". *IEEE Journal on Selected Areas in Communications, vol.7, no.1,* pp. 122-129, January 1989.

[29] **J.C.S. Cheung.** "Receiver techniques for wideband time division multiple access mobile radio systems". *PhD Mini-Thesis, Univ. of Southampton,* 1990.

[30] **J.B. Anderson, T. Aulin and C.E. Sundberg.** *Digital phase modulation,* Plenum Press, 1986.

[31] **M.R.L. Hodges, S.A. Jensen and P.R. Tattersall.** "Laboratory testing of digital cellular radio systems". *BTRL Journal, vol.8, no.1,* pp. 57-66, January 1990.

[32] **L. Hanzo, R. Steele and P.M. Fortune.** "A subband coding, BCH coding and 16-QAM system for mobile radio speech communications". *IEEE Tr. on VT., Vol. 39,* pp. 327-340, November 1990.

[33] **D.J. Targett and H.R. Rast.** "Handover-enhanced capabilities of the GSM system". *Proc. of Digital Cellular Radio Conference, Hagen, FRG,* pp. 3C/1-3C/11, October 12-14, 1988.

[34] **E. Bacs and L. Hanzo.** "A simple real-time adaptive speech detector for SCPC systems". *Proc of ICC'85, Chicago,* pp. 1208-1212, May, 1985.

[35] **J.A. Jankowski.** "A new digital voice-activated switch". *Comsat Tech. Journal, vol.6, no.1,* pp. 159-170, Spring 1976.

[36] **D.K. Freeman, G. Cosier, C.B. Southcott and I. Boyd.** "The voice activity detector for the pan-European digital cellular mobile telephone service". *Proc. of ICASSP'89, Glasgow,* pp. 369-372, 23-26. May, 1989.

[37] **S. Hansen.** "Voice activity detection (VAD) and the operation of discontinuous transmission (DTX) in the GSM system". *Proc. of Digital Cellular Radio Conference, Hagen, FRG,* pp. 2b/1-2b/14, October 12-14, 1988.

[38] **P.C.J. Arend.** "Security aspects and the implementation in the GSM system". *IBID.,* pp. 4a/1-4a/7, October 12-14, 1988.

Glossary

A3	Authentication algorithm
A5	Cyphering algorithm
A8	Confidential algorithm to compute the cyphering key
AB	Access burst
ACCH	Associated control channel
ADC	Administration centre
AGCH	Access grant control channel
AUC	Authentication centre
AWGN	Additive gaussian noise
BCCH	Broadcast control channel
BER	Bit error ratio
BFI	Bad frame indicator flag
BN	Bit number
BS	Base station
BS-PBGT	BS power budget: to be evaluated for power budget motivated handovers
BSIC	Base station identifier code
CC	Convolutional codec
CCCH	Commom control channel
CELL_BAR_ACCESS	Boolean flag to indicate, whether the MS is permitted to access the specific traffic cell
CNC	Comfort noise computation

771

CNI	Comfor noise insertion
CNU	Comfort noise update state in the DTX handler
DB	Dummy burst
DL	Down link
DSI	Digital speech interpolation to improve link efficiency
DTX	Discontinuous transmission for power consumption and interference reduction
EIR	Equipment identity register
EOS	End of speech flag in the DTX handler
FACCH	Fast associated control channel
FCB	Frequency correction burst
FCCH	Frequency correction channel
FEC	Forward error correction
FH	Frequency hopping
FN	TDMA frame number
GMSK	Gaussian minimum shift keying
GP	Guard space
HGO	Hangover in the VAD
HLR	Home location register
HO	Handover
HOCT	Hangover counter in the VAD
HO_MARGIN	Handover margin to facilitate hysteresis
HSN	Hopping sequence number: frequency hopping algorithm's input variable
IMSI	International mobile subscriber identity
ISDN	Integrated services digital network
LAI	Location area identifier
LAR	Logarithmic area ratio
LTP	Long term predictor
MA	Mobile allocation: set of legitimate RF channels, input variable in the frequency hopping algorithm

MAI	Mobile allocation index: output variable of the FH algorithm
MAIO	Mobile allocation index offset: intial RF channel offset, input variable of the FH algorithm
MS	Mobile station
MSC	Mobile switching centre
MSRN	Mobile station roaming number
MS_TXPWR_MAX	Maximum permitted MS transmitted power on a specific traffic channel in a specific traffic cell
MS_TXPWR_MAX(n)	Maximum permitted MS transmitted power on a specific traffic channel in the n-th adjacent traffic cell
NB	Normal burst
NMC	Network management centre
NUFR	Receiver noise update flag
NUFT	Noise update flag to ask for SID frame transmission
OMC	Operation and maintenance centre
PARCOR	Partial correlation
PCH	Paging channel
PCM	Pulse code modulation
PIN	Personal identity number for MSs
PLMN	Public land mobile network
PLMN_PERMITTED	Boolean flag to indicate, whether the MS is permitted to access the specific PLMN
PSTN	Public switched telephone network
QN	Quater bit number
R	Random number in the authentication process
RA	Rural area channel inpulse response
RACH	Random access channel
RF	Radio frequency
RFCH	Radio frequency channel
RFN	Reduced TDMA frame number: equivalent representation of the TDMA frame number, which is used in the synchronisation chanel

RNTABLE	Random number table utilised in the frequency hopping algoirthm
RPE	Regular pulse excited
RPE-LTP	Regular pulse excited codec with long term predictor
RS-232	Serial data transmission standard equivalent to CCITT V24. interface
RXLEV	Received signal level: parameter used in hangovers
RXQUAL	Received signal quality: parameter used in hangovers
S	Signed response in the authentication process
SACCH	Slow associated control channel
SB	Synchronisation burst
SCH	Synchronisation channel
SCPC	Single channel per carrier
SDCCH	Stand-alone dedicated control channel
SE	Speech extrapolation
SID	Silence identifier
SIM	Subscriber identity module in MSs
SPRX	Speech received flag
SPTX	Speech transmit flag in the DTX handler
STP	Short term predictor
TA	Timing advance
TB	Tailing bits
TCH	Traffic channel
TCH/F	Full-rate traffic channel
TCH/F2.4	Full-rate 2.4 kbps data traffic channel
TCH/F4.8	Full-rate 4.8 kbps data traffic channel
TCH/F9.6	Full-rate 9.6 kbps data traffic channel
TCH/FS	Full-rate speech traffic channel
TCH/H	Half-rate traffic channel
TCH/H2.4	Half-rate 2.4 kbps data traffic channel
TCH/H4.8	Half-rate 4.8 kbps data traffic channel

TDMA	Time division multiple access
TMSI	Temporary mobile subscriber identifier
TN	Time slot number
TU	Typical urban channel inpulse response
TXFL	Transmit flag in the DTX handler
UL	Up link
VAD	Voice activity detection
VE	Viterbi equaliser
VLR	Visiting location register

Chapter **9**

Wireless QAM-based Multi-media Systems: Components and Architecture

L. Hanzo[1]

9.1   Motivation and Background

Previous chapters of this book attempted to portray the state-of-the-art of various system components, such as source and channel codecs, modems, multiple access schemes, etc. used in mobile communications, leading to a detailed discussion on the Pan-European GSM system in Chapter 8, which is the most widespread operational cellular system world-wide at the time of writing. The GSM system's example was invoked, in order to amalgamate the various system components in a system design study. This chapter has a similar goal in the context of modern wireless multi-media systems and endeavours to speculate on some of the evolutionary features of TDMA-based systems, while providing a system design study. We will amalgamate a range of system components into multi-media systems [1], portraying the interconnection of system components and characterising the expected system performance. Apart from the employment of low-rate speech and video codecs, multi-level modems [2] can also substantially

---

[1]University of Southampton and Multiple Access Communications Ltd

contribute towards reducing the required user bandwidth and can accommodate either an increased number of users, or provide potentially higher user bit-rates, if their increased channel SNR and co-channel interference requirements can be satisfied. Hence, Quadrature Amplitude Modulation (QAM) schemes are introduced in this chapter and their ability to support multi-mode operation is demonstrated. Specifically, when the instantaneous channel quality or SNR is increased, an increased number of bits per symbol can be transmitted, supporting an increased bit-rate. When the channel quality degrades, a more error-resilient, but less bandwidth-efficient modem mode can be invoked. At the time of writing the feasibility of such systems is being researched, for example as a potential evolutionary path for the well-established and widespread GSM system or for other future arrangements.

While the second-generation digital mobile radio systems of Table 1.1 are now widespread across the globe, researchers endeavour world wide to define the **third-generation** personal communications network (PCN), which is referred to as a personal communications system (PCS) in North America. The European Community's Research in Advanced Communications Equipment (RACE) programme [3, 4] and the follow-up research framework referred to as Advanced Communications Technologies and Services (ACTS) programme [5] spear-headed these initiatives in Europe. Similar campaigns were also conducted in Japan and the USA. In the European RACE programme there were two dedicated projects, endeavouring to resolve the on-going debate as regards to the most appropriate multiple access scheme, studying Time Division Multiple Access (TDMA) and Code Division Multiple Access (CDMA). The basic advantages and disadvantages of these multiple access schemes were highlighted in Chapter 1 and the most prominent TDMA system, namely the GSM system, was the topic of Chapter 8. At the time of writing in the third-generation era, however, CDMA seems to be emerging as the favourite in Europe [17], Japan and the USA, although the proposed CDMA systems are different from eachother. The design aspects of CDMA systems and the emerging third-generation European, Japanese and American system proposals will be the topic of Chapter 10, while the second-generation Pan-American so-called IS-95 CDMA system [16] was highlighted in Chapter 1.

A common requirement of all modern wireless systems is the ability to support services on a more flexible basis, than the somewhat rigid second-generation standards, promising to allocate bit-rates upto 2 Mbps on a demand basis. Therefore third-generation systems are expected to be more amenable to wireless multi-media transmission, than second generation systems. A wide variety of further associated aspects of modern wireless systems were treated in references [3]- [13].

The range of existing and future systems can be characterised with the aid of Figure 9.1 in terms of their expected **grade of mobility and bit-rate**, which are the two most fundamental parameters in terms of determining the systems' potential in terms of wireless multi-media applications. Specifically, the fixed networks are evolving from the basic 2.048 Mbit/s Integrated Services Digital Network (ISDN) towards higher-rate broad-band ISDN or B-ISDN. In comparison to these fixed networks, a higher grade of mobility, which we refer to here as **portability**, is a feature of cordless telephones (CTs), such as the Digital European Cordless Telephone (DECT), the British CT2 and the Japanese Personal Handyphone (PHP) systems, although their transmission rate is more limited than that of the fixed ISDN network. Recall that these systems were characterised in Table 1.1. The DECT system is the most flexible CT amongst them, allowing the multiplexing of 23 single-user channels in one of the duplex links between the portable station (PS) and base station (BS), which provides rates up to 23×32 kbps = 736 kbps for advanced services - although this bit-rate potential is eroded to around 500 kbps due to the various control channel overheads encountered. As suggested by Figure 9.1, wireless local area networks (WLAN) can support higher bit-rates of up to 155 Mbits/s in order to extend existing Asynchronous Transfer Mode (ATM) links to portable terminals, but they usually do not support full mobility functions, such as location update or hand-over from one base station to another. Another ambitious European initiative is targeted at high-rate, high-mobility system studies hallmarked by the so-called Mobile Broadband System (MBS), which is also featured in Figure 9.1. By contrast, as seen in the Figure, contemporary second-generation Public Land Mobile Radio (PLMR) systems, such as the Pan-European GSM, the American IS-54 and the Japanese Digital Cellular (JDC) systems cannot support high bit-rate services, since they typically have to communicate over lower quality, dispersive mobile channels, but they exhibit the highest grade of mobility - in the case of GSM including also high-speed international roaming capabilities. Again, the basic features of second-generation systems were summarised in Table 1.1 of Chapter 1.

Again, in this chapter we turned our attention to specific algorithmic and system architectural aspects of complete voice/video multi-media transceivers, which may be employed as evolutionary successors of existing second-generation systems, such as GSM or IS-54. We also evaluated their expected performance. Hence the system bandwidth was assumed to be 30 kHz, as in the American IS-54 standard [14], which allowed us to assess the potential of the proposed scheme in the context of a well-known existing system. As a further system design study, we also contrived a slightly more complex intelligent multi-mode transceiver, which was studied in the context of the 200 kHz bandwidth Pan-European GSM system.

This chapter is organised as follows. Section 9.2 gives a brief overview of recent developments in speech coding and, with reference to Chapter 3, it describes the 4.8 kbit/s so-called transformed binary pulse excited (TBPE) speech codec used in our system study. This is followed by the portrayal of a bit sensitivity analysis technique invoked in order to assist in mapping the

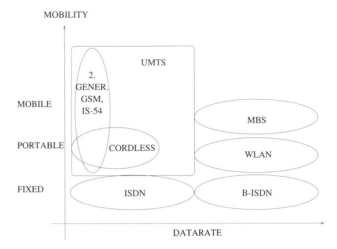

Figure 9.1: Stylised mobility versus bit-rate plane classification of existing and future wireless systems.

speech bits to different bit-protection classes, employing source-sensitivity matched error protection. A brief discussion is provided on the design of video codecs, in particular on that of the proposed fixed-rate video codec and the ITU H.263 standard video scheme in Section 9.3, while Section 9.5 is focused on the choice of modulation, in particular on 16-level quadrature amplitude modulation (16-QAM). Since the so-called full-response, linearly amplified modulation techniques have not been treated in previous chapters of this book, more attention is devoted to this topic than to other, previously considered system components. Section 9.6 highlights how packet reservation multiple access (PRMA) improves the efficiency of the TDMA radio link by surrendering passive time slots for active users contending for an available slot. Finally, two novel voice-video systems are proposed and investigated in Section 9.7 in the context of the 30 kHz bandwidth IS-54 system and the 200 kHz bandwidth GSM system, in order to be able to relate their performance to that of these well-known systems, before concluding in Section 9.8.

9.2   Speech Coding Aspects

9.2.1   Recent Speech Coding Advances

Let us commence our discourse on speech coding aspects with a brief overview of the recent speech compression literature [20]- [48], noting that Chapter 3 provided an in-depth treatment of speech coding. Following the International Telecommunications Union's (ITU) 64 kbit/s Pulse Code

Modulation (PCM) and 32 kbps Adaptive PCM (ADPCM) G.721 standards, in 1986 the 13 kbit/s Regular Pulse Excitation (RPE) [28,29] codec was selected for the Pan-European mobile system known as GSM, and more recently Vector Sum Excited Linear Prediction (VSELP) [30,31] codecs operating at 8 and 6.7 kbit/s were favoured in the American IS-54 and the JDC wireless networks. These developments were followed by the 4.8 kbit/s American Department of Defence (DoD) codec [32]. The state-of-art was documented in a range of excellent monographs by O'Shaughnessy [33], Furui [34], Anderson and Mohan [35], Kondoz [36], Kleijn and Paliwal [37] and in a tutorial review by Gersho [38]. More recently the 5.6 kbit/s half-rate GSM quadruple-mode Vector Sum Excited Linear Predictive (VSELP) speech codec standard developed by Gerson et al [39] was approved, while in Japan the 3.45 kbit/s half-rate JDC speech codec invented by Ohya, Suda and Miki [40] using the so-called Pitch Synchronous Innovation (PSI) CELP principle was standardised. Other currently investigated schemes are the Prototype Waveform Interpolation (PWI) proposed by Kleijn [41], Multi-Band Excitation (MBE) suggested by Griffin and Lim [42] and Interpolated Zinc Function Prototype Excitation (IZFPE) codecs advocated by Hiotakakos and Xydeas [43]. In the low-delay, but more error sensitive backward adaptive class the 16 kbps ITU G.728 codec [44] developed by Chen *et.al.* from the AT&T speech team hallmarks a significant step. This was followed by the equally significant development of the more robust, forward-adaptive 10 ms delay G.728 ACELP arrangement proposed by the Cherbrook team [46,47], AT&T and NTT [48]. Lastly, the standardisation of the 2.4 kbps DoD codec led to intensive research in this very low-rate range and the Mixed Excitation Linear Predicitve (MELP) codec by Texas Instrument was identified [49] in 1996 as the best overall candidate scheme.

Following the above speech coding review let us now briefly concentrate on the specific 4.8 kbps codec employed in our 30-kHz bandwidth multimedia system.

9.2.2   The 4.8 kbit/s Speech Codec   [22, 51, 55]

Again, speech codecs were discussed in depth in Chapter 3, where we have shown that in code excited linear predictive (CELP) codecs a Gaussian process with slowly varying power spectrum is used to represent the residual signal after short-term and long-term prediction, and the speech waveform is generated by filtering Gaussian distributed stochastic excitation vectors through the time-varying linear pitch and LPC synthesis filters [50]. Here we follow the approach of Section 3.4, since at bit-rates around 4.8 kbps CELP codecs and their derivatives are the most successful schemes and restrict our discussion on speech codecs to a terse summary, in order to allow readers to consult this chapter in isolation from the rest of the book.

More specifically, in the CELP codec of Figure 3.34 in Section 3.4 the Gaussian distributed excitation vectors of dimension N are either stored in a codebook or generated in real-time, in order to avoid excessive storage requirements and the optimum excitation sequence is determined by the exhaustive search of the excitation codebook. The codebook entries $c_k(n)$ of Figure 3.34, $k = 1 \ldots L$, $n = 0 \ldots N-1$, after scaling by a gain factor G_k, are filtered through the synthesis filter, in order to produce the weighted synthetic speech $\tilde{s}_w(n)$, which is compared to the weighted original speech $s_w(n)$ for finding the specific codebook entry, which results in the best possible N-sample synthetic speech segment.

However, for a typical excitation frame length of $N = 40$ and codebook size $L = 1024$, the complexity of the original CELP codec proposed by Atal and Schroeder [52, 53] becomes excessively high for real-time implementation. Hence a plethora of computationally efficient solutions have been suggested in the literature [30,31,51,54] in order to ease the computational load encountered, while still maintaining perceptually high speech quality. As mentioned above, the VSELP principle [30, 31] was favoured in the American IS-54 [14], the JDC [15] and in the Pan-European half-rate GSM standards [39], while the Algebraic CELP (ACELP) [54] excitation was incorporated in the ITU G.729 [46] and G.723 recommendations. When the bit-rate is reduced below about 4.8 kbps, other approaches, such as that employed in the Japanese 3.45 kbit/s half-rate JDC speech codec invented by Ohya, Suda and Miki [40] using the so-called Pitch Synchronous Innovation (PSI) CELP can be employed. Multi-Band Excitation (MBE) suggested by Griffin and Lim [42], the Interpolated Zinc Function Prototype Excitation (IZFPE) codecs proposed by Hiotakakos and Xydeas [43] and the DoD MELP codec [49] are also efficient at rates below 4.8 kbps.

In our experiments here we opted for the 4.8 kbps transformed binary pulse excited (TBPE) speech codec of Section 3.5.1 proposed by Salami [22, 51,55], but our system-design hints are applicable to any other 4.8 kbps codec, such as the DoD codec [32] or the ACELP scheme of Section 3.4.2.4 and reference [19]. A leading-edge half-rate system can also be contrived on the basis of the previously mentioned US DoD 2.4 kbps MELP codec, which can double the number of users supported in the 30-kHz bandwidth of the proposed system. The attraction of TBPE codecs when compared to CELP codecs accrues from the fact that the excitation optimisation can be achieved in a direct computation step [51], as it was shown in Section 3.5.1.

The TBPE algorithm of Section 3.5.1 is summarised here briefly for convenience, where the Gaussian excitation vector is assumed to take the form of:

$$\mathbf{c} = \mathbf{Ab}, \qquad (9.1)$$

and the binary vector $\mathbf{b}$ has M elements of ± 1, while the $M \times M$ matrix $\mathbf{A}$ represents an orthogonal transformation. Due to the orthogonality of $\mathbf{A}$ the binary excitation pulses of $\mathbf{b}$ are transformed into independent, unit variance Gaussian components of $\mathbf{c}$. The set of 2^M binary excitation vectors gives rise to 2^M Gaussian vectors of the original CELP codec.

Parameter	Number of Bits
10 LSFs	36
LTPD	$2 \cdot 7 + 2 \cdot 5$
LTPG	$4 \cdot 3$
GP	$4 \cdot 2$
EG	$4 \cdot 4$
Excitation	$4 \cdot 12$
Total	144/30 ms

Table 9.1: Bit-allocation scheme of the 4.8 Kbit/s TBPE codec.

The block diagram of the TBPE codec was shown in Figure 3.41 of Chapter 3. As seen in the Figure, the weighted synthetic speech is generated for all $2^M = 1024$ codebook vectors and subtracted from the weighted input speech in order to find the one resulting in the best synthesised speech quality. The synthetic speech is generated at the output of the weighted synthesis filter, which is excited by the vectors given by the superposition of the adaptive codebook vector scaled by the long term predictor gain (LTPG) - which is synonymously also referred to as the adaptive codebook gain - and that of the orthogonally transformed binary vectors output by the binary pulse generator scaled by the stochastic excitation gain.

The bit allocation of our TBPE codec is summarised in Table 9.1. We note that a similar 4.8 kbps Algebraic Code Excited Linear Predicitve (ACELP) codec exhibiting the same bit-allocation scheme - apart from the different encoding of the excitation vectors - can be designed using the excitation model of Section 3.4.2.4. This ACELP excitation model was used in our system proposed in reference [19]. Returning to the bit-allocation table, the spectral envelope is represented by ten line spectrum frequencies (LSFs), which are scalar quantised using 36 bits, as it was detailed in Chapter 3. The 30 ms long speech frames hosting 240 samples are divided into four 7.5 ms subsegments having 60 samples. The subsegment excitation vectors **b** have 12 transformed duo-binary samples with a pulse-spacing of $D = 5$. The long term predictor (LTP) delays (LTPD) are quantised with seven bits in odd and five bits in even indexed subsegments, while the LTP gain (LTPG) is quantised with three bits. The excitation gain (EG) factor is encoded with four bits, while the grid position (GP) of candidate excitation sequences by two bits. A total of 28 or 26 bits per subsegment is used for quantisation, which yields $36 + 2 \cdot 28 + 2 \cdot 26 = 144$ bits/30 ms, i.e. a bit-rate of 4.8 kbit/s. In the next subsection we will give a rudimentary introduction to objective speech quality measures, which will be used in our bit-sensitivity evaluation carried out in order to design an appropriate embedded error correction scheme. Let us now provide a brief introduction to speech quality measures, which can be used in our bit sensitivity investigations.

9.2.3 Speech Quality Measures

In general the speech quality of communications systems is difficult to assess and quantify. However, in our system performance evaluations an easily evaluated objective speech quality measure is needed. The most reliable speech quality evaluation methods are based on subjective quality assessments, such as the so-called mean opinion score (MOS), which uses a five-point scale between one and five. MOS-tests use evaluation of speech by untrained listeners, but their results depend on the test conditions. Specifically, the selection and ordering of the test material, the language, and listener expectations all influence their outcome. A variety of other subjective measures is discussed in references [56]- [58], but subjective measures are tedious to derive and difficult to quantify during system development.

Objective speech quality measures do not provide results that could be easily converted into MOS values, but they facilitate quick comparative measurements during research and development. Most objective speech quality measures quantify the distortion between the speech communications system's input and output either in the time or frequency domain. The conventional SNR can be defined as

$$\text{SNR} = \frac{\sigma_{in}^2}{\sigma_e^2} = \frac{\sum_n s_{in}^2(n)}{\sum_n [s_{out}(n) - s_{in}(n)]^2}, \tag{9.2}$$

where $s_{in}(n)$ and $s_{out}(n)$ are the sequences of input and output speech samples, while σ_{in}^2 and σ_e^2 are the variances of the input speech and that of the error signal, respectively. A major drawback of the conventional SNR is its inability to give equal weighting to high- and low-energy speech segments, because its computation will be dominated by the higher-energy voiced speech segments. Therefore the reconstruction fidelity of voiced speech is given higher priority than that of low-energy unvoiced sounds, when computing the arithmetic mean of the SNR. Hence a system optimised for maximum SNR usually is suboptimum in terms of subjective speech quality.

Some of the problems of SNR computation can be overcome by using the segmental SNR (SEGSNR)

$$\text{SEGSNR}^{dB} = \frac{1}{M} \sum_{m=1}^{M} 10 \log_{10} \frac{\sum_{n=1}^{N} s_{in}^2(n)}{\sum_{n=1}^{N} [s_{out}(n) - s_{in}(n)]^2}, \tag{9.3}$$

where N is the number of speech samples within a segment of typically 15-25 ms, while M is the number of 15-25 ms segments, over which SEGSNR^{dB} is evaluated. The advantage of using SEGSNR^{dB} over conventional SNR is that it averages the SNR^{dB} values related to 15-20 ms speech segments, giving a better weighting to low-energy unvoiced segments by effectively computing the geometric mean of the SNR values due to averaging in the logarithmic domain instead of the arithmetic mean. Hence the SEGSNR values correlate better with subjective speech quality measures, such as the MOS.

For linear predictive hybrid speech codecs, such as the 13 kbps RPE GSM codec, spectral domain measures typically have better correlation with perceptual assessments than time domain measures. The so-called cepstral distance measure physically represents the logarithmic spectral envelope distortion and is computed as [56, 57] :

$$
\check{\mathrm{CD}} = \sqrt{\left[C_0^{in} - C_0^{out}\right]^2 + 2\sum_{j=1}^{3p}\left[C_j^{in} - C_j^{out}\right]^2}, \qquad (9.4)
$$

where C_j^{in} and C_j^{out}, $j = 0 \ldots 3p$ are the cepstral coefficients of the input and output speech, respectively, and p is the order of the short-term predictor filter which is typically 8-10. These cepstral coefficients can be readily computed from the coefficients of the short-term predictor using the results of reference [57]. Using these measures, let us now briefly consider the error sensitivity of the TBPE encoded bits, which will allow us to assign the bits to appropriate bit protection classes.

9.2.4 Bit Sensitivity Analysis

In our bit sensitivity investigations [55] we systematically corrupted each bit of a 144 bit TBPE frame and evaluated the SEG-SNR and CD degradation. Our results are depicted for the first 63 bits of a TBPE frame in terms of SEGSNR (dB) in Figure 9.2, and in terms of CD (dB) in Figure 9.3. For the sake of completeness we note that we previously reported our findings on a somewhat more sophisticated sensitivity evaluation technique in reference [19], where the effects of error propagation across speech frame boundaries due to filter memories was also taken into account by integrating or summing these degradations over all consecutive frames, where the error propagation inflicted measurable SEGSNR and CD reductions. However, for simplicity, in this treatise we refrain from using this technique and demonstrate the principles of source sensitivity-matched error protection using a less complex procedure.

Specifically, we recall from Table 9.1 that the first 36 bits represent the 10 LSFs describing the speech spectral envelope. Concentrating on the LSF bits initially, the SEG-SNR degradations of Figure 9.2 indicate the most severe waveform distortions for the first 10 bits describing the first 2-3 LSFs. The CD degradation in Figure 9.3, however, was quite severe for all LSFs, particularly for the most significant bits (MSBs) of the individual parameters. This was confirmed by our informal subjective tests. Whenever possible, at least the MSBs of the LSF bits should be protected against corruption.

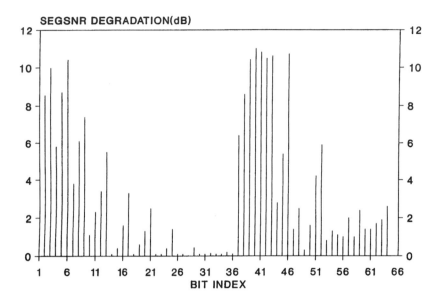

Figure 9.2: Bit sensitivities for the 4.8 Kbit/s codec expressed in terms of SEGSNR (dB).

Considering the remaining 27 bits seen in Figures 9.2 and 9.3, the parameters concerned are the LTPD, LTPG, GP, EG and Excitation pulse parameters for the first subsegment. We highlight our findings for the case of the first 27-bit subsegment only, as the other subsegments have identical behaviours. Bits 37-43 represent the LTP delays and bits 44-47 the LTP gains. Their errors are more significant in terms of SEG-SNR than in CD, as demonstrated by Figures 9.2 and 9.3. This is because the LTPD and LTPG parameters describe the spectral fine structure and do not seriously influence the spectral envelope distortion evaluated in terms CD, although they seriously degrade the recovered waveform, as indicated by the associated SEGSNR degradation. As the TBPE codec is a stochastic codec with random excitation patterns, bits 48-63 assigned to the excitations and their gains are not particularly vulnerable to transmission errors. This is because the redundancy in the signal is removed by the long-term and short-term predictors. Furthermore, the TBPE codec exhibits exceptional inherent excitation robustness, as the influence of a channel error is restricted to one component of the vector $\mathbf{b}$ and its effect in the excitation is spread and diminishes after the orthogonal transformation $\mathbf{c} = \mathbf{Ab}$. In conventional CELP codecs this is not the case, as a codebook address error causes the decoder to select a different excitation pattern from its codebook causing considerably more speech degradation than encountered by the TBPE codec.

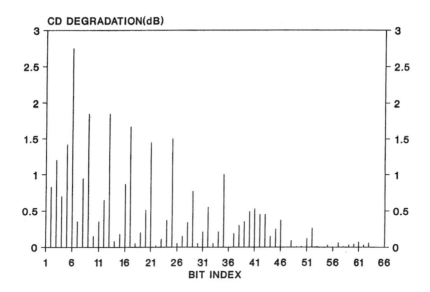

Figure 9.3: Bit sensitivities for the 4.8 Kbit/s codec expressed in terms of CD (dB).

In general, most robust performance is achieved if the bit protection is carefully matched to the bit sensitivities, but the SEG-SNR and CD sensitivity measures portrayed in Figures 9.2 and 9.3 often contradict. Therefore we combine the two measures to give a sensitivity figure S, representing the average sensitivity of a particular bit. The bits must be first ordered both according to their SEG-SNR and CD degradations portrayed in Figures 9.2 and 9.3, respectively, in order to derive their 'grade of prominence', where 1 represents the highest and 63 the lowest sensitivity. Observe that the highest CD degradation is caused by bit 6, which is the MSB of the second LSF in the speech frame, while the highest SEG-SNR degradation is due to bit 40 in the group of bits 37-43, representing the LTP delay. Furthermore, bit 6 is the seventh in terms of its SEG-SNR degradation, hence its sensitivity figure is $S = 1 + 7 = 8$, as seen in the first row of Table 9.2. On the other hand, the corruption of bit 40, the most sensitive in terms of SEG-SNR, results in a relatively low CD degradation, as it does not degrade the spectral envelope representation characterised by the CD, but spoils the pitch periodicity and hence the spectral fine-structure. This bit is the 19 th in terms of its SEG-SNR degradation, giving a sensitivity figure contribution of 19 plus 1 due to CD degradation, i.e. the combined sensitivity figure is $S = 20$, as shown by row 6 of Table 9.2. The combined sensitivity figures for all the LSFs and the first 27-bit subsegment are similarly summarised in ascending order in column 3 of Table 9.2, where column 2 represents the

Bit no. in frame	Bit index in frame	Sensitivity figure	Bit no. in frame	Bit index in frame	Sensitivity figure
1	6	8	36	57	76
2	9	14	37	10	79
3	5	16	38	28	80
4	3	16	39	19	80
5	41	19	40	61	80
6	40	20	41	59	82
7	13	21	42	62	84
8	2	23	43	15	85
9	43	24	44	60	88
10	8	25	45	34	89
11	46	25	46	50	91
12	42	26	47	31	92
13	17	27	48	55	95
14	39	31	49	27	95
15	4	31	50	23	97
16	21	32	51	14	97
17	12	37	52	47	98
18	38	38	53	58	102
19	25	43	54	54	103
20	16	44	55	53	105
21	52	45	56	56	105
22	7	45	57	18	105
23	1	45	58	33	108
24	37	48	59	49	109
25	45	49	60	26	109
26	11	55	61	30	110
27	20	58	62	22	119
28	51	60	63	36	125
29	29	60			
30	35	60			
31	44	63			
32	32	68			
33	48	69			
34	24	71			
35	63	76			

Table 9.2: Bit-sensitivity figures for the 4.8 Kbit/s TBPE codec

bit-index in the first 63-bit segment of the 144-bit TBPE frame.

On the basis of the above bit-sensitivity analysis [55] the speech bits were assigned in three sensitivity classes for embedded source-matched forward error correction to be detailed in Subsection 9.7.2.1. In closing we note that a more detailed discussion on various speech compression and transmission techniques will be provided in a forthcoming monograph [20]. Having described the proposed 4.8 kbit/s TBPE speech codec we now focus our attention on the design of the fixed, but arbitrarily programable videophone codec proposed for our multi-media communicator.

9.3   Video Coding Issues [112]

9.3.1   Recent Video Coding Advances

The theory and practice of image compression has been consolidated in a number of monographs by Netravali and Haskell [115], Jain [103], Jayant and Noll [80] and Hanzo *et. al.* [112]. Hence in this chapter we refrain from detailing the basics of video compression and concentrate mainly on the associated system aspects. A plethora of various video codecs have been proposed in the excellent special issues edited by Tzou, Mussmann and Aigawa [59], by Hubing [60], Gharavi et al [114] and Girod et al [61] for a range of bit-rates and applications, but the individual contributions by a number of renowned authors are too numerous to review.

Khansari, Jalali, Dubois and Mermelstein [62, 113] as well as Mann Pelz [63] reported promising results on adopting the International Telecommunications Union's (ITU) standardised H261 variable-rate codec [112] for wireless applications, which was originally designed for benign, low error-rate Gaussian channels. Since this codec employs so-called variable-length coding techniques, a single bit error can result in the erroneous decoding of an entire run-length coded string of bits, potentially leading to catastrophic video degradations. By invoking powerful signal processing and error-control techniques the authors succeeded in remedying the inherent source coding problems inflicted by stretching the codec's application domain to hostile wireless environments. Further important contributions in the field were due, for example, to Chen *et. al.* [64], Illgner and Lappe [65], Zhang [66], Ibaraki, Fujimoto and Nakano [67], Watanabe et al [68] etc. and the MPEG4 consortium's endeavours [71], as well as due to the efforts of the mobile audio-video terminal (MAVT) consortium. The applicability of the ITU standard H.263 codec [70, 72] to mobile videophony was investigated for example by Färber, Steinbach and Girod [69] as well as Cherriman and Hanzo [106]- [112]. A common feature of the above codecs is that unless an efficient bit-rate control mechanism, such as the adaptive packetisation algorithm of reference [110, 112] is used, the scheme has a time-variant bit-rate, which cannot be readily accommodated by contemporary second-generation wireless systems.

As a different design alternative, Streit and Hanzo offered [102] a comparative study of a range of fixed but arbitrarily programmable-rate 176×144 pixel head-and-shoulders Quarter Common Intermediate Format (QCIF) video codecs specially designed for fixed-rate videotelephony over existing and future mobile radio speech systems on the basis of a recent research programme [75,92,99–101,112]. These codecs employ novel motion-compensation and -tracking techniques as well as video 'activity' identification and tracking, which will be highlighted during our further discourse. Various motion compensated error residual coding techniques were compared, which dispensed with the self-descriptive, zig-zag scanning and run-length coding principle of the H.261 and H.263 codecs [112] for the sake of maintaining a time-invariant bit-rate and improved robustness against channel errors, while tolerating some compression ratio reduction. Within the implementational complexity and bit-rate limits always an optimum constant bit allocation was sought in order to be able to adapt the codec to the requirements imposed by existing second-generation wireless speech systems. Having reviewed some of the recent advances in video compression, let us now briefly highlight the rudimentary principles of video coding and commence our elaborations by considering the removal of temporal redundancy using motion compensation, before highlighting the concepts used by the above-mentioned fixed-rate video codecs.

9.3.2   Motion Compensation

The ultimate goal of low-rate image coding is to remove redundancy, predictability or self-similarity in both spatial and temporal domains, which correspond to the so-called intra-frame and inter-frame redundancy, manifesting themselves within a given video frame and with respect to consecutive frames, respectively. These redundancy reduction measures allow us to reduce the required transmission bit rate. The temporal correlation between successive image frames is typically removed using so-called block-based motion compensation, where each segment or block of the video frame to be encoded is assumed to be a motion-translated version of the corresponding block in the previously encoded video frame. How this can be achieved is the subject of this subsection.

The vector describing the above-mentioned motion translation is referred to as the motion vector (MV), which is typically found with the aid of correlation techniques, as it will be described during our forthcoming discourse. Specifically, as portrayed in Figure 9.4, a motion translation region or search scope is stipulated within the previous frame. To be more explicit, instead of using the previous original frame, the so-called 'locally decoded' frame is used in the motion compensation, where the phrase 'locally decoded' implies decoding it at the encoder, i.e. where it was encoded. This 'local decoding' yields an exact replica of the video frame at the distant decoder's output. This so-called local decoding operation is necessary,

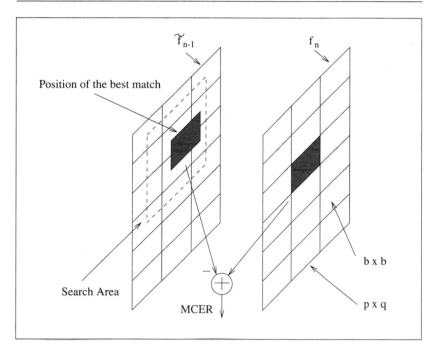

Figure 9.4: Simplified schematic of motion compensation ©J. Streit [98], 1996.

since the previous original frame is not available at the distant decoder and hence without the local decoding operation the distant decoder would have to use the reconstructed version of the previous frame in its attempt to reconstruct the current frame. The absence of the original video frame would lead to a mismatch between the operation of the encoder and decoder, a phenomenon, which will become more clear during our further elaborations.

In order to accomplish motion compensation, the current block to be encoded is slid over the previously stipulated search region of the locally decoded previous frame and the location of the highest correlation is deemed to be the destination of the motion translation. As suggested by Figure 9.4, motion compensation (MC) is then carried out by subtracting the appropriately motion translated previous 'locally decoded' block from the current block to be encoded, in order to generate the so-called motion compensated error residual (MCER). Clearly, the image is decomposed in motion translation described by means of the MVs and in the MCER. Both the MVs and the MCER have to be encoded and transmitted to the decoder for image reconstruction.

Again, the motion compensation removes the temporal inter-frame redundancy and hence the variance of the MCER becomes typically lower than that of the original image, unless there is a substantial amount of new information introduced in the current frame, which cannot be predicted on the basis of the previous frame. Hence MC typically improves the codec's bit-rate economy, although in high quality video coding, where there is a limited interframe correlation due to newly introduced picture objects, the situation is often reversed. This is mainly due to the fact that albeit the MC-engendered MCER-reduction is rather modest, a substantial fraction of the bit-rate budget must be dedicated to the encoding of the MVs. Efficient codecs can circumvent this problem by carrying out an intra/inter-coded decision on a block-by-block basis, which is signalled to the decoder using a one-bit flag. This measure prevents the codec from 'wasting' bits on encoding the inefficient motion vectors in the case of blocks, where the inter-frame correlation is low. Having considered the basic principles of motion compensation, let us now consider how the MCER can be encoded for transmission to the decoder.

The MCER frame can be represented using a range of techniques [102, 112], including subband coding [85,86] (SBC), wavelet coding [87], Discrete Cosine Transformation [78,79,99,103] (DCT), vector quantisation (VQ) [90, 91] or Quad-tree [88,89,95,100] (QT) coding. In this chapter we will restrict our treatment of video compression issues to a rudimentary overview, the interested reader is referred to the literature cited above, in order to probe further, although some aspects of DCT-based MCER-coding schemes will be discussed during our forthcoming elaborations.

When a low codec complexity and low bit-rate are required, the motion compensation technique described above can be replaced by a simple technique, often referred to as frame-differencing. In frame-differencing the whole of the previous locally decoded image frame is subtracted from the frame to be encoded without the need for the above correlation-based motion prediction, which may become very computationally intensive for high-resolution, high-quality video portraying high-dynamic scenes. The schematic of such a simple video codec based on simple frame-differencing is shown in Figure 9.5, which will be described in the next paragraph. Although the variance or energy of the MCER remains somewhat higher for frame-differencing than in case of full motion compensation, there is no pattern-matching search, which reduces the complexity and no MVs have to be encoded, which may reduce the overall bit-rate.

Returning to Figure 9.5, observe that after frame-differencing the encoded MCER is conveyed to the transceiver and also locally decoded. As mentioned earlier, this is necessary to be able to generate the locally reconstructed video signal, which is invoked by the encoder in subsequent MC steps. Again, the encoder uses the locally reconstructed, rather than the original input video frames in the MC process, since the original uncoded frames are unavailable at the decoder. Hence invoking the original previous video frame at the encoder, while the reconstructed one at the decoder, would result in 'mis-alignment' between the encoder and decoder

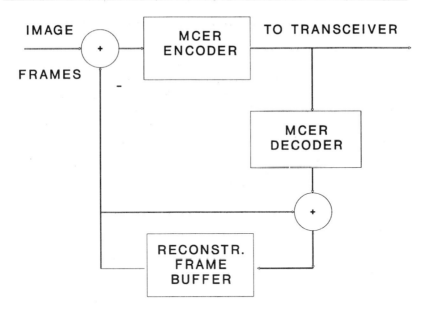

Figure 9.5: Simple video codec schematic.

due to using different frames at both ends for motion compensation. This so-called local reconstruction operation is carried out by the adder in the Figure, superimposing the decoded MCER on the previous locally decoded video frame. The philosophy of the codec's operation is similar, if full MC is used. As alluded to before, efficient codecs, such as for example the ITU H.263 scheme [112], often combine the so-called inter-frame and intra-frame coding techniques on a block-by-block basis, where MC is employed only if it was deemed advantageous in MCER reduction terms.

In case of highly correlated consecutive video-frames the MCER typically exhibits 'line-drawing' characteristics, where large sections of the frame difference signal are 'flat', characterised by low pixel magnitude values, while the motion contours, where the frame differencing has failed to predict the current pixels on the basis of the previous locally decoded frame, are represented by larger values. Consequently, efficient MCER residual coding algorithms must be able to represent such textured MCER patterns adequately. Again, some examples of encoding the MCER efficiently by subband coding [85, 86] (SBC), wavelet coding [87], Discrete Cosine Transformation [78, 79, 99, 103] (DCT), vector quantisation (VQ) [90, 91] or Quad-tree [88, 89, 95, 100] (QT) coding can be found in the literature [112]. At this point we concentrate our attention on the rudimentary portrayal of a fixed-rate DCT-based videophone codec, suitable for the proposed multimedia system.

9.3.3 A Fixed-rate Videophone Codec [22, 99][2]

In this subsection we set out to briefly describe the philosophy of fixed-rate video source coding, which was portrayed in depth in references [99–102, 112] by Streit and Hanzo. The particular codec advocated here was detailed in reference [99]. The video codec's outline is depicted in Figure 9.6, which will be detailed below. The coding algorithm was designed to produce a fixed, but programmable bit-rate, which was adjusted to 852 bits/90 ms$\approx$9.47 kbps in order to match the bit-packing requirements of the proposed 30 kHz bandwidth system. The codec's operation is initialised in the intra-frame coded mode, but once it switched to the inter-frame coded mode, any further mode switches are optional and only required if a drastic video scene change occurs.

9.3.3.1 The Intra-Frame Mode

As seen at the bottom of Figure 9.6, in the intra-frame mode the encoder transmits the coarsely quantised block averages for the current frame, which provides the low-resolution initial frame required for the operation of the inter-frame codec at the commencement of video communications. This results in a very coarse intra-frame coded initial frame, which is used by the inter-frame coded mode of operation in order to improve the video quality in successive coding steps. This initial 'warm-up' phase of the codec typically is imperceptible to the untrained viewer, since it typically takes less than 1-2 s and it is a consequence of the fixed bit-rate constraint imposed by contemporary second-generation mobile radio systems.

Furthermore, the intra-frame mode is also invoked during later coding stages in a number of blocks in order to mitigate the effect of transmission errors and hence prevent encoder/decoder misalignment, as it will be detailed during our later elaborations. In this context this operation is often referred to as partial forced updating (PFU), since the specific blocks concerned are partially updated by superimposing an attenuated version of the intra-frame coded block averages. For 176×144 pixel CCITT standard Quarter Common Intermediate Format (QCIF) images we limited the number of video encoding bits per frame to 852. In order to transmit all block averages with a 4-bit resolution, while not exceeding the 852 bits per video frame budget, the forced-update block size was fixed to 11 × 11 pixels, since there are 852/4=213 blocks that can be encoded on this basis, yielding a block size of 176×144/213 $\approx$119 or 11×11 pixels.

9.3.3.2 Cost/Gain Controlled Motion Compensation

In motion-compensation often 8×8 blocks are used, since the associated MC complexity of the correlation operation would be quadrupled due to

---

[2]This subsection is supported by a down-loadable video compression demonstration package under the WWW address http://www-mobile.ecs.soton.ac.uk

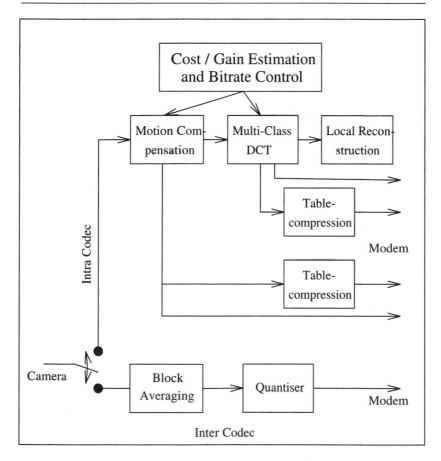

Figure 9.6: Fixed-rate DCT-based codec schematic ©Kluwer, Hanzo, Streit et al, 1995 [22].

doubling the block size, although the number of MVs could be reduced to a quarter - at the cost of some MCER reduction penalty. At the commencement of the encoding procedure the motion compensation (MC) scheme determines a MV for each of the 8×8 blocks using full-search [103, 112]. The MC search window is fixed to 4 × 4 pixels around the centre of each block and hence a total of 4 bits are required for the encoding of 16 possible positions for each MV. Although this search window is relatively small, it was found adequate in limited-motion head-and-shoulders videotelephony. Before the actual motion compensation takes place, the codec tentatively determines the potential benefit of the motion compensation in terms of motion compensated error energy reduction. Then the codec selects those blocks as 'motion-active' ones, whose MCER reduction gain exceeds a cer-

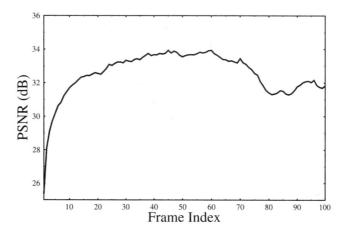

Figure 9.7: PSNR versus frame index performance of the 9.47 kbps video codec for the 'Miss America' sequence, ©Kluwer, Hanzo, Streit et al, 1995 [22].

tain threshold. This method of classifying the blocks as 'motion-active' and 'motion-passive' results in an 'active/passive table', which consists of a one bit activity flag for each block, marking it as passive or active. In case of 8×8 blocks and 176×144 pel QCIF images this table consists of 396 entries, which can be compressed using a technique reminiscent of a two stage so-called quad-tree based compression [99, 112], the details of which are beyond the scope of this chapter As a result, a typical 396-bit active/passive table containing 30 active flags can be compressed to less than 150 bits.

This implies that constrained by the extremely low bit-rate budget of 852 bits/frame, in this codec only a total of about 30 8×8 blocks can be marked as motion-active, corresponding to about 10% of the total video frame area. For the remaining motion-passive blocks simple frame-differencing is invoked, since employing full motion compensation is not justified in terms of the achievable MCER reduction, especially in the light of the required MC search complexity and the increased number of MV encoding bits.

If, however, the number of bits allocated to the compressed activity tables and active motion vectors exceeds half of the total number of available bits/frame, i.e. 852/2=426, a number of blocks satisfying the motion-active criterion will be relegated to the motion-passive class. This process takes account of the subjective importance of various blocks and does not ignore motion-active blocks in the central eye and lip regions of the image, while relegating those, which are closer to the fringes of the image. Pursuing a similar approach, gain control is also applied to the Discrete Cosine Transform (DCT) based compression [103] of the MCER. Let us however initially consider briefly the philosophy of DCT-based compression in the next section.

9.3.3.3 Transform Coding

9.3.3.3.1 One-dimensional Transform Coding As it is well-known from Fourier theory, signals are often synthesised by so-called orthogonal basis functions, a term, which will be augmented during our further discourse. Specifically, when using Fourier transforms, an analogue time-domain signal, which can be the luminance variation along a scan-line of a video frame, can be decomposed into its constituent frequencies.

For signals, such as the above-mentioned video signal representing the luminance variation along a scan-line of a video frame, orthogonal series expansions can provide a set of coefficients, which equivalently describe the signal concerned. We will make it plausible that these equivalent coefficients may become more amenable to quantisation, than the original time-domain signal.

For example, for a one-dimensional time-domain sample sequence $\{x(n), 0 \leq n \leq N - 1\}$ a so-called unitary transform is given in a vectorial form by $\underline{X} = \underline{A}\underline{x}$, which can also be expressed in a less compact scalar form as [103]:

$$X(k) = \sum_{n=0}^{N-1} a(k,n) \cdot x(n), \;\; 0 \leq k \leq N - 1 \qquad (9.5)$$

where the transform is referred to as unitary, if $\underline{A}^{-1} = \underline{A}^{*T}$ holds. The associated inverse operation requires us to invert the matrix $\underline{A}$ and due to the above-mentioned unitary property we have $\underline{x} = \underline{A}^{-1}\underline{X} = \underline{A}^{*T}\underline{X}$, yielding [103]:

$$x(n) = \sum_{k=0}^{N-1} X(k) a^*(k,n), \;\; 0 \leq k \leq N - 1 \qquad (9.6)$$

which gives a **series expansion** of the time-domain sample sequence $x(n)$ in the form of the **transform coefficients** $X(k)$. The columns of $\underline{A}^{*T}$, i.e. the vectors $\underline{a}_k^* \triangleq \{a^*(k,n), 0 \leq n \leq N - 1\}$ are the so-called basis vectors of $\underline{A}$ or the **basis vectors of the decomposition**. According to the above principles the time-domain signal $x(n)$ can be equivalently described in the form of the **decomposition** in Equation 9.6, where the **basis functions** $a^*(k,n)$ are weighted by the transform coefficients $X(k)$

and then superimposed on each other, which corresponds to their summation at each pixel position of the transformed block. The transform-domain weighting coefficients $X(k)$ can be determined from Equation 9.5.

The transform-domain coefficients $X(k); k = 0 \cdots N - 1$ often give a more 'compact' representation of the time-domain samples $x(n)$, implying that if the original time-domain samples $x(n)$ are correlated, then in the transform-domain most of the signal's energy is concentrated in a few transform-domain coefficients. To elaborate a little further - according to the Wiener-Khintshin theorem - the autocorrelation function (ACF) and the power spectral density (PSD) are Fourier transform pairs. Due to the Fourier-transformed relationship of the ACF and PSD it is readily seen that a slowly decaying autocorrelation function, which indicates a predictable signal $x(n)$ in the time-domain is associated with a PSD exhibiting a rapidly decaying low-pass nature. Therefore, in the case of correlated time-domain $x(n)$ sequences the transform-domain coefficients $X(k)$ tend to be statistically small for high frequencies, i.e. for high transform coefficient indices and exhibit large magnitudes for low-frequency transform-domain coefficients, i.e. for low transform-domain indices. This concept will be exposed in a little more depth below, but for a deeper exposure to these issues the reader is referred to Jain's excellent book [103].

9.3.3.3.2 Two-dimensional Transform Coding The above one-dimensional signal decomposition can also be extended to two-dimensional (2D) signals, such as 2D image signals of a video frame, as follows [103]:

$$X(k,l) = \sum_{m=0}^{N-1} \sum_{n=0}^{N-1} x(m,n) \cdot a_{k,l}(m,n) \quad 0 \le k,l \le N - 1 \quad (9.7)$$

$$x(m,n) = \sum_{k=0}^{N-1} \sum_{l=0}^{N-1} X(k,l) \cdot a_{k,l}^*(m,n) \quad 0 \le m_1 n \le N - 1 \quad (9.8)$$

where $\{a_{k,l}^*(m,n)\}$ is a set of discrete two-dimensional basis functions, $X(k,l)$ are the 2D transform-domain coefficients and $\underline{\underline{X}} = \{X(k,l)\}$ constitutes the transformed image.

As in the context of the one-dimensional transform, the two-dimensional (2D) time-domain signal $x(m,n)$ of a video block to be encoded can be equivalently described in the form of the decomposition in Equation 9.8, where the 2D basis functions $a_{k,l}^*(m,n)$ are weighted by the coefficients $X(k,l)$ and then superimposed on each other, which again, corresponds to their summation at each pixel position in the video frame. The transform-domain weighting coefficients $X(k,l)$ can be determined from Equation 9.7. Once a spatially correlated image block $x(m,n)$ of for example $N \times N = 8 \times 8$ pixels is orthogonally transformed using the Discrete Cosine Transform

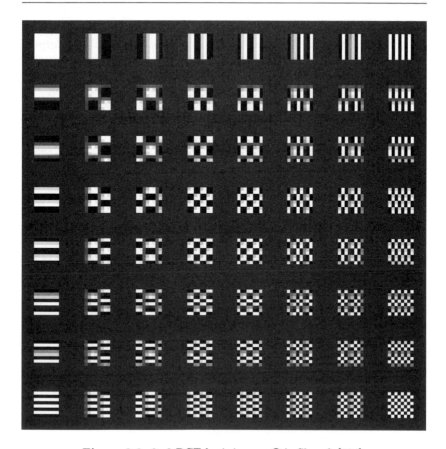

Figure 9.8: 8×8 DCT basis images ©A. Sharaf [104].

(DCT) matrix $\underline{\underline{A}}$ defined as [103]:

$$A_{mn} = \frac{2c(m)c(n)}{N} \sum_{i=0}^{N} \sum_{j=0}^{N} b(i,j) \cos \frac{(2i+1)m\pi}{2N} \cos \frac{(2j+1)n\pi}{2N}$$

$$c(m) = \begin{cases} \frac{1}{\sqrt{2}} & \text{if}(m=0) \\ 1 & \text{otherwise} \end{cases} \tag{9.9}$$

the transform-domain image described by the DCT coefficients can be quantised for transmission to the decoder. **The rationale behind invoking the DCT is that the frequency-domain coefficients $X(k,l)$ can typically be quantised using a lower number of bits, than the original image pixel values $x(m,n)$,** which will be further augmented during our forthcoming discourse.

For illustration's sake the associated two-dimensional 8×8 DCT **basis-**

images are portrayed in Figure 9.8, where for example the top left-hand corner represents the zero horizontal and vertical spatial frequency, since there is no intensity or luminance change in any direction across this basis image. Following similar arguments, the bottom right corner corresponds to the highest vertical and horizontal frequency, which can be represented using 8×8 basis images, since the luminance changes from black to white between adjacent pixels in both the vertical and horizontal directions. Similarly, the basis image in the top right-hand corner corresponds to the highest horizontal frequency, but zero vertical frequency component and by contrast, the bottom left basis image represents the highest vertical frequency, but zero horizontal frequency. **In simple terms the decomposition of Equations 9.7, 9.8 can be viewed as finding the required weighting coefficients $X(k,n)$, in order to superimpose the weighted versions of all the 64 different 'patterns' in Figure 9.8 for the sake of re-constituting the original 8×8 video block.** In other words, each original 8×8 video block is represented as the sum of the 64 appropriately weighted 8×8 basis images.

It is plausible that for blocks, over which the video luminance or gray shade does not change dramatically, i.e. at a low 'spatial frequency', most of the video frame's conveyed energy is associated with these low spatial frequencies. Hence the associated low-frequency transform-domain coefficients $X(k,n)$ are high and the high-frequency coefficients $X(k,n)$ are of low magnitude. By contrast, if there is a lot of fine-detail in a video frame, such as in a finely striped pattern or in a checker-board pattern, most of the video frame's conveyed energy is associated with high spatial frequencies. Most practical images contain more low spatial frequency energy, than high-frequency energy. This is also true for those motion-compensated video blocks, where the motion compensation was efficient and hence resulted in a 'flat' block, associated with a low spatial-frequency. For these blocks therefore most of the high-frequency DCT coefficients can be set to zero at the cost of neglecting only a small fraction of the video block's energy, residing in the high spatial-frequency DCT coefficients. In simple terms this corresponds to gentle low-pass filtering, which in perceptual terms results in a slight blurring of the high spatial-frequency image fine details.

In other words, upon exploiting that the human eye is rather insensitive to high spatial frequencies, in particular, when these appear in moving pictures, the spatial frequency-domain block is amenable to data compression. Again, this can be achieved by more accurately quantising and transmitting the high-energy, low-frequency coefficients, while typically coarsely representing or masking out the low-energy, high-frequency coefficients. We note, however that in motion-compensated codecs there may be blocks along the edges of moving objects, where the MCER does not retain the above-mentioned spatial correlation and hence the DCT does not result in significant energy compaction. Again, for a deeper exposure to the DCT the reader is referred to for example reference [103].

3	2	1	0	0	0	0	0	2	1	0	0	0	0	0	0
2	1	0	0	0	0	0	0	3	2	0	0	0	0	0	0
1	0	0	0	0	0	0	0	1	1	0	0	0	0	0	0
0	0	0	0	0	0	0	0	0	0	0	0	0	0	0	0
0	0	0	0	0	0	0	0	0	0	0	0	0	0	0	0
0	0	0	0	0	0	0	0	0	0	0	0	0	0	0	0
0	0	0	0	0	0	0	0	0	0	0	0	0	0	0	0
0	0	0	0	0	0	0	0	0	0	0	0	0	0	0	0
2	3	1	0	0	0	0	0	2	2	1	0	0	0	0	0
1	2	1	0	0	0	0	0	2	2	0	0	0	0	0	0
0	0	0	0	0	0	0	0	1	0	0	0	0	0	0	0
0	0	0	0	0	0	0	0	0	0	0	0	0	0	0	0
0	0	0	0	0	0	0	0	0	0	0	0	0	0	0	0
0	0	0	0	0	0	0	0	0	0	0	0	0	0	0	0
0	0	0	0	0	0	0	0	0	0	0	0	0	0	0	0
0	0	0	0	0	0	0	0	0	0	0	0	0	0	0	0

Table 9.3: Bit allocation tables for the four DCT quantisers, where the top left-hand corner indicates the number of bits allocated to the DC-component of the DCT ©IEEE, Hanzo, Streit, 1995 [99].

9.3.3.4 Gain Controlled Quadruple-Class DCT

Having reviewed the basics of DCT, let us now return to the DCT-based video codec schematic of Figure 9.6 and consider the specific codec design. Focusing our attention on the gain-controlled DCT-based MCER compression, every 8×8 block is tentatively DCT transformed, quantised and transformed back to the temporal domain, in order to assess the potential benefit of marking the block as DCT-active, when judged in terms of MCER reduction. In order to take account of the above-mentioned time-variant, non-stationary nature of the MCER and its time-variant frequency-domain distribution, four different sets of DCT quantisers were designed. The quantisation distortion associated with each quantiser is computed by quantising the MCER tentatively, in order to be able to choose the best quantiser. This measure improves the video quality at the cost of increased complexity. As it was shown in reference [99], a total of ten bits are allocated for each of the four quantisers, which are characterised by the bit allocation scheme of Table 9.3. The four DCT quantisers correspond to different DCT coefficient energy distributions across the spatial frequency domain, where the top left-hand corner indicates the number of bits allocated to the DC-component of the DCT.

Each quantiser is a trained Max-Lloyd quantiser [103], catering for a specific frequency-domain energy distribution class. However, a joint feature of all of them is that the high-frequency components were masked out. All DCT blocks, whose coding gain exceeds a certain threshold are marked as DCT-active, resulting in a similar 'active/passive table' as for the motion vectors. For the DCT-activity table we also applied the same run length compression technique [99, 112], as above in the context of the motion activity table. Again, if the number of bits required for the encoding of the DCT-active blocks exceeds half of the maximum allowable number of bits, i.e. 852/2=426, the blocks around the fringes of the image are considered DCT-passive, rather than those in the central eye and lip sections. If, however, the active DCT coefficients and activity-tables do not fill the 852-bit fixed-length transmission burst, the number of active DCT blocks is increased and all activity tables are recomputed.

The codec's bit-allocation scheme is summarised in Table 9.4. The so-called frame-alignment word (FAW) or unique word is used to allow the codec to re-synchronise at the beginning of each 852-bit frame in the case of transmission errors. Furthermore, 22 blocks out of the 25 384 pixels/(8 · 8)=396-block QCIF frame are partially forced up-dated using the block means, partially overlaid on the contents of the local reconstructed frame buffer in order to enhance the codec's robustness. The corresponding bit-rate contribution seen in the Table due to PFU is 22×4=96 bits. A total of 30 blocks are marked as DCT- and motion-active, yielding a total of 852 bits per frame, or a video rate of 852 bits/90 ms≈9.47 kbps, as seen in Table 9.4.

FAW	PFU	MV	DCT	Total
22	22×4	< 376	< 376	852

Table 9.4: Video codec bit allocation scheme [98] Streit, 1996.

The encoded FAW, PFU, MV and DCT parameters are then transmitted to the decoder and also locally decoded in order to be used in future motion predictions. The video codec's Peak Signal to Noise Ratio (PSNR) versus frame index performance is shown in Figure 9.7, where an average PSNR of about 33.3 dB was achieved for the widely used QCIF-resolution Miss America (MA) sequence.

The associated subjective video quality is adequate for the transmission of low-activity head-and-shoulders videophone sequences, but for high-activity sequences typically higher transmission rates are required. Consequently, the higher video-rate necessitates the allocation of more than one speech slot per transmission frame for video communications, which may compromise the voice capacity of the multi-media system. In this respect a higher flexibility can be guaranteed by the 200 kHz bandwidth system to be described in Section 9.7.3, which is capable of supporting more timeslots and users than the 30 kHz system of Section 9.7.2.

For a more detailed discourse on the proposed video codec the interested reader is referred to [99, 112]. Further fixed-rate wireless videophone systems were proposed in references [100, 101, 112], which are based on

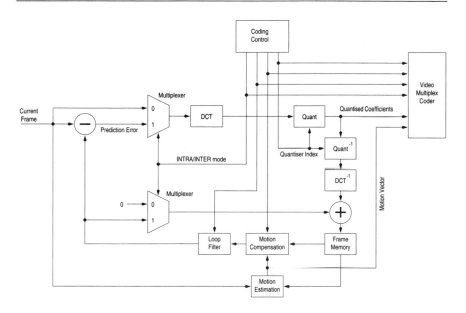

Figure 9.9: Simplified H.261/H.263 schematic©Cherriman [105] 1995.

fixed-rate QCIF codecs operating in the range of 8-13 kbps at 10 frames/s. As expected, the performance of these video codecs improves, as the affordable bit-rate increases and adequate subjective videophone quality can be maintained for moderate-activity QCIF images at bit-rates in excess of 8-15 kbps, when using a scanning rate around 10 frames/s. As mentioned before, in references [106]- [112] Cherriman and Hanzo proposed various system design alternatives based on the H.263 standard video codec, which will be briefly highlighted below.

9.3.4 The H.263 Standard Video Codec

As a potential programmable-rate design alternative to the previously proposed fixed-rate video codec, here we briefly introduce the H.261 and H.263 ITU standard codecs [105, 112], which will be employed in our 200 kHz bandwidth candidate system described in Section 9.7.3. Although the H.263 codec is flexible in terms of supporting five different video resolutions, including the 288×352 pixel ITU Common Intermediate Format (CIF), the Quarter CIF (QCIF), Sub-QCIF (SQCIF), as well as the higher resolution 4×CIF and 16×CIF video formats, for low-rate wireless videophony we are restricted to QCIF or SQCIF images. The full description of the H.263 codec can be found in reference [72], while various features of the proposed video system were discussed in [106]- [112], hence here only a brief exposure is offered.

The H.261 and H.263 codecs share the simplified schematic of Figure 9.9 and they operate under the instructions of the coding control block, selecting the required inter/intra frame mode, the quantisation and bit-allocation scheme etc. Similarly to our previously described fixed-rate codec, DCT [103, 115] is invoked also in the H.261/H.263 codecs of Figure 9.9, in order to compress either the blocks of the current original frame in the intra-frame coded mode or the motion compensated prediction error blocks in the inter-frame coded mode. This is controlled by the Multiplexer in the Figure. Whether the intra- or inter-frame coded mode is enabled by the Coding Control block, depends on a number of factors, such as the required bit-rate, robustness against channel errors, etc. As mentioned before, there may be input sequences, for which intra-frame coding is just as efficient as inter-frame coding, since due to the lack of correlation in the motion compensated error residual the DCT does not always lead to energy compaction.

A large selection of quantisers are stored and invoked in the H.261/H.263 codecs, depending on the required bit-rate and video quality, where again, the index of the quantiser to be used is selected by the Coding Control block of Figure 9.9. The quantised DCT coefficients are then transmitted to the decoder via the Video Multiplex Coder and also locally inverse-quantised by the $QUANT^{-1}$ block, before inverse-DCT (DCT^{-1}) is employed, in order to generate the locally decoded replica of the signal that was subjected to DCT. Specifically, to reconstruct either the motion compensated prediction error or the original intra-coded current frame, in the latter case the '0' signal shown in the schematic of Figure 9.9 is gated through by the Multiplexer in order to reconstruct and store the locally decoded frame in the Frame Memory. In contrast, if inter-frame coding is used, the locally decoded motion prediction error is gated through by the Multiplexer and added to the previous locally decoded frame.

Observe in the Figure that in the inter-frame coded mode both the current video frame and the previous locally decoded frame that was stored in the Frame Memory are input to the Motion Estimation block, which then generates the Motion Vectors. The MVs are transmitted to the decoder and are also employed by the Motion Compensation block in order to properly position the best matching replica of the current block to be encoded, which was identified in the previously reconstructed block. Motion compensation is then carried out by subtracting the motion translated best matching block of the previous decoded frame from the current original block. Finally, all encoded information is multiplexed for transmission by the Video Multiplex Coder.

The H.263 [72, 73, 112] ITU standard codec scheme is in many respects similar to its predecessor, the H.261 codec [74, 112], but it incorporates a number of recent advances in the field, such as for example using half-pixel resolution in the motion compensation, which improves the MC process and hence reduces the variance of the MCER. Invoking the half-pixel resolution requires an interpolation process, which generates an additional pixel amongst all the existing pixels and hence supports a potentially improved MC process at the cost of an increased complexity. Furthermore, the H.263 scheme allows configuring the codec for a lower datarate or better error resilience and supports four so-called 'negotiable coding options', which are detailed in the Recommendation. These negotiable options can be 'negotiated' by the encoder and decoder about to commence communications, in order to use the 'lowest common denominator' of their optional features in their communications.

The bitstream generated by the H.261 and H.263 codecs is structured in a number of hierarchical layers, including the so-called picture layer, group of blocks layer, macroblock layer and block layer [72, 73], each of which represents a gradually reduced-size video-frame segment, commencing with specifying the coded information of the whole video frame - down to the 8×8 block layer. In order to allow a high grade of flexibility and to adapt to various images, each of these layers has a 'self-descriptive' structure, specifying the various coding parameters of the given layer. The coded information of the upper three layers commences with a unique word, allowing the codec to re-synchronise after loss of synchronisation following transmission errors. The 'self-descriptive' received bitstream typically informs the decoder as to the inter- or intra-coded nature of a frame, the video resolution used, whether to expect any more information of a certain type, the index of the currently used quantiser, the location of encoded blocks containing large transform coefficients, etc.

The H.263 scheme achieves a high compression ratio for transmissions over channels exhibiting a low bit error rate, but since the bit stream is 'self-descriptive', any transmission error can corrupt the segments describing the coding parameters used, resulting in catastrophic error events associated with using mismatching decoders. Hence the H.263 codec is rather vulnerable to channel errors. In references [106]- [112] Cherriman and Hanzo reported on the design of a low-rate video transceiver, where the H.263 codec was constrained to operate at a near-constant bit-rate using an appropriate bit-rate control and packetisation algorithm, which adjusted the quantiser such that it would output the required number of bits per frame. Furthermore, using a low-rate feedback channel, the contents of the local and remote decoder was 'frozen', when transmission errors occurred, which prevented the propagation of errors across blocks and allowed the codec to operate over hostile wireless channels. A more detailed exposure of video compression and transmission aspects is provided in the monograph [112].

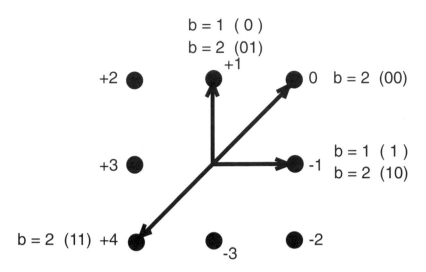

Figure 9.10: Coding ring ©ETT, Hanzo and Yuen [195].

9.4 Graphical Source Compression

9.4.1 Introduction to Graphical Communications

Telewriting is a multi-media telecommunication service enabling the bandwidth-efficient transmission of handwritten text and line graphics through fixed and wireless communication networks [196]- [201]. Differential chain coding (DCC) has been successfully used for graphical communications over E-mail networks or teletext systems [198], where bit-rate economy is achieved by exploiting the correlation between successive vectors. reference [202] addressed also some of the associated communications aspects. A plethora of further excellent treatises were contributed to the literature of chain coding by R. Prasad and his colleagues from Delft University [198]- [205].

9.4.2 Fixed-Length Differential Chain Coding

In chain coding (CC) a square-shaped coding ring is slid along the graphical trace from the current pixel, which is the origin of the legitimate motion vectors, in steps represented by the vectors portrayed in Figure 9.10. The bold dots in the Figure represent the next legitimate pixels during the graphical trace's evolution. In principle the graphical trace can evolve to any of the surrounding eight pixels and hence a three-bit codeword is required for lossless coding. Differential chain coding [199]- [201] (DCC) exploits that the most likely direction of stylus movement is a straight extension, with a diminishing chance of 180° turns. This suggests that the

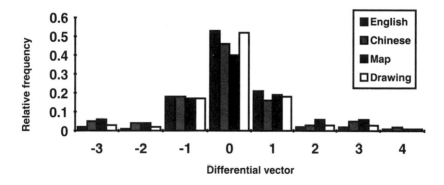

Figure 9.11: Relative frequency of differential vectors for a range of dynamo-graphical source signals ©ETT, Hanzo and Yuen [195].

coding efficiency can be improved using the principle of entropy coding by allocating shorter codewords to more likely transitions and longer ones to less likely transitions. This argument is supported by the histogram of the differential vectors of a range of graphical source signals, including English and Chinese handwriting, a Map and a technical Drawing, portrayed in Figure 9.11, where the vectors 0, +1 and -1 are seen to have the highest relative frequency.

In this section we embark on exploring the potential of a novel graphical coding scheme dispensing with the variable length coding principle of conventional DCC codecs, which we refer to therefore as fixed length differential chain coding (FL-DCC). FL-DCC was contrived in order to comply with the time-variant resolution- and/or bit-rate constraints of intelligent adaptive multi-mode terminals, which can be re-configured under network control, in order to satisfy the momentarily prevailing tele-traffic, robustness, quality, etc. system requirements. In order to maintain lossless graphics quality under lightly loaded traffic conditions, the FL-DCC codec can operate at a rate of $b = 3$ bits/vector, although it has a higher bit-rate than DCC. However, since in voice and video coding typically perceptually unimpaired lossy quantisation is used, we embark on exploring the potential of the re-configurable FL-DCC codec under $b < 3$ low-rate, lossy conditions.

Based on our findings in Figure 9.11 concerning the relative frequencies of the various differential vectors, we decided to evaluate the performance of the FL-DCC codec using the $b = 1$ and $b = 2$ bit/vector lossy schemes. As demonstrated by Figure 9.10, in the $b = 2$-bit mode the transitions to pixels -2, -3, $+2$, $+3$ are illegitimate, while vectors 0, +1, -1 and +4 are legitimate. In order to minimise the effects of transmission errors the Gray codes seen in Figure 9.10 were assigned. It will be demonstrated

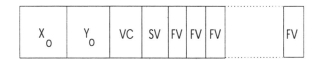

Figure 9.12: Coding syntax ©ETT, Hanzo and Yuen [195].

that, due to the low probability of occurance of the illegitimate vectors, the associated subjective coding impairment is minor. Under degrading channel conditions or higher tele-traffic load the FL-DCC coding rate has to be reduced to $b = 1$, in order to be able to invoke a less bandwidth efficient, but more robust modulation scheme or to generate less packets contending for transmission. In this case only vectors +1 and -1 of Figure 9.10 are legitimate. The subjective effects of the associated zig-zag trace will be removed by the decoder, which can detect these characteristic patterns and replace them by a fitted straight line.

In general terms the size of the coding ring is given by $2n\tau$, where $n = 1, 2, 3 \ldots$ is referred to as the order of the ring and τ is a scaling parameter, characteristic of the pixel separation distance. Hence the ring shown in Figure 9.10 is a first order one. The number of nodes in the ring is $M = 8n$.

The data syntax of the FL-DCC scheme is displayed in Figure 9.12. The beginning of a trace can be marked by a typically 8 bit long pen-down (PD) code, while the end of trace by a pen-up (PU) code. In order to ensure that these codes are not emulated by the remaining data, if this were incurred, bit stuffing must be invoked. We found however that in complexity and robustness terms using a 'vector counter' (VC) for signalling the trace-length to the decoder constituted a more attractive alternative for our system. The starting coordinates X_0, Y_0 of a trace are directly encoded using for example 10 and 9 bits in the case of a video graphics array (VGA) resolution of 640×480 pixels.

The first vector displacement along the trace is encoded by the best fitting vector defined by the coding ring as the starting vector (SV). The coding ring is then translated along this starting vector to determine the next vector. A differential approach is used for the encoding of all the following vectors along the trace, in that the differences in direction between the present vector and its predecessor are calculated and these vector differences are mapped into a set of 2^b fixed length b-bit codewords, which we refer to as 'fixed vectors' (FV). We will show that the coding rate of the proposed FL-DCC scheme is lower for $b = 2$ and $b = 1$ than that of DCC.

When a curve is encoded by FL-DCC, it is sliced by the coding ring into small segments. Consider a sampled curve segment s. Let v be the vector link produced by the coding ring. The coding rate of a chain code

	$b = 1$ bit/vector	$b = 2$ bit/vector	DCC bit/vector
English script	0.8535	1.7271	2.0216
Chinese script	0.8532	1.7554	2.0403
Map	0.8536	1.7365	2.0396
Drawing	0.8541	1.7911	1.9437
Theoretical	0.9	1.80	2.03

Table 9.5: Coding rate comparison.

is defined [198], [200] in bits per unit length of the curve segment as

$$r = \frac{E[b(s, v)]}{E[l_n(s)]} \qquad (9.10)$$

where $b(s, v)$ is the number of bits used to encode a vector link v, $l_n(s)$ is the length of the curve segment s, while $E(x)$ represents the expected value of a random variable x. It has been shown [201] that for the set of all curves, the product of a segment length $l_n(s)$ and the probability $p(\alpha)$ that this segment occurs with a direction α must be constant. Thus the expected curve segment length for a ring of order n is given by [199]:

$$E[l_n(s)] = \int_0^{\pi/4} \frac{8 \cdot n \cdot \tau}{\cos \alpha} \cdot p(\alpha) d\alpha = \frac{\pi \cdot n \cdot \tau}{2 \cdot \sqrt{2}}. \qquad (9.11)$$

Therefore, the theoretical coding rate of FL-DCC becomes:

$$r = \frac{E[b(s, v)]}{\pi \cdot n \cdot \tau/(2 \cdot \sqrt{2})}. \qquad (9.12)$$

The theoretical and experimental coding rates of the $b = 1$ and $b = 2$ FL-DCC schemes are shown in Table 9.5. In the next Section the associated transmission issues are considered.

9.4.3 FL-DCC Graphical Codec Performance

The performance of the proposed FL-DCC graphical codecs was evaluated for a range of graphical source signals, including an English script, a Chinese script, a drawing and a map using a coding ring of $M = 8$. Table 9.5 shows the associated coding rates produced by FL-DCC for $b = 1$ and $b = 2$ as well as by DCC along with the corresponding theoretical coding rates. Both FL-DCC schemes achieve a lower coding rate than DCC. The corresponding subjective quality is portrayed in Figure 9.13 for two of the input signals previously used in Table 9.5. Observe that for $b = 2$ no subjective degradation can be seen and the degradation associated with $b = 1$ is also fairly low. This is due to the fact that the typical fuzzy granular error pat-

Figure 9.13: Decoded information for FL-DCC with $b = 1$ (bottom), $b = 2$ (centre) and DCC (top).

terns inflicted by the $b = 1$ FL-DCC scheme, when a straight line section is approximated by a zig-zag pattern, can be detected and smoothed by the decoder. The overall graphical system performance will be highlighted in the forthcoming system performance section.

Following the above speech, video and graphical source coding issues, we now address the transmission aspects of the proposed multi-media systems. Let us initially consider the factors affecting the choice of modulation.

9.5 Modulation Issues

9.5.1 Choice of Modulation

The appropriate choice of the modem scheme is based on the interplay of equipment complexity, power consumption, spectral efficiency, robustness against channel errors, co-channel and adjacent channel interference as well as the propagation phenomena, which depends on the cell size. Equally important are the associated issues of linear or non-linear amplification and filtering, the applicability of non-coherent, differential detection, soft-decision detection, equalisation and other associated issues [2], most of which will be addressed in a certain depth during our further discourse. The above, often conflicting factors led to a proposal by the European Research in Advanced Communications Equipment (RACE) project, which is referred to as the Advanced Time Division Multiple Access (ATDMA) initiative [116,117]. This proposal did not become a third-generation standard. Nonetheless, the main features of the ATDMA system framework are interesting, since this proposal reflects the philosophy and spirit of discussions leading to the European UMTS standard proposals to be highlighted in Chapter 10. Furthermore, some of the ATDMA features may influence the evolution of existing second-generation systems, such as GSM. Hence the main ATDMA features are summarised in Table 9.6 [116,117], which will be described during our forthcoming discourse. Suffice to say at this stage that these features were defined on the basis of providing higher bit-rates and bandwidth-efficiency for benign indoor picocells, while ensuring backwards compatibility with existing second-generation systems, such as GSM, for example.

Specifically, the Pan-European GSM system described in Chapter 8 or the Digital European Cordless Telecommunications (DECT) scheme highlighted in Chapter 1 employ constant envelope partial response Gaussian Minimum Shift Keying (GMSK), which was the topic of Chapter 6. The main advantage of GMSK is that since it is a so-called constant envelope modulation scheme, it ignores any fading-induced or amplifier-specific amplitude fluctuations present in the received signal and hence facilitates the utilisation of power-efficient non-linear class-C amplification. In benign pico- and micro-cellular propagation conditions, however, low transmitted power and low signal dispersion are the typical characteristics. Hence the employment of more bandwidth efficient multilevel modulation schemes becomes realistic [2]. In fact the American IS-54 [14] and JDC [15] second-generation digital systems have already opted for 2 bits/symbol modulation. in the case of the so-called multi-level full-response modulation schemes the influence of each modulation symbol is restricted to its own signalling interval, as opposed to the partial response GMSK modems of Chapter 6. Since these multi-level modems have not been treated in preceeding chapters of this book, this section attempts to provide a rudimentary overview of some of the associated multi-level modulation issues in a slightly more detailed style. For a more detailed account on full-response multi-level modulation schemes the reader is referred to [2].

Returning to Table 9.6, the ATDMA European proposal identified the following cellular structures, as the most typical propagation environments: 'long' or large macro cells, 'short' or small macro cells, micro cells and pico

Cell type	Long-macro	Short-macro	Micro	Pico
Modulation	GMSK	4/16-QAM		
Baud-rate (kBd)	360	225	900	
Carrier spacing (kHz)	276.92		1107.69 = 4 × 276.92	
Bit-Rate (kbps)	360	450/900	1800/3600	
Bwidth eff. (bps/Hz)	1.3	1.625/3.25		

Table 9.6: ATDMA cell types and modulation schemes.

cells. As the propagation environment becomes more friendly, higher channel SNRs may be maintained and hence more bandwidth-efficient modem schemes can be employed, moving from the 1 bit/symbol partial-response GMSK scheme to 2 bits/symbol or even to 4 bits/symbol full-response signalling. In ATDMA parlance the latter two schemes are referred to as 1 and 2 bits/symbol Offset Quadrature Amplitude Modulation (OQAM), where the so-called inphase (I) and quadrature phase (Q) components are offset by half of the signalling interval, which will limit the encountered signal envelope swing, since only one of the I and Q components can change at any instant. Hence OQAM is less sensitive to power amplifier non-linearities [2] than conventional QAM, since the latter allows a simultaneous change of both the I and Q components. In conventional terminology, however, 1 and 2 bits/symbol OQAM corresponds to 4-level or 16-level QAM, which is our preferred terminology. Here we will restrict our treatment to a rudimentary overview of QAM techniques; for a more detailed treatise on the subject the interested reader is referred to reference [2].

Observe furthermore in Table 9.6 that the carrier spacing and signalling rate are also different for the various cell sizes, which is a consequence of the lower signal dispersion or excess delay spread of picocells due to their low transmitted power and propagation distances. This is justified for example by the fact that the large-cell GSM system at a signalling rate of 271 kbps using a bit-interval of 3.69 μs experiences dispersive long-delay multipath components and hence requires a channel equaliser, while DECT at 1152 kbps, where the bit-interval duration is reduced by a factor of four, does not. The 'long-macro' GMSK ATDMA signalling rate is then 360 kBaud, which results in a bandwidth efficiency of 360 kBaud/276.92 kHz=1.3 bps/Hz. This is very similar to the 271 kbps/200 kHz≈1.35 bps/Hz GSM bandwidth efficiency. As seen in Table 1.1 of Chapter 1, in the American IS-54 system a signalling rate of 24.3 kBaud or a bit-rate of 48.6 kbps was accommodated in a 30 kHz bandwidth, yielding a bandwidth efficiency of 1.62 bps/Hz. This was achieved using a Nyquist filter (see Section 9.5.2) with a so-called roll-off factor of $\alpha = 0.35$ and then allowing adjacent channels to slightly overlap, while tolerating the associated adjacent channel interference. Specifically, this corresponds to allowing an overlap of the adjacent channel spectra for attenuations higher than 24 dB with respect to the signal level measured at the carrier frequency. A similar philosophy was pursued in the ATDMA proposal, where using 2bits/symbol 4-QAM signalling in a bandwidth of 276.92 kHz, and assuming a Nyquist excess filtering bandwidth (see Section 9.5.2) of 35% the achievable bit-rate became 450 kbps. The corresponding bandwidth efficiency is again 1.62 bps/Hz. Lastly, for 4bits/symbol signalling under identical filtering requirements an efficiency of 3.2 bps/Hz is maintained in the more benign indoor propagation scenarios of Table 9.6, provided that the required channel SNR and Signal-to-Interference Ratio (SIR) can be maintained.

Although many of the ATDMA parameters of Table 9.6 are non-integer values, all physical layer clocks and carrier bit-rates can be derived from a single reference oscillator frequency of 14.4 MHz, which is an important practical consideration. We note furthermore that slots in picocells can be concatenated to give so-called 'double slots' with increased bit-rates, since the associated dispersion is low due to the low pathlength differences. Here we curtail our discussion on the ATDMA proposal, a more detailed discourse on the various ATDMA transport modes can be found in references [116,117].

Here we simply introduce the ATDMA framework as an example of the recently often quoted so-called 'software radio' concept, where the transceiver is designed to reconfigure itself in a number of different modes, adapting to various propagation environments, teletraffic requirements, etc., facilitated by the flexible base-band 'algorithmic tool-box', as detailed in [8]. Since the family of constant envelope partial response GMSK modems was fully characterised in Chapter 6, here we refrain from detailing partial-response modems. Let us now turn our attention to the class of multi-level full-response modems, which can exploit the higher Shannonian channel capacity of high-SNR channels by transmitting several bits per information symbols and hence ensure high bandwidth efficiency. These modems are also often referred to as Quadrature Amplitude Modulation (QAM) schemes [2].

9.5.2 Quadrature Amplitude Modulation [2]

9.5.2.1 Background

Until quite recently QAM developments were focused at the benign AWGN telephone line and point-to-point radio applications [123], which led to the definition of the CCITT telephone circuit modem standards V.29-V.33 based on various QAM constellations ranging from uncoded 16-QAM to trellis coded (TC) 128-QAM [2]. In recent years QAM research for hostile fading mobile channels has been motivated by the ever-increasing bandwidth efficiency demand for mobile telephony [124]- [133], although QAM schemes require power-inefficient class A or AB linear amplification [134]- [137]. However, the power consumption of the low-efficiency class-A amplifier [136], [137] is less critical than that of the digital speech, image and channel codecs. Out-of-band emissions due to class AB amplifier non-linearities generating adjacent channel interferences can be reduced by some 15 dB using the adaptive predistorter proposed by Stapleton *et.al.* [171, 172]. The spectral efficiency of QAM in various macro- and micro-cellular frequency re-use structures was studied in comparison to a range of other modems in Chapter 17 of reference [2], while burst-by-burst adaptive modem arrangements were proposed for example in references [144]- [169]. Let us now highlight the basic concepts of quadrature amplitude modulation.

9.5.2.2 Modem Schematic

Multi-level full-response modulation schemes have been considered in depth in reference [2]. In this chapter only a terse introduction is offered, concentrating on the fundamental modem schematic of Figure 9.14.

If an analogue source signal must be transmitted, the signal is first low-pass filtered and analogue-to-digital converted (ADC) using a sampling frequency at least twice the signal's bandwidth - hence satisfying the Nyquist criterion. The generated digital bitstream is then mapped to complex modulation symbols by the MAP block, as seen in Figure 9.15 in case of mapping 4 bits/symbol to a 16-QAM constellation.

9.5.2.2.1 Gray Mapping and Phasor Constellation

The process of mapping the information bits onto the bitstreams modulating the I and Q carriers plays a fundamental role in determining the properties of the modem, which will be elaborated on at a later stage in Section 9.5.2.3. Suffice to say here that the mapping can be represented by the so-called constellation diagram of Figure 9.15. A range of different constellation diagrams or so-called phasor diagrams was introduced in Figure 1.38 of Chapter 1, where a phasor constellation was defined as the resulting two-dimensional plot when the amplitudes of the I and Q levels of each of the points which could be transmitted (the constellation points) are drawn in a rectangular coordinate system. For a simple binary amplitude modulation scheme, the constellation diagram would be two points both on the positive x axis. For a binary PSK (BPSK) scheme the constellation diagram would consist again of two points on the x axis, but both equidistant from the origin, one to the left and one to the right. The 'negative amplitude' of the point to the left of the origin represents a phase shift of 180 degrees in the transmitted signal. If we allow phase shifts of angles other than 0 and 180 degrees, then the constellation points move off the x axis. They can be considered to possess an amplitude and phase, the amplitude representing the magnitude of the transmitted carrier, and the phase representing the phase shift of the carrier relative to the local oscillator in the transmitter. The constellation points may also be considered to have cartesian, or complex co-ordinates, which are normally referred to as inphase (I) and quadrature

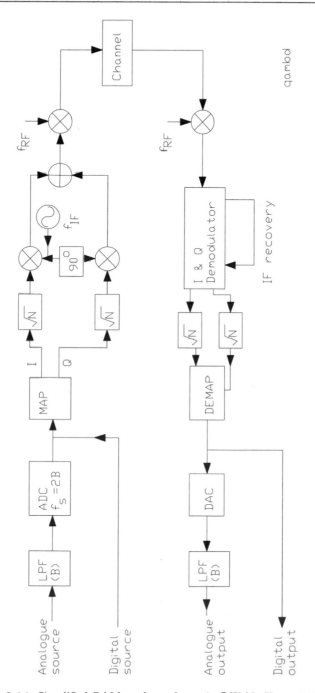

Figure 9.14: Simplified QAM modem schematic ©Webb, Hanzo 1994 [2].

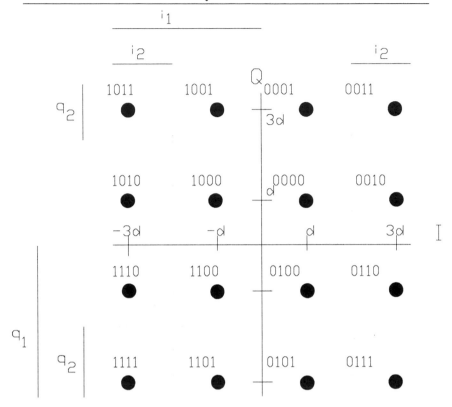

stargray

class II decision boundaries

Figure 9.15: 16-QAM square constellation ©Webb, Hanzo 1994 [2].

(Q) components corresponding to the x and y axes, respectively.

In the square-shaped 16-QAM constellation of Figure 9.15 each phasor is represented by a four-bit symbol, constituted by the in-phase bits i_1, i_2 and quadrature bits q_1, q_2, which are interleaved to yield the sequence i_1, q_1, i_2, q_2. The quaternary quadrature components I and Q are Gray encoded by assigning the bits *01, 00, 10* and *11* to the levels *3d, d, -d* and *-3d*, respectively. This constellation is widely used because it has equidistant constellation points arranged in a way that the average energy of the phasors is maximised. Using the geometry of Figure 9.15 the average energy is computed as

$$E_0 = (2d^2 + 2 \times 10d^2 + 18d^2)/4 = 10 \times d^2. \qquad (9.13)$$

For any other phasor arrangement the average energy will be less and therefore, assuming a constant noise energy, the signal to noise ratio required to achieve the same bit error rate (BER) will be higher, a topic to be studied comparatively in the context of two different 16-QAM constellations at a later stage in Section 9.5.2.3.

Notice from the mapping in Figure 9.15 that the Hamming distance amongst the constellation points, which are 'closest neighbours' with a Euclidean distance of $2d$ is always one. The Hamming distance between any two points is the difference in the mapping bits for those points, so points labelled 0101 and 0111 would have a Hamming distance of 1, and points labelled 0101 and 0011 would have a Hamming distance of 2. This is a fundamental feature of the Gray coding process and ensures that whenever a transmitted phasor is corrupted by noise sufficiently that it is incorrectly identified as a neighbouring constellation point, the demodulator will choose a phasor with a single bit error. This minimises the error probability.

It is plausible that the typical quaternary I or Q component sequence generated by the MAP block of Figure 9.14 would require an infinite transmission bandwidth due to the abrupt changes at the signalling interval boundaries. Hence these signals must be bandlimited before transmission in order to contain the spectrum within a limited band and so minimise interference with other users or systems sharing the spectrum. This filtering is indicated in Figure 9.14 by the square-root Nyquist-filter blocks denoted by $\sqrt{N}$, where the rationale behind the notation will become clear in the next section.

9.5.2.2.2 Nyquist Filtering A full theoretical treatment of Nyquist filtering was provided in reference [2], hence here we restrict our discussions to a rudimentary introduction. An ideal linear-phase low-pass filter (LPF) with a cut-off frequency of $f_N = f_s/2$, where $f_s = 1/T$ is the signalling frequency, T is the signalling interval duration and $f_N = 1/(2T)$ is the so-called Nyquist frequency, would be able to pass most of the energy of the quadrature components I and Q within a compact frequency band. Due to the linear phase response of the filter all frequency components would exhibit the same group-delay. Because such a filter has a $(\sin x)/x$ function shaped impulse response with equidistant zero-crossings at the sampling instants $n \cdot T$, this ideal low-pass filter does not result in inter-symbol-interference (ISI) between consecutive signalling symbols. After its inventor Nyquist [118] this ideal low-pass transfer function and its derivatives about to be introduced in the next paragraph are referred to as the Nyquist characteristic. However, such an ideal low-pass filter is unrealisable, as all practical low-pass (LP) filters exhibit amplitude and phase distortions, particularly towards the transition between the pass- and stop-band. Conventional Butterworth, Chebichev or inverse-Chebichev LP filters have impulse responses with non-zero values at the equi-spaced sampling instants $n \cdot T$ and hence introduce ISI. They therefore degrade the bit error rate (BER) performance.

Nyquist's fundamental theoretical work [118] suggested that special pulse shaping filters must be deployed, ensuring that the total transmission path, including the transmitter, receiver and the channel, has an impulse response with a unity value at the current signalling instant and zero-crossings at all other consecutive sampling instants $n \cdot T$. He showed that any odd-symmetric frequency-domain extension characteristic fitted to the ideal LPF amplitude spectrum yields such an impulse response, and is therefore free from ISI. Two examples of the corresponding filter characteristics are shown in Figure 9.16, which will be described during our forthcoming deliberations.

A practical odd-symmetric extension characteristic is the so-called raised-cosine (RC) characteristic fitted to the above-mentioned ideal low-pass filter characteristic [2]. The parameter controlling the bandwidth of the Nyquist filter is the so-called roll-off factor α, which is unity, if the ideal LPF bandwidth is doubled by the extension characteristic. If $\alpha = 0.5$ a total bandwidth of $1.5 \times f_N = 1.5/(2T)$ results, and so on. The lower the value of the roll-off factor, the more compact the spectrum becomes, but the higher the complexity of the required filter and other receiver circuitry, such as clock and carrier recovery [2]. The stylised frequency response of these filters is shown in Figure 9.16 for $\alpha = 0.9$ and $\alpha = 0.1$. It follows from Fourier theory that the wider the transmission band, the more sharply decaying the impulse response. A sharply decaying impulse response has a favourable effect concerning the mitigation of the potential ISI in the case of imperfect clock recovery, when there is a time-domain jitter superimposed on the optimum sampling instant. Hence in terms of system performance an $\alpha = 1$ filtering scheme is more favourable than a more sharply filtered but more bandwidth-efficient scheme.

In case of additive white Gaussian noise (AWGN) with a uniform power spectral density (PSD) the noise power admitted to the receiver is proportional to its bandwidth. Therefore it is also necessary to limit the received signal bandwidth at the receiver to a value close to the transmitter's bandwidth. Optimum detection theory [119] shows that the SNR is maximised, if so-called matched filtering is used, where the Nyquist characteristic of Figure 9.16 is divided between two identical filters, a transmitter- and a receiver-filter, each characterised by the square root of the Nyquist shape, as suggested by the filters $\sqrt{N}$ in Figure 9.14.

In conclusion of our discourse on filtering issues we note that Feher [121] proposed non-linear filtering (NLF) as a low-complexity alternative to Nyquist-filtering, which operates by simply fitting a time-domain quarter period of a sine wave between two symbols for both of the quadrature carriers. This technique can be simply implemented by using a look-up table and there is no contribution from previous symbols at any sample point, which is advantageous when complex high-level QAM constellations are transmitted. The disadvantage of this form of filtering is that it is less

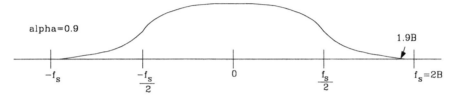

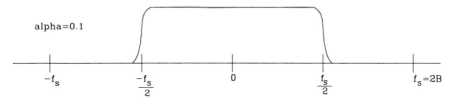

Figure 9.16: Stylised frequency response of two Nyquist filters with $\alpha = 0.9$ and 0.1 ©Webb, Hanzo 1994 [2].

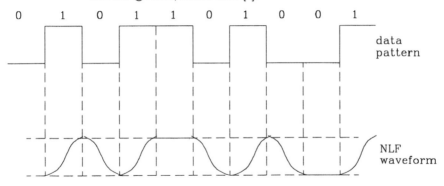

Figure 9.17: Stylised NLF waveforms ©Webb, Hanzo 1994 [2].

spectrally efficient than optimal partial-response filtering schemes. Nevertheless, its implementational advantages often render this loss of efficiency acceptable. The power spectrum of a NLF signal is given by [121]:

$$S(f) = T \left(\frac{\sin 2\pi fT}{2\pi fT} \frac{1}{1 - 4(fT)^2} \right)^2 \tag{9.14}$$

and the corresponding original and NLF waveforms are given in Figure 9.17 for the I or Q quadrature component.

9.5.2.2.3 Modulation and Demodulation   Once the analogue I and Q signals have been generated and filtered, they are modulated by an I-Q modulator as shown in Figure 9.14. This modulator essentially consists of

Transmitted RF spectrum

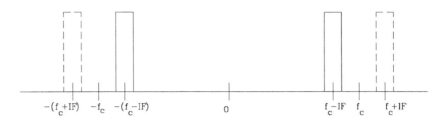

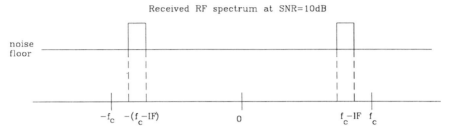

Figure 9.18: Stylised transmitted and received spectra ©Webb, Hanzo 1994 [2].

two mixers, one for the I channel and another for the Q channel. The I channel is mixed with an intermediate frequency (IF) signal that is in phase with respect to the carrier, and the Q channel is mixed with an IF that is 90 degrees out of phase. This process allows both signals to be transmitted over a single channel within the same bandwidth using quadrature carriers. In a similar fashion, the signal is demodulated at the receiver. Provided that the signal degradation is kept to a minimum, the orthogonality of the I and Q channels will be retained and their information sequences can be independently demodulated.

Following I-Q modulation, the signal is modulated by a radio frequency (RF) mixer, increasing its frequency to that used for transmission. Since the IF signal occurred at both positive and negative frequencies, it will occur at both the sum and difference frequencies when mixed up to the RF. Since there is no reason to transmit two identical sidebands, one is usually filtered out, as seen plotted in dashed lines in Figure 9.18. We also note that in theory one could dispense with the IF stage, mixing the base-band component directly to the transmission frequency, if it were possible to design the required extremely narrow-band so-called notch-filters at the RF for removing the unwanted modulation products and out-of-band spectral spillage. Since this results in filter-design problems, in practical systems the signal is converted up to the RF usually in two or more mixing stages.

The transmission channel is often the most critical factor influencing the performance of any communications system. Here we consider only the addition of noise based on the signal to noise ratio (SNR). The noise is often the major contributing factor to signal degradation and its effect exhibits itself in terms of a noise floor, as portrayed in the received RF spectrum of Figure 9.18.

The RF demodulator mixes the received signal down to the IF for the I-Q demodulator. In order to accurately mix the signal back to the appropriate intermediate frequency, the RF mixer operates at the difference between the IF and RF frequencies. Since the I-Q demodulator includes IF recovery circuits, the accuracy of the RF oscillator frequency is not critical. However, it should be stable, exhibiting a low phase noise, since any noise present in the down-conversion process will be passed on to the detected I and Q baseband signals, thereby adding to the possibility of bit errors. The recovered IF spectrum is similar to the transmitted one but with the additive noise floor seen in the RF spectrum of Figure 9.18.

Returning to Figure 9.14, I-Q demodulation takes place in the reverse order to the modulation process. The signal is split into two paths, with each path being mixed down with IFs that are 90 degrees apart. Since the exact frequency of the original reference must be known to determine the absolute phase, IF carrier recovery circuits are used to reconstruct the precise reference frequency at the receiver. The recovered I component should be almost identical to that transmitted, with the only differences being caused by noise.

9.5.2.2.4 Data Recovery Once the analogue I and Q components have been recovered, they must be digitised. This is carried out by the bit detector. The bit detector determines the most likely bit transmitted by sampling the I and Q signals at the correct sampling instants and comparing them to the legitimate I and Q values of $-3d, -d, d, 3d$ in the case of a square 16-QAM constellation. From each I and Q decision two bits are derived, leading to a 4-bit 16-QAM symbol. The four recovered bits are then passed on to the DAC. Although the process might sound simple, it is complicated by the fact that the 'right time' to sample is a function of the clock frequency at the transmitter. The data clock must be regenerated upon recovery of the carrier. Any error in clock recovery will increase the BER. Again, these issues are treated in more depth in reference [2].

If there is no channel noise or the SNR is high, the reconstructed digital signal is identical to the original input signal. Provided the DAC operates at the same frequency and with the same number of bits as the input ADC, then the analogue output signal after low-pass filtering with a cut-off frequency of B, is also identical to the output signal of the LPF at the input of the transmitter. Hence it is a close replica of the input signal. Following the above basic modem schematic description let us now consider two often used 16-QAM constellations.

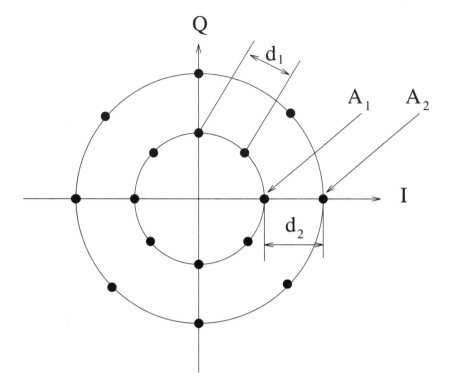

Figure 9.19: Star 16-QAM constellation.

9.5.2.3 QAM Constellations

A variety of different constellations have been proposed for QAM transmissions over Gaussian channels. However, in practice often the constellations shown in Figures 9.15 and 9.19 are preferred. The essential problem is to maintain a high minimum distance, d_{min}, between constellation points whilst keeping the average power required for the constellation to a minimum. Calculation of d_{min} and the average power is a straightforward geometric procedure, and has been performed for a range of constellations by Proakis [138]. The results show that the square constellation of Figure 9.15 is optimal for Gaussian channels. We will show that the star constellation of Figure 9.19 requires a higher energy to achieve the same minimum distance d_{min} amongst constellation points than the square constellation of Figure 9.15 and hence the latter is often preferred for Gaussian channels. However, there may be implementational reasons for favouring circular constellations over the square ones.

When designing a constellation, consideration must be given to:

1. The minimum Euclidean distance amongst phasors, which is characteristic of the noise immunity of the scheme.

2. The minimum phase rotation amongst constellation points, determining the phase jitter immunity and hence the scheme's resilience against clock recovery imperfections and channel phase rotations.

3. The ratio of the peak-to-average phasor power, which is a measure of robustness against non-linear distortions introduced by the power amplifier.

It is quite instructive to estimate the optimum ring ratio RR for the star constellation of Figure 9.19 in AWGN under the constraint of a constant average phasor energy E_0. Accordingly, a high ring ratio value implies that the Euclidian distance amongst phasors on the inner ring is reduced, while the distance amongst phasors on different rings is increased. In contrast, upon reducing the ring ratio the cross-ring distance is reduced and the distances on the inner ring become larger.

Intuitively, one expects that there will be an optimum ring ratio, where the overall bit error rate (BER) constituted by detection errors on the same ring plus errors between rings is minimised. Suffice to say here that the minimum Euclidean distance amongst phasors is maximised if $d_1 = d_2 = A_2 - A_1$ in the star constellation of Figure 9.19. Using the geometry of Figure 9.19 we can write that:

$$\cos 67.5^o = \frac{d_1}{2} \cdot \frac{1}{A_1}$$
$$d_1 = 2 \cdot A_1 \cdot \cos 67.5^o$$

and hence

$$A_2 - A_1 = d_1 = d_2 = 2 \cdot A_1 \cdot \cos 67.5^o.$$

Upon dividing both sides by A_1 and introducing the ring ratio RR we arrive at:

$$RR - 1 = 2 \cdot \cos 67.5^o$$
$$RR \approx 1.77.$$

Simulation results using a variety of ring ratios in the interval of $1.5 < RR < 3.5$ both over Rayleigh and AWGN channels showed [2, 139] that the BER does not strongly depend on the ring ratio, exhibiting a flat BER minimum for RR values in the above range.

Under the constraint of having identical distances amongst constellation points, when $d_1 = d_2 = d$, the average energy E_0 of the star constellation can be computed as follows:

$$E_0 = \frac{8 \cdot A_1^2 + 8 \cdot A_2^2}{16} = \frac{1}{2}(A_1^2 + A_2^2)$$

where

$$A_1 = \frac{d}{2 \cdot \cos 67.5^o} \approx \frac{d}{0.765} \approx 1.31d$$

and
$$A_2 \approx 1.77 \cdot A_1 \approx 2.3d$$
yielding
$$E_0 \approx 0.5 \cdot (5.3 + 1.72)d^2 \approx 3.5d^2.$$

The minimum distance of the constellation for an average energy of E_0 becomes:
$$d_{min} \approx \sqrt{E_0/3.5} \approx 0.53 \cdot \sqrt{E_0},$$
while the peak-to-average phasor energy ratio is:
$$r \approx \frac{(2.3d)^2}{3.5d^2} \approx 1.5.$$

The minimum phase rotation θ_{min}, the minimum Euclidean distance d_{min} and the peak-to-average energy ratio r are summarised in Table 9.7 for both of the above constellations.

Let us now derive the above characteristic parameters for the square constellation. Observe from Figure 9.19 that $\theta_{min} < 45^o$, while the distance between phasors is $2 \cdot d$. Hence the average phasor energy becomes:

$$
\begin{aligned}
E_0 &= \frac{1}{16}\left[4 \cdot (d^2 + d^2) + 8(9d^2 + d^2) + 4 \cdot (9d^2 + 9d^2)\right] \\
&= \frac{1}{16}(8d^2 + 80 \cdot d^2 + 72d^2) \\
&= 10d^2.
\end{aligned}
$$

Hence, assuming the same average phasor energy E_0 as for the star constellation we now have a minimum distance of
$$d_{min} = 2d = 2 \cdot \sqrt{E_0/10} = \sqrt{E_0/2.5} \approx 0.63 \cdot \sqrt{E_0}.$$

Lastly, the peak-to-average energy ratio r is given by:
$$r = \frac{18d^2}{10d^2} = 1.8.$$

The square constellation's characteristics are also summarised in Table 9.7. Observe that the star constellation has a higher jitter immunity and a slightly lower peak-to-average energy ratio than the square scheme. However, the square phasor constellation has an almost 20 % higher minimum distance at the same average phasor energy and hence it is very attractive for AWGN channels, where noise is the dominant channel impairment. Let us now consider the bit error rate (BER) versus channel Signal-to-Noise Ratio (SNR) performance of the maximum-minimum distance square-constellation 16-QAM over AWGN channels.

Type	θ_{min}	d_{min}	r
Star	45°	$0.53\sqrt{E_0}$	1.5
Square	$< 45^{o}$	$0.63 \cdot \sqrt{E_0}$	1.8

Table 9.7: Comparison of the star and square constellations.

9.5.2.4   16-QAM BER versus SNR Performance over AWGN Channels

9.5.2.4.1   Decision Theory   Before analysing the effects of errors let us briefly review the roots of decision theory in the spirit of Bayes' theorem formulated as follows:

$$P(X/Y) \cdot P(Y) = P(Y/X) \cdot P(X) = P(X,Y), \qquad (9.15)$$

where the random variables X and Y have probabilities of $P(X)$ and $P(Y)$, their joint probability is $P(X,Y)$ and their conditional probabilities are given by $P(X/Y)$ and $P(Y/X)$.

In decision theory the above theorem is invoked in order to infer from the noisy analogue received sample y, what the most likely transmitted symbol was, assuming that the so-called *a priori* probability $P(x)$ of the transmitted symbols $x_n, n = 1 \dots M$ is known. Given that the received sample y is encountered at the receiver, the conditional probability $P(x_n/y)$ quantifies the chance that x_n has been transmitted:

$$P(x_n/y) = \frac{P(y/x_n) \cdot P(x_n)}{P(y)}, \qquad n = 1 \dots N \qquad (9.16)$$

where $P(y/x_n)$ is the conditional probability of the continuous-valued noise-contaminated sample y, given that $x_n, n = 1 \dots N$ was transmitted. The probability of encountering a specific y value will be the sum of all possible combinations of receiving y, given that $x_n, n = 1 \dots N$ was transmitted, which can be written as:

$$P(y) = \sum_{n=1}^{N} P(y/x_n) \cdot P(x_n) = \sum_{n=1}^{N} P(y, x_n). \qquad (9.17)$$

Let us now consider the case of binary phase shift keying (BPSK), where there are two legitimate transmitted values, x_1 and x_2 which are contaminated by noise, as portrayed in Figure 9.20. The conditional probability of receiving any particular noise-contaminated analogue sample y, given that x_1 or x_2 was transmitted is quantified by the Gaussian probability density

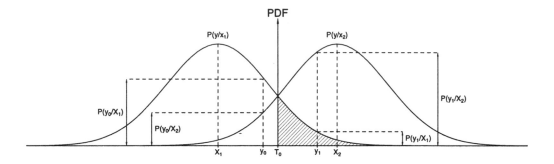

Figure 9.20: Transmitted samples and noisy received samples for BPSK.

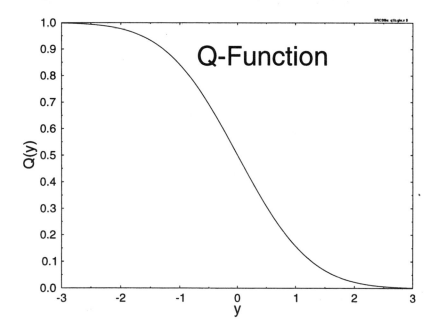

Figure 9.21: Gaussian Q-function

functions (PDFs) seen in Figure 9.20, which are described by:

$$P(y/x) = \frac{1}{\sigma\sqrt{2\pi}}e\frac{-(y-x)^2}{2\sigma^2}, \qquad (9.18)$$

where $x = x_1$ or x_2 is the mean and σ^2 is the variance.

Observe from the Figure that the shaded area represents the probability of receiving values larger than the threshold T_0, when x_1 was transmitted and this is equal to the probability of receiving a value below T_0, when x_2 was transmitted. As displayed in the Figure, when receiving a specific $y = y_0$ sample, there is an ambiguity, as to which symbol was transmitted. The corresponding conditional probabilities are given by $P(y_0/x_1)$ and $P(y_0/x_2)$ and their values are also marked on Figure 9.20. Given the knowledge that x_1 was transmitted, we are more likely to receive y_0 than with the premise that x_2 was transmitted. Hence, upon observing $y = y_0$ statistically speaking it is advisable to decide that x_1 was transmitted. Following similar logic, when receiving y_1 as seen in Figure 9.20, it is logical to conclude that x_2 was transmitted.

Indeed, according to optimum decision theory [188], the optimum decision threshold above which x_2 is inferred is given by:

$$T_0 = \frac{x_1 + x_2}{2} \tag{9.19}$$

and below this threshold x_1 is assumed to have been transmitted. If $x_1 = -x_2$ then $T_0 = 0$ is the optimum decision threshold minimising the bit error probability.

In order to compute the error probability in the case of transmitting x_1, the PFD $P(y/x_1)$ of Equation 9.18 has to be integrated from x_1 to ∞, which gives the shaded area under the curve in Figure 9.20. In other words, the probability of a zero-mean noise sample exceeding the magnitude of x_1 is sought, which is often referred to as the *noise protection distance*, given by the so-called Gaussian Q-function:

$$Q(x_1) = \frac{1}{\sigma\sqrt{2\pi}} \int_{x_1}^{\infty} e^{\frac{-y^2}{2\sigma^2}} \, dy, \tag{9.20}$$

where σ^2 is the noise variance. Notice that since $Q(x_1)$ is the probability of exceeding the value x_1, it is actually the complementary cumulative density function (CDF) of the Gaussian distribution.

Assuming that $x_1 = -x_2$, the probability that the noise can carry x_1 across $T_0 = 0$ is equal to that of x_2 being corrupted in the negative direction. Hence, assuming that $P(x_1) = P(x_2) = 0.5$, the overall error probability is given by:

$$\begin{aligned} P_e &= P(x_1) \cdot Q(x_1) + P(x_2) \cdot Q(x_2) \\ &= \frac{1}{2}Q(x_1) + \frac{1}{2}Q(x_1) = Q(x_1). \end{aligned} \tag{9.21}$$

The values of the Gaussian Q-function plotted in Figure 9.21 are tabulated in many textbooks [188], along with values of the Gaussian PDF in the case of zero-mean, unit-variance processes. For abscissa values of $y > 4$ the following approximation can be used:

$$Q(y) \approx \frac{1}{y\sqrt{2\pi}} e^{\frac{-y^2}{2}} \quad \text{for} \quad y > 4. \tag{9.22}$$

Having provided a rudimentary introduction to decision theory, let us now focus our attention on the demodulation of 16-QAM signals in AWGN.

9.5.2.4.2 QAM Modulation and Transmission In general the modulated signal can be represented by

$$s(t) = a(t)\cos[2\pi f_c t + \Theta(t)] = Re[a(t)e^{j[w_c t + \Theta(t)]}], \tag{9.23}$$

where the carrier $\cos(w_c t)$ is said to be amplitude modulated if its amplitude $a(t)$ is adjusted in accordance with the modulating signal, and is said to be phase modulated if $\Theta(t)$ is varied in accordance with the modulating signal. In QAM the amplitude of the baseband modulating signal is determined by $a(t)$ and the phase by $\Theta(t)$. The inphase component I is then given by

$$I = a(t)\cos\Theta(t) \tag{9.24}$$

and the quadrature component Q by

$$Q = a(t)\sin\Theta(t). \tag{9.25}$$

This signal is then corrupted by the channel. Here we will only consider AWGN. The received signal is then given by

$$r(t) = a(t)\cos[2\pi f_c t + \Theta(t)] + n(t) \tag{9.26}$$

where $n(t)$ represents the AWGN, which has both an inphase and quadrature component. It is this received signal which we will attempt to demodulate.

9.5.2.4.3 16-QAM Demodulation in AWGN The demodulation of the received QAM signal is achieved by performing quadrature amplitude demodulations using the decision boundaries constituted by the coordinate axes and the dotted lines portrayed in Figure 9.15 for the I and Q components, as shown below for the bits i_1 and q_1:

$$\begin{array}{llll} \text{if} & \text{I,Q} \geq 0 & \text{then} & i_1, q_1 = 0 \\ \text{if} & \text{I,Q} < 0 & \text{then} & i_1, q_1 = 1 \end{array} \tag{9.27}$$

The decision boundaries for the 3rd and 4th bits i_2 and q_2, respectively, are again shown in Figure 9.15, and thus:

$$
\begin{array}{llllll}
\text{if} & \text{I,Q} & \geq & 2d & \text{then} & i_2, q_2 = 1 \\
\text{if} & --2d \leq & \text{I,Q} & < 2d & \text{then} & i_2, q_2 = 0 \\
\text{if} & --2d > & \text{I,Q} & & \text{then} & i_2, q_2 = 1.
\end{array}
\tag{9.28}
$$

We will show that in the process of demodulation the positions of the bits in the QAM symbols associated with each point in the QAM constellation have an effect on the probability of them being in error. In the case of the two most significant bits (MSBs) of the four bit symbol i_1, q_1, i_2, q_2, i.e. i_1 and q_1, the distance from a demodulation decision boundary of each received phasor in the absence of noise is $3d$ for 50 % of the time, and d for 50 % of the time; if each phasor occurs with equal probability. The average protection distance for these bits is therefore $2d$ although the bit error probability for a protection distance of $2d$ would be dramatically different from that calculated. Indeed, the average protection distance is never encountered, we only use this term to aid our investigations. The two least significant bits (LSB), i.e. i_2 and q_2 are always at a distance of d from the decision boundary and consequently the average protection distance is d. We may consider our QAM system as a class one (C1) and as a class two (C2) subchannel, where bits transmitted via the C1 subchannel are received with a lower probability of error than those transmitted via the C2 subchannel.

Observe in the phasor diagram of Figure 9.15 that upon demodulation in the C2 subchannel, a bit error will occur if the noise exceeds d in one direction or $3d$ in the opposite direction, where the latter probability is insignificant. Hence the C2 bit error probability becomes

$$
P_{2G} = Q\left\{ \frac{d}{\sqrt{N_0/2}} \right\} = \frac{1}{\sqrt{2\pi}} \int_{\frac{d}{\sqrt{N_0/2}}}^{\infty} \exp\left(-x^2/2\right)dx
\tag{9.29}
$$

where $N_0/2$ is the double-sided spectral density of the AWGN, $\sqrt{N_0/2}$ is the corresponding noise voltage, and the $Q\{\}$ function was given in Equation 9.20 and Figure 9.21. As the average symbol energy of the 16-level QAM constellation computed for the phasors in Figure 9.15 is

$$
E_0 = 10d^2,
\tag{9.30}
$$

then we have that

$$
P_{2G} = Q\left\{ \sqrt{\frac{E_0}{5N_0}} \right\}.
\tag{9.31}
$$

For the C1 subchannel data the bits i_1 and q_1 are at a protection distance of d from the decision boundaries for half the time, and their protection distance is $3d$ for the remaining half of the time. Therefore the probability of a bit error is

$$P_{1G} = \frac{1}{2}Q\left\{\frac{d}{\sqrt{N_0/2}}\right\} + \frac{1}{2}Q\left\{\frac{3d}{\sqrt{N_0/2}}\right\} = \frac{1}{2}\left[Q\left\{\sqrt{\frac{E_0}{5N_0}}\right\} + Q\left\{3\sqrt{\frac{E_0}{5N_0}}\right\}\right].$$
$$(9.32)$$

The C1 and C2 error probabilities P_{1G} and P_{2G} as a function of E_b/N_0 are given by *Equation 9.31* and *9.32* and displayed in Figure 9.22 as a function of the channel SNR in contrast to a range of other modulation schemes. Note that for 1 bit/symbol uncoded transmissions the E_b/N_0 and SNR values are identical, but for example for 2 bit/symbol transmissions for a given signal and noise energy, i.e. channel SNR, the E_b/N_0 value must be reduced by a factor of two or 3.01 dB. Viewing this observation from a different angle, 2 bit/symbol transmissions require a 3 dB higher signal energy or SNR for maintaining a constant E_b/N_0 value. Similarly, for 4 bit/symbol transmissions a factor four E_b/N_0 reduction is necessary for a fixed SNR value, which corresponds to a 6.02 dB higher channel SNR. Returning to the Figure, the BER versus channel SNR performance of binary phase shift keying (BPSK), quaternary phase shift keying (QPSK), 16-QAM and 64-QAM BER are portrayed. For 16-QAM the two protection classes differ by a factor of two in terms of their BER. Similarly to our above deliberations, in reference [2] we also showed that 64-QAM exhibits three subchannels, whose BERs are also shown in the Figure. Observe in the Figure that given a certain channel SNR, i.e. a constant signal power, in harmony with our expectations, 2 bit/symbol transmissions require about 3 dB higher channel SNR for a given BER than binary signalling. A further 6 dB is necessitated by 16-QAM and an additional 6 dB by 64-QAM transmissions. The average probability P_{AV} of bit error for the 16-level QAM system is then computed as:

$$P_{AV} = (P_{1G} + P_{2G})/2. \qquad (9.33)$$

Our simulation results gave virtually identical curves to those in Figure 9.22, exhibiting a BER advantage in using the C1 subchannel over using the C2. The computation of the error rate over Rayleigh fading channels is more involved. For the square 16-QAM constellation Cavers [178] provided symbol error rate formulae, while for the star constellation analytical error rate formulae were disseminated by Adachi [189]. With the above considerations in mind let us now concentrate our attention on multilevel communications over Rayleigh fading channels, which were described in Chapter 2.

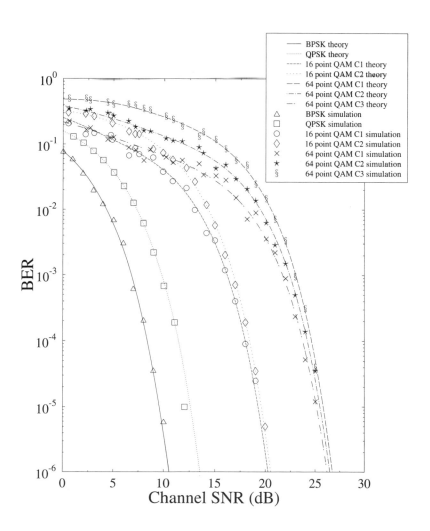

Figure 9.22: BPSK, QPSK, 16-QAM and 64-QAM BER versus channel SNR performance over AWGN channels ©Torrance, 1996 [179].

9.5.2.5 Reference Assisted Coherent QAM for Fading Channels

Over fading channels a number of additional measures have to be taken in order be able to invoke bandwidth-efficient multi-level QAM schemes. The major difficulty is that over fading channels the transmitted phasors' magnitude is attenuated and their phase is rotated by the channel, as it was shown in Figure 1.1. Two powerful methods have been proposed in order to ensure adequate QAM operation in fading environments. Both these techniques deliver channel measurement information in terms of attenuation and phase shift due to fading. The first is transparent tone in band (TTIB) assisted modulation proposed by McGeehan, Bateman *et.al* [180]- [187], where a pilot carrier is inserted typically in the centre of the modulated spectrum. At the receiver the signal is extracted and used to estimate the channel-induced attenuation and phase rotation, as it was detailed also in [2]. A disadvantage of TTIB schemes is their relatively high complexity and expanded spectral occupancy, since the pilot tone is inserted in one or several spectral gaps created by segmenting the signal spectrum and shifting the contiguous spectrum segments apart.

An alternative lower complexity technique is pilot symbol assisted modulation (PSAM) [178], where known channel sounding phasors are periodically inserted into the transmitted time-domain signal sequence. Similarly to the frequency domain pilot tone, these known symbols deliver channel measurement information. Let us here concentrate our attention on the latter, implementationally less complex technique.

9.5.2.5.1 PSAM System Description
Following Caver's approach [178], the block diagram of a general PSAM scheme is depicted in Figure 9.23, where the pilot symbols p are cyclically inserted into the data sequence prior to pulse shaping, as demonstrated by Figure 9.24.

A frame of data is constituted by M symbols, and the first symbol in every frame is assumed to be the pilot symbol $b(0)$, followed by $(M-1)$ useful data symbols $b(1), b(2) \ldots b(M-1)$.

Detection can be carried out by matched filtering, and the output of the matched filter is split into data and pilot paths, as seen in Figure 9.23. The set of pilot symbols can be extracted by decimating the matched filter's sampled output sequence using a decimation factor of M. The extracted sequence of pilot symbols must then be interpolated in order to derive a channel estimate $v(k)$ for every useful received information symbol $r(k)$. Decision is carried out against a decision level reference grid, scaled and rotated according to the instantaneous channel estimate $v(k)$.

Observe in Figure 9.23 that the received data symbols must be delayed according to the interpolation and prediction delay incurred. This delay becomes longer, if interpolation is carried out using a longer history of the received signal to yield better channel estimates. Consequently, there is a trade-off between processing delay and accuracy, an issue documented by

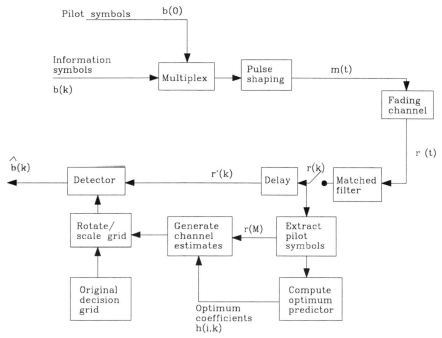

Figure 9.23: PSAM schematic © [178] ©IEEE, 1991, Cavers.

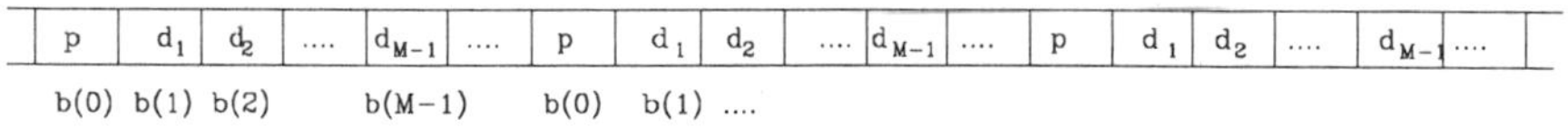

p	d$_1$	d$_2$	....	d$_{M-1}$	....	p	d$_1$	d$_2$	....	d$_{M-1}$	....	p	d$_1$	d$_2$	....	d$_{M-1}$	....

b(0) b(1) b(2)   b(M−1)   b(0)   b(1) ....

Figure 9.24: Insertion of pilot symbols in PSAM ©Webb and Hanzo 1994. [2]

Torrance and Hanzo [143] for a wide range of parameters. The interpolation coefficients can be kept constant over a whole pilot-period of length M, but better channel estimates can be obtained, if the interpolator's coefficients are optimally updated for every received symbol.

The complex envelope of the modulated signal can be formulated as:

$$m(t) = \sum_{k=-\infty}^{\infty} b(k)p(t - kT), \qquad (9.34)$$

where $b(k) = -3, -1, 1$ or 3 represents the quaternary I or Q components of the 16-QAM symbols to be transmitted, T is the symbol duration and $p(t)$ is a band-limited unit-energy signalling pulse, for which we have:

$$\int_{-\infty}^{\infty} |p(t)|^2 dt = 1. \qquad (9.35)$$

The value of the pilot symbols $b(kM)$ can be arbitrary, although sending a sequence of known pseudo-random symbols instead of using always the same phasor avoids the transmission of a periodic tone, which would increase the detrimental adjacent channel interference [140].

The narrowband Rayleigh channel is assumed to be 'flat'-fading, which implies that all frequency components of the transmitted signal suffer the same attenuation and phase shift. This condition is met, if the transmitted signal's bandwidth is much lower than the channel's coherence bandwidth. The received signal is then given by:

$$r(t) = c(t) \cdot m(t) + n(t), \tag{9.36}$$

where $n(t)$ is the AWGN and $c(t)$ is the channel's complex gain. Assuming a Rayleigh-fading envelope $\alpha(t)$, a uniformly distributed phase $\phi(t)$ and a residual frequency offset of f_0, we have:

$$c(t) = \alpha(t)e^{j\phi(t)} \cdot e^{j\omega_0 t}. \tag{9.37}$$

The matched filter's output symbols at the sampling instants kT are then given as:

$$r(k) = b(k) \cdot c(k) + n(k). \tag{9.38}$$

Without imposing limitations on the analysis, Cavers [178] assumed that in every channel sounding block $b(0)$ was the pilot symbol and considered the detection of the useful information symbols in the range $\lfloor -M/2 \rfloor \le k \le \lfloor (M-1)/2 \rfloor$, where $\lfloor \bullet \rfloor$ is the integer of $\bullet$. Optimum detection is achieved if the corresponding channel gain $c(k)$ is estimated for every received symbol $r(k)$ in the above range. The channel gain estimate $v(k)$ can be derived as a weighted sum of the surrounding K received pilot symbols $r(iM)$, $\lfloor -K/2 \rfloor \le i \le \lfloor K/2 \rfloor$, as shown below:

$$v(k) = \sum_{i=\lfloor -K/2 \rfloor}^{\lfloor K/2 \rfloor} h(i,k) \cdot r(iM), \tag{9.39}$$

and the weighting coefficients $h(i,k)$ explicitly depend on the symbol position k within the frame of M symbols.

The estimation error $e(k)$ associated with the gain estimate $v(k)$ is computed as:

$$e(k) = c(k) - v(k). \tag{9.40}$$

Let us now consider the computation of the optimum channel gains.

9.5.2.5.2 Channel Gain Estimation in PSAM While previously proposed PSAM schemes used either a low-pass interpolation filter [140] or an approximately Gaussian filter [141], Cavers employed an optimum Wiener filter [142] to minimise the channel estimation error variance

$\sigma^2{}_e(k) = E\{e^2(k)\}$, where $E\{\ \}$ represents the expectation. This well-known estimation error variance minimisation problem can be formulated as follows:

$$
\begin{aligned}
\sigma_e^2(k) &= E\{e^2(k)\} = E\{[c(k) - v(k)]^2\} \\
&= E\left\{\left[c(k) - \sum_{i=\lfloor -K/2 \rfloor}^{\lfloor K/2 \rfloor} h(i,k) \cdot r(iM)\right]^2\right\}.
\end{aligned}
\tag{9.41}
$$

In order to find the optimum interpolator coefficients $h(i,k)$, minimising the estimation error variance $\sigma^2{}_e(k)$ we consider estimating the k th sample and set:

$$
\frac{\partial \sigma_e^2(k)}{\partial h(i,k)} = 0 \quad \text{for} \quad \lfloor -K/2 \rfloor \le i \le \lfloor K/2 \rfloor.
\tag{9.42}
$$

Then using Equation 9.41 we have:

$$
\frac{\partial \sigma^2{}_e(k)}{\partial h(i,k)} = E\left\{2\left[c(k) - \sum_{i=\lfloor -K/2 \rfloor}^{\lfloor K/2 \rfloor} h(i,k) \cdot r(iM)\right] \cdot r(jM)\right\} = 0.
\tag{9.43}
$$

After multiplying both square bracketed terms with $r(jM)$, and computing the expected value of both terms separately, we arrive at

$$
E\{c(k) \cdot r(jM)\} = E\left\{\sum_{i=\lfloor -K/2 \rfloor}^{\lfloor K/2 \rfloor} h(i,k) \cdot r(iM) \cdot r(jM)\right\}.
\tag{9.44}
$$

Observe that

$$
\Phi(j) = E\{c(k) \cdot r(jM)\}
\tag{9.45}
$$

is the cross-correlation of the received pilot symbols and complex channel gain values, while

$$
R(i,j) = E\{r(iM) \cdot r(jM)\}
\tag{9.46}
$$

represents the pilot symbol autocorrelations; hence Equation 9.44 yields:

$$
\sum_{i=\lfloor -K/2 \rfloor}^{\lfloor K/2 \rfloor} h(i,k) \cdot R(i,j) = \Phi(j), \quad j = \left\lfloor -\frac{k}{2} \right\rfloor \dots \left\lfloor \frac{k}{2} \right\rfloor.
\tag{9.47}
$$

If the fading statistics can be considered stationary, the autocorrelations $R(i,j)$ will only depend on the difference $|i-j|$, giving $R(i,j) = R(|i-j|)$. Therefore Equation 9.47 can be written as:

$$
\sum_{i=\lfloor -K/2 \rfloor}^{\lfloor K/2 \rfloor} h(i,k) \cdot R(|i-j|) = \Phi(j), \quad j = \lfloor -K/2 \rfloor \dots \lfloor K/2 \rfloor,
\tag{9.48}
$$

which is a form of the well-known Wiener-Hopf equations [142], often used in estimation and prediction theory, as we have shown with reference to optimum linear prediction of speech signals in Chapter 3.

This set of K equations contains K unknown prediction coefficients $h(i,k)$, $i = \lfloor -K/2 \rfloor \ldots \lfloor K/2 \rfloor$, which must be determined in order to arrive at a minimum error variance estimate of $c(k)$ by $v(k)$. First the correlation terms $\Phi(j)$ and $R(|i-j|)$ must be computed and to do this the expectation value computations in Equations 9.45 and 9.46 need to be restricted to a finite duration window. This approach is referred to as the *autocorrelation method*, which was detailed in the context of speech coding in Chapter 3. The pilot autocorrelation, $R(i,j)$, may then be calculated from the fading estimates at the pilot positions within this window. Calculation of the received pilots' and the complex channel gains' cross correlation is less straightforward, because in order to calculate the cross-correlation the complex channel gains have to be known at the position of the data symbols as well as the pilot symbols. However, the channel gains are only known at the pilot positions, while for the data symbol positions they must be derived by interpolation. Hence in reference [143] Torrance and Hanzo proposed fitting a polynomial to the known samples of $R(|i-j|)$ and then estimated the values of $\Phi(j)$ for the unknown positions in order to provide a wide range of PSAM modem BER versus channel SNR performance figures for 1, 2 and 4 bits/symbol signalling.

The set of Equations 9.48 can also be expressed in a convenient matrix form as:

$$
\begin{bmatrix}
R(0) & R(1) & R(2) & \ldots & R(K) \\
R(1) & R(0) & R(1) & \ldots & R(K-1) \\
R(2) & R(1) & R(0) & \ldots & R(K-2) \\
\vdots & \vdots & \vdots & \ldots & \vdots \\
R(K) & R(K-1) & R(K-2) & \ldots & R(0)
\end{bmatrix}
\tag{9.49}
$$

$$
\cdot
\begin{bmatrix}
h\left(\lfloor -\frac{K}{2}\rfloor, k\right) \\
h\left(\lfloor -\frac{K}{2}+1\rfloor, k\right) \\
h\left(\lfloor -\frac{K}{2}+2\rfloor, k\right) \\
\vdots \\
h\left(\lfloor \frac{K}{2}\rfloor, k\right)
\end{bmatrix}
=
\begin{bmatrix}
\Phi\left(\lfloor -\frac{K}{2}\rfloor\right) \\
\Phi\left(\lfloor -\frac{K}{2}+1\rfloor\right) \\
\Phi\left(\lfloor -\frac{K}{2}+2\rfloor\right) \\
\vdots \\
\Phi\left(\lfloor \frac{K}{2}\rfloor\right)
\end{bmatrix},
$$

which can be solved for the optimum predictor coefficients $h(i,k)$ by matrix inversion using Gauss-Jordan elimination or the recursive Levinson-Durbin algorithm of Chapter 3. Once the optimum predictor coefficients $h(i,k)$ are known, the minimum error variance channel estimate $v(k)$ can be derived from the received pilot symbols using Equation 9.39, as also demonstrated by Figure 9.23.

9.5.2.5.3 PSAM Performance [143] Torrance *et.al.* in reference [143] also compared the performance of the above Cavers-interpolator with that of the conventional linear, low-pass and a higher-order polynomial interpolator using 1, 2 and 4 bit/symbol modems and concluded that in the fast-fading IS-54 environment investigated the highest-complexity minimum mean-squared error Cavers-interpolator did not significantly outperform the above low-complexity linear, low-pass or polynomial interpolators in terms of reduced residual BER. In these experiments the propagation frequency was increased from the 900 MHz IS-54 frequency to the 1.8 GHz propagation frequency of the next generation of systems, the vehicular speed was fixed at 50 km/h or approximately 30 mph, and the signalling rate was set to 20 kBd, which corresponded to a modulation excess bandwidth of 50 %, when using the the standard IS-54 bandwidth of 30 kHz. The corresponding Doppler frequency f_d was

$$f_d = (v \cdot f_p)/c = (13.88\text{m/s} \cdot 1.8 \cdot 10^9 \text{Hz})/(3 \cdot 10^8 \text{m/s}) \approx 83.3\text{Hz},$$

where v is the vehicular speed and f_p is the propagation frequency. The corresponding normalised Doppler frequency is

$$f_d \cdot T = 83.3\text{Hz} \cdot 1/(20 \cdot 10^3 \text{Baud}) \approx 0.0042.$$

Due to its approximately 13-times higher signalling rate of 271 kbps the GSM-like DCS1800 system under identical propagation conditions results in a relative Doppler frequency of 0.0003, which is associated with a less dramatically fading signal envelope and hence better fade tracking properties. The 1 bit/symbol, 2 bit/symbol and 4 bit/symbol modulation schemes were combined with all four interpolators and their bit error rate (BER) performance was evaluated at channel SNRs of 20, 30 and 40 dB, which yielded $3 \cdot 4 \cdot 3 = 36$ sets of results. In each set of results pilot Buffer lengths of 3, 5, 7, 9, 11 PSAM frames and pilot separation or Gap values of 10, 20, 40, 60, 80, 100, 116 were employed, leading to a plethora of performance curves, which allowed us to generate a corresponding set of 3-dimensional (3D) graphs of BER versus Buffer and Gap. These results are presented in Figure 9.25 as a set of 3-dimensional (3D) graphs of BER versus Buffer and Gap for pilot-assisted square-constellation 16-QAM. The corresponding graphs for BPSK and QPSK were presented by Torrance in reference [179], while a variety of further results can be found in [143]. As an alternative to the above coherent PSAM scheme let us now consider the advantages and disadvantages of non-coherent differential detection using the star constellation of Figure 9.19.

9.5.2.6 Differentially Detected QAM [2]

We have shown above that the so-called 'maximum minimum distance' square-shaped QAM constellation [122] is optimal for transmissions over

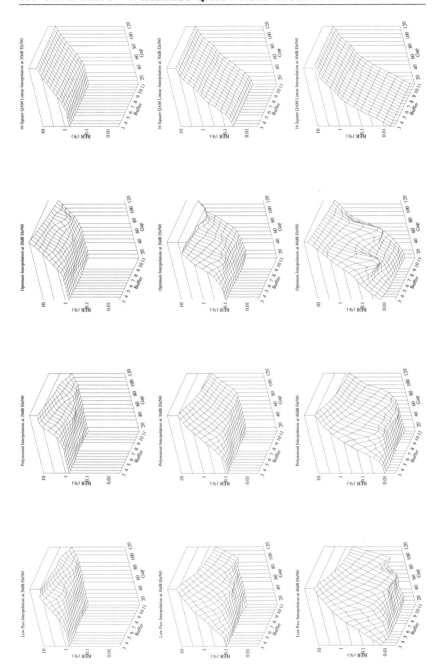

Figure 9.25: BER versus 'Buffer' length and pilot 'Gap' performance of pilot-assisted 16-QAM over Rayleigh channels at 20 kBd, 50 km/h, 1.8 GHz ©IEE 1995 Torrance, Hanzo [143].

Additive White Gaussian (AWGN) channels, since it has the highest average distance amongst its constellation points, yielding the highest noise protection distances for a given average power. We have also introduced the star QAM constellation in Figure 9.19 and compared some of its properties with those of the square constellation of Figure 9.15 in Table 9.7. In this subsection we will introduce a differentially encoded version of the star constellation shown in Figure 9.19, which can be often advantageously employed over fading channels.

When using the previously discussed square-shaped 16-QAM constellation, it is essential to be able to separate the information modulated onto the in-phase (I) and quadrature-phase (Q) carriers with the aid of coherent demodulation, invoking the Transparent-tone-in-band (TTIB) principle invented by McGeehan and Bateman [181], [187], [190] or employing the above PSAM schemes [178]. In order to achieve this, a perfectly phase-coherent replica of the transmitter's I and Q carrier has to be recovered by the carrier recovery circuitry. In contrast, in the so-called differentially encoded schemes it is not necessary to derive this phase-coherent reference carrier, an issue which will be elaborated on below.

The pivotal point of differentially encoded non-coherent QAM demodulation is that of finding a rotationally symmetric QAM constellation, where all constellation points are rotated by the same amount. Such a rotationally symmetric, differentially encoded 'star-constellation' was proposed by Webb *et.al.* in reference [133], which is similar to the star scheme shown in Figure 9.19 in terms of the location of its phasors, but differentially encoded, as it will be described below. We have seen in Table 9.7 that a disadvantage of the proposed star 16-QAM (16-StQAM) constellation is its lower average energy.

Our differential encoder obeys the following rules. The first bit $b1$ of a four-bit symbol is differentially encoded onto the phasor magnitude, yielding a ring-swap for an input logical one and maintaining the current magnitude, i.e., ring for $b1 = 0$. Bits $(b2, b3, b4)$ are then differentially Gray-coded onto the phasors of the particular ring pin-pointed by $b1$. Accordingly, $(b2, b3, b4) = (0, 0, 0)$ implies no phase change, $(0, 0, 1)$ a change of 45°, $(0, 1, 1)$ a change of 90°, etc.

The corresponding non-coherent differential 16-StQAM demodulation is equally straightforward, having decision boundaries at a concentric ring of radius $B = (A_1 + A_2)/2$ and at phase rotations of $(22.5^o + n.45^o)$ $n = 0 \ldots 7$. Assuming received phasors of P_t and P_{t+1} at consecutive sampling instants of t and $t + 1$, respectively, bit $b1$ is inferred by evaluating the condition:

$$\left| \frac{P_{t+1}}{P_t} \right| \geq (A_1 + A_2)/2. \tag{9.50}$$

If this condition is met, $b1 = 1$ is assigned, otherwise $b1 = 0$ is demodulated.

Bits $(b2, b3, b4)$ are then recovered by computing the phase difference

$$\Delta\Theta = (\Theta_{t+1} - \Theta_t) \quad (\bmod\ 2\pi) \tag{9.51}$$

and comparing it against the decision boundaries $(22.5^o + n.45^o)$ $n = 0 \ldots 7$. Having decided which rotation interval the received phase difference $\Delta\Theta$ belongs to, Gray-decoding delivers the bits $(b2, b3, b4)$.

From our previous discourse it is plausible that the less dramatic the fading envelope and phase trajectory fluctuation between adjacent signalling instants, the better this differential scheme works. This implies that lower vehicular speeds are preferred by this arrangement, if the signalling rate is fixed. Therefore the modem's performance improves for low pedestrian speeds, when compared to typical vehicular scenarios. Alternatively, for a fixed vehicular speed higher signalling rates are favourable, since the relative amplitude and phase changes introduced by the fading channel between adjacent information symbols are less drastic.

Similar differentially encoded and non-coherently detected constellations can be used in conjunction with any number bits/symbol. In reference [143] Torrance *et.al.* presented the BER of pilot-symbol assisted 1, 2 and 4 bits/symbol BPSK, QPSK and 16-QAM for channel SNRs of 20, 30 and 40 dB in contrast to their lower-complexity differentially detected counterparts. These comparative results are reproduced in Figure 9.26 [143] as a set of BER versus number of bits per symbol curves for both the pilot-assisted and differential schemes using channel SNRs of 20, 30 and 40 dB. Explicitly, the bold symbols in the Figure represent the PSAM schemes, while the hollow symbols correspond to the differentially detected schemes. Observe in the Figure that as the modulation constellation becomes less complex, ie the number of bits per symbol is reduced, the benefits of coherent modulation are reduced, although this is also a function of the channel SNR. In contrast, for higher order constellations, such as QPSK and 16Q-AM, PSAM does reduce the residual BER of the slightly less complex, differentially detected schemes while having a somewhat higher delay.

For a full treatise on various aspects of QAM the interested reader is referred to [2], where the modem performance was documented for various constellations and channel conditions. As a brief performance comparison, we remind the reader that in Figure 9.26 we portrayed the coherent and non-coherent modem's residual BER performances for 1, 2 and 4 bits/symbol signalling under identical conditions. Observe in the Figure that the differentially detected scheme has typically a factor two higher BER due to the fact that in the case of an erronnous decision errors occur in both the current and the forthcoming signalling interval, where the current phasor is used as a reference in deriving the next one. Some further BER degradation is expected due to the reduced distance of the constellation points in the star constellation, although this does not appear to be a significant factor over fading channels. In conclusion, star- and

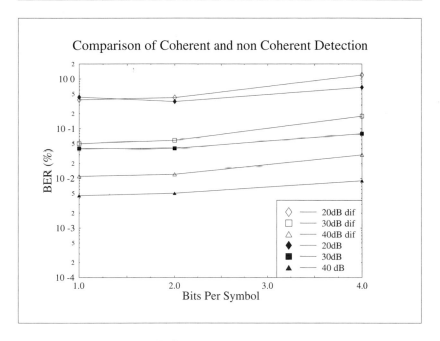

Figure 9.26: Residual BER for 1,2 and 4 bits per symbol PSAM modulation compared with equivalent differential schemes ©IEE [143] Torrance and Hanzo, 1995.

square-constellation QAM have a high bandwidth efficiency in exchange for a typically higher channel SNR and SIR requirement. The BER versus channel SNR performance of coherently detected pilot-assisted QAM is slightly higher than that of the lower complexity star-QAM, providing the system designer with a choice of implementational options. In the next section we consider burst-by-burst adaptive QAM schemes.

9.5.2.7 Burst-by-burst Adaptive Modems

Burst-by-burst adaptive multi-level modulation was first suggested by Steele and Webb in references [2,144,145] for slowly-fading wireless pedestrian channels, inspiring intensive further research in recent years [147]-[162], in particular by Kamio, Sampei, Sasaoka, Morinaga, Morimoto, Harada, Okada, Komaki and Otsuki at Osaka University and the Ministry of Post in Japan [146]- [149], as well as by Goldsmith *et.al.* [150]- [156] at Stanford University in the USA or by Pearce, Burr and Tozer [157] in the UK. The proposed schemes provide a means of realising some of the time-variant channel capacity potential of the fading wireless channel [165], invoking a more robust Transmission Scheme (TS) on a burst-by-burst basis, when the channel is of low quality and vice-versa, while maintaining

Switching levels(dB)	l_1	l_2	l_3	l_4
Mean-Speech (1%)	3.31	6.48	11.61	17.64
Mean-BER Data (0.01%)	7.98	10.42	16.76	26.33

Table 9.8: Switching levels for speech and computer data systems through a Rayleigh channel, shown in instantaneous channel SNR (dB) to achieve Mean BERs of 1×10^{-2} and 1×10^{-4}, respectively.

a certain target bit error rate (BER) performance. The most appropriate TS is dependent upon the time-variant instantaneous Signal-to-Noise Ratio (SNR) and Signal-to-Interference Ratio (SIR). The TS can be chosen according to the following regime [158]:

$$
\text{TS} = \begin{cases}
\text{No Transmission (Notx)} & \text{if } l_1 > s^2/N \\
\text{BPSK} & \text{if } l_1 \le s^2/N < l_2 \\
\text{QPSK} & \text{if } l_2 \le s^2/N < l_3 \\
\text{Square 16 Point QAM} & \text{if } l_3 \le s^2/N < l_4 \\
\text{Square 64 Point QAM} & \text{if } s^2/N \ge l_4,
\end{cases} \tag{9.52}
$$

where s is the instantaneous signal level, N is the average noise power, and l_1, l_2, l_3 and l_4, are the BER-dependent optimised switching levels. Time Division Duplex (TDD) was proposed, in order to exploit the reciprocity of the channel under high SIR conditions, which allowed us to estimate the prevalent SNR on a burst-by-burst basis [159]. The reciprocity of the up- and down-link channel conditions in the TDD frame is best approximated, if the corresponding TDD slots are adjacent. This requirement, however, imposes various practical constraints on the transceiver design, which are beyond the scope of this book.

In reference [158] the analytical upper-bound performance of such a scheme was characterised over slow Rayleigh-fading channels, while in [161] an unequal protection phasor constellation for signalling the current TS was proposed. The problem of appropriate power assignment was discussed for example in [148, 150].

In reference [158] a combined BER- and Bits per Symbol (BPS) based optimisation cost-function was defined and minimised, in order to find the required TS switching levels for maintaining average target BERs of 1×10^{-2} and 1×10^{-4}, irrespective of the instantaneous channel SNR. These BER values can then be further mitigated by forward error correction coding and in the case of the lower BER scheme can be rendered virtually error-free. The former scheme was referred to as the speech TS, and the latter as the adaptive data TS. The optimised TS switching levels l_1, l_2, l_3 and l_4 are summarised in Table 9.8 [158]. The average BPS performance B of this adaptive modem was derived for a Rayleigh fading

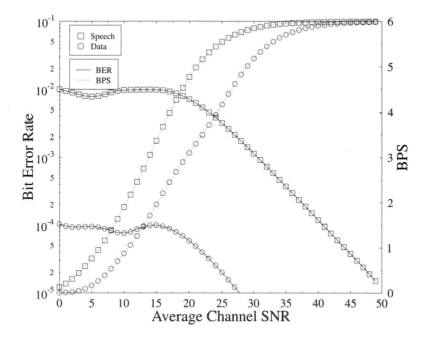

Figure 9.27: Upper bound BER and BPS performance of adaptive QAM in Rayleigh channel optimised separately for 'Speech' and 'Data' transfer ©IEE Torrance and Hanzo, 1996 [158].

channel in reference [158], which can be written as:

$$
\begin{aligned}
B \;=\; & 1 \cdot \int_{l_1}^{l_2} F(s,S)\,ds + \; 2\; \cdot \int_{l_2}^{l_3} F(s,S)\,ds \\
& + \, 4 \; \cdot \int_{l_3}^{l_4} F(s,S)\,ds + \; 6 \; \cdot \int_{l_4}^{\infty} F(s,S)\,ds,
\end{aligned}
\tag{9.53}
$$

where $F(s,S)$ is the PDF of the Rayleigh channel, which was given in Chapter 2, S is the average power and the integrals characterise the received signal level domains, where the 0, 1, 2, 4 and 6 bits/symb TSs of Equation 9.52 are used. Since the transmissions can be disabled for the duration of deep channel fades, the transmitted information has to be buffered, which results in latency. In references [162, 163] the latency performance of these schemes was quantified and frequency hopping as well as statistical multiplexing were proposed to mitigate its latency and buffer requirements.

Considering Figure 9.27, the desired BER is achieved between 0 and 50 dB for both the speech and computer data schemes, when using the switching thresholds of Table 9.8, which were optimised for maintaining

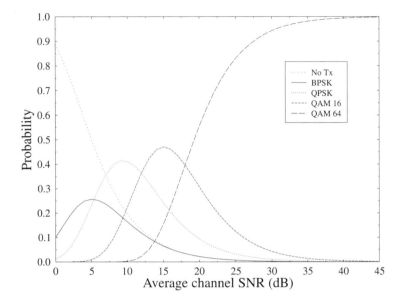

Figure 9.28: Probability of the individual AQAM modem modes optimised for speech transmission at an average BER of 1% plotted against average channel SNR for a Rayleigh channel ©Torrance, Hanzo, 1996 [176].

mean BERs of 1×10^{-2} and 1×10^{-4}, respectively. The targeted BPS performance is achieved at about 18 dB and 19 dB average channel SNRs for the speech and computer data schemes, respectively. Observe in the Figure that both the speech and data BER profiles outperform the BER requirements for average channel SNRs higher than these values, since the modem cannot switch to higher order modes than 64-QAM. The system was capable of maintaining the target BER performances at extremely low average SNR values. This robust performance was achieved at the cost of reducing the BPS throughput below that of BPSK, which was possible due to disabling transmissions for low instantaneous SNR values.

In Figure 9.28 the probability density function (PDF) of the various AQAM modem modes versus the average channel SNR performance was documented. Note in Figure 9.28 that given a certain average channel SNR, there is a finite probability that the AQAM modem assumes potentially any of its legitimate modes of operation. However, for example at an average SNR of 15 dB the most frequently invoked mode is 16-QAM, while in excess of this SNR 64-QAM is employed predominantly. Nonetheless, even the No TX mode has a finite probability of occurrence above 15 dB, which is a consequence of channel estimation errors.

A variety of decision feed-back equalised wideband burst-by-burst adaptive modem schemes were studied in references [166]- [169]. Specifically, the maximum achievable throughput or BPS performance of decision feedback equalised AQAM was compared to the Shannonian channel capacity limit, in order to gauge the potential gains due to adaptivity, while the block turbo coded performance of AQAM was the topic of [167–169], exhibiting channel SNR gains up to 20 dB in comparison to conventional non-adaptive systems. Although AQAM research is in its infancy, the initial results are encouraging and hence they are likely to stimulate further research.

9.5.2.8 Summary of Multi-level Modulation

In closing we note that in cellular frequency-reuse structures, where the co-channel interference is a dominant impairment, the bandwidth-efficiency of these schemes is often eroded due to the increased frequency reuse factor required by the higher interference sensitivity of these schemes. In Chapter 17 of reference [2] we have shown that the true spectral efficiency - which is also often referred to as area spectral efficiency - of a modulation scheme, taking into account the effect of the required frequency reuse factor is dependent on the bit error ratio (BER) targeted. The required BER in turn is dependent on the robustness of the source codecs used. However, for example in the indoor picocells of the ATDMA system of Table 9.6 the partitioning walls and floors mitigate the co-channel interference and this facilitates the employment of 16-QAM.

In this rudimentary introduction to QAM techniques we assumed perfect clock recovery and dispensed with considering a range of important aspects of the transceiver design, such as clock and carrier recovery, wideband aspects and channel equalisation, the effects of co- and adjacent-channel interference, trellis coding, etc. which are treated in depth in the corresponding chapters of reference [2]. The analytical error rate performance of square and star QAM was characterised in references [178] and [189] by Cavers and Adachi, respectively. We note that the advantages and disadvantages of the above modem schemes will be elaborated on again in Section 9.7 in the context of two different systems based on non-coherently detected star 16-QAM and coherently detected square 16-QAM. Let us now briefly consider the principles of Packet Reservation Multiple Access (PRMA) in the next section.

9.6 Packet Reservation Multiple Access

PRMA is a relative of slotted ALOHA contrived for conveying speech signals on a flexible demand basis via time division multiple access (TDMA) systems. PRMA was documented in a series of excellent treatises by Nanda, Goodman *et.al.* [191], while a PRMA-assisted adaptive differential pulse code modulation (ADPCM) transceiver was proposed in reference [26]. The voice activity detector (VAD) [26] queues the active speech spurts to contend for an up-link TDMA time-slot for transmission to the BS. Alternatively, a VAD similar to that of the GSM system described in Chapter 8 can be employed. Inactive users' TDMA time slots are offered by the BS to other users, who become active and are allowed to contend for the unused time slots with a less than unity permission probability. This measure prevents previously colliding users from consistently colliding in their further attempts to attain a time-slot reservation. For a seven-slot PRMA system the optimum permission probability allowing to support the highest number of users was found to be 0.6, as shown in Table 9.9.

If several users contend for an available slot, neither of them will be granted it, while if only one user requires the time slot, he can reserve it for future communications. When many users are contending for a reservation, the collision probability is increased and hence a speech packet might have to contend for a number of consecutive slots, until its maximum contention delay of typically 32 ms expires. In this case the speech packet must be dropped, but the packet dropping probability must be kept below 1%, a value inflicting minimal degradation in perceivable speech quality in contemporary speech codecs. As an example, the 8 kbps G.729 CCITT/ITU ACELP candidate codec's target was to inflict less than 0.5 Mean Opinion Score (MOS) degradation in the case of a speech frame error rate of 3% [46].

The performance of communications systems is often evaluated in terms of the teletraffic carried, while maintaining a set of communications quality measures. In conventional TDMA mobile systems the grade of service (GOS) degrades due to speech impairments caused by call blocking, hand-over failures and speech frame interference engendered by noise, as well as co- and adjacent-channel interference. In PRMA-assisted systems calls are not blocked due to the lack of an idle time-slot, but the packet dropping probability is increased gracefully. Hand-overs will be performed in the form of contention for an idle time slot provided by the specific BS offering the highest signal quality amongst the potential hand-over target BSs.

The specific physical up-link to the BS offering the best signal quality during decoding the packet header is not likely to substantially degrade during the life-time of an active speech spurt having a typical mean duration of 1 s or some thirty consecutive 30 ms speech frames. If, however, the link degrades before the next active spurt is due for transmission, the subsequent contention phase is likely to establish a link with another BS. Hence this process will have a favourable effect on the channel's quality, effectively simulating a diversity system having independent fading channels and limiting the time spent by the MS in deep fades, thereby avoiding channels with high noise or interference.

This advantageous property can be exploited to train a self-adjusting adaptive system using the channel segregation scheme proposed for PRMA systems in reference [170]. Accordingly, each BS evaluates and ranks the quality of its idle physical channels constituted by the unused time slots on a frame-by-frame basis and identifies a certain number of slots, N, with the highest quality, i.e. lowest noise and interference. These high-quality, low-interference channels are segregated for contention, while the lower quality idle slots contaminated by noise and interference are temporarily disabled. Hence upon a new access request the BS is likely to receive a signal having low interference, which maximizes the chances of successful packet decoding, unless a collision caused by a simultaneous MS attempt to attain a reservation has occurred. When a successful, uncontended reservation takes place, the BS promotes the highest quality disabled time slot to the set of N segregated channels, unless its quality is unacceptably low. It appears plausible that if N is high, the packet dropping probability becomes low, but the physical channels constituted by the time slots might become heavily interfered with, while if N is low, we have a packet dropping-dominated scenario, which equally limits the GOS.

Clearly, the main cause of GOS degradation in PRMA systems is limited to speech packet corruption due to noise or interference and packet dropping [192]. They both result in different subjective speech or GOS degradation, which we will attempt to quantitatively compare in terms of the objective segmental signal to noise ratio (SEGSNR) degradation. Quantifying these GOS degradations in relative terms in contrast to each other will allow us to appropriately split the acceptable overall degradation between packet dropping and packet corruption. With the system elements described in previous Sections of this chapter, in the next section we focus our attention on the amalgamated PCS transceiver proposed.

9.7 Multi-mode Multi-media Transceivers

9.7.1 Flexible Transceiver Architecture

It transpired from our previous discussions that a high-performance transceiver is expected to be reconfigurable in a number of different operational modes for reasons to be augmented below. The schematic of such a flexible, toolbox-based multi-media system is portrayed in Figure 9.29. The key optimisation criterion of such a multi-media PS is that of finding the best compromise amongst a number of contradicting design factors, such as power consumption, robustness against transmission errors, spectral efficiency, audio/video quality and so forth. As argued before, the time-variant optimisation criteria of a flexible multi-media system can only be met by an adaptive scheme, comprising the firmware of a suite of system components and invoking that combination of speech codecs, video codecs, embedded channel codecs, voice activity detector (VAD) and modems, which fulfils the prevalent one [2]. A few examples are maximising the teletraffic carried or the robustness against channel errors, while in other cases minimisation

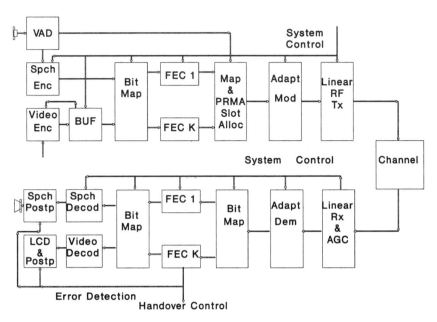

Figure 9.29: Flexible Multi-media Communicator Schematic, ©IEEE, Hanzo, 1998, [1].

of the bandwidth occupancy, the call blocking probability or the power consumption is of prime concern.

Focusing our attention on the speech and video links displayed in Figure 9.29, the voice activity detector (VAD) is employed to control the packet reservation multiple access (PRMA) slot allocator [2], multiplexing speech and video. Control traffic and system information is carried by packet headers added to the composite signal by the 'Bit Mapper' before K-class source sensitivity-matched forward error correction coding (FEC) takes place. Observe that the 'Video Encoder' supplies its bits to an adaptive buffer (BUF) having a feed-back loop. If the PRMA video packet delay becomes too high or the buffer fullness exceeds a certain threshold, the video encoder is instructed to reduce its bit-rate, implying a concomitant dropping of the image quality.

The Bit Mapper assigns the most significant source coded bits (MSB) to the input of the strongest FEC codec, FEC K, while the least significant bits (LSB) are protected by the weakest one, FEC 1. K-class FEC coding is used after mapping the speech and video bits to their appropriate bit protection classes, which ensures source sensitivity-matched transmission. 'Adaptive Modulation' originally proposed by Steele and Webb [194], which was discussed in Section 9.5.2.7 is employed [2], [146]- [175] with the number of modulation levels, the FEC coding power and the speech/video source coding algorithm adjusted by the 'System Control' according to the dominant propagation conditions, bandwidth and power efficiency requirements, channel blocking probability or PRMA packet dropping probability. If the communications quality or the prevalent system optimisation criterion cannot be improved by adaptive transceiver re-configuration, the serving BS will hand the PS over to another BS providing a better grade of service.

One of the most important and reliable parameters used to control these algorithms is the 'Error Detection' flag of the FEC decoder of the most significant bit (MSB) class of speech and video bits, namely FEC K. This flag can also be invoked to control the speech and video 'Postprocessing' algorithms. The adaptive modulator transmits the user bursts from the PS to the BS using the specific PRMA slot allocated by the BS for the PS's speech, data or video information via the linear radio frequency (RF) transmitter (Tx). Although the linear RF transmitter has a low power efficiency, its power consumption is less critical due to the low transmitted power requirement of the multi-media PCN than that of the digital signal processing (DSP) hardware.

The receiver structure essentially follows that of the transmitter. After linear class-A amplification and automatic gain control (AGC) the 'System Control' information characterising the type of modulation and the number of modulation levels must be extracted from the received signal, before demodulation can take place. This information also controls the various internal bit mapping algorithms and invokes the appropriate speech and video decoding as well as FEC decoding procedures. After 'Adaptive Demodulation' at the BS the source bits are mapped back to their original bit protection classes and FEC decoded. As mentioned, the error detection flag of the strongest FEC decoder, FEC K, is used to control hand-overs or speech and video post-processing. The FEC decoded speech and video bits are finally source decoded and the recovered speech arrives at the earpiece, while the video information is displayed on a flat liquid crystal display (LCD).

The system control algorithms of the re-configurable mobile multi-media communicator will dynamically evolve over the years. PSs of widely varying complexity will coexist, with newer ones providing backward compatibility with existing ones, while offering more intelligent new services and more convenient features. Following the above rudimentary system-level overview of multi-mode transceivers, in the next two subsections we incorporate the previously highlighted system components in two different transceivers, in order to provide guidelines for designers of novel transceivers and to characterise the expected performance of such systems. Two well-understood system design contexts were selected for hosting the system components and for evaluating their performance, namely that of a 30 kHz bandwidth and a 200 kHz bandwidth system, which are the bandwidths of the Pan-American IS-54 system and the Global System of Mobile communications, known as GSM. We note, however that the systems studied are different from the standard schemes mentioned, only their bandwidths are identical. Nonetheless, systems similar to those studied in this chapter may become evolutionary successors of the IS-54 and GSM schemes, respectively. Let us initially consider the proposed 30 kHz bandwidth system in the next subsection, commencing with a brief description of the channel coding schemes used to protect the source-coded speech and video bits.

9.7.2   A 30 kHz Bandwidth Multi-media System

9.7.2.1   Channel-coding and Bit-mapping

Both convolutional and block error correction codes were portrayed in depth in Chapter 4, hence here we retsrict our discussions to the practical aspects of the FEC scheme proposed for our multi-media system shown in Figure 9.29.

Encoding a speech packet using an integer number of FEC-coded blocks allows us to carry out direct comparisons between different sources of speech impairments on the basis of dropped PRMA packets due to network overload or corrupted speech packets due to channel impairments. Furthermore, in our proposed packet radio transceivers we opted for binary Bose-Chaudhuri-Hocquenghem (BCH) block codes (see Section 4.4.3), since we found that the subjective speech quality of BCH-coded speech was often preferable to convolutionally coded speech due to longer unimpaired speech segments, even if the objective Segmental Signal to Noise Ratio (SEGSNR) and Bit Error Rate (BER) performances of the convolutional and block codes were similar. This was because the speech quality is typically more strongly dependent on the frame error rate (FER) than on decoded BER. Furthermore, powerful block codes also have a reliable error detection capability, which can be advantageously exploited in order to invoke speech post-enhancement in the case of error events [193].

A set of appropriate FEC codes for the 4.8 kbps TBPE speech codec is constituted by the BCH5=BCH(63,36,5), BCH2=BCH(63,51,2) and BCH1=BCH(63,57,1) codes, correcting 5, 2 and 1 bits per 63-bit frame, respectively. Accordingly, the most sensitive class 1 (C1) 36 speech bits of Table 9.2 are protected by the powerful BCH(63,36,5) code, while the less vulnerable 51 class 2 (C2) and 57 class 3 (C3) bits are encoded by the BCH(63,51,2) and BCH(63,57,1) codes, respectively. The total number of protected bits is 144. The packet header, which is conveying control information, is also BCH(63,36,5) coded, hence $4 \cdot 63 = 252$ bits per 30 ms are transmitted. This corresponds to transmitting 252/4=63 four-bit 16-QAM symbols and, upon adding three so-called ramp-symbols, yields a signalling rate of 66 symbols/30 ms = 2.2 kBd. The total bit-rate became 8.4 kbit/s. Recall from Chapter 8 that the ramping symbols are included, in order to assist the transceiver to power up and down smoothly, as it was highlighted in the context of the power ramping mask of the GSM system in Figure 8.29, which was necessary for mitigating the spurious adjacent channel emissions.

The above FEC scheme has the advantage of curtailing error propagation across speech frame boundaries and over-bridging deep channel fades for typical urban vehicular speeds. For example, for a vehicular speed of 30 mph or 13.3 m/s the travelling distance is 39.9 cm/30 ms speech frame. For a propagation frequency of 1.8 GHz the wavelength is about 15 cm, and therefore interleaving over a time-domain interval corresponding to a travelled distance of about 40 cm ensures adequate error randomisation for the FEC scheme to work efficiently. However, for pedestrian speeds there is a danger of idling in deep fades, in which case the employment of a switch-diversity scheme or frequency hopping, as in Chapters 7 and 8 is essential.

The 852-bit video frame is encoded using 12 BCH(127,71,9) code-words, yielding a total of 1524 FEC-coded bits per video frame. A pair of these BCH codewords form a video packet of 254 bits or 64 four-bit 16-QAM symbols, which is expanded by two ramp symbols in order to generate a 66-symbol packet. For delivering the 1524-bit BCH-coded video frame, 1524/254=6 such 66-symbol packets are necessary, but during the 90 ms video frame repetition time there are only three 30 ms speech frames, implying that two reserved time-slots per 30 ms PRMA frame are required for video users. This is equivalent to a video signalling rate of $2 \times 2.2 = 4.4$ kBd. The video transceiver also obeys the structure of Figure 9.29.

The receiver seen in Figure 9.29 carries out the inverse functions of the transmitter. The error detection capability of the strongest BCH(63,36,5) decoder is exploited to initiate hand-overs and to invoke speech post-processing [193], if the FEC decoder's error correction capability happens to be overloaded due to interference or contention-induced PRMA packet collision. The system elements of Figure 9.29 were simulated and the key transceiver parameters of both the non-coherently detected 30-kHz bandwidth system and those of the coherently detected 200-kHz bandwidth schemes are summarised in Table 9.12, while the associated system performance will be characterised in the following two subsections.

The transmitted Baud-rate of our non-coherent transceiver was fixed to 20 kBd, in order for the PRMA signal to fit in a 30 kHz channel slot, as in the IS-54 system, when using a modem excess bandwidth of 50 % [2]. Hence our transceiver can accommodate TRUNC(20 kBd/2.2 kBd) = 9 time slots, where TRUNC represents truncation to the nearest integer. The slot duration was 30 ms/9 $\approx$ 3.33 ms and one of the PRMA users was transmitting speech signals recorded during a telephone conversation, while all the other users generated negative exponentially distributed speech spurts and speech gaps with mean durations of 1 and 1.35 s. The PRMA parameters used are summarised in Table 9.9, where it was made explicit again that two

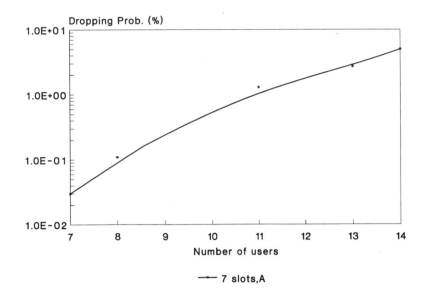

Figure 9.30: Packet dropping versus number of speech users performance ©Kluwer, 1995 Hanzo et al, [22].

PRMA parameter	30 kHz band-width system	200 kHz band-width system
Channel rate (kBd)	20	100
Speech rate (kBd)	2.2	3.1
Video rate (kBd)	4.4	16
Frame duration (ms)	30	30
Total no. of slots	9	32
No. of PRMA speech slots	7	32/42/47
No. of TDMA video slots	2	No. of users×6
Slot duration (ms)	3.33	0.9375
Packet header length (bits)	63	63
Maximum speech delay (ms)	32	30
Speech perm. prob.	0.6	0.2
Video perm. prob.	1	1

Table 9.9: Summary of PRMA/TDMA parameters.

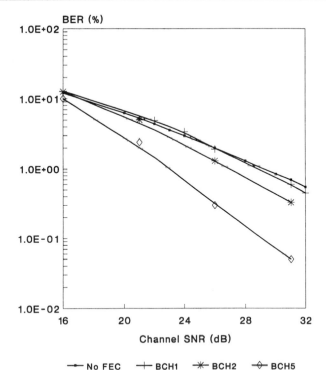

Figure 9.31: Non-coherent 16-QAM BER versus channel SNR performance over Rayleigh-fading channels for a vehicular speed of 30 mph, propagation frequency of 1.8 GHz and signalling rate of 20 kBD without diversity using BCH1, BCH2 and BCH5 coding©Kluwer, 1995 Hanzo *et al* [22].

time slots are reserved for a videophone user all the time and 7 slots are dynamically assigned to PRMA for the speech users. Let us now consider the robustness of the non-coherently detected 30-kHz bandwidth candidate system against channel effects.

9.7.2.2 Performance of a 30-kHz Bandwidth Multi-media System

In order to be able to contrast the performance of our multi-media system with that of an existing second-generation benchmarker, the system performance was evaluated for a narrowband Rayleigh-fading channel exhibiting a propagation frequency of 1.8 GHz, vehicular speed of 30 mph and signalling rate of 20 kBd, which is characteristic of an up-converted IS-54 system. The corresponding modem performance of the star and square QAM schemes was comparatively studied by Torrance and Hanzo in ref-

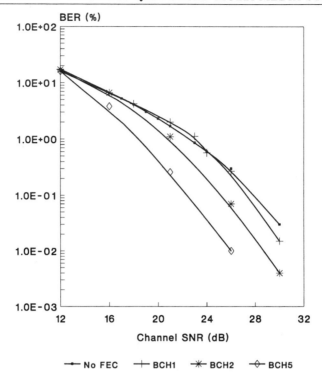

Figure 9.32: Non-coherent 16-QAM BER versus channel SNR performance over Rayleigh-fading channels for a vehicular speed of 30 mph, propagation frequency of 1.8 GHz and signalling rate of 20 kBD with diversity using BCH1, BCH2 and BCH5 coding ©Kluwer, 1995 Hanzo *et al* [22].

erence [143] and was summarised in Figure 9.26. Note that instead of the 24.3 kBaud signalling rate of the IS-54 scheme the more conservative 20 kBaud was used here, but a further potentially 20 % higher number of users can be supported, when increasing the Bd-rate to 24.3 kBd. The packet dropping probability versus number of PRMA speech users curve of the proposed system is portrayed in Figure 9.30. Observe that about 10-11 users can be supported by our 7-slot PRMA scheme with $P_{drop} < 1\%$, a value inflicting almost negligible speech degradation, while also supporting a videophone user.

The BER versus channel SNR performance of our 16-QAM transceiver with and without second-order selection diversity and BCH1, BCH2 and BCH5 FEC coding is depicted in Figures 9.7.2.1 and 9.7.2.1. When no diversity is invoked, at a channel SNR of about 28 dB the most important C1 bits protected by the BCH5 codec have a BER of about 0.1%, while the less

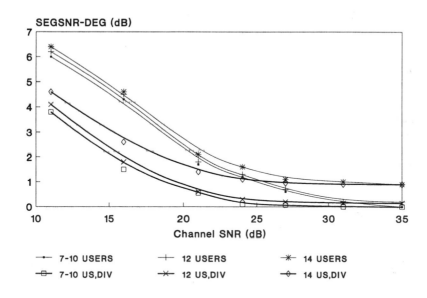

Figure 9.33: SEGSNR-DEG versus channel SNR performance of the proposed non-coherent 16-StQAM transceiver over Rayleigh-fading channels for a vehicular speed of 30 mph, propagation frequency of 1.8 GHz and signalling rate of 20 kBD with and without diversity parameterised with the number of PRMA users supported ©Kluwer, 1995 Hanzo et al, [22].

sensitive C2 and C3 bits attain BERs of about 0.5 and 1 %, respectively. These values are sufficiently low for nearly unimpaired speech transmission. When second-order diversity is used, these target BERs are achieved around 24 dB channel SNR. Although the lowest integrity C3 BCH1 codec does not provide a reduced BER for the least significant bits, since it is often overloaded, it ensures periods of unimpaired transmission for the most robust speech bits, which has a favourable subjective effect on the perceived speech quality. The 16-QAM modem is sensitive to co-channel interference, requiring SIR values similar to the minimum channel SNR necessitated [2]. Therefore it is beneficial to use the channel segregation algorithm proposed in references [170, 192] for mitigating the co-channel interference by classifying the slots on the basis of how interfered they are. Severely interfered slots can be temporarily disabled, at the cost of reducing the number of actively utilised slots, while high-quality, low-interference slots can guarantee the SIR required for unimpaired communications. No time-slot classification is necessary, if the system is used in an indoor environment, where the partitioning walls and floors naturally contribute towards the interference

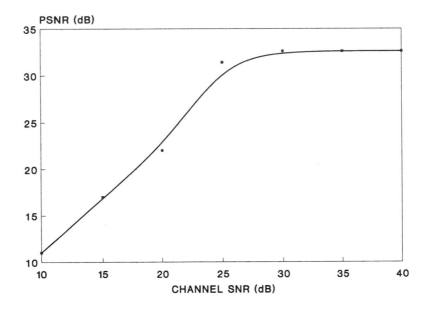

Figure 9.34: PSNR versus channel SNR performance of the proposed diversity-assisted non-coherent 16-QAM videophone scheme over Rayleigh-fading channels for a vehicular speed of 30 mph, propagation frequency of 1.8 GHz and signalling rate of 20 kBD ©Kluwer, 1995 Hanzo et al, [22].

mitigation.

The overall objective SEGSNR degradation (SEGSNR-DEG) versus channel SNR performance of our diversity-assisted PCS transceiver is displayed in Figure 9.33 parameterised with the number of PRMA users supported. While for 7-10 users no speech degradation can be observed, if the channel SNR is in excess of about 24 dB, for 12 users the SEGSNR-DEG due to PRMA packet dropping becomes noticeable, although not subjectively objectionable. in the case of 14 users, however, there is a consistent SEGSNR-DEG of about 1 dB due to the 4-5 % packet dropping probability seen in Figure 9.30. Without diversity about 5 dB higher channel SNR is necessitated in order to achieve a similar performance to that of the diversity-assisted scheme.

The PSNR versus channel SNR performance of the diversity-assisted video transceiver is portrayed in Figure 9.34, where in harmony with the voice transceiver a channel SNR of about 22-25 dB is required for near-unimpaired video quality. Without diversity the video scheme lacks robustness, since the corrupted run-length coded activity tables affect the whole of each video frame. We note that the video codec's robustness can be significantly improved upon dispensing with the quad-tree based motion- and DCT-activity table compression, as it was demonstrated in reference [99], at the cost of about 30% higher bit-rate requirement.

Let us now focus our attention on the 200-kHz bandwidth coherently detected multi-media system in the next subsection.

9.7.3 A 200 kHz Bandwidth Multi-mode, Multi-media System

In contrast to the previous subsection, where we used a non-coherent star 16-QAM multi-media scheme evaluated in the framework of a 30-kHz bandwidth system, here we portray the performance of a coherent multi-mode arrangement in the context of a GSM-like system following the approaches proposed by Hanzo and Woodard [19] as well as Cherriman *et.al.* [106]-[112]. While our discourse on the 30-kHz bandwidth system was somewhat more detailed, now we restrict our treatment to a brief summary of the rationale behind this system. These investigations will allow us to gauge the expected system performance in comparison to that of a well-known and widespread benchmarker, namely the GSM system. The basic differences between the two systems are as follows:

- The bandwidth was increased from 30 kHz to 200 kHz.

- Instead of the previous 4.8 kbps TBPE speech codec here a dual-rate 4.7/6.5 kbps Algebraic Code Excited Linear Predictive (ACELP) codec was employed [19], which was detailed in Section 3.4.2.4.

- The fixed-rate proprietary DCT video codec of Section 9.3.3 was exchanged against the programmable-rate ITU standard H.263 video codec of Section 9.3.4.

- The non-coherent star 16-QAM scheme of Section 9.5.2.6 was replaced by a coherent 4/16/64-QAM pilot-assisted scheme of Section 9.5.2.5, although for voice transmissions only the latter two modes were employed. The more vulnerable video codec necessitated the employment of the lower-rate, lower-quality but more robust 4-QAM mode.

- The system was arranged to convey the lower-quality 4.7 kbps ACELP-coded speech using the more robust 16-QAM mode, while the 6.5 kbps-coded higher-quality speech was transmitted at the same user Baud-rate within the same bandwidth, but requiring higher channel SNRs for 64-QAM transmissions.

- Similarly, the H.263 video codec was programmed to generate 1176, 2352 or 3560 bits per QCIF video frame for 4, 16 and 64-QAM transmission, respectively, which resulted in the same signalling rate, irrespective of the modem mode of operation used.

- A novel graphical source codec was introduced, in order to facilitate graphical correspondence with the aid of multiplexing speech, video and graphical signals using PRMA.

Codec Mode	4.7 kbps	6.5 kbps
LSF Parameters	34 bits/30 ms	34 bits/30 ms
Excitation and LTP	4 ·27=108 bits/30 ms	6·27=162 bits/30 ms
Total	142 bits/30 ms	196 bits/30 ms

Table 9.10: 4.7/6.5 kbps dual-rate ACELP codec bit-allocation.

The components of this scheme were described during our earlier discussions. Specifically, the principles of ACELP coding were addressed in Section 3.4.2.4, while the dual-rate codec's bit-allocation scheme and our bit sensitivity investigations were described in [19,20]. Suffice to say here that the ACELP codec uses 34 Line Spectral Frequency (LSF) bits for spectral quantisation in both of its modes and a total of 27 bits per subsegment for the LTP and excitation parameters. In its 4.7 kbps mode there are four 7.5 ms excitation optimisation subsegments per 30 ms, while in the 6.5 kbps mode there are six 5 ms subsegments. Hence the total number of bits per 30 ms frame is either $34+4\times27=142$ or $34+6\times27=196$, yielding the required rates, as it is also shown in Table 9.10. Let us now consider some of the system-level details of this intelligent dual-mode scheme.

9.7.3.1 Low-quality Speech Mode

In this subsection we highlight our code design approach using the 4.7 kbit/s codec and note that similar principles were followed in case of the 6.5 kbit/s codec. The sensitivity of the 4.7 kbps ACELP source bits was evaluated similarly to the bit-sensitivities of the TBPE codec. Our detailed bit-sensitivity analysis was portrayed in [19,20], which used the weighted SEGSNR and CD based approach introduced earlier in the context of the TBPE codec, but it also took account of the different error propagation properties of the various bits over consecutive speech frames.

Intuitively, one would expect that the more closely the FEC protection power is matched to the source sensitivity, the higher the system's robustness. In order to limit the system's complexity and the variety of candidate schemes, in the case of the 4.7 kbit/s ACELP codec we experimented with a single-class or full-class BCH codec, a twin-class and a quad-class scheme, while maintaining the same channel coding rate. We found that similar results were obtained for the twin- and quadruple-class scheme, hence we opted for the lower-complexity twin-class protection [19].

Our propagation conditions were characterised by a pedestrian speed of $v =3$ mph, propagation frequency of $f_p =1.8$ GHz, pilot symbol spacing of $P=10$ and a signalling rate of 100 kBd, which fitted in a bandwidth of 200 kHz, when using a unity Nyquist roll-off factor. The corresponding

Doppler frequency f_d was:

$$f_d = (v \cdot f_p)/c = (1.388\text{m/s} \cdot 1.8 \cdot 10^9\text{Hz})/(3 \cdot 10^8\text{Hz}) \approx 8.33\text{Hz},$$

while the corresponding normalised Doppler frequency was:

$$f_d \cdot T = 8.33\text{Hz} \cdot 1/(10^5\text{Baud}) \approx 0.0000833,$$

which is a slower fading rate than that of the previous non-coherent system due to the ten-fold reduced speed of 3 mph and the five-fold increased signalling rate.

Note however that the above conservative choice of roll-off factor and the associated bandwidth efficiency can be improved substantially, when opting for a roll-off factor of 0.5, as in the case of the previously described star 16-QAM system, or even for 0.35, as proposed for the ATDMA system, at the cost of a slight complexity increase and clock jitter sensitivity. We found that for a channel SNR of about 20 dB the pilot-assisted 16-QAM modem provided two independent QAM subchannels exhibiting different bit error rates (BERs), which was demonstrated earlier in Figure 9.22 over AWGN channels. The BER is about a factor of three times lower for the higher integrity path referred to as the Class 1 (C1) subchannel than for the C2 subchannel over Rayleigh-fading channels [2, 19]. We capitalised on this feature to provide unequal source sensitivity-matched error protection combined with different BCH codecs for our ACELP codecs.

If the ratio of the BERs of these QAM subchannels does not match the sensitivity constraints of the ACELP codec, it can be 'fine-tuned' with the aid of different BCH codecs, while maintaining the same number of BCH-coded bits in both subchannels. However, the increased number of redundancy bits of stronger BCH codecs requires that a higher number of sensitive bits are directed to the lower integrity C2 subchannel, whose channel coding power must be concurrently reduced in order to accommodate more source bits. This non-linear optimisation problem can only be solved experimentally, assuming a certain sub-division of the source bits, which would match a given pair of BCH codecs.

When designing the twin-class protection scheme and opting for the approximately half-rate BCH(127,71,9) codec in both subchannels, the ACELP source bits have to be split into two classes, each hosting 71 bits. From our bit sensitivity analysis [19,20] we observed that the more sensitive bits require almost an order of magnitude lower BER than the more robust bits, in order to inflict a similar SEGSNR penalty. Hence both classes must be protected by different codes, and after some experimentation we found that the BCH(127,57,11) and BCH(127,85,6) codes employed in the C1 and C2 16-QAM subchannels provide the required integrity. Each 142-bit ACELP frame is encoded by two BCH codewords, yielding $2 \cdot 127{=}254$ encoded bits and curtailing error propagation at the transmission packet boundaries. The FEC-coded speech bit-rate became $\approx$8.5 kbps, implying an overall coding rate of 4.7 kbps/8.8 kbps $\approx$ 0.553.

The PRMA control header was allocated a BCH(63,24,7) code and hence the total PRMA framelength became 317 bits, representing 30 ms speech and yielding a total bit-rate of $\approx$ 10.57 kbps. The 317 bits constitute 80 16-QAM symbols, and 9 pilot symbols as well as 2+2=4 ramp symbols must be added, resulting in a PRMA transmission packet-length of 93 symbols per 30 ms speech frame. Hence the signalling rate becomes 3.1 kBd. Using a PRMA bandwidth of 200 kHz, similarly to the Pan-European GSM system, and a filtering excess bandwidth of 100 % allowed us to accommodate 100 kBd/3.1 kBd $\approx$ 32 PRMA slots. When using an excess bandwidth of 50%, as in our star 16-QAM system, the signalling rate would be 133 kBd, accommodating 42 PRMA time slots. Similarly, when opting for the ATDMA excess bandwidth of 35%, the signalling rate could be increased to 148 kBd, supporting 47 PRMA slots, which in turn would further increase the PRMA gain expressed in terms of the number of users supported.

9.7.3.2   High-quality Speech Mode

Following the approach proposed in the previous subsection we designed a triple-class source-matched protection scheme for the 6.5 kbps ACELP codec. The reason for using three protection classes this time is that the 6.5 kbps ACELP codec's higher bit-rate must be accommodated by a 64-level QAM constellation, which inherently provides three different integrity subchannels, which again, was shown earlier in Figure 9.22 over AWGN channels. When using second-order switched-diversity and pilot-symbol assisted coherent square-constellation 64-QAM [2] amongst our previously stipulated propagation conditions with a pilot-spacing of $P=5$ and channel SNR of about 25 dB, the C1, C2 and C3 subchannels have BERs of about 10^{-3}, 10^{-2} and $2 \cdot 10^{-2}$, respectively [2].

The source sensitivity-matched codes for the C1, C2 and C3 subchannels are BCH(126,49,13), BCH(126,63,10) and BCH(126,84,6), while the packet header was allocated again a BCH(63,24,7) code. The total number of BCH-coded bits becomes $3 \cdot 126 + 63 = 441/30$ ms, yielding a bit rate of 14.7 kbps. The resulting 74 64-QAM symbols are amalgamated with 15 pilot and 4 ramp symbols, giving 93 symbols/30 ms, which is equivalent to a signalling rate of 3.1 kBd, as in the case of the low-quality mode of operation. Again, 32 PRMA slots can be created, as for the low-quality system, accommodating more than 50 speech users in a bandwidth of 200 kHz and yielding an equivalent speech user bandwidth of about 200 kHz/50 users = 4 kHz, while maintaining a packet dropping probability of about 1 %.

9.7.3.3 Multi-mode Video Transmission [110, 112] [3]

Similarly to the above dual-mode philosophy, we also contrived a multi-mode videophone scheme, where the H.263 video codec was used and a number of speech slots were dedicated to videophony. In this system again, we considered transmitting QCIF images, where the video codec was programmed to generate 3560, 2352 or 1176 bits per frame, which were then transmitted using 64-, 16- or 4-QAM, respectively, at a constant signalling rate, requiring the same bandwidth.

Earlier in this chapter in Figure 9.22 we have shown that in square-constellation QAM schemes the bits can be assigned to a number of different integrity classes. The number of integrity classes depends on the number of modulation levels, and in 4-QAM there is only one integrity class, in 16-QAM there are 2, while in 64-QAM there are 3 classes, often also referred to as sub-channels. By using different strength FEC codes on each QAM sub-channel it is possible to equalise the probability of errors on the QAM sub-channels for video transmission. This means that all sub-channels' FEC codes should break down at approximately the same channel SNR. This is desirable, if all bits to be transmitted are equally important. Since the H.263 datastreams are variable length coded, one error can cause a loss of synchronisation and corrupt the rest of the frame. Therefore in this case most bits are equally important, and hence 'equalisation' of the QAM sub-channels' BER is desirable. The FEC codes used in our system in order to achieve a similar BER in all QAM sub-channels are summarised in Table 9.11.

Eight of each of the codewords are required for the encoding of the generated 3560, 2352 and 1176 bits/frame in all three modes of operation. In order to generate video packets compatible with the speech packets, again, the same BCH(63,24,7) packet header was selected. In the 4-QAM mode the 255+63=318 bits constitute 159 2-bit symbols, and after adding 17 pilot symbols as well as 2+2=4 so-called ramp symbols, during which the power amplifier is smoothly ramped up and down, in order to mitigate out-of-band spectral spillage, the resulting framelength becomes 180 symbols. Hence for delivering the 180-symbol video packets we require double-length speech slots, since the speech slots were 93 symbol long. Consequently, for the transmission of an FEC-coded video frame eight such 180-symbol packets per 90 ms are necessary, which can be accommodated by reserving three double-length speech slots per 30 ms PRMA frame, capable of delivering up to nine such video packets per 90 ms frame repetition interval. The corresponding video signalling rate including the packet header becomes 8×180 symb/90 ms = 16 kBaud. It is worth noting that it is possible to packetise the FEC-coded 1176 video bits as 4-bit 16-QAM symbols, rather than 2-bit 4-QAM symbols, which halves the length of the transmission burst

---

[3]This subsection is supported by a range of video demonstrations using various channel conditions under the WWW address http://www-mobile.ecs.soton.ac.uk

Modulation scheme	FEC codes used
4 QAM	BCH(255,147,14)
16 QAM	Class 1: BCH(255,179,10)
	Class 2: BCH(255,115,21)
64 QAM	Class 1: BCH(255,199,7)
	Class 2: BCH(255,155,13)
	Class 3: BCH(255,91,25)

Table 9.11: FEC codes used for the 4, 16 and 64 QAM transmission modes in conjunction with H.263 video coding ©Cherriman, Hanzo [110], 1995.

expressed in terms of modulation symbols. Consequently, this stream then can be arranged to occupy single, rather than double slots. Therefore the signalling rate can be reduced to 8 kBd at the cost of requiring higher channel SNR and SIR values than the more robust, but higher-rate 16 kBd 4-QAM system. Alternatively, the number of video bits can be doubled, which improves the associated video quality. Clearly, these different system configuration modes highlight the underlying trade-offs that designers of such flexible systems are faced with [112].

When comparing the 16-QAM modes of operation of the coherent and non-coherent video transceivers, in the non-coherent star 16-QAM system an 9.47 kbps video codec was used, while in the coherent 16-QAM scheme a bit-rate of 2352 bits/(90 ms) $\approx$26.13 kbps was maintained. Furthermore, the coherent scheme uses a stronger FEC scheme and some channel capacity is also dedicated to pilot symbols. Similar arguments are also valid for the 64-QAM mode, which can improve the associated video quality, delivering 3560 bits/frame, at the cost of higher required channel SNRs. Again, six time slots must be dedicated to support an H.263-based videophone call, which substantially reduces the system's speech capacity.

9.7.3.4   Packet reservation multiple access assisted multi-level graphical communications [195]

9.7.3.4.1   Graphical Transmission Issues In order to complement the proposed multi-rate FL-DCC graphical source codec discussed in Section 9.4, we used the previously introduced re-configurable multi-mode QAM modem. The most robust, but least bandwidth efficient 4-level Quadrature Amplitude Modulation (4-QAM) [2] mode can be used in outdoor scenarios in conjunction with the $b = 1$ mode of the FL-DCC codec. The less robust but more bandwidth efficient 16-QAM mode may be invoked in friendly indoor cells in order to accommodate the $b = 2$ mode of operation of the FL-DCC codec. When the channel conditions are extremely favourable, the modem can also be configured as a 64-QAM scheme,

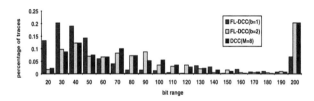

Figure 9.35: Histogram of the average trace length ©ETT, Hanzo and Yuen [195].

in which case it can deliver $b = 3$ bits per FL-DCC vector, allowing lossless coding.

As in our previous multi-mode schemes, we exploited that the BER of the lower quality class 2 (C2) 16-QAM subchannel was found to be a factor 2-3 times higher, than that of the higher integrity class 1 (C1) subchannel. Hence the more vulncrable FL-DCC coded bits were transmitted via the C1 subchannel, while the more robust bits over the C2 16-QAM subchannel. For error correction coding we used binary BCH codes, which were the topic of Chapter 4. Their specific coding parameters will be specified during our forthcoming discourse.

9.7.3.4.1.1 Graphical Packetisation Aspects In order to determine the desirable length of the transmission packets, in Figure 9.35 we evaluated the histogram of the average trace length, which exhibited a very long low-probability tail. This probability tail was represented by the bars at an encoded trace-length of 200 bits in the Figure. Observe furthermore that, as expected, the highest concentration of short traces was recorded for $b = 1$, which was followed by $b = 2$ and the DCC mode of operation. However, most traces generated less than a few hundred bits, even when $b = 2$ was used. In order to be able to use a fixed packet length, while maintaining robustness against channel errors and curtailing transmission error propagation across packets and/or traces, we decided to tailor the number of bits per trace to the packet length of 222 bits. If a longer trace was encountered, it was forcibly truncated to this length and the next packet started with a new 'artificial' pen-down code. If, however, a shorter trace was encountered, a second trace was also fitted into the current packet and eventually truncated to the required length for transmission. Unfortunately, the additional forcibly included VC code and the X_0, Y_0 coordinates portrayed in Figure 9.12 increased the number of bits generated but mitigated the error propagation effects. The proportion of the bit-rate increase evaluated in terms of percent for various packet lenghts in the case of the FL-DCC $b = 1$ scheme is portrayed in Figure 9.36.

Recall that the C1 16-QAM subchannel had a factor 2-3 times better BER than the C2 subchannel. This ratio would remain approximately

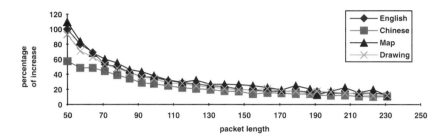

Figure 9.36: Proportion of bit-rate increase due to fixed-length trace termination versus packet length for the FL-DCC $b = 1$ scheme ©ETT, Hanzo and Yuen [195].

the same if we were to use the same FEC code in both subchannels, but the C2 BER would remain excessively high for the transmission of the starting vectors (SV) and fixed-length vectors (FV) of Figure 9.12. The BCH(255,131,18) and BCH(255,91,25) codes were found to ensure the required balance between the more and less robust FL-DCC bits, when employed in the C1 and C2 16-QAM subchannels, respectively. In other words, employing the more powerful BCH(255,91,25) codec in the higher-BER C2 16-QAM subchannel reduced the integrity difference between the subchannels and ensured the required source-sensitivity matched unequal error protection. The $2 \times 255 = 510$ BCH-coded bits constitute 128 4-bit 16-QAM symbols. After adding 14 pilot symbols according to a pilot spacing of 10 and concatenating 4 ramp symbols for smooth power amplifier ramping in order to minimise the out-of-band emissions, the resulting 146-symbol packets are queued for transmission to the BS. The same packet format can be used for the voice/video packets. Both the voice and the video codecs generate in their 8 kbps mode of operation 160 bits/20 ms frame and hence the 222-bit packet can accommodate a 62-bit signalling and control header in each 146-symbol packet. The corresponding single-user voice/video signalling rate becomes 146 symbols/20 ms = 7.3 kBd. Due to their significantly lower rates and higher delay tolerance, graphical users assemble their 146-symbol transmission packets over a longer period, before their transmission. The issue of maximum achievable graphical data rate will be addressed in the Results Section.

In the event of channel quality degradation the more robust 4-QAM mode can be invoked under BS control. However, since the 4-QAM packets can only convey half the number of bits, when compared to 16-QAM, the re-configurable graphical source codec has to halve its bit-rate, as we have seen in the case of our voice and video codecs earlier in this chapter. Explicitly, under unfavourable channel conditions the FL-DCC graphical source encoder can reduce the number of bits from $b = 2$ to 1 under the control of the BS, while providing adequate graphics quality. Again, these issues will be discussed in more depth during our further discussions. In order to maintain the same 222-bit long framing structure, as in the case of the 16-QAM mode of operation, we used two codewords of the shortened BCH(255,111,21) code in conjunction with the 4-QAM modem scheme, since the 4-QAM modem does not have different integrity sub-channels.

9.7.3.4.2 Graphics, Voice and Video Multiplexing using PRMA

Packet reservation multiple access (PRMA) was introduced earlier in this chapter as a convenient technique of surrendering passive speech or video slots, in favour of users, who are becoming active. Here PRMA is also invoked in order to multiplex the graphics source information with the voice and video packets for transmission to the base station. We have argued before that the packet dropping probability due to packet collision must remain below $P_{drop} = 1$ % [206], in order to minimise the speech degradation. Fortunately this initial speech spurt clipping is hardly perceivable, if the 1 % dropping probability requirement is not violated.

In contrast to speech, graphical data packets cannot be dropped, but tolerate longer delays and can be allocated to slots, which are not reserved by speech users in the present frame. In our exepriments the user transmitted graphical traces generated by a writing tablet. A less than unity permission probability either allowed or disabled permission to contend during any particular slot for a reservation within the current PRMA frame. Wong and Goodman noted [207] that it is advantageous to control data packet contentions on the basis of the fullness of the contention buffer. Specifically, contentions are disabled, until a certain number of packets awaits transmission, which reduces the probability of potential packet collisions due to the frequent transmission of short graphical data bursts. While speech communications stability can be defined as a dropping probability $P_{drop} < 1\%$, data transmission stability is specified in terms of maximum graphical data delay or buffer requirement. The PRMA packet multiplexer is depicted in Figure 9.37, and as an example, we indicated in the Figure a set of permission probabilities P, which could be employed by the various speech, data and video users in order to satisfy their corresponding delay and packet dropping constraints, while maximising the multiplexer's throughput capacity. The specific permission probability values indicated in the Figure were only used for the sake of illustration, they have to be optimised for the specific prevalent service quality requirements.

9.7.3.5 Performance of the 200 kHz Bandwidth Multi-mode, Multi-media System

9.7.3.5.1 Speech Performance The PRMA parameters used in our 100 kBaud, 200-kHz bandwidth multi-media system are summarised in Table 9.9 in contrast to those of the previously described 30-kHz bandwidth scheme. Observe in the Table that the most significantly different system

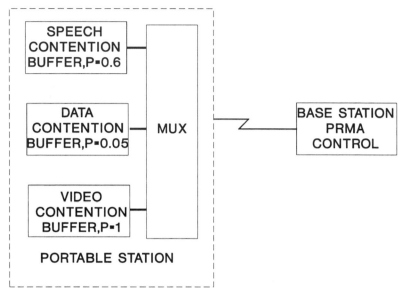

Figure 9.37: PRMA packet multiplexer ©ETT, Hanzo and Yuen [195].

parameters are the signalling rate and hence the number of PRMA slots and the number of users supported. Furthermore, the GSM-like scheme uses stronger error correction coding, which increased the non-coherent systems's speech Baud-rate from 2.2 to 3.1 kBaud. Furthermore, the optimum permission probability is reduced from 0.6 to 0.2 at a concomitant increase of the number of slots form 7 to 31, since best PRMA performance is achieved, if the number of users contending at any instant is kept approximately proportional to the total number of slots available.

The number of speech users supported by the 32-slot PRMA system becomes explicit from Figure 9.38, where the packet dropping probability versus number of users is displayed. Observe that more than 55 users can be served with a dropping probability below 1 %. The number of PRMA users per slot becomes about 1.72, which is higher than the 1.43 PRMA user/slot parameter of the 30-kHz bandwidth, 20 kBaud non-coherent multi-media scheme. This is a consequence of the higher statistical multiplexing gain associated with a higher number of slots and users. In order to restrict the subjective effects of PRMA-imposed packet dropping, according to Figure 9.38 the number of users must be below 60. As a comparative basis it is worth noting that the 8 kbps CCITT/ITU ACELP speech codec's target was to inflict less than 0.5 Mean Opinion Score (MOS) degradation in case of a speech frame error rate of 3% [46]. Again, in case of roll-off factors of

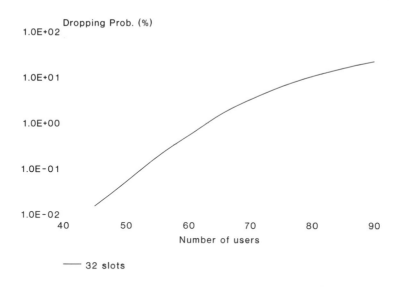

Figure 9.38: Packet dropping probability versus number of users for 32-slot PRMA ©IEEE, Hanzo, Woodard 1995, [19].

0.5 and 0.35 the 200 kHz system can accommodate 42 and 47 PRMA slots, supporting up to 70 and 80 users, respectively, when assuming 1.72 PRMA users per slot.

The SEGSNR versus channel SNR performance of the re-configurable 100 kBd transceiver using 32-slot PRMA is shown in Figure 9.39 for different number of conversations. Observe in the Figure that in contrast to the 30 kHz bandwidth system here we used SEGSNR, rather than SEGSNR degradation, as a system performance measure, since we wanted to portray the quality difference between the higher and lower quality modes of operations. We emphasise furthermore again that the number of users (us.) here is related to the 100 kBd, 100% excess bandwidth, 32-slot scenario. For the 42 and 47-slot scenarios similar tendencies can be observed. When the channel SNR was in excess of about 25 dB, the 6.5 kbps/64-QAM system outperformed the 4.7/16-QAM scheme in terms of both objective and subjective speech quality. Furthermore, at around 25 dB channel SNR, where the 16-QAM and 64-QAM SEGSNR curves cross each other in Figure 9.39 it is preferable to use the inherently lower quality but unimpaired mode of operation. When supporting more than 32 users, as in our PRMA-assisted system, speech quality degradation is experienced due to packet corruption caused by channel impairments and packet dropping caused by PRMA

Parameter	30 kHz System	200 kHz System
		Low/High Quality Mode
Speech Codec	4.8 kbps TBPE	4.7/6.5 kbps ACELP
Speech FEC	Triple-class BCH	Twin-/Triple-class Binary BCH
FEC-coded Speech Rate	8.6 kbps	8.5/12.6 kbps
Video Codec	Fixed-rate DCT	Packetised H.263
No. of Video bits/fr	852	1176, 2352 or 3560
Video Rate (kbps)	9.47	13.1, 26.1 or 39.6
Video FEC	BCH(127,71,9)	see Table 9.11
Modulation for Speech	Star 16-QAM	Square 16-QAM/64-QAM
Modulation for Video	Star 16-QAM	Square 4-/16-/64-QAM
Demodulation	Non-coh.	Coherent, Diversity, PSAM
Equaliser	No	No
Speech Signalling Rate (kBd)	2.2	3.1
Video Signalling Rate (kBd)	4.4	8/16
VAD	GSM-like (Chap 3)	GSM-like
Multiple Access	7-slot PRMA +	(32/42/47-slot PRMA)-
	2-slot TDMA	No. of Video Slots (n×6 or 3)
Speech Frame Length (ms)	30	30
Slot Length (ms)	3.33	0.94/0.71/0.64
Channel rate (kBd)	20	100-148
System Bandwidth (kHz)	30	200
No. of PRMA Speech Users	> 10	> 50-80
No. of PRMA Users/slot	1.43	> 1.72
Equiv. Speech User Bandwidth	3	2.5-4
Min. Channel SNR/SIR (dB)	24	15/25

Table 9.12: Transceiver parameters.

packet collisions. These impairments yield different subjective perceptual degradations, which we will attempt to compare in terms of the objective SEGSNR degradation. Quantifying these speech imperfections in relative terms in contrast to each other will allow system designers to adequately split the tolerable overall speech degradation between packet dropping and packet corruption. Observe in Figure 9.39 that the rate of change of the SEGSNR curves is more dramatic due to packet corruption caused by low-SNR channel conditions than due to increasing the number of users. As long as the number of users does not significantly exceed 50, the subjective effects of PRMA packet dropping show an even more benign speech quality penalty than that suggested by the objective SEGSNR degradation, because frames are typically dropped at the beginning of a speech spurt due to a failed contention, rather than during active speech spurts.

In conclusion, our re-configurable transceiver has a single-user rate of 3.1 kBd, and can accommodate 32 PRMA slots at a PRMA rate of 100 kBd in a bandwidth of 200 kHz. The number of users supported is in excess of 50 and the minimum channel SNR for the lower speech quality mode is about 15 dB, while for the higher quality mode it is about 25 dB. The number of time slots can be further increased to 42, when opting for a modulation access bandwidth of 50%, accommodating a signalling rate of 133 kBd within the 200 kHz system bandwidth. This will inflict a slight bit error rate penalty, but will pay dividends in terms of increasing the number of PRMA users by about 20. The parameters of the proposed transceiver

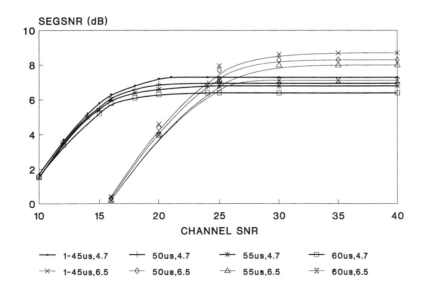

Figure 9.39: SEGSNR versus channel SNR performance of the re-configurable
100 kBd transceiver using 32-slot PRMA for different number of
conversations ©IEEE, Hanzo, Woodard 1995, [19].

are summarised in Table 9.12. In order to minimise packet corruption due
to interference, the employment of a time-slot quality ranking algorithm is
essential for invoking the appropriate mode of operation. When serving 50
users, the effective user bandwidth becomes 200 kHz/50 = 4 kHz, which
guarantees the convenience of wireless digital speech communication in a
bandwidth similar to conventional analogue telephone channels. The 4 kHz
user bandwidth can be further reduced to 200 kHz/70 users ≈ 2.9 or even
to 200 kHz/80 users = 2.5, when using a modulation roll-off factor of 0.5 or
0.35, respectively, hence accommodating more PRMA slots and therefore
supporting more PRMA users.

9.7.3.5.2 Video Performance   The corresponding H.263-based video
system performance was also evaluated under the same propagation conditions, but in addition to the previous 4.7 kbps 16-QAM and 6.5 kbps
64-QAM modes the more robust 4-QAM mode was introduced, in order
to provide higher robustness for the more error-sensitive video bits. In
the various operating modes investigated the PSNR versus channel SNR
curves of Figures 9.40 and 9.41 were obtained for AWGN and Rayleigh
channels, respectively. Since both the H.261 and H.263 source codecs have

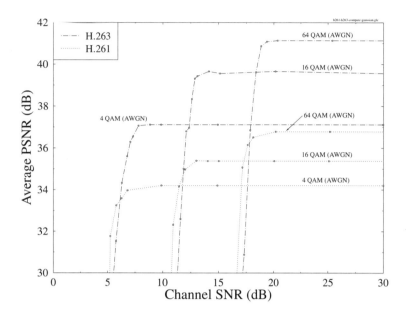

Figure 9.40: Performance comparison of the proposed adaptive H261 and H263 transceivers over AWGN channels ©Cherriman, Hanzo [110], 1995.

had similar robustness against channel errors, and their transceivers were identical, the associated 'corner SNR' values, where unimpaired communications broke down, were virtually identical for both systems over both AWGN and Rayleigh channels. However, as expected, the H.263 codec exhibited always higher video quality at the same bit-rate or system bandwidth. We note in closing that the described H.263-based video scheme reduces the speech capacity of the GSM-like system by six speech users, every time a new video user is admitted to the system. Let us now conclude our findings throughout this chapter.

9.7.3.5.3 Graphical System Performance The overall graphical system robustness is characterised below in terms Figures 9.42-9.45. The graphical representation quality was evaluated in terms of both the mean squared error (mse) and the Peak Signal to Noise Ratio (PSNR). In analogy to the previous PSNR in video telephone quality evaluation, the graphical PSNR was defined as the squared ratio of the maximum possible spatial deviation from the uncoded graphical trace across the graphical screen, related to the quantisation-induced deviation from the original graphical trace, which is the equivalent of the quantisation distortion.

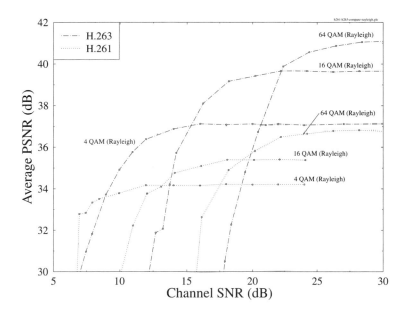

Figure 9.41: Performance comparison of the proposed adaptive H261 and H263
transceivers over Rayleigh channels ©Cherriman, Hanzo [110],
1995.

When using a resolution of 640 × 480 pixels, the maximum spatial deviation energy is $640^2 + 480^2 = 640\,000$, corresponding to a maximum diagonal spatial deviation of 800 pixels. The lossy coding energy was measured as the mean squared value of the pixel-to-pixel spatial distance between the original graphical input and the FL-DCC graphical output. For perfect channel conditions the $b = 1$ and $b = 2$ FL-DCC codec had PSNR values of 49.47 and 59.47 dB, respectively. As we showed in Figure 9.13, the subjective effects of a 10 dB PSNR degradation due to using $b = 1$ instead of $b = 2$ are not severe in terms of readability.

The system's robustness was characterised in Figures 9.42-9.45 upon transmitting the handwriting sequence of Figure 9.13 a high number of times, in order to ensure the statistical soundness of the graphical quality investigations. The best-case propagation scenario was encountering the stationary Additive White Gaussian Noise (AWGN) channel. The more bandwidth-efficient but less robust 16QAM mode of operation is characterised by Figure 9.42. Observe that transmissions over Rayleigh channels with diversity (RD) and with no diversity (RND) are portrayed using both one (TX1) and three (TX3) transmission attempts, in order to improve the system's robustness. Two-branch selection diversity using two independent

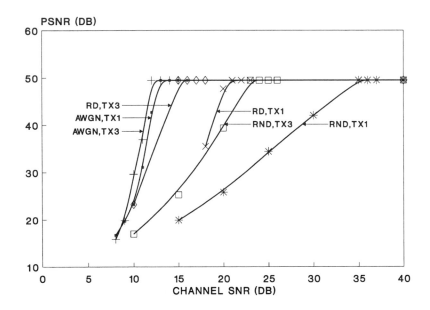

Figure 9.42: Graphical PSNR versus channel SNR performance of the $b =$ 1-bit FL-DCC/16-QAM mode of operation over various channels ©ETT, Hanzo and Yuen [195].

Rayleigh channels was studied in conjunction with various selection criteria and we found that using the channel with the minimum phase shift between pilots slightly outperformed the maximum energy criterion. Similarly, over AWGN channels one or three transmission attempts were invoked.

As seen in Figures 9.42 and 9.43 in the case of the 16-QAM mode, over AWGN channels the required channel SNR for unimpaired graphical communications is around 11 dB with ARQ and 12 dB without ARQ. This marginal improvement is attributable to the fact that the AWGN channel exhibits always a fairly constant bit error rate (BER) and hence during re-transmission attempts the chances of successful transmissions are only marginally improved. In contrast, over Rayleigh channels the received signal has a high probability of emerging from a deep fade by the time the packet is re-transmitted. This is the reason for the significantly improved robustness of the ARQ-assisted Rayleigh scenarios of Figures 9.42 and 9.43. Specifically, with diversity the ARQ attempts reduced the required channel SNR by about 5 dB to around 15 dB, while without diversity an even higher 10-12 dB ARQ gain is experienced. When using the $b = 2$-bit FL-DCC scheme, Figure 9.43 shows that the error-free PSNR was increased to nearly 60 dB, while the system's robustness against channel errors and the

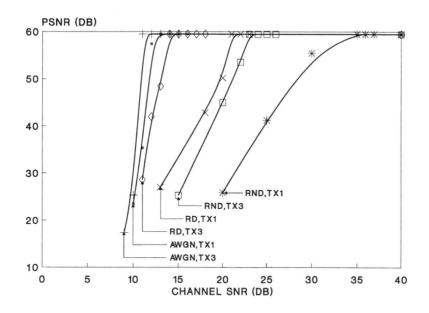

Figure 9.43: Graphical PSNR versus channel SNR performance of the $b = 2$-bit FL-DCC/16-QAM mode of operation over various channels ©ETT, Hanzo and Yuen [195].

associated 'corner SNR' values remained unchanged.

Figure 9.44 demonstrates that when the more robust 4-QAM mode was invoked, over AWGN channels SNR values of 5 and 8 dB were necessitated by the ARQ-aided TX3 scheme and the non-ARQ assisted TX1 systems, respectively. The diversity-assisted RD, TX3 and RD, TX1 systems required a minimum SNR of about 10 and 17 dB for unimpaired graphical communications, which had to be increased to 20 and 27 dB without diversity. Clearly, both the ARQ- and diversity-assistance played a crucial role in terms of improving the system's robustness. The same tendencies can be noted in Figure 9.45 as regards the $b = 2$-bit FL-DCC scheme. When using the lossless $b = 3$-bit FL-DCC or the conventional DCC scheme, the graphical PSNR becomes infinite, hence the corresponding PSNR versus channel SNR curves cannot be plotted. However, the transceiver's performance predetermines the minimum channel SNR values required for unimpaired graphical quality, which are about the same as for the above schemes. We note furthermore that since each packet commences with a pen-down code, the system can switch between the 4- and 16-QAM modes arbitrarily frequently without any objectionable perceptual quality degradation or graphical switching transients.

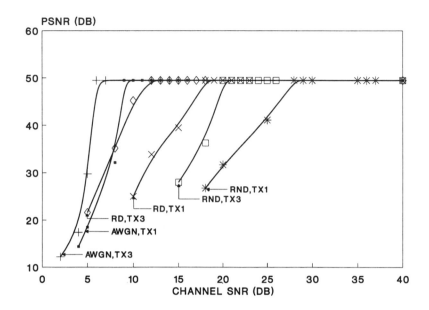

Figure 9.44: Graphical PSNR versus channel SNR performance of the $b = 1$-bit FL-DCC/4-QAM mode of operation over various channels ©ETT, Hanzo and Yuen [195].

The subjective effects of channel errors are demonstrated by Figure 9.46 in the case of the diversity- and ARQ-assisted Rayleigh 16-QAM, FL-DCC $b = 1$ scenario, where the graphical PSNR was gradually reduced from the error-free 49.47 dB at the top left hand corner to 42.57, 37.42, 32.01, 27.58 and 21.74 dB at the bottom right hand corner, respectively. Here we quoted the PSNR values, rather than the channel SNR values, since the associated graphical quality can be ensured by various system configuration modes under different channel conditions. The corresponding channel SNRs for the various system configuration modes can be inferred from the intercept points of the horizontal line corresponding to a particular PSNR value in Figures 9.42-9.45. Note that when the channel BER becomes high and hence the PSNR is degraded by more than about 5 dB, the graphical artifacts become rather objectionable. In this case it is better to disable the decoder's output by exploiting the error detection capability of the BCH decoder.

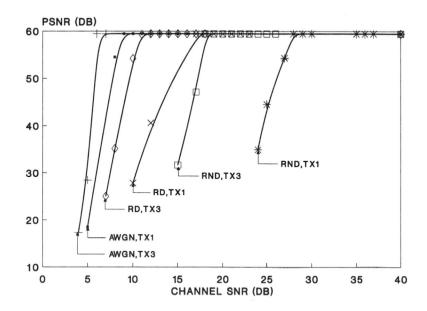

Figure 9.45: Graphical PSNR versus channel SNR performance of the $b = 2$-bit FL-DCC/4-QAM mode of operation over various channels ©ETT, Hanzo and Yuen [195].

9.8 Summary and Conclusions

In contrast to previous chapters of this book, which mainly dealt with a range of system components, ranging from speech codecs to channel codecs and modems, this chapter attempted to address system design and performance issues in the context of a 30-kHz bandwidth and a 200-kHz bandwidth system, which are the bandwidths of two well-known systems, namely the Pan-American IS-54 and the Pan-European GSM system. Following a brief overview of speech and video coding advances as well as bit sensitivity issues, a slightly deeper discussion was offered on bandwidth-efficient full-response multi-level modem schemes, since these modems were not detailed in previous chapters.

Then the potential of a bandwidth-efficient 2.2 kBd PRMA-assisted TBPE, DCT, BCH, 16-QAM scheme has been investigated under the assumption of benign channel conditions. Within the 30-kHz bandwidth about 10 voice users plus a video telephone user can be supported, if channel SNR and SIR values in excess of about 24 dB can be maintained. The main transceiver features are summarised in contrast to those of the 200-kHz bandwidth dual-mode speech system in Table 9.12.

The system performance of the 30-kHz bandwidth non-coherent

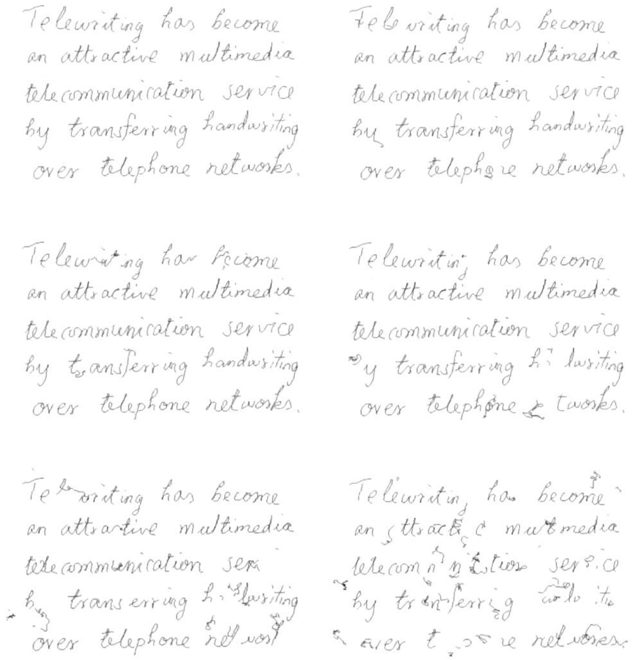

Figure 9.46: Subjective effects of transmission errors for the $b = 1$ 16-QAM, RD, TX3 scheme for PSNR values of (left to right, top to bottom) 49.47, 42.57, 37.42, 32.01, 27.58 and 21.74 dB, respectively. ©ETT, Hanzo and Yuen [195].

transceiver was further improved at the cost of slightly higher implementational complexity, when using a more sophisticated multi-mode, pilot symbol assisted, block-coded coherent square-constellation modem. The potential of this system was evaluated and the main system features were summarised also in Table 9.12 in contrast to those of the non-coherent star 16-QAM 30-kHz bandwidth system. As seen in the Table, the coherent scheme employed stronger overall error correction coding and a more robust modem, requiring lower SNR and SIR values over the more slowly-fading ($f_d \cdot T \approx 0.0000833$) DCS1800-like pedestrian channel than the lower-complexity non-coherent scheme over the slower-fading ($f_d \cdot T \approx 0.0042$) IS-54-like channel.

Furthermore, the increased system bandwidth of 200 kHz supports more time slots than the 30-kHz bandwidth system, which improves the statistical multiplexing gain of the PRMA scheme from 1.43 to 1.72 and hence increases also the number of users per PRMA slot. Lastly, due to its multi-mode nature, this scheme can also adapt to time-variant propagation environments. This multi-mode transceiver was then also exploited to transmit three different quality video signals at bit-rates of 13.1, 26.1

or 39.6 kbps in one of the 4-, 16- or 64-QAM modes, depending on the prevalent channel conditions. Overall, as expected, in performance terms we favour PSAM-assisted schemes, if their slightly higher complexity is acceptable, since it was more robust against transmission impairments and had a lower effective user bandwidth requirement than the non-coherent scheme.

An adaptive FL-DCC graphical coding scheme was also proposed for graphical communications, which has a lower coding rate and similar graphical quality to DCC in the case of $b = 1$ and 2. The codec can be adaptively re-configured to operate at $b = 1$, $b = 2$ or even at $b = 3$ in order to comply with the prevailing network loading and/or propagation conditions, as well as graphical resolution requirements. The proposed FL-DCC codec was employed in an intelligent, re-configurable wireless adaptive multimedia communicator, which was able to support a mixture of speech, video and graphical users. The minimum required channel SNR in the ARQ-assisted 16-QAM mode was 11 dB and 15 dB over AWGN and diversity-aided Rayleigh channels, respectively. The system's robustness against channel errors was improved at a concomitantly reduced graphical resolution, when 4-QAM was invoked.

In order to probe further into the field of wireless multimedia communications, the interested reader is referred to reference [2] for a range of novel transceiver components, including channel equalisers, clock and carrier recovery circuitries, orthogonal frequency division multiplex schemes and adaptive transceivers, while various multimedia systems were proposed and analysed in references [19–22, 25, 26, 55, 75, 86, 92, 99–101, 112].

9.9 Acknowledgement

Over the years I have had the privilege of collaborating with many former and current colleagues in the Department of Electronics and Computer Science at the University of Southampton, UK, while working on various research projects. These colleagues and valued friends, too numerous to mention, have influenced my views concerning various aspects of wireless multimedia communications and I am grateful for the enlightment gained form our collaborations on various projects, papers and books. I am particularly indebted to Raymond Steele, Professor of Communications and the Editor of this book, whose long-term vision motivated many of the investigations reported. I am also grateful to J. Brecht, Jon Blogh, Marco Breiling, M. del Buono, Clare Brooks, Peter Cherriman, Byoung Jo Choi, Joseph Cheung, Peter Fortune, Peter Cherriman, Joseph Cheung, Lim Dongmin, D. Didascalou, S. Ernst, David Greenwood, Hee Thong How, Thomas Keller, Xiao Lin, Chee Siong Lee, Tong-Hooi Liew, Matthias Muenster, V. Roger-Marchart, Redwan Salami, Juergen Streit, Jeff Torrance, Spiros Vlahoyiannatos, William Webb, John Williams, Jason Woodard, Choong Hin Wong,

Henry Wong, Lie-Liang Yang, Bee-Liong Yeap, Mong-Suan Yee, Kai Yen, Andy Yuen and many others, with whom I have enjoyed an association. My sincere thanks are also due to the EPSRC, UK, the Commission of the European Communities, Brussels, Belgium and Motorola ECID, Swindon, UK for sponsoring some of my recent research.

The topic of this chapter was a range of powerful multimedia systems based on bandwidth-efficient multi-level modulation, which can accommodate more users in a given bandwidth, than their second generation counterparts, such as the systems in Table 1.1 and hence these solutions may offer an evolutionary path for the existing systems. The system design principles introduced can also be invoked in the context of the forthcoming third-generation systems. In the next chapter we will concentrate on the third-generation proposals currently under consideration in Europe, the US and Japan.

Bibliography

[1] **L. Hanzo**, "Bandwidth-efficient Wireless Multimedia Communications", *Proc. of the IEEE*, July 1998, Vol. 86, No. 7, pp 1342-1382

[2] **W.T. Webb** and **L. Hanzo**, "Modern quadrature amplitude modulation: Principles and applications for fixed and wireless channels", *IEEE Press-John Wiley co-publication*, ISBN 0-7273-1701-6, p 557, 1994

[3] **J. D. Gibbson**, "The mobile communications handbook", *IEEE Press-CRC Press*, 1996

[4] **IEEE Personal Communications**, "Special Issue: The European Path Toward UMTS", *The magazine of nomadic communications and computing*, Feb. 1995, Vol. 2, No. 1

[5] *Advanced Communications Technologies and Services (ACTS), Workplan, European Commission, DGXIII-B-RA946043-WP, Aug. 1994*

[6] **IEEE Personal Communications** "Feature Topic: Wireless Personal Communications" *The magazine of nomadic communications and computing*, Apr. 1995, Vol. 2, No. 2

[7] **IEEE Communications Magazine**, "Feature Topic: Universal Telecommunications at the Beginning of the 21st Century", Nov. 1995, Vol. 33, No. 11

[8] **IEEE Communications Magazine**, "Feature Topic: "Software Radios", Vol. 33 No. 5, May 1995

[9] **IEEE Communications Magazine**, "Feature Topic: Wireless Personal Communications", Jan. 1995, Vol. 33, No. 1.

[10] **IEEE Communications Magazine**, "Feature Topic: European Research in Mobile Communications", Vol. 34, No. 2, pp 60-106, Feb. 1996

[11] **D.C. Cox**, "Wireless personal communications: A perspective", *in J.D. Gibbson (Ed.): The mobile communications handbook, IEEE Press-CRC Press*, pp 209-241, 1996

[12] **A.D. Kucar**, "Mobile radio: An overview", *in J.D. Gibbson (Ed.): The mobile communications handbook, IEEE Press-CRC Press*, pp 242-262, 1996

[13] **V.O.K. Li, X. Qiu**, "Personal Communications Systems". *Proc. of the IEEE*, Vol. 83, No. 9, pp 1210- 1243, Sept. 1995

[14] *Dual-mode subscriber equipment- Network equipment compatibility specification, Interim Standard IS-54, Telecomm. Industry Association (TIA), Washington, DC, 1989, TIA*

[15] *Public Digital Cellular (PDC) Standard, RCR STD-27, Research and Development Centre for Radio Systems, Japan, PDC*

[16] *Mobile station - Base station compatibility standard for dual-mode wideband spread spectrum cellular system, EIA/TIA Interim Standard IS-95, Telcomm. Industry Association (TIA), Washington, DC, 1993, TIA*

[17] **P.W. Baier, P. Jung** and **A. Klein**, "Taking the challenge of multiple access for third-generation cellular mobile radio systems - A European View", *IEEE Comms. Mag.*, Vol. 34, No. 2, pp 82-89, Feb. 1996

[18] **R. Prasad**, "CDMA for wireless personal communications", *Artech House*, 1996

[19] **L. Hanzo** and **J.P. Woodard**, "An intelligent multimode voice communications system for indoors communications", *IEEE Tr. on Veh. Technology*, Vol. 44, No. 4, pp 735-749, ISSN 0018-9545, Nov. 1995

[20] **L. Hanzo, J.P. Woodard**, "Modern Voice Communications", in preparation

[21] **L. Hanzo, R. Salami, R. Steele** and **P.M. Fortune**, "Transmission of Digitally Encoded Speech at 1.2 KBd for PCN", *IEE Proc-I*, Vol. 139, No 4, pp 427-447, Aug. 1992

[22] **L. Hanzo et.al.** "A low-rate multi-level voice/video transceiver for personal communications", *Wireless Personal Communications, Kluwer Academic Publishers*, Issue 2:3, pp 213-230, Nov. 1995

[23] **J Woodard** and **L. Hanzo**, "A re-configurable speech transceiver", *Proc. of 2nd International Workshop on Mobile Multimedia Communications, MoMuc-2, Bristol, UK.* 11-13 Apr, 1995

[24] **J. Williams, L. Hanzo** and **R. Steele**, "Performance comparison of wireless speech communications schemes", *Proc. of 2nd International Workshop on Mobile Multimedia Communications, MoMuc-2, Bristol, UK.* 11-13 Apr, 1995

[25] **J. Williams, L.Hanzo, R.Steele** and **J.C.S Cheung**, "A comparative study of microcellular speech transmission schemes", *IEEE Tr. on Veh. Technology*, Vol. 43, No. 4 pp 909-925, Nov. 1994

[26] **L.Hanzo, et al.** "A packet reservation multiple access assisted cordless telecommunications scheme", *IEEE Tr on VT.*, vol. 43, No. 2, pp 234-245, May 1994

[27] **J.C.S. Cheung, L. Hanzo, W.T. Webb** and **R. Steele**, "Effects of Packet Reservation Multiple Access on Objective Speech Quality", *Electr. Letters*, 21, Vol. 29, No. 2, pp 152-153, Jan. 1993

[28] **P. Kroon** and **E.F. Deprettere**, "Regular Pulse Excitation - A Novel Approach to Effective Multipulse Coding of Speech", *IEEE Trans. on Acoustics, Speech and Signal Processing*, pp. 1054–1063, 1986

[29] **P. Vary, K. Hellwig, R. Hofmann, R. Sluyter, C. Galland** and **M. Rosso**, "Speech Codec for the European Mobile Radio System", *Proc. ICASSP*, pp. 227–230, April 1988

[30] **I.A. Gerson** and **M.A. Jasuik**, "Vector Sum Excited Linear Prediction (VSELP) Speech Coding at 8 kbps," *IEEE Journal on Selected Areas in Communications*, pp. 461–464, 1990

[31] **I.A. Gerson** and **M.A. Jasuik**, "Vector Sum Excited Linear Prediction (VSELP)," in *Advances in Speech Coding* (Bishnu S. Atal, Vladimir Cuperman and Allen Gersho, ed., pp. 69–80, Kluwer Academic Publishers, 1991

[32] **J. Campbell, V. Welch** and **T. Tremain**, "An Expandable Error-protected 4800 bps CELP Coder (U.S. Federal Standard 4800 bps Voice Coder)," *Proc. ICASSP*, pp. 735–738, 1989

[33] **D. O'Shaughnessy**, *Speech Communication, Human and Machine*, Addison-Wesley Publishing Company, 1987

[34] **S. Furui**, *Digital Speech Processing, Synthesis and Recognition*, Marcel Dekker Inc., 1989

[35] **J.B. Anderson** and **S. Mohan**, *Source and Channel Coding - An Algorithmic Approach*, Kluwer Academic Publishers , 1991

[36] **A.M. Kondoz**, *Digital Speech: Coding for Low Bit-Rate Communications Systems* , John Wiley, 1994

[37] **W.B. Kleijn** and **K.K. Paliwal**, *Speech Coding and Synthesis*, Elsevier Science, 1995

[38] **Allen Gersho**, "Advances in Speech and Audio Compression", *Proceedings of the IEEE*, pp. 900–918, June 1994

[39] **I.A. Gerson, M.A. Jasuik, J-M. Muller, J.M. Nowack** and **E.H. Winter**, "Speech and Channel Coding for the half-rate GSM Channel". *Proceedings ITG-Fachbericht 130, VDE-Verlag, Berlin*, pp. 225–233, Nov 1994

[40] **T. Ohya, H. Suda** and **T. Miki**, "5.6 kbits/s PSI-CELP of the half-rate PDC speech coding standard ", *Proceeding of the IEEE Conference on Vehicular Technology*, pp. 1680–1684, June 1994

[41] **W. B. Kleijn**, "Encoding Speech Using Prototype Waveforms", *IEEE Trans. on Acoustics, Speech and Signal Processing*, pp. 386–399, Oct 1993

[42] **D. W. Griffin** and **J. S. Lim**, "Multiband Excitation Vocoder", *IEEE Trans. on Acoustics, Speech and Signal Processing*, pp. 1223–1235, Aug 1988

[43] **D.J. Hiotakakos** and **S. Xydeas**, "Low Bit-Rate Coding Using an Interpolated Zinc Excitation Model", *Proceeding of the IEEE Singapore International Conference on Communications Systems* , pp. 865–869, Nov 1994

[44] **J.-H.Chen, R.V. Cox, Y.-C. Lin, N. Jayant** and **M.J. Melchner**, "A Low-Delay CELP Coder for the CCITT 16 kb/s Speech Coding Standard", *IEEE Journal on Selected Areas in Communications*, pp. 830–849, June 1991

[45] Coding of Speech at 16 kbit/s Using Low-Delay Code Excited Linear Prediction, *CCITT Recommendation G.728*, 1992

[46] R.A.Salami, C.Laflamme, J.-P. Adoul and D. Massaloux, "A Toll Quality 8 Kb/s Speech Codec for the Personal Communications System (PCS)", *IEEE Transactions on Vehicular Technology*, pp. 808–816, August 1994

[47] Coding of Speech at 8 kbit/s Using Conjugate-Structue Algebraic Code-Excited Linear Prediction (CS-ACELP), *ITU Draft Recommendation G.729*, February 1996

[48] A. Kataoka, J-P. Adoul, P. Combescure and P. Kroon, "ITU-T 8-kbits/s Standard Speech Codec for Personal Communication Services", *Proceedings of International Conference on Universal Personal Communications, Tokyo, Japan*, pp. 818–822, Nov 1995

[49] A. McCree *et.al.*, "A 2.4 kbit/s MELP candidate for the new US federal standard", *Proc. of ICASSP'96, Atalanta, Georgia, US*, pp 200-203, 1996.

[50] M.R. Schroeder and B.S. Atal, "Code-Excited Linear Prediction (CELP): High Quality Speech at Very Low Bit-Rates", *Proc. of ICASSP'85*, pp. 937-941, 1985.

[51] R.A. Salami, "Binary pulse excitation: a novel approach to low complexity CELP coding" in *Advances in speech coding*, Kluwer Academic Publishers, pp. 145-156, 1991

[52] B.S. Atal and M.R. Schroeder. "Stochastic coding of speech signals at very low bit-rates". *IEEE Int. Conf. Commun.*, May 1984.

[53] M.R. Schroeder and B.S. Atal. "Code-Excited Linear Prediction (CELP): High quality speech at very low bit-rates". *Proc. ICASSP'85, Tampa, Florida, USA* pp. 937–940, 26-29 March, 1985.

[54] J-P. Adoul *et.al.*, "Fast CELP coding based on algebraic codes", *Proc. ICASSP'87*, pp. 1957-1960, 1987

[55] L. Hanzo, R. Salami, R. Steele and P.M. Fortune, "Transmission of Digitally Encoded Speech at 1.2 KBd for PCN", *IEE Proc.-I, vol. 139, no. 4*, pp. 437-447, Aug. 1992

[56] N. Kitawaki, M. Honda and K. Itoh, "Speech-Quality Assessment Methods for Speech-coding Systems", *IEEE Communications Magazine*, vol. 22, no. 10, pp. 26-33, Oct. 1984

[57] A.H. Gray and J.D. Markel, "Distance Measures for Speech Processing", *IEEE Tr. on Acoutics, Speech and Signal Processing*, vol. ASSP-24, no. 5, pp. 380-391, Oct. 1976

[58] N. Kitawaki, H. Nagabuchi and K. Itoh, "Objective Quality Evaluation for Low-Bit-Rate Speech Coding Systems", *IEEE Tr. JSAC*, vol. 6, no. 2, pp. 242-249, Febr. 1988

[59] Guest Editors:, K.H. Tzou, H.G. Mussmann and K. Aizawa, "Special issue on very low bit-rate video coding", *IEEE Transactions on Circuits and Systems for Video Technology*, 4(3):213–357, June 1994

[60] Guest Editor: N. Hubing, "Speech and image coding", *Special Issue of the IEEE Journal on JSAC*, 10(5):793–976, June 1992

[61] Guest Editors: **B. Girod** *et.al.*, "Special issue on image sequence compression", *IEEE Transactions on Image Compression*, 3(5):465–716, September 1994

[62] **M. Khansari, A. Jalali, E. Dubois** and **P. Mermelstein**, "Robust low bit-rate video transmission over wireless access systems", In *Proc. of International Comms. Conf. (ICC)*, pages 571–575, 1994

[63] **R. Mann Pelz**, "An unequal error protected px8 kbit/s video transmission for DECT", In *Vehicular Technology Conference*, pages 1020–1024. IEEE, 1994

[64] **T.C. Chen**, "A real-time software based end-to-end wireless visual communications simulation platform", In *Proc. of SPIE Conf. on Visual Communications and Image Processing*, pages 1068–1074, 1995

[65] **K. Illgner** and **D. Lappe**, "Mobile multimedia communications in a universal telecommunications network", In *Proc. of SPIE Conf. on Visual Communications and Image Processing*, pages 1034–1043, 1995

[66] **Y.Q. Zhang**, "Very low bit-rate video coding standards", In *Proc. of SPIE Conf. on Visual Communications and Image Processing*, pages 1016–1023, 1995

[67] **H. Ibaraki** *et.al.*, "Mobile video communication techniques and services", In *Proc. of SPIE Conf. on Visual Communications and Image Processing*, pages 1024–1033, 1995

[68] **K. Watanabe** *et.al.*, "A study on transmission of low bit-rate coded video over radio links", In *Proc. of SPIE Conf. on Visual Communications and Image Processing*, pages 1025–1029, 1995

[69] **N. Färber, E. Steinbach** and **B. Girod**, "Robust H.263 video transmission over wireless channels," *in Proc. of International Picture Coding Symposium (PCS)*, (Melborne, Australia), pp. 575–578, March 1996.

[70] **M. W. Whybray** and **W. Ellis**, "H.263 - video coding recommendation for PSTN videophone and multimedia," *in IEE Colloquium (Digest)*, pp. 6/1–6/9, IEE, England, Jun 1995.

[71] **P. Sheldon, J. Cosmas** and **A. Permain**, "Dynamically adaptive control system for MPEG-4", In *Proceedings of the 2nd International Workshop on Mobile Multimedia Communications*, 1995

[72] **ITU-T**, *Draft Recommendation H.263: Video coding for low bit-rate communication,*

[73] **Telenor Research and Development**, P.O.Box 83, N-2007 Kjeller, Norway, *H.263 Software Codec*

[74] **ITU-T**, *Recommendation H.261: Video codec for audiovisual services at px64 Kbit/s*, March 1993

[75] **J.Streit** and **L.Hanzo.** "A fractal video communicator", In *IEEE Conference on Vehicular Technology*, pages 1030–1034, Stockholm, Sweden, June 1994

[76] **N. Jayant**, "Signal compression: Technology targets and research directions", in *Speech and Image Coding, Special Issue of the IEEE JSAC*, June 1992, Vol. 10, No. 5, pp 793-976, pp 796-818

[77] **Guest Editor: N. Hubing**, "Speech and Image Coding", *Special Issue of the IEEE JSAC*, Vol. 10, No. 5, pp 793-976, June 1992

[78] **ISO/IEC 11172 MPEG 1 International Standard**, "Coding of moving pictures and associated audio for digital storage media up to about 1.5 Mbit/s", Parts 1-3

[79] **ISO/IEC CD 13818 MPEG 2 International Standard**, "Information Technology, Generic coding of moving video and associated audio information", Parts 1-3.

[80] **N.S. Jayant** and **P. Noll**, "Digital coding of waveforms", Prentice-Hall, 1984

[81] *Proceedings of the 2nd International Workshop on Mobile Multimedia Communications, MoMuC-2, Apr. 11-13, Bristol, UK, 1995*

[82] **S. Sheng, A. Chandrakashan** and **R.W. Brodersen**, "A Portable Multimedia Terminal", *IEEE Comms. Mag.*, Vol. 30. No. 12, pp 64-75, Dec 1992

[83] "Special Issue on Very Low Bit-Rate Video Coding", *IEEE Transactions on Circuits and Systems for Video Technology*, Vol 4, No. 3, pp 213-357, June 1994

[84] "Special Issue on Image Sequence Compression", *IEEE Tr. on Image processing*, Sept. 1994, Vol. 3, No. 5, Guest Editors: B. Girod et al

[85] **J.W. Woods (Ed.)**. *Subband image coding*, Kluwer Academic Publishers, 1991

[86] **R. Stedman, H. Gharavi, L. Hanzo** and **R. Steele** "Transmission of Subband-Coded Images via Mobile Channels", *IEEE Tr. on Circuits and Systems for Video Technology*, Vol.3, No. 1, pp 15-27, Feb 1993

[87] **J. Katto, J. Ohki, S. Nogaki** and **M. Ohta**, "A wavelet codec with overlapped motion compensation for very low bit-rate environment", *IEEE Tr. on Video Technology*, Vol. 4, No. 3, pp 328-338, June 1994

[88] **P. Strobach**, "Tree-Structured Scene Adaptive Coder", *IEEE Trans. Commun.*, vol COM-38, pp 477-486, 1990

[89] **J. Vaisey** and **A. Gersho**, "Image compression with variable block size segmentation", *IEEE Tr. on Signal Processing*, Vol. 40, No. 8, pp 2040-2060, Aug 1992

[90] **A. Gersho** and **R.M. Gray**, *Vector Quantization and Signal Compression*, Kluwer Academic Publishers, 1992

[91] **B. Ramamurthi** and **A. Gersho**, "Classified Vector Quantization of Images", *IEEE Tr. on Com.*, Vol 31, No 11, pp 1105-1115, Nov. 1986

[92] **L. Hanzo** *et.al.*, "A Portable Multimedia Communicator Scheme", in *R.I. Damper, W. Hall and J.W. Richards (Ed.) Multimedia Technologies and Future Applications*, pp 31-54, Pentech Press, London, 1993

[93] **S.Liu** and **F.M. Wang**, "Hybrid video coding for low bit-rate applications", *Proc. of IEEE ICASSP*, 19-22, pages 481–484, April 1994

[94] **J.F. Arnold X. Zhang** and **M.C. Cavenor**, "Adaptive quadtree coding of motion -compensated image sequences for use on the broadband ISDN", *IEEE Transactions on Circuits and Systems for Video Technology*, Vol 3,No 3, pp 222–229, June 1993

[95] **E. Shustermann** and **M. Feder**, "Image Compression via Improved Quadtree Decomposition Algorithms", *IEEE Transactions on Image Processing*, Vol 3, No 2, pp 207-215, March 1994

[96] *Proceedings of the International Workshop on Coding Techniques for Very Low Bit-rate Video, Hi-Vision Hall, Shinagawa, Tokyo, 8-10 Nov. 1995*

[97] **P. Sheldon, J. Cosmas** and **A. Permain**. "Dynamically adaptive control system for MPEG-4", *Proceedings of the 2nd International Workshop on Mobile Multimedia Communications, Apr. 11-13, Bristol, UK*

[98] **J. Streit**, "Digital Image Compression". *Phd. Thesis*, Dept. of Electr., Univ. of Southampton, 1996

[99] **L. Hanzo** and **J. Streit**, "Adaptive low-rate wireless videophone systems", *IEEE Tr. on Video Technology*, Vol. 5, No. 4, 305-319, ISSN 1051-8215, Aug 1994

[100] **J. Streit** and **L. Hanzo**. "Quadtree-based parametric wireless videophone systems", *IEEE Tr. on Circuit and Systems for Video Technology*, pp 225-237, Apr 1996

[101] **J. Streit** and **L. Hanzo**, "Vector-quantised low-rate cordless videophone systems", Submitted to IEEE Transactions Vehicular Technology

[102] **J. Streit** and **L. Hanzo**, "Comparative study of programmable-rate videophone codecs for existing and future multimode wireless systems", Vol.8, Issue No 6, 1997, pp 551-572

[103] **A.K. Jain**, *Fundamentals of Digital Image Processing*, Prentice-Hall, 1989

[104] **A. Sharaf**: Video coding at very low bit-rates using spatial transformations, *PhD Thesis*, Dept. of Electr. and Electr. Eng., King's College, London, 1997

[105] **P. Cherriman**, "Packet video communications", *Phd Thesis*, Dept. of Electr., Univ. of Southampton, UK, 1999

[106] **P. Cherriman** and **L. Hanzo**, "ARQ-assisted H.261 and H.263-based programable video transceivers", *1995 Research Journal of the Dept. of Electr., Univ. of Southampton, UK, 1996*

[107] **P. Cherriman** and **L. Hanzo**, H261 and H263-based Programable Video Transceivers, *Proc. of ICCS'96/ISPACS'96 Singapore*, 25-29. Nov. 1996, pp 1369-1373

[108] **P. Cherriman** and **L. Hanzo**, Robust H.263 Video Transmission Over Mobile Channels in Interference Limited Environments, *Proc. of First Wireless Image/Video Communications Workshop*, 4-5 Sept, 1996, Loughborough, UK, pp 1-7

[109] **P. Cherriman** and **L. Hanzo**, Power-controlled H.263-based wireless videophone performance in iterference-limited scenarios, *Proc. of Personal, Indoor and Mobile Radio Communications*, PIMRC'96, Taipei, Taiwan, 15-18 Oct., 1996, pp 158-162

[110] **P. Cherriman** and **L. Hanzo**, Programable H.263-based wireless video transceivers for interference-limited environments, *IEEE Tr. on CSVT*, June 1997, Vol. 8, No.3, pp 275-286

[111] **P. Cherriman, T. Keller** and **L. Hanzo**, Orthogonal Frequency Division Multiplex transmission of H.263 encoded video over highly frequency-selective wireless networks http://www-mobile.ecs.soton.ac.uk/peter/robust-h263/robust.html, to appear in IEEE Tr. on CSVT, 1999

[112] **L. Hanzo, P. Cherriman** and **J. Streit**,Modern Video Communications: Principles and applications for fixed and wireless channels, IEEE Press, in preparation [4]

[113] **M. Khansari, A. Jalali, E. Dubois** and **P. Mermelstein**, "Low bit-rate video transmission over fading channels fo wireless microcellular systems", *IEEE TR. on CSVT*, Vol. 6, No. 1, pp 1-11, Feb 1996

[114] **H. Gharavi, H. Yasuda** and **T. Meng**, "Special issue on wireless visual communications", *IEEE Tr. on CSVT*, Vol. 6, No. 2, Apr. 1996

[115] **A.N. Netravali** and **B.G. Haskell**, *Digital pictures: representation and compression*, Plenum Press, 1988

[116] **A. Urie, M. Streeton** and **C. Mourot**, "An advanced TDMA mobile access system for UMTS", *IEEE Comms. Mag.*, pp 38-47, Feb 1995

[117] **European RACE D731 Public Deliverable**, "Mobile communication networks, general aspects and evolution", Sept. 1995

[118] **H. Nyquist**, "Certain factors affecting telegraph speed", *Bell System Tech Jrnl*, p 617, Apr. 1928

[119] **H.R. Raemer**, "Statistical communication theory and applications", Prentice Hall, Inc., Englewood Cliffs, New Jersey, 1969

[120] **N.G.Kingsbury**, "Transmit and receive filters for QPSK signals to optimise the performance on linear and hard limited channels", *IEE Proc. Pt.F*, Vol.133, No.4, pp345-355, July 1996

[121] **K.Feher**, *Digital communications - satellite/earth station engineering*, Prentice Hall 1983

[122] **C.N. Campopiano** and **B.G. Glazer**, "A Coherent Digital Amplitude and Phase Modulation Scheme", *IRE Trans. Commun. Systems*, Vol CS-10, pp 90- 95, 1962

[123] **G.D. Forney** *et.al.*, "Efficient Modulation for Band-Limited Channels", *IEEE Journal on Selected Areas in Communications*, Vol SAC-2, No 5, pp 632-647, Sept 1984

[124] **K. Feher**, "Modems for Emerging Digital Cellular Mobile Systems", *IEEE Tr. on VT*, Vol 40, No 2, pp 355-365, May 1991

[125] **M. Iida** and **K. Sakniwa**, "Frequency Selective Compensation Technology of Digital 16-QAM for Microcellular Mobile Radio Communication Systems", *Proc. of VTC'92, Denver, Colorado*, pp 662-665

---

[4]For detailed contents please refer to http://www-mobile.ecs.soton.ac.uk

[126] **R.J. Castle** and **J.P. McGeehan**, "A Multilevel Differential Modem for Narrowband Fading Channels", *Proc. of VTC'92, Denver, Colorado*, pp 104-109

[127] **D.J. Purle, A.R. Nix, M.A. Beach** and **J.P. McGeehan**, "A Preliminary Performance Evaluation of a Linear Frequency Hopped Modem", *Proc. of VTC'92, Denver, Colorado*, pp 120-124

[128] **Y. Kamio** and **S. Sampei**, "Performance of Reduced Complexity DFE Using Bidirectional Equalizing in Land Mobile Communications", *Proc. of VTC'92, Denver, Colorado*, pp 372-376

[129] **T. Nagayasu, S. Sampei** and **Y. Kamio**, "Performance of 16-QAM with Decision Feedback Equalizer Using Interpolation for Land Mobile Communications", *Proc. of VTC'92, Denver, Colorado*, pp 384-387

[130] **E. Malkamaki**, "Binary and Multilevel Offset QAM, Spectrum Efficient Modulation Schemes for Personal Communications", *Proc. of VTC'92, Denver, Colorado*, pp 325-378

[131] **Z. Wan** and **K. Feher**, "Improved Efficiency CDMA by Constant Envelope SQAM", *Proc. of VTC'92, Denver, Colorado*, pp 51-55

[132] **H. Sasaoka**, "Block Coded 16-QAM/TDMA Cellular Radio System Using Cyclical Slow Frequency Hopping", *Proc. of VTC'92, Denver, Colorado*, pp 405-408

[133] **W.T. Webb, L. Hanzo** and **R. Steele**, "Bandwidth-Efficient QAM Schemes for Rayleigh-Fading Channels", *IEE Proceedings*, Vol 138, No 3, pp 169- 175, June 1991

[134] **A.S. Wright** and **W.G. Durtler**, "Experimental Performance of an Adaptive Digital Linearized Power Amplifier", *IEEE Tr. on VT*, Vol 41, No 4, pp 395- 400, Nov. 1992

[135] **M. Faulkner** and **T. Mattson**, "Spectral Sensitivity of Power Amplifiers to Quadrature Modulator Misalignment", *IEEE Tr. on VT*, Vol 41, No 4, pp 516-525, Nov. 1992

[136] **P.B. Kennington** *et.al.*, "Broadband Linear Amplifier Design for a PCN Base-Station", *Proc. of 41st IEEE VTC*, pp 155-160, May 1991

[137] **R.J. Wilkinson** *et.al.*, "Linear Transmitter Design for MSAT Terminals", *2nd Int. Mobile Satellite Conference*, June 1990

[138] **J.G.Proakis**, *Digital Communications*, McGraw Hill 1983

[139] **Y. C. Chow, A. R. Nix** and **J. P. McGeehan**, "Analysis of 16-APSK modulation in AWGN and Rayleigh fading channel", *Electronic Letters*, Vol 28, pp 1608-1610, Nov 1992

[140] **M.L. Moher** and **J.H. Lodge**, "TCMP – A Modulation and Coding Strategy for Rician Fading Channels", *IEEE J. Select. Areas Commun.*, Vol. 7, pp. 1347-1355, Dec. 1989

[141] **S. Sampei** and **T. Sunaga**, "Rayleigh Fading Compensation Method for 16-QAM in Digital Land Mobile Radio Channels", *Proc. IEEE Veh. Technol. Conf.*, San Francisco, CA, pp. 640-646, May 1989

[142] **S. Haykin**, *Adaptive Filter Theory*, Prentice Hall, 1991

[143] **J. Torrance** and **L. Hanzo**, "A comparative study of pilot symbol assisted modem schemes", *Proc. of IEE RRAS'95 Conference*, 26-28 Sept., 1995, Bath, UK, Conf. Public. No. 415, pp 36- 41

[144] **R. Steele** and **W.T. Webb**, "Variable rate QAM for data transmission over Rayleigh fading channels," in *Wireless '91, Calgary, Alberta*, pp. 1–14, IEEE, 1991.

[145] **W. Webb** and **R. Steele**, "Variable rate QAM for mobile radio", *IEEE Transactions on Communications*, 1995, Vol. 43, No. 7, pp 2223-2230

[146] **Y. Kamio, S. Sampei, H. Sasaoka** and **N. Morinaga**, "Performance of modulation-level-control adaptive-modulation under limited transmission delay time for land mobile communications," in 45th *Vehicular Technology Conference*, pp. 221–225, IEEE, 1995.

[147] **S. Sampei, S. Komaki** and **N. Morinaga**, Adaptive Modulation/TDMA Scheme for Large Capacity Personal Multi-Media Communication Systems, *IEICE Transactions on Communications*, 1994, Vol. 77. No. 9, pp 1096-1103

[148] **M. Morimoto, H. Harada, M. Okada** and **S. Komaki**, A study on power assignment of hierarchical modulation schemes for digital broadcasting, *IEICE Transactions on Communications*, 1994, Vol. 77, No. 12, pp 1495-1500

[149] **S. Otsuki, S. Sampei** and **N. Morinaga**, "Square-QAM adaptive modulationTDMA/TDD systems using modulation level estimation with Walsh function.*Electronics Letters*, November 1995. pp 169-171.

[150] **S.G. Chua** and **A. Goldsmith**, "Variable-rate variable-power mqam for fading channels," in 46th *Vehicular Technology Conference*, pp. 815–819, IEEE, 1996.

[151] **A. Goldsmith** and **S. Chua**, "Variable-rate variable-power MQAM for fading channels," *to appear in IEEE Tr. on Veh. Techn.*, 1997. http://www.systems.caltech.edu.

[152] **A. Goldsmith**, "The capacity of downlink fading channels with variable rate and power," *IEEE Tr. on Veh. Techn.*, vol. 46, pp. 569–580, Aug. 1997.

[153] **A. Goldsmith** and **P. Varaiya**, "Capacity of fading channels with channel side information," *to appear in IEEE Tr. in Inf. Theory*, 1998. http://www.systems.caltech.edu.

[154] **M.-S. Alouini** and **A. Goldsmith**, "Capacity of fading channels under different adaptive transmission and diversity techniques," *submitted to IEEE Tr. on Veh. Techn.*, 1998. http://www.systems.caltech.edu.

[155] **M.-S. Alouini** and **A. Goldsmith**, "Area spectral efficiency of cellular mobile radio systems," *submitted to IEEE Tr. on Veh. Techn.*, 1998. http://www.systems.caltech.edu.

[156] **A. Goldsmith** and **S. Chua**, "Adaptive coded modulation," *submitted to IEEE Tr. on Communications*, 1998. http://www.systems.caltech.edu.

[157] **D. A. Pearce, A. G. Burr** and **T. C. Tozer**, "Comparison of counter-measures against slow Rayleigh fading for TDMA systems," in *Colloquium on Advanced TDMA Techniques and Applications*, pp. 9/1–9/6, IEEE, 1996.

[158] **J. M. Torrance** and **L. Hanzo**, "Upper bound performance of adaptive modulation in a slow Rayleigh fading channel." *Electronics Letters*, April 1996. pp 169-171.

[159] **J. M. Torrance** and **L. Hanzo**, "Adaptive modulation in a slow Rayleigh fading channel," in *Proceedings of the 7th Personal, Indoor and Mobile Radio Communications (PIMRC) Conference*, pp. 497–501, IEEE, 1996.

[160] **J. M. Torrance** and **L. Hanzo**, "Optimisation of switching levels for adaptive modulation in a slow Rayleigh fading channel." *Electronics Letters*, June 1996. pp 1167 - 1169.

[161] **J.M. Torrance** and **L. Hanzo**, Demodulation level selection in adaptive modulation, *Electr. Letters*, Vol. 32, No. 19, 12th of Sept., 1996, pp 1751-1752

[162] **J. Torrance** and **L. Hanzo**, Latency considerations for adaptive modulation in slow Rayleigh fading, *Proc. of IEEE VTC'97*, Phoenix, USA, 1997, pp 1204-1209

[163] **J. Torrance** and **L. Hanzo**, Statistical Multiplexing for Mitigating Latency in Adaptive Modems, *PIMRC'97*, 1-4. Sept. 1997, Helsinki, Finland, pp 938-942

[164] **M-S Alouini** and **A. Goldsmith**, Area spectral efficiency of cellular mobile radio systems, *Proc. of IEEE VTC'97*, Phoenix, USA, pp 652-656

[165] **M-S Alouini** and **A. Goldsmith**, Capacity of Nakagami Multipath Fading Channels, *Proc. of IEEE VTC'97*, Phoenix, USA, pp 652-656

[166] **C.H. Wong** and **L. Hanzo**, Upper-bound Performance of a Wideband Burst-by-burst Adaptive Modem, *Proc. of VTC'99, 1999, Houston, USA*

[167] **T.H. Liew, C.H. Wong** and **L. Hanzo**, Block turbo coded burst-by-burst adaptive modems; *Proceeding of Microcoll'99*, Budapest, Hungary, 21-24 March, 1999

[168] **C.H. Wong, T.H. Liew** and **L. Hanzo**, Blind modem mode detection aided block turbo coded burst-by-burst wideband adaptive modulation; *Proceedings of ACTS'99*, Sorrento, Italy, June, 1999

[169] **C.H. Wong, T.H. Liew** and **L. Hanzo**, Blind-detection Assisted, Block Turbo Coded, Decision-feedback Equalised Burst-by-Burst Adaptive Modulation; *submitted to IEEE Tr. on Comms.*, 1998

[170] **M. Frullone, G. Riva, P. Grazioso** and **C. Carciofy**, "Investigation on Dynamic Channel Allocation Strategies Suitable for PRMA Schemes", *1993 IEEE Int. Symp. on Circuits and Systems, Chicago*, pp 2216-2219, May 1993

[171] **S.P. Stapleton** and **F.C. Costescu**, "An Adaptive Predistorter for a Power Amplifier Based on Adjacent Channel Emissions", *IEEE Tr. on VT*, Vol 41, No 1, pp 49-57, Febr. 1992

[172] **S.P. Stapleton, G.S. Kandola** and **J.K. Cavers**, "Simulation and Analysis of an Adaptive Predistorter Utilizing a Complex Spectral Convolution", *IEEE Tr. on VT*, Vol 41, No 4, pp 387-394, Nov. 1992

[173] **Norihiko Morinaga**, "Advanced wireless communication technologies for achieving high-speed mobile radios", *IEICE Transactions on Communications*, vol. 78, no. 8, pp. 1089–1094, 1995

[174] **J.M. Torrance** and **L. Hanzo**, "Upper bound performance of adaptive modulation in a slow Rayleigh-fading channel", *Electr. Letters*, Vol. 32, No. 8, pp 718-719, 11th Apr. 1996

[175] **J.M. Torrance** and **L. Hanzo**, "Optimisation of Switching Levels for Adaptive Modulation in Slow Rayleigh Fading", *Electr. Letters*, Vol. 32, 20th June, 1996, No. 13, pp 1167-1169

[176] **J.M. Torrance** and **L. Hanzo**, "Performance upper bound of adaptive QAM in slow Rayleigh-fading environments", *Proc. of ICCS'96/ISPAC'96 Singapore*, Nov. 1996, pp 1653-1657

[177] **J. Torrance** and **L. Hanzo**, "Adaptive modulation in a slow Rayleigh fading channel", Proc. of Personal, Indoor and Mobile Radio Communications, *PIMRC'96*, Taipei, Taiwan, 15-18 Oct., 1996, pp 1049-1053

[178] **J.K. Cavers**, "An Analysis of Pilot Symbol Assisted Modulation for Rayleigh Fading Channels", *IEEE Tr. on VT*, Vol 40, No 4, pp 686-693, Nov 1991

[179] **J.M. Torrance**, " Adaptive Full Response Digital Modulation for Wireless Communication Environments", *Phd Thesis*, Dept. of Electr., Univ. of Southampton, UK, 1997

[180] **J.P.McGeehan** and **A. Bateman**, "Phase-locked transparent tone in band (TTIB): A new spectrum configuration particularly suited to the transmission of data over SSB mobile radio networks", *IEEE Trans Comm, Vol.COM-32*, pp81-87, 1984

[181] **A.Bateman** and **J.P.McGeehan**, "Feedforward transparent tone in band for rapid fading protection in multipath fading", *IEE Int. Conf. Comms.* Vol.68, pp9-13, 1986

[182] **A.Bateman** and **J.P.McGeehan**, "The use of transparent tone in band for coherent data schemes", *IEEE Int. Conf. Comms., Boston, Mass.*, 1983

[183] **A.Bateman, G.Lightfoot, A.Lymer** and **J.P.McGeehan**, "Speech and data transmissions over a 942MHz TAB and TTIB single sideband mobile radio system", *IEEE Trans Veh. Tech., Vol.VT-34* pp13-21, 1985

[184] **A.Bateman** and **J.P.McGeehan**, "Data transmissions over UHF fading mobile radio channels", *Proc IEE Pt.F, Vol.131*, 1984, pp364-374

[185] **J.P.McGeehan** and **A.Bateman**, "A simple simultaneous carrier and bit synchronisation system for narrowband data transmissions", *Proc. IEE, Pt.F, Vol.132*, pp69-72, 1985

[186] **J.P.McGeehan** and **A.Bateman**, "Theoretical and experimental investigation of feedforward signal regeneration", *IEEE Trans. Veh. Tech., Vol.VT-32*, pp106-120, 1983

[187] **A.Bateman**, "Feedforward transparent tone in band: Its implementation and applications", *IEEE Trans. Veh. Tech.* Vol.39, No.3, pp235-243, Aug 1990

[188] **B.Sklar**, "Digital communications", *Prentice Hall*, 1988

[189] **F. Adachi**, "Error rate analysis of differentially encoded and detected 16APSK under Rician fading", *IEEE Tr. on Veh. Techn.*, Vol. 45, No. 1, pp 1-12, Feb 1996

[190] **J.K.Cavers**, "The performance of phase locked transparent tone in band with symmetric phase detection", *IEEE Trans. on Comms.*, Vol.39, No.9, pp1389-1399, Sept 1991

[191] **S. Nanda, D.J. Goodman** and **U. Timor**, "Performance of PRMA: A Packet Voice Protocol for Cellular Systems", *IEEE Tr. on VT*, Vol 40, No 3, , pp 584-598, Aug 1991

[192] **M. Frullone, G. Falciasecca, P. Grazioso, G. Riva, A. M.** and **Serra**, "On the performance of packet reservation multiple access with fixed and dynamic channel allocation", *IEEE Tr. on Veh. Techn.*, Vol. 42. No. 1, pp 78-86, Feb 1996

[193] **Goodman, D.J., Lockhart, G.B., Wasem, O.J.,** and **Wong, W.C.**, "Waveform substitution techniques for recovering missing speech segments in packet voice communications", *IEEE Tr. on ASSP*, vol. 34, pp 1440-1448, Dec. 1986

[194] **R.Steele** and **W. T. Webb**, "Variable rate QAM for data transmissions over mobile radio channels", *Keynote paper, Wireless 91, Calgary Alberta*, July 1991

[195] **L. Hanzo** and **H. Yuen**: "Robust programmable-rate wireless graphical correspondence", *European Tr. on Communications*, No. 3, Vol. 8, May-June, 1997, pp 271-283

[196] **J.C. Arnbak, J.H. Bons** and **J.W. Vieveen**: "Graphical correspondence in electronic-mail networks using personal computers," *IEEE Journal on Selected Areas in Communications*, Feb. 1989, SAC-7, pp.257-267

[197] **M. Klerk, R. Prasad, J.H. Bons** and **N.B.J. Weyland**: "Introducing high-resolution line graphics in UK teletext using differential chain coding," *IEE Proceedings*, Part I, vol. 137, pp. 325-334, Dec. 1990.

[198] **D.L. Neuhoff** and **K.G. Castor**: "A rate and distortion analysis of chain codes for line drawings, " *IEEE Tran. Information Theory*, vol. IT-31, pp. 53-68, Jan. 1985.

[199] **K. Liu** and **R. Prasad**: "Performance analysis of differential chain coding," *European Transaction on Telecommunications and Related Technology*, vol. 3, pp. 323-330, Jul-Aug. 1992

[200] **A. B. Johannessen, R. Prasad, N.B.J. Weyland** and **J.H. Bons**: "Coding efficiency of multiring differential chain coding," *IEE Proceedings*, Part I, vol. 139, pp. 224-232, April 1992.

[201] **R. Prasad, J.W. Vivien, J.H. Bons** and **J.C. Arnbak**: "Relative vector probabilities in differential chain coded line drawings", *Proc. of IEEE Pacific Rim Conference on Comms., Computers and Signal Processing*, pp 138-142, Victoria, Canada, June 1989

[202] **L. Yang, H.I. Cakil** and **R. Prasad**: "On-line handwritting processing in indoor and outdoor radio environment for multimedia", *Proc. of IEEE VTC'94*, Stockholm, Sweden, 1015-1019

[203] **R. Prasad, P.A.D. Spaargaren** and **J.H. Bons**: "Teletext reception in a mobile channel for a broadcast tele-information system", *IEEE Tr. on VT.*, Vol. 42, No. 4, Nov. 1993, pp 535-545

[204] **L-P.W. Niemel** and **R. Prasad**: "A novel description of handwritten characters for use with generalised Fourier descriptors", *ETT* Vol. 3, No. 5, Sept-Oct. 1992, pp 455-464

[205] **L-P.W. Niemel** and **R. Prasad**: "An improved character description method based on generalized Fourier descriptors", *ETT* Vol. 5, No. 3, May-June 1994, pp 371-376

[206] **D.J. Goodman** and **X.S. Wei**: Efficiency of packet reservation multiple access, *IEEE Transactions on vehicular technology*, Vol. 40, No. 1, Febr. 1991, pp. 170-176.

[207] **W. Wong** and **D. Goodman**: A packet reservation access protocol for integrated speech and data transmission, *Proc. of the IEE*, Part-I, Dec. 1992, Vol 139, No 6, pp 607-613.

Glossary

ACELP	Algebraic Code Excited Linear Prediction
ACTS	Advanced Communications Technologies and Services
ADC	Analogue Digital Converter
ADPCM	Adaptive Differential Pulse Code Modulation
ATDMA	Advanced Time Division Multiple Access
ATM	Asynchronous Transfer Mode
AWGN	Additive White Gaussian Noise
B-frames	Bi-directional-frames
B-ISDN	Broadband-ISDN
BCH	Bose-Chaudhuri-Hocquenghem, A class of forward error correcting codes (FEC)
BER	Bit Error Rate, the number of the bits received incorrectly
BPSK	Binary Phase Shift Keying
BS	Base Station
CD	Cepstral Distance
CD-DEG	Cepstral Distance Degradation
CDMA	Code Division Multiple Access
CELP	Code Excited Linear Prediction
CT2	British Cordless Telephone System
DAC	Digital Analogue Converter
DCT	The Discrete Cosine Transform, transforms data into the frequency domain. Commonly used for video compression by removing high frequency components in the video frames

DECT	Digital European Cordless Telephone
DFD	Displaced Frame Difference
DoD	US Department of Defence
EG	Excitation Gain
FAW	Frame Alignment Word
FEC	Forward Error Correction
FPLMTS	Future Public Land Mobile Telecommunications System
G.728	ITU 16 kbps speech coding standard
G.729	ITU 8 kpbs speech coding standard
GMSK	Gaussian Minimum Shift Keying
GOS	Grade of Service
GP	Grid Position
GSM	A Pan-European digital mobile radio standard, operating at 900 MHz.
H.261	A video coding standard [74], published by the ITU in 1990
H.263	A video coding standard [72], due to be published by the ITU in 1996
I-component	Inphase-component
IF	Intermediate Frequency
ISDN	Integrated Services Digital Network, digital replacement of the analogue telephone network
ISI	Inter Symbol Interference
ITU	International Telecommunications Union, formerly the CCITT, standardisation group
IZFPE	Interpolated Zinc Function Pulse Excitation
LP filtering	Low-Pass filtering
LPF	Low-Pass Filter
LSB	Least Significant Bit
LSF	Line Spectra Frequency
LTP	Long Term Prediction
LTPD	Long Term Prediction Delay
LTPG	Long Term Prediction Gain
MAVT	Mobile Audio Video Terminal
MBE	Multi Band Excitation

MC	Motion Compensation
MCER	Motion Compensated Error Residual
MELP	Mixed Excitation Linear Prediction
MOS	Mean Opinion Score
MPEG	Motion Picture Expert Group, also a video coding standard designed by this group that is widely used
MSB	Most Significant Bit
MV	Motion Vector, a vector to estimate the motion in a frame
NLF	Non-Linear Filtering
NTT	Nippon Telegraph and Telephone Company
OFDM	Orthogonal Frequency Division Multiplexing
P-frames	Predicted-frames
PCM	Pulse Code Modulation
PCN	Personal Communications Network
PDC	Personal Digital Cellular
PHP	Personal Handy Phone
PLMR	Public Land Mobile Radio
PRMA	Packet Reservation Multiple Access
PS	Portable Station
PSAM	Pilot Symbol Assisted Modulation, a technique where known symbols (pilots) are transmitted regularly. The effect of channel fading on all symbols can then be estimated by interpolating between the pilots
PSD	Power Spectral Density
PSI	Pitch Synchronous Innovation
PWI	Prototype Waveform Interpolation
Q-component	Quadrature-component
QAM	Quadrature Amplitude Modulation
QCIF	Quarter Common Intermediate Format Frames containing 176 pixels vertically and 144 pixels horizontally
QT	Quad Tree
RACE	Research in Advanced Communications Equipment

RC filtering	Raised Cosine filtering
RF	Radio Frequency
RPE	Regular Pulse Excitation
SBC	Sub Band Coding
SEGSNR	Segmental Signal-to-Noise Ratio
SEGSNR-DEG	Segmental Signal-to-Noise Ratio Degradation
SNR	Signal to Noise Ratio, noise energy compared to the signal energy
TBPE	Transform Binary Pulse Excitation
TC	Transform Coding
TCM	Trellis Coded Modulation
TDMA	Time Division Multiple Access
TTIB	Transparent Tone in Band
UMTS	Universal Mobile Telecommunications System
V.29-V.34	ITU Data Transmission Modems
VAD	Voice Activity Detection
VQ	Vector Quantisation
VSELP	Vector Sum Excited Linear Prediction
WLAN	Wireless Local Area Network

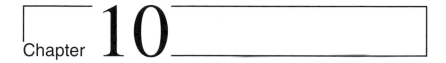

Chapter 10

Third-Generation Wireless Systems

K. Yen and L. Hanzo[1]

10.1 Introduction

The evolution of third-generation (3G) systems began in the late 1980s, when the International Telecommunication Union's - Radiocommunication Sector (ITU-R) Task Group (TG) 8/1 defined the requirements for the 3G mobile radio systems. This initiative was then known as Future Public Land Mobile Telecommunication System (FPLMTS) [1,2]. This led to the identification of the frequency spectrum for FPLMTS on a world-wide basis during the World Administrative Radio Conference (WARC) in 1992 [2], as the bands 1885-2025 MHz and 2110-2200 MHz - an issue to be detailed in the context of Figure 10.1 during our further discourse.

The tongue-twisting acronym of FPLMTS was also aptly changed to IMT-2000, which refers to the International Mobile Telecommunications system in the year 2000. Besides possessing the ability to support services from rates of a few kbps to as high as 2 Mbps in a spectrally efficient way, IMT-2000 aimed to provide a seamless global radio coverage for global roaming. This implied the ambitious goal of aiming to connect virtually any two mobile terminals world-wide. The IMT-2000 system is aiming to be flexible in order to operate in any propagation environment, such as indoor, outdoor to indoor and vehicular scenarios. It is also aiming to be sufficiently flexible to handle so-called circuit as well as packet mode

---

[1]University of Southampton and Multiple Access Communications Ltd

services and to handle services of variable data rates. In addition, these requirements must be fulfilled with a quality of service (QoS) comparable to that of the current wired network at an affordable cost.

Several regional standard organizations - led by the European Telecommunications Standards Institute (ETSI) in Europe, by the Association of Radio Industries and Businesses (ARIB) in Japan and by the Telecommunications Industry Association (TIA) in the United States - have been dedicating their efforts to specifying the standards for IMT-2000. A total of 15 Radio Transmission Technology (RTT) IMT-2000 proposals were submitted to ITU-R in June 1998, five of which are satellite based solutions, while the rest are terrestrial solutions. Table 10.1 shows a list of the terrestrial-based proposals submitted by the various organizations and their chosen radio access technology. Although Table 10.1 is very informative, reflecting a variety of views across the wireless research community, which span the range of cordless telephony-based solutions, such as the Digital European Cordless Telecommunications (DECT) system of Table 10.1, or the second-generation (2G) IS-136 system, here we will concentrate on the CDMA-based solutions. A rudimentary discussion on CDMA was provided in Chapter 1 in the context of the Pan-American IS-95 system and in this chapter a basic familiarity with CDMA principles is assumed.

It transpires from Table 10.1 that most standardization bodies have based their terrestrial oriented solutions on Wideband-CDMA (W-CDMA), due to its previously mentioned advantageous properties, which satisfy most of the requirements for 3G mobile radio systems. W-CDMA is aiming to provide improved coverage in most propagation environments in addition to an increased user capacity. Furthermore, it simplifies frequency planning due to its unity frequency reuse. As argued in Chapter 1, CDMA has the ability to combat - or in fact to benefit from - multipath fading through RAKE multipath diversity combining [3–5]. Hence, in this chapter we will concentrate on the proposed terrestrial transmission technologies advocated by the three major regional standardization bodies, namely ETSI, ARIB and TIA, whereby the access technology is based on W-CDMA. The corresponding systems have been termed UMTS Terrestrial Radio Access (UTRA), Wideband-CDMA (W-CDMA) and cdma2000, respectively. In order to avoid confusion between the Japanese W-CDMA proposal itself and the W-CDMA access technology in general, we shall refer to the ARIB's RTT as IMT-2000.

Since ETSI and ARIB have been harmonizing their standardization efforts, aiming for the same W-CDMA technology, their proposals became indeed very similar and hence we shall discuss these two RTT proposals collectively in Section 10.2. The RTT of cdma2000 will be highlighted separately in Section 10.3. At the time of writing, ITU are still deliberating on their decisions concerning a single global standard. Although the proposals have been submitted, active research is still in progress in order to improve and optimize the systems. Hence, the parameters and technolo-

Proposal	Description	Access technology	Source
DECT	Digital Enhanced Cordless Telecommunications	Multicarrier TDMA (TDD)	ETSI Project (EP) DECT
UWC-136	Universal Wireless Communications	TDMA (FDD and TDD)	USA TIA TR45.3
WIMS W-CDMA	Wireless Multimedia and Messaging Services Wideband CDMA	Wideband CDMA (FDD)	USA TIA TR46.1
TD-CDMA	Time-Division Synchronous CDMA	Hybrid with TDMA/CDMA/ SDMA (TDD)	Chinese Academy of Telecommunication Technology (CATT)
W-CDMA	Wideband CDMA	Wideband DS-CDMA (FDD and TDD)	Japan ARIB
CDMA II	Asynchronous DS-CDMA	DS-CDMA (FDD)	South Korean TTA
UTRA	UMTS Terrestrial Radio Access	Wideband DS-CDMA (FDD and TDD)	ETSI SMG2
NA: W-CDMA	North America Wideband CDMA	Wideband DS-CDMA (FDD and TDD)	USA T1P1-ATIS
cdma2000	Wideband CDMA (IS-95)	DS-CDMA (FDD and TDD)	USA TIA TR45.5
CDMA I	Multiband synchronous DS-CDMA	Multiband DS-CDMA	South Korean TTA

Table 10.1: Proposals for the radio transmission technology of terrestrial IMT-2000 (obtained from ITU's web site : http://www.itu.int/imt).

gies presented in this chapter may evolve further. It should also be noted that this chapter serves as an overview of the three main proposed systems. Readers may want to refer to a recent book by Ojanperä and Prasad [6], which addresses W-CDMA 3G mobile radio systems in more depth. Again, here we assume that the reader is familiar with the basic CDMA principles.

10.2   UMTS/IMT-2000 Terrestrial Radio Access [7]- [14]

Universal Mobile Telecommunications System (UMTS) is the term introduced by the ETSI/Special Group Mobile (SMG) for the 3G wireless mobile communication system in Europe [1, 8, 9, 15–18]. Research activities for UMTS within ETSI have been spearheaded by the European Union's (EU) sponsored programmes, such as the Research in Advanced Communication Equipment (RACE) [19, 20] and the Advanced Communications Technologies and Services (ACTS) [8, 16, 20] initiative. The RACE programme, which comprised of two phases, was started in 1988 and ended in 1995. The objective of this programme was to investigate and develop testbeds for the air interface technology candidates. The ACTS programme succeeded the RACE programme in 1995. Within the ACTS Future Radio Wideband Multiple Access System (FRAMES) project two multiple access modes have been chosen for intensive study, as the candidates for UMTS terrestrial radio access (UTRA). They are based on Time Division Multiple Access (TDMA) with and without spreading, and on W-CDMA [7, 21, 22].

As early as January 1997, ARIB decided to adopt W-CDMA as the terrestrial radio access technology for their IMT-2000 proposal and proceeded to focus their activities towards the detailed specifications of this technology [18]. Driven by a strong support behind W-CDMA worldwide and this early decision from ARIB, a consensus agreement was reached by ETSI in January 1998 to adopt W-CDMA as the terrestrial radio access technology for UMTS. Since then, ARIB and ETSI have harmonized their standards in order to aim for the same W-CDMA technology. In this section we will highlight the key features of the terrestrial RTT behind the ETSI and ARIB proposals. The descriptions that follow are applicable to both UTRA and IMT-2000, unless it is stated otherwise. Most of the material in this section is based on an amalgam of References [8]- [9].

10.2.1   Characteristics of UTRA/IMT-2000

The proposed spectrum allocations for UTRA and IMT-2000 are shown in Figures 10.1 and 10.2, respectively. As can be seen, UTRA and IMT-2000 are unable to utilize the full allocated frequency spectrum for 3G mobile radio systems, since those frequency bands have also been partially allocated to the DECT and Personal Handyphone System (PHS) , respectively, which were characterized in Table 1.1 of Chapter 1. The radio access supports both Frequency Division Duplex (FDD) and Time Division Duplex (TDD) operations. The operating principles of these two schemes were detailed in Chapter 1, which are augmented here in the context of Figure 10.3.

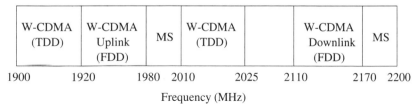

MS : Mobile satellite application

Figure 10.1: The proposed spectrum allocation in UTRA.

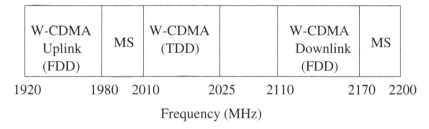

MS : Mobile satellite application

Figure 10.2: The proposed spectrum allocation in IMT-2000.

Specifically, the uplink and downlink signals are transmitted using different carrier frequencies f_1 and f_2, respectively, separated by a frequency guard band in FDD mode. On the other hand, the uplink and downlink messages in the TDD mode are transmitted using the same carrier frequency f_c, but in different time-slots, separated by a guard time. As seen from the spectrum allocation in Figures 10.1 and 10.2, the paired bands of 1920-1980 MHz and 2110-2170 MHz are allocated for FDD operation in the uplink and downlink, respectively, whereas the TDD mode is operated in the remaining unpaired bands [8]. However, in case of asymmetric services, such as for example computer file downloading or video on demand (VoD), only one of the FDD bands is required and hence the more flexible TDD link could potentially double the link's capacity by allocating all time-slots in one direction. The parameters designed for FDD and TDD operations are such that they are mutually compatible so as to ease the implementation of a dual-mode terminal capable of accessing the services offered by both FDD and TDD operators. We note furthermore that recent research advocates

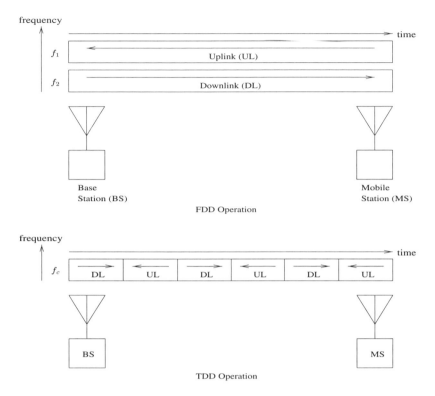

Figure 10.3: Principle of FDD and TDD operation.

the TDD mode quite strongly in the context of burst-by-burst adaptive CDMA modems [23, 24], since the uplink-downlink reciprocity can be advantageously exploited in order to adjust the modem parameters, such as the spreading factor or the number of bits per symbol on a burst-by-burst basis. This allows the system to more efficiently exploit the time variant wireless channel capacity, hence maintaining a higher bits/s/Hz bandwidth efficiency.

Table 10.2 shows the basic parameters of the UTRA/IMT-2000 proposals. Some of these parameters are discussed during our further discourse, but significantly more information can be gleaned concerning these systems by carefully studying the table. It is also informative to compare these parameters to the IS-95 CDMA system parameters of Table 1.1. Both systems are operated at a basic chip rate of 4.096 Mcps, giving a nominal bandwidth of 5 MHz, when using root-raised cosine Nyquist pulse shaping filters (see Chapter 9) with a roll-off factor of 0.22. IMT-2000 has an additional lower chip rate of 1.024 Mcps, corresponding to a bandwidth of 1.25 MHz. Increased chip rates of 8.192 Mcps and 16.384 Mcps are also specified in

Radio access technology	FDD : DS-CDMA
	TDD : TDMA/CDMA
Operating environments	Indoor/Outdoor to indoor/Vehicular
Chip rate (Mcps)	UTRA : 4.096/8.192/16.384
	IMT-2000 : 1.024/4.096/8.192/16.384
Channel bandwidth (MHz)	UTRA : 5/10/20
	IMT-2000 : 1.25/5/10/20
Nyquist roll-off factor	0.22
Duplex modes	FDD and TDD
Channel bit-rates (kbps)	FDD (UL) : 16/32/64/128/256/512/1024
	FDD (DL) : 32/64/128/256/512/1024/2048
	TDD (UL/DL) : 512/1024/2048/4096
Frame length	10 ms
Spreading factor	FDD : variable, 4 to 256
	TDD : variable, 2 to 16
Detection scheme	Coherent with time-multiplexed pilot symbols
Inter-cell operation	FDD : Asynchronous
	TDD : Synchronous
Power control	Open and closed-loop
Transmit power dynamic range	80 dB (UL), 30 dB (DL)
Handover	Soft handover
	Inter-frequency handover

Table 10.2: UTRA/IMT-2000 basic parameters.

order to cater for much higher user bit-rates ($>$ 2 Mbps).

UTRA/IMT-2000 fulfilled the requirements of 3G mobile radio systems by offering a range of user bit-rates up to 2 Mbps. Various services having different bit-rates and quality of service (QoS) can be readily supported using so-called Orthogonal Variable Spreading Factor (OVSF) codes, which will be highlighted in Section 10.2.6.1, and service multiplexing which will be discussed in Figure 10.12. As opposed to the common pilot channel of the second-generation IS-95 system, which was portrayed in Chapter 1, the third-generation UTRA / IMT-2000 systems invoked dedicated pilot symbols embedded in the users' data-stream. These can be invoked in order to support the operation of adaptive antennae at the base station, which was not facilitated by the common pilot channel of the IS-95 system.

Irrespective of whether a common pilot channel is used or dedicated pilots are embedded in the data, they facilitate the employment of coherent detection. Coherent detection is known to provide better performance, than non-coherent detection [26], a fact also argued in Chapter 9. Furthermore, the inclusion of short spreading codes enables the implementation of various performance enhancement techniques, such as interference cancellers and joint-detection algorithms. In order to support flexible system deployment in indoor and outdoor environments, **inter-cell-asynchronous operation** is used in the FDD mode. This implies that no external timing source, such as a beacon or Global Positioning System (GPS) is required. However, in the TDD mode inter-cell synchronization is required in order to

be able to seamlessly access the time-slots offered by adjacent Base Stations
(BS) during handovers. This is achieved by maintaining synchronization
between the base stations.

Radio access is concerned mainly with the physical layer of the International Standardization Organization/Open Systems Interconnection
(ISO/OSI) Reference Model. Hence, in the following sections we will mainly
concentrate on the physical layer of the UTRA/IMT-2000 proposals. We
note here, furthermore, that there have been proposals in the literature for
allowing TDD operation also in certain segments of the FDD spectrum,
since FDD is incapable of surrendering the uplink or downlink frequency
band of the duplex link, when the traffic demand is basically simplex. In
fact, segmenting the spectrum in FDD/TDD bands inevitably results in
some inefficiency in bandwidth utilization terms, especially in the case of
asymmetric or simplex traffic. Hence in reference [27] the idea of eliminating the dedicated TDD band was investigated, where TDD was invoked
within the FDD band by simply allowing TDD transmissions in either the
uplink or downlink frequency band, depending on which one was less interfered with. This flexibility is unique to CDMA, since as long as the amount
of interference is not excessive, FDD and TDD can share the same bandwidth. This would be particularly feasible in the indoor scenario of [27],
where the surrounding outdoor cell could be using FDD, while the indoor
cell would reuse the same frequency band in TDD mode. The buildings'
walls and partitions could mitigate the interference between the FDD/TDD
schemes.

10.2.2   Transport Channels

Transport channels are offered by the physical layer to the higher OSI layers
and they can be classified into two main groups, as shown in Table 10.3 [7,8].
The Dedicated transport CHannel (DCH) is related to a specific mobile
station-base station link and it is used to carry user and control information between the network and a mobile station. Hence the DCHs are bidirectional channels. There are four transport channels within the common
transport channel group, as shown in Table 10.3. The Broadcast Control
Channel (BCCH) is used to carry system- and cell-specific information on
the downlink (DL) to all mobiles over the entire cell. This channel conveys
information, such as the downlink transmit power of the base station and
the uplink (UL) interference power measured at the base station, which are
vital for the mobile station in adjusting its transmit power required for the
target Signal-to-Interference plus Noise Ratio (SINR) of the base station,
as we shall see in Section 10.2.8. The Forward Access Channel (FACH) of
Table 10.3 is a downlink common channel used for carrying control information and short user data packets to mobile stations, if the system knows
the serving base station of the mobile station. On the other hand, the
Paging Channel (PCH) of Table 10.3 is used to carry control information

Dedicated transport channel	Common transport channel
Dedicated Channel (DCH) (UL/DL)	Broadcast Control Channel (BCCH) (DL)[†]
	Forward Access Channel (FACH) (DL)
	Paging Channel (PCH) (DL)
	Random Access Channel (RACH) (UL)

[†]In IMT-2000, this is known as Broadcast Channel (BCH)

Table 10.3: UTRA/IMT-2000 transport channels.

to a mobile station, when the serving base station of the mobile station is unknown in order to page the mobile station, when there is a call for the mobile station. The Random Access Channel (RACH) of Table 10.3 is an uplink channel used by the mobile station to carry control information and short user data packets to the base station in order to support the mobile station's access to the system, when it wishes to set up a call.

The philosophy of these channels is fairly plausible and it is informative and enlightening to explore the differences between the somewhat less flexible control regime of the 2nd-generation GSM system of Chapter 8 and the more advanced 3rd-generation proposals, which we leave for the motivated reader due to lack of space. Suffice to say here that unfortunately it is unfeasible to design the control regime of a sophisticated mobile radio system by 'direct synthesis' and hence some of the solutions reviewed throughout this section in the context of the 3G proposals may appear somewhat heuristic and quite ingenious. These solutions constitute an amalgam of the wireless research community's experience in the design of the existing second-generation systems and of the lessons learned from their operation. Further contributing factors in the design of the 3G systems were based on solving the signalling problems specific to the favoured physical layer traffic channel solutions, namely CDMA. In order to mention only one of them, the TDMA-based GSM system of Chapter 8 was quite robust against power control inaccuracies, while the Pan-American IS-95 CDMA system required an accurate power control. As we will see in Section 10.2.8 during our forthcoming discourse, the power control problem was solved quite elegantly in the 3G proposals. We will also see that statistical multiplexing schemes, such as ALOHA - the original root of the recently more familiar Packet Reservation Multiple Access (PRMA) procedure highlighted for example in Chapter 9 - found their way into public mobile radio systems. A variety of further interesting solutions have also found applications in these 3G proposals, which are the results of the past decade of wireless system research. Let us now review the range of physical channels in the next section.

10.2.3 Physical Channels

The transport channels are transmitted using the physical channels. The physical channels are organized in terms of superframes, radio frames and timeslots, as shown in Figure 10.4. The philosophy of this hierarchical

Dedicated Physical Channels	Transport Channels
Dedicated Physical Data Channel (DPDCH) (UL/DL)	DCH
Dedicated Physical Control Channel (DPCCH) (UL/DL)	

Common Physical Channels	Transport Channels
Physical Random Access Channel (PRACH) (UL)	RACH
Primary Common Control Physical Channel (PCCPCH) (DL)	BCCH
Secondary Common Control Physical Channel (SCCPCH) (DL)	FACH
Synchronization Channel (SCH) (DL)	PCH

Table 10.4: Mapping the transport channels of Table 10.3 to the UTRA physical channels. The equivalent mapping of IMT-2000 is seen in Table 10.5.

frame structure is also reminiscent to a certain degree of the GSM TDMA frame hierarchy of Chapter 8. However, while in GSM each TDMA user had an exclusive slot-allocation, in W-CDMA the number of simultaneous users supported is dependent on the users' required bit-rate and their associated spreading factors. The mobile stations can transmit continuously in all slots or discontinuously, for example when invoking a voice activity detector (VAD). Some of these issues will be addressed in Section 10.2.4.

As seen in Figure 10.4, the UTRA/IMT-2000 superframe consists of 72 radio frames, with 16 timeslots within each radio frame. The duration of each timeslot is 0.625 ms, which gives a duration of 10 ms and 720 ms for the radio frame and superframe, respectively. The 10 ms frame-duration also conveniently coincides for example with the frame-length of the ITU's G729 speech codec for speech communications, while it is a 'sub-multiple' of the GSM system's various full- and half-rate speech codec's frame durations, which were detailed in Chapter 8. We also note that a convenient mapping of the video-stream of the H.263 videophone codec of Chapter 9 can be arranged on the 10 ms-duration radio frames for supporting interactive video services, while on the move. In the FDD mode, a downlink physical channel is defined by its spreading code and frequency. Furthermore, in the uplink, the modem's orthogonal in-phase and quadrature-phase channels are used to deliver the data and control information simultaneously in parallel on the modem's I and Q branches - as it will be augmented in Figure 10.18 - and hence the knowledge of the relative carrier phase, namely whether the I or Q branch is involved, constitutes part of the physical channel's identifier. On the other hand, in the TDD mode, a physical channel is defined by its spreading code, frequency and timeslot. The format of the physical channels is different for UTRA and IMT-2000. Hence, we will highlight them individually, commencing with the UTRA structure.

10.2.3.1 UTRA Physical Channels

Similarly to the transport channels of Table 10.3, the physical channels in UTRA can also be classified, as dedicated and common channels. Table 10.4 shows the type of physical channels and the corresponding mapping of transport channels on the physical channels in UTRA.

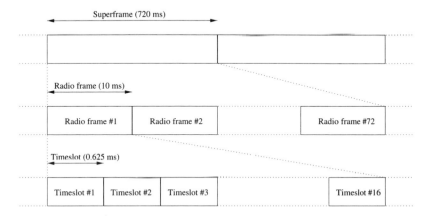

Figure 10.4: UTRA/IMT-2000 physical channel structure. On the UTRA downlink DPDCH and DPCCH are interspersed by time-multiplexing. On the uplink they are mapped to the I and Q modem branches, as it will be augmented in the context of Figure 10.18.

The configuration of the information in the timeslots of the physical channels differs from one another in the uplink and downlink, as well as in the FDD and TDD modes. Figures 10.5 and 10.6 show the structure of one timeslot for each physical channel on the downlink (DL) and uplink (UL), respectively, in the FDD mode. The timeslot structures of the TDD mode will be highlighted subsequently during our further discourse in this section. The structure of the Synchronisation Channel (SCH) will be explained in more detail in Section 10.2.9.

The dedicated physical channel of Figure 10.5 can be divided into a dedicated physical data channel (DPDCH) and a dedicated physical control channel (DPCCH), both of which are bi-directional. The DPDCH is used to transmit the DCH information between the base station and mobile station. The DPCCH is used to transmit the pilot symbols, transmit power control (TPC) commands and an optional so-called transport-format indicator (TFI). At the time of writing, the number of pilot symbols and the length of the TPC as well as TFI segments, which constitute the total overhead of the data channels, is undecided. Given that the TPC and TFI segments render the transmission packets 'self-descriptive', the system becomes very flexible, supporting burst-by-burst adaptivity, which sub-

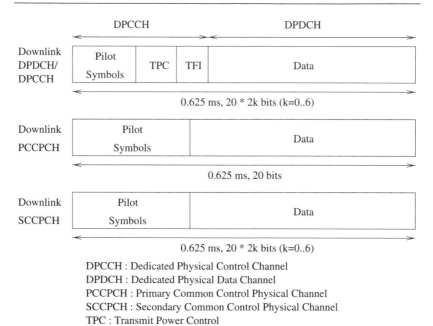

DPCCH : Dedicated Physical Control Channel
DPDCH : Dedicated Physical Data Channel
PCCPCH : Primary Common Control Physical Channel
SCCPCH : Secondary Common Control Physical Channel
TPC : Transmit Power Control
TFI : Transport Format Indicator

Figure 10.5: UTRA downlink FDD physical channel timeslot configuration,
which are mapped to the time slots of Figure 10.4. On the
UTRA downlink DPDCH and DPCCH are interspersed by time-
multiplexing. On the uplink they are mapped to the I and Q
modem branches, as it will be augmented in the context of Figure 10.18. The equivalent IMT-2000 structure is seen in Figure 10.7.

stantially improves the system's performance [23, 24], although this side-
information is vulnerable to transmission errors. The pilot symbols are used
to facilitate coherent detection on both the uplink and downlink - as it was
discussed in Chapter 9 in the context of QAM, - and also to enable the
implementation of performance enhancement techniques, such as adaptive
antennas and interference cancellation. The TPC commands provide a fast
and efficient power control scheme, which is essential in DS-CDMA using
the techniques to be highlighted in Section 10.2.8. The TFI carries information concerning the instantaneous parameters of each transport channel
multiplexed on the physical channel.

On the UTRA FDD uplink of Figure 10.6, the DPDCH and DPCCH
messages of Table 10.4 are transmitted in parallel on the in-phase (I) and
quadrature-phase (Q) branches of the modem, as it will become more explicit during our further discourse in the context of Figure 10.18 [7]. By
contrast, at the top of the downlink structure of Figure 10.5, the DPDCH

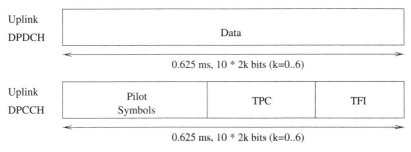

DPDCH : Dedicated Physical Data Channel
DPCCH : Dedicated Physical Control Channel
TPC : Transmit Power Control
TFI : Transport Format Indicator

Figure 10.6: UTRA uplink FDD physical channel timeslot configuration, which
is mapped to the time slots of Figure 10.4. The DPDCH and
DPCCH messages are transmitted in parallel on the I and Q
branches of the modem of Figure 10.18. The equivalent IMT-2000
uplink FDD time-slot configuration is seen in Figure 10.8. These
DPDCH and DPCCH bursts are time-multiplexed at the top of
Figure 10.5, yielding 20×2^k bits per 0.625 ms.

and DPCCH are time-multiplexed into one physical channel time-slot of
Figure 10.4. The reason for the parallel transmission on the uplink is to
avoid Electromagnetic Compatibility (EMC) problems due to discontinuous
transmission of the DPDCH of Table 10.4 [18]. Discontinuous transmission
occurs, when temporarily there are no data to transmit, but the link is still
maintained by the DPCCH. If the uplink DPCCH were time-multiplexed
with the DPDCH, as in the downlink of Figure 10.6, this could create short,
sharp energy spikes. Since the mobile station may be located near sensitive
electrical equipment, these spikes could affect these equipment.

The Primary Common Control Physical Channel (PCCPCH) of Ta-
ble 10.4 and Figure 10.5 is used by the base station in order to continuously
broadcast the BCCH information at a fixed rate of 20 bits/0.625 ms = 32
kbps to all mobiles in the cell. The Secondary Common Control Physical
Channel (SCCPCH) of Table 10.4 and Figure 10.5 carries the FACH and
PCH information on the downlink and they are transmitted only, when data
are available for transmission. Knowledge of the SCCPCH bit-rate can be
acquired from the BCCH information transmitted on the PCCPCH.

The parameter k in Figures 10.5 and 10.6 determines the spreading factor (SF) of the physical channel. The highest SF is 256 for $k = 0$, which
corresponds to the lowest channel bit-rate and the highest spreading gain,
while the highest channel bit-rate has a SF of 4, when $k = 6$. Hence the
bit-rates available for the uplink DPDCH are 16/32/64/128/256/512/1024
kbps, due to the associated 'payload' of 10×2^k bits per 0.625 ms bursts

in Figure 10.6, where $k = 0 \ldots 6$. Recall that the uplink structure of Figure 10.6 invoked I/Q multiplexing, as it will be demonstrated in Figure 10.18. By contrast, the downlink structure of Figure 10.5 refrains from I/Q-multiplexing and the timeslot payload is 20×2^k bits per 0.625 ms, but the exact data rate is unspecified.

This hierarchically structured set of legitimate rates provides a high flexibility in terms of the services supported. Notice that the channel bit-rates of the downlink dedicated physical channels are twice as high as those of the uplink dedicated physical channels. This is due to the time-multiplexed DPCCH and DPDCH on the downlink of Figure 10.5, while the DPCCH and DPDCH are transmitted in parallel on the modem's I and Q branches in the uplink of Figures 10.6 and 10.18. If higher bit-rates are required, then several DPDCHs with only one DPCCH can be transmitted in parallel, using a technique known as multicode transmission [25], which will be explained in more detail in the context of Figure 10.16 in Section 10.2.5. The SCCPCH also has a variable bit-rate, similarly to that of the downlink dedicated physical channel portrayed at the bottom of Figure 10.5. On the other hand, again, the PCCPCH has a constant bit-rate of 20 bits/0.625 ms = 32 kbps. Since the chip rate is 4.096 Mcps and each time slot has a duration of 0.625 ms, there will be 4.096 Mcps $\times$ 0.625 ms = 2560 chips per time slot.

At this stage it is worth mentioning that the available control channel rates are significantly higher in the 3G systems, than in their 2G counterparts of Table 1.1 in Chapter 1. For example, the maximum BCCH signalling rate in GSM is about an order of magnitude lower than the above mentioned 32 kbps UTRA BCCH rate. In general, this increased control channel rate will support a significantly more flexible system control than the 2G systems. For comparison, we refer to the 'Control Channel Rate' row of Table 1.1. Having highlighted the UTRA physical channels, let us now consider the corresponding IMT-2000 solutions.

10.2.3.2 IMT-2000 Physical Channels

The type of physical channels and their mapping to/from the transport channels in IMT-2000 are shown in Table 10.5. The dedicated channels of IMT-2000 are basically similar to those of Table 10.4 in UTRA. The differences are in the common physical channels. The so-called 'perch' channel has a similar function to that of the SCH in UTRA. However, as seen from the mapping in Table 10.5, the Broadcast Channel (BCH) information is also carried by the perch channel, whereas in UTRA an additional physical channel, namely the PCCPCH of Figure 10.5 is used to carry the BCCH information.

These physical channels are also arranged in terms of superframes, radio frames and timeslots, with parameters similar to those in UTRA, as it was shown in Figure 10.4. However, the configuration of the timeslots is slightly

Dedicated Physical Channels	Transport Channels
Dedicated Physical Data Channel (DPDCH) (UL/DL)	DCH
Dedicated Physical Control Channel (DPCCH) (UL/DL)	

Common Physical Channels	Transport Channels
Physical Random Access Channel (PRACH) (UL)	RACH
Perch Channel (DL)	BCH*
Common Physical Channel (CPCH) (DL)	FACH
	PCH

* BCH in IMT-2000 corresponds to BCCH in UTRA

Table 10.5: Mapping the transport channels of IMT-2000 to physical channels. The equivalent mapping of UTRA is seen in Table 10.4.

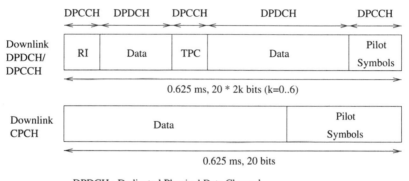

DPDCH : Dedicated Physical Data Channel
DPCCH : Dedicated Physical Control Channel
CPCH : Common Control Channel
TPC : Transmit Power Control
RI : Rate Information

Figure 10.7: IMT-2000 FDD downlink physical channel timeslot configuration, which is mapped to the time slots of Figure 10.4. On the uplink DPDCH and DPCCH are mapped to the I and Q modem branches, as it will be augmented in the context of Figure 10.18. The equivalent UTRA structure is seen in Figure 10.5.

different from those in UTRA, which is demonstrated by Figures 10.7 and 10.8 for the FDD downlink and uplink, respectively. The Rate Information (RI) at the top of Figure 10.7 has the same function as the TFI of Figure 10.5 in UTRA, rendering the transmission bursts 'self-descriptive' and hence IMT-2000 is also capable of supporting burst-by-burst adaptivity [23,24].

In contrast to the previous FDD structures of Figures 10.5-10.8, in TDD operation, the burst structure of Figure 10.9 is used, where each time-slot's

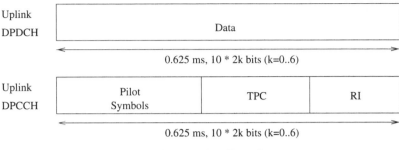

0.625 ms, 10 * 2k bits (k=0..6)

0.625 ms, 10 * 2k bits (k=0..6)

DPDCH : Dedicated Physical Data Channel
DPCCH : Dedicated Physical Control Channel
TPC : Transmit Power Control
RI : Rate Information

Figure 10.8: IMT-2000 FDD uplink physical channel timeslot configuration, which is mapped to the time slots of Figure 10.4. On the IMT-2000 downlink DPDCH and DPCCH are interspersed by time-multiplexing. On the uplink they are mapped to the I and Q modem branches, as it will be augmented in the context of Figure 10.18. The equivalent UTRA structure is seen in Figure 10.6.

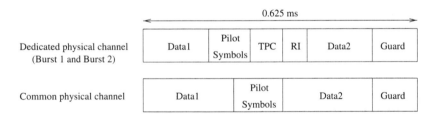

Figure 10.9: Burst configuration in the IMT-2000/UTRA TDD mode, which is augmented in Figure 10.11 for the dedicated physical channel.

transmitted information can be arbitrarily allocated to the downlink or uplink, with the exception of the first burst in the TDD frame of Figure 10.10, which is always assigned to the downlink. Hence, this flexible allocation of the uplink and downlink burst in the TDD mode enables the use of an adaptive modem [23, 24] whereby the modem parameters, such as the spreading factor or the number of bits per symbol can be adjusted on a burst-by-burst basis to optimize the link quality. This first slot, known as the 'beacon slot' only contains the downlink physical control information, such as the BCCH, PCH, SCH or FACH information of Tables 10.4 and 10.5 for UTRA and IMT-2000, respectively. Three examples of possible TDD uplink/downlink allocations are shown in Figure 10.10. A symmetric uplink/downlink allocation refers to a scenario, where an equal number of downlink and uplink bursts are allocated within a frame, while in asymmetric uplink/downlink

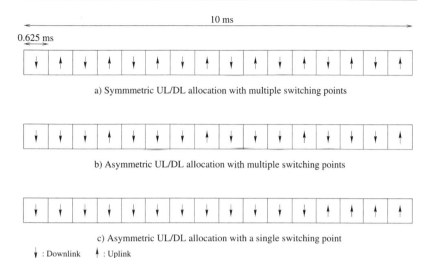

a) Symmmetric UL/DL allocation with multiple switching points

b) Asymmetric UL/DL allocation with multiple switching points

c) Asymmetric UL/DL allocation with a single switching point

↓ : Downlink   ↑ : Uplink

Figure 10.10: Uplink/downlink allocation examples for the 16 slots in IMT-2000 and UTRA TDD operation using the timeslot configurations of Figure 10.9. The first slot is always a downlink slot, providing physical control information for the mobile station, such as the BCCH, PCH, SCH or FACH information.

allocation, there is an unequal number of uplink and downlink bursts, such as for example in 'near-simplex' file download or video-on-demand. In the TDD mode, the configuration of the information in the burst differs from that in FDD mode due to the presence of a guard time. Figure 10.9 shows an example of the TDD burst configuration for the common and dedicated physical channels. In UTRA, two different traffic burst structures, known as Burst 1 and Burst 2, are defined, as shown in Figure 10.11. The parameter k, where $k = 0, 1, 2, 3$ in Figure 10.11 determines the spreading factor of the burst. Hence, the spreading factor of a TDD burst can be variable, ranging from $976/(61 \times 2^3) = 2$ to $976/(61 \times 2^0) = 16$, as derived from Burst 1. Following the same approach, it can be easily shown that the spreading factor of Burst 2 is also in the range of 2 to 16, which was stipulated earlier in Table 10.2. With these spreading factors, the channel bit-rate of a single QPSK modulated TDD burst can be 512/1024/2048/4096 kbps, as given in Table 10.2. Having highlighted the basic features of the various UTRA/IMT-2000 channels, let us now consider, how the various services are error protected, interleaved and multiplexed on the DPDCH, an issue discussed with reference to Figure 10.12 in the context of UTRA/IMT-2000.

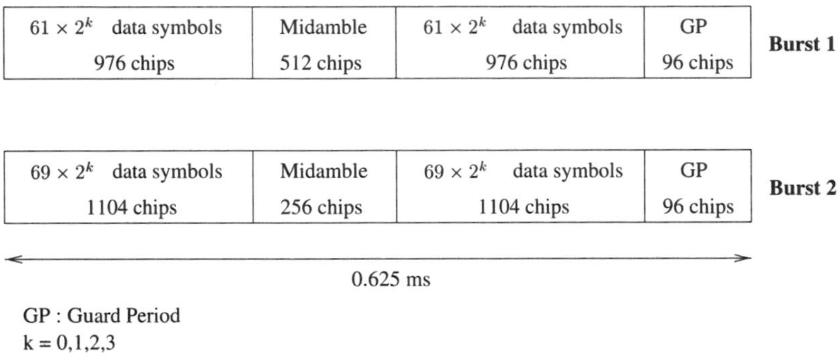

61×2^k data symbols	Midamble	61×2^k data symbols	GP	Burst 1
976 chips	512 chips	976 chips	96 chips	

69×2^k data symbols	Midamble	69×2^k data symbols	GP	Burst 2
1104 chips	256 chips	1104 chips	96 chips	

0.625 ms

GP : Guard Period
k = 0,1,2,3

Figure 10.11: Configuration of two different types of traffic bursts, as defined in UTRA, namely Burst 1 and Burst 2. The midamble contains the control symbols such as the pilot symbols, the TPC and RI, as portrayed in Figure 10.9. At the time of writing, the number of symbols in the respective fields in a TDD burst is still undecided in IMT-2000.

10.2.4   Service Multiplexing and Channel Coding in UTRA/IMT-2000

Service multiplexing is employed, when multiple services of identical and/or different bit-rates requiring different quality of service (QoS) belonging to the same user's connection are transmitted. An example would be the simultaneous transmission of voice, video, data and handwriting transmission service for a multimedia application. These issues were also addressed in Chapter 9, where PRMA was used for multiplexing the various services. Accordingly, each service is represented by its corresponding transport channels, as described in Section 10.2.2. A possible method of transmitting multiple services is by using code multiplexing with the aid of orthogonal codes. Every service could have its own DPDCH and DPCCH, each assigned to a different orthogonal code. This method is not very efficient, however, since a number of orthogonal codes would be reserved by a single user, while on the uplink it would also inflict self-interference. Alternatively, these services can be time-multiplexed into one or several DPDCHs, as shown in Figure 10.12 for UTRA/IMT-2000.

Transport channels belonging to different services with different QoS requirements are first channel coded individually, using various coding techniques. Several forward error correction (FEC) techniques are proposed for channel coding. The FEC technique used is dependent on the QoS requirement of that specific service. The potential FEC techniques are listed in Table 10.6, together with their corresponding parameters. Convolutional coding is used for services with a BER requirement on the order of 10^{-3}, for example, for voice services. For services requiring a lower BER, namely

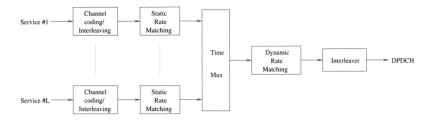

Figure 10.12: Service multiplexing in UTRA/IMT-2000.

	Convolutional	Reed-Solomon	Turbo[†]
BER requirement	10^{-3}	10^{-6}	10^{-6}
Rate	1/4 to 1	TBD[‡]	1/3 or 1/2
Constraint length	9	N/A	3

[†]Turbo coding is still under investigation in ETSI, optional in IMT-2000
[‡]TBD : To be decided

Table 10.6: UTRA and IMT-2000 channel coding parameters.

on the order of 10^{-6}, additional outer Reed-Solomon (RS) coding and outer interleaving concatenated with the inner convolutional coding is applied. These techniques were discussed in depth in Chapter 4. Instead of RS coding, turbo coding is proposed in IMT-2000 as an optional coding scheme. Turbo coding is known to guarantee a high performance [28] over AWGN channels at the cost of increased interleaving-induced latency or delay and at a high implementational complexity. At the time of writing, turbo coding is still under investigation within the ETSI. Each coded transport channel is then interleaved by the 'Channel coding/Interleaving' block, as shown in Figure 10.12. The depth of this so-called inner interleaving can range from one radio frame (10 ms) to as high as 80 ms, depending on the type of service being interleaved. For example, for a speech service, which belongs to the so-called real-time or interactive services, but can tolerate a BER of $10^{-3} to 10^{-2}$, decoding time is critical and hence a short interleaver depth is more feasible. On the other hand, for non-real-time services, such as data services, more emphasis is placed on achieving a low BER, than on fast decoding time, and hence a higher interleaver depth is more beneficial. The effect of the interleaver depth was also studied in Chapter 4.

The output of each coded transport channel of Figure 10.12 will have a different bit-rate. Hence, before time-multiplexing them on a physical channel, a so-called **static rate matching** procedure is required, as seen in Figure 10.12. Static rate matching is coordinated amongst the different coded transport channels, such that the bit-rate of each channel is adjusted to a level that fulfils its minimum QoS requirements [8]. On the downlink, the bit-rate is also adjusted so that the total instantaneous transport channel bit-rate approximately matches the defined bit-rate of the physical channel, as listed in Table 10.2. Static rate matching is based on code puncturing, which was treated in Chapter 4, and repetition.

After static rate matching, the coded transport channels are time-multiplexed, as portrayed in Figure 10.12 in order to produce the DPDCH 'payload'. The total instantaneous bit-rate of the DPDCH 'payload' may not be equal to the defined DPDCH bit-rate. Hence, a process referred to as **dynamic rate matching** is used to match the instantaneous bit-rate to one of the defined DPDCH bit-rates highlighted in Section 10.2.3. If the instantaneous bit-rate exceeds the maximum defined DPDCH bit-rate of 1.024 Mbps, then multicode transmission is invoked, which is highlighted in Section 10.2.5, whereby several DPDCHs are transmitted in parallel. After the bit-rate of the multiplexed channels is matched to that of the DPDCH, the data are interleaved, as seen in Figure 10.12.

Having highlighted the various channel coding techniques and having seen the structures of the physical channels in FDD mode and TDD mode, as illustrated by Figures 10.5-10.8 and Figure 10.11, respectively, let us now consider, how services of different bit-rates are mapped on the dedicated physical data channels (DPDCH) of Figures 10.5-10.8 and Figure 10.11 with the aid of two examples. Specifically, we consider the mapping of several speech services on the DPDCH in FDD mode and an example of the mapping of a 2.048 Mbps data service on the DPDCH in TDD mode.

10.2.4.1   Mapping Several Speech Services to the Physical Channels in FDD Mode [11]

In this example we shall assume that an 8 kbps G.729 speech codec was used to compress the speech signal, generating 80 bits/10 ms. As illustrated in Figure 10.12, each service is first channel coded, before it is time-multiplexed with other services in order to produce a single bit stream. Figure 10.13 shows the channel coding procedure of an 8 kbps speech service. Speech services usually have a moderate BER requirement, in the region of 10^{-3}. Hence, according to Table 10.6, convolutional coding will be employed. Since the duration of a radio frame in UTRA/IMT-2000 is 10 ms, the incoming 8 kbps bit stream is first split into segments of 10 ms, with each segment containing a total of 8 kbps $\times$ 10 ms = 80 bits, as demonstrated in Figure 10.13. A 16-bit CRC checksum is then attached to each 80-bit segment for the purpose of error detection. As a result, the number of bits in a segment is increased to 16 + 80 = 96 bits, as illustrated in Figure 10.13. Next, a block of 8 tail bits is concatenated to the 96-bit segment in order to flush the shift registers of the convolutional encoder, as discussed also in the GSM system of Chapter 8. Thus a total of 96 + 8 = 104 bits are conveyed to the convolutional encoder, as shown in Figure 10.13. A coding rate of $R = 1/3$ and a constraint length of $K = 9$ is used for the convolutional encoding, as listed in Table 10.6. The output of the convolutional encoder will have a total of 104 bits $\times$

3 = 312 bits per 10 ms segment. Interleaving, which is optional, can be performed across the frame after the convolutional encoding. The output of the channel coding/interleaving block in Figure 10.12 constitutes a so-called 'dedicated channel (DCH) radio unit', which is a 312-bit segment in this case, as shown in Figure 10.13. Hence, the channel coding process has increased the bit-rate of an 8 kbps speech service to 31.2 kbps.

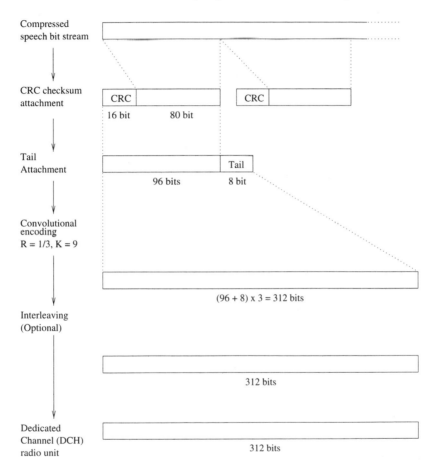

Figure 10.13: Convolutional coding of an 8 kbps speech service.

In order to illustrate the concept of service multiplexing and how these multiplexed services are eventually mapped on the DPDCH, let us assume that there are Q number of simultaneous speech services to be transmitted in the same connection in the FDD mode. These speech services are individually channel coded, as shown in Figure 10.12 and the channel coding procedure is illustrated in Figure 10.13. Hence, there will be Q separate DCH radio units at the input of the time-multiplexer of Figure 10.12, as shown in Figure 10.14. These DCHs are time-multiplexed in order to produce a single bit stream. We mentioned in Section 10.2.3, and also emphasized in Table 10.2 and in Figures 10.5-10.8 that a single dedicated physical data channel (DPDCH) can assume one of the available channel bit-rates, namely 16/32/64/128/256/512/1024 kbps. Since the bit-rate of the time-multiplexed DCHs may not be equal to any one of these bit rates, rate matching has to be employed in order to adapt the time-multiplexed bit-rate to one of the available DPDCH bit-rates within one radio frame, as shown in Figure 10.12. As mentioned previously, rate matching can be performed by bit puncturing or repetition. Hence for example, if only one DCH is present, which has a bit-rate of 31.2 kbps, then rate matching will increase this bit-rate to the nearest available DPDCH bit-rate, which is 32 kbps or 320 bits per radio frame. In the event, when the bit-rate of the time-multiplexed DCH bit stream exceeds the maximum available bit-rate, then multicode transmission is used, which will be highlighted in Section 10.2.5. In this case, the bit-rate will be matched to any of the available channel bit-rates within one radio frame, namely to $L \times$ 16/32/64/128/256/512/1024 kbps, where L denotes the number of radio frames of equal rate required to convey the information, as illustrated in Figure 10.14. After the bit-rate is matched, interleaving is performed across each of the L radio frames in order to produce the L DPDCHs. Let us now consider the channel coding and mapping procedures of a 2.048 Mbps data service.

10.2.4.2 Mapping a 2.048 Mbps Data Service to the Physical Channels in TDD Mode [10]

In contrast to 2G mobile systems, 3G mobile systems must be capable of supporting data services with rates as high as 2 Mbps. Hence in this example we will illustrate the mapping of a 2 Mbps data service to the dedicated physical data channels (DPDCH) in TDD mode using the TDD bursts shown in Figure 10.11. Unlike speech services, which have a moderate BER requirement, a low BER on the order of 10^{-6} is often required for the transmission of data services. Hence, more powerful FEC methods, such as turbo coding and concatenated inner convolutional/outer Reed-Solomon (RS) coding are needed for channel coding, as shown in Table 10.6. In this example, concatenated convolutional/RS coding will be invoked as the channel coding technique. Similarly to the channel coding of a speech service, as given in Figure 10.13, the incoming data bit stream of the 2 Mbps data service is broken down into 10 ms segments, each containing 2 Mbps × 10 ms = 20480 bits, as it can be seen in Figure 10.15. Each segment is first coded using the outer Reed-Solomon coding scheme. Since the RS coding rate is undecided at the time of writing, we will use the coding rate of 200/210 in this example, as it was given in reference [10] noting that a number of different-rate RS-coded user scenarios were also exemplified

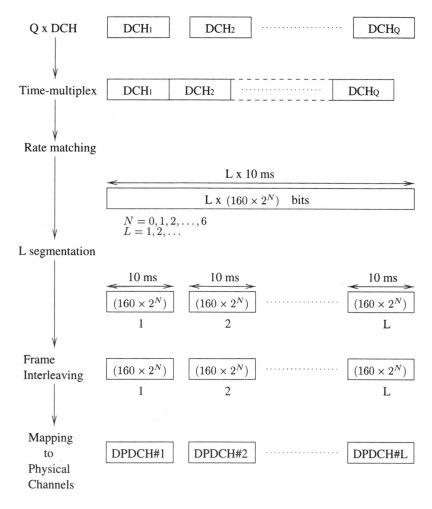

Figure 10.14: Mapping of the channel coded speech service portrayed in Figure 10.13 to the dedicated physical data channels of Figures 10.5-10.8 in FDD mode. The value L denotes the number of radio frames required to convey the information. When $L > 1$, multicode transmission, as highlighted in Section 10.2.5, is employed. When $L = 1$, single code transmission is used. The corresponding schematic is seen in Figure 10.12.

in the standard. Hence, the number of bits in a segment is increased to $20480 \times 210/200 = 21504$ bits at the output of the outer RS encoder, as displayed in Figure 10.15. After the outer interleaving, X number of signalling bits from Layer 2 of the OSI seven-layer structure and 16 blocks of 8 tail bits are then added to the 21504-bit segment, before the inner convolutional coding is applied. Following the example in reference [10], a convolutional coding rate of 2/3 is used. Hence after channel coding, the segment would contain $[21504 + X + (16 \times 8)] \times 3/2 = 32256 + 3X/2 + 192$ coded bits. These coded bits have to be mapped to the TDD physical channels of Figure 10.11, where two configurations are defined, namely Burst 1 and Burst 2. Assuming a spreading factor of 16, Burst 1 can accommodate a total of 2×976 chips/16=122 symbols, which constitute 244 bits, when QPSK modulation is used. On the other hand, Burst 2 can accommodate 2×1104 chips/16=138 symbols or 276 bits. In TDD mode, two methods can be used to transmit a block of data, either allocating several time-slots or allocating several orthogonal codes per time-slot, as in multicode transmission. Each burst must contain either 244 bits or 276 bits, again, assuming that the spreading factor is 16. Hence, either bit puncturing or repetition has to be used in order to adapt the coded bit stream such that the total number of bits in the segment becomes an integer factor of either 244 bits, as in Burst 1, or 276 bits, as in Burst 2. At the left of Figure 10.15, we see that puncturing is used in order to reduce the total number of bits to $117 \times 244 = 28548$ bits, such that 117 bursts of 'Burst 1' are used for transmission. Alternatively, the coded bit stream can be punctured in order to reduce the total number of bits in a segment to $104 \times 276 = 28704$ bits, as illustrated at the right of Figure 10.15 for transmission, using 104 'Burst 2' type packets. These bits are then QPSK modulated and mapped to the dedicated physical channels seen in Figure 10.11.

Following the above brief discussions on service multiplexing, channel coding and interleaving, let us now concentrate on the aspects of variable-rate and multicode transmission in UTRA/IMT-2000.

10.2.5 Variable Rate and Multicode Transmission in UTRA/IMT-2000

Three different techniques have been proposed for supporting variable rate transmission, namely multicode-, modulation-division multiplexing-(MDM) and multiple processing gain (MPG) based techniques [29]. UTRA and IMT-2000 employ a number of different processing gains, or variable spreading factors in order to transmit at different channel bit-rates, as highlighted previously in Section 10.2.3. We argued in Chapter 1 that the spreading factor has a direct effect on the performance and capacity of a DS-CDMA system. Since the chip rate is constant, the spreading factor - which is defined as the ratio of the spread bandwidth to the original information bandwidth - becomes lower, as the bit-rate increases. Hence, there is a limit to the value of the spreading factor used, which is 4 in FDD mode in the proposed 3G standards. Multicode transmission [25, 29, 30] is used, if the total bit rate to be transmitted exceeds the maximum bit-rate supported by a single DPDCH, which was stipulated as 1.024 Mbps. When

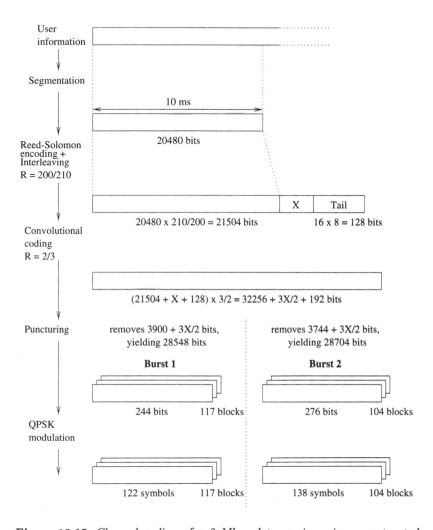

Figure 10.15: Channel coding of a 2 Mbps data service using concatenated convolutional/Reed-Solomon coding and mapping to the TDD dedicated physical channels, namely Burst 1 and Burst 2, as shown in Figure 10.11 for a spreading factor of 16. The corresponding schematic is seen in Figure 10.12.

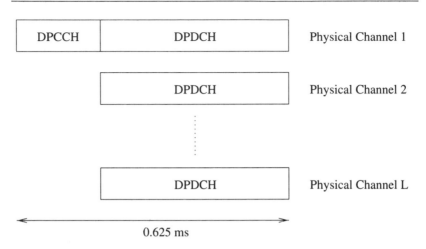

Figure 10.16: Downlink FDD slot format for multicode transmission in IMT-2000/UTRA, based on Figure 10.5, but dispensing with transmitting DPCCH over all multicode physical channels.

this happens, the bit-rate is split amongst a number of spreading codes and the information is transmitted using two or more codes. However, only one DPCCH is transmitted during this time. Hence, on the uplink, one DPCCH and several DPDCH are code-multiplexed and transmitted in parallel, as it will be augmented in the context of Figure 10.18. On the downlink, the DPDCH and DPCCH are time-multiplexed on the first physical channel associated with the first spreading code. If more physical channels are required, the DPCCH part in the slot will be left blank, as shown in Figure 10.16.

10.2.6   Spreading and Modulation

As we argued in Chapter 1, the performance of DS-CDMA is interference limited [31]. The majority of the interference originates from the transmitted signals of other users within the same cell, as well as from neighbouring cells. This interference is commonly known as multiple access interference (MAI). Another source of interference, albeit less dramatic, is a result of the wideband nature of CDMA, which causes several replicas of the transmitted signal to reach the receiver at different time instants, hence inflicting what is known as inter-path interference. However, the advantages gained from wideband transmissions, such as multipath diversity and the noise-like properties of interference, outweigh the drawbacks.

   In order to reduce the MAI and hence to improve the systems's performance and capacity, **the IMT-2000/UTRA physical channels are spread using two different codes, namely the so-called channel-**

ization code[2] and a typically longer so-called scrambling code. By contrast, recall that in the IS-95 CDMA system of Chapter 1, for example in the downlink schematic of Figure 1.42, there were three different orthogonal codes. Namely the 64-chip Walsh-codes of Figure 1.41, the in-phase and quadrature-phase pseudo-noise sequences, PNI and PNQ, which are the so-called 'short-codes' of 32768 chip-duration and the $2^{42} - 1$ chip-duration long codes. The IS-95 short codes are the same cell-specific codes in both the uplink and downlink, while the long codes are user-specific and they are also identical in the uplink and downlink. The cdma2000 system of Section 10.3 follows the IS-95 philosophy.

10.2.6.1   Orthogonal Variable Spreading Factor Codes in UTRA/ IMT-2000

The UTRA/IMT-2000 channelization codes are derived from a set of orthogonal codes known as Orthogonal Variable Spreading Factor (OVSF) codes [32]. OVSF codes are generated from a tree-structured set of orthogonal codes, such as the Walsh-Hadamard codes of Chapter 1, using the procedure shown in Figure 10.17. Each channelization code is denoted by $c_{N,n}$, where $n = 1, 2, \ldots, N$ and $N = 2^x, x = 2, 3, \ldots 8$. Each code $c_{N,n}$ is derived from the previous code $c_{(N/2),n}$ as follows [32] :

$$
\begin{bmatrix}
c_{N,1} \\
c_{N,2} \\
c_{N,3} \\
\vdots \\
c_{N,N}
\end{bmatrix}
=
\begin{bmatrix}
c_{(N/2),1} | c_{(N/2),1} \\
c_{(N/2),1} | \bar{c}_{(N/2),1} \\
c_{(N/2),2} | c_{(N/2),2} \\
\vdots \\
c_{(N/2),(N/2)} | \bar{c}_{(N/2),(N/2)}
\end{bmatrix}
, \qquad (10.1)
$$

where $[|]$ denotes an augmented matrix and $\bar{c}_{(N/2),n}$ is the binary complement of $c_{(N/2),n}$. Hence, for example, according to Equation (10.1) and Figure 10.17 $c_{N,1} = c_{8,1}$ is created by simply concatenating $c_{(N/2),1}$ and $c_{(N/2),1}$, which simply doubles the number of chips. By contrast, $c_{N,2} = c_{8,2}$ is generated by attaching $\bar{c}_{(N/2),1}$ to $c_{(N/2),1}$. From Equation (10.1) we see that, for example, $c_{N,1}$ and $c_{N,2}$ at the left-hand side of Equation (10.1) are not orthogonal to $c_{(N/2),1}$, since the first half of both was derived from $c_{(N/2),1}$ in Figure 10.17, but they are orthogonal to $c_{(N/2),n}, n = 2, 3, \ldots, (N/2)$. The code $c_{(N/2),1}$ in Figure 10.17 is known as the mother code of the codes $c_{N,1}$ and $c_{N,2}$, since these two codes are derived from $c_{(N/2),1}$. The codes on the 'highest'-order branches ($k = 6$) of the tree at the left of Figure 10.17 have a spreading factor of 4 and they are used for transmission at the highest possible bit-rate for a single channel, which is 1024 kbps. On the other hand, the codes on the 'lowest'-order branches ($k = 0$) of the tree at the right of Figure 10.17 have a spreading factor of 256 and these are used for transmission at the lowest bit-rate,

---

[2]In IMT-2000, the channelization codes are known as spreading codes.

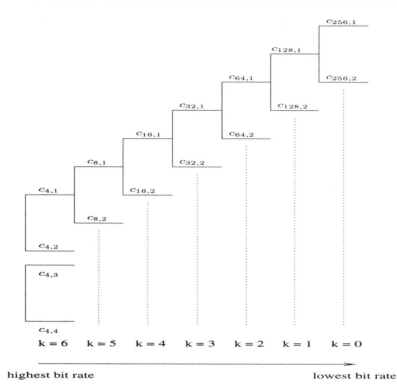

Figure 10.17: Orthogonal variable-spreading factor code tree in UTRA/IMT-2000 according to Equation 10.1. The parameter k in the figure is directly related to that found in Figures 10.5-10.8.

which is 16 kbps. Orthogonality between parallel transmitted channels of the same bit rate is preserved by assigning each channel a different orthogonal code accordingly. For channels with different bit-rates transmitting in parallel, orthogonal codes are assigned, ensuring that no code is the mother-code of the other. Hence, OVSF channelization codes provide total isolation between different users' physical channels on the downlink which transmits all codes synchronously and hence eliminates multiple access interference. OVSF channelization codes also provide orthogonality between the different DPDCHs seen in Figure 10.16 during multicode transmission.

However, since there is only a limited set of OVSF codes, which is likely to be insufficient to support a large user-population, while also allowing identification of the base stations by the mobile stations on the downlink, **each cell will reuse the same set of OVSF codes.** However, orthogonal codes, such as the orthogonal OVSF codes, in general exhibit poor asynchronous cross-correlation properties [33] and hence the cross-correlations of the OVSFs of adjacent asynchronous base stations will become unacceptably high, degrading the correlation receiver's performance at the mobile station. On the other hand, certain long codes such as Gold codes, exhibit low asynchronous cross-correlation, which is advantageous in CDMA applications [3]. Hence in UTRA/IMT-2000, cell-specific long codes are used in order to reduce the inter-cell interference on the downlink. On the uplink, MAI is reduced by assigning different scrambling codes to different users. Table 10.7 shows the parameters and techniques used for spreading and modulation in UTRA and IMT-2000, which will be discussed in depth in the next section.

	Channelization codes	Scrambling codes
Type of codes	OVSF (Section 10.2.6.1)	DL : Gold codes (UTRA/IMT-2000), Extended very large (VL) Kasami codes (UTRA) UL : Gold codes
Code length	Variable	DL : 10 ms of $(2^{18} - 1)$-chip Gold code UL : 10 ms of $(2^{41} - 1)$-chip Gold code, 256-chip Kasami code (UTRA), 737.28 s of $(2^{41} - 1)$-chip Gold code (IMT-2000)
Type of spreading	DL : BPSK UL : BPSK	DL : BPSK (UTRA), QPSK (IMT-2000) UL : QPSK
Data Modulation	DL : QPSK (FDD and TDD) UL : BPSK (FDD), QPSK (TDD)	

Table 10.7: UL/DL and FDD/TDD spreading and modulation parameters in UTRA/IMT-2000.

10.2.6.2 Uplink Spreading and Modulation

Let us commence our discourse with a brief note concerning the choice of spreading codes in general [34, 35]. Suffice to say that the traditional measures used in comparing different codes are their cross-correlations (CCL) and auto-correlation (ACL). If the CCL of the channelization codes of different users is non-zero, this will increase their interference, as perceived by the receiver. Hence a low CCL reduces the MAI. The so-called out-of-phase ACL of the codes, on the other hand, plays an important role during the initial synchronization between the base station and mobile station, which has to be sufficiently low in order to minimize the probability of synchronizing to the wrong ACL peak. Let us now continue our discourse with the uplink spreading issues with reference to Table 10.7. A model of the uplink transmitter for a single DPDCH is shown in Figure 10.18 [7]. We have seen in Figure 10.6 that the DPDCH and DPCCH are transmitted in parallel on the I and Q branches of the uplink, respectively. Hence, to avoid I/Q channel interference, different orthogonal spreading codes are assigned to the DPDCH and DPCCH on the I and Q branches, respectively. The technique is referred to **as dual-channel spreading**. These two channelization codes for DPDCH and DPCCH, denoted by c_D and c_C in Figure 10.18, respectively, are allocated in a pre-defined order. Hence, the base station and mobile station only need to know the spreading factor of the channelization codes, but not the code itself. After spreading, the

BPSK modulated I and Q branch signals are summed in order to produce a complex signal, where G in Figure 10.18 is a power gain adjustment for the DPCCH. In the event of multi-code transmission, different orthogonal spreading codes are assigned to each DPDCH for the sake of maintaining orthogonality and they can be transmitted on either the I or Q branch. In this case, the base station and mobile station have to agree on the number of channelization codes to be used.

The complex signal is then scrambled by a user-specific complex scrambling code, denoted by c_{scramb} in Figure 10.18 [7]. This scrambling code is a complex Gold code constructed from two m-sequences using the polynomials of $1 + X^3 + X^{41}$ and $1 + X^{20} + X^{41}$, following the procedure highlighted by Proakis in reference [26]. This code is also shown in Table 10.7. The Q-branch Gold code is a shifted version of the I-branch Gold code, where a shift of 1024 chips was recommended. Each Gold code is rendered different from one another by assigning a unique initial state to one of the shift registers of the m-sequence. The initial state of the other shift register is a continuous sequence of logical '1'. The base station will inform the mobile station about the specific initial sequence used via the access grant message. Complex-valued scrambling balances the power on the I and Q branches. This can be shown by letting c_s^I and c_s^Q be the I and Q branch scrambling codes, respectively. The spread data of Figure 10.18 [7] can be written as:

$$d(t) = c_D \cdot b_{\mathrm{DPDCH}} + jG \cdot c_C \cdot b_{\mathrm{DPCCH}}, \qquad (10.2)$$

where b_{DPDCH} and b_{DPCCH} represent the DPDCH message and the DPCCH message, respectively. Assuming that the power level in the I and Q branches of Figure 10.18 is unbalanced due to their different bit-rates or different QoS requirements on DPDCH and DPCCH and if only real-valued scrambling is used, then the output becomes:

$$s(t) = c_s^I \left(c_D \cdot b_{\mathrm{DPDCH}} + jG \cdot c_C \cdot b_{\mathrm{DPCCH}} \right), \qquad (10.3)$$

which is also associated with an unbalanced power level on the I and Q branches. By contrast, if complex-valued scrambling is used, then the output of Figure 10.18 [7] becomes:

$$
\begin{aligned}
s(t) &= \left(c_D \cdot b_{\mathrm{DPDCH}} + jG \cdot c_C \cdot b_{\mathrm{DPCCH}} \right) \cdot \left(c_s^I + j c_s^Q \right) & (10.4) \\
&= c_s^I \cdot c_D \cdot b_{\mathrm{DPDCH}} - G \cdot c_s^Q \cdot c_C \cdot b_{\mathrm{DPCCH}} & (10.5) \\
&\quad + j \left(c_s^Q \cdot c_D \cdot b_{\mathrm{DPDCH}} + G \cdot c_s^I \cdot c_C \cdot b_{\mathrm{DPCCH}} \right). & (10.6)
\end{aligned}
$$

As it can be seen, the power on the I and Q branches is the same, regardless of the power level of the DPDCH and DPCCH. Hence complex scrambling improves the power efficiency by reducing the peak-to-average power fluctuation. This also relaxes the linearity requirements of the up-link power amplifier used. The whole process of spreading using orthogonal codes and complex-valued scrambling codes is known in this context as Orthogonal

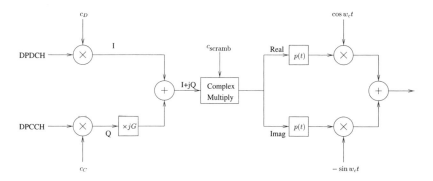

Figure 10.18: FDD uplink transmitter in UTRA/IMT-2000.

Complex QPSK (OCQPSK) modulation[3]. The pulse shaping filters, $p(t)$, are root-raised cosine Nyquist filters with a roll-off factor of 0.22, which were introduced in Chapter 9.

Because of the asynchronous nature of the mobile stations' uplink transmissions, every user can employ the same set of channelization codes. In UTRA, instead of the long Gold scrambling code of $(2^{41} - 1)$ chip duration, short scrambling codes such as extended very large (VL)-Kasami codes of length 256 chips can be used. This code was introduced in order to ease the implementation of multiuser detection at the base station [18]. Explicitly, the multiuser detector has to invert the so-called system matrix [36], the dimension of which is proportional to the sum of the channel impulse response duration and the spreading code duration. Hence using a relatively short spreading code is an important practical consideration. Let us now consider the downlink modulation and spreading.

10.2.6.3 Downlink Spreading and Modulation

The time-multiplexed DPDCH and DPCCH burst at the top of Figure 10.5 are first QPSK modulated in order to form the I and Q channels, before spreading to the chip rate using the OVSF channelization code c_{ch} of Section 10.2.6.1. Different users are assigned different channelization codes for maintaining their orthogonality. The base station will inform the users about their corresponding channelization codes via the Access Grant Message.

The resulting signal in Figure 10.19 is then scrambled by a cell-specific scrambling code c_{scramb}. As seen in Table 10.7, this scrambling code is in a complex, i.e. QPSK form in IMT-2000, while in UTRA, it is in a real or BPSK form. The scrambling code is selected from one of 32 groups of scrambling codes, each group containing 16 different scrambling codes, giving a total of 512 available scrambling codes [12]. The reason for catego-

---

[3]OCQPSK is also known as Hybrid PSK (HPSK)

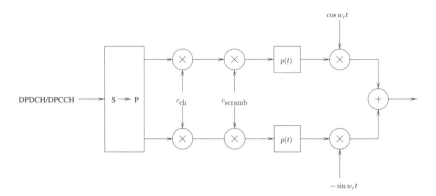

Figure 10.19: FDD/TDD downlink transmitter in UTRA/IMT-2000.

rizing the scrambling codes into groups is to facilitate fast cell identification, which will be augmented in Section 10.2.9. Similarly to the downlink, the pulse shaping filters are root-raised cosine Nyquist filters with a roll-off factor of 0.22, which were discussed in Chapter 9. Figure 10.19 shows the model of a UTRA/IMT-2000 downlink transmitter for one user.

10.2.7   Random Access

If data transmission is initiated by a mobile station, it is required to send a random access request to the base station. Since such requests can occur at any time, collisions may result, when two or more mobile stations attempt to access the network simultaneously. Hence in order to reduce the probability of a collision, the random access procedure in UTRA is based on the slotted ALOHA technique [8], which is a statistical multiplexing procedure similar to the PRMA technique discussed in Chapter 9.

Random access requests are transmitted to the base station via the Random Access Channel (RACH). UTRA and IMT-2000 have different RACH burst structures. The burst structure of the RACH in UTRA is shown in Figure 10.20 which is transmitted according to the regime of Figure 10.22, as it will be described during our forthcoming discourse. The duration of one random access burst is 11.25 ms. It consists of a preamble and a message part, while between the preamble and message portion of the burst there is an idle period of 0.25 ms duration. The purpose of the idle period is to allow the base station to detect the preamble and then to prepare subsequently for receiving the message itself [8]. The preamble carries a signature, which is a complex orthogonal Gold code of length 16, spread by a cell-specific 256-chip orthogonal Gold code. The structure of the message part is shown in Figure 10.21, whereby the data and control information are transmitted in parallel, again, by mapping them on the I and Q modem branches, as seen in Figure 10.18. The control part seen at the bottom of

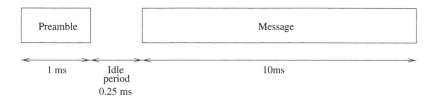

Figure 10.20: Structure of the UTRA random access burst, which is detailed in Figure 10.21.

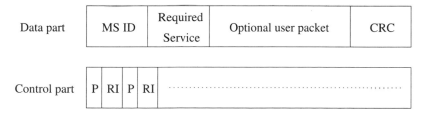

Figure 10.21: Structure of the 10 ms-duration UTRA random access burst message part seen in Figure 10.20. The data and control parts are mapped on the I and Q modem branches in Figure 10.18. RI is the spreading-factor-related rate information, while P represents the pilots.

Figure 10.21 simply contains the pilot symbols (P) and rate information (RI), which contains information about the spreading factor used by the data part. Hence this control part must be detected first in order to infer the spreading factor of the data part at the top of Figure 10.21. The control part has a fixed bit-rate with a spreading factor of 256. On the other hand, the data part of Figure 10.21 can have a spreading factor ranging from 32 to 256, which is communicated to the base station with the aid of the RI in the control part, as mentioned previously. The data part consists of a random mobile station identification (MS ID), a 'Required Service' field and Cyclic Redundancy Checking (CRC) for error detection. The required service indicates the function of the random access burst. If the actual information packets to be transmitted from the mobile station are short and infrequent, then these packets can be transmitted in the 1'Optional user packet' field of Figure 10.21. However, if the packets to be transmitted are long and frequent, then the mobile station will request a dedicated physical channel to be set up in order to transmit those packets. Whether a short packet is transmitted or a request is made for allocating a dedicated physical channel, is indicated in the required service field.

Before any random access burst can be transmitted, the mobile station has to obtain certain information via the downlink BCCH transmitted on the Primary CCPCH according to the format of Figure 10.5. The informa-

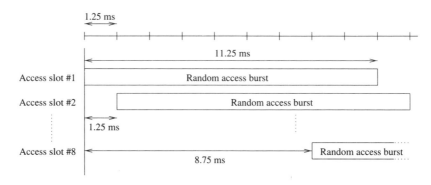

Figure 10.22: ALOHA-based physical uplink random access bursts in UTRA, which are transmitted using the RACH burst format of Figures 10.21 and 10.21.

tion includes the cell-specific spreading codes for the preamble and the message of Figures 10.20 and 10.21 itself, the available signatures, the uplink access slots of Figure 10.22, which can be contended for in ALOHA mode and the spreading factors for the message, the interference level measured at the base station, and the Primary CCPCH transmit power level. All this information can be readily available, once synchronization is achieved, as it will be discussed in Section 10.2.9.

The random access slot starting instants are spaced 1.25 ms apart in Figure 10.22. The random access burst can only be transmitted in one of these access slots. Hence, the physical random access channel scheduling takes place, as shown in Figure 10.22, with 8 available access slots in a 10 ms frame duration. After acquiring all the necessary information via the BCCH which is mapped on the PCCPCH according to the format of Figure 10.5, the mobile station will randomly select a signature from the available signatures and will commence transmitting its uplink RACH bursts according to the formats of Figures 10.20 and 10.21 on a randomly selected access slot chosen from the set of available access slots as seen in Figure 10.22. The transmit power is adjusted via an open-loop power control scheme, which will be highlighted in Section 10.2.8.2 since at this stage of the mobile station's access no closed-loop power-control is possible.

After the RACH burst is transmitted, the mobile station will listen for the acknowledgement of reception transmitted from the base station on the FACH of Table 10.4. If the mobile station fails to receive any acknowledgement after some pre-defined time, it will retransmit the RACH burst in another randomly selected slot of Figure 10.22 and the procedure is repeated.

In IMT-2000 a different random access burst structure is used, which is shown in Figure 10.23. The data are carried on the I branch, while the signature is repeatedly transmitted on the Q branch of the modem, as seen

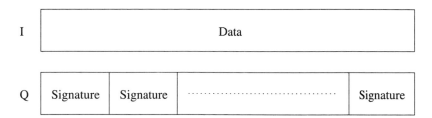

Figure 10.23: Structure of the 10 ms-duration IMT-2000 uplink random access burst. The corresponding UTRA-burst was shown in Figure 10.21, while the ALOHA procedure obeys Figure 10.22. The data and the signature are mapped to the I and Q branches of Figure 10.18.

in Figure 10.18. The procedure of access request is the same as in UTRA. Let us now consider the issues of power control in the next section.

10.2.8   Power Control

Accurate power control is essential in CDMA in order to mitigate the so-called near-far problem [37, 39], as we argued in Chapter 1. Furthermore, power control has a dramatic effect on the coverage and capacity of the system.

10.2.8.1   Closed-Loop Power Control in UTRA/IMT-2000

Closed-loop power control is employed on both the uplink and downlink. Since the power control procedure is the same on both links, we will only elaborate on the uplink procedure, noting that the TPC commands are conveyed in the downlink and uplink directions according to the format of Figures 10.5 and 10.6, respectively.

The base station measures the received power of the desired uplink DPCCH transmitted using the schematic of Figure 10.18 after RAKE combining and also estimates the total received interference power in order to obtain the estimated Signal-to-Interference Ratio (SIR). This SIR estimation process is performed every 0.625 ms, or a timeslot duration, in which the SIR estimate is compared to a target SIR. The value of the target SIR depends on the required quality of the connection. According to the values of the measured and required SIRs, the base station will generate a transmit power control (TPC) command, which is conveyed to the mobile station using the burst at the top of Figure 10.5. If the estimated SIR is higher than the target SIR, the TPC command will instruct the mobile station to lower the transmit power of the DPDCH and DPCCH by a step size of Δ_{TPC} dB. Otherwise, the TPC command will instruct the mobile station to increase the transmit power by the same step size. The step size may differ from cell to cell and it is typically 0.25-1.5 dB[4]. Explicitly, transmitting at an unnecessarily high power reduces the battery life, while degrading other users' reception quality, who - as a consequence - may request a power increment, ultimately resulting in an unstable overall system operation.

On the downlink, so-called base station-diversity combining may take place, whereby two or more base stations transmit the same information to the mobile station in order to enhance its reception. These base stations are known as the active base station set of the mobile station. Power control is performed by all the base stations independently. Hence, the mobile station may receive different TPC commands from its active set of base stations. In this case, the mobile station will adjust its transmit power according to a simple algorithm, increasing the transmit power only, if the TPC commands from all the base stations indicate an 'increase power' instruction. If one of the base stations issued a 'decrease power' TPC command, then the mobile station will decrease its transmit power according to the required step size. If the mobile station received more than one 'decrease power' TPC command, then it will decrease its transmit power according to the largest step size indicated. In this way, the multiuser interference will be kept to a minimum without significant deterioration of the performance, since at least one base station has a good reception. Again, the uplink and downlink procedures are identical, obeying the TPC transmission formats of Figures 10.5 and 10.6, respectively.

10.2.8.2 Open-Loop Power Control During the Mobile Station's Access

As mentioned in Section 10.2.7, open-loop power control is used to adjust the transmit power for the random access burst of Figures 10.20-10.23, since no closed-loop operation is feasible at this stage of the mobile station's access request. Prior to any data burst transmission, the mobile station would have acquired information about the interference level measured at the base station and also about the base station's Primary CCPCH transmitted signal level, which are conveyed to the mobile station via the BCCH according to the format of Figure 10.5. At the same time, the mobile station would also measure the power of the received Primary CCPCH. Hence, with the knowledge of the transmitted and received power of the Primary CCPCH, the downlink path loss can be found. The random access burst of Figures 10.20-10.23 should be received by the base station under all practical conditions. Since the interference level and the path loss are now known, the required transmitted power of the random access burst can be readily calculated.

---

[4]In IMT-2000, the step size is fixed at 1dB

10.2.9 Cell Identification

System- and cell-specific information are conveyed via the BCCH transmitted by the Primary CCPCH of Table 10.4 and Figure 10.5 in UTRA or over the so-called Perch Channel of Table 10.5 in IMT-2000, which is transmitted according to the format of Figure 10.24 from the corresponding base station to the mobile station, as mentioned in Section 10.2.3. Figure 10.24 will be detailed in our forthcoming discourse. Before a mobile station can access the network, a variety of system- and cell-specific information has to be obtained. The Primary CCPCH of Figure 10.5 is also spread by a cell-specific scrambling code, identical to the scrambling code c_{scramb} shown in Figure 10.19, which minimises the inter-cell interference and assists in identifying the cells. Hence, the first step for the mobile station is to recognize this scrambling code and to synchronize with the corresponding base station.

As specified in Section 10.2.6.3, there are a total of $32\times16{=}512$ downlink scrambling codes available in the network. Theoretically it is possible to achieve scrambling code identification by cross-correlating the Primary CCPCH broadcast signal with all the possible 512 scrambling codes. However, this would be an extremely tedious and slow process, unduly delaying the mobile station's access to the network. Furthermore, synchronization between the base station and the mobile station has to be established. In order to achieve a fast cell identification by the mobile station, UTRA and IMT-2000 adopted a 3-step approach [38], which invoked the synchronization channel (SCH) and the perch channel broadcast from all the base stations in the network, respectively. The perch channel of IMT-2000 basically carries out the functions of the SCH in UTRA. The SCH is transmitted during frame synchronization along with all the downlink physical channels, as it will be highlighted in the context of Figure 10.24. The concept behind this 3-step approach is to divide the total number of possible scrambling codes into groups, in this case into 32 groups, each containing a smaller set of scrambling codes, namely 16 codes, yielding a total of 512 codes. Once the knowledge of which group the scrambling code belongs to is acquired, the mobile station can proceed to search for the correct scrambling code from a smaller subset of the possible codes.

The frame structure of the downlink synchronization channel is shown in Figure 10.24, where the slots correspond to those shown in Figure 10.4. It consists of two sub-channels, the so-called Primary SCH and Secondary SCH, transmitted in parallel using orthogonal code multiplexing. As seen in Figure 10.24, in the Primary SCH an unmodulated orthogonal Gold code, known as the Primary Synchronization Code (PSC), of length 256 chips is transmitted periodically at the beginning of each slot, which is denoted by c_p in Figure 10.24. The same PSC is used by all the base stations in the network. This allows the mobile station to establish slot-synchronization and to proceed to the frame-synchronization phase with the aid of the secondary SCH. On the Secondary SCH, a sequence of 16 different consecutive unmodulated orthogonal Gold codes, each of length 256 chips, are transmitted with a period of one radio frame duration, ie 10 ms, as seen at the bottom of Figure 10.24. An example of this sequence would be

$$c_1\ c_1\ c_2\ c_{11}\ c_6\ c_3\ c_{15}\ c_7\ c_8\ c_8\ c_7\ c_{15}\ c_3\ c_6\ c_{11}\ c_2, \qquad (10.7)$$

where each of these 16 orthogonal Gold codes are selected from a set of 17 different orthogonal Gold codes, known as Secondary Synchronization Codes (SSC). These 17 SSCs are also orthogonal to the PSC. The specific sequence of 16 SSCs, denoted by $c_i^1, \ldots, c_i^{16}$ where $i = 1, \ldots, 17$ in Figure 10.24 is used as a code in order to identify and signal to the mobile station, which of the 32 scrambling code groups the scrambling code - which is used by the particular base station concerned - belongs to. Specifically, when each of the 17 legitimate 256-chip sequences can be picked for any of the 16 positions in Figure 10.24 and with no other further constraints, one could construct

$$
\begin{aligned}
c_{i,j}^{\text{repeated}} &= \binom{i + j - 1}{j} \\
&= \frac{(i + j - 1)!}{j!(i - 1)!} \\
&= \frac{32!}{16! \cdot 16!} \\
&= 601,080,390 \qquad (10.8)
\end{aligned}
$$

different such sequences, where $i = 17$ and $j = 16$. However, the 16 different 256-chip sequences seen at the bottom of Figure 10.24 must be constructed such that their cyclic shifts are also unique. In other words, none of the cyclic shifts of the 32 required $16 \times 256 = 4096$-chip sequences can be identical to any of the other sequences' cyclic shifts. Provided that these conditions are satisfied, the 16 specific 256-chip secondary SCH sequences can be recognized within one 10 ms-radio frame-duration of 16 slots and hence both slot and frame synchronization can be established within the particular frame received. The BCH data of Table 10.5 in IMT-2000 and the pilot symbols following the PSC only appeared in the perch channel, as illustrated by the dotted slot in Figure 10.24. By contrast, these two types of information are transmitted in the PCCPCH in UTRA, as shown in Figure 10.5. Using this technique, initial cell identification and synchronization can be carried out in three basic steps. **Step one:** The mobile station uses the 256-chip PSC of Figure 10.24 to perform cross-correlation with all the received Primary SCHs of the base stations in its vicinity. The base station with the highest correlator output is then chosen, which constitutes the best cell site with the lowest path loss. Several periodic correlator output peaks have to be identified in order to achieve a high detection re-

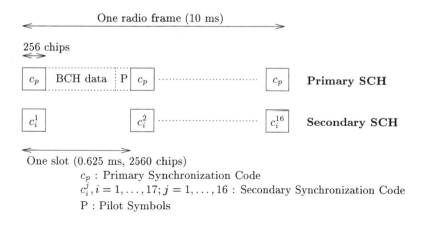

c_p : Primary Synchronization Code
$c_i^j, i = 1, \ldots, 17; j = 1, \ldots, 16$: Secondary Synchronization Code
P : Pilot Symbols

Figure 10.24: Frame structure of the UTRA/IMT-2000 downlink synchronization channel, which is mapped on the slots of Figure 10.4. The primary and secondary SCH are transmitted in parallel using orthogonal codes. In IMT-2000, the BCH data are correlated with the possible scrambling codes [8].

liability. Slot synchronization is also achieved in this step by recognizing the 16 consecutive c_p sequences, providing 16 periodic correlation peaks.

Step two: Once the best cell site is identified, the scrambling code group of that cell site is found by cross-correlating the Secondary SCH with the 17 possible SSCs in each of the 16 timeslots of Figure 10.24. This can be easily implemented using 17 correlators, since the timing of the SSCs is known from Step 1. Hence, there are a total of $16 \times 17 = 272$ correlator outputs. From these outputs, a total of $32 \times 16 = 512$ decision variables corresponding to the 32 possible sequences and 16 cyclic shifts of each $16 \times 256 = 4096$-chip sequence are obtained. The highest decision variable determines the scrambling code group. Consequently, frame synchronization is also achieved.

Step three: With the scrambling code group identified and frame synchronization achieved, the scrambling code is acquired in UTRA by cross-correlating the received Primary CCPCH signal symbol-by-symbol with the 16 possible scrambling codes belonging to the identified group. In IMT-2000, the cross-correlation is performed on the BCH data symbol-by-symbol. Once the exact scrambling code is identified, the BCCH information of Table 10.4, which is conveyed by the PCCPCH of Figure 10.5, can be detected. Let us now consider some of the associated handover issues.

10.2.10 Handover

Theoretically, DS-CDMA has a frequency reuse factor of one [40]. This implies that neighbouring cells can use the same carrier frequency without interfering with each other, unlike in TDMA or FDMA. Hence, seamless uninterrupted handover can be achieved, when mobile users move between cells, since no switching of carrier frequency and synthesizer re-tuning is required. However, in hierarchical cell structures (HCS)[5], using a different carrier frequency is necessary in order to reduce the inter-cell interference. In this case, interfrequency handover is required. Furthermore, since the various operational GSM systems used different carrier frequencies, handover from UTRA systems to GSM systems will have to be supported during the transitory migration phase, while these systems will co-exist. Hence, handovers in terrestrial UMTSs can be classified into inter-frequency and intra-frequency handovers.

10.2.10.1 Intra-frequency Handover or Soft Handover

Soft handover [41, 42] involves no frequency switching, since the new and old cells use the same carrier frequency. The mobile station will continuously monitor the received signal levels from the neighbouring cells and compares them against a set of thresholds. This information is fed back to the network. Based on this information, if a weak or strong cell is detected, the network will instruct the mobile station to drop or add the cell from/to its active base station set. In order to ensure a seamless handover, a new link will be established before relinquishing the old link, using the so-called 'make before break' approach.

10.2.10.2 Inter-frequency Handover or Hard Handover

In order to achieve handovers between different carrier frequencies without affecting the data flow, a technique known as slotted mode[6] can be used [43]. According to this technique, the downlink data, which normally occupy the entire 10 ms frame of Figure 10.25 are 'time-compressed', such that they only occupy a portion of the frame, while no data are transmitted during the remaining portion, as shown in Figure 10.25. The latter portion is known as the idle period and it has a variable duration. The idle period can occur at the beginning, at the centre, or at the end of the frame. The compression of data can be achieved by channel-code puncturing, a procedure, which obliterates some of the coded parity bits, thereby slightly reducing the code's error correcting power (see Chapter 4), or by adjusting the coding rate. In order to maintain the quality of the link, the instantaneous power is also increased during the slotted mode operation. After receiving the data, the mobile station can use this idle period in the frame in order to switch

---

[5]Microcells overlaid by a macrocell.
[6]Slotted mode is also known as Compressed Mode

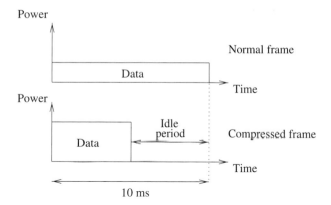

Figure 10.25: Downlink frame structure during slotted mode of operation during UMTS/IMT-2000 handovers.

to other carrier frequencies of other cells and to perform the necessary link-quality measurements for handover.

Alternatively, a dual receiver can be used in order to perform inter-frequency handovers. One receiver can be tuned to the desired carrier frequency for reception, while the other receiver can be used to perform handover link-quality measurements at other carrier frequencies. This method results in a higher hardware complexity at the mobile station.

The 10 ms frame length of UTRA/IMT-2000 was chosen such that it is compatible with the multiframe length of 120 ms in GSM, as seen in Chapter 8. Hence, the mobile station is capable of receiving the Frequency Correction Channel (FCCH) and Synchronization Channel (SCH) in the GSM frame using slotted mode transmission and to perform the necessary handover link-quality measurements [6].

10.2.11 Inter-cell Time Synchronization in the UTRA/ IMT-2000 TDD mode

Time-synchronization between base stations is required, when operating in TDD mode in order to support seamless handovers. A simple method of maintaining inter-cell synchronization is by periodically broadcasting a so-called beacon code from a source to all the base stations. The propagation delay can be easily calculated from the fixed distance between the source and the receiving base stations. There are three possible ways of transmitting this beacon code: via the terrestrial radio link, via the physical wired network, or via the Global Positioning System (GPS).

Global time-synchronization in 3G mobile radio systems is achieved by dividing the synchronous coverage region into three areas, namely the so-called sub-area, main-area and coverage area, as shown in Figure 10.26.

Inter-cell synchronization within a sub-area is provided by a sub-area bea-

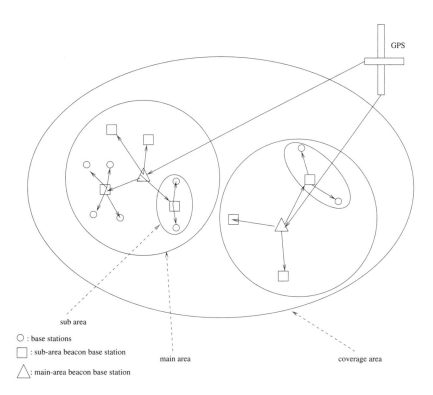

Figure 10.26: Inter-cell time-synchronization in UTRA/IMT-2000 TDD mode.

con base station. Since the sub-area of Figure 10.26 is smaller than the
main-area, transmitting the beacon code via the terrestrial radio link or
the physical wired network is more feasible. All the sub-area beacon base
stations in a main-area are in turn synchronized by a main-area beacon
base station. Similarly, the beacon code can be transmitted via the terrestrial radio link or the physical wired network. Finally, all the main-area
beacon base stations are synchronised using the GPS system. The main
advantage of dividing the coverage regions into smaller areas is that each
lower hierarchical area can still operate on its own, even if the synchronization link with the higher hierarchical areas is lost. Having reviewed the
basic features of UTRA/IMT-2000, let us now consider the Pan-American
cdma2000 system.

Radio Access Technology	DS-CDMA, Multicarrier CDMA
Operating environments	Indoor/Outdoor to indoor/Vehicular
Chip rate (Mcps)	1.2288/3.6864/7.3728/11.0592/14.7456
Channel bandwidth (MHz)	1.25/3.75/7.5/11.25/15
Duplex modes	FDD and TDD
Frame length	5 and 20 ms
Spreading factor	variable, 4 to 256
Detection scheme	Coherent with common pilot channel
Inter-cell operation	FDD : Synchronous
	TDD : Synchronous
Power control	Open- and closed-loop
Handover	Soft-handover
	Inter-frequency handover

Table 10.8: The cdma2000 basic parameters.

10.3 The cdma2000 Terrestrial Radio Access [44]- [46]

The current 2G mobile radio systems standardised by TIA in the United States are IS-95-A and IS-95-B [44]. The radio access technology of both systems is based on narrowband DS-CDMA with a chip rate of 1.2288 Mcps, which gives a bandwidth of 1.25 MHz. The basic features of this system were summarized in Table 1.1 of Chapter 1. IS-95-A was commercially launched in 1995, supporting circuit and packet mode transmissions at a maximum bit-rate of only 14.4 kbps [44]. An enhancement to the IS-95-A standards, known as IS-95-B, was developed and introduced in 1998 in order to provide higher data rates, on the order of 115.2 kbps [18]. This was feasible without changing the physical layer of IS-95-A. However, this still falls short of the 3G mobile radio system requirements. Hence the technical committee TR45.5 within TIA has proposed cdma2000, a 3G mobile radio system that is able to meet all the requirements laid down by ITU. One of the problems faced by TIA is that the frequency bands allocated for the 3G mobile radio system, identified during WARC'92 to be 1885-2025 MHz and 2110-2200 MHz, has already been allocated for Personal Communications Services (PCS) in the United States from 1.8 GHz to 2.2 GHz. In particular, the CDMA PCS based on the IS-95 standards has been allocated the frequency bands of 1850-1910 MHz and 1930-1990 GHz. Hence, the 3G mobile radio systems have to fit into the allocated bandwidth without imposing significant interference on the existing applications. Thus, the framework for cdma2000 was designed such that it can be overlaid on IS-95 and it is backwards compatible with IS-95. Most of this section is based on references [44]- [46].

10.3.1 Characteristics of cdma2000

The basic parameters of cdma2000 are shown in Table 10.8. The cdma2000

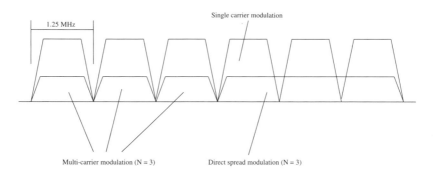

Figure 10.27: Example of an overlay deployment in cdma2000. The multi-carrier mode is only used in the downlink.

system has a basic chip rate of 3.6864 Mcps, which is accommodated in a bandwidth of 3.75 MHz. This chip rate is in fact three times the chip rate used in the IS-95 standards, which is 1.2288 Mcps. Accordingly, the bandwidth was also trebled. Hence, the existing IS-95 networks can also be used to support the operation of cdma2000. Higher chip rates on the order of $N \times 1.2288$ Mcps, $N = 6, 9, 12$ are also supported. These are used to enable higher bit-rate transmission. The value of N is an important parameter in determining the channel coding rate and the channel bit-rate. In order to transmit the high chip-rate signals ($N > 1$), two modulation techniques are employed. In the **direct-spread modulation mode**, the symbols are spread according to the chip-rate and transmitted using a single carrier, giving a bandwidth of $N \times 1.25$ MHz. This method is used on both the uplink and downlink. In **multicarrier (MC) modulation**, the symbols to be transmitted are de-multiplexed into separate signals, each of which is then spread at a chip rate of 1.2288 Mcps. N different carrier frequencies are used to transmit these spread signals, each of which has a bandwidth of 1.25 MHz. This method is used for the downlink only, since in this case, transmit diversity can be achieved by transmitting the different carrier frequencies over spatially separated antennas.

By using multiple carriers, cdma2000 is capable of overlaying its signals on the existing IS-95 1.25 MHz channels and its own channels, while maintaining orthogonality. An example of an overlay scenario is shown in Figure 10.27. Higher chip rates are transmitted at a lower power, than lower chip rates. Hence, the interferences are kept to a minimum.

Similarly to UTRA and IMT-2000, cdma2000 also supports TDD operation in unpaired frequency bands. In order to ease the implementation of a dual-mode FDD/TDD terminal, most of the techniques used for FDD operation can also be applied in TDD operation. The difference between these two modes is in the frame structure, whereby an additional guard time has to be included for TDD operation.

Dedicated Physical Channels (DPHCH)	Common Physical Channels (CPHCH)
Fundamental Channel (FCH) (UL/DL)	Pilot Channel (PICH) (DL)
Supplemental Channel (SCH) (UL/DL)	Common Auxiliary Pilot Channel (CAPICH) (DL)
Dedicated Control Channel (DCCH) (UL/DL)	Forward Paging Channel (PCH) (DL)
Dedicated Auxiliary Pilot Channel (DAPICH) (DL)	Sync Channel (SYNC) (DL)
Pilot Channel (PICH) (UL)	Access Channel (ACH) (UL)
	Common Control Channel (CCCH) (UL/DL)

Table 10.9: The cdma2000 physical channels.

In contrast to UTRA and IMT-2000, where the pilot symbols of Figures 10.5 and 10.7 are time-multiplexed with the dedicated data channel on the downlink, cdma2000 employs a common code multiplexed continuous pilot channel on the downlink, as in the IS-95 system of Chapter 1. The advantage of a common downlink pilot channel is that no additional overhead is incurred for each user. However, if adaptive antennas are used, then additional pilot channels have to be transmitted from each antenna.

Another difference with respect to UTRA and IMT-2000 is that the base stations are operated in synchronous mode in cdma2000. As a result of this, the same PN code but with different phase offsets can be used to distinguish the base stations. Using one common PN sequence can expedite cell acquisition as compared to a set of PN sequences, as we have seen in Section 10.2.9 for IMT-2000/UTRA. Let us now consider the cdma2000 physical channels.

10.3.2   Physical Channels in cdma2000

The physical channels (PHCH) in cdma2000 can be classified into two groups: Dedicated Physical Channels (DPHCH) and Common Physical Channels (CPHCH). DPHCHs carry information between the base station and a single mobile station, while CPHCHs carry information between the base station and several mobile stations. Table 10.9 shows the collection of physical channels in each group. These channels will be elaborated on during our further discourse. Typically, all physical channels are transmitted using a frame length of 20 ms. However, the control information on the so-called Fundamental Channel (FCH) and Dedicated Control Channel (DCCH) can also be transmitted in 5 ms frames.

Each base station transmits its own downlink Pilot Channel (PICH), which is shared by all the mobile stations within the coverage area of the base station. Mobile stations can use this common downlink PICH in order to perform channel estimation for coherent detection, soft handover and for fast acquisition of strong multipath rays for RAKE combining. The PICH is transmitted orthogonally along with all the other downlink physical channels from the base station by using a unique orthogonal code (Walsh code 0) as in the IS-95 system of Table 1.1 in Chapter 1. The optional Common Auxiliary Pilot Channels (CAPICH) and Dedicated Auxiliary Pilot Channels (DAPICH) are used to support the implementation of antenna arrays. CAPICHs provide spot coverage shared amongst a group of mobile

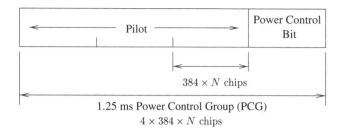

Figure 10.28: Uplink pilot channel structure in cdma2000 for a 1.25 ms duration PCG, where $N = 1, 3, 6, 9, 12$ is the rate-control parameter.

stations, while a DAPICH is directed towards a particular mobile station. Every mobile station also transmits an orthogonal code-multiplexed uplink pilot channel (PICH), which enables the base station to perform coherent detection in the uplink as well as to detect strong multipaths and to invoke power control measurements. This differs from IS-95, which supports only non-coherent detection in the uplink due to the absence of a coherent uplink reference. In addition to the pilot symbols, the uplink PICH also contains time-multiplexed power control bits assisting in downlink power control. A power control bit is multiplexed onto the 20 ms frame every 1.25 ms, giving a total of 16 power control bits per 20 ms frame or 800 power updates per second, implying a very agile, fast response power control regime. Each 1.25 ms duration is referred to as a Power Control Group, as shown in Figure 10.28.

The use of two dedicated data physical channels, namely the so-called Fundamental (FCH) and Supplemental (SCH) channels, optimizes the system during multiple simultaneous service transmissions. Each channel carries a different type of service and is coded and interleaved independently. However, in any connection, there can be only one FCH, but several SCHs can be supported. For a FCH transmitted in a 20 ms frame, two sets of uncoded data rates, denoted as Rate Set 1 (RS1) and Rate Set 2 (RS2), are supported. The data rates in RS1 and RS2 are 9.6/4.8/2.7/1.5 kbps and 14.4/7.2/3.6/1.8 kbps, respectively. Regardless of the uncoded data rates, the coded data rate is 19.2 kbps and 38.4 kbps for RS1 and RS2, respectively, when the rate-control parameter is $N = 1$. The 5 ms frame only supports one data rate, which is 9.6 kbps. The SCH is capable of transmitting higher data rates, than the FCH. The SCH supports variable data rates ranging from 1.5 kbps for $N=1$ to as high as 2073.6 kbps, when $N=12$. Blind rate detection [47] is used for SCHs not exceeding 14.4 kbps, while rate information is explicitly provided for higher data rates. The dedicated control physical channel has a fixed uncoded data rate of 9.6 kbps on both 5ms and 20 ms frames. This control channel rate is more than an order of magnitude higher than that of the IS-95 system in Table 1.1, and hence supports a substantially enhanced system control.

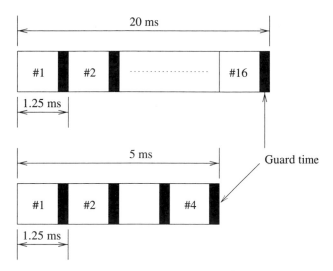

Figure 10.29: The cdma2000 TDD frame structure.

The Sync Channel (SYCH) - note the different acronym in comparison to the SCH abbreviation in UTRA/IMT-2000 - is used to aid the initial synchronization of a mobile station to the base station and to provide the mobile station with system-related information, including the Pseudo Noise (PN) sequence offset, which is used to identify the base stations and the long code mask, which will be defined explicitly in Section 10.3.4. The SYCH has an uncoded data rate of 1.2 kbps and a coded data rate of 4.8 kbps.

Paging functions and packet data transmission are handled by the downlink Paging Channel (PCH) and the downlink Common Control Channel (CCCH). The uncoded data rate of the PCH can be either 4.8 kbps or 9.6 kbps. The CCCH is an improved version of the PCH, which can support additional higher data rates, such as 19.2 and 38.4 kbps. In this case, a 5 ms or 10 ms frame length will be used. The PCH is included in cdma2000 in order to provide IS-95-B functionality.

In TDD mode, the 20 ms and 5 ms frames are divided into 16 and 4 timeslots, respectively. This gives a duration of 1.25 ms per timeslot, as shown in Figure 10.29. A guard time of 52.08 μs and 67.44 μs is used for the downlink in multicarrier modulation and for direct spread modulation, respectively. In the uplink, the guard time is 52.08 μs. Having described the cdma2000 physical channels of Table 10.9, let us now consider the service multiplexing and channel coding aspects.

10.3.3   Service Multiplexing and Channel Coding

Services of different data rates and different QoS requirements are carried by different physical channels, namely by the FCH and SCH of Table 10.9. This differs from UTRA and IMT-2000, whereby different services were time multiplexed onto one or more physical channels, as seen in Figure 10.12. These channels in cdma2000 are code-multiplexed using Walsh codes. Two types of coding schemes are used in cdma2000, as shown in Table 10.10. Basically, all channels use convolutional codes for forward error correction which were discussed in Chapter 4. However, for SCHs at rates higher than 14.4 kbps, turbo coding [28] is preferable. The rate of the input data stream is matched to the given channel rate by either adjusting the coding rate, or using symbol repetition with and without symbol puncturing, or alternatively, by sequence repetition. Tables 10.11 and 10.12 show the coding rate and the associated rate matching procedures for the various downlink and uplink physical channels, respectively, when $N = 1$. Following the above brief notes on the cdma2000 channel coding and service multiplexing issues, let us now focus our attention on the spreading and modulation processes.

	Convolutional	Turbo
Rate	1/2 or 1/3 or 1/4	1/2 or 1/3 or 1/4
Constraint length	9	4

Table 10.10: The cdma2000 channel coding parameters.

10.3.4   Spreading and Modulation

There are generally three layers of spreading in cdma2000, as shown in Table 10.13. Each user's uplink signal is identified by different offsets of a long code, a procedure which is similar to that of the IS-95 system portrayed in Chapter 1. As seen in Table 10.13, this long code is an m-sequence with a period of $2^{42} - 1$ chips. The construction of m-sequences was highlighted by Proakis in reference [26]. Different user offsets are obtained using a long code mask. Orthogonality between the different physical channels of the same user belonging to the same connection in the uplink is maintained by spreading using Walsh codes, which were introduced in Chapter 1.

In contrast to the IS-95 downlink of Figure 1.42, whereby Walsh code spreading is performed prior to QPSK modulation, the data in cdma2000 are first QPSK modulated, before spreading the resultant I and Q branches with the same Walsh code. In this way, the number of Walsh codes available is increased twofold due to the orthogonality of the I and Q carriers. The length of the uplink/downlink (UL/DL) channelization Walsh codes of Table 10.13 varies according to the data rates. All the base stations in the system are distinguished by different offsets of the same complex downlink m-sequence, as indicated by Table 10.13. This downlink m-sequence code

Physical Channel	Data Rate	Code Rate	Repetition	Puncturing	Channel rate
SYCH	1.2 kbps	1/2	×2	0	4.8 ksps
PCH	4.8 kbps	1/2	×2	0	19.2 ksps
	9.6 kbps	1/2	×1	0	19.2 ksps
CCCH	9.6 kbps	1/2	×1	0	19.2 ksps
	19.2 kbps	1/2	×1	0	38.4 ksps
	38.4 kbps	1/2	×1	0	76.8 ksps
FCH	1.5 kbps	1/2	×8	1 of 5	19.2 ksps
	2.7 kbps	1/2	×4	1 of 9	19.2 ksps
	4.8 kbps	1/2	×2	0	19.2 ksps
	9.6 kbps	1/2	×1	0	19.2 ksps
	1.8 kbps	1/3	×8	1 of 9	38.4 ksps
	3.6 kbps	1/3	×4	1 of 9	38.4 ksps
	7.2 kbps	1/3	×2	1 of 9	38.4 ksps
	14.4 kbps	1/3	×1	1 of 9	38.4 ksps
SCH	9.6 kbps	1/2	×1	0	19.2 ksps
	19.2 kbps	1/2	×1	0	38.4 ksps
	38.4 kbps	1/2	×1	0	76.8 ksps
	76.8 kbps	1/2	×1	0	153.6 ksps
	153.6 kbps	1/2	×1	0	307.2 ksps
	307.2 kbps	1/2	×1	0	614.4 ksps
	14.4 kbps	1/3	×1	1 of 9	38.4 ksps
	28.8 kbps	1/3	×1	1 of 9	76.8 ksps
	57.6 kbps	1/3	×1	1 of 9	153.6 ksps
	115.2 kbps	1/3	×1	1 of 9	307.2 ksps
	230.4 kbps	1/3	×1	1 of 9	614.4 ksps
DCCH	9.6 kbps	1/2	×1	0	19.2 ksps

Table 10.11: The cdma2000 downlink physical channel (see Table 10.9) coding parameters for $N = 1$, where Repetition ×2 implies transmitting a total of two copies.

is the same as that used in IS-95, which has a period of $2^{15} = 32768$ and it is derived from m-sequences. The feedback polynomials of the shift registers for the I and Q sequences are $X^{15} + X^{13} + X^9 + X^8 + X^7 + X^5 + 1$ and $X^{15} + X^{12} + X^{11} + X^{10} + X^6 + X^5 + X^4 + X^3 + 1$, respectively. The offset of these codes must satisfy a minimum value, which is equal to $N \times 64 \times$Pilot_Inc, where Pilot_Inc is a so-called code reuse parameter, which depends on the topology of the system, analogously to the frequency reuse factor in FDMA. Let us now focus our attention on downlink spreading issues more closely.

10.3.4.1   Downlink Spreading and Modulation

Figure 10.30 shows the structure of a downlink transmitter for a physical channel. In contrast to the IS-95 downlink transmitter shown in Figure 1.42 of Chapter 1, the data in the cdma2000 downlink transmitter shown in Figure 10.30 are first QPSK modulated before spreading using Walsh codes. As a result, the number of Walsh codes available is increased twofold due to the orthogonality of the I and Q carriers, as mentioned previously. The

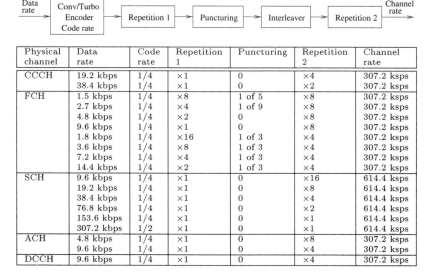

Physical channel	Data rate	Code rate	Repetition 1	Puncturing	Repetition 2	Channel rate
CCCH	19.2 kbps	1/4	×1	0	×4	307.2 ksps
	38.4 kbps	1/4	×1	0	×2	307.2 ksps
FCH	1.5 kbps	1/4	×8	1 of 5	×8	307.2 ksps
	2.7 kbps	1/4	×4	1 of 9	×8	307.2 ksps
	4.8 kbps	1/4	×2	0	×8	307.2 ksps
	9.6 kbps	1/4	×1	0	×8	307.2 ksps
	1.8 kbps	1/4	×16	1 of 3	×4	307.2 ksps
	3.6 kbps	1/4	×8	1 of 3	×4	307.2 ksps
	7.2 kbps	1/4	×4	1 of 3	×4	307.2 ksps
	14.4 kbps	1/4	×2	1 of 3	×4	307.2 ksps
SCH	9.6 kbps	1/4	×1	0	×16	614.4 ksps
	19.2 kbps	1/4	×1	0	×8	614.4 ksps
	38.4 kbps	1/4	×1	0	×4	614.4 ksps
	76.8 kbps	1/4	×1	0	×2	614.4 ksps
	153.6 kbps	1/4	×1	0	×1	614.4 ksps
	307.2 kbps	1/2	×1	0	×1	614.4 ksps
ACH	4.8 kbps	1/4	×1	0	×8	307.2 ksps
	9.6 kbps	1/4	×1	0	×4	307.2 ksps
DCCH	9.6 kbps	1/4	×1	0	×4	307.2 ksps

Table 10.12: The cdma2000 uplink physical channel (see Table 10.9) coding parameters for $N = 1$, where Repetition ×2 implies transmitting a total of two copies.

	Channelization Codes (UL/DL)	User-specific scrambling-codes (UL)	Cell-specific scrambling codes (DL)
Type of codes	Walsh codes	Different offsets of a real m-sequence	Different offsets of a complex m-sequence
Code length	Variable	$2^{42} - 1$ chips	2^{15} chips
Type of Spreading	BPSK	BPSK	QPSK
Data Modulation	DL : QPSK UL : BPSK		

Table 10.13: Spreading parameters in cdma2000.

user data are first scrambled by the long scrambling code by assigning a different offset to different users for the purpose of improving user privacy, which is then mapped to the I and Q channels. This long scrambling code is identical to the uplink user-specific scrambling code given in Table 10.13. The downlink pilot channels of Table 10.9 (PICH, CAPICH, DAPICH) and the SYNC channel are not scrambled with a long code since there is no need for user-specificity. The uplink power control symbols are inserted into the FCH at a rate of 800 Hz, as it was shown in Figure 10.30. The I and Q channels are then spread using a Walsh code and complex multiplied with the cell-specific complex PN sequence of Table 10.13, as portrayed in Figure 10.30. Each base station's downlink channel is assigned a different Walsh code in order to eliminate any intra-cell interference since all Walsh codes transmitted by the serving base station are received synchronously.

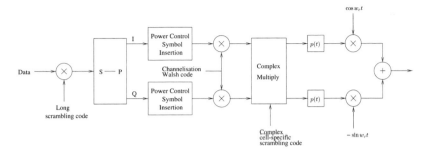

Figure 10.30: The cdma2000 downlink transmitter. The long scrambling code
is used for the purpose of improving user privacy. Hence, only the
paging channels and the traffic channels are scrambled with the
long code. The common pilot channel and the SYNC channel are
not scrambled by this long code (the terminology of Table 10.13
is used).

The length of the downlink channelization Walsh code of Table 10.13 is
determined by the type of physical channel and its data rate. Typically
for $N = 1$, downlink FCHs with data rates belonging to RS1, i.e. those
transmitting at $9.6/4.8/2.7/1.5$ kbps, use a 128-chip Walsh code and those
in RS2, transmitting at $14.4/7.2/3.6/1.8$ kbps use a 64-chip Walsh code.
Walsh codes for downlink SCHs can range from 4-chip to 128-chip Walsh
codes. The downlink PICH is an unmodulated sequence (all 0s) spread by
Walsh code 0. Finally, the complex spread data in Figure 10.30 is baseband
filtered using the Nyquist filter impulse responses $p(t)$ in Figure 10.30 and
modulated on a carrier frequency.

For the case of multi-carrier modulation, the data are split into N
branches immediately after the long code scrambling of Figure 10.30 which
was omitted in the figure for the sake of simplicitly. Each of the N branches
is then treated as a separate transmitter and modulated using different carrier frequencies.

10.3.4.2   Uplink Spreading and Modulation

The uplink cdma2000 transmitter is shown in Figure 10.31. The uplink
PICH and DCCH of Table 10.9 are mapped to the I data channel, while
the uplink FCH and SCH of Table 10.9 are mapped to the Q channel in
Figure 10.31. Each of these uplink physical channels belonging to the same
user is assigned different Walsh channelization codes in order to maintain
orthogonality, with higher rate channels using shorter Walsh codes. The
I and Q data channels are then spread by complex multiplication with
the user-specifically offset real m-sequence based scrambling code of Table 10.13 and a complex scrambling code, which is the same for all the
mobile stations in the system, as seen at the top of Figure 10.31. However,

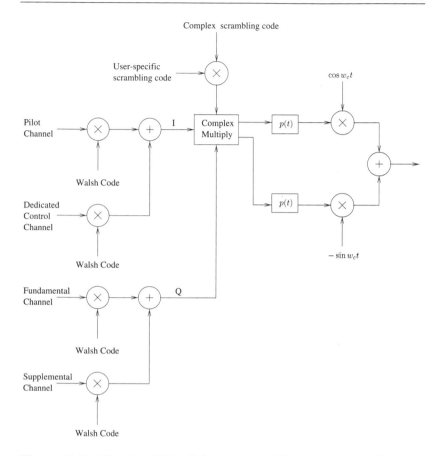

Figure 10.31: The cdma2000 uplink transmitter. The complex scrambling code
is identical to the downlink cell-specific complex scrambling code
of Table 10.13 used by all the base stations in the system (the
terminology of Table 10.13 is used).

this latter complex scrambling code is not explicity shown in Table 10.13,
since it is identical to the downlink cell-specific scrambling code. This complex scrambling code is only used for the purpose of quadrature spreading.
Hence, in order to reduce the complexity of the base station receiver, this
complex scrambling code is identical to the cell-specific scrambling code
of Table 10.13 used on the downlink by all the base stations. Let us now
consider the cdma2000 random access process.

10.3.5 Random Access

The mobile station initiates an access request to the network by repeatedly transmitting a so-called 'access probe', until a request acknowledgement is received. This entire process of sending a request is known as an 'access attempt'. Within a single access attempt, the request may be sent to several base stations. An access attempt addressed to a specific base station is known as a 'sub-attempt'. Within a sub-attempt, several access probes with increasing power can be sent. Figure 10.32 shows an example of an access attempt. The access probe transmission follows the slotted ALOHA

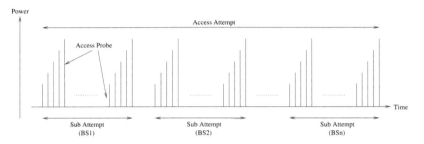

Figure 10.32: An access attempt by a mobile station in cdma2000 using the access probe of Figure 10.33.

algorithm, which is a relative of PRMA portrayed in Chapter 9. An access probe can be divided into two parts, as shown in Figure 10.33 The access

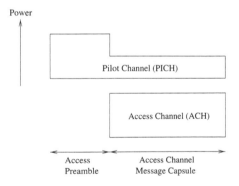

Figure 10.33: A cdma2000 access probe transmitted using the regime of Figure 10.32.

preamble carries a non-data bearing pilot channel, at an increased power level. The so-called 'access channel message capsule' carries the data bearing Access Channel (ACH) or uplink Common Control Channel (CCCH) messages of Table 10.9 and the associated non-data bearing pilot channel. The structure of the pilot channel is similar to that of the uplink pilot channel (PICH) of Figure 10.28 except that in this case, there are no time-multiplexed power control bits. The preamble length in Figure 10.33 is an integer multiple of the 1.25 ms slot intervals. The specific access preamble length is indicated by the base station, which depends on how fast the base station can search the PN code space in order to recognize an access attempt. The ACH is transmitted at a fixed rate of either 9.6 or 4.8 kbps, as seen in Table 10.12. This rate is constant for the duration of the access probe of Figure 10.32. The ACH or CCCH and their associated pilot channel are spread by the spreading codes of Table 10.13, as shown in Figure 10.34. Different ACHs or CCCHs and their associated pilot channels are spread by different long codes.

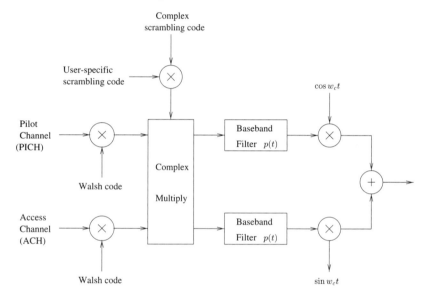

Figure 10.34: The cdma2000 access channel modulation and spreading. The complex scrambling code is identical to the downlink cell-specific complex scrambling code of Table 10.13 used by all the base stations in the system (the terminology of Table 10.13 is used).

The access probes of Figure 10.33 are transmitted in pre-defined slots, where the slot length is indicated by the base station. Each slot is sufficiently long in order to accommodate the preamble and the longest message of Figure 10.33. The transmission must begin at the start of each 1.25 ms slot. If an acknowledgement of the most recently transmitted probe is not received by the mobile station after a time-out period, another probe is transmitted in another randomly chosen slot, obeying the regime of Figure 10.32.

Within a sub-attempt of Figure 10.32, a sequence of access probes is transmitted, until an acknowledgement is received from the base station.

Each successive access probe is transmitted at a higher power compared to the previous access probe, as shown in Figure 10.35. The initial power (IP) of the first probe is determined by the open-loop power control plus a nominal offset power that corrects for the open-loop power control imbalance between uplink and downlink. Subsequent probes are transmitted at a power level higher than the previous probe. This increased level is indicated by the Power Increment (PI). Following the above discussions on the mobile station's random access procedures, let us now highlight some of the cdma2000 handover issues.

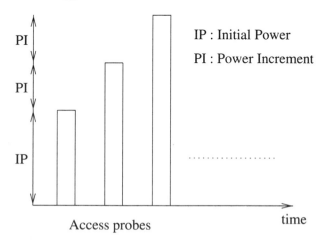

Figure 10.35: Access probes within a sub-attempt of Figure 10.32.

10.3.6 Handover

Intra-frequency or soft-handover is initiated by the mobile station. While communicating, the mobile station may receive the same signal from several base stations. These base stations constitute the so-called 'Active Set' of the mobile station. The mobile station will continuously monitor the power level of the received pilot channels (PICH) transmitted from neighbouring base stations, including those from the mobile station's active set. The power levels of these base stations are then compared against a set of thresholds according to an algorithm, which will be highlighted during our further discourse. The set of thresholds consists of the so-called static thresholds, which are maintained at a fixed level, and the so-called dynamic thresholds, which are dynamically adjusted based on the total received power. Subsequently, the mobile station will inform the network, when any of the monitored power levels exceeds the thresholds.

Whenever the mobile station detects a PICH, whose power level exceeds a given static threshold, denoted as T_1, this PICH will be moved to a so-called candidate set and will be searched and compared more frequently against a dynamically adjusted threshold denoted as T_2. This value of T_2 is a function of the received power levels of the PICHs of the base stations in the active set. This process will determine, whether the candidate base station is worth adding to the active set. If the overall power level in the active set is weak, then adding a base station of higher power would improve the reception. By contrast, if the overall power level in the active set is relatively high, then adding another high-powered base station may not only be unnecessary, it will in fact utilize more network resources.

For the base stations that are already in the active set, the power level of their corresponding PICH is compared against a dynamically adjusted threshold, denoted as T_3, which is also a function of the total power of the PICH in the active set, similar to T_2. This is to ensure that each base station in the active set is contributing sufficiently to the overall power level. If any of the PICH's power level dropped below T_3, after a specified period of time allowed in order to eliminate any uncertainties due to fading which may have caused fluctuations in the power level, the base station will be, again, moved to the candidate set where it will be compared with a static threshold T_4. At the same time, the mobile station will report to the network the identity of the low-powered base station in order to allow the corresponding base station to increase its transmit power. If the power level decreases further below a static threshold, denoted as T_4, then the mobile station will again report this to the network and the base station is subsequently dropped from the candidate set.

Inter-frequency or hard-handovers can be supported between cells having different carrier frequencies. Here we conclude our discussions on the cdma2000 features and provide some rudimentary notes on a number of advanced techniques, which can be invoked in order to improve the performance of the 3G W-CDMA systems.

10.4 Performance Enhancement Features

The treatment of adaptive antennas, multiuser detection, interference cancellation or the portrayal of transmit diversity techniques is beyond the scope of this chapter. Here we simply provide a few pointers to the associated literature.

10.4.1 Adaptive Antennas

The transmission of time-multiplexed pilot symbols on both the uplink and downlink as seen for UTRA and IMT-2000 in Figures 10.5-10.8 facilitates the use of adaptive antennas. Adaptive antennas are known to enhance the capacity and coverage of the system [48, 49].

10.4.2   Multiuser Detection/Interference Cancellation

Following Verdu's seminal paper [50], extensive research has shown that Multiuser Detection (MUD) [36,51–57] and Interference Cancellation techniques [58–69] can substantially improve the performance of the CDMA link in comparison to conventional RAKE receivers. However, using long scrambling codes increases the complexity of the MUD [18]. As a result, UTRA introduced an optional short scrambling code namely the extended Kasami code of Table 10.7, as mentioned in Section 10.2.6.2 in order to reduce the complexity of MUD [8]. Another powerful technique is invoking burst-by-burst adaptive CDMA [23,24] in conjunction with MUD.

On the other hand, interference cancellation schemes require accurate channel estimation in order to reproduce and deduct or cancel the interference. Several stages of cancellation are required in order to achieve a good performance, which in turn increases the canceller's complexity. It was shown that recursive channel estimation in a multistage interference canceller improved the accuracy of the channel estimation and hence gave better BER performance [70].

Due to the complexity of the multiuser or interference canceller detectors, it is not feasible to implement them in the mobile station. Hence, MUD or interference cancellation are optionally proposed for the uplink in the base station at the time of writing.

10.4.3   Transmit Diversity

10.4.3.1   Time Division Transmit Diversity

In Time Division Transmit Diversity (TDTD), the dedicated downlink transmit signal is switched between base stations. The transmitting base station can be selected according to either a fixed pattern or based on the quality of the signal received by the mobile station. The latter technique is known as Selection Transmit Diversity (STD) [71], while the former is known as Time Switched Transmission Diversity (TSTD) [72].

In TSTD, suppose there are several separate antennas involved in the transmission. Each antenna transmits one timeslot of the downlink dedicated channel frame in turn, in a fixed pattern. Signals for other users may have a different pattern. In STD, the transmitting antenna is dynamically selected based on a control signal transmitted from the mobile station periodically, indicating the mobile station's perceived preference. Each antenna involved in the diversity will transmit a Primary CCPCH. The control signal transmitted by the mobile station carries information about the quality of the received Primary CCPCH signal at the mobile station. The best received Primary CCPCH from the corresponding antenna will then be invoked in order to transmit the user's signal.

10.4.3.2 Orthogonal Transmit Diversity [6]

In Orthogonal Transmit Diversity (OTD), the signal is transmitted using two separate antennas. These signals are spread using different orthogonal channelization codes, so that self-interference is eliminated in flat fading.

$$*\qquad\qquad\qquad*$$

We have presented an overview of the terrestrial radio transmission technology of 3G mobile radio systems proposed by ETSI, ARIB and TIA. All three proposed systems are based on Wideband-CDMA. Despite the call for a common global standard, there are some differences in the proposed technologies, notably the chip rates and inter-cell operation. These differences are partly due to the existing 2G infrastructure already in use all over the world, specifically due to the heritage of the GSM and the IS-95 systems. Huge capital has been invested in these current 2G mobile radio systems. Hence, the respective regional standard bodies have endeavoured to ensure that the 3G systems are compatible with the 2G systems. Due to the diversified nature of these 2G mobile radio systems, it is not an easy task to reach a common 3G standard that can maintain perfect backwards compatibility.

Due to changes in the UMTS/IMT2000 specifications some of the information in Chapter 10 became obsolete. An updated version of this chapter can be downloaded in both .pdf and .ps format from

http://www-mobile.ecs.soton.ac.uk

Bibliography

[1] **J. Rapeli**, "UMTS:Targets, System Concept, and Standardization in a Global Framework," *IEEE Personal Communications*, vol. 2, no. 1, pp. 20–28, February 1995.

[2] **M.H. Callendar**, "Future Public Land Mobile Telecommunication Systems," *IEEE Personal Communications*, vol. 12, no. 4, pp. 18–22, 1994.

[3] **A.J. Viterbi**, *CDMA-Principles of Spread Spectrum Communication*, Addison Wesley, 1995.

[4] **S.G. Glisic** and **B. Vucetic**, *Spread Spectrum CDMA Systems for Wireless Communications*, Artech House Publishers, 1997.

[5] **R. Prasad**, *CDMA for Wireless Personal Communications*, Artech House, 1996.

[6] **T. Ojanperä** and **R. Prasad**, *Wideband CDMA for Third Generation Mobile Communications*, Artech House, 1998.

[7] **A. Toskala**, **J.P. Castro**, **E. Dahlman**, **Matti Latva-aho**, and **Tero Ojanperä**, "FRAMES FMA2 Wideband-CDMA for UMTS," *European Transactions on Telecommunications*, vol. 9, no. 4, pp. 325–336, July-August 1998.

[8] **E. Dahlman**, **B. Gudmundson**, **M. Nilsson**, and **Johan Sköld**, "UMTS/IMT-2000 Based on Wideband CDMA," *IEEE Communications Magazine*, vol. 36, no. 9, pp. 70–80, September 1998.

[9] **T. Ojanperä**, "Overview of Research Activities for Third Generation Mobile Communication," in *Wireless Communications TDMA versus CDMA*, Savo G. Glisic and Pentti A. Leppänen, Eds., pp. 415–446. Kluwer Academic Publishers, 1997.

[10] **ETSI/SMG/SMG2**, "The ETSI UMTS Terrestrial Radio Access (UTRA) ITU-R RTT Candidate Submission," Tech. Rep., European Telecommunications Standards Institute, June 1998.

[11] **ARIB**, "Japan's Proposal for Candidate Radio Transmission Technology on IMT-2000 : W-CDMA," Tech. Rep., Association of Radio Industries and Businesses, June 1998.

[12] **F. Adachi, M. Sawahashi**, and **H. Suda,** "Wideband DS-CDMA for Next-generation Mobile Communications Systems," *IEEE Communications Magazine*, vol. 36, no. 9, pp. 56–69, September 1998.

[13] **F. Adachi** and **M. Sawahashi,** "Wideband Wireless Access Based on DS-CDMA," *IEICE Transactions on Communications*, vol. E81-B, no. 7, pp. 1305–1316, July 1998.

[14] **A. Sasaki,** "Current Situation of IMT-2000 Radio Transmission Technology Study in Japan," *IEICE Transactions on Communications*, vol. E81-B, no. 7, pp. 1299–1304, July 1998.

[15] **P.-G. Andermo** and **L.-M. Ewerbring,** "A CDMA-Based Radio Access Design for UMTS," *IEEE Personal Communications*, vol. 2, no. 1, pp. 48–53, February 1995.

[16] **P.W. Baier, P. Jung**, and **A. Klein,** "Taking the Challenge of Multiple Access for Third-Generation Cellular Mobile Radio Systems - A European View," *IEEE Communications Magazine*, vol. 34, no. 2, pp. 82–89, February 1996.

[17] **E. Berruto, M. Gudmundson, R. Menolascino, W. Mohr**, and **M. Pizarroso,** "Research Activities on UMTS Radio Interface, Network Architectures, and Planning," *IEEE Communications Magazine*, vol. 36, no. 2, pp. 82–95, February 1998.

[18] **T. Ojanperä** and **R. Prasad,** "An Overview of Air Interface Multiple Access for IMT-2000/UMTS," *IEEE Communications Magazine*, vol. 36, no. 9, pp. 82–95, September 1998.

[19] **A. Baier, U.-C. Fiebig, W. Granzow, Wolfgang Koch, Paul Teder,** and **Jörn Thielecke,** "Design Study for a CDMA-Based Third-Generation Mobile System," *IEEE Journal on Selected Areas in Communications*, vol. 12, no. 4, pp. 733–743, May 1994.

[20] **J. Schwarz da Silva, Bernard Barani**, and **Bartolomé Arroyo-Fernández,** "European Mobile Communications on the Move," *IEEE Communications Magazine*, vol. 34, no. 2, pp. 60–69, February 1996.

[21] **E. Nikula, A. Toskala, E. Dahlman, Laurent Girard**, and **Anja Klein,** "FRAMES Multiple Access for UMTS and IMT-2000," *IEEE Personal Communications*, vol. 5, no. 2, pp. 16–24, April 1998.

[22] **F. Ovesjö, E. Dahlman, T. Ojanperä, Antti Toskala**, and **Anja Klein,** "FRAMES Multiple Access Mode 2 - Wideband CDMA," in *IEEE International Conference on Personal, Indoor and Mobile Radio Communication*, Helsinki, Finland, pp. 42–46, September 1997, PIMRC'97.

[23] **E. L. Kuan, C. H. Wong** and **L. Hanzo,** "Burst-by-burst Adaptive Joint Detection CDMA," *to appear in Proc. of VTC'99*, Houston, USA, May, 1999.

[24] **E. L. Kuan, C. H. Wong** and **L. Hanzo,** "Upper-bound Performance of Burst-by-burst Adaptive Joint Detection CDMA," *submitted to IEEE Communications Letters*, 1998.

[25] **F. Adachi, K. Ohno, A. Higashi, Tomohiro Dohi**, and **Yukihiko Okumura,** "Coherent multicode DS-CDMA mobile Radio Access," *IEICE Transactions on Communications*, vol. E79-B, no. 9, pp. 1316–1324, September 1996.

[26] **J.G. Proakis,** *Digital Communications,* McGraw-Hill, 1995.

[27] **M. O. Sunay, Z. -C. Honkasalo, A. Hottinen, H. Honkasalo** and **L. Ma,** "A Dynamic Channel Allocation Based TDD DS CDMA Residential Indoor System," *IEEE 6th International Conference on Universal Personal Communications, ICUPC'97,* San Diego, pp. 228–234, October 1997.

[28] **A. Fujiwara, H. Suda,** and **F. Adachi,** "Turbo Codes Application to DS-CDMA Mobile Radio," *IEICE Transactions on Communications,* vol. E81A, no. 11, pp. 2269–2273, November 1998.

[29] **M. J. Juntti,** "System Concept Comparison for Multirate CDMA with Multiuser Detection," in *Proceedings of the IEEE Vehicular Technology Conference,* Ottawa, Canada, May 1998, pp. 18–21.

[30] **S. Ramakrishna** and **J. M. Holtzman,** "A Comparison between Single Code and Mulitple Code Transmission Schemes in a CDMA System," in *Proceedings of the IEEE Vehicular Technology Conference,* Ottawa, Canada, May 1998, pp. 791–795.

[31] **M. B. Pursley,** "Performance Evaluation for Phase-Coded Spread-Spectrum Multiple-Access Communication-Part I: System Analysis," *IEEE Transactions on Communications,* vol. COM-25, no. 8, pp. 795–799, August 1977.

[32] **F. Adachi, M. Sawahashi,** and **K. Okawa,** "Tree-structured Generation of Orthogonal Spreading Codes with Different Lengths for Forward Link of DS-CDMA Mobile Radio," *Electronic Letters,* vol. 33, no. 1, pp. 27–28, January 1997.

[33] **R. F. Ormondroyd** and **J. J. Maxey,** "Performance of Low Rate Orthogonal Convolutional Codes in DS-CDMA," *IEEE Transactions on Vehicular Technology,* vol. 46, no. 2, pp. 320–328, May 1997.

[34] **M. K. Simon, J. K. Omura, R. A. Scholtz** and **B. K. Levitt,** "Spread Spectrum Communications Handbook," McGraw-Hill, 1994.

[35] **T. Kasami,** "Combinational Mathematics and its Applications," University of North Carolina Press, 1969.

[36] **E. L. Kuan** and **L. Hanzo,** "Joint Detection CDMA Techniques for Third-generation Transceivers," *Proceedings of ACTS'98,* Rhodos, Greece, June 1998, pp 727-732.

[37] **A. Chockalingam, Paul Dietrich, Laurence B. Milstein,** and **Ramesh R. Rao,** "Performance of Closed-Loop Power Control in DS-CDMA Cellular Systems," *IEEE Transactions on Vehicular Technology,* vol. 47, no. 3, pp. 774–789, August 1998.

[38] **K Higuchi, M Sawahashi,** and **F Adachi,** "Fast Cell Search Algorithm in DS-CDMA Mobile Radio using Long Spreading Codes," in *Proceedings of the IEEE Vehicular Technology Conference,* Phoenix, U.S.A., May 1997, vol. 3, pp. 1430–1434.

[39] **R. R. Gejji,** "Forward-Link-Power Control in CDMA Cellular-Systems," *IEEE Transactions on Vehicular Technology,* vol. 41, no. 4, pp. 532–536, November 1992.

[40] **W.C.Y. Lee,** "Overview of Cellular CDMA," *IEEE Transactions on Vehicular Technology,* vol. 40, no. 2, pp. 291–302, May 1991.

[41] **D. Wong** and **T. J. Lim,** "Soft Handoffs in CDMA Mobile Systems," *IEEE Personal Communications,* vol. 4, no. 6, pp. 6–17, December 1997.

[42] **C. C. Lee** and **R. Steele,** "Effects of Soft and Softer Handoffs on CDMA System Capacity," *IEEE Transactions on Vehicular Technology,* vol. 47, no. 3, pp. 830–841, August 1998.

[43] **M Gustafsson**, **K Jamal**, and **E Dahlman,** "Compressed Mode Techniques for Inter-frequency measurements in a wide-band DS-CDMA system," in *IEEE International Conference on Personal, Indoor and Mobile Radio Communication,* Helsinki, Finland, September 1997, PIMRC'97, pp. 231–235.

[44] **D.N. Knisely**, **S. Kumar**, **S. Laha**, and **S. Nanda,** "Evolution of Wireless Data Services : IS-95 to cdma2000," *IEEE Communications Magazine,* vol. 36, no. 10, pp. 140–149, October 1998.

[45] **TIA,** "The cdma2000 ITU-R RTT Candidate Submission," Tech. Rep., Telecommunications Industry Association, 1998.

[46] **D. N. Knisely**, **Q. Li**, and **N. S. Rames,** "cdma2000 : A Third Generation Radio Transmission Technology," *Bell Labs Technical Journal,* vol. 3, no. 3, pp. 63–78, July-September 1998.

[47] **Y. Okumura** and **F. Adachi,** "Variable-Rate Data Transmission with Blind Rate Detection for Coherent DS-CDMA Mobile Radio," *IEICE Transactions on Communications,* vol. E81B, no. 7, pp. 1365–1373, July 1998.

[48] **J. C. Liberti, Jr.** and **T. S. Rappaport,** "Analytical Results for Capacity Improvements in CDMA," *IEEE Transactions on Vehicular Technology,* vol. 43, no. 3, pp. 680–690, August 1994.

[49] **J. H. Winters,** "Smart Antennas for Wireless Systems," *IEEE Personal Communications,* vol. 5, no. 1, pp. 23–27, February 1998.

[50] **S. Verdu,** "Minimum Probability of Error for Asynchronous Gaussian Multiple-Access Channel," *IEEE Transactions on Communications,* vol. 32, no. 1, pp. 85–96, January 1986.

[51] **S. Moshavi,** "Multi-User Detection for DS-CDMA Communications," *IEEE Communications Magazine,* vol. 34, no. 10, pp. 124–136, October 1996.

[52] **T. J. Lim** and **L. K. Rasmussen,** "Adaptive Symbol and Parameter Estimation in Asynchronous Multiuser CDMA Detectors," *IEEE Transactions on Communications,* vol. 45, no. 2, pp. 213–220, February 1997.

[53] **T. J. Lim** and **S. Roy,** "Adaptive Filters in Multiuser (MU) CDMA Detection," *Wireless Networks,* vol. 4, no. 4, pp. 307–318, June 1998.

[54] **L. Wei**, **L. K. Rasmussen** and **R. Wyrwas,** "Near Optimum Tree-search Detection Schemes for Bit-synchronous Multiuser CDMA Systems over Gaussian and Two-path Rayleigh Fading Channels," *IEEE Transactions on Communications,* vol. 45, no. 6, pp. 691–700, June 1997.

[55] **T. J. Lim** and **M. H. Ho,** "LMS-Based Simplifications to the Kalman Filter Multiuser CDMA Detector," *Proceedings of IEEE Asia-Pacific Conference on Communications/International Conference on Communication Systems,* Singapore, November, 1998.

[56] **D. You** and **T. J. Lim,** "A Modified Blind Adaptive Multiuser CDMA Detector," *Proceedings of IEEE International Symposium on Spread Spectrum Techniques and Application,* Sun City, South Africa, pp. 878–882, September 1998.

[57] **S. M. Sun, L. K. Rasmussen, T. J. Lim** and **H. Sugimoto,** "Impact of Estimation Errors on Multiuser Detection in CDMA," *Proceedings of IEEE Vehicular Technology Conference,* Ottawa, Canada, pp. 1844–1848, May 1998.

[58] **Y. Sanada** and **Q. Wang,** "A Co-Channel Interference Cancellation Technique Using Orthogonal Convolutional Codes on Multipath Rayleigh Fading Channel," *IEEE Transactions on Vehicular Technology,* vol. 46, no. 1, pp. 114–128, February 1997.

[59] **M.K. Varanasi** and **B. Aazhang,** "Multistage Detection in Asynchronous Code-Division Multiple-Access Communications," *IEEE Transactions on Communications,* vol. 38, no. 4, pp. 509–519, April 1990.

[60] **P. Patel** and **J. Holtzman,** "Analysis of a Simple Successive Interference Cancellation Scheme in a DS/CDMA System," *IEEE Journal on Selected Areas in Communications,* vol. 12, no. 5, pp. 796–807, June 1994.

[61] **P. H. Tan** and **L. K. Rasmussen,** "Subtractive Interference Cancellation for DS-CDMA Systems," *Proceedings of IEEE Asia-Pacific Conference on Communications/International Conference on Communication Systems,* Singapore, November, 1998.

[62] **K. L. Cheah, H. Sugimoto, T. J. Lim, L. K. Rasmussen** and **S. M. Sun,** "Performance of Hybrid Interference Canceller with Zero-Delay Channel Estimation for CDMA," *Proceedings of IEEE Global Communications Conference '98,* Australia, pp. 265–270, November 1998.

[63] **S. M. Sun, L. K. Rasmussen** and **T.J. Lim,** "A Matrix-Algebraic Approach to Hybrid Interference Cancellation in CDMA," *Proceedings of IEEE International Conference on Universal Personal Communications '98,* Florence, Italy, pp. 1319–1323, October 1998.

[64] **A. L. Johansson** and **L. K. Rasmussen,** "Linear Group-wise Successive Interference Cancellation in CDMA," *Proceedings of IEEE International Symposium on Spread Spectrum Techniques and Application,* Sun City, South Africa, pp. 121–126, September 1998.

[65] **S. M. Sun, L. K. Rasmussen, H. Sugimoto** and **T. J. Lim,** "A Hybrid Interference Canceller in CDMA," *Proceedings of IEEE International Symposium on Spread Spectrum Techniques and Application,* Sun City, South Africa, pp. 150–154, September 1998.

[66] **D. Guo, L. K. Rasmussen, S. M. Sun, T. J. Lim** and **C. Cheah,** "MMSE-Based Linear Parallel Interference Cancellation in CDMA," *Proceedings of IEEE International Symposium on Spread Spectrum Techniques and Application,* Sun City, South Africa, pp. 917–921, September 1998.

[67] **L. K. Rasmussen, D. Guo, Y. Ma** and **T. J. Lim,** "Aspects on Linear Parallel Interference Cancellation in CDMA," *Proceedings of IEEE International Symposium on Information Theory '98,* Cambridge, US, pp. 37, August 1998.

[68] **L. K. Rasmussen, T. J. Lim, H. Sugimoto** and **T. Oyama,** "Mapping Functions for Successive Interference Cancellation in CDMA," *Proceedings*

of IEEE Vehicular Technology Conference, Ottawa, Canada, pp. 2301–2305, May 1998.

[69] **S. M. Sun, T. J. Lim, L. K. Rasmussen, T. Oyama, H. Sugimoto and Y. Matsumoto,** "Performance Comparison of Multi-stage SIC and Limited Tree-Search Detection in CDMA," *Proceedings of IEEE Vehicular Technology Conference*, Ottawa, Canada, pp. 1854–1858, May 1998.

[70] **Mamoru Sawahashi, Yoshinori Miki, Hidehiro Andoh,** and **Kenichi Higuchi,** "Pilot Symbol-Assisted Coherent Multistage Interference Canceller Using Recursive Channel Estimation for DS-CDMA Mobile Radio," *IEICE Transactions on Communications*, vol. E79-B, no. 9, pp. 1262–1269, September 1996.

[71] **A. Wittneben** and **T. Kaltenschnee,** "TX Selection Diversity with Prediction: Systematic Nonadaptive Predictor Design," *IEEE 44th Vehicular Technology Conference, 1994*, pp. 1246–1250, June 1994.

[72] **A. Hottinen** and **R. Wichman,** "Transmit Diversity by Antenna Selection in CDMA Downlink," in *Proceedings of ISSSTA '98*, Sun City, South Africa, September 1998.

Glossary

2G	Second Generation
3G	Third Generation
ACL	Auto Correlation
ACTS	Advanced Communications Technology and Services
ARIB	Association of Radio Industries and Businesses
AWGN	Additive White Gaussian Noise
BCCH	Broadcast Control Channel
BER	Bit Error Rate
BPSK	Binary Phase Shift Keying
BS	Base Station
CAPICH	Common Auxiliary Pilot Channel
CCCH	Common Control Channel
CCL	Cross Correlation
CDMA	Code Division Multiple Access
CPHCH	Common Physical Channel
CRC	Cyclic Redundancy Check
DAPICH	Dedicated Auxiliary Pilot Channel
DCCH	Dedicated Control Channel
DCH	Dedicated Channel
DECT	Digital Enhanced Cordless Telecommunications
DL	Downlink
DPCCH	Dedicated Physical Control Channel
DPDCH	Dedicated Physical Data Channel
DPHCH	Dedicated Physical Channel
DS-CDMA	Direct Sequence Code Division Multiple Access
EMC	Electromagnetic Compatibility

ETSI	European Telecommunications Standards Institute
EU	European Union
FACH	Forward Access Channel
FCCH	Frequency Correction Channel
FCH	Fundamental Channel
FDD	Frequency Division Duplex
FDMA	Frequency Division Multiple Access
FEC	Forward Error Correction
FPLMTS	Future Public Land Mobile Telecommunication System
FRAMES	Future Radio Wideband Multiple Access System
GPS	Global Positioning System
HCS	Hierarchical Cell Structure
IMT-2000	International Mobile Telecommunications 2000
ISO/OSI	International Standardization Organization/Open Systems Interconnection
ITU	International Telecommunication Union
ITU-R	International Telecommunication Union - Radiocommunication Sector
MAI	Multiple Access Interference
MC	Multicarrier
MDM	Modulation Division Multiplexing
MPG	Multiple Processing Gain
MS	Mobile Station
OCQPSK	Orthogonal Complex Quadrature Phase Shift Keying
OVSF	Orthogonal Variable Spreading Factor
PCCPCH	Primary Common Control Physical Channel
PCH	Paging Channel
PCS	Personal Communications Services
PHCH	Physical Channel
PHS	Personal Handyphone System
PICH	Pilot Channel
PN	Pseudo Noise
PRMA	Packet Reservation Multiple Access
PSC	Primary Synchronization Code
QoS	Quality of Service
QPSK	Quadrature Phase Shift Keying

RACE	Research in Advanced Communication Equipment
RACH	Random Access Channel
RI	Rate Information
RS	Reed-Solomon
RTT	Radio Transmission Technology
SCCPCH	Secondary Common Control Physical Channel
SCH	Synchronisation Channel
SF	Spreading Factor
SIR	Signal-to-Interference Ratio
SSC	Secondary Synchronization Code
SYCH	Sync Channel
TDD	Time Division Duplex
TDMA	Time Division Multiple Access
TFI	Transport Format Indicator
TIA	Telecommunications Industry Association
TPC	Transmit Power Control
UL	Uplink
UMTS	Universal Mobile Telecommunications System
UTRA	Universal Mobile Telecommunications System Terrestrial Radio Access
VoD	Video on Demand
W-CDMA	Wideband Code Division Multiple Access
WARC	World Administrative Radio Conference

Chapter 11

Wireless ATM

P. Pattullo[1] and R. Steele[2]

11.1 Introduction

During the 1990s there has been an explosive growth in the usage of mobile communications. World-wide, there are now hundreds of millions of subscribers, using a wealth of first and second generation networks. The vast majority of teletraffic on these networks is speech communications. There are messaging and data services but their variety has been limited because they must be transmitted over channels that normally transport speech communications.

In the fixed networks there has been a rapid growth in the demand for multimedia communications. The range of multimedia services is wide and may be broadly classified into dialogue, messaging, information retrieval and distribution. This classification includes video, speech, and data services, as well as web browsing, file transfer, and email. Historically, either separate networks were required to support the different communications services, or complex interworking procedures were necessary. Over the past decade, asynchronous transfer mode (ATM) systems have been developed as a solution for the transport of multi-service communications traffic in fixed networks. As fixed network multimedia demands continue to grow, a concomitant increase in the demand for *mobile* multimedia traffic is forecast. Transport solutions for the efficient provision of mobile multimedia are necessary, and one of these is wireless ATM (WATM).

This chapter is intended to introduce the reader to the enhancements that are necessary to efficiently transport mixed multimedia traffic within

---

[1]Multiple Access Communications Ltd
[2]University of Southampton and Multiple Access Communications Ltd

mobile radio communications networks. It focuses on techniques based around the ATM concepts. Section 11.2 provides an overview of ATM. This is a pre-requisite to the understanding of WATM and the concepts of multimedia traffic transport. The enhancements that must be made to ATM to support wireless traffic are discussed in Section 11.3. Dynamic radio resource control techniques that ensure a high utilisation of the radio spectrum are examined in Section 11.4. Options for the architecture of the supporting network infrastructure are discussed in Section 11.5, while Section 11.6 describes the transport performance of mobile multimedia traffic in microcellular networks.

11.2   Overview of ATM

Because this book is concerned with mobile radio communications, we will provide some background on ATM before considering the introduction of ATM in wireless mobile networks. ATM has been conceived to radicalise the fixed networks by replacing the conventional isochronous and asynchronous methods of transmission by one based on fixed length packet transmissions that support multimedia services in an efficient way. In conventional telephony, users are connected using a series of switched links, and the call is said to be circuit-switched. Multiple access links are also used whereby many users simultaneously use the same physical media. Each user is assigned a channel in the form of a regular timeslot (TDMA), a particular frequency (FDMA), or a unique access code (CDMA), for the duration of the call. Circuit-switching is suitable for continuous real-time communications. During a call, all information is delivered with a constant transmission delay and bit rate. However, the resources remain reserved whether or not the connection is in active use. For example, in a telephone conversation, each party typically only speaks for 40% of the time, so for 60% of the time when the user is listening the link is only transmitting background noise. As the speed and variety of real-time services grow, simple circuit-switching becomes less efficient.

Data communications, such as the transfer of computer files, or world wide web browsing, are more efficiently conveyed using packet-switching techniques rather than circuit switched ones. Packet-switching provides each connection with a large transport resource for short periods of time as and when required. Store and forward techniques, such as those innate in the Internet Protocol (IP), are common, and can achieve high utilisation of the network links. However, packet-switching networks require switching servers with large information buffers. The route that the information packets take to their destination is referred to as connectionless, i.e. it is not established prior to the initial transmission of the data, and individual packets may take different routes between the same endpoints. Therefore, delivery delays are inconsistent, packets transmitted in a sequence may

arrive in a different order, some packets may not arrive at all or may have
an unacceptable bit error rate (BER), and some packets are so delayed that
they are discarded. Packet-switching is therefore not well suited for real-
time communications, such as video-links, unless careful control procedures
are implemented.

Individual services have different requirements. Computer data may
require very low bit error rates (BER) and be tolerant to long delays, while
voice links may tolerate a relatively large BER as long as the delivery delay
is short. Similarly, a high-quality interactive video-link requires a variable
high data rate, low BER and low delay; a messaging service may need
a low data rate and perhaps a zero BER, and so on. Such individuality
makes it difficult to efficiently cater for multiple services within a single
network. As the number and variety of communications services grow,
it becomes desirable to flexibly and efficiently transport both real-time
and non-real-time services over the same network, providing an individual
quality-of-service (QoS) to every connection.

ATM has been designed with the specific aim of forming a common
networking standard that provides flexible support for the transmission of
all established communications services, while allowing for the introduction of future services. ATM permits high-speed transmission of real-time
and non-real-time services simultaneously. A high bandwidth is offered
with seamless integration between wide area public and private ATM networks [1,2]. ATM is a combined switching and multiplexing technique based
upon asynchronous time division multiplexing (ATDM). A hybrid of circuit
and packet-switching processes are used to transport fixed length packets
across pre-defined routes. Efficiency is achieved through statistical multiplexing of the multimedia services, whilst QoS is guaranteed on a per connection basis that is agreed at call establishment. The standards continue
to be specified. The main controllers are the European Telecommunication
Standards Institute (ETSI), the International Telecommunications Union
(ITU), and the American National Standards Institute (ANSI). General
support for ATM is increasing, as the number of ATM network installations grow. The ITU has selected ATM as the preferred transfer mechanism
for Broadband-ISDN.

11.2.1   ATM Cell

ATM communications are based on the transport of short, fixed length
packets, commonly referred to as cells. This terminology is unfortunate for
cellular radio engineers who must now be conscious as to whether we are
talking about a radio cell or an ATM cell. Nevertheless we are stuck with
the definition that an ATM packet is a cell. The word asynchronous in ATM
relates to the fact that cells are only transmitted when required, i.e. the
network resources are allocated on demand. The cells are routed across a
path known as a virtual circuit, which is established at call set-up. The cell

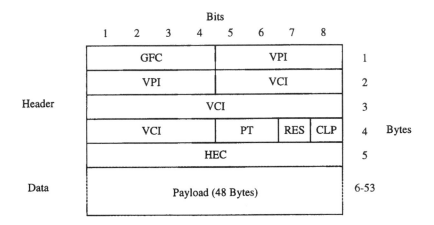

Figure 11.1: ATM cell (UNI format).

length is 53 bytes. The first 5 bytes are the cell header, and the remaining 48 bytes contain the information payload. The cell length was derived as a compromise between link efficiency and packetisation delay. A larger cell would have a greater payload-to-header ratio, and would be more efficient for the transport of computer data. However, a larger cell would have a longer packetisation delay (the time taken to fill the cell with data), which would make it unsuitable for real-time communications requiring low delay, e.g. telephony. The 5 byte header places approximately a 10% overhead on the payload. The cell header provides route identification information, error correction data, and network control information [1]. There are two cell formats, one specified for user-network interface (UNI) links, and the other for network-network interface (NNI) links. The UNI cell format is represented diagrammatically in Figure 11.1, where the terms in Figure 11.1 have the following meaning.

- Generic flow control (GFC) - 4 bits: controls traffic flow between the user and the network, allowing identification of multiple terminals connected to the same access link.

- Virtual channel identification (VCI) - 12 bits: identifies the virtual routing channel for the cell. The VCI is updated when routed by virtual channel crossconnects, i.e. switches.

- Virtual path identifier (VPI) - 8 bits: identifies the virtual routing path for the cell. A virtual path is a bundle of virtual channels that can be switched as a unit, assisting control and efficiency. The VPI is updated at virtual path crossconnects.

- Payload type (PT) - 2 bits: indicates whether the cell payload contains

Characteristic	Service			
	CBR	VBR	ABR	UBR
User Bit	Constant	Variable		
Data Type	Stream		Message/Pack	
	Connection			Connectionless

Figure 11.2: ATM service classes.

user data or network management information.

- Cell loss priority (CLP) - 1 bit: used to reduce congestion in the net-work, by marking low priority cells, or cells from users exceeding their agreed data rate, and allowing deletion of these cells when necessary.

- Header error control (HEC) - 8 bits: one bit of error correction or multiple bit error detection code, only for the cell header information. The HEC field can also be used to determine the position of the cell boundaries within continuous data streams.

- Reserved (RES) - 1 bit: available for future development.

- Payload - 384 bits: contains the user data that is to be transmitted over the network. Some bytes of the payload may be used for protocol control information to assist the terminal equipment with data processing, e.g. cell sequence numbers. Protocol control information is controlled by the application operating at the users' terminal equipment. The payload cannot be altered by the network, except in the special case of network management cells.

In the NNI cell format the four bits of GFC information are replaced and used to extend the VPI information to 12 bits. ATM switches contain translation tables to route cells onward according to their VPI and VCI information, and to update this routing information for the next leg of the journey. ATM switches require a high cell throughput rate and low queueing delay, and must cope with priority and broadcast functions.

11.2.2   Service Classes

In order to classify the general type of communications traffic to be trans-ported in each connection, the following four broad service classes are iden-tified. The characteristics of these classes are shown diagrammatically in Figure 11.2

- Constant bit rate (CBR): circuit-switched emulation for real-time traf-fic, cells are generated by the source at a constant rate, e.g. pulse code modulated (PCM) speech.

- Variable bit rate (VBR): either real-time or non-real-time traffic, where the sources generate cells at a non-constant rate, eg, video, video-mail, silence detected speech.

- Available bit rate (ABR): non-real-time traffic, such as computer or message data, connection-oriented, utilising the bandwidth remaining after CBR and VBR allocations.

- Unspecified bit rate (UBR): packet-oriented, connection-less data, without delay or delivery guarantees.

VBR and ABR should be the most abundant forms of ATM traffic, providing statistical multiplexing gains. The network must support CBR traffic, as this is a common existing traffic format, although little multiplexing gain is achievable with this traffic. Header generation, header error correction, cell routing, and cell multiplexing are controlled by the ATM layer (AL) [2,3]. The ATM adaptation layer (AAL) controls segmentation and reassembly of the user data for the cell payloads. There are five recommended AAL types, optimised for different service types. These steal between zero and four bytes of the payload for protocol control information, such as cell sequence information, time-stamps, error control, packet boundary identifiers, etc. The AAL type used by each connection is agreed between the users' terminal equipment, and is not affected by the ATM network.

11.2.3   Statistical Multiplexing

Asynchronous multiplexing provides an increase in the number of user transmissions that can be supported over a network, known as the statistical gain [2]. This gain is achieved because the probability of all connections transmitting at their peak rate at the same time is very small. For example, consider n sources placing cells onto a single transmission link to n sinks. The probability of a particular source generating a cell within a frame is p_s. The capacity of the link is that all n sources deliver a cell within a frame, i.e. there are n cells carried per frame. The link utilisation is the average ratio of the number of cells carried to the total number of slots in each frame, ie:

$$\text{Link utilisation} = E\left[\frac{\text{Total number of cells in frame}}{\text{Total number of slots per frame}}\right] = \frac{\sum_{s=0}^{n} p_s}{n}, \quad (11.1)$$

where $E[(\bullet)]$ means the expectation of $(\bullet)$. If p_s is the same for all sources, then p_s equals p, and the link utilisation equals p. Either additional sources may be served with the existing link capacity, or the link capacity may be reduced and still serve the existing number of sources. However, this introduces the possibility of more cells being generated than can be transported within a single time-frame, and hence the possibility of cells being deleted. The probability of cell loss (deletion) conforms to a binomial distribution. The more sources there are, the smaller the deviation of the average cell generation rate, and the lower the probability of cell loss. The probability of cell loss, $P_{cell\ loss}$, is

$$P_{cell\ loss} = \sum_{i}^{n} \binom{i}{n} P^i (1-p)^{n-i}, \qquad (11.2)$$

where I is the link capacity (number of time-slots per frame) and n is the number of sources. The statistical gain is the achievable reduction in the link capacity that can provide an acceptable probability of cell loss, compared to the link capacity required for zero cell loss probability. Statistical analysis of the multiplexing gain rapidly becomes extremely complex when multiple services are generating cells with independent probabilities and individual QoS parameters. Intelligent resource management schemes are critical to the success of ATM networks.

11.2.4 Virtual Connections

We have already seen that ATM is a hybrid of circuit and packet-switching techniques. Like circuit-switching, the path over which the data will travel is established at call set-up, and remains fixed for the duration of the call. However, like packet-switching, network resources are not reserved, and are allocated on demand as cells arrive at the switch input buffers, thus providing statistical multiplexing. These connection-oriented routes without reserved network resources are called *virtual connections*.

The route of a virtual connection is described by virtual channel identifiers (VCIs) and virtual path identifiers (VPIs). Taken together, the VCI and VPI uniquely identify a single connection at the input to a switch. The identification is split into two parts so that connections that arrive on the same input port and leave by the same output port can be grouped together and switched as a single entity, without the need for the switch to read the full identification. The VPI identifies the group to which a connection belongs, whilst the VCI provides the remaining information to uniquely identify a connection. The VPI/VCI information can be changed by each switch for the next leg of the route, thus ensuring that the connection has a unique identification when it arrives at the input buffer of the next switch. There are in fact two classes of ATM switches: high-level switches which only change the VPI field, and those that can change both the VPI and VCI fields. Every ATM switch contains a translation table, in which an entry exists for every virtual channel connection passing through the switch. Each incoming cell has its VCI and/or VPI information read, and, according to the translation table, is switched to the appropriate output port. Its cell header is then updated with the appropriate VPI/VCI

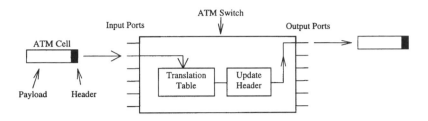

Figure 11.3: An ATM switch using the cell header and a translation table.

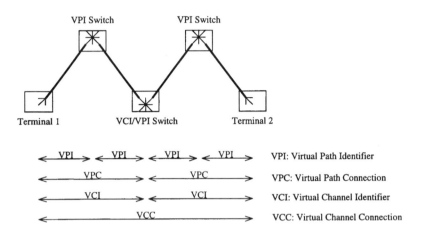

Figure 11.4: Virtual paths and channels.

for the next leg of the journey. Figure 11.3 shows an example ATM switch architecture.

An example of a virtual connection is shown in Figure 11.4. A virtual connection is shown connecting Terminal 1 with Terminal 2. The virtual connection is routed through three ATM switches that form part of an ATM network, and each switch is connected to many other switches (not shown). The figure shows that a virtual channel connection (VCC) is composed of two concatenated virtual path connections (VPCs) [4]. The virtual channel identifier (VCI) remains constant over the route of a virtual path connection, and is only altered at the VCI/VPI switch. In the example, each VPC is composed of two switching legs, each with different virtual path identifiers (VPIs), that are changed at both the VPI switches and the VCI/VPI switches.

In general, a group of connections following the same route for two or more legs is given the same VPI and can be switched as a group. Switching a group of connections together using only the VPI reduces the amount of processing required at the VPI switches. Connection routing basically consists of selecting a route that has suitable resources available, and setting up the translation table entry at each switch on the route. Concepts from both circuit and packet-switched routing are utilised. The QoS implications of adding a new connection to a path must be considered before a route is accepted.

Little control can be exerted after a new connection has been accepted onto the network, so connection admission control (CAC) is the primary mechanism for controlling the network loading. CAC is carried out at call establishment, and must consider the requested service parameters, investigate the network for a suitable route, and set up the virtual connection if possible. Control of the network relies on statistics collected from the nodes and complex resource allocation algorithms. The network control processing may be distributed amongst many network nodes or located at a few powerful nodes - typically, the first node that a new connection arrives at will carry out CAC for that connection. There is a necessary trade-off between the network control complexity, the amount of control information that must be transmitted across the network, and the link utilisation efficiency.

Policing of the network ensures excess loading does not unduly degrade the QoS and keeps each connection within its agreed traffic load limits. To maintain QoS levels, violating cells and low-priority cells may be deleted. A counter is incremented each time a cell arrives at a network interface. The counter is decremented at an agreed average cell rate, and provided that the counter stays below an agreed limit to allow for cell clustering, no cell deletion occurs.

11.2.5 Service Parameters

ATM agrees QoS parameters on a per call connection basis. This means that every connection request can define its individual characteristics, and the call admission control (CAC) algorithms must decide whether the necessary resources to service the connection are available. To initiate a call the user-terminal requests a connection by placing a single cell on a predefined virtual channel reserved for signalling. A specific virtual signalling channel is then assigned to the terminal by the network. The user-terminal provides information as to the QoS parameters required and the destination address. The network must investigate a suitable route and, if one is found, set up a virtual channel. The QoS parameters used to describe a connection are shown in Table 11.1 as specified by the ATM Forum.

Cell transfer delay (CTD) over the network is an important service parameter, and if too large, will prevent the establishment of real-time services. Delay originates from three sources: physical propagation delay, delay in forming an ATM cell, and queueing delay at the switches. Consecutive cells of a connection will experience different switching delays dependent upon the instantaneous traffic load of the switches they pass

Parameter	Acronym	Description
Cell Transfer Delay	CTD	Maximum allowable delay to cell delivery
Cell Delay Variation	CDV	Acceptable variance of cell transfer delay
Peak Cell Rate	PCR	Maximum cell transmission rate
Sustainable Cell Rate	SCR	Average cell transmission rate
Minimum Cell Rate	MCR	Minimum cell transmission rate
Burst Tolerance	BT	Max. number of cells in a burst at the PCR
Cell Loss Ratio	CLR	Acceptable ratio of lost cells to total cells

Table 11.1: ATM service parameters.

through. These different switching delays result in a variable CTD, the variance of which is identified by the cell delay variation (CDV) parameter. Some services are sensitive to CDV, especially if the cell arrival rate is used as part of a clock recovery process. CDV can be minimised by using output buffering at the final switch before cell delivery to the terminal. In order to keep delays similar for services requiring, say both voice and video, multiple connections to the same user (e.g. combined voice and video) may be forced to follow the same route over the network. Current telephone speech transmission typically uses 64 kbit/s logarithmic pulse code modulation (PCM) coding, with a 3.1 kHz speech bandwidth, from 300 Hz to 3.4 kHz [5, 6]. Packetisation of 8 bits/sample at 64 kbit/s for 47 samples per cell payload takes 5.9 ms, which is fairly slow, i.e. the packetisation delay is 5.9 ms. If a lower bit rate is used then this delay increases. It is assumed that one byte of the payload is used for protocol control information, such as a time-stamp. Fixed network ATM transmission and switching delay is likely to be in the order of a few milliseconds - less than the existing public switched telephone network (PSTN). Packetisation delay can be decreased by only partially filling cells with data and padding the remaining payload with dummy bytes, or by multiplexing samples of several voice channels into a single packet [7], but this adds processing complexity. The ATM Forum has recently specified a new ATM adaptation layer that allows data from several users to be multiplexed into a single ATM cell [8, 9].

Cell transfer delay also affects call echo. People like to hear an amount of echo in speech communications. A few milliseconds of delay is acceptable, but 20-30 ms of delay sounds reverberant, and over 50 ms sounds like real echo [5]. Guidelines in the UK suggest end-to-end delay should be below 23 ms for speech circuits that do not have echo control. As 5 ms of this end-to-end delay is allowed in the user-equipment at each end, the maximum allowable delay over the network is only 13 ms. If echo cancellation is used then longer delays can be tolerated. Hence, echo cancellation equipment is often required in the fixed network, particularly for international calls. Note that a delivery delay of 30 ms is significantly shorter than that typically found in cellular networks, which may also require echo cancellation.

The acceptable peak cell rate (PCR) is dependent on the resources available on each link of a virtual circuit and the maximum possible transmission rate of each link. Whilst operators may be able to provide high-rate national links, the peak rates of links near to the end terminals, such as the local loop to subscriber buildings, may be more restricted. The communications wiring in most buildings is made from copper, as is most wiring to the local exchange [10]. Wiring is often treated as a long-term investment by businesses. Therefore, adequate provision of ATM over copper, to the desk, should be ensured. In company premises this copper cabling is normally unscreened twisted pair (UTP). This contains relatively inexpensive 4-pair cables emanating from a wiring closet, with a normal maximum distance of 100 m to each data jack unit, e.g. desk. The minimum quality specified for data transfer is category-3 UTP, typically providing up to 16 Mbit/s data transmission, e.g. in Ethernet and token ring networks. Most new installations use category-5 UTP, which typically support transmissions up to 100 Mbit/s. ATM Forum specifications exist for ATM over copper UTP; providing 51 Mbit/s for up to 100 m with category-3, and 155 Mbit/s with category-5. A lower cost alternative, at 25.6 Mbit/s, is being defined.

The remaining parameters of Table 11.1 are needed to assist in call admission control and network resourcing calculations. The sustainable cell rate (SCR) is used to define the average rate at which cells are being generated by a connection. The minimum cell rate (MCR) defines the minimum rate at which a connection will be placing cells onto the network. The burst tolerance (BT) defines the maximum number of cells that will be added to the network in a single burst at the peak cell rate. The BT value is necessary to evaluate the amount of input buffering required at the switches in order to minimise the probability of buffer overload and hence cell loss. The cell loss ratio (CLR) defines the proportion of cells that the connection is willing to accept not being delivered. Some services require a very low CLR, whilst other services are quite tolerant of cell loss. The required probability of cell loss directly affects the statistical multiplexing gain that can be achieved, so is important to resource dimensioning. The CLR can also be used to select cells that can be deleted at times of traffic overload.

Providing service parameters on an individual connection basis makes call charging a complex issue, involving many more variables than traditional circuit-switched technologies. Charging regimes are likely to be based on the agreed QoS parameters, duration, distance, and time of day, etc. ATM may also be vulnerable to security attacks, e.g. eavesdropping, service interruption and unauthorised access, although security mechanisms are developing [11].

Service	Bit Rate	Max. Delay	Max. Cell Loss
Voice audio	8-128 kbit/s	Low	High/Medium
Video Telephony	64-356 kbit/s	Medium	Medium
Video	0.5-5.0 Mbit/s	Medium/Low	Medium/Low
Digital data	10-1000 kbit/s	High/Medium	Low
File transfer	10-1000 kbit/s	High	Low

Table 11.2: Typical wireless multimedia service requirements.

11.3   Wireless ATM Mobility

ATM is receiving increasing acceptance as the standard system for broadband wide area and local area networking. In cellular communications, multimedia services are increasing in second generation systems, such as GSM and IS-95, while the third generation (3G) cellular networks will be rich in multimedia services. As ATM is particularly suited to multimedia service provision it is appropriate that serious consideration be given to allowing ATM to be introduced into the radio interface of 3G cellular networks [12–15]. By doing so the cellular base stations become another ATM node, and ATM communications with mobile users becomes seamless. However, the broadband optical fibre fixed network enjoys very low bit error rates, and, as ATM was conceived to function in this environment, no forward error correction (FEC) codes have been specified. By contrast, mobile radio channels suffer from severe degradation which can only be overcome by FEC coding.

The bandwidth available for mobile systems is very limited compared to that in wired networks. For example, fixed network ATM links may transport 155 Mbit/s supporting a huge range of data rates, whilst typical user data rates for the universal mobile telecommunications system (UMTS) are between 384 kbit/s and 8 kbit/s, although the maximum data rate per connection is around 2 Mbit/s. Ensuring efficient use of the radio spectrum in mobile communications, whilst meeting QoS requirements with the relatively small allotted bandwidth, is difficult. If ATM is used over the radio interface these difficulties are predominantly due to the multimedia traffic requirements, rather than to the ATM structure itself. Dynamic radio control protocols are necessary to deliver asymmetric multi-rate traffic, and the use of small radio-cells is necessary for the delivery of high data rates. Unless large amounts of spectrum are made available to network operators, high-rate services may only be supportable in limited locations, or for users with limited mobility. Example bit rates, acceptable delivery delays and acceptable cell loss rates of typical wireless services are shown in Table 11.2.

Wireless ATM (WATM) is applicable for many forms of wireless data delivery, e.g. in cellular communications, wireless LAN, wireless local loop

(WLL), and cordless applications. Discussions in this book are predominantly based on cellular communications, which encompass many of the parameters that need consideration for other WATM applications.

As we have pointed out, the original ATM specifications were not designed for wireless access. To achieve seamless delivery of ATM to mobile users, enhancements are required in two main areas:

- *ATM mobility*, which involves enhancements to the fixed ATM network infrastructure to ensure that it is able to support calls with mobile users. ATM mobility is primarily concerned with call handover, location management, routing, and network management [13, 14, 16]. Providing the fixed ATM network with the functionality required to support mobility should be achievable independently of the radio access technology, enabling support of both end-to-end WATM systems and interworking with other cellular and wireless LAN applications.

- *Radio access layer infrastructure*, which involves enhancing the air interface protocols of cellular systems to support the requirements of ATM. The radio access layer infrastructure has a medium access control (MAC) layer to support the sharing of radio resources between multiple terminals and services, a data link control (DLC) layer to mitigate the effect of radio channel errors, and a wireless physical layer (PHY) for transmission over the air interface.

The remainder of this section discusses the enhancements necessary to the fixed network to support ATM mobility, whilst Section 11.4 concentrates on the radio access layer infrastructure. After an overview of possible fixed network architectures that can support ATM mobility for cellular networks in Section 11.3.1, some of the important aspects are considered in more detail. Section 11.3.2 examines the complexities of supporting handover of ATM connections as mobile users move between different radio-cells. Section 11.3.3 considers the difficulties of supporting the ATM quality-of-service parameters when using WATM, and Section 11.3.4 discusses location management and call routing issues.

11.3.1 Network Architectures for ATM Mobility

An ATM system has a predominantly flat network architecture, sometimes called a cloud network, or mesh network, i.e. there is no strict, hierarchical call routing tree and calls do not have to pass through a centralised call switching centre. Most ATM nodes are connected to many other nodes, and many possible routes exist between any two endpoints. Existing cellular networks share control amongst different levels of a hierarchical tree structure, e.g. mobile switching centres (MSCs), base station controllers (BSCs), home location registers (HLRs), etc. For WATM, some form of mobility servers must be connected to the ATM network to cater for the

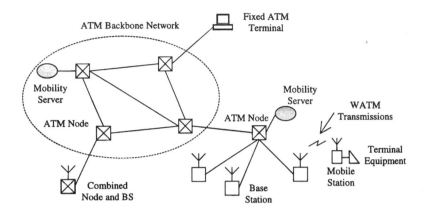

Figure 11.5: Example WATM architecture.

usual cellular control layers of MSC/BSC/HLR, and so forth. These mobility servers take advantage of the flat ATM network structure as much as possible, so that a local call is routed locally, whilst authentication, billing information, etc. are all individually routed to the nearest relevant database. Alternatively, less sophisticated mobility servers could overlay a traditional hierarchical cellular switching architecture onto the ATM network. Figure 11.5 shows an example of a WATM architecture in which it has been assumed that the mobility servers cater for the needs of ATM mobility within the fixed network, replacing the individual MSC/BSC/BS approach. Base stations control access to the continuously changing population of mobile stations. For now, we assume that all the functionality that is required to control the radio access layer (RAL) is included in the base stations.

The enhancements necessary to provide ATM mobility in the fixed network could be included in the functionality of new ATM node equipment, but to enable existing ATM networks to support WATM it is necessary to develop mobility servers that can be attached to the existing nodes. A two-stage migration path could be employed to utilise ATM within cellular networks. The first stage utilises the fixed ATM network in conjunction with mobility servers to replace the traditional hierarchical cellular architecture. This provides efficiency in the utilisation of the wired network, and would require the base stations to act as gateways between the traditional air interface and the ATM backhaul network. Stage two is concerned with directly conveying ATM cells over the air interface to the mobile station, thus forming a complete WATM network.

11.3.2   Handover Schemes

As a mobile station (MS) roams, or the network conditions fluctuate, it may be necessary for the radio link to handover from one base station (BS) to another. There must be an associated handover in the routing of the fixed network connection [16,17]. Handover control is therefore required at both the fixed network and radio levels. We are not concerned with the radio level handover at this stage, as the techniques for this are relatively well known. Unfortunately, ATM does not support re-routing of virtual circuits after establishment, so special techniques are required for network level handover. It may be arranged for base station controllers (BSCs) to control groups of BSs, and that all connections for a BS group will pass through a single BSC. A handover between BSs in the same group is relatively simple in this case. The BSC controls the cell flow and routing to the BSs. However, network level handover is necessary for handover between BSs controlled by different BSCs, and is considered below.

As ATM cells are non-periodically transported, it is not known when cells on a particular path will pass through each node. During handover, delays involved in setting up new paths, emptying buffers, forwarding or re-routing cells, and aligning the fixed and wireless networks can lead to cell loss or to cells being delivered in the wrong order, or, if the handover procedure involves cell cloning, to cells being delivered more than once. These phenomena affect different services in different ways, but generally degrade the quality-of-service (QoS) and are undesirable. As mobile communications traffic increases, the radio-cells are made smaller to increase the network capacity, and the number of network level handovers increases. Handover techniques must be rapid and robust to minimise QoS degradation.

The WATM network level handover techniques are based around three main re-routing techniques. Cell forwarding involves concatenating a new connection to the target BS onto the route of the existing connection. Virtual connection tree handover overlays a virtual hierarchy onto the ATM network, routing connections through central points, while dynamic re-routing establishes a new optimum path for the call on handover. The following subsections discuss these three handover techniques further.

11.3.2.1   Cell Forwarding

On handover initiation, the cell forwarding technique establishes a new path from the current BS to the BS to be used after handover completion, which is referred to as the target BS. On handover completion, the path to the target BS is concatenated onto the virtual connection already in use, i.e. the existing virtual connection is elongated to reach the target BS. As all cells continue to traverse the original fixed network path, when the radio link with the current BS ends, new cells destined for transmission to the MS simply continue to the target BS along the newly elongated path. New

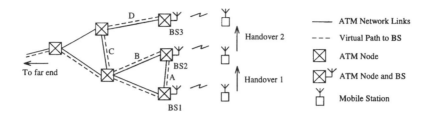

Figure 11.6: Handover using cell forwarding as a MS moves from BS1 to BS3.

uplink cells received at the target BS from the MS follow the elongated path, passing through the node at the previous BS. As transmission and queueing delays at the BSs cannot be guaranteed, uplink cells from the new BS are only allowed through the previous BS's node after all queued cells have been sent, ensuring that the correct cell sequence is maintained. Therefore, the handover procedure involves cell queueing, which places an additional load on the network's buffering resources. The affect of such loading must be carefully considered as it is important that buffering, and hence delay, is minimised for successful re-routing of real-time traffic.

Several handovers may occur over the duration of a call, sometimes between the same two BSs. Long, linked chains of connections can build up, with cells following arduous routes, and possibly traversing the same link more than once. Figure 11.6 shows a MS travelling through two handovers, firstly between BS1 and BS2 where a single connection is added to the route (link A). The second handover, from BS2 to BS3, requires three additional connections (links B, C and D) including returning to a node that is already in the path. The utilisation of the fixed network is inefficient, especially if a hierarchical structure exists. Due to the delays involved in establishing new connections to the handover BSs it may be necessary to have pre-established virtual connections between all the handover candidate BSs.

To improve link utilisation the Bahama handover scheme [15] suggests that the concatenated route should be periodically replaced with an optimised route. The Bahama scheme also suggests using the VPI as the actual BS address rather than for identifying the individual forwarding links. This is contrary to the ATM specifications, and is not appropriate for large networks due to limitations in the maximum number of addresses. The cell forwarding process can be implemented without requiring a specific mobility server to be added to the network, but this relies on there being distributed intelligence at the BSs.

11.3.2.2   Virtual Connection Tree

A virtual connection tree is overlaid onto the network to control handover for a group of BSs [13, 18]. An ATM mobility server (AMS) at one node

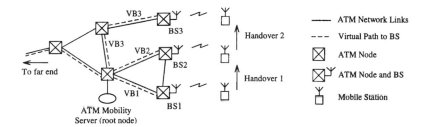

Figure 11.7: Virtual connection tree handovers (VB stands for virtual branch).

acts as a root node through which all connections to BSs in its domain must pass. Handover is controlled and executed by the MS, whilst the AMS is responsible for service continuation during network handover. When a MS requests a connection, the AMS establishes links to every BS within its tree, called virtual branches (VBs). Only one VB in the set carries the data for a particular MS at any time. A MS selects which BS to handover to, and handover is recognised in real-time by the AMS when an uplink cell arrives on a previously idle VB. The downlink cells are then switched onto the appropriate VB and so to the new BS. Virtual connection tree handover is shown in Figure 11.7, where VB1 is used initially, then VB2 after handover one, and finally VB3 after handover two.

Unfortunately, uplink cells on the VB from the new BS may arrive at the root node before all the cells from the old BS arrive, and downlink cells may have already been routed towards the old BS but not transmitted before the MS had performed a handover, causing either re-ordering of the cell sequence or cell loss. Handover delimiters can be used to surmount these problems. On handover, the MS sends a start delimiter cell on the new VB and an end delimiter on the old VB; if the AMS receives the start delimiter before the end delimiter then uplink cells are queued until the end delimiter is received. If the old BS receives downlink cells after the end delimiter has been sent it must forward these to the new BS via a specific handover channel, or return them to the AMS to do so. Thus, to achieve smooth handover, limited buffering and special cell control protocols are required.

As every connection requires a VB to be set-up to every BS in the tree, a very large number of paths must be reserved. This is inefficient as QoS parameters must be agreed at call set-up. Handover is sometimes necessary outside of the virtual tree, for which other handover techniques must be used. If the virtual tree is small the number of VBs can be reduced, but the number of inter-tree handovers will increase.

Source routing mobile circuit handover [19] improves the virtual connection tree approach by making the virtual tree dynamic. VBs are only found for BSs to which a MS is likely to handover, and these paths are de-

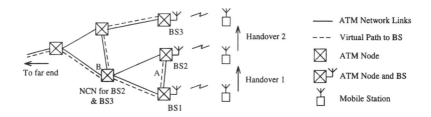

Figure 11.8: Nearest common node handover.

termined but not actually reserved. Upon a handover request, the new VB resources are reserved, and, if possible, the handover is completed. Every node has the capability of acting as the root node for the tree, locating the root node as far down the network hierarchy as possible. The root node can change as the tree alters. Efficiency is improved as fewer branches are required and the resource is not actually reserved; however, rapid assignment of the resources on handover can be troublesome. A hierarchical tree must still be overlaid on the network.

11.3.2.3 Dynamic Re-routing

On handover, dynamic re-routing attempts to transfer the connection to form a new optimum path to the new BS. This is usually carried out at the nearest ATM node to the radio-end that is common to both paths. Determining the optimal path may be complex and it can be difficult to achieve rapid handover. Avoiding cell loss and cell re-ordering is difficult as the path may change at any node.

Nearest common node (NCN) handover [16] is a dynamic re-routing technique based on determining the closest common node between the existing and desired links, and switching to the new connection through this node. This aims to minimise the resources required and preserves network bandwidth. An example of NCN handover is shown in Figure 11.8. In handover one there is a direct link from BS1 to BS2 so Link A is added to the path with BS1 acting as the NCN. For handover two, Node B acts as the NCN, and the path is re-routed. NCN re-routing uses zone managers (ZMs) that control a cluster of one or more BSs, and have look-up tables containing the addresses of neighbouring zones. Real-time traffic is not suitable for prolonged buffering but can tolerate some loss, whereas data traffic is less tolerant of cell loss but can accept increased buffering delay. NCN routing alters the handover protocol slightly according to the traffic type. Importantly, the NCN scheme assumes that the transmission delays and latency incurred during handover are negligible in comparison to the delays in the air interface. Such assumptions require confirmation in real networks.

On receiving a handover request for a real-time connection, the existing ZM checks to see if there is a direct connection to the desired ZM. If so, one ZM acts as an anchor and the connection is made. However, both routes are held until the handover stabilises. If a direct link exists, this method follows a cell forwarding approach as in handover one of Figure 11.8. If there is no direct link, the ZM sends a handover initiation message containing the addresses of the two ZMs to the far terminal. Each node along the way checks to see if it is the NCN by checking if the routes to the two ZMs and the far terminal all use different output ports. When the NCN is found, a route from the NCN to the new BS is established and the radio handover completed, but downlink information is routed to both BSs until stability is reported. This information is transmitted by whichever BS is actually in contact with the MS. The MS uses time sequence information to discard any duplicate data it receives. On the uplink, the MS sends cells to either BS which, in turn, forwards them to the ZM.

The procedure is slightly different for non-real-time traffic. When radio handover begins, both BSs buffer the downlink information until the handover is successfully completed (or fails). If the old BS's buffer is not empty, then the data are sent to the new BS and re-sequenced onto the new path. Uplink data are buffered at the MS until the handover has stabilised, ensuring cell integrity and cell sequence. After the handover is completed the cell delivery rate over the air interface is increased so that the handover buffers can be emptied. There are timing concerns, and the overall transfer speed requires confirmation. The NCN routing technique offers acceptable network complexity and efficient use of link resources. The suitability of NCN handover to different network topographies, the robustness against handover instabilities, and the service-dependent procedures are appealing.

In summary, of the handover schemes considered, the handover requirements that an ATM network must support are difficult to achieve because they require extensive control procedures and copious amounts of signalling data. Pre-establishment of new paths can decrease delay but reduces link efficiency. If the candidate BSs for handover can be predicted, then the number of paths involved can be minimised. The schemes discussed are in development at the time of writing, and it is likely that adaptive solutions will be utilised for different service types to provide rapid and efficient handovers. To attain an efficient WATM air interface, increased signalling must be carried by the fixed ATM network.

11.3.3 Quality-of-Service

Achieving individual quality-of-service (QoS) targets for each call connection is extremely challenging. In a cellular radio environment, the quality of the radio channels, the number of active connections on each BS, and the cell generation rate of the connections are all time-variant. Similarly, the loading and routing of the wired network links are also time-variant. The radio link normally causes the most serious limitations, and hence attracts the highest QoS effort [20]. Cell scheduling over the radio link, and routing in the wired link are the primary controllers. Cell delivery probability and cell transfer delay are the most important QoS statistics in WATM. In addition to the fixed network delays, cells are queued at the BS on the downlink, and the MS on the uplink, before being transported over the limited capacity radio-link.

The cell scheduler, typically located in the BS, determines priorities of queued cells and correspondingly assigns the radio resources. Determining cell priority is particularly difficult on the uplink as the scheduler must either obtain knowledge of, or predict, the state of cell queues of all the MSs in its domain. Usually, the BS transmits a broadcast message to all MSs, informing them when downlink cells will be delivered to them, and when they can transmit uplink cells. Cell scheduling within the BS is dealt with in more depth in Section 11.4.1. If both real-time and non-real-time traffic are being transported, real-time cells can be rapidly delivered and non-real-time cells can be queued until sufficient capacity is available. The QoS protocols must be dynamic to maximise link efficiency under the time-varying conditions.

A single high-rate connection may consume a large proportion of a BS's radio capacity. When the high-rate connection requires handover there is a strong possibility that the new BS will not have sufficient capacity to accept it. In some systems, eg GSM, it is common for channels to be reserved specifically for handover, but it is inefficient to reserve a large amount of capacity at every candidate BS in case a high-rate handover request occurs. The problem is exacerbated as MSs travel between picocells, microcells, and macrocells during a call, which have varying capacity and service provisions. To achieve good resource management with multimedia traffic requires a more global view of the network and traffic profiles than is normal in existing cellular networks.

Prediction of handover times and the identity of candidate BSs allow the network to provision allocate more economically. Handover can be anticipated using neighbour lists, received signal reports, positioning systems, personal usage profiles, etc. A simple form of prediction can be based on radio-cell profiles, where a statistical handover probability and call request probability is assigned to every BS according to the time of day and current loading. The profiles would be generated by analysing traffic characteristics over a period of time. Radio-cell profiles may be more ethically acceptable than recording individual mobile users' profiles. Knowledge of the urgency of a handover request could also assist the network and allow handover prioritisation. Accurate handover prediction involves intensive processing, so it should be concentrated on connections that require a high QoS.

If a BS predicts QoS degradation due to overload, it may request additional resources and/or reduce its loading by forcing handovers of some connections to nearby BSs. To provide additional resources requires dynamic capacity assignment, and the BS needs spare transceiver hardware to provide peak capacity. Dynamic capacity may use real-time frequency planning, or a set of frequency plans that can be automatically deployed as required, or dynamic channel allocation schemes, etc. Spare channels could be reserved specifically for dynamic allocation, and shared between a cluster of BSs. Inter-BS time division multiple access could allow neighbouring BSs to use the same carrier on a flexible time-share basis. This may be particularly useful if wide channel bandwidths are used in order to provide high peak data rates. As synchronisation is important this may only be possible in microcellular environments. Bandwidth reservation should include both physically reserved capacity and rapidly assignable dynamic capacity. Dynamic capacity allocation is dependent upon network-wide traffic conditions and so involves much intra-network signalling.

QoS parameters are agreed at call establishment, and are fixed for the call duration. It would be beneficial if the terminal and network could re-negotiate QoS parameters during connections, so that links can be maintained at a service grade appropriate to the traffic conditions. In this manner, rather than a call being dropped, it could be downgraded when the MS moves into an area with less resources. Similarly, if more resources become available the QoS parameters could be upgraded from those agreed at call establishment. Users may not desire connection re-grading, but it may be a necessary price to pay to achieve mobility with high-rate services. A control flag indicating whether re-negotiation is acceptable could be included at call set-up, and the service charged accordingly. QoS re-negotiation is not possible in current ATM standards, but will be considered for inclusion in mobility enhancements.

11.3.4 Location Management and Routing

Location management is concerned with knowing the approximate location of MSs as they roam within a network. It provides location information when queried by network control for call set-up, roaming, authentication, billing, and so forth. A two-tier database architecture is commonly used in wide area networks, involving home and visitor location registers. BSs are grouped into zones known as location areas. A home location register (HLR) permanently keeps details of MSs which are normally registered in its location area. A visitor location register (VLR) is attached to every location area, and keeps a copy of the details of every MS that is currently visiting the area. When the network receives an enquiry regarding a particular mobile, the enquiry is directed to the mobile's HLR. If the mobile is within it home location area it is paged and transactions commence to set up the call. Should the mobile be elsewhere the HLR knows the current location area as details of the mobile have been copied from the HLR to the VLR responsible for this location area. The call is routed to the mobile switching centre (MSC) that has an interface to the VLR. All the

BSs within the location area of the VLR page the MS. Notice that a location area may include many BSs or just a single BS. In the former, the paging load is large, whereas in the latter it is small. However, when there are many BSs within a location area there is less housekeeping to do in the fixed network. This is because each time a mobile changes its location area its details must be removed from the old VLR and then placed in the new VLR, and the HLR is involved in this process as it must always know where to find its mobiles. Thus, small location areas use small amounts of radio resources, but impose significant signalling on the back-haul network, and vice versa. GSM uses such a two-tier location management architecture, more details of which can be found in Chapter 8, and the signalling requirements for a WATM architecture are discussed in reference [21].

In a wireless LAN environment, as distinct from a cellular network, the HLR/VLR technique is rather cumbersome. As the number of BSs in a wireless LAN is relatively small, every BS can record every MS currently within its domain, and a MS is located by querying every BS until it is found. Thus, large, centralised databases do not need to be maintained, and no paging on the air interface is required to determine the precise BS that the MS is camped on. This scheme requires a large amount of signalling over the fixed network, and is not suitable for large networks. To connect the wireless LAN with public WATM networks requires significant interworking. An example of location management in a WATM-LAN is given in reference [22].

Routing in any network is complex, and no less so in ATM where it is closely related to ensuring QoS. ATM mobility requires additional functionality for re-routing of connections to deal with handover, as discussed in Section 11.3.2. Rapid and efficient algorithms that operate in both the local and wide area are necessary. Routing functionality is developing in wired ATM implementations, along with intelligent network procedures, and mobility enhancements. If a fixed link fails, automatic re-routing may take place. After a path is identified, every node along the route is investigated to see if they have the necessary resources; if not, another route is tried, or the call is rejected. The virtual path concept of ATM may be used to establish separate paths for control signalling, call information, and user-data, so that each can be routed optimally. Hence, local calls may be routed locally, whilst billing information is sent to a central database, and mobile location queried via separate paths. Compared to traditional cellular architectures where almost all traffic passes through the mobile switching centre, optimal routing techniques can greatly reduce the loading on the fixed network.

11.4   Radio Access Infrastructure

As ATM was not originally designed for wireless transmission, integration is required to make the radio access layer (RAL) function in series with the

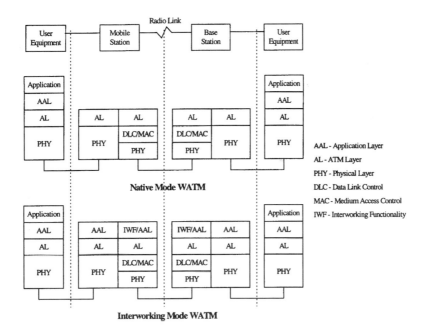

Figure 11.9: Example user plane protocol stacks for native and interworking ATM.

fixed ATM network. There are two options for integrating the WATM air interface with the fixed networks, namely the native mode and the interworking mode, as indicated in Figure 11.9. The protocol stack at the user equipment is the same in either mode, whilst the functionality at the interface between the fixed ATM network and the MS or BS equipment differs. The ATM adaptation layer (AAL) segments and re-assembles the user data for the cell payloads. The ATM layer (AL) carries out cell multiplexing, cell routing, header generation and header error correction. Over the radio link, the protocols below the AAL are referred to as the radio access layer (RAL). The RAL controls the flow of information over the air interface. It consists of medium access control (MAC) to control the sharing of radio resources amongst the multiple terminals, data link control (DLC) to cater for the effects of radio channel errors, and a physical layer (PHY) for dealing with the actual radio transmission and reception protocol [13, 18, 23]. Basic RAL technology is well known from the first and second generation cellular networks, but improved intelligent and dynamic protocols are necessary to deal efficiently with the characteristics of WATM multimedia traffic.

- *Native mode*, entailing transparent transport of the cell payloads. Specific ATM adaptation layer (AAL) processing is not necessary at the

MS or BS, as shown in Figure 11.9. Only ATM layer and physical
layer protocols are affected at the MS and BS. Native mode WATM
minimises the interworking functionality, but may not make the most
efficient use of the radio spectrum.

- *Interworking mode*, where the AAL is terminated at the BS and MS,
  and hence the cell payloads are affected by the interworking functionality (IWF). Wireless-specific protocols transport the user data and re-
  assemble the ATM payloads at the receiving end. Interworking mode
  can utilise the radio spectrum more efficiently, but adds significant
  functionality requirements to the MS and BS. Also, the MS/BS-specific
  functionality may require modification if new services are introduced
  that can be handled by the existing ATM network protocols but not
  by the interworking functionality.

Native mode WATM transport is desirable for inter-network operability
and future adaptability, and although an interworking approach is feasible,
many of the benefits of ATM may be lost during the interfacing transitions.
Radio channel errors, limited bandwidth and high data rates necessitate
powerful adaptive control protocols for the air interface to be able to deliver
the high QoS requirements of ATM to the MSs. Therefore, even native
mode WATM requires some interworking for radio transmission, but this
can be successfully kept below the AAL protocol level, i.e. in the radio
access layer.

The RAL functionality can be distributed over several elements of a
network architecture, but we assume for now that the RAL functionality is
located in each base station and mobile station. In order to allow radio resource management the BS requires knowledge of some of the ATM control
information, e.g. QoS parameters for determining cell transmission priorities. However, ATM control signalling and connection admission control
(CAC) is normally terminated at the ATM node, so special techniques are
needed for providing the BS with the required information. There are two
ATM node architectural options for this.

- *Remote node*; the ATM node continues to terminate the ATM signalling
  connection and passes the necessary information onto the BS. This
  minimises the BS complexity but needs extra processing in the node
  and a specific control link between the node and BS. Existing ATM
  nodes would require upgrading to permit implementation of this option.

- *Private node*; the BS includes basic ATM node functionality, and terminates both the wired and wireless signalling. This requires more
  complexity at the BS which also acts as a small ATM node, but allows
  superior integration into existing ATM networks.

The following subsections examine the difficulties and current solutions
for providing a RAL for WATM. We are primarily concerned with native

mode WATM, where the ATM cell payload is directly transmitted over the air interface (along with radio control information), although most of the discussions are also relevant to the interworking mode in which the user data are removed from the ATM cell structure and re-assembled according to some wireless protocol before transmission over the air interface. In either case multimedia traffic still has to be dealt with efficiently, and the radio interface must have the flexibility to do this.

11.4.1 Medium Access Control

The medium access control (MAC) layer controls the sharing of radio channels amongst multiple mobile users, ensuring efficient use of spectrum and meeting the QoS requirements [24]. Circuit-switched synchronous time division multiplexing (STDM), used in second generation cellular networks, is simple and efficient for constant bit rate (CBR) services, but inefficient for multimedia traffic. Packet-switching is efficient, but cannot support the QoS guarantees required. As in a wired ATM network, a hybrid solution is necessary, but this must also cater for distributed mobile nodes (the MSs) accessing a common medium, namely the radio spectrum. Various packet reservation multiple access (PRMA) methods and dynamic slot assignment (DSA) schemes have been investigated [24–26]. Implementation of the MAC layer is dependent on the following system choices.

- *Multiple access technique.* TDMA, FDMA, and CDMA are all feasible. As ATM is packet-based, TDMA schemes offer the most intuitive relationship, although any of the techniques are appropriate. For simplicity we refer to slots of a TDMA system; however, the codes of a CDMA system, or the frequencies of an FDMA system can be treated as analogous to these slots.

- *Duplexing scheme.* Frequency division duplex (FDD) eases the transceiver design and is suitable for large radio-cells, but time division duplex (TDD) allows reconfigurable asymmetric links that may be efficient for multimedia transport.

- *Capacity allocation.* Allocation of circuit-switched (constant bit rate) channels to calls is simple and robust, but dynamic allocation schemes are more efficient and applicable to the traffic principles of ATM. Dynamic schemes require additional control information to be passed over the air interface, which can itself be inefficient, and introduce delay.

- *Slot assignment signalling.* MSs must be informed of slot assignments. Continuous slot assignment signalling may be appropriate for variable bit rate (VBR) traffic, whilst schemes that only inform the MSs when the slot assignments need to be altered may be more efficient for constant bit rate (CBR) traffic. In practice, a hybrid format is likely to be implemented.

- *Capacity requirement signalling.* The MS must provide the controller (eg, the base station) with details of its current uplink requirements, allowing the controller to determine resource allocation and whether to grant access to the MS. The MS may concatenate this information with uplink data packets, place it in dedicated slots, or send it on a separate channel.

- *ATM cell scheduling.* A controller determines the optimal transmission order of cells in both the downlink and uplink queues, with particular attention to probable delay and cell loss. This ordering is dependent on information regarding the real-time traffic parameters of each connection.

- *Data link control and physical layers.* The MAC layer must co-operate with and support the lower radio access layers, to provide functions such as error control, power control and authentication information.

Dynamic MAC schemes offer flexible, real-time, resource assignment per connection, and, if necessary, can assign the entire transmission resources of a BS to a single MS. By this procedure the peak cell rate of a connection may be proportionate to maximum throughput of the transmitter, as long as the average cell rate is much lower. In a time-slot structure, the capacity allocation techniques can be represented as varying between two extremes.

- *Vertical reservation*; set length frames are split into fixed time-slots and each call is assigned a particular fixed slot (or slots) in each frame, as employed in GSM.

- *Horizontal reservation*; the frame structure is removed and every time-slot is allocated in turn, as necessary, to realise complete multiplexing.

Pure vertical reservation is not suitable for variable bit rate (VBR) services, and is only suitable for CBR services if the cell rate and frame rate are integer factors, otherwise many slots may be allocated and left unused, and slot utilisation will fall rapidly. Pure horizontal reservation requires a reservation message to be exchanged for every single slot, in turn, and hence commands a large signalling overhead. This allows dynamic priorities to be assigned, and is robust to errors as signalling is only valid for the following slot. However, the large control overheads are probably unacceptable. Therefore, it is necessary to utilise some frame structuring, with reservation indications being transmitted for a group of slots at a time. As scheduling takes place at the start of each frame there is a minimum average delay of half a frame length between scheduling and transmission. It is possible to minimise delay for high priority data by scheduling multiple slots near the start of the frame.

The MAC scheme is of vital importance to the success of WATM, especially in relatively low bandwidth implementations, such as wide-area cellular networks. A range of MAC schemes currently under development is examined below, and references are provided for further details. Uplink slot assignments are much more difficult than downlink slot assignments as the controller must either predict or obtain knowledge of the current status of mobiles within its area. The descriptions below concentrate on the uplink processes, with the assumption that a subset of these is used for the downlink.

11.4.1.1 Adaptive PRMA

As part of the European ACTS programme, the MEDIAN project is developing adaptive packet reservation multiple access (PRMA) [25, 27, 28], which is biased towards vertical reservation. Real-time control of the PRMA parameters is proposed to cope with a variety of QoS demands; MSs reserve slots for uplink transmissions within frames. The BS transmits a broadcast signal informing all MSs which downlink slots are relevant to them, which uplink slots have been reserved for them, and which uplink slots are available for random access attempts. Collisions occur if MSs attempt to access the same available slot, so a permission probability controls whether a MS has the right to attempt access at any instant. The BS also controls the number of unreserved slots, the number of slots in each direction, and the access permission probabilities. Each time-slot can carry one ATM cell plus its associated error coding and signalling needs.

The controller determines assignments and permission probabilities using algorithms accounting for QoS agreements, buffer delays, etc. As broadcast slots are issued a frame in advance of transmission, the controller must perform rapid processing; however, processing requirements can be reduced by bundling calls into sets that are handled in a particular manner. After a successful access, the BS places an acknowledgement in the next broadcast slot, and reserves a slot in each frame until the MS does not send any more cells. If no acknowledgement is received, the MS re-transmits the cell on the next permitted access slot, and so on until the cell is successfully transmitted or expires. When the MS stops sending on reserved slots, the BS sends a negative acknowledgement to it and re-assigns the slot.

The dynamic adaptability of adaptive PRMA is apparent and seems well suited to the traffic mix. Implementation is relatively simple. Inefficiencies arise from the number of contentions and slots that have been reserved but are not used, i.e. every time a MS does not wish to transmit on a consecutive frame once established. Enhancements may be achievable for particular service classes, such as CBR services needing one slot every n frames where the slots could be reserved without access requests. Also, one bit could be used to indicate whether or not the mobile desires its reservation to be held in the subsequent frame.

A scheme called statistical packet assignment multiple access (SPAMA) [29, 30] uses a similar frame structure. SPAMA uses statistical non-deterministic bandwidth assignment performed by a BS on the basis of the average cell rate required by a service. The probability of a connection requiring a slot in a frame is calculated and used to allocate uplink slots. Initial results indicate relatively high throughput efficiency, with low delay, and minimal processing complexity.

11.4.1.2 Dynamic Slot Assignment

The dynamic slot assignment++ (DSA++) scheme [26] tends towards horizontal reservation. The concept uses a distributed queueing system to transmit cells according to a residual lifetime and cell loss rate. A broadcast burst is used, as in PRMA, including feedback messages for fast collision resolution. On the uplink slots, MSs transmit parameters that inform the BS of their current capacity requirements - such as the residual lifetime of the most critical queued cell, the mean residual lifetime of queued cells, and the length of their queues. These are used to determine uplink slot reservations.

Random access is only required for call set-up, or if a MS needs to rapidly update its dynamic parameters. Random access subslots are used to reduce contention, with four subslots taking the period of one normal slot. For random access, a MS sends its identification and dynamic parameters, but no data. The frame length can be suddenly shortened to reduce delay if contentions were received in the preceding frame. If a MS has not requested access for a long period, it can be polled to send its parameters on a reserved subslot.

11.4.1.3 Distributed Queueing Request Update Multiple Access

The distributed queueing request update multiple access DQRUMA [23,31] scheme uses physically separate links (eg, FDD), imposing a similar time-slot structure to that used by DSA++. A MS uses random access to send a transmit request. The BS sends an acknowledgement if the transmit request is received correctly. The MS then waits for a transmit permission signal from the BS before sending its cell. Within the transmission of this cell is information regarding further slot requests, thereby reducing random access requirements. CBR service connections are programmed to automatically generate transmit permissions, when necessary.

The DQRUMA scheme is quite general as it does not specify the algorithms for generating permissions, nor for dealing with contention. It is based upon horizontal reservation on a slot-by-slot basis. This may cause problems with MSs having to decode the transmit permissions on all slots after requesting access, wasting power and preventing received power measurements from neighbouring BSs being taken. DQRUMA proposes sending very limited information to the BS regarding capacity requirements, i.e. buffer full or empty, although this could be extended.

11.4.2   Polling Scheme for Adaptive Antenna Arrays

This polling scheme is designed to support an air interface utilising adaptive antenna arrays at the BS [14, 18]. Adaptive antenna arrays use antenna beam forming techniques to maximise the gain in the direction of the wanted signal, and minimise the gain in the direction of interfering signals, thus providing high-quality radio channels. To maintain a high-quality channel it is important to periodically contact MSs to update the antenna weightings, otherwise the channel will degrade beyond recovery and channel re-establishment procedures will need to be performed.

A token is sent to each MS in turn informing it how many cells it can transmit. The MS replies instantly. If the MS has nothing to transmit, it returns a pilot tone so that the antenna weightings can be updated. The BS then transmits downlink cells using the appropriate antenna weightings. A time-out period is required to terminate the downlink transmission process so that re-polling and adjustment of weightings can be carried out before the channel degenerates beyond recovery. This polling scheme is efficient if the time-out interval is long enough and the number of MSs is small. The time-out period depends on the rate of change of the radio channel, which is primarily dependent upon the MS velocity and carrier frequency, e.g. 50 ms time-out for pedestrians at a 1 GHz carrier frequency, falling to 1 ms time-out at 35 mph at a 5 GHz carrier frequency, which is rather short. Power efficiency and decoding of other BS data may be problematic.

An alternative proposal that can also cater for high velocity MSs is to use a broadband polling cycle [18]. At the start of its slot, before the antenna beam is formed, the MS sends a broadband tone so that the BS can rapidly calculate the antenna weightings and form the antenna beam with which it receives the payload data from the MS. On the downlink a broadband signal is first sent to the MS, to which it replies with a tone so that the antenna weightings can be calculated and the downlink information is then sent. This removes the time-out restraint and, in effect, allows PRMA to be used in conjunction with high-quality channels. Very fast signal processing is required, but the link efficiency is increased by decreasing channel errors.

All the medium access control schemes described above are similar in principle. The differences tend to reside in the adaptability of the BS, the contention resolution schemes, and the amount of information returned to the controller for management of cell assignments. There are various trade-offs between signalling overheads, delivery delay, channel utilisation, and processing requirements. Throughput efficiencies from 70% to 90% have been reported. Downlink traffic is relatively simple to control, requiring only good scheduling algorithms. Traffic that can be anticipated should be carried in some form of reservation mode, whilst contention modes are required for some relatively random-type traffic and for random access attempts. The control must be carried out by the scheduler based upon the information it can ascertain regarding priorities and parameters.

In this chapter we focus on a suitable MAC scheme for WATM, assumed to utilise a TDMA slot structure with a broadcast message indicating reservations. To minimise delay, the broadcast message must be rapidly constructed and transmitted, and must refer to the earliest slot available after transmission, allowing for demodulation delays [25].

11.4.3   Data Link Control Layer

ATM assumes a high-quality transmission channel, which is often not the case in cellular radio [32]. The error controls cannot cope with the errors introduced over the air interface, and so additional techniques are required. The wireless data link control (DLC) layer removes these errors before cells are released from the air interface control. For efficiency, the error detection and correction techniques used should correlate with the required QoS. Real-time services are more sensitive to cell delay than non-real-time services, which require low bit error rates (BERs). The radio channel can be enhanced using complex and costly radio physical layer techniques, but this is unlikely to achieve fixed line quality [16, 33].

Forward error control (FEC) is used to decrease the BER by allowing a number of errors to be corrected. FEC increases the number of bits transmitted. The quality of the radio channel is time-variant, so the optimum amount and type of FEC is difficult to define. Additional protocols are necessary when the channel is particularly poor, or when the BER for a service must be minimal. Automatic repeat request (ARQ) techniques are particularly suitable for non-real-time services. The MAC layer must allow acknowledgements (positive or negative) to be sent in both the uplink and downlink directions. ARQ can only be used for real-time links if re-transmission is fast enough, possibly requiring special procedures. It may be better to use an estimate of the correct cell rather than risk deletion whilst awaiting re-transmission. The protocol should be aware of the current channel load, adjusting control so that all connections meet their QoS.

Interleaving is a powerful technique for overcoming burst error impairments, but large interleaving depths increase delay and are difficult to implement with varying rate services. Interleaving of bits within a single timeslot can spread very short error bursts. If a MS requests very high data rates and is suitably equipped, it may be able to access multiple carriers on a single BS, or multiple BSs. DLC protocol would require enhancements to cope with this type of transmission.

ATM trades an increase in bandwidth for simpler switching, placing a 10% header overhead on every cell. Such a large overhead is undesirable on the air interface. Exactly the same header is attached to many cells for delivery to one MS, and therefore the header could be compressed and only transmitted in full when it changes. Alternatively, several headers could be registered with the MS and a few data bits used to indicate the relevant one for each cell delivery. Also, transmission of the ATM header error correction code over the air interface may not be specifically required as sufficient error correction may be carried out by the air interface channel coding.

11.4.4 Radio Physical Layer

The radio physical layer (PHY) provides the wireless transmissions in conjunction with radio transceivers at the BS and the MS. The physical layer must provide the highest possible link quality to minimise the DLC requirements. A wide variety of schemes exist, and more are emerging, attempting to optimise the data transmission capacity, and maximise the spectrum efficiency [12, 13, 23, 34]. In keeping with the theme of this chapter, only a brief summary of the radio physical layer requirements is presented here.

A major challenge is overcoming multipath reception characteristics, fast fading, and slow fading of the mobile radio channel. Equalisation techniques are often used in combination with known training sequences to determine the instantaneous characteristics of the radio channel. The modulation scheme must provide efficient throughput with the highest possible bit rate over time-varying radio channels. Binary and quaternary phase shift keying (PSK), quadrature amplitude modulation (QAM), and minimum shift keying (MSK) techniques are commonly used. A wide range of operating frequencies are allocated for mobile communications around the world, typically ranging from 500 MHz to 5 GHz. The most common second generation cellular systems, such as GSM, DAMPS (or IS-136), and IS-95, operate in the 800-900 MHz and 1800-1900 MHz bands. The third generation cellular systems of the international mobile telecommunications system for the year 2000 (IMT-2000) family mainly operate in the 1900-2200 MHz bands.

11.5 Microcellular Architecture

The architecture required for a microcellular WATM network is now considered. An implementation based on the dynamic concepts proposed in Section 11.3 will be considered. In microcellular networks radio-cells are small and the number of handovers is large. It is desirable for every radio access to have dynamic and asymmetric capacity, so the maximum transmission capabilities of each base station (BS) must be far in excess of the normal requirements. The wired links to the BSs must be dimensioned around the peak traffic load. The utilisation of the wired links will be low if each BS is connected using a dedicated link. The utilisation may be increased by equipping BSs with ATM switching functionality and operating them as fully interconnected ATM switching nodes; however, the cost of equipping BSs with such functionality may be excessive. Trade-offs clearly

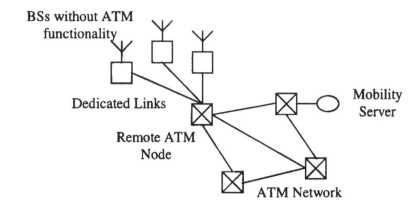

Figure 11.10: An example WATM network architecture for a location area using dedicated links from a remote ATM node to BSs without
ATM functionality. The remote node is part of the ATM network, which has mobility servers providing the ATM mobility
functionality.

exist, and a number of possible architectures are discussed below.

11.5.1   Dedicated Link to BSs from a Remote ATM Node

Every BS contains all the functionality for controlling the air interface
(MAC, DLC, PHY, buffering, etc.) and has a dedicated link to the nearest
ATM node (the remote node), but is not an ATM switching node itself.
The remote node terminates the ATM signalling and carries out the connection admission control (CAC). Much information must be exchanged
between the node and the BS regarding QoS and CAC. An example of this
architecture for a small part of a WATM network, such as a single location
area, is represented in Figure 11.10. Three BSs without ATM functionality
are shown connected by dedicated links to a remote ATM node in the fixed
ATM network. A mobility server provides the ATM mobility functionality
required in the fixed network.

   Some wireless functionality is required in the remote node as well as a
control link to the BS, the utilisation of which may be low. For handover,
the node must retrieve handover cells from the BS buffers, or the node must
contain duplicate buffering and exchange information with the BS regarding
the state of its buffers. Both these cases complicate the handover process,
and increase signalling. It would be beneficial to have at least basic ATM
node functionality at the BS to minimise the need for specialised signalling
between the ATM node and the BSs, even if the BS remains on a dedicated

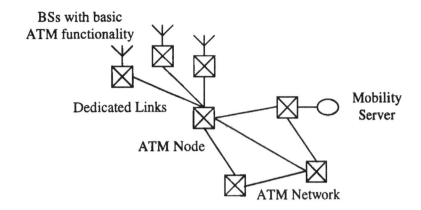

Figure 11.11: An example WATM network architecture for a location area using BSs with basic ATM functionality as simple private ATM
nodes on dedicated links to a full ATM node.

link to the full ATM node.

11.5.2   BSs as Simple Private ATM Nodes

The BSs could operate as simple private ATM nodes connected by dedicated links to a full ATM switching node, as shown in Figure 11.10. The
BSs terminate control signalling and CAC, forming radio channel control
information and sending set-up messages onto the full network. No additional functionality or control links are required at the full ATM node, so
the BSs can be added to any ATM network. No statistical gains are made
as the BSs are still on dedicated links.

11.5.3   BSs as Full ATM Nodes

The BSs are full ATM nodes with ATM connections to other nodes, as
represented in Figure 11.12. The BSs carry out their own control as part
of the general ATM network. Non-wireless traffic may also be switched via
the BSs acting as ATM switching nodes. Statistical multiplexing gains are
apparent, dependent on the switching speed, link capacity, and network
structure.

11.5.4   BSC for Semi-intelligent BSs

All handovers in the above schemes involve the network, burdening call
admission control (CAC) and signalling. Use of a base station controller
(BSC) to control a location area only requires network control for handovers

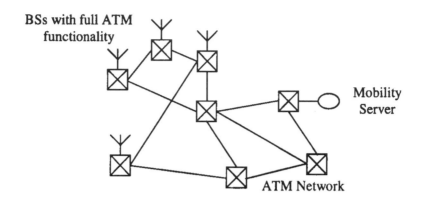

BSs with full ATM functionality

Mobility Server

ATM Network

Figure 11.12: An example WATM network architecture for a location area using BSs as full ATM nodes that can be connected to other BSs and ATM nodes.

between different location areas. Resource control intelligence can therefore be centralised at the BSC, removing the need for the BSs to communicate directly. Each BSC must act as an ATM switch, either as a full node, or on a dedicated line to a main node. Statistical gain is obtained at the BSC inputs. Figure 11.13 shows an example of this architecture. The BSs are regarded as semi-intelligent as they operate their own MAC and DLC, but are operated in conjunction with a BSC that is in charge of CAC, handover, routing, and resource control. The benefit of this is that semi-intelligent BSs are only required to communicate directly with the BSC. Instead of ATM functionality being added to every BS, a single BSC is attached at a convenient point for a number of BSs. All connections are controlled by the BSC, which finds a route to the appropriate BSs. The BSC effectively terminates incoming connections and establishes another connection to the intended terminal, concatenating the two connections to complete the path. BSC-BS signalling is over ATM virtual paths.

Handovers between location areas must use a network handover scheme (see Section 11.2.2). The majority of handovers are within the location area of the BSC. Due to the need for queueing buffers at the BSs, cell retrieval from the BSs to the BSC will be required at times. The BSC, BS, and MS could keep a count of cells received on each connection. When a handover is received at the BSC, it includes a field from the MS that identifies the number of the last downlink cell that the MS correctly received. The BSC must then retrieve the appropriate cells from the old BS and forward these to the new BS. The MS also indicates from which cell number the uplink transmissions commence, so that the BSC can verify when the uplink cells from the old BS have been received. If a BS loses communications with a

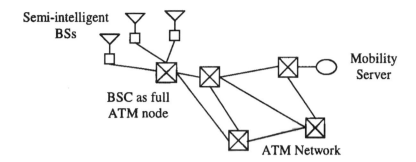

Figure 11.13: An example WATM network architecture for a location area using a BSC as a full ATM node controlling semi-intelligent BSs on dedicated links.

MS, it immediately sends the cell number of the last acknowledged downlink cell, correctly received uplink cell, and intended BS (if known) to the BSC. The BSC can ready itself to reinstate the call on another BS if necessary. Should an ATM cell be lost during handover, the MS is informed of the cell number it will receive first. To increase handover speed and help prevent cell loss the BSC could keep a short copy buffer of recent downlink cells. To minimise buffer size, copies of downlink cells should only be stored when a handover is believed to be imminent.

The above discussion assumes that cell counting can be implemented per connection without the addition of sequencing bits in every cell. The DLC already performs sequencing between the BS and MS. On the BSC-BS link, cell sequencing should not be necessary as cell losses are expected to be negligible. Periodic checking of cell counters could be performed. Although relatively simple, such a scheme may only be efficient for high QoS connections. It could also assist in reducing retrieval delays for network handovers.

11.5.5   BSC for Dumb BSs

The above architectures require intelligent BSs that control their own radio functionality. BS complexity can be drastically reduced by placing RAL control in the BSC. We refer to this 'dumb' BS as a radio port (RP), i.e. a BS with only rudimentary functionality. A RP receives formatted data from the BSC ready for transmission, along with control information for transmit powers and channel allocation. The link to the BSC is critical to ensure transmission synchronisation - limited buffering is also necessary at the RP. The data are no longer in pure ATM cell format as RP control information has been added. The simplest architectures utilise a dedicated star topology link to each RP or a ring bus structure, as shown in Figure

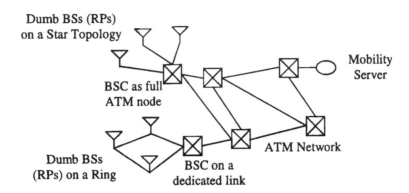

Figure 11.14: An example WATM network architecture for a location area using a BSC as a full ATM node controlling dumb BSs (RPs) on either a star or ring topology.

11.14. Even in a microcellular network, dedicated links are undesirable as costs may be prohibitive and peak bandwidth provisioning is necessary. The use of simple RPs may reduce equipment cost and can allow the amount of equipment required at the transmitter site to be much smaller than when traditional BSs are installed.

Data could be transported between a BSC and its RPs using ATM. The RP is dumb in wireless control terms, but it does perform AAL segmentation and reassembly processes, being either an ATM switching node or attached directly to a nearby node. Downlink data are formed at the BSC, segmented into ATM cells, and routed to the RP. The RP extracts the timing and control information, adds FEC coding as required, and then transmits the ATM cells over the radio interface. A similar process occurs on the uplink. This architecture would allow much of the BS functionality to be centralised at BSCs, and simple RPs to be positioned anywhere that they can be attached to the ATM network. However, as most of the radio access layer functionality is separated from the RP it is necessary to provide tight timing controls between the BSCs and RPs, which may not be possible if the link between them utilises the general ATM network.

11.5.6 Plug-in BSs

A new BS may be added to a cellular network simply by connecting it into an ATM network that has ATM mobility functionality. Two architectures in which such 'plug-in BSs' may be attached to an ATM network are shown in Figure 11.15. Firstly, a plug-in BS may also form a new ATM switching node, or secondly, it may be placed on a dedicated link to the nearest ATM node. A plug-in BS has all the ATM functionality required so that it may

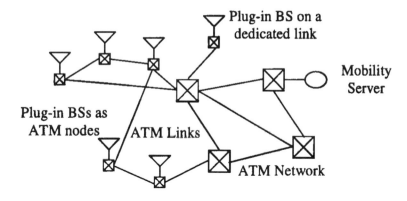

Figure 11.15: An example WATM network architecture for a location area using plug-in BSs that can act as ATM nodes or can be placed on
dedicated links attached to the nearest ATM node.

be integrated into an ATM network. The newly deployed BS would have
to inform the network of its existence, and the network would update its
databases appropriately.

The use of ATM can build redundancy into the network. ATM networks provide many different paths to a single location, and techniques for
automatic re-routing in case of equipment failure exist. The network may
cater for failure of a BS by enlarging coverage areas of neighbouring BSs,
re-assigning carrier frequencies, and re-routing connections via an appropriate BS. If a BSC should fail, the network can re-route all connections
for affected BSs via a backup BSC or via another, lightly loaded, BSC.

11.6   WATM Network Teletraffic Simulation

Current areas of research into WATM techniques tend to focus on specific
MAC and DLC protocols to achieve optimum performances for a range of
conditions. These are often based on wireless LAN scenarios rather than
the cellular networks with which this book is concerned. The consideration
of one communications link or protocol aspect in isolation is not a representative guide to a cellular network implementation. The simulation of
WATM traffic in cellular networks allows us to examine different service
types that can be supported, and the effect of different parameters and
algorithms on performance. High-level investigation of multimedia cellular
teletraffic may be useful to both network operators and system develop-
ers. The implementation of low-level protocols from other research could
enhance particular network aspects. Simulation allows analysis of the service types, user-mix, throughput delays, buffering, handovers, and QoS

guarantees, whilst accounting for path-loss and interference. The network
performance and requirements can be ascertained for various traffic loads.

This section is concerned with simulations of a microcellular WATM
network. Section 11.6.1 provides an overview of the simulation tool, Section
11.6.2 presents the results of simulations based on an urban rectilinear grid
pattern of roads with square city blocks when users are allowed to have
a mixture of voice and video services. Section 11.6.3 is concerned with
simulations of a microcellular campus network, where users are allowed to
have a combination of voice, video, and data services.

11.6.1   WATM Simulation Tool

A flexible simulator has been developed as an enhancement to Multiple
Access Communications Ltd's existing teletraffic simulation tool, Telsim. [3]
The tool loads a map with path-loss coverage profiles for a number of
BSs, generated by the NP WorkPlace[3] network planning tool. The predic-
tions provide a path-loss figure for each 5 m square bin of the map area.
Modelled mobile users roam around the map area attempting to make
calls, whilst signal-to-interference ratios (SIR) and signal-to-noise ratios
(SNR) are calculated, and channels assigned when they can be supported.
There are user-definable parameters for call initiation thresholds, handover
thresholds, frame length, slots per frame, number of users, call length,
user velocity, etc. Models are used for the call attempt distribution, call
length distribution, user movement, and so forth. Alternative procedures
and models can be added easily. Only the relevant WATM features are
presented here, other references include further details of the basic Telsim
simulator [35, 36].

11.6.1.1   Medium Access Control

The simulation works on a time division slot basis; we assume one ATM
cell is carried per slot along with all the required coding and signalling
information. A slot assignment scheme is used to dynamically assign every
slot within every frame, assuming a controlling BSC. This is referred to as
the dynamic slot assignment (DSA) scheme. Using the path-loss profiles
and the position of each MS and BS, signal-to-interference ratios (SIR)
and signal-to-noise ratios (SNR) are calculated. An ATM cell can only
be transported between a MS and BS if particular SIR and SNR criteria
are met. Modelling of the control channels is not explicitly included. The
BSC controls all the BSs in the simulation and determines the order in
which each uplink and downlink slot is assigned. Each active connection
generates cells which are queued at the BS or MS as appropriate until a
slot is allocated or the cell is deleted due to exceeding the delivery delay

---

[3]Telsim and NP WorkPlace are proprietary products of Multiple Access Communications Ltd

threshold or buffer length, according to the parameter-set for the service-type of the connection.

An alternative slot assignment scheme is used to simulate the slot assignments of a typical circuit-switched, second generation cellular network, such as GSM. This scheme uses synchronous time division multiplexing and is referred to as the fixed slot assignment (FSA) scheme. When a call is accepted by the system it is assigned a particular slot allocation within the frame structure, with one uplink and one downlink slot. The slot allocation is fixed so that the same uplink and downlink slot is reserved for a particular call in every frame until the call is terminated or handed over to another BS. High-rate services are modelled in the FSA scheme as having more than one slot per frame, where the number of slots is fixed for a particular high-rate service, and the slots are permanently assigned for the call duration, or until handover.

11.6.1.2 Service Characteristics

The characteristics of different services in the simulator are modelled by the cell generation rate, maximum delivery delay, maximum buffer length, service priority, and the type of source model. The parameters used to define a service are detailed in Table 11.3. The table shows that three different source models are available for generating the ATM cells.

The CBR source model generates cells at a constant bit rate according to the specified peak cell rate. Data are assumed to be generated continuously and segmented to form ATM cells. At each frame boundary the newly filled cells are added to the appropriate uplink or downlink queues. Notice that although cells are generated at a constant rate they must pass through the queues, and that they are not necessarily delivered over the radio interface at a constant rate. For speech services, the voice activity detection (VAD) model is placed on top of the CBR cell generation model, so that, in the periods when a user is speaking, cells are generated at a constant rate, but when the user is deemed to be silent, no cells are generated. A four-state Markov machine is used to model the voice activity, using the Brady model [37]. Normally, either one party or the other is speaking, but at times both are speaking, and at other times both are silent. If the parties were continuously talking, the cell generation rate would equal the specified peak cell rate. On average, the model generates cells at 40% of the specified peak cell rate.

The video source model uses an eight state Markov model [38], in order to generate cells at a variable rate. Eight different generation rates are used, varying from 30% to 100% of the specified peak cell rate. Twenty-five video frames are generated per second and the data are packed into ATM cells and then added to the cell queues at the video frame intervals, i.e. 25 cell generation bursts per second. A sample of the cell generation output rate for 800 video frames (32 s) is shown in Figure 11.16. The average cell

Parameter	Description	Unit
Fraction of users	Proportion of total user population for this service type.	%
Priority	Servicing order priority (1 is highest).	
Maximum delay	Period by which a cell may be delayed by before it is deleted.	s
Maximum uplink queue	Number of cells that may be stored in MS transmission buffer.	cells
Maximum downlink queue	Number of cells that may be stored in BS transmission buffer for downlink to one MS.	cells
Dropping Counter	Number of cells consecutively deleted that will cause the call to be dropped.	cells
Peak cell rate	Maximum cell generation rate for the source model.	cell/s
Source model		
Constant bit rate (CBR)	Continuously generates cells at the peak cell rate.	
Voice activity detection (VAD)	Four-state Markov model for voice call cell generation with neither, one, or both parties speaking at a time.	
Video	Eight-state Markov model of a variable rate video source, with 25 video frames per second.	

Table 11.3: WATM simulation service-type parameters.

generation rate is approximately 60% of the peak cell rate specified for the service.

Up to four service types may be defined in a simulation, along with the relative proportion of users of each type, and the total user population. Connections are prioritised according to their service type; real-time services with a low delay threshold typically have the highest priority (priority-one). Newly generated cells are added to the queues at the start of each frame, at a position dependent upon their priority, following all existing delayed cells of an equal or higher priority. Cells are thus assigned slots in the order of their delay on a first-in-first-out (FIFO) basis, until all cells of each service type have been serviced, for each priority in turn.

11.6.1.3   Call Admission Control

Call attempts are distributed randomly around the map, on roads and in open spaces. A density grid can be defined to model user hot-spots and quiet areas. Calls are attempted according to a Poisson process, spread

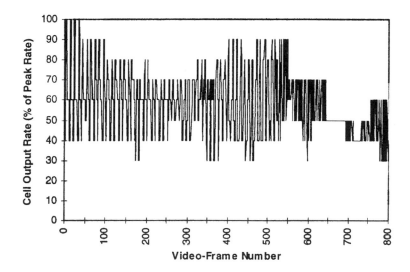

Figure 11.16: Sample cell generation rate of the eight-state Markov video model.

between the service classes in their assigned proportions. The average call duration is defined and the call length varied according to a negative exponential distribution. For a call attempt to be accepted, call initiation SIR and SNR thresholds must be exceeded, and one uplink and one downlink slot must be available in the current frame. This is referred to as the standard call admission procedure.

The standard call admission procedure can introduce unnecessary call blocking in the WATM simulation. This is because, during slot assignment, low-priority cells are buffered and assigned to unused slots after all other cells have been assigned, but before call attempts are received. Therefore, call attempts are sometimes blocked as all slots have been assigned in the current frame, but, in reality, a new call could have been accepted as a low-priority cell could have been held in the buffer longer and delivered in a later frame. Therefore, an alternative call acceptance protocol is available which is referred to as the 'accept all calls' (AAC) algorithm. In this protocol, if the SNR and SIR criteria are met, the call will be automatically accepted onto the BS, as long as the number of active calls on the BS is below a certain limit. The latter protocol effectively excludes the call set-up slots from the simulation and provides a simple call admission technique. In reality, random access slots would be required (see Section 11.4.1), and complex call admission control algorithms are used.

11.6.1.4   Handover

During a call, if a superior signal is available to a MS from an alternative BS, a handover timer is initiated. This models the time required for handover set-up and prevents handover due to a single signal report. On time-out, the call will attempt to handover to the new BS, which will accept the handover call if it has slots available. A handover hysteresis margin is included to reduce the possibility of rapidly repeated handovers. The scheme is prioritised, so that a high-priority user will be allowed to handover to the target BS before a low-priority user that is already on the target BS is serviced. Conversely, high-priority users already on the target BS will be assigned slots before a low-priority user can attempt to handover to it. An alternative handover algorithm may be selected, whereby the handover timer is started whenever a connection suffers cell deletion, regardless of signal quality. In this case, on timer expiry, the MS will handover to the BS that offers the next best signal connection and has slots available. The timer is reset if any cells are successfully delivered before the timer expires. The latter handover algorithm is referred to as the 'handover on cell loss' (HOL) algorithm.

11.6.2   Rectilinear Grid Network Simulations

A hypothetical microcellular network was generated for a rectilinear 'Manhattan' street pattern, consisting of streets running North-South and East-West. The streets were 20 m wide, separated by square building blocks of side 50 m. Antennas were placed at the street intersections, and the NP WorkPlace was used to predict the coverage of each microcell in bins of 5 m square for a receiver height of 2 m, at a carrier frequency of 1800 MHz. The propagation model assumes that antenna heights are low compared to the building heights and therefore diffraction over the building roofs into adjacent streets is negligible. The path-loss predictions for four such BSs on the rectilinear grid loaded into Telsim are shown in Figure 11.17. A frequency assignment of four carriers has been applied to the full network using 21 BSs, as shown in Figure 11.18. To limit edge effects, a central area of this rectilinear network, measuring 500 m by 500 m, was used in the simulations.

All the simulations on the rectilinear grid used a frame length of 10 ms with 8 slots per frame. The mean call length was set as 60 s, and the mean user velocity was set to be 2 m/s. Each simulation ran for 100,000 frames, with the total number of users in the network being increased in steps of 400 between runs. Each user had an average offered traffic rate of 0.025 Erlangs. A network load of 4000 users therefore corresponds to a mean of 100 active connections.

The following sections discuss the performance of various simulations run on the rectilinear grid network. Firstly, Section 11.6.2.1 compares the performance achieved by simulations using the fixed slot assignment (FSA)

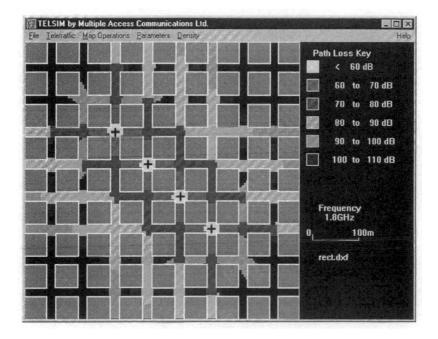

Figure 11.17: Rectilinear network coverage cluster of four microcells.

and dynamic slot assignment (DSA) schemes for transporting circuit-switched voice calls, based on voice calls carried by GSM. Next, Section 11.6.2.2 reports on the effect of giving the voice calls WATM characteristics and transporting them using the DSA scheme. In Section 11.6.2.3 a mixture of voice and video calls are applied to the DSA scheme. Section 11.6.2.4 then compares the DSA and FSA schemes for transporting these voice and video services. Returning to the DSA scheme, the effects on performance of several alternative algorithms are investigated. In Section 11.6.2.5, call attempts are allowed to be sent to a secondary BS, Section 11.6.2.6 considers the HOL algorithm, and Section 11.6.2.7 analyses the performance achieved when the AAC algorithm was used. Finally, Section 11.6.2.8 reports on the performance achieved when the HOL and AAC algorithms were used in conjunction.

11.6.2.1 Dynamic versus Fixed Slot Assignment Schemes Transporting GSM-based Voice Traffic

The performance of the DSA and FSA schemes for transporting typical second generation voice services was simulated. This provided knowledge of the performance of existing second generation cellular networks (using FSA), such as GSM, and investigated whether the voice services of such

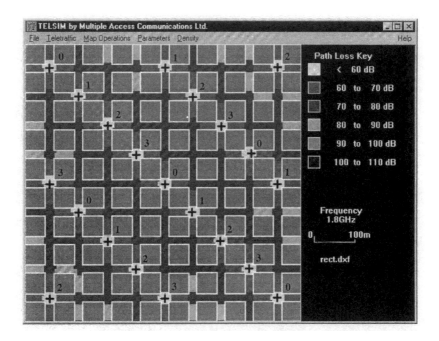

Figure 11.18: Full rectilinear network with 21 microcellular BSs, showing the
frequency plan assignment of the four carrier frequencies.

networks could be successfully transported by the DSA scheme. Second
generation cellular schemes are primarily designed to transport a single
service type, namely circuit-switched voice.

For the FSA scheme one slot was reserved in each direction per frame
per active connection. Similarly, for the DSA scheme one cell (i.e. one
slot) was added to the queue in each direction per frame per connection,
and zero delivery delay was allowed, i.e. the cell had to be delivered in the
subsequent frame or else it was deleted from the queue. Thus, the DSA
scheme was used but the benefits of a queueing cell were not available, thus
simulating the application of a GSM-based voice service to WATM.

The cell delivery probability and call blocking probability achieved in
the simulations against the number of users in the network are presented in
Figure 11.19 (see curves marked FSA and DSA). Cell delivery probability is
an important ATM statistic and is often used to compare the performance
of systems. It can be seen that the FSA scheme displayed a significantly
better cell delivery probability (CDP), particularly as the traffic load increased. The FSA scheme achieved a CDP of over 99.9% (see curve marked
FSA), whilst the DSA scheme delivered 99.9% of cells at 2000 users, decreasing steadily to only 99.2% at 6000 users, as shown by the curve marked
DSA. Call dropping also rose faster with the DSA scheme, as may be ex-

pected with a lower CDP. The lower CDP of DSA can be attributed to the assignment and handover schemes employed. DSA is more likely to accept handover requests than FSA as slot assignments are cleared at the end of each frame, whilst the FSA scheme only clears slot reservations when a call ends or is handed over. With the simple voice service the DSA scheme therefore readily hand calls over to a new BS, forcing existing users of that BS to be delayed, and as no queueing is allowed, causing cell deletion and QoS degradation. It is clear that second generation systems are well suited to carry the basic voice service.

The advantages of DSA and hence WATM start becoming apparent when voice activity detection (VAD) is introduced. As fixed slot schemes assign paired slots, the benefit of VAD is a reduction in interference. Conversely, DSA only assigns the slots that are required by each talking party, and uses the spare slots to carry more calls. The penalty of filling all the slots made available by VAD is that interference is not reduced, and the operator must ensure that an adequate call quality is maintained. Figure 11.19 shows that DSA-VAD still presented a lower CDP than FSA-VAD, but displayed a much lower call blocking probability, and actually carried significantly more successful calls.

The simulations showed that existing cellular networks are better suited to the transport of simple CBR voice traffic than the more complex DSA system. However, the increased capacity of DSA-type systems is apparent when VAD is introduced. By using DSA with VAD the user capacity increased from 2000 to 4400 users for 2% call blocking probability. However, the CDP dropped from 99.95% to 99.40%. The dynamic slot assignment and handover schemes require precise and fast control algorithms to ensure optimum cell delivery and QoS guarantees. DSA schemes will become more beneficial multimedia services, delivery delay parameters, and service prioritisation are introduced.

11.6.2.2 DSA Scheme Transporting Voice Traffic With WATM Characteristics

The voice traffic of the previous section was now given WATM characteristics in that queueing was allowed, and DSA was used. WATM characteristics required a service parameter set to be defined, with cell delivery delay being a particularly important parameter as voice is a real-time service. A voice service was defined to operate at a nominal GSM data rate of 22.4 kbit/s, which, given that each ATM cell carries 384 bits, corresponds to a cell rate of 58.3 cells/s. Note that the frame rate remained at 100 frames per second, with eight slots per frame. The maximum cell delivery delay was set as 30 ms, i.e. a cell could be delayed for up to three frames. If the cell had not been delivered after three frames it was deleted from the queue. A strict QoS parameter was used for call dropping, namely if 58 cells were consecutively deleted, the call was dropped. This equated to ap-

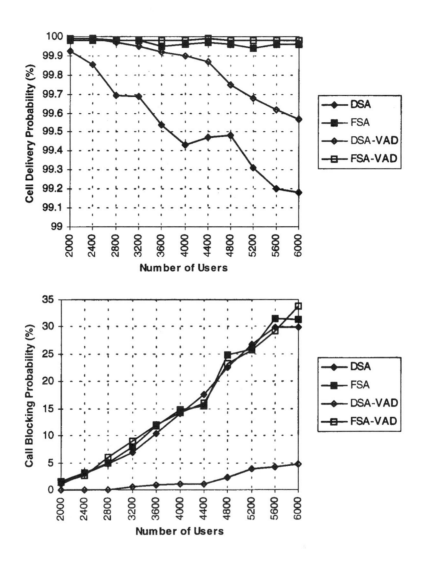

Figure 11.19: Cell delivery and call blocking probabilities for GSM-type voice traffic using the dynamic (DSA) and fixed (FSA) slot assignment schemes, with and without VAD. No queues are used.

proximately half a second of cell loss in both directions with a CBR service, or one second with VAD. Simulations were carried out with CBR and VAD speech.

Figure 11.20 shows the cell delivery and call blocking probabilities as a function of the numbers of users in the simulation. VAD speech resulted in a much lower call blocking probability than for CBR speech. Call blocking exceeded 2% with 3600 users for CBR services, but this value was not reached until there were 8000 users when VAD was used. The FSA scheme reported in Section 11.6.2.1 exceeded 2% call blocking with only 2400 users (shown in Figure 11.19). The vast majority of cells were delivered in the first frame, but the introduction of a short permissible delivery delay markedly improved the CDP. The CBR service provided a CDP of over 99.99% to 5200 users, and 99.90% to 8000 users. The VAD service presented over 99.99% CDP in all simulations - better than the FSA scheme reported above. Call dropping was negligible throughout. These simulations were re-run with the call dropping counter decreased to 15 cells. This made little difference, suggesting that when cell loss did occur it was usually only for short periods.

Allowing a short delivery delay and not requiring the cell generation rate to match the frame rate resulted in a significant performance improvement compared to transporting GSM-based voice traffic using the DSA scheme without the benefits of a small queue. This improvement was particularly apparent when VAD was used (i.e. a variable bit rate service). This simulation also showed that the DSA scheme could deliver CBR services with a reasonable quality without needing special handling techniques, such as pre-determined slot reservations in each frame, as long as a short delivery delay was acceptable.

11.6.2.3   DSA With A Mixture of Voice and Video Services

A cardinal attribute of ATM is the transport of multiple service types, so a combination of voice and video services was examined, both being real-time services. The VAD voice service with 58.3 cells per second was again used, as described in Section 11.6.2.2. A mid-rate video service with a peak rate of 128 kbit/s (333 cells per second) was defined, generating new cells at a burst rate of 25 video-frames per second. The eight-state Markov video model was used, with a minimum cell rate of 30% of the peak rate. A dropping counter of 333 cells was set, and the relative priority of the services was altered to observe the effect. The population consisted of 80% voice users and 20% video users.

The cell delivery probability (CDP) and the probability of call dropping using a 30 ms delay threshold, for video services as priority-one (P1) and voice as priority-two (P2), and vice versa, are shown in Figure 11.21. Note that priority-one corresponds to the high-priority, and priority-two corresponds to the low-priority service, i.e. the priority-one cells are ser-

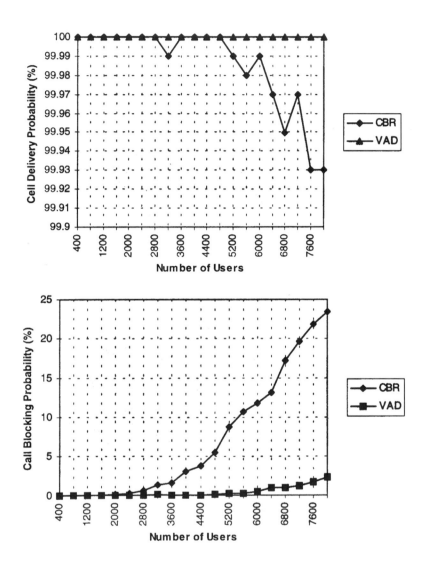

Figure 11.20: Cell delivery and call dropping probabilities for 22.4 kbit/s WATM-based voice traffic using DSA where queueing is allowed, with cell generation either at a constant bit rate (CBR) or using voice activity detection (VAD).

viced before any of the priority-two cells. CDP was good at low loads, but fell rapidly for low-priority voice (P2), remaining at over 99.9% for high-priority voice (P1). The CDP for high-priority video varied between 100% and 99.6%, falling slowly, and dropped away more rapidly when specified as low-priority. The curves show that whichever service was priority-one achieved the highest CDP. The relative degradation in the CDP for voice calls when changed from priority-one to priority-two was much greater than the degradation of video calls. Low-priority speech suffered from a large probability of call dropping, which remained below 0.5% for all other services and priorities. This indicated a need for voice to be a higher priority than video. This was due to the relatively high rate of the video services which generated many cells at a time in video-frame bursts, causing voice cells to be delayed for several frames at a time, thereby to expire and to be deleted from the queue.

In all the simulations there was a relatively large probability of new call attempts being blocked (not shown). The call blocking probability on all services increased linearly, to 30% for 5200 users. Such high call blocking was due to delayed cells filling the spare slots in each frame so that no slots were available for new call attempts. This indicates the need for special call admission control procedures for WATM, controlling the resources reserved for random access attempts.

Cell loss for video services is often regarded as being more critical than cell loss for voice services, suggesting that video services should remain as high-priority. Alternatively, the delay threshold for video users could be increased to 100 ms, thus attempting to improve video CDP whilst giving priority to voice, and still providing a reasonable QoS for video signals. Simulation results for this approach are shown in Figure 11.22 for voice as priority-one and video as priority-two. The cell delivery and call dropping probabilities when voice had a delay threshold of 30 ms and video had a 100 ms threshold, compared to the case where both voice and video connections had a delivery delay threshold of 30 ms, are displayed. CDP and call dropping were improved for 100 ms delay video, particularly below 3200 users where CDP remained over 99.95%. At high loads, the cell delivery probabilities still fell faster than with video as priority-one. The increased delay threshold improved the QoS for video without degrading the speech quality. This demonstrated how various service parameters may be altered to support different QoS requirements.

11.6.2.4   Dynamic versus Fixed Slot Assignment with Voice and Video Traffic

The voice and video traffic carried on the DSA scheme in the previous section was also simulated using the FSA scheme in order to examine how typical second generation cellular networks may perform with such traffic mixtures. In the FSA scheme, video calls were allocated three slots per

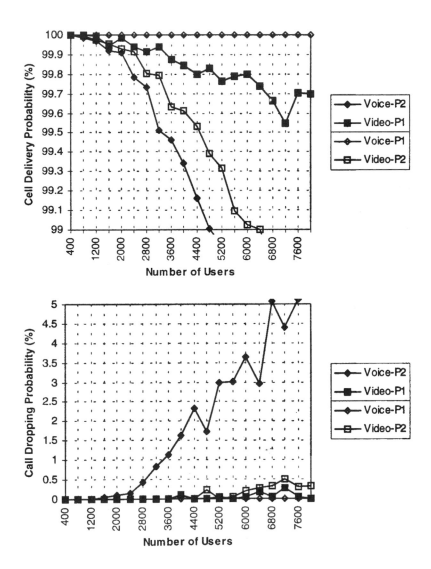

Figure 11.21: Cell delivery and call dropping probabilities for 22.4 kbit/s VAD
voice and 128 kbit/s VBR video, using DSA with 30 ms delay
thresholds, with video as priority-one (P1) and voice as priority-
two (P2), and vice versa.

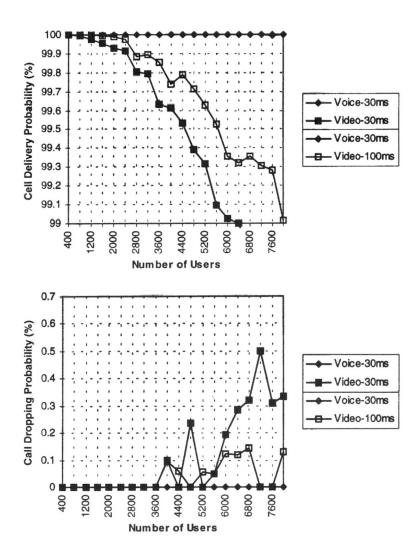

Figure 11.22: Cell delivery and call dropping probabilities for 22.4 kbit/s VAD voice and 128 kbit/s VBR video with voice as priority-one and video as priority-two, comparing 30 ms and 100 ms delay thresholds for video, whilst using a 30 ms delay threshold for voice.

frame and voice calls were allocated one slot per frame. Figure 11.23 shows the call blocking and dropping probabilities achieved relative to the results for the DSA scheme presented in the previous section, which assigned voice calls as priority-one and video calls as priority-two with delivery delay thresholds of 30 ms and 100 ms, respectively. The FSA scheme carried voice services with only slightly less success than was achieved by the DSA scheme. However, the video calls on the FSA scheme (which required 3 slots per frame) suffered from excessive call blocking and dropping. These simulations clearly demonstrate the inadequacy of FSA for transporting multimedia traffic.

11.6.2.5   Allowing Call Attempts on a Secondary BS

Let us return to the results of the DSA scheme of Section 11.6.2.3, which had a mixture of voice and video users. It was found that call blocking probability severely limited the QoS. Consequently, an alternative call attempt scheme was simulated, in which, if a call attempt was blocked, the MS was allowed to re-direct the call attempt to an alternative BS - the BS that offered the second-best signal level. Figure 11.24 shows graphs of the changes in CDP and call blocking for different loads compared to the earlier simulations which did not allow call attempt redirection. The cell delay thresholds were 30 ms for voice and 100 ms for video. Voice was the priority-one service. The call attempt redirection technique increased interference, but decreased call blocking. Two thousand users could now be supported with 2% blocking probability or 3200 users supported with a 5% blocking probability. However, an increase in video call dropping was introduced for more than approximately 3200 users, commensurate with a severe degradation in the CDP. The priority-one users (voice) were essentially unaffected by the new call attempt procedure. This technique successfully decreased new call blocking and allowed more calls to be carried, but the increased interference and channel utilisation incurred additional cell loss and call dropping. Nevertheless there was a net gain in performance.

11.6.2.6   Allowing Handover on Cell Loss

The previous simulations have shown that an increase in network load resulted in greater cell loss and call dropping. In an attempt to decrease such degradation in performance an alternative handover algorithm was tested, whereby handover was initiated if a connection suffered from cell deletion. As explained in Section 11.6.1.4, this is referred to as the 'handover on cell loss' (HOL) algorithm. The HOL algorithm was again compared to the results of Section 11.6.2.3 where voice and video services were mixed on the DSA scheme, with voice services as priority-one with a 30 ms delay allowed, and video services as priority-two with a permitted delivery delay of 100 ms.

The graphs with and without handover on cell loss (HOL), presented

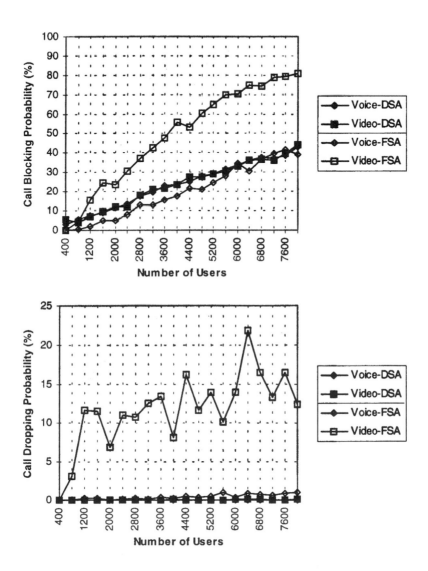

Figure 11.23: Call blocking and dropping probabilities for voice and video connections, comparing the FSA and DSA schemes.

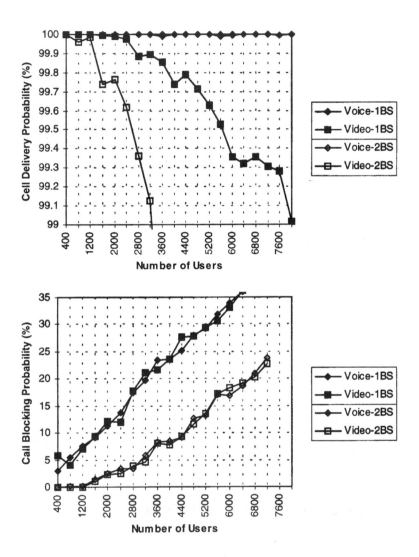

Figure 11.24: Cell delivery and call blocking probabilities for 22.4 kbit/s VAD
voice and 128 kbit/s VBR video with voice as priority-one and
video as priority-two, with blocked call attempts being redirected
to a second BS (2BS) compared to no redirection being permitted
(1BS).

in Figure 11.25, show that a significant performance improvement was achieved by implementing handover on cell loss, particularly at high user loads. There was little difference in call blocking or the total number of calls carried, and call dropping remained negligible, but, as traffic increased, the video CDP was substantially improved, remaining above 99.9% with up to 6000 users, without affecting speech calls. This performance enhancement was attained because large numbers of handovers were instigated. The graph shows that as loading passed 2800 users, video handovers rose from the normal average of 0.7 handovers per call, to 3.5 handovers per call when there were 6000 users. Voice calls displayed only a marginal increase in handover rate. Therefore, a significant QoS improvement was achieved by introducing the HOL algorithm, provided that the associated increase in the number of handovers, and hence control signalling, was acceptable. It is likely that the handover algorithm could be optimised, reducing the number of handovers required but still providing a significant performance improvement.

11.6.2.7 Accept All Calls Algorithm

Call admission procedures were removed from the simulation by use of the 'accept all calls' (AAC) procedure, as explained in Section 11.6.1.3. When a request for service was made, it was accepted by a BS if the SNR and SIR criteria were satisfied. Consequently, a channel overhead must be allowed for the omission of a CAC scheme. Such a scheme may have periodic slots reserved for random access attempts, or a specific call attempt and paging channel may be provided, or call attempts may be placed on a macrocell layer which hands suitable calls onto the microcells. See Section 11.4.1 for further details.

With the AAC algorithm a large number of active calls could be assigned to a single BS, so a limit of 100 active calls were allowed per BS. Call blocking is normally zero with such a high limit. Comparison of these results with the standard call acceptance algorithm must be treated with caution as the CAC overhead provisions were not simulated. During simulation with 4800 users, 1880 call attempts were made in 100,000 frames, spread amongst 21 BSs, each supporting eight uplink slots. On average, 90 call attempts were made per BS, the highest receiving 174 call attempts. On average, only one slot in 7500 slots was actually needed for call initiation requests. Consequently, although CAC schemes must allow for the distribution of call attempts, contention, etc. the CAC radio resource requirements are relatively small, and the lack of CAC protocol should only introduce a small error into the results.

Comparison of the results with and without the AAC algorithm (without the HOL algorithm), can be made with reference to Figure 11.26. Again, the scheme of Section 11.6.2.3 was used with voice as priority-one and video as priority-two, with 30 ms and 100 ms delivery delay thresh-

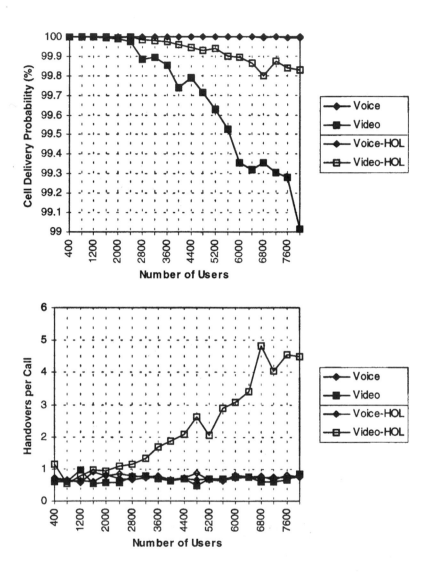

Figure 11.25: Cell delivery probability and handover rate for voice and video connections, with handover on cell loss (HOL) compared to the standard handover algorithm which only operated according to received signal strength.

olds. Observe that there is a significant rise in the number of successfully carried calls with increasing load. However, rapidly falling CDP occurred for video signals when the number of users exceeded 1600, and call dropping was incurred for more than 2400 users. There was negligible voice service degradation. As more users were accepted on already loaded BSs, the existing low-priority users (video) got excessively delayed, and hence dropped in favour of the high-priority calls (voice). While accepting more calls can provide a significant capacity increase, it is clearly necessary for the CAC to be intelligent to prevent an excessive number of calls being allowed onto the network, thus degrading existing calls beyond acceptable limits. Rapidly re-assignable capacity would also be advantageous to help meet localised, instantaneous traffic peaks.

11.6.2.8 Accept All Calls Algorithm Combined with the Handover on Cell Loss Algorithm

The possibility of further improvement was examined by combining the handover on cell loss (HOL) algorithm with the accept all calls (AAC) algorithm that was utilised in the previous section. The results of the cell delivery probability and call dropping probability are shown in Figure 11.27, comparing the performance of both the AAC and HOL algorithms with that of the AAC algorithm on its own. When both algorithms were used there was a significant improvement in the probability of call dropping, which became negligible. The video CDP was also improved, remaining at over 99.9% with 2400 users, but still tailing off fairly rapidly, concomitant with a large increase in the handover rate. The handover rate was significantly increased by the HOL algorithm; for example, for 2400 users there were 2.5 handovers per video call - an increase of over 300% on the value without HOL. At this loading, the average channel utilisation was 23%, whilst the most heavily loaded BS had an average utilisation of 39%. Channel utilisation was well distributed amongst the BSs. The handovers per BS were also fairly evenly distributed, indicating that all the BSs benefit from the HOL procedure at times. In comparison with the results without either AAC or HOL (Figure 11.22), the video CDP was slightly decreased. However, many more calls were successfully carried, i.e. when using both algorithms, the capacity of the network was significantly greater, concomitant with only a slight reduction in the cell delivery probability.

The above simulations indicate that CAC procedures are extremely important for traffic queueing schemes such as DSA, and handover on cell loss can improve the probability of cell delivery and decrease call dropping. However, the CAC and handover schemes can induce problems of cell loss and multiple handovers, so it is essential that the schemes are carefully controlled. These algorithms could be significantly improved if they utilised more variables, such as the surrounding network conditions, in order to further enhance the QoS. The number of active transceivers at a BS could

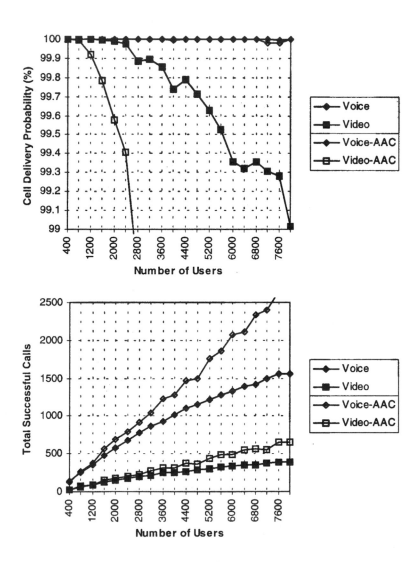

Figure 11.26: Cell delivery probability and total successful calls for voice and video connections, comparing the acceptance of all calls (AAC) algorithm with the standard call admission procedure.

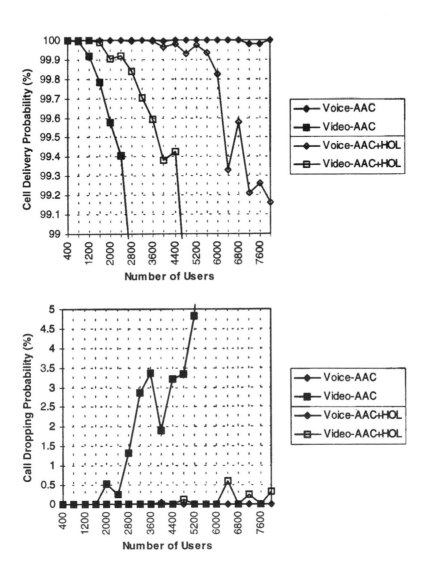

Figure 11.27: Cell delivery and call dropping probabilities for voice and video connections, accepting all call attempts (AAC), with and without handover on cell loss (HOL).

be made a function of traffic load, although this would require additional transmission hardware at the BSs, in addition to sophisticated network control procedures.

11.6.3   Campus Network Simulations

We now progress to simulating the application of WATM to a real microcellular network as opposed to a network operating on a theoretical rectilinear grid. Due to the irregularities of street layouts and the difficulties in obtaining the desired antenna positions, the coverage profiles of BSs in real networks have a complex shape. The network used for these simulations is the DOLPHIN network designed for operation on the campus of Southampton University. The network consists of 16 distributed microcellular base stations, located no more than 10 m above ground level, and two minicellular base stations placed on two of the tallest buildings on the campus. Radio propagation coverage information for this network was generated using the NP WorkPlace. The path-loss profiles for MSs at 2 m above ground level were loaded into the Telsim simulator with the appropriate three-dimensional map data for the area. The simulation area was divided into a grid of 25 squares - users were not allowed to attempt calls in the four most north-westerly squares as these were off-campus. Eight carrier frequencies were used, with one frequency reserved for each of the two minicells, and the remaining six shared between the 16 microcells. The area used and the associated frequency plan are shown loaded into Telsim in Figure 11.28, where one of the microcell BSs is off the map, although its coverage is included. The location of the other microcell BSs are indicated by '+' symbols, and the two minicells (carriers 7 and 8) are indicated by 'x' symbols.

Multimedia traffic includes both real-time and non-real-time services, so a mixture of these was simulated, namely a non-real-time data service, and voice and video real-time services. The data service generated ATM cells at a constant bit rate of 14.4 kbit/s, the voice service was 12 kbit/s and used VAD, whilst the video service was 64 kbit/s when operating at its peak rate and used the VBR video model (see Section 11.6.1.2). The full parameter set of the three services is shown in Table 11.4. Note that the data service had a maximum delivery delay of ten seconds, was the lowest priority, and desired a particularly high CDP. Voice services were priority-one, video services were priority-two, and data services were priority-three. The effect of this is that data cells were held in queues until the queued voice and video cells had been delivered. The dropping counter was set as the equivalent to five second's worth of contiguously deleted cells at the peak rate for both directions. Half of the user population were voice users, with video and data users forming 25% each. The system parameters used in the simulation were based on those of the DOLPHIN network. The frame length was 11.75 ms, with six slots in each direction per frame. Frequency

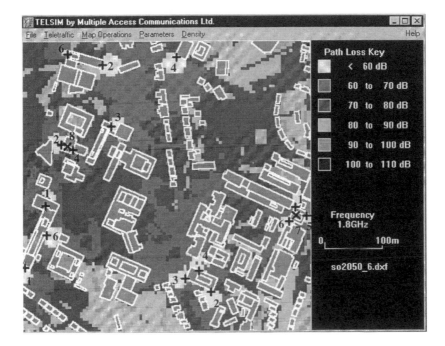

Figure 11.28: Campus network path-loss profile at 1800 MHz, based on the DOLPHIN network on the campus of Southampton University, showing the frequency plan carrier assignments, with microcellular BSs marked as '+' and the two minicellular BSs marked as 'x'.

division duplex was used. The average call length was 60 s, and each user offered an average of 0.025 Erlangs of traffic. Each simulation was run for 100,000 frames.

The following sections present the results of various WATM simulations of the DOLPHIN network. First, in Section 11.6.3.1, we study the performance achieved using the DSA scheme with a combination of voice, video and data services. In the subsequent sections the performance of further simulations, in which a number of parameter values and network procedures have been altered, are compared to these first results. Section 11.6.3.2 compares the performance of the fixed and dynamic slot assignment schemes for transporting this multimedia traffic. Returning to the DSA scheme, Section 11.6.3.3 investigates the importance of the HOL algorithm, Section 11.6.3.4 and Section 11.6.3.5 examine the effect of altering the relative priorities of the three services, whilst Section 11.6.3.6 studies how important cell buffering is to the DSA scheme. The effect of handover delay is investigated in Section 11.6.3.7, the effect of altering the handover hysteresis margin is examined in Section 11.6.3.8, and the need

Parameter	Speech 12 kbit/s VAD	Video 64 kbit/s VBR	Data 14.4 kbit/s CBR	Unit
Fraction of users	50	25	25	%
Priority	1	2	3	
Maximum delay	0.035	0.106	10	s
Maximum uplink queue	10	100	1000	cells
Maximum down-link queue	10	100	1000	cells
Dropping count	312	1667	375	cells
Peak cell rate	31.2	166.7	37.5 cells/s	
Source model	VAD	Video	CBR	

Table 11.4: WATM service parameter set for DOLPHIN network simulations.

for oversailing minicellular coverage is discussed in Section 11.6.3.9.

11.6.3.1   Combined Voice, Video and Data Services

As described in the previous sections, the DSA scheme was used in combination with the accept all calls (AAL) admission control algorithm and the handover on cell loss (HOL) handover algorithm. Using the parameter set shown in Table 11.4, voice, video and data traffic was offered to the network, with voice as priority-one, video as priority-two, and data as priority-three. Half the users made voice calls, 25% made video calls, and the remaining 25% made data calls. Data cells had a maximum delivery delay of 10 s, which in effect meant the buffer size was infinite. As the AAC algorithm was used for call admission control the probability of call blocking was negligible. Also, for up to 4800 users, the probability of call dropping was negligible. Figure 11.29 shows the cell delivery probability (CDP) as a function of the number of users in the simulation.

Because of the large delivery delay allowed for data traffic an excellent CDP was achieved, averaging over 99.99% for up to 3200 users. Video and voice services had similar CDPs, averaging at 99.8% for up to 3000 users, beyond which there was some degradation. There was some variation in the results for video and voice services, which may be due partly to the duration of each simulation, but the lack of improvement of the CDP, even for very low numbers of users, indicated gaps in signal coverage or poor SIR areas. For example, users moving round corners may experience rapid signal loss or high levels of interference, or when they moved into some areas close to irregular buildings. Often, cells with low delay thresholds

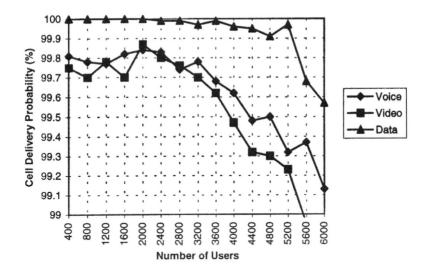

Figure 11.29: Simulation results of the DOLPHIN network on the campus of
Southampton University. The curves show the cell delivery probability for 12 kbit/s VAD voice, 64 kbit/s VBR video, and 14.4
kbit/s CBR data in a user mix of 2:1:1, using HOL and AAC
algorithms.

were deleted before the user returned to an area with sufficiently high SIR.
The cells with large delay thresholds were often held in queues until the
user returned to an area with good coverage or was handed over to a better
BS, when they were successfully delivered. The number and irregularity
of buildings was much higher than in the rectilinear network simulations
reported earlier in this chapter. It was found that only 99.5% of the area
was able to satisfy the call initiation thresholds for SIR and SNR, compared
to 100% for the rectilinear grid network simulations.

The channel utilisation per base station is shown in Figure 11.30. A
large variation in utilisation was found between the different base stations.
This variation decreased slightly as the traffic load grew, more BSs became
overloaded, and more calls were handed off to other BSs due to cell loss.
The areas of the radio-cells were radically different for each BS, with a
few dominant BSs having large radio-cells, and hence the highest channel
utilisation. The Maths and Faraday minicells, and the PhysicsII microcell
were the three most utilised BSs. Utilisation of the Maths minicell was
high, at 91% with 5200 users, at which point the average BS utilisation was
only 55%, and the Boldrewood site was at only 6% utilisation. Handovers
commenced at an average of three per call, increasing slowly as the load
increased, mainly for voice and video users, to six per call at 3600 users.

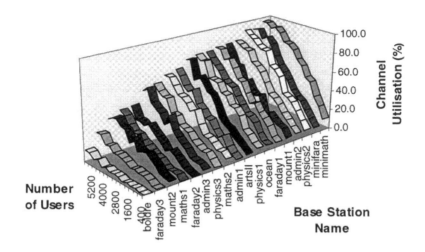

Figure 11.30: Channel utilisation per BS for 12 kbit/s VAD voice, 64 kbit/s VBR video, and 14.4 kbit/s CBR data in a user mix of 2:1:1, using HOL and AAC algorithms.

The voice and video handovers increased because they had short delay thresholds and were therefore forced into cell deletion as the load increased, which initiated the HOL algorithm. Handovers increased even more rapidly with further load, particularly for the real-time services, indicating that the network was becoming overloaded, and calls were commonly being forced into cell deletion. Note that this simulation is used as the general reference for the following sections, with the parameters that were altered stated in *each* case.

11.6.3.2 Dynamic versus Fixed Slot Assignment Scheme with Voice, Video, and Data Traffic

The previous simulations applied a mixture of voice, video, and data traffic to the dynamic slot assignment scheme, similar to schemes that are likely to be used in WATM networks. It is important to compare the performance of the dynamic slot assignment scheme with the fixed slot assignment scheme typical of second generation cellular networks (see Section 11.6.1.1 for further information).

Simulations similar to those of the previous section were run, but the FSA scheme was used in place of the DSA scheme. The FSA scheme assigned calls to a certain number of slots in each frame, in a fixed position within the frame, and for the duration of the call. The FSA scheme had no capability for cell queueing, priority assignment or handover on cell loss, and treated all services in the same manner. Voice calls were assigned one

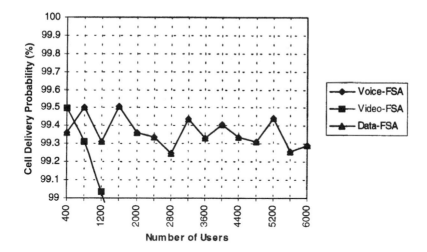

Figure 11.31: Cell delivery probability for voice, video and data connections, when transported using the fixed slot assignment (FSA) scheme.

slot per frame, and when VAD was used the unused slots could not be re-assigned, so the effect of VAD was to decrease interference. Data calls were also assigned one slot per frame. Notice that the FSA scheme cannot tell the difference between a data call and a voice call and simply reserves one slot per frame for a call of either service-type. The higher data rate of the video calls meant that they were assigned two slots per frame.

Figure 11.31 displays the CDP achieved for the voice, video, and data calls using the FSA scheme. Voice calls had a reasonable CDP of around 99.5% at all traffic loads. As data calls were serviced in exactly the same manner as voice calls, they achieved exactly the same CDP as the voice calls. Whilst a 99.5% CDP is acceptable for voice calls it is not acceptable for data calls which generally require very low error rates. The figure also shows that video connections suffered from high loss at all but the lightest loading. High levels of call blocking and call dropping were also reported, particularly for the video service, and, in comparison with the DSA scheme, comparatively few calls were successfully completed. The traffic carried per user for the DSA and FSA schemes as a function of the number of users, where each user was offered 25 mErlangs of traffic, are displayed in Figure 11.32. Observe that the DSA scheme successfully carried significantly more traffic, particularly at high traffic loads. The traffic carried per user on the FSA scheme dropped rapidly as the load increased, whilst there was only a marginal decrease when the DSA scheme was used.

These results demonstrate that the FSA scheme is not suitable for the transportation of multimedia traffic. Some sort of dynamic slot assignment

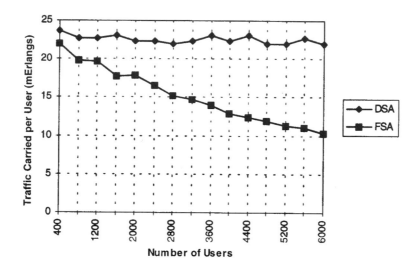

Figure 11.32: The average traffic per user for voice, video and data connections, when transported using the fixed slot assignment (FSA) scheme or the dynamic slot assignment (DSA) scheme.

(DSA) scheme, such as may be used with WATM, is much better suited to the transport of this traffic. Having confirmed the superiority of dynamic slot assignment over fixed slot assignment, the following sections investigate the benefits, or otherwise, of particular aspects of the DSA scheme.

11.6.3.3 The Absence of Handover on Cell Loss

Let us return to the dynamic slot assignment scheme to test the usefulness of the handover on cell loss (HOL) algorithm. The same set-up employed in Section 11.6.3.1 was again used, but with the handover on cell loss algorithm disabled, so that handover was only initiated according to signal quality measurements. It was found that removing the HOL algorithm introduced call dropping. The call dropping for data services rapidly increased as the traffic load increased. Figure 11.33 displays the CDP performance achieved when the HOL algorithm was removed. Comparing this with the curves in Figure 11.29 reveals that removing the HOL algorithm provided a small decrease in the CDP at low traffic loads, but as the population exceeded 2000 users, the CDP fell rapidly for all services. Notice that the CDP fell most rapidly for the data services. This was because the data services were the lowest priority, so when a BS became overloaded they suffered the most. When the HOL algorithm was available, BS overloading forced some calls to handover, thus decreasing the overload. In the simulation without the HOL algorithm, handovers could only be invoked

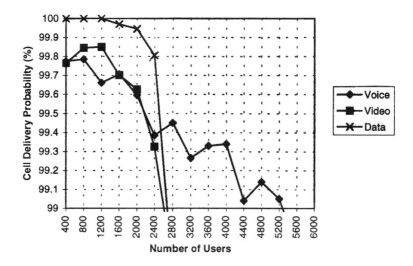

Figure 11.33: Cell delivery probability for voice, video and data connections, with the handover on cell loss (HOL) algorithm disabled.

according to signal quality measurements, and they were stable at around three handovers per call, regardless of the traffic load. This simulation revealed that the HOL algorithm provided a slight improvement in QoS at low loads, and a significant improvement as the traffic level increased.

11.6.3.4 High-Priority Video

The priority assigned to a service affects the QoS available within the dynamic slot assignment scheme. Returning to the network set-up described in Section 11.6.3.1, the service priorities were altered from voice as priority-one and video as priority-two, to video as priority-one and voice as priority-two. The data service remained as priority-three. The video CDP improved at the expense of voice calls, as shown in Figure 11.34 when compared to the curves in Figure 11.29. As the traffic increased, the voice CDP fell rapidly, passing 99% at 3200 users, whilst handover of voice calls swiftly increased - by 300% at this loading. Slots were assigned to video calls in preference to voice calls so that the voice calls, which could only tolerate three frames of delivery delay, were commonly forced into cell deletion. The HOL algorithm then attempted to handover the voice calls, thereby raising the handover rate. As a consequence, the HOL scheme successfully prevented the voice calls from suffering a large probability of call dropping. Notice that the CDP of the data calls was similar to that shown in Figure 11.29. It is clear that the QoS given to each service can be altered by changing the relative priority of the services. The effects of changing

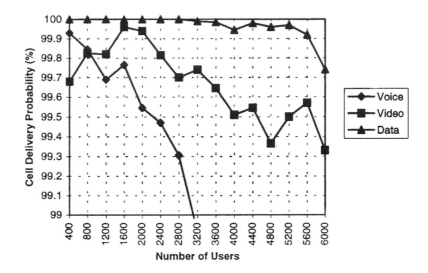

Figure 11.34: Cell delivery probability for voice, video and data connections, with video as priority-one, voice priority-two, and data priority-three (with HOL and AAC).

such priorities must be carefully considered before implementation on an established cellular network.

11.6.3.5   Equal Priority Services

The advantages of the prioritisation levels used in the dynamic slot assignment scheme were further examined by running simulations in which the different services were not prioritised. Again, the same set-up as used in Section 11.6.3.1 was simulated, but with all three services assigned an equal priority so that the longest delayed cells were assigned slots first, regardless of the service to which they belonged. Compared to our findings in Section 11.6.3.1, this approach resulted in almost complete cell delivery for data connections, but increased cell loss for video and voice users. This was because data cells had the longest delivery delay threshold, so they could stay in the queue for a long time and, as all services had equal priority, were not overtaken by prioritised cells. As the voice and video services had only short delivery delay thresholds, their QoS was severely degraded when they were not prioritised within the queues. This verified the benefits of using different priorities for different services. Further, it may be beneficial to dynamically adjust the priorities as the traffic profiles alter. For example, if voice connections were priority-one, then delayed video or data cells approaching deletion could be promoted to top priority as long as the other services still achieved acceptable service grades. Another simulation was run in which voice and video calls were assigned an equal priority, but data calls were assigned a lower priority. This achieved results similar to those with video as priority-one, voice as priority-two, and data as priority-three (Section 11.6.3.4) but with a slightly improved speech delivery performance.

11.6.3.6   Delay Buffering

To establish the benefits of buffering ATM cells whilst suitable air interface resources are sought, a simulation was run without buffering. If a cell was not delivered in the first frame after generation, it was deleted. All other parameters were the same as those used in Section 11.6.3.1. Voice calls were hardly affected by this change, but delivery of video and data cells was poor as shown by the curves in Figure 11.35 (note the change of scale of the cell delivery probability axis). As data cells were priority-three they were commonly not serviced in the frame in which they were generated and thus received a high rate of deletion. As the video cells were priority-two, they had to wait until after the voice cells had been serviced, resulting in a lower CDP than the data cells. This was because it was difficult for all the cells in a video-frame burst to be transmitted within one frame. The voice calls suffered little from the removal of buffering as they were serviced first in each frame. We observed that large numbers of handovers due to cell loss were initiated. Our simulations confirm that cell buffering is advantageous in WATM networks.

11.6.3.7   Speed of Handover

There are numerous handovers in microcellular networks, and even more if handover on cell loss is used. The effect of changing the speed at which handover was completed was investigated. The set-up of Section 11.6.3.1 was again used, except that the handover timer was increased from three frames (35 ms) to 85 frames (one second). The consequence was a vastly degraded cell delivery probability on all the services, as displayed in Figure 11.36 (note the change of scale of the cell delivery probability axis). CDP of the real-time services fell past 99% with only 800 users. On average, the handover rate decreased by two-thirds compared to when fast handover was used. This was because a cell was often successfully delivered before the handover timer expired, resetting the timer. These simulations confirmed that fast handover is indispensable for achieving low cell loss rates in microcellular networks. It is possible that different timers could be utilised depending on whether a handover was initiated for signal quality or for cell delivery reasons.

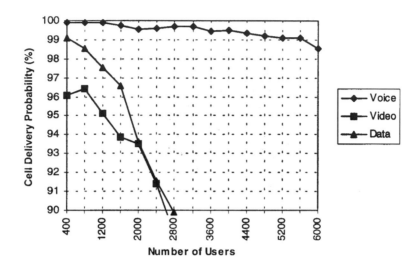

Figure 11.35: Cell delivery probability for voice, video and data connections, with minimal cell delay buffering (note change of scale of the cell delivery probability axis).

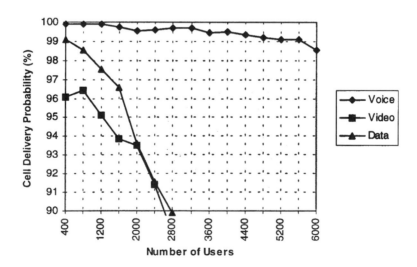

Figure 11.36: Cell delivery probability for voice, video and data connections, with the handover timer increased from 35 ms to 1 second (note the y-axis scale change).

11.6.3.8 Increased Handover Hysteresis

Several of the preceding simulations reported high rates of handover, typically three or more handovers per call. Although the HOL algorithm causes additional handovers, it is also likely that multiple rapid handovers are being caused at the irregular overlaps at the radio-cell boundaries. One method of reducing the number of handovers is through the use of a handover hysteresis margin. Handover due to signal strength is only initiated if the signal strength of another BS is better than the signal strength of the existing BS by at least a certain amount, the handover hysteresis margin. The simulations so far have operated with a handover hysteresis margin of 4 dB; the effect of increasing this margin is now investigated.

The simulation set-up of Section 11.6.3.1 was again used, but with a handover hysteresis margin of 12 dB. Figure 11.37 shows the number of handovers per call with the 4 dB and 12 dB hysteresis margins. Increasing the handover hysteresis margin from 4 dB to 12 dB reduced the average handover rate by a factor of approximately 2.5. At low traffic loads, there was approximately one handover per call. As the load grew, the numbers of handovers increased, but at a slower rate than when the 4 dB margin was used. Overall, the stability of the network was improved. At low loads it was found that the cell delivery probability was unaffected, but at high loads there was a slight decrease in the CDP. Another consequence of the handover margin increase was that a MS that was handed over to another BS due to cell deletion was less likely to immediately seek handover back to the original BS which still offered a superior signal quality.

11.6.3.9 Absence of Minicell Coverage

The DOLPHIN network used two oversailing minicells that covered a large proportion of the campus area, including areas already covered by the microcellular BSs. Simulations were run with the two minicellular BSs disabled, otherwise the set-up was the same as that used in Section 11.6.3.1. When the two minicells were removed, significantly worse QoS was encountered. Figure 11.38 shows that the CDP dropped fairly rapidly as the traffic load increased, particularly for real-time services. The handover rate increased, and so did the probability of call dropping. These effects occurred when users entered areas where the microcells did not offer adequate coverage, or suffered from too much interference, i.e. the continuity of the network coverage was poor without the oversailing coverage provided by the minicells. Also, the minicells normally provided the second-best signal to a MS operating on a microcellular BS, so the minicell was often used for handovers initiated due to cell loss. Oversailing coverage by minicells is important for the successful operation of a microcellular network, and, for optimum efficiency, special procedures should be used to ensure that the oversailing coverage layer is available to MSs that need it. Nevertheless, as much traffic as possible should be directed towards the microcellular BSs.

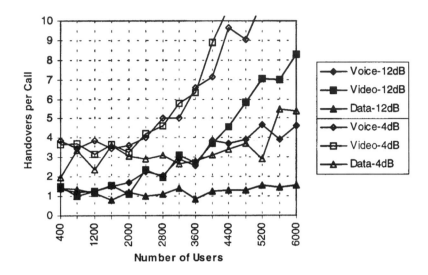

Figure 11.37: Handover rate for voice, video and data connections, with handover hysteresis margins of 4 dB or 12 B.

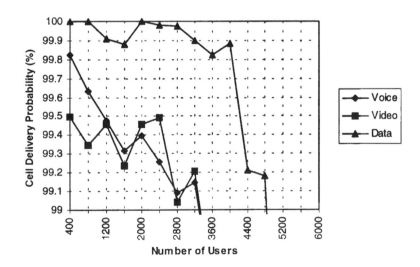

Figure 11.38: Cell delivery probability for voice, video and data connections, on the DOLPHIN network when the coverage of the two oversailing minicells was removed.

11.7 Summary of WATM Simulations

The simulations revealed that multiple-service transport is realisable in a WATM microcellular network. The fixed slot assignment schemes typical of second generation cellular networks may be optimal for the transport of a single, circuit-switched, service, but are not suited to multimedia traffic which requires dynamic slot assignment schemes.

The performances of DSA schemes have been found to be vastly improved if cell queueing and service prioritisation are included. Transporting variable bit rate traffic provides significant statistical multiplexing gain, while constant bit rate services are also accommodated. Non-real-time services can be delivered with very low cell loss if large delay thresholds are acceptable and adequate buffer sizes are provided. Non-real-time services should generally be assigned a low delivery priority so that real-time services can be delivered first, although this requires substantial buffers for holding delayed cells. Even real-time services can accept some delivery delay, which helps in achieving an acceptable QoS. Altering the priorities and delivery delays of different services can have profound effects on the QoS achieved by each service. The probability of cell loss can be decreased by increasing the delivery delay threshold. The dynamic slot assignment scheme used in the simulation could be optimised further. Optimisation should take account of more network statistics, and could treat connections differently according to the instantaneous traffic load, e.g. if a low-priority call was suffering from a particularly poor QoS, a high-priority call could be intentionally degraded to just meet its minimum QoS agreements, thus allowing the QoS for the low-priority call to be improved. In general, achieving a high channel utilisation efficiency when multimedia traffic is transported requires the use of dynamic slot assignment schemes.

It has been found that selective handover of calls to non-optimum BSs (in terms of signal coverage) can be used to greatly enhance the cell delivery probability and to reduce the probability of call dropping. Such intelligent handover techniques may increase the handover rate substantially, so careful control is required in selecting which MSs should be forced to handover, and when. In order to prevent QoS degradation it is necessary for the handover operation to be fast and reliable. Similarly, intelligent call admission control algorithms are necessary to manage the access of new calls and handover requests, thus controlling the mix of services carried by each BS.

In general, it is desirable for all the radio access layer protocols to be dynamic and intelligent in order to take account of the ongoing traffic demands on each BS and in the surrounding network. Rapidly re-assignable capacity would reduce the need for forced handovers, call blocking, and call dropping, particularly for high-rate users. Variable rate codecs and error control schemes could be utilised, varying the user rate according to the current loading and channel quality, thus preserving network resources. However, call dropping and call blocking may be unavoidable at times, and selective call dropping and blocking according to a user's service agreements could be used to control peak traffic loads.

Although multimedia traffic transport can be supported, the current bandwidth limitations, delivery delay, and cell loss rates of wireless networks make it extremely difficult to meet the ATM QoS requirements of the fixed network. Normally, mobile ATM connections will have lower QoS guarantees than fixed ATM connections.

11.8 WATM Conclusions

The WATM concepts have been presented in this chapter. The advent of multimedia traffic is a major challenge for cellular network design, and ATM offers a sensible (but not obligatory) means of transport, allowing seamless integration of fixed and radio links. WATM requires enhancements to the fixed ATM network infrastructure to cope with the mobility of users. It is essential that reliable and rapid handover control schemes are developed for the fixed ATM network. Mobile location management, QoS control, and routing management for mobile users are important associated issues. The functionality required at a BS depends upon the surrounding network architecture and may or may not include full ATM switching functionality. Different architectural options will be optimal in different scenarios. Full ATM/WATM network integration is beneficial and could facilitate plug-in BSs, operating over public and private networks.

In this discourse, a WATM microcellular network teletraffic simulation tool has been used to study many scenarios and algorithms that may be involved in the transport of multimedia traffic over cellular networks. The simulations have verified the effectiveness of the WATM concept, successfully mixing real-time, non-real-time, constant bit rate, and variable bit rate services. A number of network control enhancements have been suggested. The simulations confirmed that the medium access control protocols, data link control protocols, and network management schemes must be dynamic and intelligent, and should take into account the instantaneous traffic loading on each BS and in the surrounding network. Intelligent handover and call admission schemes can provide vast improvements to the QoS. The rapid re-assignment of capacity over a wide area would be beneficial. It must be emphasised that, given current bandwidth availabilities, satisfying the QoS expected in the fixed ATM network is economically impractical in wireless networks. Therefore, acceptable mobile service grades should be defined, or the available radio spectrum increased.

Bibliography

[1] **J. L. Adams,** "Asynchronous transfer mode - an overview," *BT Technology Journal*, vol. 13, pp. 9–14, July 1995.

[2] **S. Bates,** "The ATM - is it a waste of space?," *IEE Electronics and Communication Journal*, vol. 8, pp. 225–233, October 1996.

[3] **M. Jeffrey,** "Asynchronous transfer mode: the ultimate broadband solution," *IEE Electronics & Communications Journal*, vol. 6, pp. 143–151, June 1994.

[4] **V. J. Friesen, J. J. Harms** and **J. W. Wong,** "Resource management with virtual paths in ATM networks," *IEEE Network*, vol. 10, pp. 10–20, September/October 1996.

[5] **D. M. Alley, Y. I. Kim,** and **A. Atkinson,** "Audio services for an ATM network," *BT Technology Journal*, vol. 13, pp. 80–91, July 1995.

[6] **D. J. Wright,** "Voice over ATM: an evaluation of network architecture alternatives," *IEEE Network*, vol. 10, pp. 22–27, September/October 1996.

[7] **J. H. Baldwin, B. H. Bharucha, B. T. Doshi, S. Dravida, and S. Nanda,** "AAL-2, a new ATM adaption layer for small packet encapsulation and multiplexing," *Bell Laboratory Technical Journal*, vol. 2, pp. 111–131, Spring 1997.

[8] **M. Bergenwall, T. Kaaresoja, Y. Raivio, and S. Uskela,** "CATMES - Cellular ATM evaluation system," in *Proceeding of ACTS Mobile Communication Summit '98*, (Rhodes, Greece), ACTS, 8-11 June 1998.

[9] **Recommendation,** "I363.2. Broadband-ISDN ATM adaption layer specification: Type 2 AAL," *International Telecommunications Union*, September 1997.

[10] **D. J. Thorne,** "ATM over copper," *BT Technology Journal*, vol. 13, pp. 15–25, July 1995.

[11] **Chen,** "Monitoring and control of ATM networks using special cells," *IEEE Network*, vol. 10, September/October 1996.

[12] **D. Raychaudhuri,** "Wireless ATM networks: architecture, system design and prototyping," *IEEE Personal Communications*, vol. 3, pp. 42–49, August 1996.

[13] **B. Walke, D. Petras, and D. Plassmann,** "Wireless ATM: air interface and network protocols of the mobile broadband system," *IEEE Personal Communications*, vol. 3, pp. 50–56, August 1996.

[14] **M. Umehira, M. Nakura, H. Sato, and A. Hashimoto,** "ATM wireless access for mobile multimedia: concept and architecture," *IEEE Personal Communications*, vol. 3, pp. 39–48, October 1996.

[15] **H. Armbruster,** "The flexibility of ATM: Supporting future multimedia and mobile communications," *IEEE Communications Magazine*, vol. 33, pp. 76–84, February 1995.

[16] **B. A. Ayko and D. C. Cox,** "Re-routing for handoff in a wireless ATM network," *IEEE Personal Communications*, vol. 3, pp. 26–35, October 1996.

[17] **D.Plassmann and A. Kadelka,** "Network handover for wireless ATM systems," in *ACTS conference*, November 1996.

[18] **A. Acampora,** "Wireless ATM: a perspective on issues and prospects," *IEEE Personal Communications*, vol. 3, pp. 8–17, August 1996.

[19] **O. Yu and V. Leung,** "B-ISDN architectures and methodology for mobile executed hadoff in wireless ATM networks," in *6th IEEE International Symposium on personal, Indoor and Mobile Radio Communications*, (Toronto, Canada), September 1995.

[20] **A. Kassler, A. Lupper, and D. Sun,** "A QoS aware audio/video communication subsystem for wireless ATM networks," in *ACTS Mobile Communications Summit 98*, (Rhodes, Greece), 8-11 June 1998.

[21] **B. Aykol and C. Cox,** "Signalling alternatives in a wireless ATM network," *IEEE Journal on Selected Areas in Communications*, vol. 15, pp. 35–49, January 1997.

[22] **M. Veeraraghavan, M. Karol, and K. Eng,** "Mobility and connection management in wireless ATM LAN," *IEEE Journal on Selected Areas in Communications*, vol. 15, pp. 50–68, January 1997.

[23] **E. A. Ayanoglu, K. Y. Eng, and M. J. Karol,** "wireless ATM: Limits, challenges and proposals," *IEEE Personal Communications*, vol. 3, pp. 18–34, August 1996.

[24] **D. Petras, A. Kramling, and A. Hettich,** "Design principles for a MAC protocol of an ATM air interface," in *1996 IEEE International Symposium on Personal and Mobile Radio Communications*, (Granada, Spain), November 1996.

[25] **F. D. Priscoli,** "Adaptive parameter computation in a PRMA, TDD based medium access control for ATM wireless networks," *IEEE Global Telecommunications,* pp. 1779–1783, November 1996. Globecom 96.

[26] **D. Petras,** "Medium access control protocol for wireless, transparent ATM access," in *IEEE Wireless Communications Systems Symposium,* (Long Island, New York), November 1995.

[27] **S. Nanda, D. J. Goodman, and U. Timor,** "Performance of PRMA: a packet voice protocol for cellular systems," *IEEE Transactions on Vehicular Technology,* vol. 40, pp. 584–598, August 1991.

[28] **D. J. Goodman and X. W. Sherry,** "Efficiency of packet reservation multiple access," *IEEE Transactions on Vehicular Technology,* vol. 40, pp. 170–176, February 1991.

[29] **J. Brecht, L. Hanzo and M. D. Buono,** "Multi-frame packet reservation multiple access for variable-rate users," in *IEEE International Symposium on Personal, Indoor and Mobile Radio Communications,* (Helsinki, Finland), pp. 430–438, 1-4. September 1997.

[30] **J. Brecht, L. Hanzo,** "Statistical packet assignment multiple access for wireless asynchronous transfer mode systems", *Proceedings of ACTS Summit'97,* Aalborg, Denmark, October, 1997, pp 734-738

[31] **M. J. Karol, Z. Liu, and K. Y. Eng,** "Distributed-queueing request multiple access DQRUMA for wireless packet ATM networks," in *Proceedings ICC '95,* pp. 1224–31, June 1995.

[32] **W. A. H. Berkvens and P. F. M. Smulders,** "ARQ for real-time services in wireless ATM networks," in *ACTS mobile communications Summit 98,* (Rhodes, Greece), 8-11 June 1998.

[33] **D. Petras, A. Hettich, and A. Kramling,** "Performance evaluation of a logical link control protocol for and ATM air interface," in *7th IEEE International symposium on Personal, Indoor and Mobile Radio Communications,* (Taipei, Taiwan), October 1996.

[34] **J. Aldis,** "On the choice of physical layer techniques for wireless ATM," in *ACTS Mobile Communications Summit 98,* (Rhodes, Greece), 8-11 June 1998.

[35] **R. Steele, J. Williams, D. Chandler, S. Dehghan, and A. Collard,** "Teletraffic performance of GSM-900/DCS-1800 in street microcells," *IEEE Communications Magazine,* vol. 33, pp. 102–108, March 1995.

[36] **W. T. Webb, J. E. B. Williams, and R. Steele,** "Microcellular teletraffic levels," in *7th IEE European conference on Mobile and Personal Communications,* (Brighton), pp. 125–130, 13-15 December 1993. Pub No 387.

[37] **P. T. Brady,** "A model for generating on-off speech patterns in two-way conversation," *Bell System Technical Journal,* vol. 48, pp. 2445–2472, September 1969.

[38] **P. Skelly, M. Schwartz, and S. Dixit,** "A histogram-based model for video traffic in an ATM multiplexer," *IEEE Transactions on Networking,* vol. 1, pp. 446–459, August 1993.

Index

Author Index